Handbook of Research on Green Technologies for Sustainable Management of Agricultural Resources

R.S. Sengar
Sardar Vallabhbhai Patel University of Agriculture and Technology, India

Reshu Chaudhary
Sardar Vallabhbhai Patel University of Agriculture and Technology, India

H.S. Bhadauriya
Jiwaji University, Gwalior, India

A volume in the Advances in Environmental Engineering and Green Technologies (AEEGT) Book Series

IGI Global
PUBLISHER of TIMELY KNOWLEDGE

Published in the United States of America by
IGI Global
Engineering Science Reference (an imprint of IGI Global)
701 E. Chocolate Avenue
Hershey PA, USA 17033
Tel: 717-533-8845
Fax: 717-533-8661
E-mail: cust@igi-global.com
Web site: http://www.igi-global.com

Library of Congress Cataloging-in-Publication Data

Names: Sengar, Rakesh S. editor. | Chaudhary, Reshu, editor. | Bhadauria, H., 1961- editor.
Title: Handbook of research on green technologies for sustainable management of agricultural resources / R. Sengar, Reshu Chaudhary, H. Bhadauria, editors.
Description: Hershey, PA : Engineering Science Reference, [2021] | Includes bibliographical references and index. | Summary: "As agricultural productivity has been gaining prominence in recent years and is now become the important points for the modern agriculture, this book covers different eco-friendly technologies for the sustainable management of natural resources"-- Provided by publisher.
Identifiers: LCCN 2021017624 (print) | LCCN 2021017625 (ebook) | ISBN 9781799884347 (hardcover) | ISBN 9781799884354 (ebook)
Subjects: LCSH: Sustainable agriculture.
Classification: LCC S494.5.S86 H365 2021 (print) | LCC S494.5.S86 (ebook) | DDC 338.1--dc23
LC record available at https://lccn.loc.gov/2021017624
LC ebook record available at https://lccn.loc.gov/2021017625

This book is published in the IGI Global book series Advances in Environmental Engineering and Green Technologies (AEEGT) (ISSN: 2326-9162; eISSN: 2326-9170)

British Cataloguing in Publication Data
A Cataloguing in Publication record for this book is available from the British Library.

For electronic access to this publication, please contact: eresources@igi-global.com.

Advances in Environmental Engineering and Green Technologies (AEEGT) Book Series

Sang-Bing Tsai
Zhongshan Institute, University of Electronic Science and Technology of China, China & Wuyi University, China
Ming-Lang Tseng
Lunghwa University of Science and Technology, Taiwan
Yuchi Wang
University of Electronic Science and Technology of China Zhongshan Institute, China

ISSN:2326-9162
EISSN:2326-9170

Mission

Growing awareness and an increased focus on environmental issues such as climate change, energy use, and loss of non-renewable resources have brought about a greater need for research that provides potential solutions to these problems. Research in environmental science and engineering continues to play a vital role in uncovering new opportunities for a "green" future.

The **Advances in Environmental Engineering and Green Technologies (AEEGT)** book series is a mouthpiece for research in all aspects of environmental science, earth science, and green initiatives. This series supports the ongoing research in this field through publishing books that discuss topics within environmental engineering or that deal with the interdisciplinary field of green technologies.

Coverage

- Cleantech
- Electric Vehicles
- Contaminated Site Remediation
- Green Transportation
- Pollution Management
- Biofilters and Biofiltration
- Waste Management
- Industrial Waste Management and Minimization
- Radioactive Waste Treatment
- Alternative Power Sources

IGI Global is currently accepting manuscripts for publication within this series. To submit a proposal for a volume in this series, please contact our Acquisition Editors at Acquisitions@igi-global.com or visit: http://www.igi-global.com/publish/.

The Advances in Environmental Engineering and Green Technologies (AEEGT) Book Series (ISSN 2326-9162) is published by IGI Global, 701 E. Chocolate Avenue, Hershey, PA 17033-1240, USA, www.igi-global.com. This series is composed of titles available for purchase individually; each title is edited to be contextually exclusive from any other title within the series. For pricing and ordering information please visit http://www.igi-global.com/book-series/advances-environmental-engineering-green-technologies/73679. Postmaster: Send all address changes to above address.

Titles in this Series

For a list of additional titles in this series, please visit: www.igi-global.com/book-series

Handbook of Research on Water Sciences and Society
Ashok Vaseashta (International Clean Water Institute, USA & Transylvania University of Brasov, Romania & Academy of Sciences of Moldova, Moldova) Gheorghe Duca (Institute of Chemistry, Moldova State University, Moldova & Academy of Science of Moldova and Romanian Academy, Romania.) and Sergey Travin (Federal Research Center for Chemical Physics and N. N. Semenov of the Russian Academy of Sciences, Russia)
Engineering Science Reference • © 2022 • 728pp • H/C (ISBN: 9781799873563) • US $465.00

Optimization and Decision-Making in the Renewable Energy Industry
Figen Balo (Fırat University, Turkey) Ayşe Topal (Nigde Omer Halisdemir University, Turkey) Ezgi Demir (Sumer Robotics Engineering and Consultancy Ltd., UK) and Alptekin Ulutaş (Sivas Cumhuriyet University, Turkey)
Engineering Science Reference • © 2022 • 304pp • H/C (ISBN: 9781668424728) • US $245.00

Prevention and Management of Soil Erosion and Torrential Floods
Slobodan Milutinović (University of Niš, Serbia) and Snežana Živković (Faculty of Occupational Safety, University of Niš, Serbia)
Engineering Science Reference • © 2022 • 335pp • H/C (ISBN: 9781799884590) • US $195.00

Applications of Nature-Inspired Computing in Renewable Energy Systems
Mohamed Arezki Mellal (M'Hamed Bougara University, Algeria)
Engineering Science Reference • © 2022 • 326pp • H/C (ISBN: 9781799885610) • US $225.00

Physiology, Genomics, and Biotechnological Applications of Extremophiles
Aparna B. Gunjal (Dr. D.Y. Patil Arts, Commerce, and Science College, Pune, India) Rebecca Thombre (The University of Kent, UK & Blue Marble Space Institute of Science, USA) and Javid A. Parray (Department of Higher Education, Government Degree College, Srinagar, India)
Medical Information Science Reference • © 2022 • 473pp • H/C (ISBN: 9781799891444) • US $345.00

Optimal Planning of Smart Grid With Renewable Energy Resources
Naveen Jain (College of Technology and Engineering, Udaipur, India) Jai Kumar Maherchandani (College of Technology and Engineering, Udaipur, India) Navneet Kumar Agrawal (College of Technology and Engineering, Udaipur, India) and Trilok Gupta (College of Technology and Engineering, Udaipur, India)
Engineering Science Reference • © 2022 • 294pp • H/C (ISBN: 9781668440124) • US $225.00

IGI Global
PUBLISHER of TIMELY KNOWLEDGE

701 East Chocolate Avenue, Hershey, PA 17033, USA
Tel: 717-533-8845 x100 • Fax: 717-533-8661
E-Mail: cust@igi-global.com • www.igi-global.com

List of Contributors

Table of Contents

Detailed Table of Contents

Green energy use is motivated by its environmentally friendly behaviour, provided green jobs, renewability, equal repartition, and economic profitability. The Anthropocene is characterized by dramatic increasing pollutants amounts and earth temperature. Not only industrial activities are responsible for the setting of the Anthropocene. Instead, agriculture byproducts were the main resource of pollution and global warming. So, with increasing agricultural activities using traditional methods, the Anthropocene has been going toward a more polluted and warmer Earth. Nonetheless, using green technologies for agriculture is a candidate to reshape the destiny of the Anthropocene. The use of green energy in agriculture will reduce greenhouse gases and polluting byproducts. Accordingly, the scenario of an Anthropocene marked by an apocalyptic end will change. Global warming and its repercussions including sea level rise, decreasing ice surface, and extreme climatic events will be toned down.

Climatic conditions have been changed drastically, which has become a big challenge to sustain agricultural production. Stagnation in production has been observed in the recent past due to different types of biotic and abiotic stresses. The present available varieties do not have resilience to climate change. To sustain the productivity and minimize the yield losses, the best way is to develop climate resilient cultivars, which can produce in abundance even under abrupt climate changes in a sustainable manner. For achieving this, a better understanding of crop response to the newly developed techniques is also required in order to satisfy the needs of the growing population. Further, improving the resilience

Increasing use of agro chemicals in crop production for management of different types of biotic stresses has become the serious environmental and health concern. The unmindful use of these chemicals is the reason for increasing greenhouse gas emission, water body pollution, and toxic residue in the food items. There are good agronomic practices that can minimize biotic and also abiotic stresses in crop production, if followed properly. The crop diversification, balanced use of plant nutrition, protecting the friendly bio agents naturally occurring in the crop field are some of the effective non-chemical methods for pest management. Herbicides increasingly provide a suitable shift for the weed control function of tillage, but the overall effective strategy is to adopt good crop husbandry. The refinement of the seeding and crop establishment methods and standardization of conservation agriculture practices will make such systems progressively striking economic alternates for non-chemical pest management.

In India, the field of protected cultivation is still evolving. Methods of more efficient input application and protection technology are being devised. Low-cost protected technology like plastic low tunnels or walk in tunnels can be used for off-season vegetable cultivation for getting high returns. Similarly, insect-proof net houses can be used on a large scale for safe vegetable cultivation to minimize use of pesticides in vegetable cultivation and virus-free crop production. More eco-friendly plastics are being produced for making the processes more efficient. As a result, protected cultivation could bring prosperity to horticultural crop growers in hilly and plateau regions of India and ensure nutritional security for everyone.

It is perceived that contemporary agricultural practices are impairing natural nutrient cycling, leading to unsustainable soils. The effects include loss of nutrients from the soil and loss due to erosion. Significant public concern has been raised during the last decade over the products from conventional agriculture. Researchers, governments, farmers, and people are constantly in search of healthier methods for producing food. Organic farming has been gaining attention in these lines as it is perceived to be good for the health of humans and the environment along with providing quality produce. Organic farming still has to emerge from the substantial challenges it is facing towards becoming a prime source for food and fiber in the long term. The chapter aims to identify the challenges of organic farming and review the available techniques towards transforming current farming methods to organic farming, making them resilient along with protecting the environment and human health.

Our material lives and comfort have increased as a result of industrialization and urbanization, but changes in land use and land cover pattern in metropolitan areas have diminished vegetative covers. Many problems arise due to decline in vegetation such as urban heat islands, global warming, air pollution, and so on. Several efforts are in vogue to address these issues, but adoption of green roofs is one of the most eco-friendly concepts in India. This natural solution plays a key role in urban transformation towards a sustainable environment by improving the urban environment and lowering the urban heat island effect. In this chapter, prefabricated layers with the growth medium were explained to build the green roof. Both soil (sandy or clayey) and soilless (cocopeat or compost) media were utilized to raise the vegetative layers. The depth of the substrate determines whether the green roof system is intensive or extensive, and based on this, the crop for growing on the green roof is chosen.

By 2025, two-thirds of the world's population may be facing water shortages, according to the World Wildlife Federation. By 2030, water demand is forecast to increase by 40%. Over 1.2 billion are basically living in areas of physical water scarcity. And almost 1.6 billion face economic water shortage. And as our population continues to grow, there's just going to be more problems. So, we are in extreme need of that creation which can treat used water and can also provide us electricity so that we can use that water for irrigation purposes, toilet flush, and industrial purposes. In recent years the research work reported about an innovative constructed wetland microbial fuel cell (CW-MFCs) for more deduction of pollutants and instantaneous bioelectricity generation. A microbial fuel cell joined with constructed wetland (CW-MFC) is a novel device to treat the wastewater and create power that has more volume for wastewater treatment, is easy to repair, environmentally friendly, requires less space, and is also most cost-efficient than other devices.

Due to the exploding world population, there is a tremendous increase in demand for agricultural output, which has resulted in the pressure for large-scale crop production. Therefore, chemical fertilizers are used extensively to increase the agricultural output, which has caused hazardous effects on human health,

More than five million tons of chemical pesticides are applied annually throughout the world. Out of this, about 65% are used for agriculture, and the remainder by public health agencies and government agencies. The primary source of pesticides in ecosystems is agriculture and forestry. Pesticide poisoning is a major public health problem. These chemical pesticides affect nonpathogenic organisms, and due to that, environmental problems arise. To avoid the use of these hazardous chemicals, integrated pest management strategies are beneficial in which cultural, physical and mechanical, biological and chemical control measures are included. IPM strongly focuses on supporting natural mortality factors like pests and natural enemies. IPM requires a great deal of sampling, monitoring, and decision making to be effective.

In this era, agriculture is one of the central issue. In the world where scarcity and ailments associated with malnourishment is high and anticipated to extend within the future, new approaches have to be necessitated. Traditional techniques alone cannot meet the obligatory measures hence alternative measures need to be created. Thus, green biotechnology is the crucial requirement. Therefore, this paper narrates the significance of green biotechnology to cope with future challenges, emphasizing the applications of the techniques and the necessity for its implementation.

Over the past few decades, the man-induced changes in the climate of the earth have become the focus of scientific and social attention. The most imminent of these is the increased levels of greenhouse gases (GHGs) such as carbon dioxide, methane, and nitrous oxide in the atmosphere leading to global warming. The recent report of the Intergovernmental Panel on Climate Change (IPCC) has reconfirmed the increasingly strong evidence of global climate change and projected that the global average temperature of the air above the earth's surface would rise by 1.8-6.4oC by the end of the century depending upon the developmental pathways of the countries. For the Indian region (South Asia), the IPCC has projected 0.5-1.2oC rise in temperature by 2020, 0.88-3.16oC by 2050, and 1.56-5.4 oC by 2080, depending on the scenario of future development. Overall, the temperature increase is likely to be much higher in winter (rabi) than in the rainy (kharif) season. Precipitation is likely to increase by 15-40% by the end of the century.

Soil pollution by heavy metals and metalloids has led to significant environmental and human health hazards. Conventional remediation technologies for trace elements are cost-intensive and degrade soil quality. Phytoremediation offers an in-situ, ecological, and economic technology for remediation of metals from soil employing hyperaccumulators. However, selection of phytoremediation technology for a given site depends on soil conditions, contaminant nature and concentration, coverage and depth of polluted area, nature of plant used, etc. As phytoremediation is a slower process, it alone becomes inefficient and thus requires integration of various agronomic practices such as plant genotypes, soil properties, irrigation, fertilizers and chelators addition, crop rotation, planting methods, weed and pest control, and inoculation of plant growth promoting rhizobacteria and mycorrhizae to up-scaling phytoremediation technology for reclaiming and restoring contaminated sites effectively in a short time span.

In the present time, polyhydroxyalkanoates have established themselve as the alternatives of petroleum-based synthetic polymers due to their biodegradability and eco-friendly nature. In the last two decades, several scientists have engaged in search of cost-effective alternatives of producing PHA at larger scale. Therefore, many plant species have been genetically engineered for this purpose. The major obstacles in producing PHA polymers in transgenic plants are the regulation of the appropriate monomer's composition and ratio synthesized in their cells. Efforts are on the way to encounter these difficulties as soon as possible. The research is also going on to enhance the production and accumulation of these biopolymers in transgenic plants. Among the targeted cell organelles, plastids have been considered as the best sites for higher production of polyhydroxyalkanoates at cheaper costs because of its maternal inheritance and its being unaffected by gene silencing. Polyhydroxyalkanoates production technologies are still costly, but these could be cost effective in the near future.

The development of synthetic chemical products has increased in the last few decades. Some synthetic chemicals are very vital (e.g., pharmaceutical products, fertilizers, textiles, etc.), and the dependency of humans on these chemicals has led to an increase in their production. Green chemistry covers the entire life cycle of a chemical product and aims to reduce hazardous substances from the design process, manufacturing, use, and final disposal. Green chemistry is also used for the removal of waste or harmful metals from water sources or for the treatment of wastewater. In this context, the elevated level of Arsenic concentrations in natural waters now becomes a global problem. It has a lethal effect on human health. The threshold concentration of Arsenic has been re-established at 0.01mg/L by WHO. The chapter focuses on green chemistry, and it describes the preparation of economic bio-sorbent for the removal of Arsenic using wood pulp and MGA-NCC bio-nanosorbent synthesis process.

Crop residues are nutrient-rich precious bio-resources having huge potential of enhancing crop productivity and maintaining soil health. The potential benefits of on-farm crop residue utilization have been well-established and well-realized, especially after the popularization of conservation agriculture. The soil health in terms of physico-chemical and biological properties significantly improves both under crop residues incorporation or retention. Yet, huge quantities of crop residues are burnt on-farm to clear the fields. Efficient management of crop residues as a component of conservation agriculture could be a non-intrusive approach for providing economically and environmentally sustainable solutions for resource-scarce and fragile agro-ecologies. This chapter is a compilation of the concerted efforts made by several researchers in understanding the role of crop residues for soil health improvement, standardizing the on-farm crop residue recycling under varied farm situations, and appropriate mechanization required for their successful use.

In the agriculture system, weed emergence is going to a major constraint to cropped as well as non-cropped areas. They compete with crop plants and hinder their uptake of nutrients, water, and space by suppressing their growth, which ultimately reduces crop yield as well as quality of produce. Due to the dynamic nature of weeds, adoption attains a presence of serious magnitude in different parts of the country. Thus, the prior need is to adopt practices that lead to prevention of weed establishment within the cultivated crop and also protect the environment from these hazards. In this chapter, the authors discuss the management strategies to mitigate the adverse effect of weeds on agricultural productivity and environmental health using green technology. Green technology is a combination of management techniques to keep the soil alive and in good health by the adoption of eco-friendly management practices such as soil solarization, plant arrangement, crop rotation, cultural practices, etc. Some future prospects of weeds management are also discussed here.

The chapter deals with the concept of organic farming as an effective way to promote sustainable agriculture worldwide. The mordent approaches of conventional farming uses chemical pesticides and fertilizers that affect the environment adversely by increasing population and diminishing soil fertility and quality. In such a situation, organic farming may be proved as a significant approach to restore the environmental condition and nutrition security for the purpose of survival. There is a need of raking awareness and research in the area of organic farming to develop a cheap and eco-friendly approach for sustainable agriculture.

The use of microbial inoculants as a source of bio-fertilizers has become a hope in the present time. It is an eco-friendly way to fertilize the crop plants; it can solve the problem of high coat of fertilizers and help in saving the economy of the world, resulting in sustainable agricultural development. These biofertilizers use living microorganisms that establish symbiotic and asymbiotic relationships with the plants or are an inoculation of microorganisms which promotes the plant growth by increasing the primary nutrient supply to the host plant and also retains the soil fertility. This chapter describes the facts of experimental results of the use of biofertilizers on different crops in India and as well as in other countries. Biofertilizers are classified on the basis of their association with plants. Generally nitrogenous and phosphatic fertilizers are used in crops plant for higher yield. The Rhizobium, Azotobacter, BGA, Frankia, Bacillus, and Mycorrhiza fulfil the need of fertilizers as microbial inoculants of N and P.

Chapter 23
Eco-Toxicology: Duckweed Growth Inhibition Toxicity Test 354
Vinay Panwar, Swami Vivekanand Subharti University, India
Laxman Singh Gautam, SSV PG College, Hapur, India

The duckweed test measures the inhibition of growth of the freshwater macrophyte, Lemma sp. This test is based on the OECD Test Guideline 221 and involves exposing Lemna sp. to contaminated water for 2-d. Two toxicity test endpoints are generally measured in this toxicity test: inhibition in growth rate and changes in biomass. The aquatic plant Lemna sp. belongs to the family Lemnacae, which has a number of worldwide species and is commonly referred to as duckweed. Lemna glbba and Lemna minor are species representative of temperature areas and are the most commonly used species in toxicity tests. Both species have a floating or submerged disc old stem (frond), and a very thin root emanates from the centre of the lower surface of each frond. Lemna sp reproduces by vegetative producing new fronds. Young plants have shorter roots and consist of two to three fronds of different sizes. The small size of Lemna, its simple structure, asexual reproduction, and generation time makes plants of this genus very suitable for laboratory testing.

Chapter 24
Organic Farming: A Key to Sustainable Development 370
Nisha Malik, Swami Vivekanand Subharti University, India
Devendra Kumar, Swami Vivekanand Subharti University, Meerut, India
Manjul Mungali, Swami Vivekanand Subharti University, India
Prateek Tripathi, The Scripps Research Institute, La Jolla, USA

The first aim in agriculture has always been to maximize the production and minimize the cost of cultivation without disturbing the environment as well as services it provides to the society. Different types of hazardous chemicals are being used in order to increase farm yield due to which agricultural challenges are also increasing day by day, and they ultimately result in health and environmental problems. Therefore, we need a better and holistic approach towards sustainable development (i.e. organic farming). It is a farming system that avoids the use of synthetic fertilizers and pesticides, thereby minimizing their deleterious effect on human and animal health and natural resources. According to the global organic organization (IFOAM), four main principles, namely the principles of health, ecology, fairness, and care, are the roots from which organic agriculture grows and develops. Thus, this chapter will be an attempt to explain the significance of organic farming that will provide not only health benefits but also environmental benefits.

Farmers' efficiency in maize production resulting from adoption of technological innovations has food security implications; low productivity from maize farms has been attributable to non-adoption of improved technologies. Multistage sampling procedure was employed for this study. Primary data collected was analyzed using descriptive statistics. The study revealed that most (78.2%) of the respondents got information on production technologies through their membership of social groups or cooperatives. Production technologies were prevalent among maize farmers in the study area; however, the level of adoption was relatively low. In addition, the constraints identified significantly affected maize farmers' adoption decisions in the study area. Formulation and implementation of policies are strongly recommended to ameliorate adoption constraints.

The extensive use of high-yielding varieties with high demand for fertilizers and pesticides leads to the unjudicious application of agri-inputs that leads to deterioration of precious natural resources. The world over, consumers are becoming serious about food quality, especially chemical residues. Organic farming is rising as a dynamic 'alternate farming system' to produce safe and quality food by environmental equilibrium. The growing demand for organic food is creating new market opportunities for farmers with premium prices. Potato can also play a potential role in improving farm income through organic production. Globally, India is the second-largest potato producer producing around 53 million tonnes from 2.2 million hectares. An organic farming system consisting of an integrated approach is the best technique for achieving higher individual crop and system productivity. The current potato scenario in India suggests its diversified production and utilization in the domestic and export market leading to higher profits by way of organic potato production.

Climate change is not very well understood including how these changes will affect potato crop, insect, disease, natural enemies, and the complex relationship among them. If the climate is warmer, will it increase yield offset losses to pests or will losses to pests outweigh yield advantage from warmer temperatures? It is likely that new insect pests of potato will stabilize in narrower areas and be able attack plants in new regions. It is likely that plants in some regions will be attacked more frequently by certain pests. This chapter explores the effect of climate change on potato insects in India.

Plant breeding and genetic engineering have made significant contributions in improving crops and thus addressing the problem of food security over the last few decades. However, due to continuously growing population, increasing urbanization, environmental pollution, land degradation, and global climate change, food security is still a major concern worldwide. More attention needs to be paid to sustainable crop production for ensuring enough food for current and future generations. Emergences of genomic technologies holds promise for improving crop productivity and provide solutions to the food shortage. Recent advances in high throughput technologies have led to generation of a tremendous amount of genomics data from plants that can be put to a variety of uses. Computational tools are essential to analyze this complex data generated from sequencing of genomes, determine complete set of genes, analyze transcripts and proteins encoded by an organism, and understand their interactions that bring out changes in phenotypes.

The RNAi-mediated technology has emerged as a powerful platform for developing eco-friendly and stable molecular tools for crop improvement. Both siRNA and miRNA are found in all the kingdoms of eukaryotic life and have been employed for crop improvement by enhancing plant nutritional value, development of seedless fruits, and enhancement of shelf life in various plants. It has also been exploited for elimination of anti-nutritional factors and allergens in the crop plants and in floricultural plants to improve and modify the flower color. RNAi-based host plant-mediated gene silencing technique can be used to combat plant pathogens such as viruses, bacteria, nematodes, fungal, and oomycetes plant pathogens of various agronomically important crop plants. RNA silencing technologies have been used in vegetable plants to enhance pest and disease resistance, alter architectural changes in plant and flowering time, improve commercial traits of fruits and flowers, improve nutritional content, eliminate toxic compounds and allergens, and promote a greater industrial product.

Sustainable crop production is a way of growing or raising food in an ecologically and ethically responsible manner. Sustainable crop production is in contrast to industrial crop production, which generally relies upon monocropping (growing only one crop in a large area of land), intensive application of commercial fertilizers, heavy use of pesticides, and other inputs that are damaging to the environment, to communities, and to farm workers. Organic farming is one of the several approaches found to meet the objectives of sustainable agriculture. Organic vegetable cultivation offers one of the most sustainable farming systems with the recurring benefits to long term soil health and provide a lasting stability in production by importing better resistance against various biotic and abiotic stresses. Most of organic products are growing at a faster rate (20%) as compared to conventional ones (5%). This growth rate is highest in Japan, USA, Australia, and EU.

Green technologies can operate in diversified areas, and at the same time, these technologies contribute to produce more healthy and nutritious food. Biotechnology, the greenest tool of green technology, has the potential to increase the agricultural production, as well as to genetically modify food or its nutritional components to support health and to reduce the risk of diet-related diseases. Recent advances in molecular biology coupled with the wealth of information from 'omics technologies' have fostered the emergence of a multidisciplinary field of science 'Nuturigenomics' that tends to unfold the role of nutrition on gene expression. Amalgamation of both green technologies and nutrigenomics will help to solve problems arising due to negative effects of food, nutrition, and environment. The chapter provides an overview of green technology and its impact on nutrition and health through the study of nutrigenomics.

Foreword

Sustainability is generally defined as "meeting the present needs without compromising the future generation to meet their own needs". It is connected to the continuity of human society's economic, social, institutional, and environmental components, as well as the non-human environment. According to present scenario, sustainable agriculture implies that current and future food demands can be met without unnecessarily compromising economic, ecological, and social/ political needs. This goes beyond simple ecological and farming-system-based interpretations of sustainability.

Sustainability of agriculture has direct implication on food security. India's population is growing at the rate of 1.9% per annum, that is, we are adding an Australia to India every year. The per capita arable land is steadily declining as a result of industrialization, urbanisation, and other development projects like as roads and railways, and it is expected to reach 0.10 ha by 2025. The need for cereals, legumes, oils, vegetables, fruits, spices and condiments, milk, meat, eggs, and fish will rise as the population grows. There is no choice except to enhance agricultural activity, putting a greater strain on the few natural resources available. It's also critical to recognise that we won't be able to attain peace and prosperity or realise our goal of making India a developed nation unless we ensure food security for all.

I am happy to know that Dr. R.S. Sengar from Sardar Vallabhbhai Patel University of Agriculture & Technology, Dr. Reshu Chaudhary from Swami Vivekanand Subharti University, Meerut and Dr. H.S. Bhadauriya from Jiwaji University, Gwalior are editing a book on an important topic related to sustainability and stability in food production. As it is a critical feature of agriculture in the twenty-first century, I hoped that this publication which has contributions from renowned experts, will definitely be useful for agricultural scientists, officers, policy makers, students and farmers for enlightening their views about green technology and sustainable agriculture by discovering solutions to serious problems in the 21st century. The editors should be commended for putting this book together.

A. K. Singh
IARI, New Delhi, India

Preface

Development and promotion of high yielding crop cultivars responsive to high input was the cornerstone of green revolution. It became imperative to expand the area under cropping, make huge investments in development of infrastructure and input supplies to augment crop production. Green revolution though adequately boosted our Indian Agriculture to greater heights, leading to self-sufficiency in food grain production, as of now the cause of concern is food quality, safety and security. Such problems evolved in modern-day agriculture, unleashed organic farming. Organic agriculture has grown out of conscious efforts inspired in people towards environmental preservation and assured food quality and is gaining huge momentum socially, environmentally, and commercially.

A holistic compilation on organic farming, intersecting the points of view of students, Scientists, policy makers and farmers is needed. This is an earnest attempt to publish a Comprehensive book on organic farming to meet out the demands and expectations of all stakeholders. This book is a ready reckoner to bring out the Organic farming - concepts, constraints, prospects, processes, procedures, and policies. It would definitely help the readers to acquire insightfulness into various aspects of organic farming starting from Introduction, marketing, certification and exporting. The book deals elaborately on various Topics including the present scenario of organic farming, biodiversity, composting methods and techniques, nutrient, weed, pest and disease management under organic farming, Conversion to organic farming, organic animal husbandry, biodynamic farming, certification Producer's and policies, export avenues in organic farming. The topics are written in a cogent, logical and systematic way for better understandability.

The authors invite valuable suggestions and feedback for further improvisation of the manuscript. The literature support, data and information obtained from various authors and articles are acknowledged profusely. We immensely thank the publishers for their praiseworthy efforts to publish this book in a readable style and format.

R. S. Sengar
Sardar Vallabhbhai Patel University of Agriculture and Technology, India

Reshu Chaudhary
Sardar Vallabhbhai Patel University of Agriculture and Technology, India

H. S. Bhadauriya
Jiwaji University, Gwalior, India

Chapter 1
Role of Green Technology Use for Agriculture in Shaping the Destiny of the Anthropocene

Elhoucine Essefi
University of Gabes, Tunisia

Soumaya Hajji
https://orcid.org/0000-0002-6371-7223
Faculty of Sciences of Sfax, Tunisia

ABSTRACT

Green energy use is motivated by its environmentally friendly behaviour, provided green jobs, renewability, equal repartition, and economic profitability. The Anthropocene is characterized by dramatic increasing pollutants amounts and earth temperature. Not only industrial activities are responsible for the setting of the Anthropocene. Instead, agriculture byproducts were the main resource of pollution and global warming. So, with increasing agricultural activities using traditional methods, the Anthropocene has been going toward a more polluted and warmer Earth. Nonetheless, using green technologies for agriculture is a candidate to reshape the destiny of the Anthropocene. The use of green energy in agriculture will reduce greenhouse gases and polluting byproducts. Accordingly, the scenario of an Anthropocene marked by an apocalyptic end will change. Global warming and its repercussions including sea level rise, decreasing ice surface, and extreme climatic events will be toned down.

INTRODUCTION

Green energy, which is any type of energy produced from natural resources such as sunlight, wind or water, has been a fertile field of study (Khoie et al., 2020; Li et al., 2020). Although there are some differences between renewable energy and green energy (Khan et al., 2020), the later usually comes from renewable energy sources, such as solar energy, wind energy, geothermal energy, biomass energy and hydropower. Each of these technologies works in a different way, either absorbing solar energy like solar

DOI: 10.4018/978-1-7998-8434-7.ch001

panels (Dixit, 2020), or using wind turbines (Hübler et al., 2021; Shaterabadi et al., 2020; Voigt, 2021) or water currents to generate electricity (Chowdhury et al., 2020; Pamungkas et al., 2020). Regarded as green energy, a resource does not cause pollution like fossil fuels. This means that not all energy used in renewable energy is environmentally friendly. For example, power generation using the burning of organic materials in sustainable forests can be renewable, but not necessarily environmentally friendly (Almsatar, 2020). Green energy is usually replenished naturally, unlike fossil fuels such as natural gas or coal that require millions of years of development time. The main sources are wind energy, solar energy and hydroelectric power including tidal and waves energy, which uses ocean and sea energy generated by tides and waves (Afsharian et al., 2020). On the other hand, during the Anthropocene, (e.g., Essefi, 2020; Essefi, 2021; Gharsalli et al., 2020), consumption of the traditional energy has caused atmospheric pollution by greenhouse gases (Lee & Lee, 2021). Energy consumption affects directly and indirectly air quality. Air pollution is the result of multiple factors including the direct effect of energy consumption growth and the indirect effect related to the development of mining, metallurgical and chemical industries, road and air traffic, the incineration of household waste, industrial and agricultural waste, etc. Environmental studies argued that, due industrial and agricultural human activities, the trajectory of the Anthropocene is toward a Great Acceleration of curves of all environmental parameters including greenhouse gases, pesticides, organic pollutants, heavy metals, nuclear by-products. Nonetheless, this trajectory may be modified if the production and the use of environmentally friendly energies increase at the expense of the traditional energies that are causing the setting of the Anthropocene conditions and threatening the humanity by apocalyptic scenario. This chapter is meant to follow the evolution of green energy use in agriculture and the potential repercussions on the trajectory of the Anthropocene.

GREEN ENERGY EVOLUTION DURING THE LAST THIRTY YEARS

As a matter of fact, industrial and agricultural uses of green are interrelated. Because some industrial uses may be oriented towards agricultural purposes. This used is for environmental goals by emission green house gases or for economic purposes by reducing coast of energy consumption. Several studies focused on the agricultural use as such to reduce CO_2 emission (Xu et al., 2020). Other proved that, in Controlled Environment Agriculture (CEA), green energy can reduce consumption of electricity by 75% (Engler & Krarti, 2021). As it is shown in Fig.1, the use of green (renewable) energy is marked by an obvious increase during the last thirty years to reach more than one tenth of the total energy. In terms countries Germany recorded highest values of green energy sharing (Kaiser et al., 2020). From 1990 to 2005, the increase started timid with a slight modification from 7% to 8%. From 2005 onward, a dramatic increase is recorded. This increase is related to a global awareness (Skordoulis et al., 2020) of the utility of the green energy in terms of economic efficiency (Gao et al., 2020) and environmental protection (Musibau et al., 2020). The six most common forms are solar power, wind power, hydropower, geothermal energy, biomass and biofuels. According to the equation (Fig.1), renewable, as it is considered clean energy, is candidate to be more produced and consumed during the few decades of the Anthropocene. Such increase will reduce the emission of greenhouse gases and polluting products resulting from the production and use of fossil energies.

Figure 1. Evolution of the green energy during the last thirty years

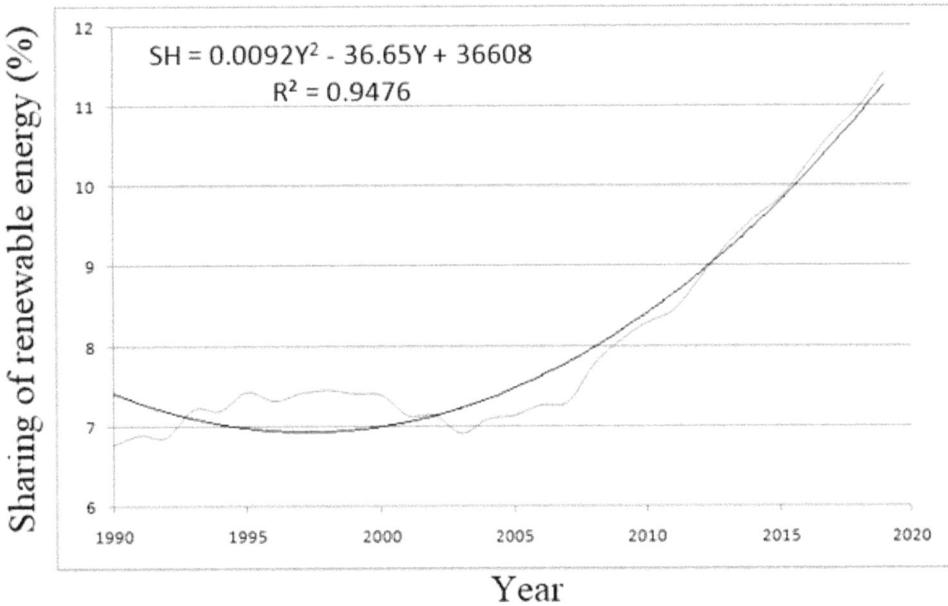

Solar Energy

This common renewable green energy source is usually made of photovoltaic cells (Teodor and Bruce, 2020), which capture sunlight and convert it into electrical energy. In agricultural regions, where there is a luck of domestic services, it is used for heating and hot water supply of buildings (Valančius & Mikučionienė, 2020; Wei &Skye, 2021), as well as for cooking (Singh, 2021) and lighting. Solar energy has become quite cheap and can be used for home uses (Alfaverh et al., 2020; Mondal & Banerjee, 2020; ALiero et al., 2021), including garden lighting, although it is also used on a large scale to power entire communities. In terms of agricultural use of the solar energy as such, several studies have proven the performance of photovoltaic power plants used to supply irrigation systems and the efficiency of photovoltaic systems in giving electricity at lower prices (Hoarcă et al., 2020). As a substitution of the fossil fuel, solar energy was proposed as a clean alternative for machine functioning, greenhouse heating and irrigation pumps (Aroonsrimorakot et al., 2020).

Wind Power

It is particularly suitable for offshore and alpine systems (Vuichard et al., 2020; Pickering et al., 2020). Wind power uses the energy of air currents around the world (Bessette & Mills, 2021). Wind energy production is related to variability in topography and seasonal to daily meteorological variability (Pickering et al., 2020). Combined use of solar and wind energy in agriculture allows to reach energy storage (Palys et al., 2021). In spite of higher production of wind energies, both wind and solar energies show an exponential increase of their production and consumption (Fig.2). In regards of their exponential evolution (Fig.2), these energies are of great potentialities in the near future, decelerating hence the abrupt great acceleration taking place during the last period of the Anthropocene.

Figure 2. Parallel evolution of solar and wind energies during the last thirty years

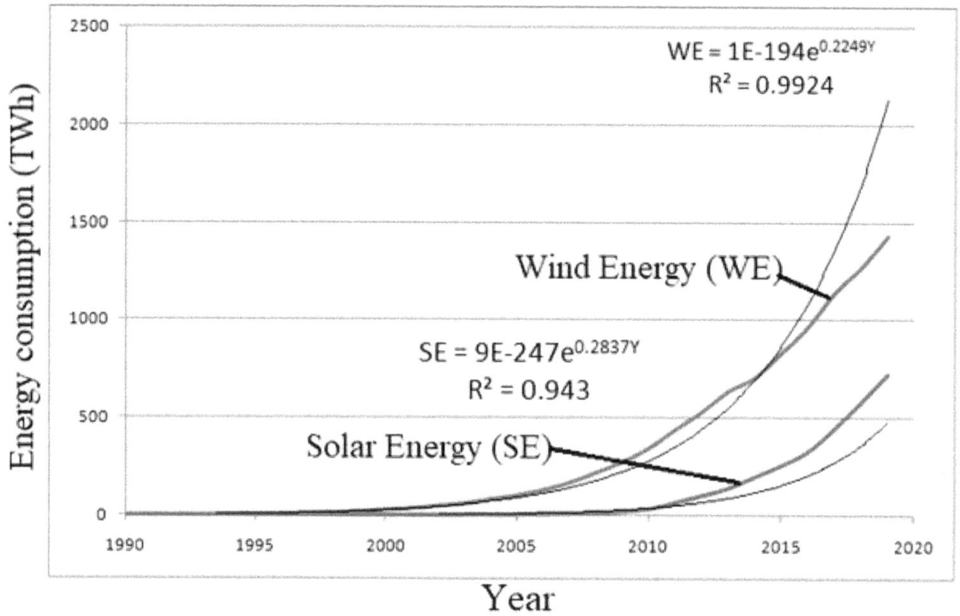

Hydroelectric Power Plant

This kind of green energy is also called hydroelectric power plant (Kumar et al., 2020; Güven & Aydemir, 2020). It uses water from rivers, streams, dams (Santos et al., 2020) or other hydrographical networks to generate energy. Hydroelectric power can also be operated on a small scale through water flow through pipes in houses (Suman et al., 2020), or it can be produced by evaporation in the ocean, rainfall or tides (Thomas, 2020). As a renewable energy, hydroelectric production does have environmental and ecological repercussions threatening the downstream of the drainage basin of hydrographical network (Kuriqi et al., 2021). Taking into account their possible effect of the environment, this recently flourishing energy (Fig.3) does have an economic and environmental interests related to their green behaviour and low coast. Had they been well exploited, they would reshape the trajectory of the Anthropocene.

Geothermal Energy

The thermal energy used by geothermal, a green energy source, is stored under the earth's crust (Davídková et al., 2020). Although drilling will require drilling to access the resource, which casts doubt on its resource environment and imposes a carefully control drilling operations to be fully environmentally friendly (Mburu et al., 2021), it is used to be a huge resource. Geothermal energy has been bathed in hot springs for thousands of years, and the same resources can be used to generate steam to turn turbines and generate electricity (Zhang et al., 2020). During the last, we noticed a linear evolution related to an increasing consumption of geothermal energy (Fig.4), which is motivated by environmental and economic consideration. Using the thermal energy encompassed within the internal layers of Earth, geothermal energy is undoubtedly one the promising green energies.

Figure 3. Hydropower consumption evolution during the last thirty year

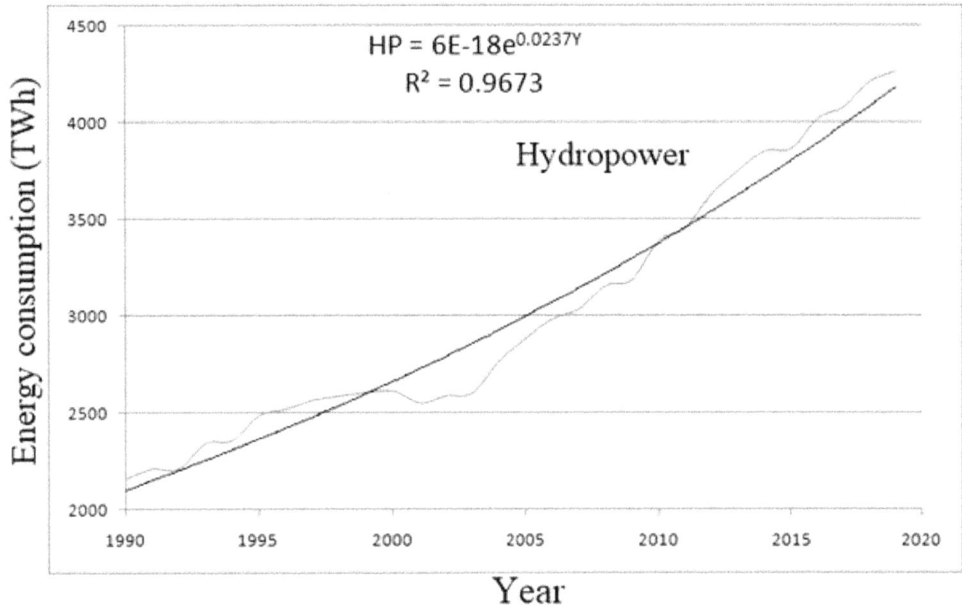

Figure 4. Evolution of the geothermal capacity during the last thirty years

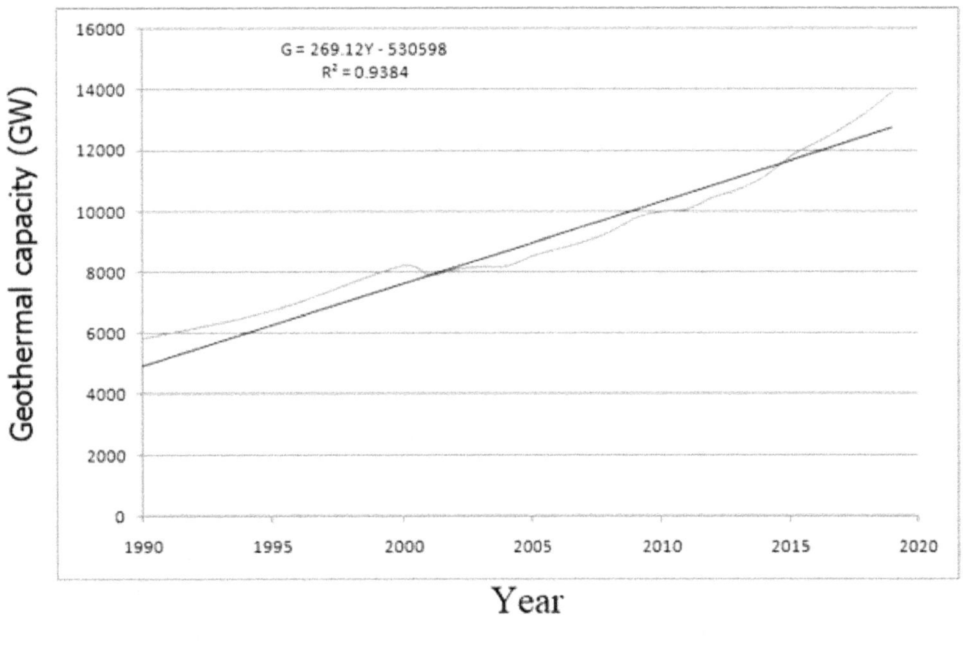

Biomass and Biofuel

These renewable resources must also be carefully managed in order to be truly labeled as "green energy." Biomass power plants use combustible wood waste (Arun et al., 2020), sawdust and organic agricultural waste to generate electricity (Honorato-Salazar & Sadhukhan, 2020). Lower than petroleum-based fuels. Biofuels do not burn biomass as mentioned above, but use this fuel Materials can be converted into fuels such as ethanol and biodiesel (Kumar et al., 2020). As it is shown in Fig.5, biofuel consumption is exponentially increasing and it is estimated that by 2050, biofuels will account for 25% of the world's transportation fuels. Taking into account its environmental effects that should be taken with serious caution, biofuel production and consumption are candidate to replace the fossil fuels use.

Figure 5. Evolution of biofuel consumption during the last thirty years

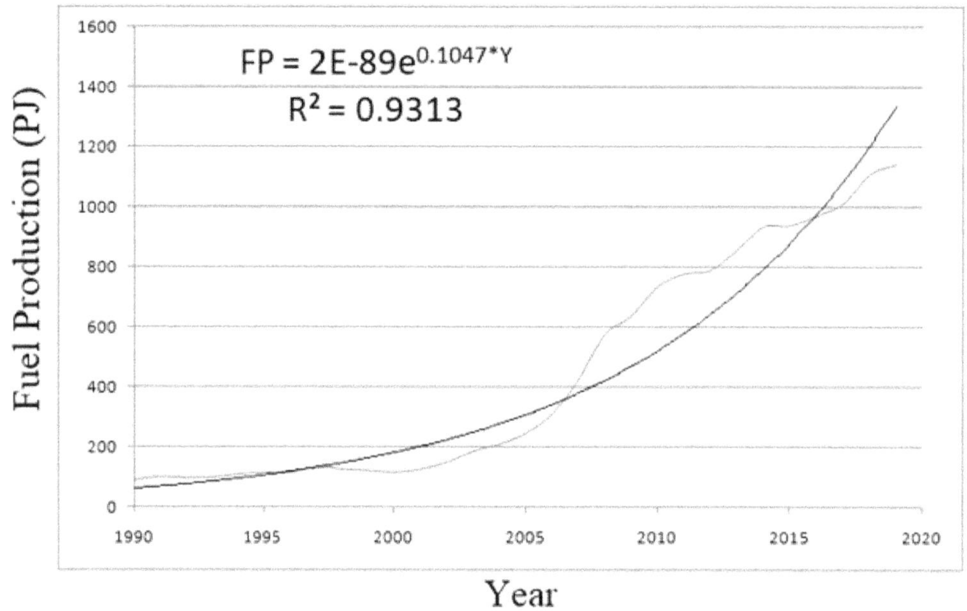

NEXT TO ABSENT GREENHOUSE GASES EMISSION

In spite of the use of logarithmic scale (Fig. 6), green energy shows no greenhouse gases comparably to traditional fossil energies. Even if combined with the emissions in the "life cycle" of clean energy (emissions at each stage of the technology life cycle: production, installation, operation, and scrap), the global warming emissions associated with renewable energy are nominal (Prasad et al., 2020). Renewable electricity generated by biomass power generation can cause various global warming emissions, depending on the resource and its sustainable supply. Increasing the supply of renewable energy would allow replacing carbon-intensive energy sources and significantly reduce world global warming emissions.

Figure 6. Carbon dioxide emission for wind, solar, hydroelectric, biogas, solar thermal, biomass, geothermal, diesel, heavy oil, coal with scrubbing, coal without scrubbing, natural gas, nuclear and fuel cell crude data were taken from Ftehenakis et al. (2008)

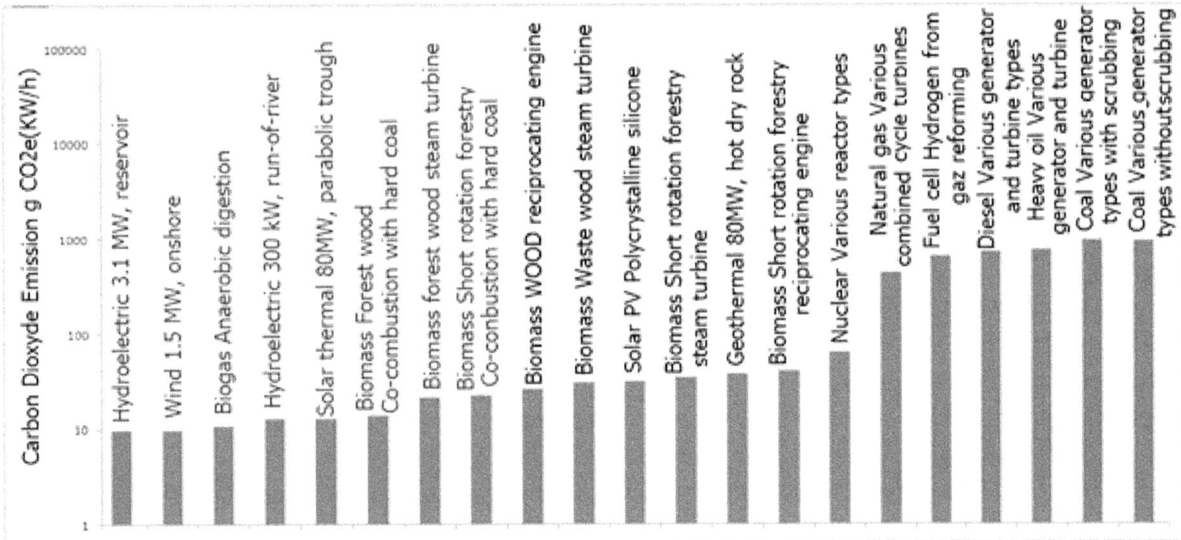

GREEN ENERGY ERA

The clustering of world green energy data during the last thirty years (Fig.7) shows the individualisation of the green era along the years 2012 to 2019. During the last three decades, energy green and consumption had started timid between 1990 and 2003. From 2004 and 2012, a transition period had taken place with an increase. From 2012 onward, a dramatic increase is still currently running to the reach the near and far future related to the flourishing of green energy production and consumption. It is indeed the green energy era that would modify the trajectory of the Anthropocene.

Figure 7. Clustering of green energy data during the last thirty years: the green energy era

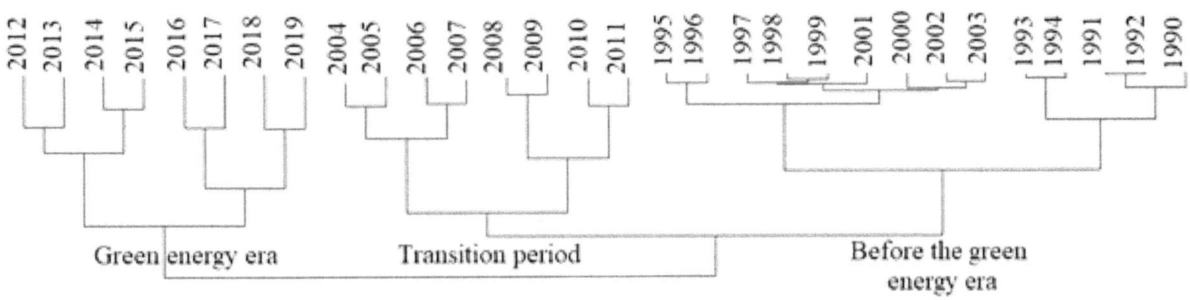

MOTIVATING FACTORS OF GREEN ENERGY SECTOR FLOURISHING DURING THE ANTHROPOCENE

Green energy from natural resources is generally renewable and clean, which means that it emits almost no greenhouse gas, and is generally easy to obtain. Even taking into account the entire life cycle of green energy, they emit much less greenhouse gases than fossil fuels, and they emit very little or only a small amount of air pollutants (Miśkiewicz, 2021). This is not only good for the Earth, but also good for the health of people and animals who need to breathe air. Green energy can also lead to stable energy prices (Anser et al., 2020) because these energy sources are usually produced locally and are not severely affected by geopolitical crises, price spikes or supply chain disruptions that may affect the global economy (Dagoumas et al., 2020). In addition, green jobs have been worldwide created (Dell'Anna et al., 2020). In 2018, the renewable energy sector created 11 million jobs, and this number will grow as we strive to achieve goals such as net zero emissions. Similar to solar and wind energy, energy infrastructure is more flexible and less dependent on central resources, which may lead to operational interruptions and poor adaptability to climate-induced climate change. Green energy also provides a cost-effective solution that can meet the energy needs of many parts of the world. This will only improve if costs continue to fall and the availability of green energy (especially in developing countries) continues to increase. Hydrogen is currently the main supplier of renewable energy in the cement, steel and chemical industries. Green energy can replace fossil fuels in the future, but this may require various means of production. For example, when the resources are easy to use, geothermal energy is particularly effective, while wind or solar energy can be. However, combining a variety of green energy sources to meet our needs and the progress in the production and development of these resources, we have reason to believe that fossil fuels may be eliminated. It will take several years to achieve this goal, but the focus is on the need to reduce climate change, improve the environment and work towards a more sustainable future.

Whatever the environmental (reducing pollution), social (providing with green jobs) political (decreasing conflict and wars upon fossil energy) repercussion, only economic benefits could motivate counties and individuals to invest in green energy. In doing so, using green energy will not be a choice between other better choices. Instead it becomes the best and the only choice to make. To understand the economics of green energy, you must compare it with fossil fuels (Kuzemko et at., 2020). The fact is that as readily available fossil resources deplete, the cost of this energy will only increase as its scarcity increases. Fuel is becoming more and more expensive, and the cost of more environmentally friendly energy is falling. Green energy, such as the ability to generate relatively inexpensive local energy solutions (such as solar parks). As we continue to accumulate knowledge and benefit from past achievements, interest, investment and development of green energy solutions have reduced costs. As a result, green energy can not only be economically viable, but can also become the first choice. Green energy efficiency varies by location: Under appropriate conditions, such as intense and frequent sunlight, it is easy to create fast and effective energy solutions. However, in order to truly be able to compare different types of energy, the entire life cycle of energy must be analyzed. This includes energy assessment. It is used to create green energy, calculate the amount of energy that can be converted into electrical energy, and any environmental remedial measures needed to create energy solutions. Environmental damage can prevent the source from truly becoming "green", but if all these factors together lead to the so-called present value of energy. For example, wind farms are considered the most effective green energy source because they require less processing and recycling than solar panels. Technology and testing have helped extend the life of wind turbines, thereby extending the life of present value of energy. However, the same applies to solar modules, which are also being vigorously developed. Green energy solutions also have the advantage of not requiring a large amount of additional energy costs. After they build, they tend to use easily renewable energy sources, such as wind. In fact, the total net energy efficiency of coal is only 29% of its original energy value, while wind energy provides high values of its original energy return. The current low efficiency ratings for them are as follows (although they may change as development progresses): How will solar energy help the environment? Green energy brings tangible benefits to the environment, because energy comes from natural resources such as sunlight, wind and water. These constantly replenished energy sources are the complete opposite of carbon emissions and volatile fossil fuels. More than a century. Generating carbon dioxide-free energy is a big step towards a cleaner future. If we can use it to meet our energy, industrial and transportation needs, then we can greatly reduce our environmental impact. What is the difference between green energy and clean energy and renewable energy? As mentioned above, there is a difference between green, clean and renewable energy. For people who often use these terms interchangeably, this is a bit confusing. A resource can be used at the same time, it can also be renewable, but it cannot be green or clean (for example) (Containing some form of biomass energy) Green energy refers to energy derived from natural resources (such as the sun). Clean energy is energy that does not emit pollutants into the air. Renewable energy comes from continuously replenished resources, such as: B. Water power, wind power or solar energy. Renewable energy is usually regarded as the same thing, but this is still controversial. For example, a hydroelectric power plant Are changes in waterways and environmental impacts beneficial to the environment? However, resources like wind energy are renewable, environmentally friendly and environmentally friendly because it comes from environmentally friendly resources that can repair themselves and will not pollute the environment. As part of the future world, it provides cleaner alternatives to many of today's energy sources. These fast renewable energy sources are not only good for the environment, but also create jobs and seem to be economically viable at any time. Development continues. The fact is that fossil fuels need to be turned into pasta, and they cannot provide a sustainable solution to our energy needs. By developing various clean energy solutions, we can create a completely sustainable future for energy supply without harming the world we live in.

CONCLUSION

Green energy has become an ultimate reality at the international level. Its use is motivated by numerous reasons:

1. Green energy is marked by weak greenhouse gases emissions and polluting products.
2. Green jobs are recently the fashion and millions of the planet inhabitants enjoy the availability of these new kinds of jobs.
3. Green energy is really renewable; the concept of reserves is indeed absent, because reserves of green energy are endless.
4. Contrary to fossil energies which are localized in some sites of the planet, motivating hence wars and political conflicts, green energy as a whole is generally equally divided upon countries allowing hence equal worldwide sustainable development.
5. Regarding the economics of green energy, the coast of green energy is lower in the long term perspectives.

Regarding these advantages, green energy will be the best choice in the coming few decades. In doing so, it will revamp all the ecological, environmental and economic equations. For instance, the Anthropocene which is basically caused by the intensive use of fossil energy may change its trajectory. Currently, the trajectory of the Anthropocene is toward the Great Acceleration. Under the current conditions all climatic and environmental parameters tend toward an apocalyptic scenario. Only an increasing of green energy use would modify this trajectory toward preindustrial conditions. It is worth noting that the green energy use had already started since the turning period (2004-2011), but this increase in not enough to revamp conditions that were already set hundreds years ago. The situation needs more time and intensive use of green energy.

REFERENCES

Afsharian, S., & Shiea, B. A. M. (2020). A Review of Tidal and Wave Energy in Southern Waters of Iran. *Journal of Marine Science*, *2*(3). doi:10.30564/jms.v2i3.1761

Alfaverh, F., Denai, M., & Sun, Y. (2020). Demand response strategy based on reinforcement learning and fuzzy reasoning for home energy management. *IEEE Access: Practical Innovations, Open Solutions*, *8*, 39310–39321. doi:10.1109/ACCESS.2020.2974286

Aliero, M. S., Qureshi, K. N., Pasha, M. F., & Jeon, G. (2021). Smart Home Energy Management Systems in Internet of Things networks for green cities demands and services. *Environmental Technology & Innovation*, 101443.

Almsatar, T. (2020). Environmental issues of biomass-burning in sub-Saharan African countries. In *Biomass Burning in Sub-Saharan Africa* (pp. 1–14). Springer. doi:10.1007/978-94-007-0808-2_1

Anser, M. K., Yousaf, Z., Zaman, K., Nassani, A. A., Alotaibi, S. M., Jambari, H., Khan, A., & Kabbani, A. (2020). Determination of resource curse hypothesis in mediation of financial development and clean energy sources: Go-for-green resource policies. *Resources Policy*, *66*, 101640. doi:10.1016/j.resourpol.2020.101640

Aroonsrimorakot, S., Laiphrakpam, M., & Paisantanakij, W. (2020). Solar panel energy technology for sustainable agriculture farming: A review. *Agricultural Technology (Thailand)*, *16*(3), 553–562.

Arun, J., Gopinath, K. P., Sivaramakrishnan, R., SundarRajan, P., Malolan, R., & Pugazhendhi, A. (2020). Technical insights into the production of green fuel from CO2 sequestered algal biomass: A conceptual review on green energy. *The Science of the Total Environment*, 142636. PMID:33065504

Bessette, D. L., & Mills, S. B. (2021). Farmers vs. lakers: Agriculture, amenity, and community in predicting opposition to United States wind energy development. *Energy Research & Social Science*, *72*, 101873. doi:10.1016/j.erss.2020.101873

Chowdhury, M. S., Rahman, K. S., Selvanathan, V., Nuthammachot, N., Suklueng, M., Mostafaeipour, A., ... Techato, K. (2020). Current trends and prospects of tidal energy technology. *Environment, Development and Sustainability*, 1–16. PMID:33041645

Dagoumas, A. S., Polemis, M. L., & Soursou, S. E. (2020). Revisiting the impact of energy prices on economic growth: Lessons learned from the European Union. *Economic Analysis and Policy*, *66*, 85–95. doi:10.1016/j.eap.2020.02.013

Davídková, V. (2020, December). Geothermal Potential of Mining Water in the Bílina Mines. *IOP Conference Series. Earth and Environmental Science*, *609*(1), 012063. doi:10.1088/1755-1315/609/1/012063

Dell'Anna, F. (2021). Green jobs and energy efficiency as strategies for economic growth and the reduction of environmental impacts. *Energy Policy*, *149*, 112031. doi:10.1016/j.enpol.2020.112031

Dixit, S. (2020). Solar technologies and their implementations: A review. *Materials Today: Proceedings*, *28*, 2137–2148. doi:10.1016/j.matpr.2020.04.134

Engler, N., & Krarti, M. (2021). Review of energy efficiency in controlled environment agriculture. *Renewable & Sustainable Energy Reviews*, *141*, 110786. doi:10.1016/j.rser.2021.110786

Essefi, E. (2020). On the Geochemistry and Mineralogy of the Anthropocene. *Int J Water Wastewater Treat*, *6*(2).

Essefi, E. (2021). Record of the Anthropocene-Great Acceleration along a core from the coast of Sfax, southeastern Tunisia. Turkish Journal of Earth Sciences Fthenakis, V. M., Kim, H. C., & Alsema, E. (2008). Emissions from photovoltaic life cycles. *Environmental Science & Technology*, *42*(6), 2168–2174.

Gao, J. (2020). Green Energy Strategic Management for Service of Quality Composition in the Internet of Things Environment. *Complexity*, *2020*, 2020. doi:10.1155/2020/6678612

Gharsalli, N., Essefi, E., Baydoun, R., & Yaich, C. (2020). The Anthropocene and great acceleration as controversial epoch of human-induced activities: case study of the Halk El Menjel Wetland, Eastern Tunisia. *Applied Ecology and Environmental Research*, *18*(3), 4137–4166. doi:10.15666/aeer/1803_41374166

Güven, A., & Aydemir, A. (2020). Dams. In *Risk Assessment of Dams* (pp. 1–14). Springer. doi:10.1007/978-3-030-47139-2_1

Hoarcă, I. C., Enescu, F. M., & Bizon, N. (2020). Comparative study regarding the integration of photovoltaic sources in agriculture. In *2020 12th International Conference on Electronics, Computers and Artificial Intelligence (ECAI)* (pp. 1-6). IEEE. 10.1109/ECAI50035.2020.9223176

Honorato-Salazar, J. A., & Sadhukhan, J. (2020). Annual biomass variation of agriculture crops and forestry residues, and seasonality of crop residues for energy production in Mexico. *Food and Bioproducts Processing*, *119*, 1–19. doi:10.1016/j.fbp.2019.10.005

Hübler, C., & Rolfes, R. (2021). Analysis of the influence of climate change on the fatigue lifetime of offshore wind turbines using imprecise probabilities. *Wind Energy (Chichester, England)*, *24*(3), 275–289. doi:10.1002/we.2572

Javed, M. S., Ma, T., Jurasz, J., & Amin, M. Y. (2020). Solar and wind power generation systems with pumped hydro storage: Review and future perspectives. *Renewable Energy*, *148*, 176–192. doi:10.1016/j.renene.2019.11.157

Kaiser, M., Bernauer, M., Sunstein, C. R., & Reisch, L. A. (2020). The power of green defaults: The impact of regional variation of opt-out tariffs on green energy demand in Germany. *Ecological Economics*, *174*, 106685. doi:10.1016/j.ecolecon.2020.106685

Khan, S. A. R., Yu, Z., Belhadi, A., & Mardani, A. (2020). Investigating the effects of renewable energy on international trade and environmental quality. *Journal of Environmental Management*, *272*, 111089. doi:10.1016/j.jenvman.2020.111089 PMID:32854892

Khoie, R., Bose, A., & Saltsman, J. (2020). The carbon emissions of wind power; a study of emissions of windmill in the panhandle of Texas. In *Conference proceedings of SOLAR* (pp. 24-25). 10.18086olar.2020.01.14

Kumar, K. A., Saket, R. K., Rajendran, R., & Eslamian, S. (2020). Design and development of green energy conversion system using waste water and organic kitchen wastes. *International Journal of Hydrology Science and Technology*, *10*(4), 315–333. doi:10.1504/IJHST.2020.108292

Kumar, N. (Ed.). (2020). *Biotechnology for Biofuels: A Sustainable Green Energy Solution*. Springer Nature. doi:10.1007/978-981-15-3761-5

Kuriqi, A., Pinheiro, A. N., Sordo-Ward, A., Bejarano, M. D., & Garrote, L. (2021). Ecological impacts of run-of-river hydropower plants—Current status and future prospects on the brink of energy transition. *Renewable & Sustainable Energy Reviews*, *142*, 110833. doi:10.1016/j.rser.2021.110833

Kuzemko, C., Bradshaw, M., Bridge, G., Goldthau, A., Jewell, J., Overland, I., & Westphal, K. (2020). Covid-19 and the politics of sustainable energy transitions. *Energy Research & Social Science*, *68*, 101685. doi:10.1016/j.erss.2020.101685 PMID:32839704

Lee, G. E., & Lee, J. H. (2021). Spatial correlation analysis using the indicators of the anthropocene focusing on atmospheric pollution: A case study of Seoul. *Ecological Indicators*, *125*, 107535. doi:10.1016/j.ecolind.2021.107535

Li, X., Zhao, L., Yu, J., Liu, X., Zhang, X., Liu, H., & Zhou, W. (2020). Water splitting: From electrode to green energy system. *Nano-Micro Letters*, *12*(1), 1–29. doi:10.100740820-020-00469-3 PMID:34138146

Mburu, S. M., Wakiru, J. M., Muchiri, P. N., & Pintelon, L. (2021). Maintenance Impact on Geothermal Drilling Operations. A Case Study Approach. *American Journal of Industrial and Business Management*, *11*(2), 183–201. doi:10.4236/ajibm.2021.112012

Miśkiewicz, R. (2021). The Impact of Innovation and Information Technology on Greenhouse Gas Emissions: A Case of the Visegrád Countries. *Journal of Risk and Financial Management*, *14*(2), 59. doi:10.3390/jrfm14020059

Mondal, S. S., & Banerjee, W. (2020). Design of Green Energy Based Charge Controller for Domestic Uses. In *Energy Systems, Drives and Automations* (pp. 173–185). Springer. doi:10.1007/978-981-15-5089-8_16

Musibau, H. O., Nepal, R., Vespignani, J., & Yanotti, M. (2020). The moderating role of green energy and energy-innovation in environmental Kuznets: Insights from quantile-quantile analysis. *Globalization and Monetary Policy Institute Working Paper*, (385).

Palys, M. J., Wang, H., Zhang, Q., & Daoutidis, P. (2021). Renewable ammonia for sustainable energy and agriculture: Vision and systems engineering opportunities. *Current Opinion in Chemical Engineering*, *31*, 100667. doi:10.1016/j.coche.2020.100667

Pamungkas, A., & Jumnahdi, M. (2020, November). The Study of Tidal Currents as Renewable Energy in Kelabat Bay, Bangka Belitung. *IOP Conference Series. Earth and Environmental Science*, *599*(1), 012001. doi:10.1088/1755-1315/599/1/012001

Pickering, B., Grams, C. M., & Pfenninger, S. (2020). Sub-national variability of wind power generation in complex terrain and its correlation with large-scale meteorology. *Environmental Research Letters*, *15*(4), 044025. doi:10.1088/1748-9326/ab70bd

Prasad, S., Singh, A., Korres, N. E., Rathore, D., Sevda, S., & Pant, D. (2020). Sustainable utilization of crop residues for energy generation: A life cycle assessment (LCA) perspective. *Bioresource Technology*, *303*, 122964. doi:10.1016/j.biortech.2020.122964 PMID:32061494

Santos, R. E., Pinto-Coelho, R. M., Drumond, M. A., Fonseca, R., & Zanchi, F. B. (2020). Damming Amazon Rivers: Environmental impacts of hydroelectric dams on Brazil's Madeira River according to local fishers' perception. *Ambio*, *49*(10), 1612–1628. doi:10.100713280-020-01316-w PMID:31994028

Shaterabadi, M., Jirdehi, M. A., Amiri, N., & Omidi, S. (2020). Enhancement the economical and environmental aspects of plus-zero energy buildings integrated with INVELOX turbines. *Renewable Energy*, *153*, 1355–1367. doi:10.1016/j.renene.2020.02.089

Singh, O. K. (2021). Development of a solar cooking system suitable for indoor cooking and its exergy and enviroeconomic analyses. *Solar Energy*, *217*, 223–234. doi:10.1016/j.solener.2021.02.007

Skordoulis, M., Ntanos, S., & Arabatzis, G. (2020). Socioeconomic evaluation of green energy investments. *International Journal of Energy Sector Management*.

Suman, R., Javaid, M., Nandan, D., Bahl, S., & Haleem, A. (2020). Electricity Generation Through Water Supply Pipes in High Rise Buildings. *Journal of Industrial Integration and Management*, 1-20.

Teodor, A. H., & Bruce, B. D. (2020). Putting photosystem I to work: Truly green energy. *Trends in Biotechnology*, *38*(12), 1329–1342. doi:10.1016/j.tibtech.2020.04.004 PMID:32448469

Thomas, O. E. (2021). Mathematical Modelling and Simulation of Tidal Energy. *International Journal of Engineering Science and Application*, *5*(1), 38–46.

Tsang, C. H. A., Huang, H., Xuan, J., Wang, H., & Leung, D. Y. C. (2020). Graphene materials in green energy applications: Recent development and future perspective. *Renewable & Sustainable Energy Reviews*, *120*, 109656. doi:10.1016/j.rser.2019.109656

Valančius, K., & Mikučionienė, R. (2020). Solar energy as a tool of renovating soviet-type multi apartment buildings. *Solar Energy*, *198*, 93–100. doi:10.1016/j.solener.2020.01.046

Voigt, C. C. (2021). Insect fatalities at wind turbines as biodiversity sinks. *Conservation Science and Practice*, *3*(5), 366. doi:10.1111/csp2.366

Vuichard, P., Stauch, A., & Wüstenhagen, R. (2020). Keep it local and low-key: Social acceptance of alpine solar power projects. *Renewable & Sustainable Energy Reviews*, 110516.

Wei, W., & Skye, H. M. (2021). Residential net-zero energy buildings: Review and perspective. *Renewable & Sustainable Energy Reviews*, *142*, 110859. doi:10.1016/j.rser.2021.110859 PMID:34413697

Xu, B., Chen, W., Zhang, G., Wang, J., Ping, W., Luo, L., & Chen, J. (2020). How to achieve green growth in China's agricultural sector. *Journal of Cleaner Production*, *271*, 122770. doi:10.1016/j.jclepro.2020.122770

Zhang, H., Meng, W., & Zhao, F. (2020, April). Study on the exploitation, utilization and cooperation of geothermal energy between Iceland and China. *IOP Conference Series. Earth and Environmental Science*, *480*(1), 012010. doi:10.1088/1755-1315/480/1/012010

Chapter 2
Effect of Climate Change on Crop Productivity:
Current Status and Steps Towards Fulfilling the Yield Gaps

Sundeep Kumar
ICAR-NBPGR, India

Ruchi Bansal
ICAR-NBPGR, India

Amit Kumar Singh
ICAR-NBPGR, India

Jyoti Kumari
ICAR-NBPGR, India

Sunish Kumar Sehgal
Dakota State University, USA

ABSTRACT

Climatic conditions have been changed drastically, which has become a big challenge to sustain agricultural production. Stagnation in production has been observed in the recent past due to different types of biotic and abiotic stresses. The present available varieties do not have resilience to climate change. To sustain the productivity and minimize the yield losses, the best way is to develop climate resilient cultivars, which can produce in abundance even under abrupt climate changes in a sustainable manner. For achieving this, a better understanding of crop response to the newly developed techniques is also required in order to satisfy the needs of the growing population. Further, improving the resilience of food production and minimizing risks against weather variability are essential for agriculture that can be achieved by broadening the genetic base of the varieties through introducing virgin germplasm as potential sources of genes for ensuring food security and sustainability in agriculture in developing countries and protecting natural resources.

DOI: 10.4018/978-1-7998-8434-7.ch002

INTRODUCTION

In the last three decades, we have experienced drastic climatic changes especially increased atmospheric temperature, enhanced level of CO_2, precipitation, high level of greenhouse gases, depicted ozone levels, UV-B radiation, drought, salinity and toxicity due to micro-nutrients etc., which have great impacts on world agricultural production (Houghton et al., 1990). Further, due to the occurrence of different abiotic and biotic stresses during crop season, we have experienced stagnation in production in past few years. Further, the ever increasing population has exerted immense pressure on agriculturists, farmers, and scientists to produce food crops in sufficient quantity. In order to meet the demands, the impact of the climate must be understood and integrated in future policy planning. Due to drastic changes in climatic conditions, we have to focus our attention on way to minimize yield losses, avoid wide fluctuations in production and sustain the higher level of productivity with limited resources. The present world cereal production as an estimate is about 2302.1 million tones which are vis-a-vis the current utilization as shown in Table 1 & Fig 1. This has created an alarming situation for planners and forced them to redesign their strategies.

Table 1. Year-wise statistics of production, supply, utilization, trade and ending stocks at the global level

World cereal market						
	2011/12	2012/13	2013/14	2014/15 estimate	2015/16 forecast	
					Previous (04 June 2015)	**Current (09 July 2015)**
	(............ *million tonnes* ....)					
Production[1]	2 349.6	2 299.8	2 524.4	2 553.6	2 524.3	**2 526.5**
Supply[2]	2 887.3	2 862.9	3 068.8	3 156.6	3 170.8	**3 171.5**
Utilization	2 313.9	2 325.8	2 433.9	2 503.9	2 524.7	**2 534.2**
Trade[3]	322.4	315.0	359.3	362.4	351.0	**358.1**
Ending Stocks[4]	563.1	544.4	603.0	645.0	634.3	**631.3**
	(............ *percent* ....)					
World stock-to-use ratio	24.2	22.4	24.1	25.5	24.7	**24.6**
Major exporters' stock-to-disappearance ratio[5]	18.1	16.9	17.3	17.8	16.5	**16.4**

Source: FAO, Rome, Italy.

The present global production of rice and wheat-the major contributor in the cereals production is about 486.9 and 662.0 m tones, respectively and to feed our growing population in 2020, we need to produce 650 m tones rice and 850 m tones wheat with annual growth rate 1.6 to 2.5per cent to the present production level, which seems difficult to achieve in a short span of time. Thus an alarming situation has arisen due to high frequency of erratic weather patterns indicating towards a climate change. To meet this challenge, agriculture scientists require developing appropriate technologies for the development of varieties, which can produce in abundance even under abrupt climate changes in a sustainable manner. It is also essential to understand the relationship between the crop and climate to develop newer technologies for augmenting the crop production in order to satisfy the needs of growing population.

Figure 1. Graphical presentation of cereal production, utilization and stocks

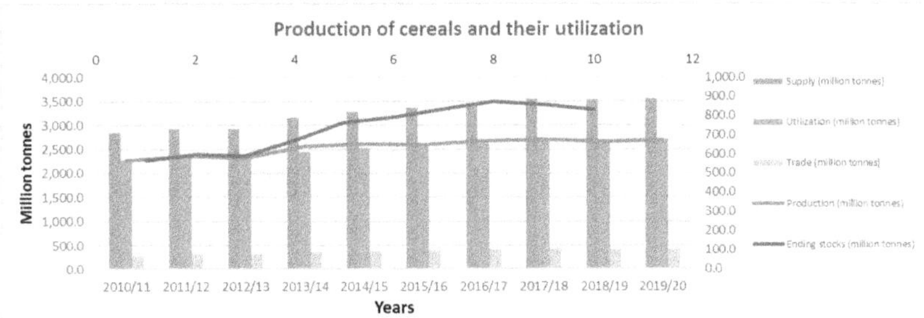

In the recent past, we have all witnessed the serious threats posed by the sudden weather fluctuations, in the form of heat stress, drought, salinity, toxicity due to micro-nutrients. Additionally evolution of new races of various diseases in almost all crops due to changing climatic conditions has taken a toll on the productivity of cereals depending on their adaptability to changed climate. It is therefore, imperative to look for tools not only to increase the crop productivity but also ensure protection against loss of potential productivity due to the environmental vagaries. During last few decades, considerable efforts have been made to improve yield performance through breeding, and to understand mechanisms involved in abiotic stress tolerance. Progress in breeding for enhanced heat, drought tolerance/resistance, and major biotic stresses is still essential for ensuring food security to the poor people in the developing world.

In Indian scenario, agriculture is highly dependent on monsoon rainfall (Kumar *et al.*, 2004). Analysis of the relationship between district level crop yield data and precipitation during 1960-2000 has shown that different regions show different responses to precipitation extremes (Asada and Matsumoto, 2009). Yield in upper Ganges basin was dependent on total precipitation and was sensitive to drought, while in lower Ganges basin, it was sensitive to flooding. There was variation among the districts due to social factors and irrigation techniques.

Rising atmospheric CO2 concentration and climate change may also affect the crop productivity through effects on pest and diseases. Pests such as aphid and weevil larvae responded positively to high CO2 concentration (Newman, 2004; Stanley and Johnson, 2008). Cheke and Tratalos (2007) observed that rainfall patterns influenced migration patterns of locust in sub-Saharan Africa. It was also predicted that changing climate can affect the resistance of crops to different diseases and abiotic stress conditions can induce the pathogenicity of the organisms by mutations (Gregory et al., 2009). Climatic variability may adversely affect the prediction and amplitude of outbreaks.

Plant genetic resources (PGR) of a crop is the sum total of genes in a crop species. PGR includes current cultivars, obsolete cultivars, parental lines, advanced breeding material, mapping populations and explored germplasm lines. Virgin germplasm is always considered as the raw material in any breeding program and a potential source of novel genes. Tropical countries have been very rich in bio-diversity due to presence of favorable environmental conditions, which favor all kinds of vegetation to grow. Germplasm lines are the potential source of genes leading to abiotic and biotic stress resistance. Due to fast-growing demands, breeders are looking forward for the significant genetic variation for biotic, abiotic and quality traits. The information on, genetic variation for the traits and their association with a suitable marker can be effectively used in marker-assisted breeding. Breeding strategies based on mega-environmentsare usually not so effective. Therefore, there is a need to assist the breeders to redefine their selection strategy for manipulating traits associated with complicated traits like heat & drought stress tolerance and incorporate resistance against prevailing biotic stresses at micro-environmental level to bring multiple resistances by integrating conventional breeding and non-conventional biotechnological approaches in the future research programmes. Further implementation of genome-wide selection strategies coupled with high throughput phenotypic technologies can help in developing selection models that could help in accelerating the genetic gain per breeding cycle leading to a rapid genetic improvement.

GLOBAL POVERTY AND HUMAN DEVELOPMENT

The global population is continuously increasing with the annual growth rate of 1.8%. As per the survey of UN Population Division and Population Reference Bureau, more than 80% of the world population is residing in developing countries while, they are sharing hardly 20% of the total natural resources during 2000. In 2050, as an estimate, more than 90% of the population would be living in developing countries and will share only 10% of the total natural resources (Fig. 2). This will be the major reason of growing poverty and starvation in the developing countries. Therefore, a collaborative efforts of developed and developing countries are required to increase the awareness, interest and research about these important issues.

Effect of Climate Change on Global Food Security

Due to a drastic change in climatic conditions, the food security of more than 40% world population is on stake. Some 800 million people still experience chronic and transitory hunger each year. Supplying of quality, healthy food in developing countries in Asia and Africa is a big challenge before the government systems in these countries. It is speculated that over the next 50 years, we face the challenge of feeding 3.5 billion additional people, most of whom will begin life in poverty. Hence, the battle to alleviate poverty and improve human health and productivity will require a huge investment and dynamic agricultural development. As an estimate, by 2050, wheat production must increase by 40% to meet the global demand (Pinstrup Andersen et al., 1999). The drastic change in climatic conditions has adversely affected the production of major cereal crops. In many parts of the world, farmers have adopted new models of cropping system which can sustain the production even under changing climatic conditions. If we take wheat as an example, the present wheat production and demand moves vis-à-vis however, this gap would be more widening during 2020 and 2050 (Fig.3).

Figure 2. Population and resource sharing between developed and developing countries after every fifty years

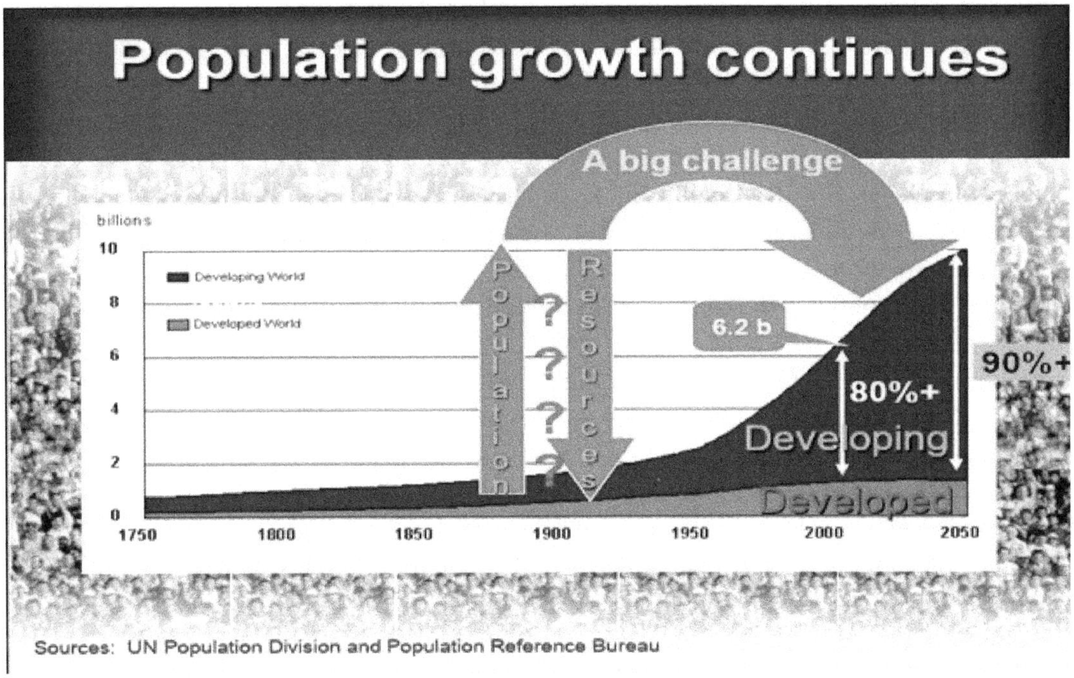

Figure 3. The gap between demands of wheat consumed at present consumption rate and at 20% more consumption in different years

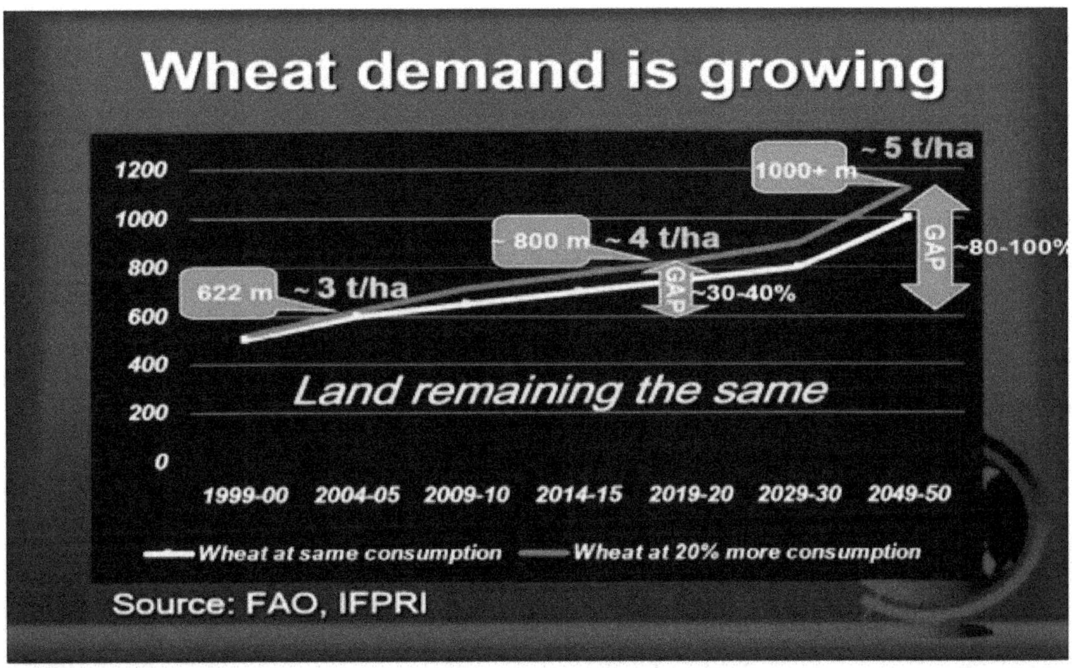

Steps Towards Fulfilling the Gaps

A Consultation 'Global climatic change and agricultural production: direct effects of hydrological and plant physiological processes' was held from 7 to 10 December 1993 at FAO Headquarters in Rome, with the support of the United Nations Environment Programme in Nairobi. The objectives of the Consultation were to analyze and assess the effects of higher temperatures and changing precipitation/evapotranspiration ratios on plant growth and food production, higher atmospheric carbon dioxide (CO_2) levels, higher ultraviolet (UV-B) radiation, higher near-surface ozone concentrations and to identify gaps in our present knowledge and to focus on critical research needs. Numbers of research institutes spread across the world have taken various steps to fill the gap between current production and growing demand for food grains. Some of the important steps towards sustainability are:

- Parental material having the potential of high production
- Novel source of resistance against various notorious diseases and abiotic stresses
- Screening of germplasm against various stresses including biotic and abiotic using conventional and non-conventional marker-assisted selection (MAS) approach
- Conservation of germplasm to be used as a source of gene(s) for future use
- Exchange of useful germplasm for its better utilization in breeding programs
- Conservation of biodiversity in the form of core/mini-core
- Development of transgenics which can produce in abundance under stress conditions
- Application of genomic selection (GS) strategies coupled with high-throughput phenotyping (HTP) to enhance the rate of genetic improvement

Expected Changes Due to the Climate in Coming Years

Following changes are expected due to abrupt climate changes:

- Due to abrupt changes in climatic conditions, agro-ecological zones would shift that will bring significant differences in crop rotation thus, would affect the society.
- Due to change in climatic conditions, food habit is also expected to changes according to those climatic conditions.
- Crops grown at high-latitude regions would improve because of increased photosynthesis, longer growing periods and extension of frost-free growing regions that provide optimum growth conditions.
- There will be crop failure at mass scale in developing countries that would result inriseofprices of agricultural products, more cereal import thus, their economy would depend upon developed countries.
- There can be a riskof complete eradication of some useful species in tropical and subtropical regions, and the greatest risk to food security.
- Due to shifting in climatic zones, the micro-environment of the soil may be unsuitable for the sustainable crop production. Therefore, new crop varieties will have to be developed.
- The confounding effect of greenhouse gases due to depletion of ozone layer, volcanoes, oceans and terrestrial ecosystems need to be predicted. Soil warming is a new source of atmospheric carbon (Oechel *et al.,* 1994).

- Deforestation, intensification of land use etc. are extremely important determinants of the terrestrial carbon fluxes into the atmosphere. At the same time, since elevated CO_2 concentrations can increase net primary biomass production, some enhancement of the terrestrial sink (i.e., more carbon storage in vegetation and soils), is to be expected, at least in the short term.
- Important parameters, such as cloudiness, atmospheric aerosols, ocean-atmosphere-terrestrial linkages, cost of technological and social adaptations, extreme climatic events, lateral water transfers, availability of water resources and soil nutrients, and multiple and inter-cropping systems are the other important factors.

Effect of Climate Change on Plant Physiology and Soil Health

Increase in the atmospheric temperature, elevated carbon dioxide concentration, high frequency of drought and cyclones and unpredicted rainfall conditions will significantly affect the agricultural productivity, farm income, and food security. Therefore, it is important to understand the changes in the physiology of plants with the onset of environmental changes. Different crop species have different responses to temperature, precipitation, CO_2 concentration, pollutants, ultraviolet radiation (UV-B). For example, cotton is highly responsive to elevated CO_2in comparison to wheat (Allen *et al.,*1987).

CO_2 is a key factor in photosynthesis and in plant growth. After diffusion into the plant through stomata, it is transformed by photosynthesis into carbohydrates. A large number of water molecules are lost by transpiration through the stomata for every CO_2 molecule entering the leaf. In a CO_2-rich environment, the larger concentration gradient forces more CO_2 into the plant, while partial closure of the stomata will reduce water losses from the leaf. Therefore reduced stomatal opening under high CO_2 environment can reduce water loss from the plant thus increasing water-use efficiency. For example, there are indications that doubling present CO_2 concentrations reduces stomatal conductance (opening) by 30-60% depending on the species (Fajer and Bazzaz, 1992). The reduction in water consumption is called the CO_2 anti-transpiration effect. The water-use efficiency (WUE) of the plant improves, since less water is used for equal or more CO_2transformed into dry matter. At the same time, net photosynthesis might increase, because photorespiration, which reduces carbon gain, is less at high CO_2 concentrations. In optimal conditions of light, moisture,and availability of nutrients, this fertilization effect could increase above- and below-ground biomass production by 10 to 40% depending on crop type and even to higher levels such as the case in cotton. Earlier and more rapid leaf production is expected, and the incremental increase in biomass can benefit the root system even more. The earlier establishment of ground cover, because of the early canopy development, may also limit water loss by direct soil evaporation. This response is even greater at high temperatures since the optimum temperature for photosynthesis increases with high atmospheric CO_2. In contrast, combined low temperatures and elevated CO_2 concentration could reduce plant growth.

On the other hand, independently of CO_2, the physiology of C_4 plants permits a generally higher photosynthetic capacity than in C_3 plants. This efficiency, however, is quickly saturated at increased CO_2 concentration. Therefore, in a CO_2-rich environment, the net improvement in photosynthesis of C_4 plants is proportionally small (about 10%, mostly in the stem), although the WUE might significantly improve (by about 40%). This category not only includes crops of major importance for food production, such as maize, sorghum, sugar cane and millet, but also tropical grasses, pasture, forage and some weed species that are critical for agricultural production. A 2°C warming in the mid-latitude could increase wheat production by nearly 10 percentwhereas at low latitude yield may reduce by the sameamount under similar conditions. So the change in climate will affect the different crop species differently (Easterling *et al.*, 2007). Global comparisons of CO_2 fertigation effects suggest that the strength of CO_2 fertigation effect will determine the resultant yield. If CO_2 fertigation is strong, North America and Europe may get benefitted. On the other hand India and Africa have been projected to face up to 5 percent yield losses despite strong CO_2 fertigation. Without CO_2 fertigation effect, all regions are projected to experience losses in productivity (Parry *et al.*, 2004; Nelson *et al.*, 2009). Projected effects of climate change on the yield of major cereals under CO_2 fertigation and without CO_2 fertigation effects are summarized in Table 2.

Impact of climate change induced high temperatures may be detrimental if they coincide with the key stages of plant growth and development. The high temperature at flowering stage may reduce the yield significantly (Wheeler *et al.*, 2000). Increase in temperature for short duration can affect enzymatic reactions and gene expression, while, in long-term, it may affect photosynthesis, growth and further yield. Drastic reduction in yield was recorded in groundnut when exposed to the temperature exceeding 42°C after flowering (Prasad *et al.*, 2003). At a temperature above 36°C, maize showed a reduction in pollen viability. Increase in temperature above 29°C for corn, 30° C for soybean and 32°C for cotton was found to reduce the yield significantly. In rice, grain sterility was observed above 35°C, while pollen viability was reduced significantly in wheat under terminal heat conditions.

Table 2. Projections of climate change induced yield effects from 2000 to 2050 by crop and management system under National Centre for Atmospheric Research, US model (NCAR) and Commonwealth Scientific and Industrial Research Organization, Australia (CSIRO) model

Region	Without CO_2 Fertigation		With CO_2 fertigation	
	CSIRO	NCAR	CSIRO	NCAR
Maize (Irrigated)				
Developing Countries	-2.0	-2.8	-1.4	-2.1
Developed Countries	-1.2	-8.7	-1.2	-8.6
Maize (Rainfed)				
Developing Countries	0.2	-2.9	2.6	-0.8
Developed Countries	0.6	-5.7	9.5	2.5
Rice (Irrigated)				
Developing Countries	-14.4	-18.5	2.4	-0.5
Developed Countries	-3.5	-5.5	10.5	9.0
Rice (Rainfed)				
Developing Countries	-1.3	-1.4	6.5	6.4
Developed Countries	17.3	10.3	23.4	17.8
Wheat (Irrigated)				
Developing Countries	-28.3	-34.3	-20.8	-27.2
Developed Countries	-5.7	-4.9	-1.3	-0.1
Wheat (Rainfed)				
Developing Countries	-1.4	-1.1	9.3	8.5
Developed Countries	3.1	2.4	9.7	9.5
			Source: Nelson *et al.*, 2009	

Effects on Soil Fertility

A sudden doubling of the atmospheric CO_2 concentration and associated higher temperatures - as in most experimental set-ups, both enclosed and under free-air enrichment may result in soil degradation including nutrient depletion. A potential increase in initial soil fertility under elevated atmospheric CO_2 can be expected if the increase in CO_2 occurs gradually, as in practice and as is the case in crop models that take into account the gradual transient increases. Additional no till practices and litter can likely to raise soil organic matter content unless litter chemistry drastically changes which would cause a decline in litter decomposition rate. However, the higher soil temperature can stimulate microbial respiration and the decomposition of the organic matter (i.e., mineralization) and cause the release of nutrients which become available for plant uptake through the root system unless microbial competition with plants for the available nutrients is intensified. Furthermore, since both the nutrient uptake efficiency and the structure (length and density) of root systems improve under elevated CO_2 concentrations, overall plant nutrient uptake can also increase. Furthermore, there are indications that the expanded root system can penetrate more deeply into the soil and reach extra sources of moisture and nutrients.

An additional driving force is an enhancement of the symbiotic association between root systems and the fungi and bacteria of rhizosphere under high CO_2 concentrations. If these concentrations do not greatly alter the existing gas composition in the rhizosphere, the colonization of the root system by mycorrhizae (fungi in symbiosis with roots, facilitating plant uptake of occluded phosphate) and nitrogen-fixing bacteria (bacteria with the ability to incorporate atmospheric nitrogen into nitrogenous compounds which can be utilized by living organisms) is likely to improve the nutrient uptake by the host plants. However, it is possible that this process will ultimately slow down if there are changes in litter chemistry, e.g., an increase in C/N ratios and an increase in tannins, occur. In such a case there will be a greater demand for the use of additional fertilizers.

CONCLUSION

There is tremendous effect of climate change on agriculture productivity due to shift in agro-climate zones, cropping system, soil micro-climate, high temperature, gaseous imbalance in environment particularly higher CO_2 concentration.The changing climatic conditions can have serious implications on agriculture and on natural ecosystems increasing pressure on farmers as well as scientists to develop climate-resilientsustainable varieties that can fulfill the requirements of the growing population. Whether or not climate change takes place, improving the resilience of food production and minimizing risks against weather variability are essential for agriculture is to meet the challenges of ensuring food security, raising rural employment in developing countries and protecting natural resources and the environment. This can prevent or mitigate the negative impacts of climate change at the same time. Integration crop livestock systems and abetter agronomic management adapted to variable conditions will enhance the sustainability of agro ecosystems and increase its resilience to face higher inter-annual weather variability and the more modest and gradual global climate change.

REFERENCES

Allen, L. H. Jr, Boote, K. J., Jones, J. W., Jones, P. H., Valle, R. R., Acock, B., Rogers, H. H., & Dahlman, R. C. (1987). Response of vegetation to rising carbon dioxide: Photosynthesis, biomass, and seed yield of soybean. *Global Biogeochemical Cycles*, *1*(1), 1–14. doi:10.1029/GB001i001p00001

Asada, H., & Matsumoto, J. (2009). Effects of rainfallvariation on rice production in the Ganges-Brahmaputra Basin. *Climate Research*, *38*, 249–260. doi:10.3354/cr00785

Cheke, R. A., & Tratalos, J. A. (2007). Migration, patchiness and population processes illustrated by two migrant pests. *Bioscience*, *57*(2), 145–154. doi:10.1641/B570209

Easterling, W. E., Aggarwal, P. K., Batima, P., Brander, K. M., Erda, L., Howden, S. M., Kirilenko, A., Morton, J., Soussana, J. F., Schmidhuber, J., & Tubiello, F. N. (2007). Food, fibre and forest products. Climate Change 2007: Impacts, adaptation and vulnerability. Contribution of working group II to the fourth assessment report of Intergovernmental Panel on Climate Change, 273-313.

Fajer, E. D., & Bazzaz, F. A. (1992). Is carbon dioxide a 'good' greenhouse gas? Effects of increasing carbon dioxide on ecological systems. *Global Environmental Change*, *2*(4), 301–310. doi:10.1016/0959-3780(92)90047-B

FAO Study-World Agriculture. (2015). *Towards 2015/2030 Summary Report*. www.fao.org

Gregory, P. J., Johnson, S. N., Newton, A. C., & Ingram, J. S. I. (2009). Integrating pests and pathogens into the climate change/food security debate. *Journal of Experimental Botany*, *60*, 2827–2838.

Houghton, J. T., Jenkins, G. J., & Ephramus, J. J. (Eds.). (1990). *Climate Change: The IPCC Scientific Assessment*. Cambridge University Press.

Kumar, K. K., Kumar, K. R., Ashrit, R. G., Deshpande, N. R., & Hansen, J. W. (2004). Climate impacts on Indian agriculture. *International Journal of Climatology*, *24*, 1375–1393.

Nelson, G. C., Rosegrant, M. W., Koo, J., Robertson, R., Sulser, T., Zhu, T., Ringler, C., & ... . (2009). *Climate change impact on agriculture and costs of adaptation*. International Food Policy Research Institute.

Newman, J. A. (2004). Climate change and cereal aphids: The relative effects of increasing CO2 and temperature on aphid population dynamics. *Global Change Biology*, *10*, 5–15.

Oechel, W. C., Cowles, S., Gurlke, N., Hastings, S. J., Lawrence, B., Prudhomme, T., Reichers, G., Strain, B., Tissue, D., & Vourlitis, G. L. (1994). Transient nature of CO_2 fertilization in Arctic tundra. *Nature*, *371*, 500–503.

Parry, M. L., Rosenzweig, C., Iglesias, A., Livermore, M., & Fischer, G. (2004). Effects of climate change on global food production under SRES emissions and socio-economic scenarios. *Global Environmental Change*, *14*, 53–67.

Prasad, P. V. V., Boote, K. J., Allen, L. H., & Thomas, J. M. G. (2003). Super-optimal temperatures are detrimental to peanut (*Arachis hypogea* L.) reproductive processes and yield at both ambient and elevated carbon dioxide. *Global Change Biology*, *9*, 1775–1787.

Stanley, J. T., & Johnson, S. N. (2008). Climate change impacts on root herbivores. In S. N. Johnson & P. J. Murray (Eds.), *Root Feeders: an ecosystem perspective*. CABI.

Wheeler, T. R., Craufurd, P. Q., Ellis, R. H., Porter, J. R., & Prasad, P. V. V. (2000). Temperature variability and the yield of annual crops. *Agriculture, Ecosystems & Environment*, *82*, 159–167.

Chapter 3
Use of Nano-Bio-Chemicals in Modern Agriculture to Accelerate Sustainable Growth

Garvita Singh
University of Delhi, New Delhi, India

Divya Singh
Banaras Hindu University, Varanasi, India

Jay Kumar
Banaras Hindu University, Varanasi, India

Ashok Kumar
Banaras Hindu University, Varanasi, India

ABSTRACT

Nanotechnology is an emerging field having diverse applications and potential in providing new strategies to many challenges agriculture. Nanotechnology in the agricultural sector focuses on sustainable food production. This nascent technology provides novel opportunities to enhance crop production and reduce environmental issues. Sustainable agriculture must be taken as an ecological method, where abiotic-biotic-living beings live together with coordinated stability of food chains and their related energy balances. The use of a network of sensors, a global positioning system (GPS), global information system, and actuators throughout an agricultural area could measure and report on a number of environmental, crop, and pest variables at a time. These reports would further support the farmers and allow choices for intervention during agricultural practices. By employing robust, portable, or remote in-situ nanotechnology-based sensing and monitoring systems in the agricultural sector, farmers can begin to market their own choices and will apply agrochemicals as and when required.

DOI: 10.4018/978-1-7998-8434-7.ch003

INTRODUCTION

Nanotechnology has emerged as one of the ''Key Enabling Technologies'' that contributes to sustainable growth in several fields of industrial applications and environmental benefits by European Commission (2012). The leading areas of nanotechnology include agriculture sector wherein continuous improvement with positive outputs is strongly needed. Technical innovation for agricultural production is of extreme importance, because of increasing global food security and climate change challenges and the limited availability of certain important plant nutrients such as phosphorus and potassium (Mukhopadhyay 2014; Yunlong & Smit 1994). Application of nanoparticles in agriculture in various forms is expected to reduce harmful plant protection tools, minimize nutrient losses, and may result in high yield due to increase in the soil fertility (Campbell et al. 2014; Paramo et al. 2020).

Nanotechnology involves the use of particles with one or more dimensions in the ranges of 100 nm or even less. These nanoparticles are quite promising in plant growth promotion and protection *vis-a-vis* management of eco-friendly agricultural practices because of their physical properties (Auffan et al. 2009; Ghormade et al. 2011). Various types of substrate with different properties are used for the synthesis of efficient nanoparticles such as metal oxides, ceramics, magnetic and ferromagnetic nanoparticles, lipids, quantum dots, polymers, dendrimers and emulsions (Puoci et al. 2008). Such nanoparticles with varied properties are widely used in different sectors as eco-friendly resources. In agriculture, they play a significant role in growth and metabolic activity of different plant species. Impact of different concentrations on important processes like seed germination, growth and reproduction in various economically important plant species has been tested (De et al. 2014; Zheng et al. 2005). It has been reported that the uptake efficiency and effects of nanoparticles on growth and metabolic functions vary among different plants. As such with a view to control pests and insects, different types of pesticides and insecticides available in the market are used in unplanned manner which result in the development of pest resistant species and killing of useful microbes leading to major changes in soil biodiversity. Additionally, biomagnification of such toxicants may also cause a sharp decline in bees, wasps and birds owing to the loss of their natural habitats (Ragaei & Sabry 2014; Tilman et al. 2002). In this scenario, application of nanotechnology seems quite beneficial and promising. Notably, the nano-formulation of pesticides and insecticides are not only non-toxic but also increases the agricultural yield. The nano-encapsulated fertilizers are well tested for their slow and specific release of nutrients. Nowadays, for precision agriculture nanoparticles are used as tools for transfer of genetic material in plant species for the development of better varieties showing resistance to insects and pests, a step towards agro-nanotechnology. Advances in nanotechnology have simplified the operation and detection of viruses and other pathogens at low cost. The detection limits of these new nanotools are very high and being used in diagnosis and treatment of various diseases (Cheng et al. 2009). Nanocapsules are not only used in the treatment of diseases but also for increasing nutrient absorption by plants. Furthermore, target-specific nanoparticles can overcome the problem of damage to non-target plant tissues and the amount of chemicals discharged into the environment (Paramo et al. 2020).

Unfortunately, in spite of all the above benefits, nanotechnology in the agricultural sector has not yet gained enough market value compared to other sectors where the usage of nanotechnology-based products is quite high. The potential of nanotechnology in agriculture is indeed enormous but certain issues such as increasing the scale of production and minimising the cost as well as risk assessment issues need utmost priority. To this effect, biologically-synthesized nanoparticles from biomolecules such as carbohydrates, proteins, amino acids and even waste biological materials can be very effective.

For instance, chitosan nanoparticles act as biopesticides against fungal infection from seedling stage. Similarly, non-toxic starch-based nanoparticles act as biostimulants and are well practiced in the agricultural sector. Advancement in nanomaterials has led to the fabrication of extremely sensitive sensors for varied applications including soil analysis, biochemical sensing, nutrient delivery and agriculture. Several other production processes of nanoparticles are being developed for a variety of applications using agricultural waste products as a source of renewable materials (raw materials) in order to replace fossil substrate. For example, a variety of useful thermoplastic nanocomposites with enhanced thermal properties are produced from agricultural wastes such as wheat straw, rice husk and soy hulls. An overview of the applications of nanotechnology in agriculture is presented in Figure 1.

Figure 1. An overview of the nanotechnology applications in sustainable agriculture

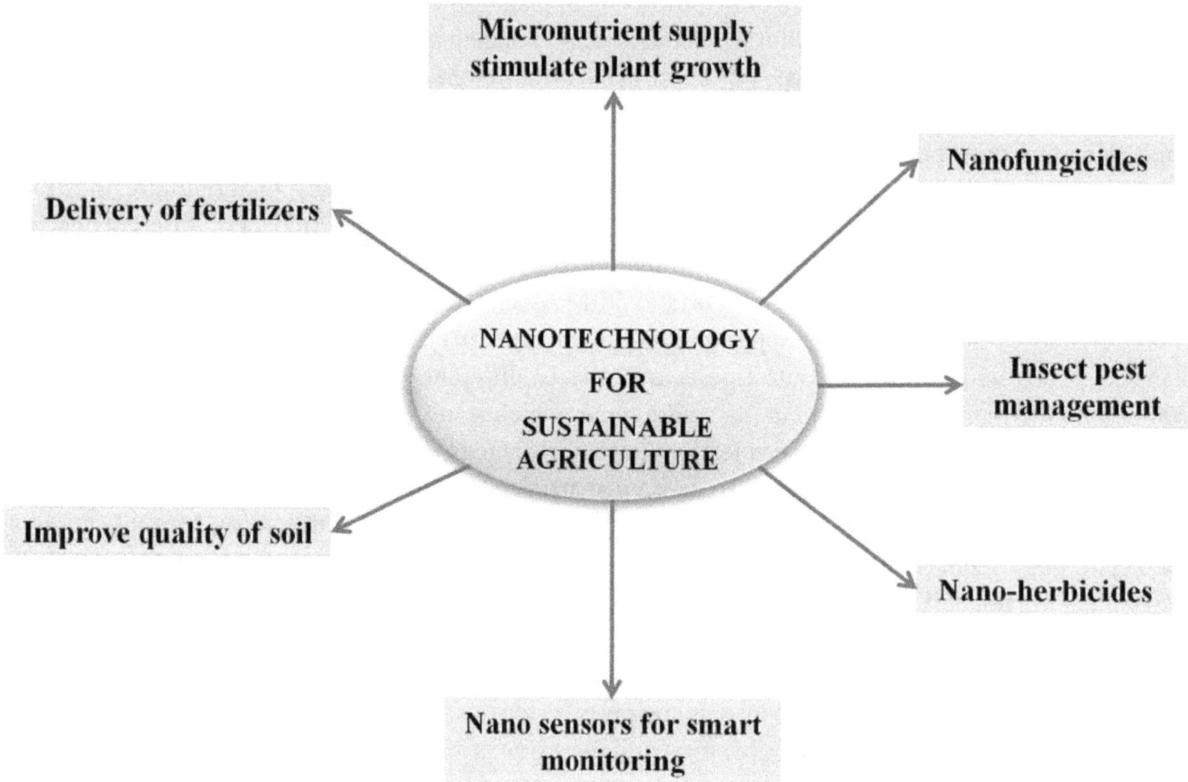

In modern agriculture, nanotechnology defines the true concept of what is precision farming which involves farm management to give maximum output from the available remains in a sustainable way. This chapter highlights basic steps with emphasis on technical innovation in agriculture elaborating the current challenges of sustainable development, and food security issues under the realm of changing climate.

NANOTECHNOLOGY IN SUSTAINABLE AGRICULTURE

An increasing global population is a major threat to agricultural food production. Food security is the basic need for the future human generations and in order to have reliable access affordable, healthy food is recognized as human right (Scott & Chen 2013). With a perspective to attain global food and nutrition security for about 9.2 billion people by the end of 2050, food production needs to be increased by 70% from now and that too in a sustainable way (Chen & Yada 2011). To achieve this challenging target, farmers should have access to scientific and innovative technologies and adopt intellectual farming practices to meet yield.

Nanotechnology has not only increased the yield but has promising application such as in controlled drug delivery, monitoring of water quality via biosensors, targeted delivery of pesticides and nanofiltration (Ion et al. 2010; Sabir et al. 2014). It has not only proven beneficial for the nutrient supply but is also important in resource management in the agricultural sector especially for maintaining the soil fertility (Gruère 2012; Mukhopadhyay 2014) and would lead to innovative applications for bringing agricultural revolution (Khot et al. 2012). Moreover, nanosensors are also being evaluated for the detection of heavy metals, even if present in traces in the soil and water samples (Ion et al. 2010). As such estimation of toxicants and their presence/activity has become much faster, highly sensitive and more flexible after the use of nanoscale particles (Kandasamy & Prema 2015; Vidotti et al. 2011). Additionally, they are well known for their efficiency of catalytic degradation of toxicants in the presence of microbes and thus act as tools for bio-remediation (Dixit et al. 2015).

World is switching towards the sustainable bio-based economy exploiting the renewable bio-resources. The method of smart agriculture is expected to substitute the toxic non eco-friendly materials in near future (Marzbani et al. 2015; Prasad et al. 2014). Important applications of the nanotechnology in agriculture are described in Table 1.

Table 1. Nano-based products and their applications in sustainable agriculture

Process/products	Nano-products	Applications	References
Fertilizers	Macronutrient fertilizers coated with nanoparticles (Zinc oxide; nanocoating of fertilizers)	Enhancement of nutrient absorption by plants and delivery of nutrient to specific sites	Milani et al. 2012
Plant protection products	Nanoemulsions and nanocapsules	Used for plant protection against pathogens e.g. neem oil nanoemulsions used as larvicidal agent	Anjali et al. 2012
Water/liquid retention	Nanoemulsions (zeolite and nano-clays)	Increased water retention and slow release of agrochemicals for proper absorption by the plants	Duhan et al. 2017
Nano sensors and diagnostic devices	Nanomaterials and nanostructures (e.g. electrochemically active nanotubes and nanofibers)	Pesticide detection with liposome- based nano-biosensors	Vamvakaki et al. 2007
Plant genetic modifications	Nanoparticles carrying DNA and RNA such as gold and silica nanomaterials	Transport DNA to transform the plant cells	Torney et al. 2007;Vijaykumar et al. 2010
Water purification and pollutant remediation	Nanoparticles e.g. nanoclays, TiO_2 nanoparticles etc	Filtering and binding to a variety of toxic substances and their removal from the environment	McMurray et al. 2006

NANOFERTILIZERS

Nanofertilizers developed with multidisciplinary technology includes nano zinc, silica, iron, titanium dioxide, ZnCdSe/ZnS core shell QDs, InP/ZnS core-shell QDs, Mn/ZnSe QDs, gold nanorods, core-shell QDs, etc. Studies pertaining to their uptake, biological role and toxicity have been critically analyzed by various researchers. All these materials are readily available in the market and replacing the traditional ones (Dimkpa 2014; Zhang et al. 2016). Several metal oxide nanoparticles such as aluminium trioxide (Al_2O_3), titanium dioxide (TiO_2) and zinc oxide (ZnO) are currently in use for better agricultural production. Several reports suggest the deficiency of zinc as one of the main factors limiting agricultural productivity in the alkaline soils (Sadeghzadeh 2013).

Nano-fertilizer technology is helpful in releasing nutrients into the soil in a planned way. This property prevents eutrophication and pollution of aquatic systems which is of major concern (Naderi & Abedi 2012; Moaveni & Kheiri 2011). In nano-fertilizers, micro and macronutrients may be encapsulated in nanomaterials in a form of thin film or as emulsion (DeRosa et al. 2010). This encapsulation of fertilizers inside a nanoparticle is carried out by three methods: a) encapsulation inside nano-porous material, b) coating with thin polymer, and c) in the form of nanoparticles or as emulsion in nanometer range (Rai et al. 2012). El-Ramady (2014) has reported the impact of nano-fertilizers on sustainable agriculture, food security and other functional aspects necessary for all-round agricultural development in any country.

Nano-Based Chemical Fertilizers

After the green revolution, amount of urea consumption or usage has increased several folds (29%) in India. In order to achieve higher production, excess fixation of nitrogen is essential which leads to the aggregation of the atmospheric N_2O (a greenhouse gas). The excessive release of such hazardous greenhouse gas results in increased atmospheric temperature and thus contributes to global warming (Trenkel 1997). Not only urea but chemical fertilizers like diammonium phosphate (DAP) and single superphosphate (SSP) are widely used in agricultural land to overcome the shortage of N, P and K in the soil.

Several reports suggest that approx. 40–70% of nitrogen, 80–90% of phosphorus, and 50–70% of potassium applied in the form of fertilizers are lost and not utilised by plants leading to excessive waste and become major cause of environmental pollution affecting the entire ecosystem (Ombodi & Saigusa 2000; Trenkel 1997). For smart agricultural practices, nanocoated fertilizers are being employed in the agriculture fields. Nanocoating provides stability and also reduces the rate of dissolution of the fertilizer and allows slow, steady release of coated fertilizer which is more effectively absorbed by plant roots. In order to reduce the rate of fertilizer consumption and risk of environmental pollution, slow releasing fertilizers are quite efficient and eco-friendly. Nano-coating of fertilizer has become more popular especially for sulphur deficient soil where fertilizer with sulphur nanocoating emerged as a useful tool. This technology facilitates the controlled and steady release of fertilizers.

In a country like India where most of the soil is deficient in macronutrients like nitrogen, phosphorus and potassium, nanocoated fertilizers can be a better option for enhancing soil fertility. Nanocoated urea and phosphates are quite beneficial because of their effective steady release in the soil vicinity. Similarly, preparation of biodegradable polymeric chitosan nanoparticles with polymethacrylic acid (PMAA) for loading NPK fertilizer is quite promising for their steady releasing property (Corradini et al. 2010). The colloidal suspension of the chitosan-polymethacrylic acid (CS-PMAA) is much more stable in the presence of nitrogen, potassium and phosphorus, because of higher charge (anionic) from the calcium phosphate than from potassium chloride and urea (Hasaneen et al. 2014). Similarly, Kaolin and biocompatible nanoparticles find much more applications as fertilizer because of their sustained releasing property (Wilson et al. 2008). Fertilizers based on silicon helps to boost plant resistance via improving seedling growth and root initiation. Several additives such as titanium dioxide can be used in order to increase water retention. As such higher solubility of several fertilizers rich in nitrogen can cause leaching and denitrification. Obviously, nanofertilizers may check the releasing pattern of nitrogen and phosphorus and also uptake by the plants, thereby avoiding any nutrient loss and preventing unnecessary nutrient interaction with microbes, air and water (Emadian 2017). Different types of controlled release fertilizers (CRFs) and slow release fertilizers (SRFs) are useful in the achievement of sustainable agriculture (Wilson et al. 2008). The use of SRFs has been reported to minimize environmental pollution by saving fertilizer consumption (Wu & Liu 2008). A wide range of SRFs and CRFs have been developed using synthetic materials or biopolymers (Subramanian & Tarafdar 2011). Various adsorbents as zeolite, halloysite, and bentonite nanoclays have become a part of nitrogen controlled release (Sharmila 2010). Nutrients absorption from soil can be maximized using nanofertilizer for promoting plant growth.

Nano-Biofertilizers

Generally, biofertilizers are living microbes which are beneficial to plants as they promote growth by enhancing the supply of nutrients to the host plant. Biofertilizer includes mycorrhizae, nodule-forming *Rhizobium*, nitrogen-fixing cyanobacterium, *Azotobacter*, *Azospirrilum*and phosphate solubilising bacterial strains such as *Pseudomonas* and *Bacillus* (Singh et al. 2020; Mokrani & El-hafid Nabti 2021). These microbes are helpful in converting complex organic matter into simple compounds that provide essential nutrients to plants, improve soil fertility, maintain the natural habitat of the soil and increase agricultural productivity. Preparation of biofertilizer, its storage and application for the best output is very important (Jha & Prasad 2006). However, there are a few drawbacks such as short shelf life, temperature sensitivity, storage, and desiccation problems. In order to overcome such drawbacks, polymeric nanoparticles have emerged as better tool where biofertilizers are coated to give formulations resistant to desiccation. One such technique includes water-in-oil emulsion which helps in storage and disperses microbes through liquid formulations and oil lowers evaporation acting as a shield around the microbes (Vandergheynst et al. 2006). This technique is specifically favourable for microbes, susceptible for desiccation. Further, water-in-oil emulsions can uphold the release kinetics by constant addition of support substances either to oil and/or water phases. In order to avoid sedimentation at the time of storage, hydrophobic silica nanoparticles are employed as they not only reduce the cell sedimentation rate but also help to increase cell viability by thickened oil phase at the time of storage (Vandergheynst et al. 2007).

Certain microorganisms present in the soil including plant growth promoting rhizobacteria (PGPRs) play an important role in growth promoting activities. Biofertilizers rich in microorganisms like *Pseudomonas fluorescens*, *Bacillus subtilis* and *Paenibacilluselgii* are very promising along with gold and silver nanoparticles for growth and development of crops of fabaceae family (Dikshit 2013). These nanoparticles show significant effects if applied even in a very minute amount in comparison to other fertilizers which make it cost effective and can be used in several hectares of crops (Shukla et al. 2015).

NANOPESTICIDES

Application of pesticides in agriculture for controlling pests or weeds in order to increase crop yield is a common practise. The worldwide consumption of pesticides is about two million tonnes per year; out of which 45% is used by Europe alone, 25% is consumed in the USA and 25% in the rest of the world (De et al. 2014). Careless and extensive use of pesticide stimulate pathogen and pest resistance, reduces soil biodiversity, kills useful soil microbes, causes biomagnification of pesticides and destroys natural habitat of farmer friends like birds (Tilman et al. 2002). In recent years, nanopesticides are emerging as potent agrochemical formulations to resolve the problems caused by the use of conventional pesticides (Grillo et al. 2021). Nanopesticides have novel characteristics in terms of their minute size containing several kinds of components such as surfactants, organic polymers and mineral nanoparticles. Water insolubility is the biggest concern that limits the production of chemicals used in crop protection. Nanoformulations are usually expected to promote the solubility of fairly soluble active ingredients to release slowly the active components in a planned way. This further avoids early degradation and facilitates higher impacts with lower pesticide dosage (Grillo et al. 2021; Lodriche et al. 2013). Nano-silicon carriers consisting of diatom frustules (pore size in the range 1 to 100 nm) are quite effective in pesticide delivery.

In modern agriculture, role of nanoparticles in pest management is exemplary. Polyethylene glycol-coated nanoparticles show increased insecticidal activity of garlic essential oil against *Triboliumcastaneum* (red flour beetle) (Yang et al. 2009). Above mentioned formulation is effective upto 80% because of its slow and sustained release of active constituents in the presence of nanoparticles. Various types of nanoparticles such as silver nanoparticles, aluminium oxide, zinc oxide and titanium dioxide are helpful in controlling grasserie disease in silkworm (caused by *Bombyxmori* and baculovirusBmNPV (*B. mori nuclear polyhedrosis* virus) and are also effective against rice weevil (caused by *Sitophilus oryzae*) (Goswami et al. 2010). A study conducted by Das et al. (2013) showed transformation of *Bombyx mori* nucleopolyhedrovirus by lipophilic coating of silica and alumina nanoparticles in hexagonal close-packed structure. Similarly capping of aspartate on gold nanoparticles when applied on *B. mori* cell line caused cytopathic effect. Ethanolic suspension of hydrophobic alumino-silicate nanoparticles are quite effective in decreasing the viral load. Alumina nanostructures when applied to wheat plant express insecticidal activity against *Sitophilus oryzae* L. and *Rhyzoperthadominica* and caused mortality after 3 days of continuous exposure (Debnath et al. 2011). Debnath et al. (2011) have tested the entomotoxicity of silica nanoparticles against rice weevil *Sitophilus oryzae* and compared the efficiency with bulk-sized silica particles larger than 1.0 µm. Silica nanoparticles have been tested against rice weevil *Sitophilus oryzae*, results showed much more entomotoxic than the bulk sized silica particles. These are highly effective against pest and cause 90% mortality. They are cheaper and more reliable alternative to control insect pest (Teodoro 2010).

Nanoencapsulated Pesticides

In nanoencapsulation, active constituents in the nanometer range are sealed in a protective coating. Nanoencapsulated insecticides, fungicides and nematicides formulations not only control pests effectively but also help in avoiding accumulation of toxic matter in the soil environment. Nanoencapsulated pesticides offer slow release, better solubility and specificity followed by permeability (Bhattacharyya et al. 2016). Nanoencapsulation provides protection to active constituents from severe environmental issues and improve persistence which in turn enhances insecticidal potential (Green & Beestman 2007).

Recently, some companies are promoting nanopesticides for sale purpose as "microencapsulated pesticides." Products such as Karate ZEON, Subdue MAXX, Ospray's Chyella, Penncap-M of Syngenta (Switzerland), and BASF's new Fastac® CS (BASF Corpn., USA) microencapsulated insecticide have been released in the market. Fastac CS can be used in more than 150 crops, including corn, soybeans, leafy vegetables, and tree nuts. It can control more than 100 insect pests such as caterpillars, aphids, stink bugs and thrips [Environmental Protection Agency (EPA-2016)].

Nanofungicides

The threat of fungal infections has emerged as one of the major concerns in agriculture as it causes severe loss in crop productivity. Fungicides are frequently applied to reduce such economic loss however this has raised another challenge related to toxicity in plants. Nanotechnology plays an important role where nanoparticles act against fungal infections. Nanoparticles of zinc oxide, silver and titanium dioxide have been tested for their antifungal role against *Macrophominaphaseolina*, a major soil borne pathogen reported from pulses and oilseed crops. Silver nanoparticles elicited higher antifungal action even at lower concentrations as compared to zinc oxide and titanium dioxide nanoparticles (Shyla et al. 2014). High response of silver nanoparticles is due to silver ions which inactivates the thiol group of cell walls causing dismantled transmembrane, altered energy metabolism and electron transport chain. Nanoparticles cause mutation in fungal genetic material, dissociating the enzyme complex leading to cell lysis (Velmurugan et al. 2009).

Nanoherbicides

Different types of wild species and weeds have emerged as the biggest threat in agriculture as they compete with the crop plants for nutrition and various other factors. Thus, they affect growth of crops mainly due to nutrients deficiency which otherwise were available to agricultural species. Eradication of weeds by traditional means is quite time consuming and tedious. However, numerous herbicides are available commercially that directly affects the life cycle of weed by killing them but they are equally harmful to the agricultural crops. They not only affect the crop species but also decrease the soil fertility and growth of microbes which are otherwise beneficial for succeeding crops (Chinnamuthu & Boopathi 2009). To overcome these problems, nanoherbicides have been formulated which showed potential role in controlling weeds in an environmental friendly manner as they do not release any harmful or toxic substrate which can enter either in soil or environment (Pérez-de-Luque & Rubiales 2009).

Effectiveness of nano zerovalent iron (nano ZVI) has been tested for dechlorination of water/soil polluted with atrazine (2-chloro-4ethylamino-6-isopropylamino-1, 3, 5-triazine) (Satapanajaru et al. 2008). Toxicity of ametryn and atrazine coated poly(ε-caprolactone) nanocapsules was found effective against green alga *Pseudokirchneriella subcapitata* and the microcrustacean *Daphnia similis.* Encapsulated herbicides showed lower toxicity against alga (*Pseudokirchneriella subcapitata*) but higher toxicity against microcrustacean (*Daphnia similis*) as compared to other herbicides (Clemente et al. 2014; Singh et al. 2021). Considering above attributes of nano-herbicides, nanosurfactant-based on soybean micelles has been developed for glyphosate-resistant crops. Interestingly, such formulation not only changes the release profile of the herbicides but also alter the interaction with soil. This study highlights the role of encapsulation as it is capable to reduce the harmful impacts of respective herbicides.

NANOPARTICLE-MEDIATED MICRONUTRIENT SUPPLY

Various micronutrients such as manganese, copper, boron, iron, molybdenum, zinc etc. are essential for growth and development of plants. New farming practices aimed to increase the crop productivity have negatively affected the presence of micronutrients in soil. In order to increase the micronutrient supply, foliar spray is done for the uptake of micronutrients (Martens & Westermann 1991). Nanotechnology can help plant growth and development by making micronutrient available in the form of nanoformulations. Nano-micronutrients when sprayed on plants or applied to the soil are rapidly taken up by roots and also support soil health. Such formulation has been reported to have potential for the slow release of agrochemicals (Mittal et al. 2020; Tao et al. 2012). Different types of nanoparticles have been used for the supply of micronutrients to the plants. Deficiency of iron is one of the major problems in plants growing in the adverse conditions like high pH and calcareous soils. Different concentrations of iron-nano oxide solution has been used to assess the effects on spike weight, grain weight, grain yield and grain (Bakhtiari et al. 2015). An increase in chlorophyll content was also observed in sub-apical leaves of soybeans under greenhouse test during hydroponic conditions in the presence of low concentration of iron oxide nanoparticles. This study clearly pointed to the beneficial role of iron oxide nanoparticles as an iron source for reducing chlorotic symptoms in soybean plants. Similarly, manganese and zinc oxide nanoparticles have been reported to support growth of mung bean (*Vigna radiata*) and chickpea (*Cicer arietinum*) seedlings (Pradhan et al. 2013).

NANOPARTICLE-BASED SMART DELIVERY SYSTEMS: APPLICATIONS AND ADVANTAGES

Delivery of Fertilizers

Huge amounts of fertilizer are being added to the soil in order to increase the food production. The addition of fertilizers in the form of ammonium, urea, nitrate, and phosphate compounds have increased the food production considerably. Several studies have shown that these fertilizers are not totally taken up by the plants because of run-off and also affect the soil microflora and cause pollution (Wilson et al. 2008).

Nanotechnology has provided tools to identify the cause of infection and management accordingly, resolve nutrient deficiency or any such problems much before the deteriorating symptoms in plants are visible. Smart agriculture promises efficient delivery and continuous monitoring of the delivery pattern which play important role in application of pesticides and fertilizers and also in genetic material transfer for the benefit of the plant.

Application systems for pesticides need to focus on efficiency enhancement and spray drift management while fertilizers face problems of bioavailability due to soil chelation, over-application and run-offs. A feasible alternative for all above mentioned problems can be attained by implementing improved delivery systems for the pesticides and fertilizers via controlled and measurable release of any necessary and limited amount of agrochemical at one point of time in order to get complete biological efficiency and to minimize the risk of pollution. To resolve the above purpose nanoparticles are explored as agronomic delivery vehicles. They are much more advantageous because of effective loading as they have larger surface area, easy attachment and fast mass transfer and controlled release of active constituent. Nanotechnology has proven its efficiency in case of genetic material transfer where delivery system faces severe challenge in terms of limited host, transportation via cell membrane and nucleus trafficking. Several insect resistant plant varieties have been developed by the application of nanoparticles assisted delivery of genetic material (Vijayakumar et al. 2010).

Precision Agriculture

The main purposes of precision in agriculture is to enhance crop yields and to cut short the role of the pesticides, fungicides by the surveillance of climatic factors and taking some already proposed actions. Surveillance is pursued by using nanosensors and monitoring systems. These sensors measure the percentage of water content and climatic factors that affect crop water use efficiency enabling the farmers to be sure that their crops are growing at their utmost efficiency. Addition of required fertilizers in order to supplement the natural soil fertility is essential for food crop production and precise management for sustainable agriculture production (Barker & Pilbeam 2006; Zhang et al. 2021). The adoption of sensors attached to a GPS system for real time surveillance of locations that have adverse conditions within the farm followed by passing the problems to the surveillance system is expected to solve the issues within no time. If implemented, these measures will surely help to achieve significant agricultural production at low cost and reduce agricultural waste and environmental pollution (Rai & Ingle 2012). Nanotechnology can also be applicable for the bioprocessing of some by-products having low economic importance (Olson et al. 2012). It can also help in controlling insects and crop plant diseases by modifying genes in the so-called nanogenentic approach or by altering the type of pesticides which is much more effective and least harmful to the environment (Philip 2011). Application of nanotechnology to overcome climate change problems especially high temperature followed by its negative effects on plant growth is completely new aspect in agriculture especially in areas close to equator. By breeding plants that can tolerate temperature stress or assist plants to absorb only useful sun rays or the development of cooling system in the form of nano-capsules in order to maintain high water content in plant tissues for longer duration are quite remarkable. Nanotechnology promises a better option of consuming rain water in the soil especially in desert areas (Sekhon 2014).

Quantum Dots (QDs)

QDs are being widely used for staining bacteria which are the most primitive life forms on the Earth. Organic dyes such as biolabels are frequently used to stain and study bacterial strains; however, these dyes are expensive and their fluorescence disappears with time. QDs are better option because of their high luminescence efficiency, narrow emission spectra, excellent photostability, symmetry and tunability depending on the particle size and material composition. Due to their broad absorption spectra, a single excitation light source can give all colours (Warad et al. 2004). Bio-labelled *Bacillus* with nanoparticle consisting of ZnS and Mn^{2+} capped with biocompatible 'chitosan' gave an orange glow when viewed under a fluorescence microscope. For the detection of *E. coli* O157:H7, QDs were used as a fluorescence marker coupled with immune magnetic separation (Su & Li 2004). For the same, magnetic beads coated with anti- *E. coli* O157 antibodies were selectively attached to the target bacteria, and biotin-conjugated anti- *E. coli* antibodies in order to form sandwich immune complexes. QDs were labelled with the immune complexes via biotin-streptavidin conjugation after magnetic separation.

QDs have also been used to detect pathogens associated with several plant diseases. QDs-fluorescence resonance energy transfer based sensors have been developed to detect witch's broom disease of lime caused by *Phytoplasmaaurantifolia. Polymyxabetae* (Keskin), the only known transmission vector of beet necrotic yellow vein virus has been successfully detected byQDs-fluorescence resonance energy transfer-based sensor (Safarpour et al. 2012). For detecting organophosphorus pesticides (OPs) in vegetables and fruits, the optical transducer of cadmium telluride (CdTe) semiconductor QDs integrated with acetylcholinesterase enzyme utilizinglayer-by-layer assembly technique has been developed. This proved to be highly sensitive biosensor for paraoxon and parathion detection and superior than the traditional gas chromatography or amperometric methods of detection (Zheng et al. 2011). Thiol-stabilized luminescent CdTeQDs which have unique optical property is used to detect methyl parathion (Chouhan et al. 2010).

Nanobarcodes

Advancements in the field of nanotechnology have led to the improvement in the resistance of plants and microbes against the environmental stresses viz. drought, salinity and diseases. Nanotechnology-based gene sequencing is capable of rapid and cost effective identification and characterization of plant gene trait resources (Branton et al. 2008). These nanobarcodes are used as the identification tags for multiplexed analysis of gene expression. The optical microscope and UV lamps are used for the identification of micrometer-sized glass barcodes which are prepared by doping with the rare earth containing different fluorescent materials (Mathew et al. 2009). Nanobarcodes are synthesized by the particles that are easily encodable, machine-readable, durable, and sub-micron sized. For the synthesis of nanobarcodes, the process is semi-automated and highly scalable, involving the electroplating of inert metals (gold, silver, etc.) into templates defining particle diameter, and then the resulting striped nanorods from the templates are released.

Sustainable Green Nanotechnology

According to the definition given by Directorate-General for internal policies of the European Union, precision agriculture is a farming management concept that measures and responds to inter and intra-field variations in crops to create a decision support system for whole farm management and to gain maximum output from the available resources. Nowadays, nanotechnology is extensively used in modern agriculture to validate the concept of precision agriculture (Auffan et al. 2009).

In the present scenario, there is a huge demand for fast, reliable, and low-cost systems for the detection, monitoring, and diagnosis of biological host molecules in agricultural sectors (Chugh et al. 2021; Vidotti et al. 2011,). Planning of modern agriculture for sustainable development without use of agrochemicals such as fertilizer and pesticide is impossible. However, every agrochemical has some potential issues including water contamination or residues on food products that threat the human being and environmental health. This invites smart nanotechnology to regulate such risks and enhance both the quality and quantity of yield (Kah 2015; Liu & Lal 2015; Sekhon 2014).

In 2016, approximately 14,000 documents were listed and about 2707 patents were reported in the world patent database in which nanotechnology is the criteria in food or agriculture sector. Further chemically synthesized nanomaterials are being replaced by biologically synthesized nonmaterials promoting green nanotechnology based on plant system, microbes and biomolecules (Prasad 2014). Advancement in nanomaterials for better crop protection along with detection of pathogens and pesticide residues were recently studied and it was concluded that nanotechnology could provide the best opportunities for agriculture in near future if adequate safety precautions are taken (Chen & Yada 2011).

Nano-Composites as Nano-Biocomposites

Composites made from particles of nanoscale ceramics or metals that are smaller than 100 nm may be greatly stronger than what is estimated by the present models of materials science. The metals that have a grain size of about 10 nm are as much as seven times harder and tougher than their normal counterparts with grain sizes of about 100 nm. A nano-composite is a solid multiphase material where one of its phases may have one, two, or three dimensions of less than 100 nm (Ajayan et al. 2003; Ke & Stroeve 2005). Nano-composites are polymers fortified with small quantities (up to 5% by weight) of nanosize particles which could upgrade the characteristics and performance of the polymer. In packaging of food, nano-composites are developed with high hindrance characteristics against the escaping of oxygen, carbon dioxide, flavor ingredients as well as water vapour. Bio-nanocomposites suitable for the use in packaging usually include derivatives of starch and cellulose, poly- butylene succinate, poly-lactic acid and others (Ayana et al. 2014). The post-harvest and shelf life of food items can be prolonged by the use of nanocomposites that possess new thermal and gas barrier characteristics and its application could safeguard transportation and storage of food commodities (Cortes-Lobos 2013).

Natural fibres from plant and wood are also used as materials for bio-composite production since the beginning of the 20th century. Being renewable together with their relative high strength, high stiffness and low density, these biocomposites are of interest as a replacement for synthetic fibre reinforced composites in a number of industrial sectors including the automobile industry, packaging, construction, and consumer product industries. Cellulose being the main constituent of plant materials is most abundant biopolymer on Earth. Cellulose chains are packed in an ordered manner to form compact microfibrils, formed by elementary fibrils (nanofibres) that are 8-50 nm in diameter and length of a few microns. Because of their crystal structure, nanofibres give strength to the plant stem. In recent years, significant research has been carried out on the production of high strength bio-composites using nanofibres from pulp, plants and agricultural crops. Recently, nanosized lignocellulosic materials have been obtained from crops and trees which had opened a new market for innovative and value-added nano-sized materials and products. Nano-sized cellulosic crystals have been used as lightweight reinforcement in polymeric matrix (Laborie 2009; Mathew et al. 2009). These can be applied in food and other packaging, construction, and transportation body structures.

Starch-Based Nanoparticles

Agricultural nanotechnology can be applied to sustainable production methods such as organic agriculture. Indeed, the Department of Agriculture, Forest, Nature and Energy (DAFNE) at the University of Tuscia in Italy is carrying out a research project for the development of starch-based nanocontainers for the delivery of nutrients, biostimulants and crop protection molecules into the plants tissues. The clear advantage of this approach is that starch is biocompatible, biodegradable and non-toxic for plants, animals and the environment. The first step to develop starch nanocontainers for sustainable and organic agriculture is to produce starch with improved content of amylose, which is the linear fraction composing starch and determines its functional properties. This is possible through improved wheat varieties having higher amylose content employing molecular mutagenesis techniques. The second step is the sustainable preparation, functionalization and characterization of starch nanoparticles suitable to be used as nanocontainers. The main approach adopted to produce nanoparticles from starch is based on the acid hydrolysis of starch granules, but these methods have several drawbacks that include long duration, low yield and environmental concerns about the production of toxic waste. To overcome these problems, it is possible to successfully apply ultrasounds as an eco-friendly approach for the production of wheat starch nanoparticles, without the need of any additional chemical reagent. They may deliver nutrients directly into the plant tissues at slow release rates and may also provide long-term feeding by protecting phosphorus and other micronutrients. Biostimulant compounds can also be slowly released through nanocontainers depending on the plant needs. Furthermore, starch nanocontainers can be used for the delivery of plant protection products, e.g. antibacterial active principles, most suitable for organic agriculture (e.g. vegetal extracts, copper etc) (Alfadul et al. 2017). In horticultural sector, recent successful experiments (in greenhouse as well as in open field) revealed the great potentiality of these nanocontainers to protect the plants along the time against harmful pathogens.

Overall, the advantages of starch nanoparticles application in agriculture are the following:

- Reduced production costs
- Reduction of harmful residues in soils
- Absence of phytotoxicity
- Decreased use of chemical compounds for nutrition/ protection of crops
- High selectivity towards the crops treated

Application of Nanotechnology in Seed Science

Seed is nature's self-perpetuating biological entity that is able to survive in harsh environmental conditions on its own. Nanotechnology can be helpful in utilizing the full potential of seed. Seed production is a tedious process especially in wind pollinated crops. Detecting pollen load that will cause contamination is a reliable method to ensure genetic purity. Pollen flight is determined by air, temperature, humidity, wind velocity, and pollen production of the crop. Use of bionanosensors specific to contaminating pollen can help alert the possible contamination and thus limit contamination. Similar types of method can be also used to prevent pollen from contaminating field crops to other genetically modified crop. Novel genes are transformed into the seeds and made available in the market. Tracking of these seeds could be done through nanobarcodes (Nicewarner-Pena et al. 2001). Basically, pathogens spread through seeds and many times destroy the stored seeds. Coating of seeds with nano structured forms of Zn, Mn, Pa, Pt, Au and Ag will not only protect seeds but also reduce the quantities of the elements used. QDs coupled with immuno-magnetic separation for *E. coli* 0157: H7 could act as a flourescence marker and will be useful to separate unviable and infected seeds (Su and Li 2004). Seeds can also be soaked with nanoencapsulations and specific bacterial strain termed as smart seed. It will thus reduce seed rate, encourage right field stand, and improve crop productivity. Smart seeds can be programmed to germinate in the presence of adequate moisture and can be dispersed over a mountain range for reforestation (Natarajan & Sivasubramanian 2007). It may sense the availability of water thereby allowing the seeds to grow only when time is right for germination. Currently metal oxide nano-particles and carbon nanotube are being tested to escalate the germination of rain-fed crops. Khodakovskaya et al. (2009) have reported improvement in the germination of tomato seeds through the use of carbon nanotube most probably due to better permeation of moisture. Results suggest that carbon nanotubes (CNTs) function as new pores for water permeation by penetrating seed coat and also act as a passage to channelize the water flow from the substrate into the seeds. All these mechanisms support germination under adverse climatic conditions which can be exploited in agricultural system.

Ecotoxicological Aspects of Nanoparticles

Advancement in the field of nanotechnology has exhibited significant output of manufactured NPs into the environment. An increasing number of research have concentrated on assessing the toxicity of NPs normally used in the industry with a view to protect human health (Rana & Kalaichelvan 2013; Du et al. 2017; Tripathi et al. 2017a, b, c). Several factors such as solubility, binding specificity to a biological site are responsible for the toxicity of a metal. Metal NPs exhibit antibacterial, anticandidal, and antifungal activities (Aziz et al. 2016; Patra & Baek 2017). Nanotoxicity might occur because of certain electrostatic interactions of NPs with membrane followed by their accumulation in cytoplasm (Rana and Kalaichelvan 2011; Aziz et al. 2015, 2016). Several NPs (TiO_2, ZnO, SiO_2, and Fullerenes) are photochemically active which means if they are exposed to light, excited electrons may be generated which may form superoxide radicals resulting in oxidative stress (Hoffmann et al. 2007). The cells respond to oxidative stress by increasing number of protective enzymes or genetic constitutions that can easily be measured (Kovochich et al. 2005; Vannini et al. 2014). This generation of reactive oxygen species (ROS) due to oxidativestress can be helpful in determining the toxicity.

Plants are the chief element in food chain and so any mishandling could hinder the ecosystem. Nowadays, plants not only grow on agricultural lands but are also developed in aqueous medium. Naturally, several NPs of iron oxide (magnetite), a magnetic form of iron ore can deposit in the plant host. It is interesting to note that the iron (II, III) oxide NPs (Fe_3O_4 -NPs) have the ability to accumulate in *Lepidiumsativum* and *Pisumsativum*. Such observations clearly suggest the fate and roles of NPs in the natural environment (Bystrzejewska-Piotrowska et al. 2012; Abbas et al. 2016). Additionally, the usage of polymeric NPs in the agricultural field supplemented with insecticides of plant origin is unique and increasingly permeated (Perlatti et al. 2013). There is no doubt that microorganisms play crucial role in maintaining ecosystem, soil health, and crop productivity but it is essential to identify the ecotoxicological aspects of the agricultural field. The introduction of engineered (either chemical or green) NPs in the agricultural field must be always followed by a routine check-up so as to maintain an eco-friendly atmosphere in the agricultural field (Prasad et al. 2017).Extensive research would be necessary to determine the delayed impacts of environmental exposure to NPs, possible adaptive mechanisms and its bioaccumulation in the food chain along with its interaction with other pollutants in the environment (Cox et al. 2017; Singh et al. 2017).

CONCLUSION

Nanotechnology is gaining increased attention for its contribution in the agricultural sector. Promising results and applications are evident in the areas of delivery of pesticides, biopesticides, fertilizers and genetic material for plant transformation. The use of nanomaterials for delivery of pesticides and fertilizers is expected to reduce the dosage and ensure controlled slow release. One of the major contributions of nanoparticles is to stabilize biocontrol preparations that will further assist in reducing the environmental hazard. Based on the unique properties of nanomaterials, nanotechnology has led to the development of nanosensors that are capable of detecting pathogens and have potential for degrading toxic chemicals into harmless and sometimes useful components. Agricultural technology should take advantage of nanotechnology to remove various contaminants from soil. Importantly, vast information is available about the individual nanomaterials, but toxicity level of many NPs is still unexplored. Hence, the application of such materials is limited due to the lack of knowledge of risk assessment and its effects on human health. Development of comprehensive database and alarm system as well as international cooperation for regulation, monitoring and legislation are necessary for exploitation of nanotechnology for use in sustainable agriculture.

ACKNOWLEDGMENT

GS is thankful to Department of Botany, Gargi College, University of Delhi, for providing constant support and encouragement. DS is grateful to DST-INSPIRE, Department of Science and Technology, New Delhi, for the award of Junior Research Fellowship (DST/INSPIRE Fellowship/2014/296, IF140707). JK is thankful to University Grants Commission (UGC), New Delhi, for providing financial assistance (23/06/2013 (I) EU-V). AK is thankful to Department of Biotechnology and Indian Council of Agricultural Research, New Delhi, for continuously supporting works on plant growth promoting bacteria by providing financial supports in the form of research projects.

REFERENCES

Abbas, S. S., Haneef, M., & Lohani, M. (2016). Nanomaterials used as a plants growth enhancer: An update. *International Journal of Pharmaceutical Sciences Review and Research*, *5*, 17–23.

Ajayan, P. M., Schadler, L. S., & Braun, P. V. (2003). Nanocomposite Science and Technology. Wiley-VCH VerlagGmbHand Co. doi:10.1002/3527602127

Alfadul, S. M., Altahir, O. S., & Khan, M. A. (2017). Application of nanotechnology in the field of food production. *Academia Journal of Scientific Research*, *5*, 143–154.

Anjali, C. H., Sharma, Y., Mukherjee, A., & Chandrasekaran, N. (2012). Neem oil (*Azadirachtaindica*) nanoemulsion a potent larvicidal agent against *Culexquinquefasciatus*. *Pest Management Science*, *68*(2), 158–163. doi:10.1002/ps.2233 PMID:21726037

Auffan, M., Rose, J., Bottero, J. Y., Lowry, G. V., Jolivet, J.-P., & Wiesner, M. R. (2009). Towards a definition of inorganic nanoparticles from an environmental, health and safety perspective. *Nature Nanotechnology*, *4*(10), 634–664. doi:10.1038/nnano.2009.242 PMID:19809453

Ayana, B., Suin, S., & Khatua, B. B. (2014). Highly exfoliated eco-friendly thermoplastic starch (TPS)/Poly(lactic acid) (PLA)/clay nanocomposites using unmodified nanoclay. *Carbohydrate Polymers*, *110*, 430–439. doi:10.1016/j.carbpol.2014.04.024 PMID:24906776

Aziz, N., Faraz, M., Pandey, R., Shakir, M., Fatma, T., Varma, A., Barman, I., & Prasad, R. (2015). Facile algae-derived route to biogenic silver nanoparticles: Synthesis, antibacterial and photocatalytic properties. *Langmuir*, *31*(42), 11605–11612. doi:10.1021/acs.langmuir.5b03081 PMID:26447769

Aziz, N., Pandey, R., Barman, I., & Prasad, R. (2016). Leveraging the attributes of *Mucorhiemalis*-derived silver nanoparticles for a synergistic broad-spectrum antimicrobial platform. *Frontiers in Microbiology*, *7*, 1984. doi:10.3389/fmicb.2016.01984 PMID:28018316

Bakhtiari, M., Moaveni, P., & Sani, B. (2015). The effect of iron nanoparticles spraying time and concentration on wheat. *Biological Forum: An International Journal*, *7*, 679–683.

Barker, A. V., & Pilbeam, D. J. (2006). *Handbook of plant nutrition*. CRC Press.

Bhattacharyya, A., Duraisamy, P., & Govindarajan, M. (2016). Nano-biofungicides: emerging trend in insect pest control. In R. Prasad (Ed.), *Advances & Applications through Fungal Nanobiotechnology* (pp. 307–319). Springer. doi:10.1007/978-3-319-42990-8_15

Branton, D., Deamer, D. W., Marziali, A., Bayley, H., Benner, S. A., Butler, T., Di Ventra, M., Garaj, S., Hibbs, A., Huang, X., Jovanovich, S. B., Krstic, P. S., Lindsay, S., Ling, X. S., Mastrangelo, C. H., Meller, A., Oliver, J. S., Pershin, Y. V., Ramsey, J. M., ... Schloss, J. A. (2008). The potential and challenges of nanopore sequencing. *Nature Biotechnology*, *10*(10), 1146–1153. doi:10.1038/nbt.1495 PMID:18846088

Bystrzejewska-Piotrowska, G., Asztemborska, M., & Steborowski, R. (2012). Application of neutron activation for investigation of Fe_3O_4 nanoparticles accumulation by plants. *Nukleonika*, *57*, 427–430.

Campbell, B. M., Thornton, P., Zougmoré, R., van Asten, P., & Lipper, L. (2014). Sustainable intensification: What is its role in climate smart agriculture? *Current Opinion in Environmental Sustainability*, *8*, 39–43. doi:10.1016/j.cosust.2014.07.002

Chen, H. D., & Yada, R. (2011). Nanotechnologies in agriculture: New tools for sustainable development. *Trends in Food Science & Technology*, *22*(11), 585–594. doi:10.1016/j.tifs.2011.09.004

Cheng, X., Chen, G., & Rodrigue, W. R. (2009). Micro- and nanotechnology for viral detection. *Analytical and Bioanalytical Chemistry*, *393*(2), 487–501. doi:10.100700216-008-2514-x PMID:19052733

Chinnamuthu, C., & Boopathi, P. M. (2009). Nanotechnology and agroecosystem. *Madras Agricultural Journal*, *96*, 17–31.

Chouhan, R. S., Vinayaka, A. C., & Thakur, M. S. (2010). Thiol-stabilized luminescent CdTe quantum dot as biological fluorescent probe for sensitive detection of methyl parathion by a fluoro-immune chromatographic technique. *Analytical and Bioanalytical Chemistry*, *397*(4), 1467–1475. doi:10.100700216-009-3433-1 PMID:20101498

Chugh, G., Siddique, K. H. M., & Solaiman, Z. M. (2021). Nanobiotechnology for agriculture: Smart technology for combating nutrient deficiencies with nanotoxicity challenges. *Sustainability*, *13*(4), 1781. doi:10.3390u13041781

Clemente, Z., Grillo, R., Jonsson, M., Santos, N. Z. P., Feitosa, L. O., Lima, R., & Fraceto, L. F. (2014). Ecotoxicological evaluation of poly(ε-caprolactone) nanocapsules containing triazine herbicides. *Journal of Nanoscience and Nanotechnology*, *14*(7), 4911–4917. doi:10.1166/jnn.2014.8681 PMID:24757962

Corradini, E., Moura, M. R., & Mattoso, L. H. C. (2010). A preliminary study of the incorporation of NPK fertilizer into chitosan nanoparticles. *Express Polymer Letters*, *4*(8), 509–515. doi:10.3144/expresspolymlett.2010.64

Cortes-Lobos, R. (2013). *Nanotechnology research in the US agri-food sectoral system of innovation: towards sustainable development* [Dissertation]. Georgia Institute of Technology.

Cox, A., Venkatachalam, P., Sahi, S., & Sharma, N. (2017). Reprint of: silver and titanium dioxide nanoparticle toxicity in plants: a review of current research. *Plant Physiology and Biochemistry*, *110*, 33–49. doi:10.1016/j.plaphy.2016.08.007 PMID:27569179

Das, S., Bhattacharya, A., Debnath, N., Datta, A., & Goswami, A. (2013). Nanoparticle-induced morphological transition of *Bombyxmori* nucleopolyhedrovirus: A novel method to treat silkworm grasserie disease. *Applied Microbiology and Biotechnology*, *97*(13), 6019–6030. doi:10.100700253-013-4868-z PMID:23588933

De, A., Bose, R., Kumar, A., & Mozumdar, S. (2014). Worldwide pesticide use. In *Targeted delivery of pesticides using biodegradable polymeric nanoparticles* (pp. 5–6). Springer. doi:10.1007/978-81-322-1689-6_2

Debnath, N., Das, S., Seth, D., Chandra, R., Bhattacharya, S. C., & Goswami, A. (2011). Entomotoxic effect of silica nanoparticles against *Sitophilus oryzae* (L.). *Journal of Pest Science*, *84*(1), 99–105. doi:10.100710340-010-0332-3

DeRosa, M. C., Monreal, C., Schnitzer, M., Walsh, R., & Sultan, Y. (2010). Nanotechnology in fertilizers. *Nature Nanotechnology*, *5*(2), 91. doi:10.1038/nnano.2010.2 PMID:20130583

Dikshit, A., Shukla, S. K., & Mishra, R. K. (2013). *Exploring nanomaterials with PGPR in current agricultural scenario*. Lap Lambert Academic Publishing.

Dimkpa, C. O. (2014). Can nanotechnology deliver the promised benefits without negatively impacting soil microbial life? *Journal of Basic Microbiology*, *54*(9), 889–904. doi:10.1002/jobm.201400298 PMID:24913194

Dixit, R., Wasiullah, Malaviya, D., Pandiyan, K., Singh, U., Sahu, A., Shukla, R., Singh, B., Rai, J., Sharma, P., Lade, H., & Paul, D. (2015). Bioremediation of heavy metals from soil and aquatic environment: An overview of principles and criteria of fundamental processes. *Sustainability*, *7*(2), 2189–2212. doi:10.3390u7022189

Du, W., Tan, W., Peralta-Videa, J. R., Gardea-Torresdey, J. L., Ji, R., Yin, Y., & Guo, H. (2017). Interaction of metal oxide nanoparticles with higher terrestrial plants: Physiological and biochemical aspects. *Plant Physiology and Biochemistry*, *110*, 210–225. doi:10.1016/j.plaphy.2016.04.024 PMID:27137632

Duhan, J. S., Kumar, R., Kumar, N., Kaur, P., Nehra, K., & Duhan, S. (2017). Nanotechnology: The new perspective in precision agriculture. *Biotechnology Reports (Amsterdam, Netherlands)*, *15*, 11–23. doi:10.1016/j.btre.2017.03.002 PMID:28603692

EC. (2012). *Communication from the Commission to the European Parliament, the Council, the European Economic and Social Committee and the Committee of the Regions. 'A European strategy for key enabling technologies-A bridge to growth and jobs'*. https://eur-lex.europa.eu/LexUriServ/ LexUriServ.do?uri=COM: 2012:0341:FIN:EN:PDF

El-Ramady, H. R. (2014). Integrated nutrient management and postharvest of crops. *Sustainable Agriculture Research*, *13*, 163–274.

Emadian, S. E. (2017). *Physiological responses of loblolly pine (Finustaeda L.) to silicon and water stress* [PhD thesis]. Texas A & M Univ, College Station, TX, USA.

Ghormade, V., Deshpande, M. V., & Paknikar, K. M. (2011). Perspectives for nano-biotechnology enabled protection and nutrition of plants. *Biotechnology Advances*, *29*(6), 792–803. doi:10.1016/j.biotechadv.2011.06.007 PMID:21729746

Goswami, A., Roy, I., Sengupta, S., & Debnath, N. (2010). Novel applications of solid and liquid formulations of nanoparticles against insect pests and pathogens. *Thin Solid Films*, *519*(3), 1252–1257. doi:10.1016/j.tsf.2010.08.079

Green, J. M., & Beestman, G. B. (2007). Recently patented and commercialized formulation and adjuvant technology. *Crop Protection (Guildford, Surrey)*, *26*(3), 320–327. doi:10.1016/j.cropro.2005.04.018

Grillo, R., Mattos, B. D., & Antunes, D. R. (2021). Foliage adhesion and interactions with particulate delivery systems for plant nanobionics and intelligent agriculture. Vol. 37, Nano Today. *Elsevier B.V.*, *2021*, 101078.

Gruère, G. P. (2012). Implications of nanotechnology growth in food and agriculture in OECD countries. *Food Policy*, *37*(2), 191–198. doi:10.1016/j.foodpol.2012.01.001

Hasaneen, M. N. A., Abdel-Aziz, H. M. M., El-Bialy, D. M. A., & Omer, A. M. (2014). Preparation of chitosan nanoparticles for loading with NPK fertilizer. *African Journal of Biotechnology*, *13*(31), 3158–3164. doi:10.5897/AJB2014.13699

Hoffmann, M., Holtze, E. M., & Wiesner, M. R. (2007). Reactive oxygen species generation on nanoparticulate material. In M. R. Wiesner & J. Y. Bottero (Eds.), *Environmental Nanotechnology. Applications & Impacts of Nanomaterials* (pp. 155–203). McGraw Hill.

Ion, A. C., Ion, I., & Culetu, A. (2010). Carbon-based nanomaterials: Environmental applications. *University Politehnica of Bucharest*, *38*, 129–132.

Jha, M. N., & Prasad, A. N. (2006). Efficacy of new inexpensive cyanobacterial biofertilizer including its shelf-life. *World Journal of Microbiology & Biotechnology*, *22*(1), 73–79. doi:10.100711274-005-7024-9

Kah, M. (2015). Nanopesticides and nanofertilizers: Emerging contaminants or opportunities for risk mitigation? *Frontiers in Chemistry*, *3*, 64. doi:10.3389/fchem.2015.00064 PMID:26636068

Kandasamy, S., & Prema, R. S. (2015). Methods of synthesis of nanoparticles and its applications. *Journal of Chemical and Pharmaceutical Research*, *7*, 278–285.

Ke, Y. C., & Stroeve, P. (2005). *Polymer-layered silicate and silica nanocomposites* (1st ed.). Elsevier. doi:10.1016/B978-044451570-4/50004-5

Khodakovskaya, M., Dervishi, E., Mahmood, M., Xu, Y., Li, Z., Watanabe, F., & Biris, A. S. (2009). Carbon nanotubes are able to penetrate plant seed coat and dramatically affect seed germination and plant growth. *ACS Nano*, *3*(10), 3221–3227. doi:10.1021/nn900887m PMID:19772305

Khot, L. R., Sankaran, S., Maja, J. M., Ehsani, R., & Schuster, E. W. (2012). Applications of nanomaterials in agricultural production and crop protection: A review. *Crop Protection (Guildford, Surrey), 35*, 64–70. doi:10.1016/j.cropro.2012.01.007

Kovochich, M., Xia, T., & Xu, J. (2005). Principles and procedures to assess nanoparticles. *Environmental Science & Technology, 39*, 1250–1256.

Laborie, M. P. G. (2009). Bacterial cellulose and its polymeric nanocomposites. In L. A. Lucia & O. J. Rojas (Eds.), *The Nanoscience & Technology ofRenewable Biomaterials*. Wiley. doi:10.1002/9781444307474.ch9

Liu, R., & Lal, R. (2015). Potentials of engineered nanoparticles as fertilizers for increasing agronomic productions. *The Science of the Total Environment, 514*, 131–139. doi:10.1016/j.scitotenv.2015.01.104 PMID:25659311

Lodriche, S. S., Soltani, S., & Mirzazadeh, R. (2013). *Silicon nanocarrier for delivery of drug, pesticides and herbicides, and for waste water treatment.* United States Patent US 20130225412 A1.

Martens, D. C., & Westermann, D. T. (1991). Fertilizer applications for correcting micronutrient deficiencies. In B. J. Alloway (Ed.), *Micronutrient Deficiencies in Global Crop Production* (pp. 549–553). Springer.

Marzbani, P., Afrouzi, Y. M., & Omidvar, A. (2015). The effect of nano-zinc oxide on particleboard decay resistance. *Maderas. Ciencia y Tecnología, 17*(ahead), 63–68. doi:10.4067/S0718-221X2015005000007

Mathew, A. P., Laborie, M. P., & Oksman, K. (2009). Cross-linked chitosan/chitin crystal nanocomposites with improved permeation selectivity and pH stability. *Biomacromolecules, 10*(6), 1627–1632. doi:10.1021/bm9002199 PMID:19388678

McMurray, T. A., Dunlop, P. S. M., & Byrne, J. A. (2006). The photocatalytic degradation of atrazine on nanoparticulate TiO2 films. *Journal of Photochemistry and Photobiology A Chemistry, 182*(1), 43–51. doi:10.1016/j.jphotochem.2006.01.010

Milani, N., McLaughlin, M. J., Stacey, S. P., Kirby, J. K., Hettiarachchi, G. M., Beak, D. G., & Cornelis, G. (2012). Dissolution kinetics of macronutrient fertilizers coated with manufactured zinc oxide nanoparticles. *Journal of Agricultural and Food Chemistry, 60*(16), 3991–3998. doi:10.1021/jf205191y PMID:22480134

Mittal, D., Kaur, G., Singh, P., Yadav, K., & Ali, S. A. (2020). Nanoparticle-based sustainable agriculture and food science: Recent advances and future outlook. *Front. Nanotechnol., 2*, 579954. doi:10.3389/fnano.2020.579954

Moaveni, P., & Kheiri, T. (2011). TiO2nano particles affected on maize (*Zea mays* L). In 2nd *International Conference on Agricultural & Animal Science*. IACSIT Press.

Mokrani, S., & El-hafid, N. (2021) Microbes Associated with Crops: Functional attributes for crop productivity. Sustainable Development and Biodiversity, 27. doi:10.1007/978-3-030-73507-4_2

Mukhopadhyay, S. S. (2014). Nanotechnology in agriculture: Prospects and constraints. *Nanotechnology, Science and Applications, 7*, 63–71. doi:10.2147/NSA.S39409 PMID:25187699

Naderi, M. R., & Abedi, A. (2012). Application of nanotechnology in agriculture and refinement of environmental pollutants. *Journal of Nanotechnology*, *11*(1), 18–26.

Natarajan, N., & Sivasubramaniam, K. (2007). Nanotechnology application in seed management. In *Application of Nanotechnology in Agriculture*. Tamil Nadu Agricultural University.

Nicewarner-Pena, S. R., Freeman, R. G., Reiss, B. D. H., He, L., Peña, D. J., Walton, I. D., Cromer, R., Keating, C. D., & Natan, M. J. (2001). Submicrometer metallic barcodes. *Science*, *294*(5540), 137–141. doi:10.1126cience.294.5540.137 PMID:11588257

Olson, D. G., McBride, J. E., Shaw, J. A., & Lynd, L. R. (2012). Recent progress in consolidated bioprocessing. *Current Opinion in Biotechnology*, *23*(3), 396–405. doi:10.1016/j.copbio.2011.11.026 PMID:22176748

Ombodi, A., & Saigusa, M. (2000). Broadcast application versus band application of polyolefin coated fertilizer on green peppers grown on andisol. *Journal of Plant Nutrition*, *23*(10), 1485–1493. doi:10.1080/01904160009382116

Paramo, L. A., Feregrino-Pérez, A. A., Guevara, R., Mendoza, S., & Esquivel, K. (2020). Nanoparticles in agroindustry: Applications, toxicity, challenges, and trends. *Nanomaterials (Basel, Switzerland)*, *10*(9), 1654. doi:10.3390/nano10091654 PMID:32842495

Patra, J. K., & Baek, K. H. (2017). Antibacterial activity and synergistic antibacterial potential of bio-synthesized silver nanoparticles against foodborne pathogenic bacteria along with its anticandidal and antioxidant effects. *Frontiers in Microbiology*, *8*, 167. doi:10.3389/fmicb.2017.00167 PMID:28261161

Pérez-de-Luque, A., & Rubiales, D. (2009). Nanotechnology for parasitic plant control. *Pest Management Science*, *65*(5), 540–545. doi:10.1002/ps.1732 PMID:19255973

Perlatti, B., Bergo, P. L. S., & Silva, M. F. G. (2013). Polymeric nanoparticle-based insecticides: a controlled release purpose for agrochemicals. In S. Trdan (Ed.), *Insecticides-Development of Safer & More Effective Technologies* (pp. 523–550). InTech Open. doi:10.5772/53355

Philip, D. (2011). *Mangiferaindica* leaf-assisted biosynthesis of well dispersed silver nanoparticles. *SpectrochimicaActa Part A*, *78*(1), 327–331. doi:10.1016/j.saa.2010.10.015 PMID:21030295

Pradhan, S., Patra, P., Das, S., Chandra, S., Mitra, S., Dey, K. K., Akbar, S., Palit, P., & Goswami, A. (2013). Photochemical modulation of biosafe manganese nanoparticles on *Vignaradiata*: A detailed molecular, biochemical, and biophysical study. *Environmental Science & Technology*, *47*(22), 13122–13131. doi:10.1021/es402659t PMID:24144189

Prasad, R. (2014). Synthesis of silver nanoparticles in photosynthetic plants. *Journal of Nanoparticles*, *2014*, 963961. doi:10.1155/2014/963961

Prasad, R., Bhattacharyya, A., & Nguyen, Q. D. (2017). Nanotechnology in sustainable agriculture: Recent developments, challenges, and perspectives. *Frontiers in Microbiology*, *8*, 1014. doi:10.3389/fmicb.2017.01014 PMID:28676790

Prasad, R., Kumar, V., & Prasad, K. S. (2014). Nanotechnology in sustainable agriculture: Present concerns and future aspects. *African Journal of Biotechnology*, *13*(6), 705–713. doi:10.5897/AJBX2013.13554

Puoci, F., Lemma, F., & Spizzirri, U. G. (2008). Polymer in agriculture: A review. *American Journal of Agricultural and Biological Sciences*, *3*(1), 299–314. doi:10.3844/ajabssp.2008.299.314

Ragaei, M., & Sabry, A. H. (2014). Nanotechnology for insect pest control. *International Journal of Science, Environment and Technology*, *3*, 528–545.

Rai, M., & Ingle, A. (2012). Role of nanotechnology in agriculture with special reference to management of insect pests. *Applied Microbiology and Biotechnology*, *94*(2), 287–293. doi:10.100700253-012-3969-4 PMID:22388570

Rai, V., Acharya, S., & Dey, N. (2012). Implications of nanobiosensors in agriculture. *Journal of Biomaterials and Nanobiotechnology*, *3*(02), 315–324. doi:10.4236/jbnb.2012.322039

Rana, S., & Kalaichelvan, P. T. (2011). Antibacterial effects of metal nanoparticles. *Advanced Biotech*, *2*, 21–23.

Rana, S., & Kalaichelvan, P. T. (2013). Ecotoxicity of nanoparticles. *ISRN Toxicology*, 574648. PMID:23724300

Sabir, S., Arshad, M., & Chaudhari, S. K. (2014). Zinc oxide nanoparticles for revolutionizing agriculture: Synthesis and applications. *TheScientificWorldJournal*, *2014*, 8. doi:10.1155/2014/925494 PMID:25436235

Sadeghzadeh, B. (2013). A review of zinc nutrition and plant breeding. *Journal of Soil Science and Plant Nutrition*, *13*(ahead), 905–927. doi:10.4067/S0718-95162013005000072

Safarpour, H., Safarnejad, M. R., Tabatabaei, M., Mohsenifar, A., Rad, F., Basirat, M., Shahryari, F., & Hasanzadeh, F. (2012). Development of a quantum dots FRET-based biosensor for efficient detection of *Polymyxabetae*. *Canadian Journal of Plant Pathology*, *34*(4), 507–515. doi:10.1080/07060661.2012.709885

Satapanajaru, T., Anurakpongsatorn, P., Pengthamkeerati, P., & Boparai, H. (2008). Remediation of atrazine-contaminated soil and water by nanozerovalent iron. *Water, Air, and Soil Pollution*, *192*(1-4), 349–359. doi:10.100711270-008-9661-8

Scott, N., & Chen, H. (2013). Nanoscale science and engineering for agriculture and food systems. *Industrial Biotechnology (New Rochelle, N.Y.)*, *9*(1), 17–18. doi:10.1089/ind.2013.1555

Sekhon, B. S. (2014). Nanotechnology in agri-food production: An overview. *Nanotechnology, Science and Applications*, *7*, 31–53. doi:10.2147/NSA.S39406 PMID:24966671

Sharmila, R. C. (2010). *Nutrient release pattern of nano-fertilizer formulations* (PhD thesis). Tamil Nadu Agricultural University, Coimbatore, Tamil Nadu, India.

Shukla, S. K., Kumar, R., & Mishra, R. K. (2015). Prediction and validation of gold nanoparticles (GNPs) on plant growth promoting rhizobacteria (PGPR): A step toward development of nano-biofertilizers. *Nano Reviews*, *4*, 439–448.

Shyla, K. K., Natarajan, N., & Nakkeeran, S. (2014). Antifungal activity of zinc oxide, silver and titanium dioxide nanoparticles against *Macrophominaphaseolina. Journal of Mycology and Plant Pathology*, *44*, 268–273.

Singh, D., Ghosh, P., Kumar, J., & Kumar, A. (2019). Plant Growth-Promoting Rhizobacteria (PGPRs): functions and benefits. In *Microbial Interventions in Agriculture and Environment* (pp. 205–227). doi:10.1007/978-981-13-8383-0_7

Singh, R. P., Handa, R., & Manchanda, G. (2021). Nanoparticles in sustainable agriculture: An emerging opportunity. *Journal of Controlled Release*, *329*, 1234–1248. doi:10.1016/j.jconrel.2020.10.051 PMID:33122001

Singh, S., Vishwakarma, K., Singh, S., Sharma, S., Dubey, N. K., Singh, V. K., Liu, S., Tripathi, D. K., & Chauhan, D. K. (2017). Understanding the plant and nanoparticle interface at transcriptomic and proteomic level: A concentric overview. *Plant Gene*, *11*, 265–272. Advance online publication. doi:10.1016/j.plgene.2017.03.006

Su, X. L., & Li, Y. (2004). Quantum dot biolabelling coupled with immunomagnetic separation for detection of *Escherichia coli* 0157:H7.2004. *Analytical Chemistry*, *76*(16), 4806–4810. doi:10.1021/ac049442+ PMID:15307792

Subramanian, K. S., & Tarafdar, J. C. (2011). Prospects of nanotechnology in Indian farming. *Indian Journal of Agricultural Sciences*, *81*, 887–893.

Tao, S., Pang, R., Chen, C., Ren, X., & Hu, S. (2012). Synthesis, characterization and slow release properties of O-naphthylacetyl chitosan. *Carbohydrate Polymers*, *88*(4), 1189–1194. doi:10.1016/j.carbpol.2012.01.076

Teodoro, S., Micaela, B., & David, K. W. (2010). Novel use of nano-structured alumina as an insecticide. *Pest Management Science*, *66*(6), 577–579. doi:10.1002/ps.1915 PMID:20127753

Tilman, D., Cassman, K. G., Matson, P. A., Naylor, R., & Polasky, S. (2002). Agricultural sustainability and intensive production practices. *Nature*, *418*(6898), 671–677. doi:10.1038/nature01014 PMID:12167873

Torney, F., Trewyn, B. G., Lin, V. S. Y., & Wang, K. (2007). Mesoporous Silica Nanoparticles Deliver DNA and Chemicals into Plants. *Nature Nanotechnology*, *2*(5), 295–300. doi:10.1038/nnano.2007.108 PMID:18654287

Trenkel, M. E. (1997). *Controlled-release and stabilized fertilizers in agriculture*. International Fertilizer Industry Association.

Tripathi, D. K., Mishra, R. K., Singh, S., Singh, S., Vishwakarma, K., Sharma, S., Singh, V. P., Singh, P. K., Prasad, S. M., Dubey, N. K., Pandey, A. C., Sahi, S., & Chauhan, D. K. (2017a). Nitric oxide ameliorates zinc oxide nanoparticles phytotoxicity in wheat seedlings: Implication of the ascorbate-glutathione cycle. *Frontiers in Plant Science*, *8*, 1. doi:10.3389/fpls.2017.00001 PMID:28220127

Tripathi, D. K., Singh, S., Singh, S., Srivastava, P. K., Singh, V. P., Singh, S., Prasad, S. M., Singh, P. K., Dubey, N. K., Pandey, A. C., & Chauhan, D. K. (2017b). Nitric oxide alleviates silver nanoparticles (AgNps)-induced phytotoxicity in *Pisumsativum* seedlings. *Plant Physiology and Biochemistry*, *110*, 167–177. doi:10.1016/j.plaphy.2016.06.015 PMID:27449300

Tripathi, D.K., Tripathi, A., & Shweta, S.S. (2017c). Uptake, accumulation and toxicity of silver nanoparticle in autotrophic plants, and heterotrophic microbes: a concentric review. *Frontiers in Microbiology, 8*, 7.

Vamvakaki, V., & Chaniotakis, N. A. (2007). Pesticide detection with a liposome-based nanobiosensor. *Biosensors & Bioelectronics*, *22*(12), 2848–2853. doi:10.1016/j.bios.2006.11.024 PMID:17223333

Vandergheynst, J. S., Scher, H., & Hong-Yun, G. (2006). Design of formulations for improved biological control agent viability and sequestration during storage. *Industrial Biotechnology (New Rochelle, N.Y.)*, *2*(3), 213–219. doi:10.1089/ind.2006.2.213

Vandergheynst, J. S., Scher, H. B., Gou, H. Y., & Schultz, D. L. (2007). Water in oil emulsions that improve the storage and delivery of the biolarvacide*Lagenidiumgiganteum*. *Biological Control*, *52*, 207–229.

Vannini, C., Domingo, G., Onelli, E., De Mattia, F., Bruni, I., Marsoni, M., & Bracale, M. (2014). Phytotoxic and genotoxic effects of silver nanoparticles exposure on germinating wheat seedlings. *Journal of Plant Physiology*, *171*(13), 1142–1148. doi:10.1016/j.jplph.2014.05.002 PMID:24973586

Velmurugan, N., Gnana Kumar, G., & Sub Han, S. (2009). Synthesis and characterization of potential fungicidal silver nano-sized particles and chitosan membrane containing silver particles. *Iranian Polymer Journal*, *18*, 383–392.

Vidotti, M., Carvalhal, R. F., Mendes, R. K., Ferreira, D. C. M., & Kubota, L. T. (2011). Biosensors based on gold nanostructures. *Journal of the Brazilian Chemical Society*, *22*(1), 3–20. doi:10.1590/S0103-50532011000100002

Vijayakumar, P. S., Abhilash, O. U., Khan, B. M., & Prasad, B. L. V. (2010). Nanogold-loaded sharp-edged carbon bullets as plant gene carriers. *Advanced Functional Materials*, *20*(15), 2416–2423. doi:10.1002/adfm.200901883

Warad, H. C., Ghosh, S. C., & Thanachayanont, C. (2004). Highly luminescent manganese doped ZnS quantum dots for biological labelling. *Proceeding of International Conference on Smart Materials/Intelligent Materials*, 203-206.

Wilson, M. A., Tran, N. H., Milev, A. S., Kannangara, G. S. K., Volk, H., & Lu, G. Q. M. (2008). Nanomaterials in soils. *Geoderma*, *146*(1-2), 291–302. doi:10.1016/j.geoderma.2008.06.004

Wu, L., & Liu, M. (2008). Preparation and properties of chitosan coated NPK compound fertilizer with controlled release and water-retention. *Carbohydrate Polymers*, *72*(2), 240–247. doi:10.1016/j.carbpol.2007.08.020

Yang, F. L., Li, S. G., Zhu, F., & Lei, C. L. (2009). Structural characterization of nanoparticles loaded with garlic essential oil and their insecticidal activity against *Triboliumcastaneum* (Herbst) (*Coleoptera: Tenebrionidae*). *Journal of Agricultural and Food Chemistry*, *57*(21), 10156–10162. doi:10.1021/jf9023118 PMID:19835357

Yunlong, C., & Smit, B. (1994). Sustainability in agriculture: A general review. *Agriculture, Ecosystems & Environment*, *49*(3), 299–307. doi:10.1016/0167-8809(94)90059-0

Zhang, P., Guo, Z., Ullah, S., Melagraki, G., Afantitis, A., & Lynch, I. (2021). Nanotechnology and artificial intelligence to enable sustainable and precision agriculture. *Nature Plants*, *7*(7), 864–876. Advance online publication. doi:10.103841477-021-00946-6 PMID:34168318

Zhang, Q., Han, L., Jing, H., Blom, D. A., Lin, Y., Xin, H. L., & Wang, H. (2016). Facet control of gold nanorods. *ACS Nano*, *10*(2), 2960–2974. doi:10.1021/acsnano.6b00258 PMID:26795706

Zheng, L., Hong, F., Lu, S., & Liu, C. (2005). Effect of nano-$TiO_{(2)}$ on strength of naturally aged seeds and growth of spinach. *Biological Trace Element Research*, *104*(1), 83–91. doi:10.1385/BTER:104:1:083 PMID:15851835

Zheng, Z., Zhou, Y., Li, X., Liu, S., & Tang, Z. (2011). Highly-sensitive organophosphorous pesticide biosensors based on nanostructured films of acetylcholinesterase and CdTe quantum dots. *Biosensors & Bioelectronics*, *26*(6), 3081–3085. doi:10.1016/j.bios.2010.12.021 PMID:21196108

Chapter 4
Sustainable Management and Production of Biofortified Crops Through Green Technology

Arun Kumar
Sardar Vallabhbhai Patel University of Agriculture and Technology, India

Rakesh Singh Sengar
https://orcid.org/0000-0002-7282-4135
Sardar Vallabhbhai Patel University of Agriculture and Technology, India

Surya Sudheer
Tallin University of Technology, Estonia

ABSTRACT

Hidden hunger is a term often used to describe malnutrition caused due to non/poor availability of minerals in the diet due to deficiency of micronutrients in staple food diet. Biofortification is an important, sustainable, and cost-effective technique of delivering micronutrients to a population that is not able to access diverse diets and other micronutrient interventions. This chapter deals with the aspects of crop biofortification, which includes breeding, agronomy, and genetic modification approaches. Here, the authors summarize the biofortification research that has been conducted on disparate crops. Success stories of biofortification include lysine and tryptophan-rich quality protein maize, Vitamin A-rich rice, Fe- and Zn-rich sorghum and pearl millet, anthocyanin- and β carotene-rich sweet potato, and several other crops for different micronutrients. Besides the challenges, the biofortified crops can fulfill the malnutrition challenge and have a bright future.

DOI: 10.4018/978-1-7998-8434-7.ch004

INTRODUCTION

The term hidden hunger is often used to describe malnutrition which caused due to non/poor availability of minerals in the diet, due to deficiency of micronutrients in staple food diet (Burchi *et. al.*, 2011). Nation's social and economic development and the human potential especially women and pre-school childrens are affected by malnutrition (Chilton *et. al.*, 2007). About 11% of all deaths globally under the age of five years, caused by micronutrient deficiencies. It has been estimated that the world's two-thirds population have a deficiency of one or more essential mineral elements in their diets. With the help of diversification of diet, supplementation of mineral, fortification of food or increasing the concentrations and/or bioavailability of mineral elements in produce, micronutrient deficiency problem can be cured in population (Dwyer *et al.*, 2015). A human requires at least 22 mineral elements for their well living and these can be provided by a specific diet. However, 6 billion people are iron (Fe) deficient, over 30% are zinc (Zn) deficient, 30% are iodine (I) deficient and 15% are selenium (Se) deficient, it has been estimated over 60% of the world's population. In addition, calcium (Ca), magnesium (Mg) and copper (Cu) deficiencies are common in many developed and developing countries (Welch and Graham, 2002). There are more than two billion people affected by deficiency of iron and zinc worldwide and mostly victims are children below the age of five and pregnant women (WHO, 2012). The annual deficiency of Fe, Zn and vitamin A in India is estimated to be 4.0milion, 2.8million and 2.3million disability-adjusted life years (DALYs) respectively. When observed with food intake, micronutrient malnutrition is more widespread compared to low energy uptake (Stewart *et al.,* 2010). Micronutrient deficiency is often referred to as "hidden hunger" as its manifestation is not obvious except in extreme cases unlike observed in insufficient energy intake (Meenakshi *et. al.,* 2010).

Micronutrient deficiency is the result of monotonous dependency on cereal based food especially in developing countries. It is also observed that 57% of pre-schoolers and mothers have vitamin A deficiency and over 330,000 child deaths every year are the result of vitamin A deficiency (VAD) at the chronic level in developing countries. Currently, the most serious global challenges to humankind are mineral malnutrition and it is avoidable (Copenhagen Consensus 2004; http://wwwcopenhagenconsensus.com). However, strategies to increase dietary diversification, mineral supplementation and food fortification have not always been successful. For this reason, the application of mineral fertilizers, combined with breeding varieties with an increased ability to acquire mineral elements, used for biofortification of crops. This is an immediate strategy not only to increase mineral concentrations in edible crops but also to improve yields on infertile soils (Graham *et. al.*, 2001).

The potential methodologies for enhanced nutrient supplementation, diversification of diets and nutritionally rich food crops have not been yet fully successful. Therefore, crop biofortification can be done by using several strategies (Garcia-Casal *et. al.*, 2016). The two most widespread nutritional disorders are Fe and Zn deficiency. It is estimated that Fe deficiency leads to reduce the work performance and also effects on body temperature regulation, development of psychomotor and intellectual performance, detrimental behavioral changes and increase the susceptibility to diseases and lead (Pb) poisoning in two billion of the world's population (Khush, 2008). Women and children have the most chances to affect by Fe deficiency because of their urgent need for child- birth and growth respectively. It has been estimated that 58% of women of developing countries are anaemic, so they bear a baby with low weight (Harding *et. al.*, 2017). The deficiency of Zn is widespread and shows its symptom in deficient condition. The symptoms of Zn deficiency are dwarfism, hypogonadism, anaemia and geophagia, heptosplenomegaly and mortality during childhood if it is prolonged (Prasad, 2013). Furthermore, in man, the Zn deficiency has been found to be associated with less amount utilization of vitamin A. Even higher number of population affected by micronutrient deficiencies in the developed countries. Taken together, a far greater number of the world's population affected by micronutrient deficiencies than protein-energy malnutrition.

Agriculture scientists are attempting to improve the nutrient content of food grains and vegetables through conventional breeding, marker-driven molecular breeding and genetic engineering, although the latter has become a subject of controversy. While ample care is needed for ensuring safety to health and the environment (biodiversity), misguided opposition needs to be countered. The consultative group on International Agricultural Research (CGIAR) has initiated the biofortification of seven food crops, including wheat, rice, pearl millet, and maize through its HarvestPlus initiative (harvestplus@cgiar.org, www.harvestplus.org, Accessed Jan 3, 2011) involving conventional breeding (Chakraborty *et. al.*,2010). India is an active partner in this initiative. In 2012, the iron rich pearl millet variety is commonly known as dhanashakti, which means prosperity and strength, and was commercialized by HarvestPlus partener Nirmal Seeds in Maharashtra, India. This firm is also trying to develop high-zinc-wheat. Bangladesh Rice Research Institute has developed zinc-rich-rice. Sweet potato varieties enriched with β carotene have also been developed by HarvestPlus. Work at the National Institute of Nutrition (NIN) has shown that about 25-50 g of the richer varieties can meet the preschool child's requirement of vitamin A. Bioavailability of β carotene from orange-fleshed sweet potato has been found to be good. Agriculture scientists need to work closely with nutrition scientists to promote such varieties. Biofortified varieties do not pose the problem of vitamin A toxicity that chemical supplements do, since there is regulation in absorption. Golden rice, enriched with β carotene has been developed by genetic engineering involving the introduction of three genes. Its safety, bioavailability and acceptability need to be tested. Bioavailability of micronutrients from biofortified crops should be the constant agenda for research since access of one micronutrient, particularly trace elements and minerals is known to inhibit the absorbance of other micronutrients competitively.

Consequences of Micronutrient Malnutrition

1. Higher disease rate.
2. Higher deathrate.
3. Lower cognitive ability.
4. Lower work productivity.
5. Effect on growth.

Possible Solutions to Micronutrient Deficiencies

Micronutrient malnutrition can be diminished either through food diversification, supplementation, fortification or biofortification. Following strategies offer a major role and are suggested-

DIETARY DIVERSIFICATION

Leafy vegetables, nuts, fresh fruits, milk, meat etc. can be included in the foods which are rich in micronutrients. Rural poor mostly depend on cereals or millets for consumption and they cannot afford different foods compared to urban. In India, 35% of total intake of calories, protein, irons and zinc is from the sorghum consumption in sorghum growing regions whereas more than 20-40% of total intake of calories, protein and micronutrients from bajra consumption in bajra predominant regions (Rao *et. al.*, 2006). Dry land India has infertile lands, where only millets can grow and hence the poor farmers entirely depend on the food which they grown in their field. So, they mostly depend on their accessible and affordable single source of food. To solve this issue, micro nutrient rich foods should be supplied to the poors.

SUPPLEMENTATION

Supplementation is a process when the micronutrients are provided orally either in the form of pills or syrup. This approach is useful for nutrients like vitamin A and also can be used for rapid improvement in Fe status in anaemic individuals. As early as 1970, the government started taking some steps especially in providing vitamin A tablets but it is not much successful. In some cases supplementation with micronutrients is the best method, like the persons who are suffering from chronic deficiency, and also the iron pills are to be given to pregnant women. It can be used as short term measures and is to be replaced with long term food based measures.

FORTIFICATION

External addition of nutrients to food at the time of food processing is known as fortification. Most commonly known example is the addition of iodine to salt and iron with wheat flour. Fortification is not an easy task, there are some problems with the fortification of food. Fe oxidizes rapidly so it is difficult to fortify in foods. The stability of the micronutrients which added to the food is only for a certain period and most of the micronutrients are lost during processing shows another problem with fortified food. The fortified foods with vitamins and minerals having a higher cost, which is not affordable by everyone.

Biofortification

The term 'Biofortification' is coined at the International Center for Tropical Agriculture in 2001 by Steve Beebe. The process biofortification is used for improving the nutritional quality of food crops through agronomic practices, conventional plant breeding or modern biotechnology methods. Biofortification increases nutrient level in crops during plant growth.

APPROACHES USED TO ENHANCE THE NUTRITION VALUE

Agronomic Approach

The micronutrient fertilizers can also be applied along with NPK fertilizers to the plants in their readily available forms which are taken up by plant roots and finally will be stored in the grains. These fertilizers can be applied to soil or through the foliar spray or can be delivered through irrigation water. The uptake of minerals from soil by plant roots influenced by several factors such as soil pH, temperature, rainfall, texture and structure of the soil (Kumar *et. al.*, 2016). Agronomic and genetic methods of biofortification are compatible and synergistic methods and both are needed for fighting malnutrition.

Transgenic Approach

Biofortification by inserting the new beneficial gene into the plants can be done with the help of the genetic engineering process. This new inserted gene can enhance the expression of storage proteins or other proteins (Garg et. al., 2018). These proteins help in the uptake of elements or can suppress some anti-nutritional factors like phytate. Biofortification through transgenic approach has been done in "Golden Rice" for enhancing vitamin A content and in Quality Protein Maize (QPM Maize) for high lysine content (Lee et. al., 2017). Mutation in opaque-2 was combined with genetic endosperm modifiers led to the formation of QPM Maize (Tripathy et. al., 2017). Several QPM cultivars have been released by National Agriculture Research System (NARS) in India. Iron biofortification in rice was done by ferritin gene transformation, metallothionin hyper-expression and phytase gene transformation (Wirth et. al., 2009). A zinc transporter gene (from Arabidopsis thaliana) over expressed in grain and it has found to increase in grain zinc content (Ramesh et. al., 2004). Few things are to be kept in mind while targeting biofortification of micro nutrients in crops, such as the plants are to be enhanced for uptake of minerals from soil by increased solubilization, increased transport of these micronutrients to grains rather than accumulating in leaves, increased sequestration of minerals in endosperm rather than husk or aluerone, increasing promoters of mineral bioavailability in grains and reduction in anti nutritional factors in grain.

Breeding Approach

Genetic variability is found in the crops naturally. We can select those lines which exhibit more micro-nutrient content by a screening of the germplasm and cultivars for micronutrient density and these can exploit in further breeding into elite lines through simple crossing or through genetic enhancement. To develop mineral dense varieties the genetic methods offers a sustainable solution by exploring natural genetic variation. Whenever enough variation does not exist in lines then mutation breeding approach is used for the creation of artificial variation. As early as in 1920, a natural mutant (opaque-2), which was rich in Lysine (twice as compared to normal) was identified in Maize. The other mutants such as fluory-2, mucronate (mc) and defective endosperm B30 (DEB30) were identified which lead to improvement of maize nutrition, later on with some transgenic interventions lead to QPM (Tripathy et. al., 2017). The target for breeding micronutrient varieties should be based on the bioavailability percentage, daily intake of food per capita, the form in which consumed and estimated average requirements (EAR). Based on these criteria, the level of concentration needed in varieties differs for different nutrients, for example, in case of Zn 10mg/Kg would be sufficient for varieties to be released for Northern India, whereas the concentration of Fe should be more as its bioavailability is less. Detection of markers which are tightly linked to Fe and Zn content enhancing genes or the genes responsible for this biofortification of micro-nutrients would hasten up the breeding process of selection. Transgenic approach has major limitations so that the transgenic is not preferred by the public and also there are several regulatory issues or lack of biosafety regulations observed, that's why the vitamin A enriched golden rice and Bt brinjal is not cultivated in India though they are released a long back (Dubock, 2014). Methods of breeding are cheap-est and sustained for a longer period, easily accepted both by farmers and rural poor consumers and provided them with a variety which is enriched with micronutrients. Plant breeding is most sustainable as farmers do not need to buy the seeds again and again unlike transgenic crop and they can reuse the same seed every year.

RATIONALE FOR BIOFORTIFICATION

In developing countries modern agriculture has been largely successful in meeting the energy needs of poor populations. Agricultural research in developing countries increased cereal production at its center in the past forty years. Nowadays agriculture is focusing on not to only produce more food for the population, but also deliver better quality food for malnutrition poor peoples (Webb *et. al.,* 2018). Biofortification approach will maintain improved nutritional status in people and will also help to severely malnourished people who require treatment by complementary interventions. Moreover, the rural poor populations who may not be able to afford commercially marketed fortified foods and supplements, biofortification provides benefits to them (Bouis *et. al.*, 2017). Traditional supplementation required for continual financial outlays whereas one-time investment in plant breeding can yield micronutrient-rich plants for farmers to grow around the world for years to come by the fortification programs. It is this multiplier aspect of biofortification that makes it more beneficial.

Advantages of Biofortification

Reaching the Malnourished in Rural Areas

The strategy of biofortification seeks to put the micronutrient-rich trait in the most profitable, highest-yielding varieties targeted to farmers and to place these traits in as many released varieties as is possible. Moreover, these crops as marketed surpluses should be reached in consumers of both rural and urban areas. The concept was clearly intervened in urban centers but the flow remained in the direction from the rural area to urban. The world cannot survive on low cost biofortified foods as they are unable to deliver a high level of nutrition but they can definitely fight against the malnourished and hidden hunger strike population to fulfill micronutrient sufficiency.

Sustainability of Biofortification

The nutritionally improved varieties will continue to be grown and consumed year after year and if once it is highly sustainable in a place, hence, recurrent expenditures are low.

Rice

Foods like fish, meat and milk are a rich source of vitamin A while vegetables like carrot, green leafy vegetables are the significant source of carotene (provitaminA). However, the poor masses of the country cannot afford varied diet as mentioned above and most of the poor people in our country depends on rice based diet, which is low in several essential micronutrients and vitamins (particularly pro-vitamin A) because rice varieties are not containing natural provitaminA. The Indian societies which depend only on rice food, the absence of β carotene in rice grains can leads to blindness, susceptibility to diseases, and increased incidence of premature death of small children. Rice plants produce β carotene (provi-

Figure 1. Agronomic and human nutritional benefits resulting from Zn enriched seeds

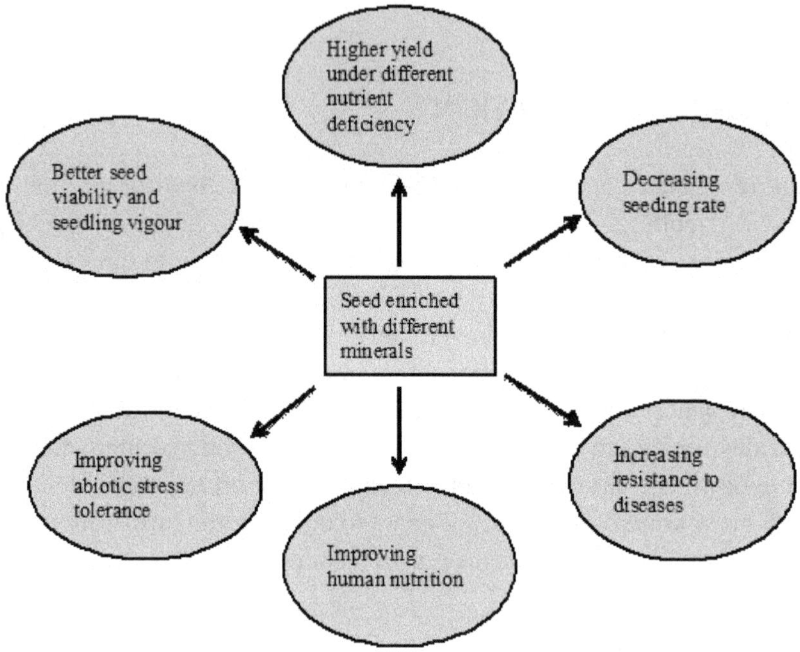

taminA) in the green tissues (not in the endosperm). The aleurone layer, the outer coat of the dehusked grains contains a number of valuable nutrients, e.g. vitamin B and nutritious fats but not provitaminA. At the time of milling and polishing, these nutrients are lost with the bran fraction. While if the fatty components are desirable to keep with the grain, these are affected by oxidative processes when exposed to air that makes the grain turn rancid. Thus, unprocessed rice, i.e. brown rice cannot be stored for a long time. Even though rice embryo consists all required genes to produce provitaminA but some of them are turned off during development. So the rice embryo cannot synthesize the β carotene (Sommer, 2008 and Mayer *et. al.*, 2011).

Golden rice is a rice variety which was developed by European scientists Ingo Potrykus and Peter Beyer. The project was started in year 1992 and published in the year 2000. In comparison with normal rice, it contain higher amount of β carotene, a source of Vitamin A. They introduced three genes for biosynthesis of β carotene in the rice embryo through genetic engineering (Potrykus, 2017). This was considered a significant breakthrough in area of biotechnology as the scientists had engineered an whole biosynthetic pathway. The original Golden rice produced 1.6 μg/g of carotenoids under greenhouse conditions and known as SGR1. In year 2004, The Louisiana State University Agricultural Center was conducted the first field trials of the golden rice cultivars. A team of biotechnology researchers of syngenta company produced another variety of golden rice known as "Golden Rice 2" in year 2005. They taken the phytoene synthase gene from maize and inserted it into the orginal golden rice for making the Golden rice 2. It is more beneficial for human being and produces 23 times more carotenoids than golden rice (up to 37 μg/g). It is estimated that, eaten 144 g of this rice would fulfill the Recommended Dietary Allowance (RDA) (Dubock, 2014).

Figure 2. The flow of different nutrients from soil through human food chains and their ecological cycling

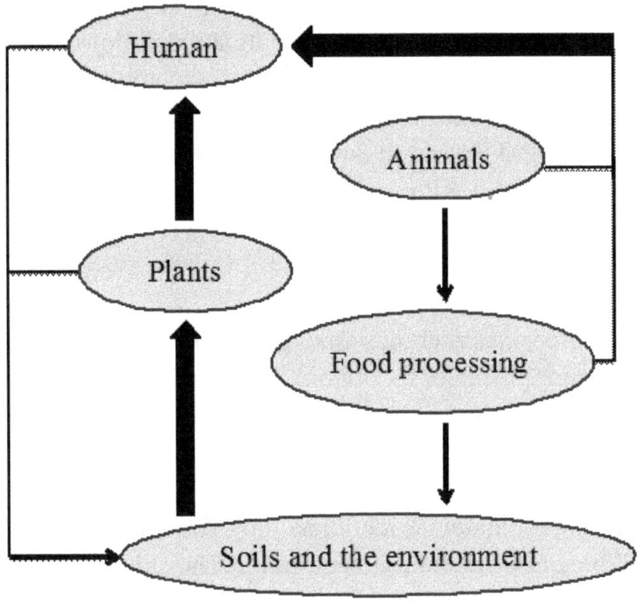

Maize

Quality Protein Maize (QPM)

A bulk of proteins (80%) are present in the endosperm of a mature maize kernel and the endosperm protein is deficient in two essential amino acids viz. lysine and tryptophan. This is due to the dominance of zein protein, which imposed unbalanced amino acid content in endosperm and results in the formation of poor quality protein in maize which have low biological value in humans. The discovery of *opaque2* mutant in maize found significant nutritional value, it enhanced the concentration of lysine and tryptophan amino acids in maize. *Opaque2 is* homozygous recessive allele which leads to the formation of double lysine and tryptophan amino acids in comparison to wild type, because it has good balancing property of all essential amino acids and it also reduces the accumulation of zein proteins and enhances the level of non-zein proteins in maize. However, *Opaque2 mutants* have pleiotropic effects such as soft texture, brittleness and insect susceptibility, lower seed density, breakage of the kernel during mechanical threshing, which decrease the value of the grain. These negative features of the mutant could be improved by the introduction of *Opaque2* modifier genes which have complex phenotypic effects and located throughout the genome. The modified- *Opaque2* genotypes which commonly known as Quality Protein Maize were developed by breeders at International Maize and Wheat Improvement Center through introgression of genetic modifier genes. These are rich in lysine and tryptophan and resemble normal maize both in kernel phenotype and agronomic performance (Babu *et. al.*, 2013). CIMMYT has played a vital role in the release of QPM varieties/synthetics and hybrids in many countries.

In 1971, the first series of biofortified maize composites were released in India such as Shakti, Rattan and Protina. These could not become popular because these were rich in protein quality but were having soft endosperm. In 1997, first hard endosperm-based opaque 2 composite/ QPM composite, Shakti-1 were released from CIMMYT germplasm. 'Shaktimaan- 1 (a white kernel-based three way cross) was the first QPM hybrid in India; it was released in 2001, followed by the release of 'Shaktimaan-2' (a white kernel-based single-cross hybrid) in 2004. `Shaktimaan-3' and `Shaktimaan-4' (first yellow-kernel based single cross hybrids) were developed together during 2006, followed by 'Shaktimaan-5' in 2013. These QPM hybrids are specifically adapted to Bihar. Later, a number of single cross QPM hybrids were released with wider adaptability to different agro-ecologies of the country viz. HQPM-1, HQPM-4, HQPM-5, HQPM-7 and Pratap QPM Hybrid-1. Several of QPM hybrids under different maturity groups are in experimental progress under AICRP programme (Muthusamy, 2015).

Provitamin A

Although maize is a major source of carotenoids among cereals, traditional maize cultivars possess less proportion of β carotene, a major precursor to vitamin A in humans. In the year 2012, Zambia released provitaminA enriched hybrids that contain 5-9 μg/g of provitaminA. With the help of MAS technology, the Scientists of IARI inserted the favourable*crt*RB1 allele to enhance Beeta-carotene concentration in seven elite parentalsinbreeds viz. VQL1, VQL2, V335, V345, HICI1105, HKI323 and HKI161. The target level set for alleviating vitamin A deficiency by HarvestPlus is 15 μg /g and the improved inbreedsprovides 8.6 to 17.5 μg /g β carotene concentration.The enhanced kernel β carotene as high as 21.7 μg/g, compared to 2.6 μg/g in the original hybrid found in reconstituted hybrids developed from improved parental genotypes (Muthusamy *et. al.,* 2014). The improved version of Vivek QPM 9 provides multi-nutrients in the maize- based diet because it contains high β carotene coupled with higher lysine and tryptophan amino acids.

Fe and Zn

To develop multi-nutrient rich maize through a systematic breeding approach, the reported Quality Protein Maize (QPM) genotypes which have a higher concentration of Fe and Zn can be explored QPM genotypes with high Fe and Zn has been identified by Mallikarjuna *et. al.,* (2014). Although the primary function of the phytate is to store phosphorus in seeds as energy source and antioxidants essentially required for the germinating seeds, the negative charge of phytic acids significantly chelate positively charged minerals and forms insoluble complexes in the gut. Hence breeding for low phytic acid assumes significance. Research efforts at the introgression of low phytate mutants (*lpa2-2*) into elite normal maize inbreds (UMI395 and UMI285) successfully done by TNAU, Coimbatore. VPKAS, Almora has been recently carried out marker-assisted introgression of *lpa1* and *lpa2* mutants in early maturing genotypes viz. CM145 and V334 respectively. These newly derived low phytate maize genotypes hold promise as they increase the bioavailability of Fe and Zn in maize, and enhance phosphorous availability for growth in monogastric animals such as poultry that digest phytate poorly (Sureshkumar *et. al.*, 2014). Further, phosphorous level in the environment and water is increased due to a higher concentration of undigested phytate eliminated by the monogastric animals. Thus newly developed genotypes of maize with low phytic acid can effectively reduce water and environmental pollution.

Pearl Millet

Pearl millet known with several names (as bajra, bajri, sajja and cumbu) in different states of India, and is a highly nutritious cereal grain. It contains higher protein with better amino acid balance, higher fat content, dietary fibre, and several minerals, including iron and zinc (Kulkarni *et. al.*, 2018). Conventional breeding approaches are most effective than genomic approaches due to former's greater cost effectiveness and short time frame for biofortified cultivar development and delivery. Fe and Zn density have quantitative inheritance hence the partnership with the public and private sector breeding programs was combined developed a multi-location testing process to accelerate the biofortification breeding process. Both private and public sector used **Iniadi germplasm** from West Africa having variability for high Fe and Zn.

Efforts from IARI

- Variability screening for the density of zinc and iron.
- Development of mapping population to map QTL for Fe and Zn density.
- These 2 lines have been identified for registration at NBPGR having high Fe and Zn density
- Pusa pearl millet inbred (PPMI 903)-Zn (72ppm) and Pusa pearl millet inbred (PPMI 904)-Fe (91ppm).
- Association mapping work is also going on in Pearl millet for grain iron and zinc content.

Variety Release

The International Crops Research Institute for Semi-Arid Tropics (ICRISAT) in collaboration with HarvestPlus (a Washington based NGO) developed the new biofortified bajra variety, called Dhanshakti. The nutritional value of staple crops can be increased by biofortification process through selective breeding or genetic modification methods.

Sorghum

Sorghum is the fourth most important consumed cereal in India. Sorghum is the staple of central and western regions of Maharashtra and the northern regions of Karnataka and Andhra Pradesh. As the whole consumption of sorghum has declined throughout India but it is still an important crop in inland regions of Central, Eastern and Western Maharashtra and Northern Karnataka with 31.8 to 54.2 kg annual per capita consumption in rural areas and 34.0 kg in urban areas. Maharashtra (47%) and Karnataka (20%) grows a large proportion of sorghum. In terms of nutrient intake, sorghum accounts for about 35% of the total intake of calories, protein, iron and zinc in the dominant production/consumption areas (Parthasarathy *et. al.*, 2006). Sorghum is a cheap source of energy, protein, iron and zinc after pearl millet among all cereals and pulses.

In the last few years large genetic variability for grain iron and zinc concentrations has been reported in sorghum by Harvest Plus for sorghum bio-fortification research. Over the years a large number of landraces (2246), hybrid parents (>500 B-lines and 100 R-lines), breeding lines and commercial sorghum cultivars (67) for assessing grain iron and zinc concentrations and important agronomic traits have been evaluated by International Crops Research Institute for the Semi-Arid Tropics (ICRISAT). The observed vast variability has been kept in the public domain in the form of a database. Major sorghum growing states indicated sufficient variability for grain iron (12-83 mg/ kg) and zinc (6-51 mg/kg) after characterization of popular Indian cultivars, hybrid parents, elite breeding lines and some selected germplasm. The base level of iron and zinc concentrations in grain observed 30 mg/kg for iron and 20 mg/kg for zinc respectively in the commercial cultivars and in the parental lines made by ICRISAT. The target level for genetic enhancement has been put at 60 mg/ kg for grain iron and 32 mg/kg for zinc at ICRISAT. To develop improved varieties and hybrid parents for grain iron and zinc, the identified landraces can be used in the crossing programmes as donors (Are *et. al.*, 2015).

Programs of Different Countries for Biofortification

Several staple food crops are being biofortify by a number of country programs running in Brazil, China, and India. The "food basket approach", favored particularly in Brazil, is useful to assess the deficiency of micronutrient in populations that consume smaller amounts of several staple crops (Saltzman *et. al.*, 2014). Brazilian Agricultural Research Corporation (Embrapa) coordinated BIOFORT Brazil and it developed nutrient-rich varieties of eight crops: cassava, sweet potato, rice, cowpea, bean, maize, pumpkin and wheat. Ten varieties have been released: three each of cassava (up to 9 ppm provitaminA); beans (up to 80 ppm iron and 50 ppm zinc); and cowpeas (up to 77 ppm iron and 53 ppm zinc); and one sweet potato variety (up to 115 ppm provitaminA). Furthermore, a provitaminA maize variety (up to 7.5 ppm) was released in 2012.

The biofortification program was started in 2005 in China and research focused on enhancing the iron, zinc, and provitaminA contents in rice, maize, wheat, and sweet potato. Released varieties include: a wheat cultivar, "Zhongmai 175", with 44 ppm zinc and 38 ppm iron; a rice variety, "Zhongguangxiang", with 6.5 ppm iron; and a sweet potato variety, "Nanshu 0101", with 93 ppm provitamin A. Additionally promising provitamin A-rich sweet potato and maize varieties are in multilocational trials. Department of Biotechnology (DBT) and the Indian Council of Agricultural Research (ICAR) have joined efforts to increase the development of biofortified varieties in India. The Indian biofortification programfocuses on rice, wheat and maize, is a long-term project of the DBT. HarvestPlusalso focuses on the development of biofortified crops and started the program to biofortify pearl millet and sorghum in collaboration with ICRISAT (Saltzman *et. al.*, 2014).

BioCassava Plus

Genetically engineered cassava with increased levels of iron and provitamin A, made by the BioCassavaPlus (BC^{+}) program. Increased shelf life, reduced level of cyanide, and improved disease resistance were the additional traits addressed by BC+ program. The first field trial for biofortified cassava for provitamin A was started in 2009, and it followed by high-iron cassava trials (Sayre *et. al.,* 2011).

Banana Biofortification

The Queensland University of Technology and the National Agricultural Research Organization of Uganda are developing transgenic provitamin A and iron bananas for Uganda. Bananas with up to 20 ppm provitamin A have been developed and trials have commenced in Uganda (Namanya, 2011). ProvitaminA bananas are expected to be released in 2019. High-iron bananas are not yet ready for human use.

African Biofortified Sorghum

A consortium of institutions for genetically modifying sorghum leaded by the Africa Harvest. Transgenic sorghum has an improved protein profile, enhanced level of provitamin A (up to 21 ppm) and reduced phytate (35-80 percent). Sorghum has lower relative bioaccessibility in transgenic varieties, but the higher levels of carotenoids in the grain increased the overall levels of accessible total and provitaminA carotenoids. Additional bioavailability studies have shown increased zinc absorption of 30-40 percent and increased iron absorption of 20-30 percent when phytate levels are reduced (Saltzman *et. al.*, 2014).

The National Agriculture Research System (NARS) has developed and released a series of biofortified varieties for ten different cereal, oilseed, vegetable and fruit crops (Yadava *et. al.,* 2018) as given in Table 1. With wide adoption and adaptation of three varieties, agriculture-based alleviation of malnutrition, especially undernutrition will become a reality. Several studies on the impact of feeding products from such fortified varieties in India and elsewhere especially in Africa, have confirmed the efficacy of such varieties in improving human nutrition and child health. All crop improvement programs must include nutritional quality traits in their genetic improvement objectives and ensure adequate availability of certified seeds of such varieties. Assured premium remunerative price for biofortified products will help faster adoption of such varieties for realizing the widespread impact.

Figure 3. Images of some biofortified crops

CONCLUSION

- In society, there are knowledge gap and lack of awareness among peoples towards the importance of biofortified crops. This could be encountered through implementing more effected field trials, knowledge exploration and extension ofpeople's awareness of the health benefits of biofortified crops and their products.
- There is also a need to redefine the potential indicators of nutrient status and for developing better strategies to understand the importance of cross nutrient synergies.

- To improve the adoption and consumption of biofortified crops, efficient delivery and market research strategies should be developed and implemented. The developmental strategies of biofortified crops should also be economic and cost effective.
- Maximum adoption and consumption of biofortified crops can be improved by the effectiveness of delivery and marketing research and strategies.
- Marker-assisted selection and transgenic methods may prove to be more effective than conventional breeding methods to enhance the level of several minerals and vitamins in a single variety.
- Agricultural research centers must adopt breeding for nutrient density as a core activity to mainstream biofortified traits.
- Only agronomic standards are considered currently to release a variety, whereas the minimum nutrient density standards should be set by the National varietal release committees to release a variety.

Table 1. Details of the baseline level of nutrients in targeted crops and levels achieved through biofortification

Crop	Nutrients	Baseline	Levels achieved
Nutritional factor			
Maize	Lysine	1.5-2.0%	> 2.5%
	Trytophan	0.3-0.4%	> 0.6%
	Provitamin A	1-2 ppm	> 8.0 ppm
Pearl millet	Zinc	30.0-35.0 ppm	> 40.0 ppm
	Iron	45.0-50.0 ppm	> 70.0 ppm
Wheat	Iron	28.0-32.0 ppm	> 38.0 ppm
	Zinc	30.0-32.0 ppm	> 40.0 ppm
	Protein	8-10%	> 12.0%
Rice	Zinc	12.0-16.0 ppm	> 20.0 ppm
	Protein	7.0-8.0%	> 10.0%
Lentil	Zinc	35-40 ppm	> 50.0 ppm
	Iron	45-50 ppm	> 62.0 ppm
Cauliflower	Bcarotene	Negligible	> 8.0 ppm
Sweet potato	Anthocyanin	Negligible	> 80.0 mg/100 g
	Bcarotene	2.0-3.0	> 13.0 mg/100 g
Pomgranate	Vitamin C	14.2-14.6 mg/100 g	> 19.0 mg/100 g
	Iron	2.7-3.2 mg/100 g	> 5.0 mg/100 g
	Zinc	0.50-0.54 mg/100 g	> 0.6 mg/100 g
Anti-nutritional factor			
Soybean	Kunitz trypsin inhibitor	30-45 mg/g of seed meal	Negligible
Mustard	Glucosinolates	> 120.0 ppm	<30.0 ppm
	Erusic acid	> 40%	<2.0 ppm

REFERENCES

Are, A. K., Anuradha, K., Ramaiah, B., & Grando, S. (2015). Recent advances in Sorghum biofortification research. *In book. Plant Breeding Reviews*, *39*, 89–124.

Babu, R., Rojas, N. P., & Prasanna, B. M. (2015). Biofortified Maize - A Genetic Avenue for Nutritional Security. *In book. Translational Genomics for Crop Breeding*, *2*, 161–176.

Bouis, H. E., & Saltzman, A. (2017). Improving nutrition through biofortification: A review of evidence from HarvestPlus, 2003 through 2016. *Global Food Security*, *12*, 49–58. doi:10.1016/j.gfs.2017.01.009 PMID:28580239

Burchi, F., Fanzo, J., & Frison, E. (2011). The role of food and nutrition system approaches in tackling Hidden Hunger. *International Journal of Environmental Research and Public Health*, *8*(2), 358–373. doi:10.3390/ijerph8020358 PMID:21556191

Chakraborty, S. N., Chakraborty, L., Agrawal, S., & Narula, G. K. (2010). Next generation protein rich potato by expressing a seed protein gene AmA1 as a result of proteome rebalancing in transgenic tuber. *Proceedings of the National Academy of Sciences of the United States of America*, *107*(41), 17533–17538. doi:10.1073/pnas.1006265107 PMID:20855595

Chilton, M., Chyatte, M., & Breaux, J. (2007). The negative effects of poverty & food insecurity on child development. *The Indian Journal of Medical Research*, *126*, 262–272. PMID:18032801

Dubock, A. (2014). The politics of Golden Rice. *GM Crops and Food: Biotechnology in Agriculture and the Food Chain*, *5*(3), 210–222. doi:10.4161/21645698.2014.967570 PMID:25437240

Dubock, A. (2014). The present status of Golden Rice. *Huazhong Nongye Daxue Xuebao*, *33*(6).

Dwyer, J. T., Wiemer, K. L., Dary, O., Keen, C. L., King, J. C., Miller, K. B., Philbert, M. A., Tarasuk, V., Taylor, C. L., Gaine, P. C., Jarvis, A. B., & Bailey, R. L. (2015). Fortification and Health: Challenges and Opportunities. *Advances in Nutrition*, *6*(1), 124–131. doi:10.3945/an.114.007443 PMID:25593151

Garcia-Casal, M. N., Pena-Rosas, J. P., & Giyose, B. (2016). Staple crops biofortified with increased vitamins and minerals: Considerations for a public health strategy. *Annals of the New York Academy of Sciences*, *1390*(1), 3–13. doi:10.1111/nyas.13293 PMID:27936288

Garg, M., Sharma, N., Sharma, S., Kapoor, P., Kumar, A., Chunduri, V., & Arora, P. (2018). Biofortified crops generated by Breeding, Agronomy, and Transgenic approaches are improving lives of millions of people around the world. *Frontiers in Nutrition*, *5*, 12. doi:10.3389/fnut.2018.00012 PMID:29492405

Graham, R. D., Welch, R. M., & Bouis, H. E. (2001). Addressing micronutrient malnutrition through enhancing the nutritional quality of staple foods: Principles, perspectives and knowledge gaps. *Advances in Agronomy*, *70*, 77–142. doi:10.1016/S0065-2113(01)70004-1

Harding, K. L., Aguayo, V. M., Namirembe, G., & Webb, P. (2017). Determinants of anemia among women and children in Nepal and Pakistan: An analysis of recent national survey data. *Maternal and Child Nutrition*, *4*(1), e12478. PMID:28857410

Khush, G. S. (2008). Biofortification of Crops for Reducing Malnutrition. *Proc. Indian Natn. Sci. Acad.*, *74*, 21–25.

Kulkarni, D. B., Sakhale, B. K., & Giri, N. A. (2018). A potential review on millet grain processing. *International Journal of Nursing Sciences*, *3*(1), 1018.

Kumar, A., Choudhary, A. K., Pooniya, V., Suri, V. K., & Singh, U. (2016). Soil Factors Associated with Micronutrient Acquisition in Crops- Biofortification Perspective. *Biofortification of Food Crops*, 158-176.

Lee, H. (2017). Transgenic Pro-Vitamin A Biofortified Crops for Improving Vitamin A Deficiency and Their Challenges. *The Open Agriculture Journal*, *11*(1), 11–23. doi:10.2174/1874331501711010011

Mallikarjuna, M. G., & Nepolian, T. (2014). Genetic variability and correlation of kernel micronutrients among exotic quality protein maize inbreads and their utility in breeding programme. *Indian Journal of Genetics and Plant Breeding*, *74*(2), 166–173. doi:10.5958/0975-6906.2014.00152.7

Mayer, J. E., & Potrykus, I. (2011). 'Golden Rice' and Biofortification - Their Potential to Save Lives Is Being Hampered by Overzealous Regulation. *Acta Horticulturae*, (941), 21–34. doi:10.17660/Acta-Hortic.2012.941.1

Meenakshi, J. V., Johnson, N. L., Manyong, V. M., Groote, H. D., Javelosa, J., & Yanggen, D. R. (2010). How cost-effective is biofortification in combating micronutrient malnutrition? An ex ante assessment. *World Development*, *38*(1), 64–75. doi:10.1016/j.worlddev.2009.03.014

Muthusamy, V. (2015). Biofortification of maize: An Indian perspective. *Indian J. Genet.*, *75*(1), 1–22. doi:10.5958/0975-6906.2015.00001.2

Muthusamy, V., Hossain, F., Thirunavukkarasu, N., Choudhary, M., Saha, S., Bhat, J. S., Prasanna, B. M., & Gupta, H. S. (2014). Development of β-Carotene rich Maize hybrids through Marker-Assisted Introgression of *β-carotene hydroxylase* Allele. *PLoS One*, *9*(12), e113583. doi:10.1371/journal.pone.0113583 PMID:25486271

Namaya, P. (2011). *Towards the biofortification of banana fruit for enhanced micronutrient content* (PhD thesis). Queensland University of Technology.

Parthasarathy, R. P., Birthal, P. S., Reddy, B. V. S., Rai, K. N., & Ramesh, S. (2006). Diagnostics of Sorghum and Pearl Millet Grains-based Nutrition in India. *International Sorghum and Millets Newsletter*, *47*, 93–96.

Potrykus, I. (2017). The GMO-crop potential for more, and more nutritious food is blocked by unjustified regulation. *Journal of Innovation & Knowledge*, *2*(2), 90–96. doi:10.1016/j.jik.2017.03.003

Prasad, A. S. (2013). Discovery of Human Zinc Deficiency: Its Impact on Human Health and Disease. *Advances in Nutrition*, *4*(2), 176–190. doi:10.3945/an.112.003210 PMID:23493534

Ramesh, S. A., Choimes, S., & Schachtman, D. P. (2004). Over-Expression of an Arabidopsis Zinc Transporter in Hordeum Vulgare Increases Short-Term Zinc Uptake after Zinc Deprivation and Seed Zinc Content. *Plant Molecular Biology*, *54*(3), 373–385. doi:10.1023/B:PLAN.0000036370.70912.34 PMID:15284493

Rao, P. P., Birthal, P. S., Reddy, B. V., Rai, K. N., & Ramesh, S. (2006). Diagnostics of sorghum and pearl millet grainsbased nutrition in India. *Int. Sorghum Millets News*, *47*, 93–96.

Saltzman, A., Birol, E., Bouis, H. E., Boy, E., De Moura, F. F., Islam, Y., & Pfeiffer, W. (2013). Biofortification: Progress toward a more nourishing future. *Global Food Security*, *2*(1), 9–17. doi:10.1016/j.gfs.2012.12.003

Sommer, A. (2008). Vitamin A deficiency and clinical disease: An historical overview. *Clinical Nutrition (Edinburgh, Lothian)*, *138*, 1835–1839. PMID:18806089

Stewart, C. P., Dewey, K. G., & Ashoran, P. (2010). The undernutrition epidemic: An urgent health priority. *Lancet*, *375*(9711), 282. doi:10.1016/S0140-6736(10)60132-8 PMID:20109955

Sureshkumar, S., Tamilkumar, P., Senthil, N., Nagarajan, P., Thangavelu, A. U., Raveendran, M., Vellaikumar, S., Ganesan, K. N., Balagopal, R., Vijayalakshmi, G., & Shobana, V. (2014). Marker assisted selection of low phytic acid trait in maize (*Zea mays* L.). *Hereditas*, *151*(1), 20–27. doi:10.1111/j.1601-5223.2013.00030.x PMID:24627970

Tripathy, S. K., & Maharana, M. (2017). Quality protein maize (QPM): Genetic basis and breeding perspective. *Tropical Plant Research*, *4*(1), 145–152. doi:10.22271/tpr.2017.v4.i1.021

Webb, P., Stordalen, G. A., Singh, S., Wijesinha-Betton, R., Shetty, P., & Lartey, A. (2018). Hunger and malnutrition in the 21st century. *BMJ (Clinical Research Ed.)*, *361*, k2238. doi:10.1136/bmj.k2238 PMID:29898884

Welch, R. M., & Graham, R. D. (2002). Breeding crops for enhanced micronutrient content. *Plant and Soil*, *245*(1), 205–214. doi:10.1023/A:1020668100330

Wirth, J., Wirth, J., Aeschlimann, B., & Yakandawala, N. (2009). Rice endosperm iron biofortification by targeted and synergistic action of nicotianamine synthase and ferritin. *Plant Biotechnology Journal*, *7*(7), 631–644. doi:10.1111/j.1467-7652.2009.00430.x PMID:19702755

Yadava, D. K., Hossain, F., & Mohapatra, T. (2018). Nutritional security through crop biofortification in India: Status & future prospects. *The Indian Journal of Medical Research*, *148*, 1–11. PMID:30666987

Chapter 5
Eco-Friendly, Non-Conventional Approaches for Sustaining Agriculture in an Organic Perspective

S. S. Rathore
Indian Agricultural Research Institute, India

Kapila Shekhawat
Indian Agricultural Research Institute, India

ABSTRACT

Increasing use of agro chemicals in crop production for management of different types of biotic stresses has become the serious environmental and health concern. The unmindful use of these chemicals is the reason for increasing greenhouse gas emission, water body pollution, and toxic residue in the food items. There are good agronomic practices that can minimize biotic and also abiotic stresses in crop production, if followed properly. The crop diversification, balanced use of plant nutrition, protecting the friendly bio agents naturally occurring in the crop field are some of the effective non-chemical methods for pest management. Herbicides increasingly provide a suitable shift for the weed control function of tillage, but the overall effective strategy is to adopt good crop husbandry. The refinement of the seeding and crop establishment methods and standardization of conservation agriculture practices will make such systems progressively striking economic alternates for non-chemical pest management.

INTRODUCTION

The increasing trend in management of agricultural affairs through use of chemicals and totally ignoring the importance of non-chemical components is responsible for present mess. Unmindful levels of commercialisation in agriculture had a devastating effect on whole agri-ecosystem. With the sole reliance on the chemicals, the people started realising the ill-effects of pesticides. The publication of 'Silent Spring'

DOI: 10.4018/978-1-7998-8434-7.ch005

by Carson 1962, created a good amount of awareness regarding the ill-effects of chemicals. These chemicals are mainly used to supplement the nutrient demand of the crop and to management different types of biotic stresses caused by pathogens and insect-pests etc. In this context managing the crop demand through eco-friendly non-chemical methods is very pertinent for reducing the cost of cultivation and to ensure environmental safety. Since the excessive use of synthetic chemicals is harmful for the environment, the good crop husbandry or good agronomic practices in predominant cropping system should have more focus on the prevention rather than curative measures against the insect –pests, pathogens. In nature the friendly insect pest and other microbes are available which manages the enemy's pest in the crop if utilized properly. The use of natural microbes and insect bio-agents for pest management needs to be promoted, the crop agronomic practices like crop rotation, diversification are also very effective in minimising the pest menace in the crops. The best management of arable and non-arable lands through right selection of crops and their varieties, intercropping and crop sequences to maintain a rich diversity of beneficial microbes by inclusion of competitors, parasites and predators of harmful pests. Planning must involve the complete farm operations by judiciously using available data and resources. The approach for non-conventional disease, insect pest management should include avoidance, varying ambience for the crops and control methods. Though, no single technique is adequate for all pests and pathogens hence, effective management of biotic stresses in crops (insect, disease and weeds) must involve the integrated management strategies, which basically means good crop husbandry.

The good soil fertility and higher crop productivity are largely determined by soil organic matter. Therefore, chemical fertilizers based soil health management is a flawed exercise as this in long run results in decline in soil organic matter. The optimum crop growth and productivity is prominently depends on soil microbes especially for supply of adequate plant nutrient. Fertile soils are being maintained with larger population of microbes. Organic matter in soil is one of the most deciding factors in maintaining optimum soil health on long term basis and dead or alive soils are defined on the basis of SOM content. The humus provides stability even in most unhealthy soil like heavy clay or sandy soils. Organic matter contains carbon compounds that were formed by living organisms. Organic agriculture builds the health of the soil, providing the foundation for healthy crops and a livelihood for good stewards of the land for the farmers. Thus, SOM is lifeline to keep soil as living and dynamic agri-ecosystem. The success of the technology in the India and other countries has generated great interest adoption of such conservation farming systems elsewhere. This is more important in present context of declining organic matter content in the soils and increasing soil degradation.

GOOD AGRONOMIC PRACTICES FOR NON-CHEMICAL MANAGEMENT IN AGRICULTURE

The good agronomic practices maintain minimum levels of pest-pathogens in the field and make crop plant competitive against the biotic and abiotic stresses. These are also known as good crop husbandry. The prevention is always better than doing any control measures for managing the stresses in crop plants. In this regards, the practices for pest avoidance, site selection, exclusion and most importantly ensuring the adequate crop diversification in the field are some of the effective non chemical measures for biotic and abiotic stress management in crop fields.

Good Agronomic Practices for Pest Avoidance

This is important to apprehend the pest and crop biology and growth habits and this is vital for effective management of biotic stresses. The type of pest infestation, its level, previous crop history, life cycles of pathogens and pest, along with soil conditions, prevailing weather conditions are significant to plan any strategy for management. This is again important to adopt the strategies which can avoid the pest-pathogen entry in the system. accordingly modified the sowing time and cultural practices to escape the crop plants from infestation.

Site Selection

Before sowing, a farmer should carefully plan out planting and crop rotation strategies to avoid any known problems. Plant pathogenic fungi such as *Armillaria*, *Fusarium* (the wilt causing fungi), *Plasmodiophora*, *Sclerotium* and *Verticillium* are true soil inhabitants and will persist in the soil for many years even in the absence of a plant host. Soil borne fungi, like *Pythium* and *Rhizoctonia*are very widespread and are hence very difficult to avoid in the fields. The well managed recorded help a lot in appropriate site selection. A farmer should note on previous soil borne disease problem associated with certain field in relation to other key areas like weed reservoir, environmental characteristics like location, and nature of the soil, water and other physical features of each site.

Exclusion

The practice of keeping out any material or object that are contaminated with pathogens or diseased plant and preventing them from entering the production system is known as exclusion. For some diseases the seed borne pathogens are the primary source of pathogen dissemination. Farmers should purchase seeds that has been tested and certified to be below a threshold infestation level and has been treated to reduce pathogen infestation level. Soil and water can harbour pathogens as well. Care should be taken that no infested soil/water is introduced to non-infested field. For example, tomato bushy stunt virus of lettuce, tomato and other crops contaminate the river, flood and runoff water and use of this contaminated water as irrigation to healthy crop led to appearance of disease. The farmers who have dredged up soil from the river in to the field have found that infestation has reached their fields. The water draining from the field also carries lots of pathogens. It can also spread after adhering to the farm implements and machinery.

Crop Rotation/Cropping Systems

Suitable cropping system is an important approach to curb on pests menace in agriculture. It is an extremely effective way to minimize most pest's problems while maintaining and enhancing soil structure and fertility. The well planned cropping systems involving suitable crop rotations are very useful in for managing diseases (soil and stubble-borne). The plant disease prevention through crop rotations depends largely upon the factors, like pathogen to survive devoid of its host and the pathogen's host plant range. Those with a wide range of hosts will be controlled less successfully (Table 1). For example, *sclerotinia* stem-rot is a common disease in canola, but it can also infect at least a half dozen other field crops. Rotations will shorten the life span of pathogens that can survive only brief periods apart from their hosts. Other situations that limit the benefit of crop rotations include: the transmission of pathogens via seed, the presence of susceptible weeds and volunteer crops that harbour pathogens, and the invasion of pathogens by wind and other means. Rotations should be used with other cultural practices to achieve the greatest benefit.

Table 1. Diseases controlled entirely or in part by rotation (Jensen, 1971)

S. N	Disease	Major Crops Attacked	Best Control Methods
1.	Ergot	rye, wheat, grasses	Rotation, tillage
2.	Bacterial blights	wheat, barley, grasses, rye	Rotation
3.	Net blotch	barley	Rotation, fungicide
4.	*Septoria* (different species)	wheat, barley	Rotation, fungicide
5.	*Septoria* (different species)	wheat	Rotation, fungicide
6.	Seedling blight	wheat, barley, corn, oats, rye	Seed treatment
7.	Smut (corn)	corn	Rotation
8.	*Verticillium* wilt	potato, sunflower, safflower	Rotation, variety
9.	Rust (sunflower)	sunflower	Variety, rotation
10.	*Sclerotinia* (white mold)	sunflower, dry beans soybean, potato, canola	Rotation 4 to 5 years

For the management of cereal cyst nematode (*Heteroderaavenae*), the crop rotation has been considered as one of the best method. In this context, *Brassica compestis var. dichotoma, B. campestris var. toria and Erucasativa* has been recommended for cultivation in one season in the infested fields. The cultivation of above crops in infested fields reduced the cyst population by 87 to 100 per cent. In addition, the paddy-wheat rotation also reduced the population of *Meloidogyne incognita*, a root-knot nematode of vegetable crop, to a zero level within three years (Sukhuja and Jain 2001).

Crop Residue Management

Reducing or removing crop residues and alternate host sites can be used to control some diseases. Incorporating the residue into the soil hastens the destruction of disease pathogens by beneficial fungi and bacteria. Burying diseased plant material in this manner also reduces the movement of spores by wind. Reduced or zero-tillage may also reduce the damage by certain pests, as the crop residue creates a micro-climate less preferred by some insects like flea beetles (Bockus et al., 1979). Pathogens survive on some of the plants as alternate habitat until an appropriate crop occurs in a nearby field. However, these crop residues may also host many beneficial insects and predators therefore; the grower must carefully assess the potential threat and benefits from pest insects in these areas before moving or removing any plants. The severity of cabbage yellowness disease and populations of *Fusarium oxysporum f. sp. conglutinans* in soil were markedly reduced by amendments of nine species including alfalfa hay and wheat straw. The effective control of cabbage yellowness disease through amendment was increased by drying the residue before incorporation. Soil solarisation through transparent polythene mulch and soils amendment with residue practically eliminated the propagules of *F. oxysporum f. sp. conglutinans* and cabbage yellowness was undetected in cabbage growing in the soil (Ramirez and Munnecke 1988). Thus, it is the most important practice to maintain a balance between crop sanitation and soil conservation.

Seed Quality

The seed lots posses high genetic purity and high germination percentage and a minimum of inert, weed and other crop seeds and are free from diseases, is known as high quality seed. It implies that if a seed material meets the certification standard, it is good quality seed and if does not meet the certification standards, it is obviously of a lower seed quality. Thus, use of high-quality seed is especially important in preventing diseases and other pests. The seed supply should be free of smut, ergot bodies or others *clerotia*, and free of kernels showing symptoms of *Fusarium* head blight infection. Seed analysis by a reputable seed testing laboratory will help to determine specific diseases in the seed supply. Relatively few diseases are exclusively seed-borne, and it is more common for pathogens to be transmitted from soil, stubble, or wind, as well as with the seed. In crops such as flax, rye and pulses, a crack in the seed coat may serve as an entry point for soil-borne micro-organisms that rot the seed once it is planted.

Weed Management

Although weeds need to be controlled to reduce their impact on crop yield and quality, a field completely free of weeds is not necessarily but the best objective should be keep weed population below economic threshold level. In many cases, weeds provide food and shelter for beneficial insects. Parasitic wasps, for example, are attracted to certain weeds with smell of flowers. Field experience has shown that the number of predators attacking pathogens increases and the number of aphids and leaf hoppers decreases on certain crops as the diversity of weeds increases.

Forecasting

Producers should pay attention to the forecasts for various pest and disease infestations for each crop in a year. Maps of these forecasts are usually available for many of the major destructive insects such as grasshoppers and wheat midge, as well as some diseases. Agro-meteorological warning and forecast can help in this way.

Altering the Growth Environment

Any crop management technique that contributes to a vigorous and competitive crop, is a tool for insects and diseases management. Producers must also be mindful that many practices that work well in conventional systems may not benefit organic systems. Certain crop species, crop varieties and equipment may work well in one system but not in the other.

Healthy Soil

Maintaining favourable soil conditions is the first line of defence against pests. A biologically active soil with good drainage supports vigorous crop growth, allowing a higher level of crop competition with weeds. Adequate and balanced soil nutrition is essential for crop quality, yield and moisture-use efficiency. The application of nutrients should be based on a sound soil testing program, accompanied by plant tissue analysis when diagnosing problems. High levels of nitrogen can occur after a high-nitrogen plough down, such as sweet clover. This results in lush leaf tissue and a dense plant canopy that provides an ideal environment for plant pathogens. However, a lush crop may also help disperse the damage by a given number of insects, so astute observations by the producer are necessary at all times. In contrast, inadequate soil phosphorous can pre-dispose to certain root diseases. A shortage of micronutrients such as zinc or copper can result in disease-like symptoms on crops, while too much of any one micronutrient may be toxic. The addition of well decomposed livestock waste manure improves soil quality and increasing the population of soil micro-organisms that compete with soil-borne plant pathogens.

Crop Competition to Minimize the Chemical Use

Producer awareness of diseases in the proximity of the farm is very important and should influence the crop and the variety of crop to be grown. These choices must also fit in with the crop rotation plan that has been developed. Some diseases are specific to certain crops. The selection of disease-resistant cultivars can be a useful tool. Certain wheat species may avoid diseases such as *Fusarium* head blight, but often agronomic factors, such as time of seeding or choice of winter versus spring wheat, have more of an influence on the incidence of disease. Plants also vary in their degree of attractiveness to diseases and vectors transmitting disease. Factors such as leaf and stem toughness, pubescence, nutrient content, plant architecture, growth habit and differences in maturity between crops and varieties can influence pest growth, reproduction and host preference.

Intercropping

The practice of intercropping (where two crops are grown at the same time) can reduce disease problems by making it more difficult for the pathogens to find a host crop. This technique also provides habitat for beneficial organisms. Strip-cropping of crops in rows with perennial legumes often leads to better disease control.

Seeding Date

Planting should be scheduled at optimum favourable conditions so that the most susceptible crop growth stage of plant does not coincide with the peak of pest cycles. Early seeding reduces crop damage caused by grasshoppers, aphids in cereal crops, wheat midge in spring wheat, barley yellow dwarf virus in barley and wheat, powdery mildew in peas and flax. Delayed seeding can be effective in avoiding wireworms and cutworms in cereal crops, Hessian fly in winter wheat, barley thrips, *ascochyta* in lentils and wheat streak mosaic virus in winter wheat. However, experience on the Prairies has generally shown that delaying seeding too long can also reduce a crop's potential yield.

Seeding Rate

Using a higher seeding rate to affect insect or disease infestations may have different results. More plants in a field may reduce the impact of a given aphid population on individual plants, but they may create a more favourable habitat for insects that prefer a dense canopy, such as true armyworm. A dense leaf canopy can also create a moist soil surface and elevated humidity within the crop, conditions favourable to certain leaf disease pathogens. Reducing the seeding rate may decrease the severity of take-all in spring wheat, but the reduced canopy may also allow weeds to invade. In other crops, reduced seeding may also produce more insect damage, as in the case of aphids, flea beetles and leafhoppers, which are attracted to the contrast between a green host and a dark soil background.

Depth and Timing of Seeding

Optimum seeding depth is also important. Deep seeding in cold soils may result in seedling blights and damping-off, especially in pulses and small seeded crops. Seeding depth should generally be no deeper than required for quick germination and even emergence. Variables include seed size, soil type and moisture conditions. If the soil is loose before seeding, a packing operation will firm up the soil and bring moisture closer to the surface. For most crops, seeding should ideally be done when the soil is warm enough for rapid germination. Seeds that remain un-germinated in cool soil are more susceptible to damage by pathogens.

Trap Strips

Trap crops are known to interfere with the visitation and oviposition by key insect pests and have been successfully utilised in soybean, cotton, okra, groundnut, potato and cauliflower. It has been demonstrated that when okra was grown as a border crop around cotton field it trapped large number of eggs of spotted bollworm and the periodic removal of eggs significantly reduced the incidence of bollworm in cotton (Thontadaryaand and Jairao 1980). Trap crops of susceptible plants are grown on land known to contain pathogens. They become infected and are then destroyed before the pathogens' life cycles are complete, thus reducing the amount of inoculums in the area. Decoy crops stimulate the hatching or germination of pathogen eggs or propagules, but the pathogens are unable to establish and infection of the decoy host and die, again reducing the amount of available inoculum. The ideal decoy crop is one that has economic value and can be used in a routine crop rotation

Tillage

Tillage can be properly timed before seeding, after harvesting and during summer fallow to reduce populations of pathogens take shelter in the soil or stubble. It is beyond doubt that ploughing (deep/shallow), raking, earthing up and hoeing influence the pest fauna considerably in agro-ecosystem. Not only that, zero tillage operation too determines the abundance of pest populations. Among these practices, the type and the time of tilling of soil affect the pest populations through direct interference as well as exposing their over-wintering stages to natural enemies/adverse weather. Deep ploughing buries insects too deep to emerge from the soil.

Roguing

The practice of roguing removes diseased/insect-pests infested crop plants. Roguing may not be practical for large fields, but could be suitable for seed plots or crops having highly infectious and destructive diseases (for example, bacterial blackleg in potatoes and certain viruses in other crops).

Direct Treatment

At a time, the organic producer will find that despite all the best efforts, an insect or disease pest will grow to levels that cause substantial crop damage. At this point, direct treatment may be necessary. Once when the crops are in the field, sometimes it becomes necessary to apply some sort of protectant or eradiant spray or dust material for disease control. In organic crop production, among disease control materials primarily copper and sulphur based fungicides are most commonly used. Products such as oil, plant extracts and other natural plant products are also compatible with organic production practices made by microorganisms are popularizing in recent times. There are a lot of commercially available biological control materials which are economically feasible and very effective also. The disease management through direct treatment can be understood under following sub-heads:

Biological Control

In a healthy, balanced ecosystem, biological control by natural predators is constantly occurring. The more diverse a cropping system, there will be greater the spectrum of pathogens, insect species and micro-organisms within it. This leads to the development of more natural predators within the ecosystem. Very few biological controls are available to reduce the effects of plant diseases, as most of the commercial products do not perform well if the disease is already established in the crop. Mycoparasitism is a form of bio-control where one fungus parasitizes another. Although this process occurs with many fungi under laboratory conditions but it hasn't been successful under field conditions.

Trichoderma is present all over the world in the soil and decomposing plant material. It does not attack and kill living plants. It helps to decompose organic matter, releasing fertiliser for the crops. It also protects crop plants from soil-borne diseases and animals that attack plant roots. It produces chemicals (antibiotics) that are toxic to other micro-organisms. Some naturally occurring fungi can be a great help in controlling diseases. The well decomposed compost, rich in *Trichoderma* can be put directly onto the soil or can put it in water for some hours and spray the water on the soil containing a disease.

Mineral Nutrition for Disease Management

All essential plant nutrients influence the health of plants and their susceptibility to disease and disease severity (Huber and Graham, 1999). Mineral nutrition affect two primary resistance mechanisms, formation of mechanical barriers (e.g. thickness of cell walls) and the synthesis of natural defence compounds (e.g. antioxidants, phytoalexins and flavenoids). Plants suffering a nutrient stress will be more susceptible to diseases, while adequate crop nutrition makes plants more tolerant or resistant to disease. Different pathogens have different infection mechanisms. The important pathogens, their mode of infection and the possible control through mineral nutrition is given as under

Fungi

Fungi penetrate the surface cells (epidermis), by passing between the cells or through them. The cell walls present a physical resistance to the fungus and stronger cell walls can prevent the infection. Certain nutrients, like Calcium, play a major role in the ability of the plant to develop stronger cell walls and tissues. Plants exude various compounds which also prevent the infection of pathogens. When some nutrients are below a certain level, the released compounds contain higher amounts of sugars and amino acids that stimulate the establishment of the fungus. Mineral nutrition also affects the production of antifungal compounds in plants.

Bacteria

Bacteria invade the plant tissues through wounds, sucking insects and through the stomata. Then they spread within the intercellular spaces and release enzymes that dissolve the plant tissue. Calcium is known in its ability to inhibit such enzymes. The ability of bacteria to spread within the plant tissues depends on the strength of the internal cells, which is greatly affected by mineral nutrition.

Nitrogen Application

Adequate nitrogen levels are necessary for disease resistance. Huber and McCay-Buis (1993) reported that when N was applied to cereal crops, take-all (*Gaeumannomycesgraminis*) disease was reduced. Also, P reduced both take-all and pythium root rot infection in cereal crops (Kiraly, 1976). However, excess nitrogen may also promote favourable conditions to plant disease. Some of the reasons for this are:

- Excess nitrogen promotes thinner and weaker cell walls
- Increased plants density and therefore high humidity and low light conditions.
- Excess nitrogen delays maturity of plant tissues and therefore increases the risk of disease infection and development.
- Unbalanced N: K ratio affects both yield and disease resistance.

Potassium and K: Ca Ratio

Potassium is a very important nutrient in plant disease prevention, as it is involved in many cellular processes that influence disease severity. It was also found effective in the prevention of bacteria, fungi and nematodes. However, excess potassium and high K: Ca ratio might result in calcium deficiency and reduced resistance to diseases. K application inhibits the panicle blast of rice (Prabhu and Silva, 2004). The K application suppresses the disease through alterations in protein or amino acid, decreasing cell permeability and the susceptibility of tissues to maceration and penetration. K reduces the incidence of various diseases such as bacterial leaf blight, sheath blight, stem rot, sesamum leaf spot in rice, black rust in wheat, sugary disease in sorghum, bacterial leaf blight in cotton, *cercospora* leaf spot in cassava, tikka leaf spot in peanut, red rust in tea, *cercospora* leaf spot in mungbean (Sharma et al., 2005)

Sulphur

Sulphur is a very important element which if supplied in sufficient quantities can prevent disease spread. The sulphur compounds are applied as foliage, fruit, seed and soil treatments. The sprays and fumigants in the form of inorganic sulphur compounds like sulphur, sulphur oxides bordeaux mixture copper sulphate sulphides (NH_4S_x, CS_2, K_2S, H_2S, P_2S_5) thiosulphates ($NH_4S_2O_3$) and organic sulphur compounds like dithiocarbamates are in use. There are various reports of disease control through sulphur application which is explained as under:

Table 2. Crop diseases which are suppressed by sulphur application

S. No.	Plant diseases	Host
1	*Fusarium* wilt, *Verticillium* wilt	Cotton, tomato
2	Club root	Crucifers
3	Black spot, black leg, late leaf spot, *Sclerotinia* stem rot, *Verticilium* wilt	Rape
4	Common scab, late blight, stem canker	Potato
5	Powdery mildew, sharp eye-spot	Wheat
6	*Rhizoctonia* root rot	Soybeans
7	Downy mildew, powdery mildew	Grape
8	Leaf blight, Stewarts wilt	Maize
9	Tobacco Mosaic Virus	Tobacco
10	*Armillaria* root rot	Peach
11	*Cercospora* leaf spot	Peanut
12	Needle blight	Pine
13	*Ramularia* leaf spot, Turfgrass*Fusarium* patch	Sugarbeets

Among the micronutrients, manganese (Mn) can control a number of diseases as Mn has an important role in lignin biosynthesis, phenol biosynthesis, photosynthesis and several other functions. Zinc (Zn) was found to have a number of different effects as in some cases it decreased, in others increased, and in others had no effect on plant susceptibility to disease. B was found to reduce the severity of many diseases because of the function that B has on cell wall structure, plant membranes and plant metabolism. Chlorine application can enhance host plants' resistance to disease. Silicon (Si) has been shown to control a number of diseases and it is believed that Si creates a physical barrier which can restrict fungal hyphae penetration, or it may induce accumulation of antifungal compounds. Also, it is important to find alternative measures to control plant diseases which do not harm the environment and at the same time increase yield and improve product quality (Atkinson and McKinlay, 1997; Batish et al., 2007; Camprubí et al., 2007).

Natural Disease Control Methods

Permitted disease-management products as non-conventional tools of disease control include copper (fixed copper and copper sulphate), lime-sulphur mixes, elemental sulphur, vinegar, soap and silica. Bordeaux mixture is considered a restricted substance, and farmers should contact their certifying body before using it. Scientific evidence on product efficacy should be researched before using them. The antifungal effect of tea tree (*Melaleuca alternifoilia*) essential oil against *Alternaria*sp. of tomato has been reported (Abbo et. al., 2009). The high risk of phytotoxicity should also be considered when using these products on certain plants; often the margin of error between benefit and damage to the plant is very small. In addition, there are environmental and ecological concerns surrounding some of these products. But, these products are not very cost-effective or feasible to apply them to field crops. Some manures are also very useful for disease management among various crops (Table 3).

Table 3. Liquid compost/manure extracts for disease management

S. No.	Disease	Type of compost
1	Late blight of potato and tomato	Horse compost extract
2	Graymold on beans and strawberries	Cattle compost extract
3	Downy and powdery mildew of grapes	Animal manure-straw compost extract
4	Powdery mildew on cucumbers	Animal manure-straw compost extract
5	Graymold on tomato and pepper	Cattle and chicken manure compost extract
6	Apple scab	Spent mushroom and compost extract

Optimum Soil Health

The sustainability of crop production largely is determined by the soil health. In general, soil organic carbon (or organic matter) is considered to be the universal indicator of optimum soil health (Rasmussen and Collins, 1991). Quality soil not only produces good crop yield, but also maintains environmental quality and consequently plant, animal and human health. In comparison with chemical intensive farming, organic farming has potential benefits in promoting soil structure formation (Pulleman et al., 2003) and enhancing soil biodiversity. Soil organic matter is the most fundamental source of fertility in organic agriculture and it is important for producers to understand the basics of organic matter cycling in the soil. Soil organic matter is that portion of the soil that consists of biological residues, from plant to animal to microorganism. Organic residues supply not only readily available nutrient sources but also the building blocks of humus. Humus is the product that is left over after decomposition has ended, and it is extremely important in increasing and maintaining soil fertility. It possesses an overall negative charge, which means that positively-charged nutrients are attracted to it. Humus can be thought of as a bank which holds nutrients and can release them in response to plant or microorganism needs.

Diverse residue sources sustain a microbial community that is efficient and has more diversity. For instance, bacteria are associated with high nitrogen materials, while the abundance of fungi increases in relation to high carbon materials. It is important to note that if the C: N ratio of the soil is too high, nitrogen will be in short supply and will be used up by microorganisms before it is available to the crop. Over time, and with good soil management, the cycling of nutrients through microbial biomass will reach equilibrium. At that point nutrients will be readily available for crop needs.

Organic Matter Status

It is estimated that around 700 MT of agricultural waste is available in the India every year, but it is not properly utilized. This implies a theoretical availability of 5 tonnes of organic manure/hectare arable land/year, which is equivalent to about 100 kg NPK/ha/yr. The improvement of soil quality through increase in soil organic carbon pool has numerous subsidiary benefits. The available data show that crop yields can be enhanced by 20–70 kg/ ha for wheat, 10–50 kg/ha for rice, and 30–300 kg/ha for maize with every 1 Mg/ha increase in soil organic carbon pool in the root zone. Adoption of recommended management practices on agricultural lands and degraded soils would enhance soil quality including the available water holding capacity, cation exchange capacity, soil aggregation, and susceptibility to crusting and erosion. Increase in soil organic carbon pool by 1 Mg/ha/year can increase food grain production by 32 million Mg/year in developing countries. This strategy would also off-set fossil fuel emissions at the rate of 0.5 Pg C/year through carbons sequestration in agricultural soils of developing countries (Lal, 2006).

Figure 1. Higher seed productivity of mustard drop under organic nutrient management compared to only chemical nutrient management over the time
(Premi et al 2012)

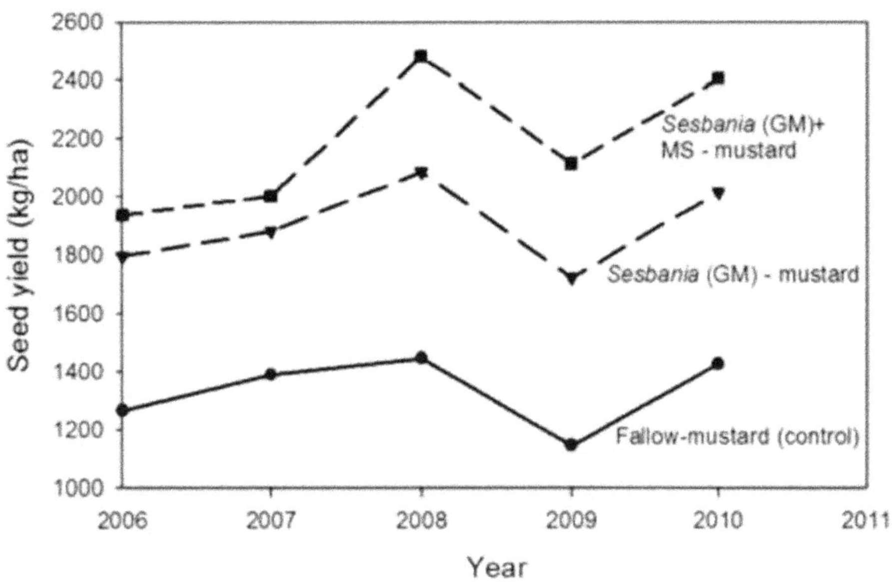

In long run, the soil fertility management through organic sources like *sesbania* green manuring and crop residue management led to higher seed yield (Fig. 1). Organic matter has usually partitioned into to two major parts, active (labile) and stable. Labile soil C pools are especially important because they are more vulnerable to climatic change and disturbance and play vital roles in nutrient cycling. Active part of organic matter is plant litter, the light fraction of the biomass and non-humic substances not integrated to mineral constitutes and active pool of organic matter is ready source of nutrients to the plants, where as stable pool of organic matter is reservoir of nutrients and it is important from long term balance of the soil. Active fraction is particularly important in maintaining the soil productivity under sustainable agriculture.

It is well known fact that the earthworms are an important indicator of soil fertility. The earthworm biomass, density and diversity were found to be highest on the organic soils compared with the conventional fertilizer fed soils (Pfiffner and Mader 1997). Several studies have shown that a relationship between earthworms and soil properties or land-use, respectively, exists. It is widely known that many of the factors influencing the incidence of pathogenic organisms in soil are directly or indirectly influenced by organic matter. For example, a plentiful supply of organic matter may favour the growth of saprophytic organisms relative to parasitic ones and thereby reduce populations of the latter. Biologically active compounds in soil, such as antibiotics and certain phenolic acids, may enhance the ability of certain plants to resist attack by pathogens.

Table 4. The living soil flora and fauna and their relationship with soil organic matter

S.No.	Organisms involved	Functions
1	Bioturbating by invertebrates and plant roots, mycorrhizae and some other micro-organisms	Maintenance of soil structure
2	Most bioturbating invertebrates and plant roots	Regulation of soil hydrological processes
3	Mostly micro-organisms and plant roots, some C protected in large compact biogenic invertebrate aggregates	Gas exchange and carbon sequestration (accumulation in soil)
4	Mostly micro-organisms	Soil detoxification
5	Mostly micro-organisms and plant roots, some soil and litter-feeding invertebrates	Nutrient cycling
6	Various saprophytic and litter-feeding invertebrates (detritivores), fungi, bacteria, actinomycetes and other micro-organisms	Decomposition of organic matter
7	Rhizobia, mycorrhizae, actinomycetes, diazotrophic bacteria and various other rhizosphere micro-organisms, ants	Symbiotic and asymbiotic relationships with plants and their roots

Table 5. Soil organic properties and their impact on soil quality

	Soil quality	Particular	Impact on the soil
1	Colour	Organic matter caused dark colour of the soil	Absorb more solar radiation
2	Water retention	OM can retain more than 20 times of its weight in water	This improve soil water holding capacity
3	With clay material	Make stable soil aggregates	Stabilises soil structures, better aeration
4	Buffer action	OM exhibits pH buffering	Maintains uniform reaction in the soil
5	Cation exchange	The CEC of humus range from 350–1400 meq/100 gm of soil	High CEC, means better availability of nutrients to crop
6	Mineralization	Decomposition of OM resulted CO2, NH4, NO3, H2PO4 and SO4	Increase plant nutrient availability
7	Solubility in water	Because of association of OM with clay, it is insoluble	Meagre amount of OM lost by leaching
8	Chelation	Form stable complexes with element and cations	It may enhance nutrient availability to the plants

The long term studies indicates that, the organic farming system was more effective than the conventional farming system in reducing soil erosion and, therefore, in maintaining soil productivity. The organic matter minimise the effect of drought on soils and on the crops. The poor utilization of rainfall is more due to inadequate land management practices (i.e. burning of crop residues, excessive tillage, eliminating hedges, etc.) than because of natural phenomena (relief, slope, rainfall intensity) that reduce organic matter levels, destroy soil structure, eliminate beneficial soil fauna and do not favour water infiltration. To minimize the impact of drought, soil needs to capture the rainwater that falls on it, store as much of that water as possible for future plant use, and allow for plant roots to penetrate and proliferate. The capacity of soil to retain and release water depends on a broad range of factors such as soil texture, soil depth, soil architecture (physical structure including pores), organic matter content and biological activity. The addition of organic matter to the soil usually increases the water holding capacity of the soil. This is because the addition of organic matter augments the number of micropores and macropores in the soil. Certain types of soil organic matter can hold up to 20 times their weight in water.

CONCLUSION

Organic matter overall improve soil through its positive impact on physical, chemical and biological properties of the soil. The maintenance of soil organic matter levels and the optimization of nutrient recycling are essential to the sustained productivity of agricultural systems. Conventional, intensive agricultural practices have significantly increased crop productivity. But, their adverse effects on soil productivity and environmental quality are also felt. This concern has led to an increasing interest in organic production systems. The long-term effects of organic on soil properties showed significantly higher organic matter content, thicker topsoil depth, higher polysaccharide content, lower modulus of rupture and less soil erosion than the conventionally-farmed soil. It is proved beyond doubt that in the long term, the organic farming system was more effective than the conventional farming system in reducing soil erosion and, therefore, in maintaining soil productivity. The natural enemies are much more vulnerable to pesticides than the targeted pests and pathogens. So when pesticides are used, natural enemies are killed and pests remained uncheck. It is therefore, essential to use non-chemical crop protection methods instead of pesticides. Non-chemical methods of crop protection have always been practised, but the introduction of chemical pesticides a few decades ago seemed to make crop protection a lot easier. Chemical pesticides are often very effective against the target pests, but sometimes they do not work at all because the pest has become resistant to the pesticide or because of unfavourable weather conditions. Non-chemical crop protection is often less effective than chemical crop protection, but it is usually less expensive and is based on locally available inputs and interventions. Therefore, green agro-techniques of organic plant nutrient and for management of abiotic stresses are of immense use for sustainable high agricultural productivity in run.

REFERENCES

Abbo, A. S. H., Idris, M. O., & ElBalla, M. M. A. (2009). The response of tea tree oil as a biofungicide against early blight disease in tomato crop (Solanum lycopersicum) in Sudan. In *Conference on International Research on Food Security, Natural Resource Management and Rural Development* (pp. 1-9). Academic Press.

Atkinson, D., & McKinlay, R. G. (1997). Crop protection and its integration within sustainable farming systems. *Agriculture, Ecosystems & Environment*, *64*(2), 87–93. doi:10.1016/S0167-8809(97)00026-1

Batish, D. R., Singh, H. P., Setia, N., Kohli, R. K., Kaur, S., & Yadav, S. S. (2007). Alternative control of littleseed canary grass using eucalypt oil. *Agronomy for Sustainable Development*, *27*(3), 171–177. doi:10.1051/agro:2007008

Bockus, W. W., Webster, R. K., Wick, C. M., & Jackson, L. F. (1979). Rice residue disposal influences overwintering inoculum level of *Sclerotiumoryzae* and stem rot severity. *Phytopathology*, *69*(8), 862–865. doi:10.1094/Phyto-69-862

Camprubí, A., Estaún, V., El Bakali, M. A., Garcia-Figueres, F., & Calvet, C. (2007). Alternative strawberry production using solarization, metham sodium and beneficial soil microbes as plant protection methods. *Agronomy for Sustainable Development*, *27*(3), 179–184. doi:10.1051/agro:2007007

Huber, D. M., & Graham, R. D. (1999). The role of nutrition in crop resistance and tolerance to disease. In Z. Rengel (Ed.), *Mineral Nutrition of Crops Fundamental Mechanisms and Implications* (pp. 205–226). Food Product Press.

Huber, D. M., & McCay-Buis, T. S. (1993). A multiple component analysis of the take-all disease of cereals. *Plant Disease*, *77*(5), 437–447. doi:10.1094/PD-77-0437

Jensen & Weiser. (1971). *Crop Rotations for Profit*. Extension Bulletin, No. 14. North Dakota State University.

Kiraly, Z. (1976). Plant disease resistance as influenced by biochemical effects of nutrients and fertilizers. *Fertilizer Use and Plant Health, Proceedings of Colloquium 12*.

Lal, R. (2006). Enhancing crop yields in the developing countries through restoration of the soil organic carbon pool in agricultural lands. *Land Degradation & Development*, *17*(2), 197–209. doi:10.1002/ldr.696

Pfiffner, L., & Mäder, P. (1997). Effects of biodynamic, organic and conventional production systems on earthworm populations. Entomological Research in Organic Agriculture, 15, 3–10.

Premi, O. P., Rathore, S. S., Shekhawat, K., Kandpal, B. K., & Chauhan, J. S. (2012). Sustainability of fallow – Indian mustard system as influenced by green manure, mustard-straw recycling and fertilizer application. *Indian Journal of Agronomy*, *57*(3), 229–234.

Pulleman, M., Jongmans, A., Marinissen, J., & Bouma, J. (2003). Effects of organic versus conventional arable farming on soil structure and organic matter dynamics in a marine loam in the Netherlands. *Soil Use and Management*, *19*(2), 157–165. doi:10.1079/SUM2003186

Ramirez, V. J., & Munnecke, D. E. (1988). Effect of solar heating and soil amendments of cruciferous residues on *Fusarium oxysporum* sp. conglutinans and other organisms. *Phytopathology*, *78*(3), 289–295. doi:10.1094/Phyto-78-289

Rasmussen, P. E., & Collins, H. P. (1991). Long-term impacts of tillage, fertilizer and crop residue on soil organic matter in temperate semi arid regions. *Advances in Agronomy*, *45*, 93–134. doi:10.1016/S0065-2113(08)60039-5

Sakhuja, P. K., & Jain, R. K. (2001). Nematode diseases of vegetable crops and their management. In T. S. Thind (Ed.), *Diseases of Fruits and Vegetables and their Management* (pp. 439–459). Kalayani Pub.

Sharma, S., Duveiller, E., Basnet, R., Karki, C. B., & Sharma, R. C. (2005). Effect of potash fertilization on helminthosporium leaf blight severity in wheat, and associated increases in grain yield and kernel weight. *Field Crops Research*, *93*(2-3), 142–150. doi:10.1016/j.fcr.2004.09.016

Thontadarya, T. S., & Jairao, K. (1980). Bhendi as trap crop in the integrated control of cotton bollworms. *Andhra Agriculture Journal*, *27*, 100–101.

Chapter 6
Role of Protected Cultivation as Green Technology for Sustainable Environment:
Future, Prospect, Potential, Production, Protection, and Profit

Awani Kumar Singh
Indian Agricultural Research Institute, India

Naved Sabir
Indian Agricultural Research Institute, India

ABSTRACT

In India, the field of protected cultivation is still evolving. Methods of more efficient input application and protection technology are being devised. Low-cost protected technology like plastic low tunnels or walk in tunnels can be used for off-season vegetable cultivation for getting high returns. Similarly, insect-proof net houses can be used on a large scale for safe vegetable cultivation to minimize use of pesticides in vegetable cultivation and virus-free crop production. More eco-friendly plastics are being produced for making the processes more efficient. As a result, protected cultivation could bring prosperity to horticultural crop growers in hilly and plateau regions of India and ensure nutritional security for everyone.

INTRODUCTION

Agriculture in India often finds itself confronted to various constraints under changing dimensions that are currently manifest in rapid climate change, soil degradation, population pressure, land fragmentation, water scarcity, marketing and entrepreneurship etc. (Saver et al. and Hasan et. al. 2021 and Singh 2004).

DOI: 10.4018/978-1-7998-8434-7.ch006

Figure 1. Plasticulture-cum-protected cultivation

Horticulture is fast emerging as one of the most dominant faces of Indian agriculture. India is second largest producer of the vegetable crops after China estimated to 146 mt from 8.4 mh area and highest producer of the fruit crops. There is little possibility of addition to the cultivable area though area under horticulture has been increased during the last two decades largely at the cost of other crops. Of all the horticultural technologies, Plasticulture-cum-protected cultivation, is emerging as one of the most promising fields particularly in terms of higher productivity, quality, increased income, entrepreneurship etc. though the area coverage of about 60,000 ha alone appears abysmal. But the progress of area coverage under protected technologies in terms of percent coverage is highly satisfying. The other major reasons for higher adoption include promotion of hi-tech horticulture by public financial institutions. Although, despite indubitable potential of protected cultivation, its spread has not been linked concomitantly with dissemination of technology, quality inputs, post-harvest management and processing units besides overall market linkages. Net result is that despite the technology performing exceedingly well by up to date farmers, majority of them keep facing challenges to cope up with the demands of hi-end horticulture. This article deals with overall gamut of technology, its potential, suitable areas and crops, tackling the challenges of biotic and abiotic stresses, market linkages and overall empowerment of farmers in terms of knowledge, input availability and use-efficiency, plant protection, post-harvest loss management, and market linkages etc. It can be safely presumed, that, if adopted, in right earnest, Plasticulture in horticulture has immense potential in realizing the sustainable solutions for higher productivity, better livelihood and nutritional security to the farmers who adopt the technology, particularly as clusters of hi-tech farming cooperates (Singh A.K. 2021 and Akshay et al 2021)

Figure 2. Hi-tech farming

Plasticulture in horticulture specially protected cultivation-cum-covered cultivation under greenhouses, polyhouses, net-houses, tunnels, mulches etc. all of which include intelligent and exhaustive use of plastics in various forms depending upon the requirements, suitability of regions, climates, crops and technologies infrastructure or paraphernalia e.g. UV stabilized polythene sheets, insect-proof nets, shade-nets, mulches, drip irrigation pipes, filters, tanks, plastic-lined water harvesting, etc. all are discretely made up of plastics. *(Singh A.K. 2021 and Singh 2004)*

Table 1. Current status of protected agriculture in world

S. No.	Country	Area under protected agriculture
1	China	331+ Lakh ha
2	Japan, Spain, Italy	60000 - 1. 0 Lakh ha
3	Netherlands, Israel, Turkey, Morocco	45000 – 85000 ha
4	USA, Canada, South Korea, UK, France and other EU nations	50000 – 110 Lakh ha
5	India	108+ Lakh ha

(*Saver et al, 2021*)

WHAT AGRICULTURE ASPIRES, PROTECTED HORTICULTURE DELIVERS

Policy initiatives for doubling the farmer income,'*PradhanMantriSinchaiYojana*' invoking 'more crop per drop' and '*Pradhan MantriRozgarYojana*' are some of the encores of the Government assertions in the quest for overall development of the nation including agriculture. Incidentally, plasticulture or protected cultivation or covered cultivation cansmake a cognizable and tangible contribution in all of these policy initiatives of the Govt., both directly and indirectly, in horticultural sector (Singh 2004).

It is interesting to note that protected cultivation including plasticultureis apparently the only sector of agriculture which is a solution for most of the its inherent constraints into an ideal system which could address them as modules of systematic and sustainable farming including healthy planting material, quality seed production, conservation and optimization and input use efficiency of water and other resources, precision farming, GAP, IPM, INM, Grafting, soilless cultivation besides minimized impaction of climate change aberrations etc. which all find convergence to strengthen farming into a hub of profitable horticulture. Apart from this, the hi-tech agriculture encourages better yields from fragmented land, entrepreneurship potential, GAP produce, export potential etc. It is by virtue of these strengths that protected cultivation technology can play a lead role in overall gamut of horticulture. Protected cultivation technology is, thus, the future of horticulture. The potential of addressing soil erosion, recycling abilities, as well as amelioration of soil and water pollution are additional boons to overall ecosystem besides the economy of agriculture. Also, protected cultivation is most amenable to hi-tech supply chain networks in horticulture which ensures higher income to growers. Over and above of these advantages, protected cultivation is very amenable to women workforce. Basket of technologies covered under protected cultivation technologies are being discussed below with relevant details (Saver et al, 2021 and Singh 2004):

Various Structures Under Protected Technology

The protected horticulture includes various structures such as polyhouse, poly-tunnel, shading net house, floating cover, bird proof net house, cloches, insect proof net house, mulching and anti-hail net house etc. Details classification of protected technology given in below (Singh et al. 2021)

Scope and Advantages of Polyhouse Technology

A polyhouse is permanent structure, covered with a transparent or translucent material. Materials used to construct a polyhouse frame may be wood, bamboo, and steel or even aluminum. Coverings material is generally plastic film (Saver et al, 2021 and Singh 2004)

Two types of polyhouses are designed in Indian condition, one is called naturally ventilated polyhouse and the other is called force ventilated (fan-pad) polyhouse. Naturally ventilated polyhouses are used mainly in temperate and moderate climate conditions while force ventilated (fan-pad) polyhousesare suitable for tropical, sub-tropical and arid zone farming systems. Polyhouse is main structure of protected -cum-covered cultivation groups. It is made by GI pipe and covered by ultra-violet (UV) stabilized plastic sheet with a service life of 3-4 years. These sheets are generally available in 7 and 9 meter widths with 200 micron thickness. Polyhouse technology is the most intensive form of commercial cultivation. Presently, various state Govtsextend financial support in the form of subsidies to the growers limited to a ceiling of 1 acre of structures costing approximately Rs 38-40 lakh of which about 50% of the cost is borne by Central Govt. with sizable and additional subsidies from State Govts which include the cost of structure, drip irrigation and fertigation system and seed/planting material etc. Polyhouse structures provide favorable growing conditions to crop plants and protect them from extreme fluctuations in weather, especially temperature, rainfall and wind and also minimizing evaporation losses besides protection from birds, animals and insect pests and diseases to quite an extent. The polyhouse structures provide 4-5 time higher yields with export quality produce; high water and fertilizer use efficiency and round the year production facility in any part of the country. Drip irrigation is mandatory for polyhouse crop cultivation along with fertigation technique because drip technology enhances water and crop productivity and minimized diseases, weeds, moisture fluctuations, irrigation frequency and judicial use of water and fertilizer. Thus, polyhouse and drip-irrigation technologies are components of horticulture based-precision farming system for better livelihood and nutritional security in the farming society. (Savir et al, 2021 and Singh 2004, 2007, 2021 and Singh, 2018).

Figure 3. Polyhouse1

Figure 4. Polyhouse 2

Greenhouse Effect and Environment under Protected Structures

A plant grows best when exposed to an environment that is optimal for that particular plant species. The aerial environment for the plant growth can be specified by the following factors and can see in figures. *(Savir et al, 2021 and Singh 2004, 2007 and Singh, 2018).*

Figure 5. Greenhouse effect and Environment

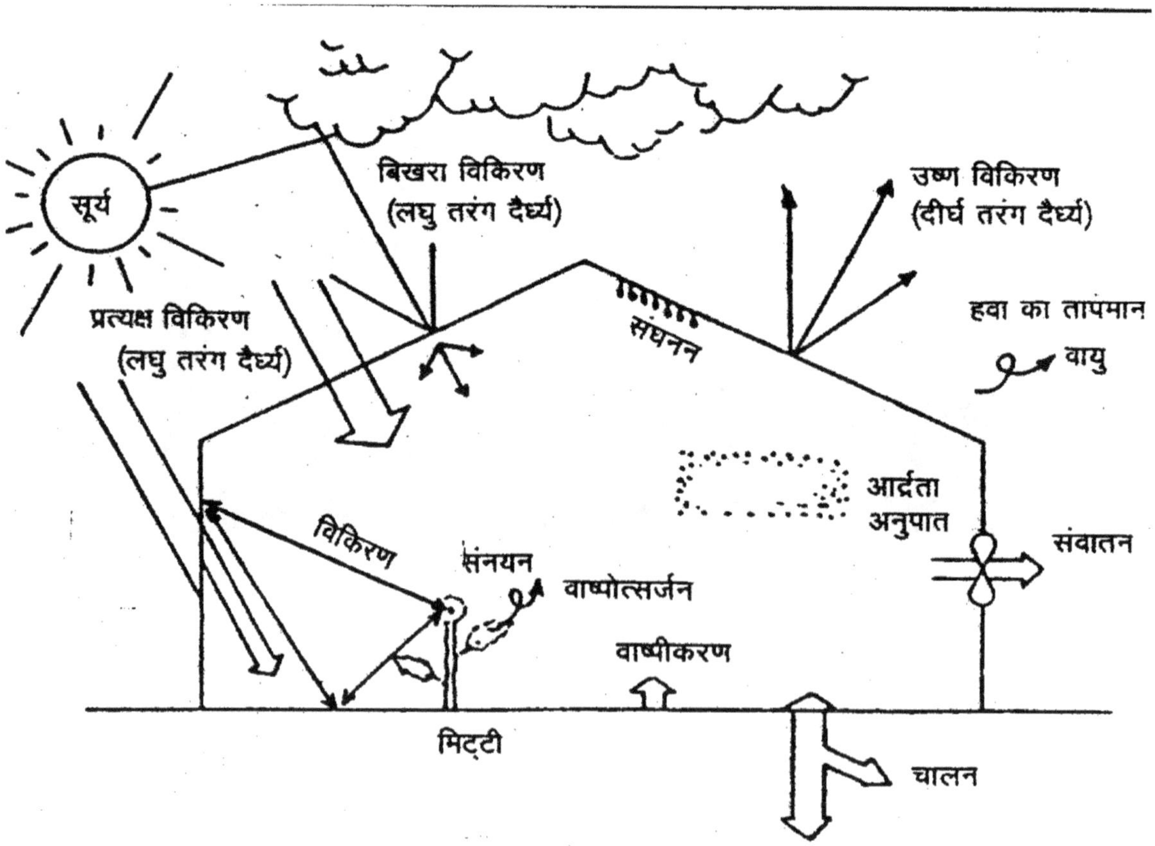

1. Heat or temperature
2. Light
3. Relative humidity
4. Carbon dioxide
5. Ventilation etc.

While plants have precise optimum environmental conditions for best growth, most are tolerant to variations in these conditions within some limits. However, permanent damage would occur when they are exposed to conditions outside these limits. At the same time, plants are subject to attack by pests and diseases. (Saver et al, 2021 and Singh 2004):

- Off-season cultivation of vegetables for getting high yield and more income.
- Year-round quality production of high value vegetable crops.
- Permanent plastic structures are suitable for high value off-season cultivation.
- Minimize crop stresses during growing season.

Suitable Horticultural Crops for Polyhouse

Horticulture-based farming systems are highly diverse numbering more than 300 species, depending upon the climate, culture, season, size of landholdings, knowledge, market etc. Being high value crops with higher nutritional contents, besides with a higher demand in urban and peri-urban areas, horticultural crops provide a paradise for the progressive grower on one end and marginal farmer. Major crops for these systems are indeterminate tomato, cherry tomato, cucumber, capsicum, bitter gourd, musk melon, summer squash, vegetable nursery etc. in vegetable groups. Among flower crops, rose, gerbera, lilium, tulip, anthurium, carnations, orchids, chrysanthemum and their nursery. Among, fruit crops group, strawberry, papaya and all fruit-nursery propagation or hardening are used for economically profitable horticulture based farming systems. (Savir et al, 2021 and Singh 2004, 2007 and *Singh, 2018*).

Crop Diversity and Rotation:

Vegetable Crops

After years of field experiences and feedback from the interaction with farmers following crops appear to give higher yield and better quality that ensure more income for improving livelihood and nutritional security. Wherein, indeterminate tomato hybrids (GS-600, ID-32 and ID-37, Rakshita, Himsona, Himsikhar, Snehlata, Naveen, Feto-1,2 etc.) on an average give quality yield of 80-100 t/acre/crop with net returns nearing Rs 3.0-4.0 lakh/acre/yr which is about 2-3 times higher yield and income as compared to traditional open farming systems. In the case of coloured capsicum (known as solanaceous apple) hybrids like *Natasha, Swarna, Indra, Bombi, Orobelle, Bachata, Inspiration* etc. in green, red and yellow colours produce 40-50 t/acre/crop yielding 6-7 times higher yield and income as compared to open cultivation of capsicum. Because the coloured capsicum cannot be grown properly in open condition. (Savir et al, 2021 and Singh 2004, 2007 and Singh, 2018, 2021).

This crop, therefore, demands to be produced under protected technology. Among vegetables, cucurbits appear most suitable for protected cultivation both in terms of season and number of crops for better utilization of structures. 2-3 crops per year of parthenocarpic cucumber (cv. Isatis, Kian, Hilton, Sun Star, Multistar, Fadia, Mini Angel etc), summer squash (cv PusaAlankar, PusaPasand, Australian Green, Seoul Green, Kora, Yellow Zucchini, Himanshu, etc), bitter gourd (PusaRasdar) or musk melon (PusaSarda) and yellow musk melon can be taken in polyhouses during early and off-seasons. Average production of 120-150 t/acre/year can be obtained in cucumber. These vegetables are transplanted during July to Aug. and harvested from May to June in the plain conditions. However, in hill areas, transplanting is done in the month of Feb.to March and harvesting during Oct. These vegetables result in better livelihood and nutritional security as compared to traditional vegetable farming in the open. (Singh et al. 2021).

Figure 6. Vegetable crops

Figure 7.

Figure 8.

Figure 9. Vegetables, cucurbits under protected cultivation

Table 2. Cropping-sequence cum rotation cum cropping-intensity of vegetable crops for polyhouse farming

Different crop cycles of polyhouse vegetables in a year	Returns/1000 m² (Rs.in lakhs)
• Year round Veg./fruits/flower crops nursery	5-10
• Veg. Nursery + Cucumber or Capsicum or Tomato or bitter gourd or Musk melon or long melon	3-4
• Spinach or Coriander leaf + Tomato (or Capsicum or Cucumber)	2-3
• Cucumber + Tomato (or Capsicum)	2-3
• Spinach or Coriander leaf + Cucumber + Cucumber + Cucumber	3-4
• Cucumber + Strawberry + Cucumber (or Bitter gourd/ Muskmelon/ Leafy vegetable)	4-5
• Veg. nursery or Spinach or Coriander leaf + Cucumber + Lettuce + Musk Melon	3-4
• Spinach or Coriander leaf + Bitter gourd + Lettuce + Musk melon	3-4
• Spinach or Coriander leaf + Bitter gourd + Lettuce + Bitter gourd	3-4
• Spinach or Coriander leaf + Musk melon + Lettuce + Musk Melon	4-5
• Spinach or Coriander leaf + Kakri + Lettuce + Musk melon	3-4
• Spinach or Coriander leaf + Summer squash + Lettuce + Musk Melon or Cucumber	3-4
• Nursery + Spinach or Coriander Leaf + Chilli or Brinjal	2-3
• Veg. Nursery + Lettuce or Spinach or Coriander leaf + Cucumber or Musk melon	2-3
• Vegetable Nursery + leafy vegetable + French bean (or okra) + Cucumber (or Musk Melon)	2-3

Integrated Nutrient Management (INM) for Polyhouse

The integrated nutrient management (INM) for cultivation of cucurbits in polyhouse requires green manure and FYM/Compost @ 200q/ha mixed in the soil at the time of field preparation. The dose of NPK @ 40, 20, 20 kg/ha given three times a month for one crop duration through drip irrigation. The integrated nutrient management (INM) combination for cultivation of capsicum in polyhouse requires green manure and FYM/Compost @ 400q/ha and it is mixed in the soil at the time of field preparation. The dose of NPK @ 140, 100, 120 kg/ha three times a month given through drip irrigation. Calcium is applied @ 5kg/ha. The integrated nutrient management (INM) combination for cultivation of tomato in polyhouse requires green manure and FYM/Compost @ 500q/ha and it is mixed in the soil at the time of field preparation. The dose of NPK @ 150, 120, 130 kg/ha three times a month through drip irrigation should be given. The application of calcium @ 5 kg/ha and boron @ 3-4 sprays should be done. The integrated nutrient management (INM) combination for cultivation of cucumber in polyhouse is FYM/Compost @ 300q/ha and it is mixed in the soil at the time of field preparation. The dose of NPK @ 50, 30, 40 kg/ha through drip irrigation should be given. The application of calcium @ 3 kg/ha should be done. The integrated nutrient management (INM) for cultivation of summer squash in polyhouse require FYM/Compost @ 200q/ha and it is mixed in the soil at the time of field preparation. The dose of NPK @ 40, 20, 30 kg/ha through drip irrigation should be given. (Savir et al, 2021 and Singh 2004, 2007 and Singh, 2018, 2021).

Flowers for Polyhouse

Polyhouse flower crops like rose, gerbera, tulip, Anthurium, carnations, orchids, lilium, and chrysanthemum and their nursery are mostly annual or perennial crops and hence seasonal crop rotations are not required. Most of these cut flowers required round the year as per market demand can be cut. Floriculture crops are highly remunerative as the quality, colour and shelf-life as well as freshness is very important and can give annual returns of 8-10 times more than the open field conditions.

Figure 10. Polyhouse flower crops

They are also more important from export perspective, so critically important in the country. Floriculture industry is led by produce from polyhouse growers mainly in Europe and developed nations. Thus, floriculture can ensure more income which in turn promotes better livelihood and nutritional security of farming families undertaking horticulture-based farming. (Savir et al, 2021 and Singh 2004, 2007 and Singh, 2018, 2021 and Prakash.et al. 2019.)

Fruits for Polyhouse

Major fruit crops recommended under polyhouse cultivation include papaya, strawberry, mango, guava, pomegranate and propagation materials of high value orchards or breeding material. These fruits crop can be cultivated for fresh and disease-free seedlings or plant production. Papaya fruits and seeds are in high demand in the market. Best quality papaya fruits yield on average 45-50kg/plant with net returns of about to Rs. 5-6lakh /acre. However, strawberry fruit yield of15-20 tons/acre are harvested in the 10 plants per sq. meter. All fruit crops are more effective in minimizing malnourishment being nutrient rich. Strawberry runners, virus-free papaya seed, budded and grafted plant sand meadow orchard in fruits are more economically and safe produced for livelihood and nutritional security. papaya can be cultivated in polyhouse as inter-cropping with leafy vegetable crops like spinach, coriander, lettuce, mustard or French been and or strawberry crops. Strawberry transplanted in the August-September and harvested up in April followed by papaya transplanted in twice, either in July to August or later in Jan-Feb month per year. Crop rotation can be used with strawberry crop i.e. leafy vegetable or vegetables nursery or cucumber + Strawberry + cucumber (or bitter gourd or muskmelon or leafy vegetable) can be taken. (Savir et al, 2021 and Singh 2004, 2007 and Singh, 2018).

Figure 11. Fruits for polyhouse

Crop Diversity and Rotations Under Net-built Structures

Generally two types of net-built structures are used *viz.*, shade-net and insect-proof net houses often termed synonymously, but indeed, a shade net is primarily a covering of UV stabilized coloured plastic net where main objective is to provide protection to the horticultural nurseries and plants from harsh sunlight and hot winds during summer seasons. On the other-hand, insect proof net houses are built for the primary objective of screening out the entry of insects or other pests by the physical barrier of nets of various mesh sizes depending upon crop and season. Both the net houses can be made by iron or GI pipe or bamboo or wooden or stone pillars etc. and are one of the most feasible forms of protected cultivation. Basic cost of fabrication of these structures may range from Rs 8-10 lakhs/acre. Details have discussed are bellow (Savir et al, 2021 and Singh 2004, 2007 and Singh, 2018).

Shade Net House Protected Structures

Shade nets generally use two colour nets (green and black) in horticulture-based farming systems. Both colour hade nets are available in 30, 50, 75 and 90% capacity of shading protection, wherein this percentage means the percentage of reduction in sunlight.

In vegetable crops 30-50% capacity of either colour or 50-90% shade nets are used for fruit-propagation in horticultural crops. Shade nets also cut down the sunlight, in turn minimizing evapotranspiration losses of moisture in general as well as reduced transpiration from plants. The total effect creates an altered relative humidity and temperature regime of microclimate favoring the plant growth. That is why these structures are especially useful for dry land and arid areas with water scarcity by increasing water use efficiency when used in combination with drip irrigation. Shading nets are the perforated plastic materials used to provide relief to the plants from the scorching sunlight, hot winds, direct rainfall, hails, birds and animals as well as insects-pests. Nets of 25% to 75% shading capacity are being used for raisingnursery,indoor plants,hardening of tissuecultured plants and growing of vegetables. During high light intensities, leafy vegetables and ornamental greens are recommended to be grown under shade nets. This structure enhances 25-35% more yield as compared to traditional methods during peak summer season. Suitable crops identified for shade net houses from economic viability include horticulture nurseries, leafy vegetables, summer tomato and capsicum. Because these systems are mainly used to grow off-season cultivation of crops, their produce fetches much higher prices in the market, hence providing good economic returns to growers, thereby improving livelihood and nutritional security to the farming families. Ornamental crops like lilium, tulip, anthurium and orchids can also be grown economically depending upon suitability of different agro-climatic zones. Therefore, shade nets have come to occupy a prime importance in horticultural promotion in the country especially vide propagation of quality planting material. *(Savir et al, 2021 and Singh 2004, 2007 and Singh, 2018).*

Figure 12. Shade nets

Figure 13. Shade nets

Insect Proof Net-house Protected Structures

Insect-proof nylon net based cultivationis newer concept more suitable for rainfed areas with much fewer days of precipitation. These structures make use of insect proof screens of different intensities of perforations, ranging from 25 mesh to 60 mesh. IPP nets of 40 or higher mesh are effective means to control entry of most flying insects and save crop from diseases.

Figure 14. Insect Proof Net-house Protected Structures

कीट अवरोधी जाली घर

Figure 15.

As compared to polyhouses, these structures are almost equally capable of screening out the insects and birds thereby avoiding the use of pesticides. These structures permit early planting of tomatoes and cucumber, bitter gourd, musk melon, summer squash, brinjal and chilli etc. without the risk of vectors. Higher mesh size (i.e. above 50), however, reduces the air exchange of the structure. UV- stabilized white coloured nets are now available which have a longer life. This structure enhances 25-35% more yield as compared to traditional methods during peak summer season.IP net structures can be used successfully for seed production in rainfed areas. Crops suitable for these structures include tomato, cherry tomato, cucumber, bitter gourd, musk melon, summer squash, papaya, meadow and high density orcharding of mango, guava and pomegranate and strawberry etc. besides very few flower crops. In irrigated areas, these structures may face problems during rainy season as it tends to flood the crop inside. Hence it is advisable to cover its roof by polythene sheets during rainy period. But this is possible only with small, half-moon type or gable type structures. Insect-proof covered cultivation technique give 2-3 times more yields and hence increase the income by about 30% normally but even more at times where insect damage is more. Thus, farmers are able to save the crop hence being safer for farmers. (Savir et al, 2021 and Singh 2004, 2007 and Singh, 2018 and Kumar et al.202 and Singh et, a.l 2021).

POLY-TUNNEL TECHNOLOGY

Poly-tunnels are flexible transparent coverings that are installed over single or multiple rows of vegetables to enhance the plant growth by warming the air around the plants in the open field during winter season when the temperature is below 8^0C. Plastic low tunnels are often used to promote the growth of plants during the period of winter season. Low tunnels are supported above the plants by using hoops of GI wire and a clear or transparent plastic of 20-30 micron is covered/stretched over the hoops and the sides are secured by placing in soil. Plastic is vented or slitted during the growing season as the temperature increase within the tunnels. The farmers can grow different varieties of summer squash (round-fruited, long-fruited), which is a emerging crop along with cultivation of netted muskmelon varieties in place of traditional varieties. Bitter gourd and round melon are two other crops with increasing demand and which usually fetches very high price during off-season and can be grown successfully by using the plastic low tunnel technology. This technology is highly suitable and profitable for the farmers.Poly-tunnel is also known as temporary or mobile micro-polyhouse or row-cover and classified as low poly-tunnel, high poly-tunnel and trench poly-tunnel. (Savir et al, 2021 and Singh 2004, 2007 and Singh, 2018).

Objectives and Utility of Poly-tunnels

The poly-tunnels are of same importance as poly-house and are used for:

- Off-season cultivation of vegetables for earning more income
- Walk-in tunnels are only erected over the crop during the peak winter months of December-mid February and thereafter, structure is removed from the crop.
- Since, the plastic is used only for two months (December- mid February) therefore; life of the plastic can be 8-10 years.
- These temporary plastic structures are suitable for off-season vegetable cultivation in northern plains and food hills. (Savir et al, 2021 and Singh 2004, 2007 and Singh, 2018).

Figure 16. Poly-tunnels

Figure 17. Poly-tunnels

Protected Surface Cover: Plastic Mulching

Plastic mulching technology, also called surface covered cultivation, largely involves mulching with polythene sheets for addressing three major issues *viz.*, soil and water conservation, leaching of nutrients as well as reduction of weeds and to some extent insect pest damage.

Figure 18. Drip irrigation

Drip irrigation is an integral part of mulching technology. Mulching involves covering the soil around the plant bases with an organic or inorganic material which makes condition more favorable for plant growth and development. Organic mulches, like leaves, straw, sawdust etc. add nutrients and humus to the soil as they decompose, improving its tilth and moisture holding capacity. Synthetic or plastic mulches have various beneficial effects on crop production. Plastic mulch accelerates plant growth by increasing soil temperature, conserving soil moisture, weed control, production of quality produce and reduction in leaching of nutrients. The plastic mulch is available in different colours. Each colour has its own significance. Transparent polyethylene mulch raises the soil temperature by soil solarization used mainly for managing soil-borne diseases and nematodes. This effect derives mostly from the suppression of latent heat loss through evaporation. Black polyethylene film also gives effective weed control by cutting down solar radiation by more than 90%, resulting in etiolated growth and the eventual death of weeds under the film. The yellow plastic mulch attracts insects so it can be used to attract and kill insects. The silver-reflective type of plastic mulch associated with higher reflectance causes insect disorientation and repels aphids. The different types of mulches are reflective plastic mulches, infra-red transmitting mulches and biodegradable plastic mulches. There are different types of laying methods for plastic mulch but large area usage has to involve the help of mechanized and tractor operated mulch layers. Similarly, several low-cost hole cutting devices and transplanting devices are now pouring in market. Suitable horticultural crops that can take advantage of mulching include plants growing vertically by nature and not the creeping types e.g. solanaceous crops andcole crops. Although, cucurbits like summer squash can also be grown on mulching but with proper staking systems only for taking better advantage of the strengths of this type of cultivation. In fruit crops also the mulching is very important especially in strawberry and gooseberry. Other fruit crops like papaya, grapes, pomegranate, guava, mango etc. under high density and meadow planting methods do use the mulching technology strategically. This technology is again useful for areas with water scarcity or rainfed ecosystems, and can conserve moisture to the extent of 50% water saving and 30% nutrient saving. Thus, we can conserve resources, economies input use and efficiency and harvest 25-30% higher yields as also the quality, playing a significant role increase in income and overall upliftment of livelihood of farming families. Details about advantage of mulching technology are given below: (Singh et.al 2013 and Singh 2018 and Saver et al 2022).

Figure 19. Mulching technology

QUALITY SEEDLINGS PRODUCTION BY PROTECTED TECHNOLOGY

Plasticulture based healthy planting material is the backbone of overall gamut of horticultural production systems. Nursery raising requires hi-end greenhouses with boom irrigation, glazing material, seeders, trays and cell pack for propagation, container carrying trays, poly-tube, bags, cups and labels for proper identification, ground cover and packing of planting material for long distance transportation.

As it is highly important to grow disease and pest free healthy planting material, soilless production of nurseries under greenhouses is very important. Generally plastic nursery trays or pro-trays having different cell sizes are used for raising vegetable and fruit seedlings. Different kinds of plastic pro-trays are used in raising flower seedlings. These trays help in proper germination of seeds and cuttingsand provide independent area for each seed and cutting to germinate, reduce the mortality rate, maintain uniform and healthy growth of seedlings, are easy in handling and storing, reliable and economical in transportation.

These plastic trays may be fixed in thermocol base trays having the same number and size of cavities before filling the media. Media used are mainly two types, *viz.*, soilless media (3 parts of coco peat + 1 part each of vermiculite and perlite) which costs about Rs 6000-7000 per 100 kg or indigenous media (2 parts cocopeat+1 part vermicompost/leaf compost and the other combination may include 2 parts vermicompost/leaf compost and one part each of sand and ash. These media may be enriched with *Trichoderma*). The cost of these media is only Rs 500-1000 per 100 kg. This media is not only cheaper but also has base nutrients available.Overall, hi-tech nursery is a little capital and labourintensive or automation based yet highly remunerative to the extent that it can be adopted in entrepreneurial model by unemployed educated youth. (Savir et al, 2021 and Singh 2004, 2007 2018, 2021).

Figure 20. Plastic trays

RAINWATER HARVESTING TECHNOLOGY FOR PROTECTED CULTIVATION

Rainwater Harvesting of Polyhouse Roofs for storing in plastic lined tanks at proper place for long duration so that even a single drop of water can be utilized preciously. This scheme is being subsidy and training from Govt. of India under scheme of '*Pradhan MantriSinchaeeYojana*'. In present scenario, concept of poly-tank cum micro-irrigation system can be employed in protected cultivation. Special polythene sheets of 300-400 gsm thickness, UV stabilized, cross-laminated in blue or black colour are utilized as standards. Life of water storage tanks so devised extends beyond 10 years easily. Additional advantage of these systems is their ability to withstand earthquakes. Besides compensating the water scarcity, rain water harvesting from polyhouse roofs can remediate water quality in areas where soil salinity and pH are inextricable issues. Such system is also the best way to provide irrigation, under low water availability areas like lower to high hills with high water facing acute water shortage in spite of rainfall. Rainwater so harvested can be best used for life-saving irrigation strategies. The micro-irrigation system saves water (40-50%), fertilizer (25-30%), labour expenditure (50%), soil-erosion cum fertility (90%), minimizes plant mortality (80%), enhances crop yield (25-75%) and improves the quality of produce as compared to traditional pattern specially when employed under protected environment.

Keeping this in view, storage of rainwater in low-cost adequate capacity poly-tank and use of this water for irrigation in crop grown in protected environment (Polyhouse or Tunnels) through gravity based drip irrigation system. Cost of installation of this type of poly-tank make is Rs 120-150/m^3 including roof top, lateral pipes and storage tanks. If drip irrigation is integrated with these structures which in turn give additional advantage to the water-scarce production system. This can even lead to increase in area under cultivation for these areas. Water storage tanks can be additionally used for fish culture or duck-fish or hen-fish farming in integration with horticultural production system as these cultures add nutrients to the water, minimizing nutrient requirements besides getting extra income from meat, fish, eggs etc. for livelihood betterment. Thus, this technology can play good role in livelihood cum nutritional security for difficult areas through horticulture based cropping system. This system has been amply demonstrated at KVK, Lohaghat, Champawat, Uttarakhand and CPCT, IARI, New Delhi. (Singh, et al. 2017, Singh 2007, and Singh et al. 2012).

Figure 21. Rainwater harvesting technology

Micro Irrigation-cum-fertigation System for Protected Cultivation

Micro-irrigation is classified mainly into two types') drip irrigation and ii) sprinkler irrigation. Drip irrigation is one of the best available tools made out of plastics in agriculture for the judicious use of water for growing various crops in large scale on sustainable basis. Drip irrigation is a low, labour intensive and highly efficient system of irrigation, which is also amenable to use in difficult situations and problematic soils, even with poor quality water. This technology is the practical example of "*More Crop PerDrop*" theme of agriculture in PMSY and has provision for 50-60% subsidy for the adoption of technology (Jitendra et al.2021, Savir et al, 2021 and Singh 2004, 2007 and Singh, 2018).

Irrigation water savings through drip irrigation can be affected by adopting a suitable drip irrigation system ranging from 36-79%. In drip irrigation, water is supplied through a network of plastic pipes using dripper/emitters. Water is supplied at a regular interval and at a required time/ quantity. Productivity gains vary from 20% to 50% depending upon the crop. Drip irrigation works by applying water slowly, directly to the soil. Drip irrigation has high efficiency as there is no loss of water in form of evaporation or run- off and water is applied directly to the plant roots. This is one of the major tools of precision farming which effectively increase water and nutrient use efficiency thus contributing significantly in water scare areas, increasing production and overall livelihood upliftment through horticulture based crops. Similarly, sprinkler irrigation is one of the more effective tools of water conservation. It is used mainly in production of horticultural crops in the open type of cultivation mainly orchards. This is used mainly in carpet natured crops and high-density crops and leafy vegetables like beans, peas, root based crops like radish, carrot etc. and ornamental landscapes and lawns including early transplanted stages of fruit crops. This irrigation technique has given a flip to the intercropping system in horticultural crops. Currently, about 69.5 mha area is under micro-irrigation, of which drip irrigation covers about 27 mha. Lately, other cropping systems have also started to adopt the drip irrigation to finally taking agriculture to overall aim of rainbow revolution. (Hasan, et.al.2018 and Singh 2007)

Figure 22. Micro irrigation-cum-fertigation system

Figure 23. Drip irrigation

FARM VALUE ADDITION

In vegetable crops, around 28-30% losses take place during harvesting, grading, packaging, transportation and marketing. The existing post-harvest loss of vegetables can be considerably reduced by adopting improved packaging, handling and efficient system of transport. Plastic is used for packaging and storage of vegetable crops for extending storage life. The plastic bags, plastic sacks, plastic blow molded containers are used for handling and storage. The cellulose film (cellophane), cellulose acetate, polyethylene (HDPE and LDPE), polystyrene, ethyl cellulose, polypropylene, PVC, edible plastic are used for packaging of vegetables. These technologies, minimize losses and add value, fetch higher price and also enhance the shelf-life of horticultural produce. (P.K. Kannaujia et al., 2019. and Saver et al., 2021)

Figure 24. Farm value addition

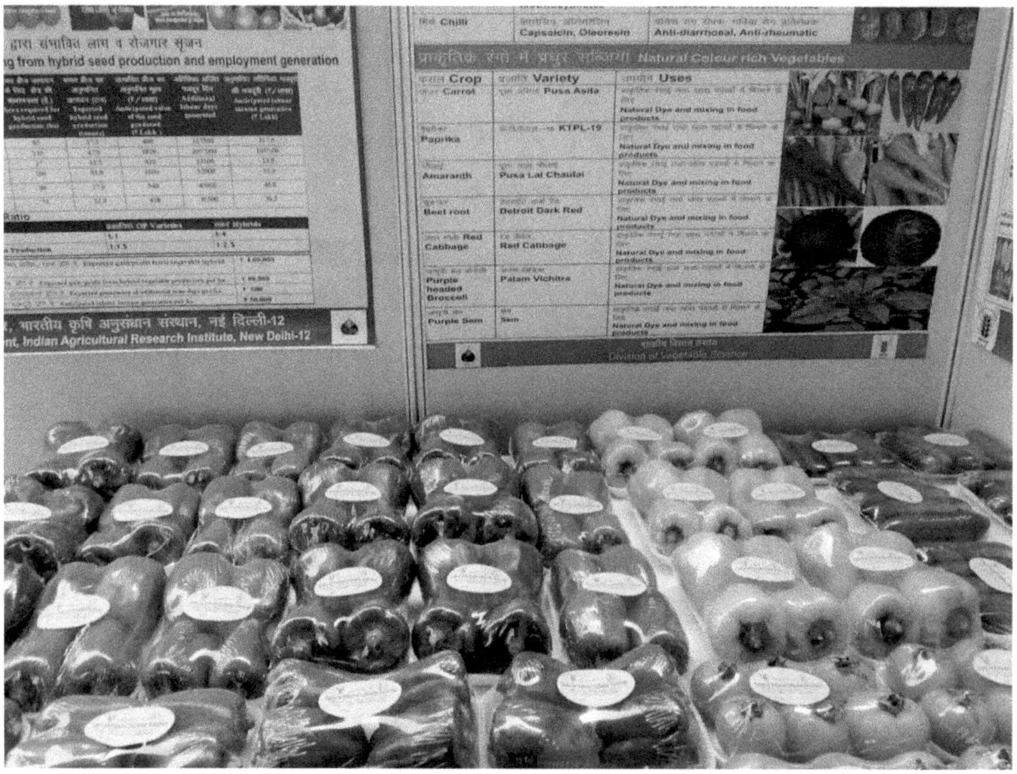

Protected cultivation based horticultural production systems, thus, have immense uses in order to conserve the water, nutrient and energy resources besides increasing the production, generating additional income for the farmers, creating more employment opportunities in rural areas for the unemployed youth and women folk. (Singh et al. 2012 and 2004)

In future, there is need to encourage environmentally degradable material, developing plastic based system for storage and handling of vegetable produce, besides adopting good agricultural practices and IPM. There is need of developing technology for production of eco-friendly plastic and eco-friendly disposal of waste plastic. (Prajapati et al. 2021 and Saver et al 2021).

Figure 25. Protected farming in horticulture

Figure 26.

Constraints for Success in Protected Farming in Horticulture

There are certain constraints for success of managing resources in protected agriculture such as infestations of soil & air borne diseases, poor quality of soil health, water and traditional methods of hill and plateau agricultural practices. It requires high initial input cost to make use for protected cultivation. (Sharma et al., 2014 and Mahawer et al., 2020).

Figure 27. A view of farm for off-season vegetable production at KVK, Lohaghat, Champawat, Uttarakhand

CONCLUSION

The protected cultivation technology, due to higher yield per unit area along with export quality produce, can prove useful in raising the socio-economic status of small and marginal farmers of the hilly and plateau areas. It is also important in present scenario of shrinking cultivable land, depleting water and other natural resources. Keeping in view the increasing demand of off-season and high value vegetables, fruits and flowers in metropolitan cities of the country, there is urgent need for diversification of traditional agriculture by production of high value horticultural crops under different protected structures for increasing the productivity and quality produce for getting high returns. Nursery raising under protected cultivation can be adopted well as a agri-entrepreneur in major vegetable growing areas of the country by unemployed youths who are graduate in agriculture or post graduate in horticulture. Low-cost protected technology like plastic low tunnels or walk in tunnels can be used for off-season vegetable cultivation for getting high returns. Similarly insect-proof net houses can be used on a large scale for safe vegetable cultivation to minimize use of pesticides in vegetable cultivation and virus free crop production. Protected cultivation of horticultural crops is reckoned to be the next logical step to open field agriculture. One million hectare under protected vegetable cultivation can easily increase the production of vegetables by about 40 million tones. This will also provide ample opportunity for the creation of additional jobs in rural sector. (Singh, Awani Kumar 2017 and (Prakash et al., 2019 and Singh, et al., 2012, 2004 and Sabir et al., 2020).

In India, the field of protected cultivation is still evolving. Methods of more efficient input application and protection technology are being devised. More eco-friendly plastics are being produced for making the processes more efficient. As a result, protected cultivation could bring prosperity to horticultural crop growers in hilly and plateau regions of India and ensure nutritional security for everyone.

Figure 28.

Figure 29. A view of polyhouse complex for off-season vegetable production at KVK, Lohaghat, Champawat, Uttarakhand

REFERENCES

Akshay, S. M., Kushwaha, H. L., Kumar, A., Bhowmik, A., & Singh, A. K. (2021). Pulsating air pollinator for greenhouse cultivation. *JSIR*, (80), 477–485.

Gupta, N., Jain, S. K., Tomar, B. S., Anand, A., Singh, J., & Singh, A. K. (2021). Influence of fruit load per vine on seed quality in cucumber (Cucumis sativus L.) grown under open field and protected environments. *Indian Journal of Horticulture, 78*(1), 93-100.

Hasan, M., Saver, N., & Singh, A. K. (2018). *Hydroponics Technology for Horticultural Crops. TB-ICN No.188/2018. IARI.*

Jitendra, K., Patel, N., Singh, R., Sahoo, P. K., Susama, S., Sehgal, V. K., Marwaha, S., & Singh, A. K. (2021). Development and evaluation of automation system for irrigation scheduling in Broccoli crop. *Indian Journal of Agricultural Sciences*, *91*(5), 221–225.

Kannaujia, P. K., Asrey, R., Singh, A. K., & Varghese, E. (2019). Postharvest treatments to reduce chilling injury in summer squash (Cucurbita Pepo) fruits during storage. *International Journal of Agricultural Sustainability*, *89*(10), 1633–1637.

Kumar, M., Shirohi, H. S., Singh, B., Tomer, B. S., & Singh, A. K. (2020). Effect of insect-proof net house on flowering and seeds quality of cucumber (Cucumis sativus L.). *Indian Journal of Agricultural Sciences*, *90*(6), 1202–1204.

Mahawer, H., Presanna, R., Gogoi, R., & Singh, A. K. (2020). *Differential modes of disease suppression elicited by silver nanoparticles alone and augmented with calothrix elenkinii against leaf blight in tomato*. European Jr. of Plant Pathology., doi:10.100710658-02021-w

Prajapati, U., Asrey, R., Varghese, E., Singh, A. K., & Singh, M. P. (2021). Effects of Post harvest ultraviolet-C treatment on shelf-life and quality of bitter gourd fruit during storage. *Elsevier Journal on Food Packaging and Self Life*, *28*(100665), 1–8.

Prakash, P., Kumar, P., Kar, A., Singh, A.K., & Anbukani, P. (2019). Progress and performance of protected cultivation in Maharashtra. *Indian Journal of Economics and Development, 15*(4), 555-563.

Prakash, P., Kumar, P., Kar, A., & Singh, A. K. (2020). Status and impact of protected, cultivation of horticultural crops in, Maharashtra. *Indian Journal of Horticulture*, *77*(3), 518–526.

Sabir, N., & Singh, A.K. (2020). Home and Smart Grading for Health and Environmental Improvement. In *Gardening for Health and Environment.* India Kitchens Garden Association.

Saver, N., Singh, A. K., & Hasan, M. (2021). Greenhouse Agriculture: Production and Protections. Pentimer Publications.

Sharma, H.K., Singh, A.K., & Lal, J. (2014). Effect of Methom Sodium on Root-knot nematode Infestation in tomato under protected Cultivation. *I.J. of Nimatology*, *43*(2), 236–237.

Singh, A.K. (2019). Producing Long Melon in Off-Season. *Indian Horticulture*, 11-13.

Singh, A. K. (2013). Traditional knowledge based adaptations of bio-mulching for ginger crop management: Empirical learning with resource-poor farmers under rain-fed agro-eco-system of western Himalaya. *Indian Journal of Traditional Knowledge*.

Singh, A. K., Shrivastva, R., & Mathala, M. J. (2004). Use of plastic in agriculture. A Hindi-Bulitin-IARI, TB-ICN:H-6/2004.

Singh, A. K. (2007a). *Drip-irrigation system for hilly areas. KVK.*

Singh, A. K. (2007b). *Polyhouse Cultivation in hilly areas. KVK.*

Singh, A. K. (2018). *Mulching Technology for Vegetables production in Hilly Area. UDWDP.*

Singh, A.K. (2021). *Protected Cultivation of Vegetables under Polyhouse structure*. UDWDP.

Sing, A. K., Shri Dhar Saber, N., Patel, N., Singh, V., Singh, A., & Kumar, A. (2021). Response of Long melon on spacing and pruning methods under insect proof net house during off-season in the plain reason of India. *Indian Journal of Agriculture Sciences, 91*(5), 226-30.

Singh, A. K., Saver, N., Jat, G. S., Singh, J., Singh, V., Singh, A., Kumar, A., & Kumar, J. (2021). Effect of different spacing and pruning methods on growth, yield and economics of Long melon during off-season under naturally ventilated polyhouse. *Indian Journal of Agriculture Sciences, 91*(6), 421-24.

Singh, R. K., Singh, A. K., & Gautam, U. S. (2021). Protected cultivation for an increase in farmer's income in Bundelkhand region of UP. New Age Protected Cultivation Magazine, 7(2), 44-48.

Singh, A. K., & Sabir, N. (2019). Design and size of Greenhouse Structure in India: Crops and Regional Suitability. Sping the Future of Horticulture, 371-392.

Singh, A. K. (2017). Horticulture based Plasticulture cum covered cultivation for livelihood security. *Indian Horticulture*, *62*(5), 68–74.

Singh, A. K. (2017). Jal-Sanchyan and Sichai Pravendhan me Sanrakkchhit Kheti ka Yogdan. *Kuroochhetra*, *64*(1), 61–63.

Singh, A. K. (2018). Protected Cultivation Technologies: A Sustainable Solution. *ISRAEL: An Indian Prospective in Agriculture Word*, *4*(5), 60–68.

Singh, A. K., Rawat, S. S., Singh, V. K., Mehta, K. S., Singh, B., Sindhu, S. S., Hasan, M., Rakesh, K., & Mukul, K. (2012). Rain Water Harvesting: A Boon for Protected Horticulture in the Peri -Urban Area of Hills. *1st Biennial International Congress on Urban Green Spaces*, 151-152.

Singh, A. K., Singh, V. K., Mehta, K. S., Singh, B., Sinthu, S. S., Kumar, R., Kumar, M., & Singh, V. (2012). Socio-Economic Empowerment of Hill Women through Protected Cultivation Technology. *Global conference on women in Agriculture*, 21-22.

Chapter 7
Sustainable Agriculture and Organic Farming

Saritha Vara
https://orcid.org/0000-0003-0473-993X
GITAM University (Deemed), India

Bhavya Kavitha Dwarapureddi
GITAM University (Deemed), India

Swathi Dash
GITAM University (Deemed), India

ABSTRACT

It is perceived that contemporary agricultural practices are impairing natural nutrient cycling, leading to unsustainable soils. The effects include loss of nutrients from the soil and loss due to erosion. Significant public concern has been raised during the last decade over the products from conventional agriculture. Researchers, governments, farmers, and people are constantly in search of healthier methods for producing food. Organic farming has been gaining attention in these lines as it is perceived to be good for the health of humans and the environment along with providing quality produce. Organic farming still has to emerge from the substantial challenges it is facing towards becoming a prime source for food and fiber in the long term. The chapter aims to identify the challenges of organic farming and review the available techniques towards transforming current farming methods to organic farming, making them resilient along with protecting the environment and human health.

INTRODUCTION

Demand for food is ever increasing with population explosion, and modern agriculture has been depending on chemical fertilizers to meet the demand. As countries are developing so the awareness among people is enhancing and thus demand for chemical free or organic food is increasing. These pressures are multiplied with issues like rising labour cost, shrinking farmland and shortage of farm workers. Though

DOI: 10.4018/978-1-7998-8434-7.ch007

intensive conventional agricultural practices depending on inorganic fertilizers, chemicals and pesticides have enhanced yield, these have also contributed to degradation of soil, biodiversity loss, enhanced crop susceptibility to pathogens and pests, negative environmental impacts etc., which have together resulted in significant consequences in food security and human health.

While there is structural decline in productivity of farm in developed countries, chief challenges in developing countries are due to substantial increase in yield with reference to quantity and quality, without further enhancement in farming costs and harmful environmental impacts. Subsequently, burdens on global environment in 21st century not only originate from population explosion but also from altering consumption forms. In order to support ever growing human demands there was an increase in food production through conventional farming practices. However, in several instances, these practices have resulted in consumption of excessive water, agrochemicals, energy, release of greenhouse emissions and also in reduced of soil productivity and fertility (Gomiero et al. 2011).

Contemporary agricultural practices implemented across the globe have adopted various approaches towards addressing sustainable environment and economic development aiming at maintaining yields along with preserving the biosphere. The concept of sustainable development intersects three primary factors – environment, society and economy which in turn interact among each two of them. Hence, interaction among environment and economy in terms of agro-ecology, environment and society in terms of environmental awareness and society and economy in terms of life standards determine sustainable development concept. From context of sustainability, one of the chief issues is maintaining soil quality, being a non-renewable resource, soil exerts several social and environmental functions. One of the target in sustainability is to search for effective methods towards nutrient recycling, pathogen and pest control, alleviation of negative impact of abiotic stress factors etc which can be enhanced through appropriate organic farming techniques.

5 principles for sustainable agriculture which captures the three pillars are proposed by FAO, 1988 these are:

1. Enhancing resource use efficiency
2. Protecting, conserving and improving natural ecosystems
3. Protecting and enhancing livelihoods of rural people and their social wellbeing.
4. Improving resilience of population, societies and ecosystems
5. Encouraging better governance of both man made and natural systems

Moreover, some studies have concluded that all SDGs either indirectly or directly are linked to sustainable and nutritious food (Rockström & Sukhdev, 2016).

In terms of IFOAM (2017), the features of organic agriculture comprise of:

- Maximising use of natural fertilizers and pesticides
- Enhancing and protecting soil health
- Supporting and enhancing diverse species and breeds
- Establishment of partially natural habitats
- Livestock management with an objective to protect and enhance animal welfare.

APPROACHES TO SUSTAINABLE AGRICULTURE

Ahirwar et al., 2020 identified some recognized terms which were precisely developed for attaining sustainable agriculture having distinct focus on environment. As these terms have robust philosophy committed to them, they mostly had:

- Vibrant philosophies;
- Objectives related to environment, economic and societal;
- Have grew as methods in their individual right over the period

In general, all these methods might be valid within assortment of production types and environments. In several cases market, practitioners are associated with them. Even though, there is a variation in scope, the chief unifying characteristic is that all the choices are made by farmers which govern how they manage the cropland over the years. Some of the approaches presented by them include:

1. Organic farming
2. Agroecology
3. Conservation agriculture
4. Nature-inclusive agriculture
5. Regenerative agriculture
6. Permaculture
7. Carbon farming
8. Biodynamic agriculture
9. High nature value farming
10. Climate-smart agriculture
11. Low external input agriculture
12. Ecological intensification
13. Circular agriculture
14. Sustainable intensification

The above list intends to clarify the landscape of different terminologies that are associated with sustainable agriculture. These are discussed in the following sections:

Organic Farming

Organic farming comprises of a series of practices for enhancing energy and nutrient use along with reducing pollution like maintaining crop diversity, crop rotations and having various combinations of plants and livestock, using amendments that are organic in nature (Nandwani and Nwosisi 2016), manure obtaining through green synthesis (de Jesus Souza et al., 2019), and fixing nitrogen through symbiosis using legumes (Romanyà and Casals, 2019). Practices of these kind are understood to consequence in enhanced degrees of organic carbon and nitrogen in soil over the years, even though initially there is a low nitrogen availability owing to insufficient nutrients in early growing seasons attributed to lower biological activity (Yang et al. 2019; Abdelrahman et al, 2020). Further, it was reported that individual soil enzymatic activity was enhanced after six months of switch to organic management even though

18 to 24 months were required towards reaching a steady state of different soil enzymatic activities (Karasawa et al., 2015). Hence, it is to be noted that introducing organic practices would require specific duration for acclimatisation of all means in use i.e at least 2 years prior to sowing for perennial crops and a minimum of three years for organic products.

Agriculture production through organic means is a way of sustaining health of ecosystems, soil and mankind (Halberg, 2012), hence, good soil quality is understood to be the cornerstone for sustainable agricultural systems like organic farming. Moreover, organic farming tends to simulate natural nutrient cycling reducing off-site losses. Effectiveness of nutrient cycling can be enhanced via numerous practices which include, cover crops, use of mulch, no or minimum tillage etc.

Agroecology

The widely accepted definition of agroecology is application of the concepts of ecology and principles for designing and managing sustainable agroecosystems (Gliessman, 2013). Agroecology is an integrated approach comprised of education, action, research and change aiming to bring sustainability to the entire food system including the concepts of ecology, economy and social. The transdisciplinary nature of agroecology stems from its values to all kinds of experience and knowledge in food system change. Agroecology provides a different path forward through provision of principles on how to design and manage agricultural systems which can withstand future crises may those be pandemics, pest outbreaks, financial meltdowns or climate disruptions through exhibiting greater levels of resilience and diversity both of which are recognized for their potential to decrease risks without compromising on yields and providing chief ecosystem services (Nicholls et al., 2018; Altieri and Nicholls, 2020).

Figure 1. Schematic representation of agroecology

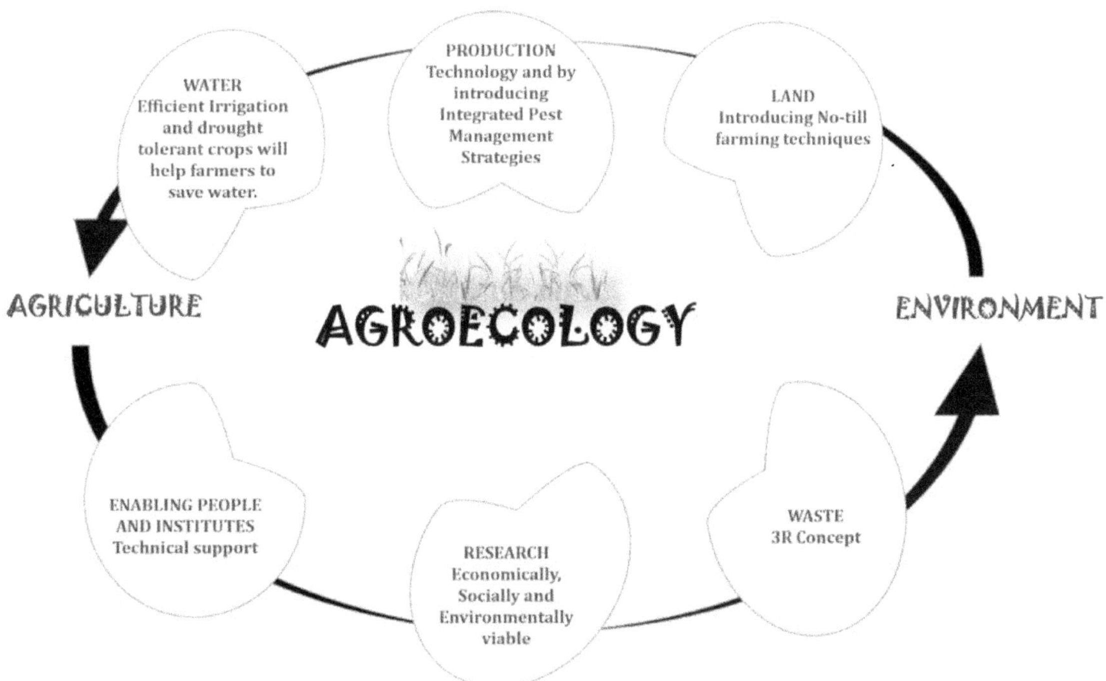

The process of transforming agriculture by leading it on sustainable path will be a tremendous challenge which can be met only when approached with 3 varied fronts concurrently: first, greater and improved knowledge on ecological relationships within different tamed agricultural species and physical environment and natural systems is required. This can be met by understanding the scientific aspects of agroecology that derives knowledge on contemporary ecological methods in deriving principles which can be used for designing and managing sustainable agroecosystems. Second, an efficient and advanced real time agricultural practices which work successfully in the contemporary conditions satisfying food needs along with paving path for enhanced sustainable systems for future. This practical aspect of agroecology is met through valuing local, indigenous and empirical knowledge that farmers have and also sharing this knowledge that can lead to reduction in distinction among production and application.

Finally, there should be a fundamental change in how humans relate to food, social and economic systems which govern the dispersal of food and also in ways which food intercedes relations of power among classes, countries and populations. Being a social alteration feature of agroecology, this not only promotes for variations leading to food security for all, but will also seek information of the ways through which these variations can be sustained and activated (Gliessman, 2018). The analysis of Miery et al., 2018 identified eight chief factors of the procedure of taking agroecology to scale: recognizing crisis which motivate search for alternatives, constructive process of learning, social organization, mobilizing discourses, efficient practices of agroecology, favourable markets, external allies and favourable policies.

Conservation Agriculture

Conservation agriculture encompasses real-world application of 3 interwoven principles which are nil or least mechanical soil disorder, diversified crops and biomass mulch soil cover in combination with other harmonizing good agricultural methods of integrated crop and production management (Valkama et al., 2020; Hunt et al., 2020; Kassam et al., 2019). Many studies have indicated that conservation agriculture leads to enhanced soil quality along with a range of soil biota (Briones and Schmidt, 2017; Knapp and Heijden, 2018). Conservation agriculture is a collection of water and soil conservation and agricultural management techniques which are intended at sustainable use of natural resources without compromising cost-effectiveness of farms and yields. Further, conservation agriculture has been anticipated as a beneficial strategy for supportive mitigation of climate change and acclimatization targets through the phenomenon of biophysical and biogeochemical pathways.

An increase in soil organic matter in conservation agricultural practices has led few researchers to propose extensive carbon sequestration potential, apart from reducing GHG emissions owing to reduced agricultural machinery inputs along with alteration in surface albedo and evapotranspiration. This is specifically seen in conditions of extreme temperatures (Seneviratne et al., 2018) and availability of soil moisture (Wilhelm et al., 2015). Such biophysical effects add towards enhancing resilience of agricultural structures to externalities like famines and heat waves (Seneviratne et al., 2018) that are anticipated to develop more frequent in future (Prestele et al., 2018)

Figure 2. Schematic representation of conservative agriculture

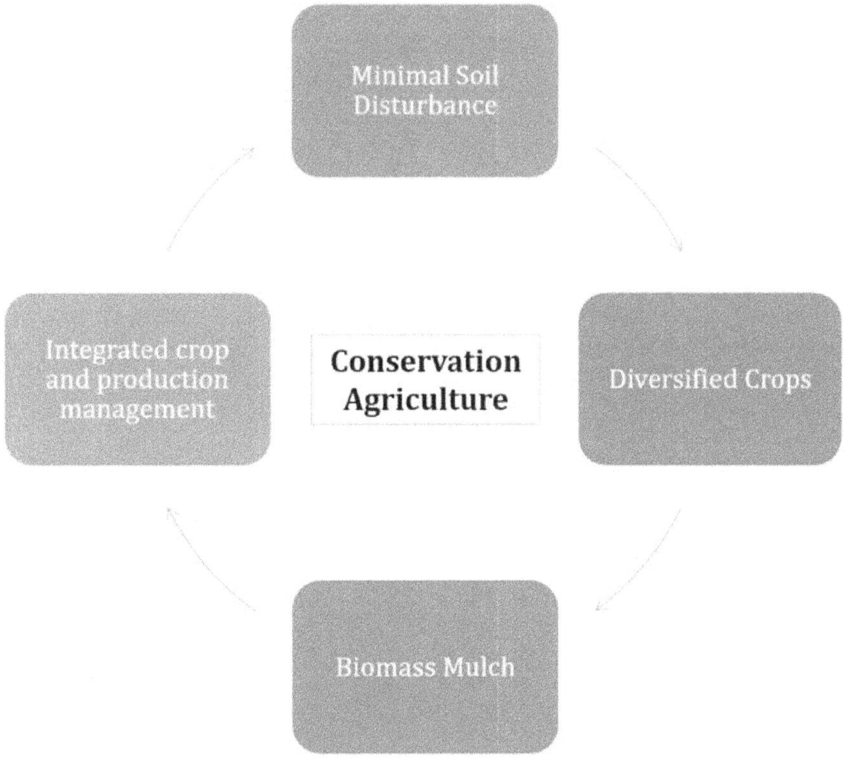

Nature-Inclusive Agriculture

A sustainable farming method which is based on best use of natural processes which are integrated in agricultural management is Nature-inclusive farming (Erisman et al., 2017). This method of farming makes farmer self-sufficient with reference to feed and manure and also works with maximum levels of water, additional landscape components (Westerink et al., 2018).

Nature-inclusive agriculture cannot be executed instinctively as there are several blockades to overcome. Hence, more knowledge on making agriculture more nature-inclusive without compromising farmer's income at the same time fitting in agricultural production styles is to generated. Several studies have contributed to the same to quote Delbaere et al., 2014; Erisman et al., 2016. One of the drawback of nature-inclusive farming is that its benefits are less visible to farmers and citizens in comparison to conventional farming. Moreover, transformation to nature-inclusive agriculture is a subject of governance (Runhaar, 2017).

Regenerative Agriculture

Perceived to be an attitude and a group of practices towards restoring and maintaining fertility and health of soil, supporting biodiversity, protecting watershed and enhancing economic and ecological resilience.

Figure 3. Schematic representation of nature inclusive agriculture

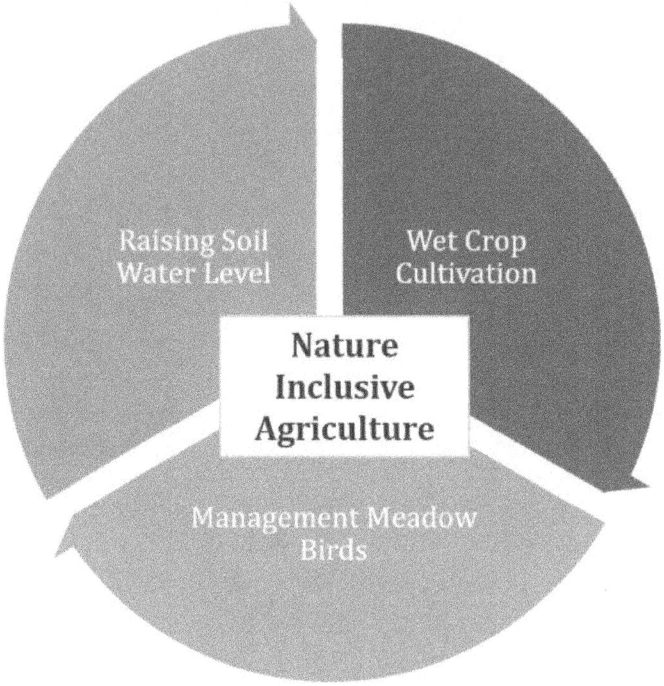

Regenerative agriculture emphases on creating circumstances for life both below and above ground, taking indications from nature, having a extended track path of growing things successfully. It can also sequester greater amounts of atmospheric carbon below ground, making it economic solution to climate change through re-carbonizing soils through photosynthesis and biology, precisely on degraded land. Multiple co-benefits of regenerative agriculture include producing healthy and nutritious food, a grave component to our response to increasing climate instability (White, 2020). Regenerative agriculture is understood to be soil centrical rather than seed centrical, constructed on the evidence that "health of soil, animals, plant and humans is one and indivisible" (Lal, 2020)

Elevitch et al, 2018 presented unifying principles which are steady across re-forming farming systems as: avoiding tillage, eradicating spatio-temporal trials of bare soil, nurturing farm plant diversity and participating cropping practices and livestock on the land. Regenerative agriculture has been strongly encouraged by NGOs, civil societies and many multi-national food companies (Giller et al., 2021)

Permaculture

Permaculture concept came from mixture of two words "permanent" and "agriculture", which is described as a strategy system and as finest practices agenda towards making and administration of sustainable and resilient agroecosystems.

Permaculture is recognized internationally as sustainable agriculture model that tactically integrates both unconventional and traditional management methods towards optimizing distribution of ecosystem services. It associates the phenomenon of perennialization, crop divergence and nature-sparing which

Figure 4. Schematic representation of regenerative agriculture

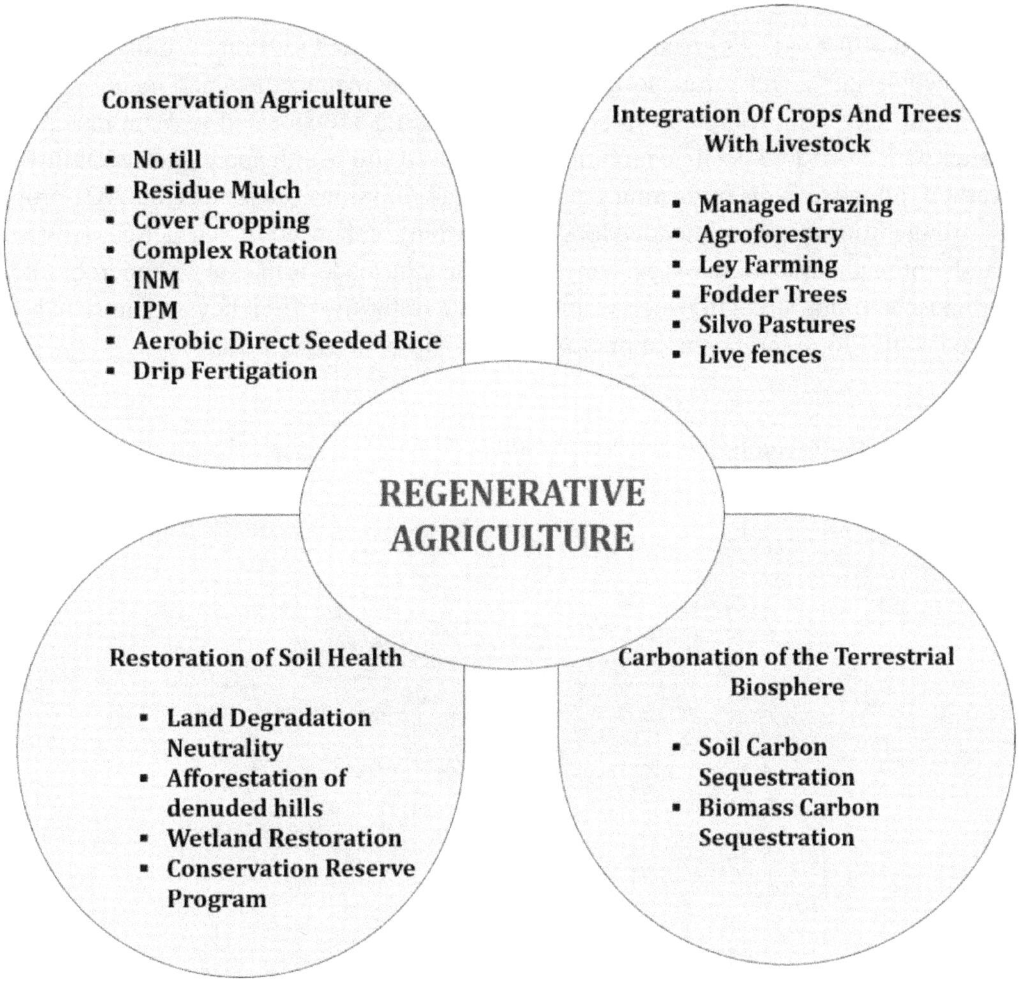

independently target various yet overlapping valuable ecosystem characteristics and processes, resulting in greater holistic ecosystem service delivery and resilience. This may be specifically suitable towards fostering ecosystem amenities in low-input, marginal, degraded and least-profitable agricultural zones (Hirschfeld and Van Acker, 2021).

Permaculture is perceived to be concept for planning sustainable socio-ecological land use system, identifying that land use systems are at no time detached from social systems. Krebs and Bach, 2018 formulated three basic ethical norms towards considering in design and administration of permaculture systems: care for earth, people and setting bounds to consumption and reproduction, and redistributing excess. According to them, vital features of permaculture towards development of agroecosystems are: characteristics of site; interactions among discrete elements on different levels, ranging from assorted cultures at field to assortment of land use at level of agroecosystem; and spatial arrangement of elements as governing factors for manifold operations.

Carbon Farming

Agricultural practice that reduces greenhouse gas emissions is called Carbon Farming. Practices in carbon farming include stubble retention, no-till cropping, grazing management and manure management. It aims to enhance CO_2 removal rate from atmosphere which is transformed to plant material and soil organic matter with a two-fold positive ambition of improved soil health and enhanced ability for long-standing carbon storage towards mitigating greenhouse gas emissions. Jansson et al., 2021 proposed the following attributes towards crops intended for carbon farming: enhanced subsurface carbon distribution for larger and profounder root biomass; exchanges with custom-made, artificial soil microbiome towards amplified rhizosphere sink strength which would facilitate water-use efficiency and nutrient acquisition and; enhanced source forte for increased photosynthesis and biomass build-up.

Figure 5. Schematic representation of carbon farming

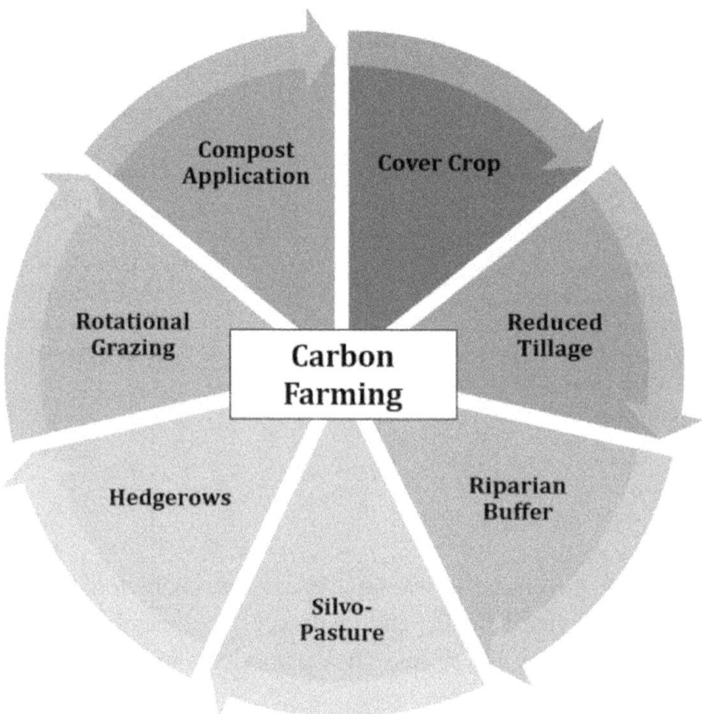

Carbon sequestration via agricultural soil is being encouraged as a win-win approach towards vindicating climate change and enhancing farmer's livelihoods, specifically in dry land. Kragt et al., 2014 presented some of the factors that hinder farmers from adopting carbon farming which are carbon price ambiguity and uncertain influence of these practices on profitability and productivity. Farmers expressed strong favourites for stubble retaining and no-tilling methods as strategies for carbon farming.

Biodynamic Agriculture

Biodynamic farming is understood as biological dynamics, a practice of organic agriculture that contemplates farm as living organization, where one action affects other. It is considered as first agricultural system that focuses on soil, ecological health and healthy food (Hakan Karadağ, 2018). Biodynamic agriculture relies mostly on composted farmyard manure (FYM) as key fertilizer, emphasizing on integration of livestock and crops, maintenance of soil, health and well-being of crops and animals. In addition, biodynamic farming utilizes compost preparations and field sprays that consist of specific minerals/plants fermented with animal manure, water and soil along with practicing crop rotations and cover crops.

Biodynamic agriculture treats plant growth, livestock care and soil fertility as ecologically interrelated tasks stressing on mystical and spiritual perspectives (Khadse et al., 2021). Biological practices in biodynamic agriculture include applying a sequence of well-known organic farming procedures to enhance soil health, while energetic practices intend to effect both metaphysical and biological aspects of the farm, adapting the farm to natural rhythms like sowing seeds during specific lunar phases (Ponzio et al., 2013). Hence, it is supposed that planting that planting crop on a day will be completely dissimilar than planting it other day. As suggested by Nabi et al., 2017, there are two characteristics of biodynamic farming: One way is using specific farming inputs that are produced from variety of mineral, herbal and physical raw materials that are treated in complex ways before applying to soil and crops in minimal doses. While other is observing the tempos in nature that go beyond effects of weather, season, sun including planetary, lunar and stellar constellations.

High Nature Value Farming

A farming system which delivers broad sets of ecosystem services, proposed as a solution towards minimizing ecosystem trade-offs is called High nature value farming. High nature value farmlands are zones where agriculture is a leading land use, that promotes high species and habitat diversity. These are seen in a multitude of environments, economic contexts, climatic conditions and production systems comprising of arable lands, grasslands, permanent croplands etc. Chief determinants of greater diversity and diverse ecosystem services delivered by high nature value agricultural processes are application of lower concentrations of pesticides and fertilizers, lesser use of machinery, substantial coverage of semi-natural vegetation along with land cover diversity.

However, high nature value farming possesses potential to deliver a variety of ecosystem services including high quality food, genetic resources maintenance, regulating soil quality, water purification, pollination, ecotourism, recreation as well as heritage. Yet, the services of high nature value are underrepresented in most of the literature. Owing to the social and ecological phenomena of ecosystem services from agriculture these are to be identified and managed with integrative and broad approaches (Plieninger et al., 2019).

Climate-Smart Agriculture

It is an approach leading to expansion in agricultural systems that aim to address the entangled threats of climate change and food security (Kurgat et al., 2020). Climate-Smart agriculture provides framework that interactions among mitigation, adaptation and enhanced food security for precisely limited farmers.

Nevertheless, advanced agricultural systems are required to find interactions among these three pillars (Andrieu et al., 2019).

Much attention is drawn towards promotion of climate-smart agricultural practices attributed to Climate Action of Sustainable development goal 13. Practices of Climate-smart agriculture contribute global food security enhancement, increased farmers' ability towards adaptation to changing climatic conditions as well as greenhouse gases mitigation. Climate-Smart agriculture not only contributes to SDG 13 but also to is inherently related to other SDGs like SDG1 – No Poverty and SDG 2 – Zero Hunger (Hellin, and Fisher, 2019). There is need for farmers to recognize what should be deemed climate-smart in context of their agricultural, biophysical and socio-economic and hence, the approach of climate-smart agriculture is perceived to be knowledge-intensive requiring significant support from institutions. Owing to these difficulties, Chandra et al., 2018 illustrated that climate-smart agriculture is "a popular scholarly solution" which is experiencing difficulties in deciphering to smaller farmers, actions of civil society and policy directions (De Pinto et al., 2020).

Low External Input Agriculture

Low external input agriculture is one that optimally utilizes locally available human and natural resources like water, vegetation, soil, local animals and plants, skill and knowledge, which are ecologically feasible, cultural adapted, ecologically sound and socially justice (Ibeawuchi et al., 2015; Firth et al., 2020). This practice is understood to be a prominent feature regarding its role in agricultural technology for reduction in rural poverty, it is perceived that this practice is more accessible to resource-poor households and would form basis for social and human capital formation.

Some prominent examples of low external input agriculture are: conservation tillage; Hedgerows and living barriers; Contour planting; cover crops; Physical structures like terraces to prevent soil erosion; Mulches; Manures and composts; Soil and water management; Controlling weeds and pests; Integrated pest management; Soil fertility enhancement; Biomass transfer and green manures; crop establishment; Planting pits; Fater cropping; Intercrops and rotations.

Ecological Intensification

Bommarco et al., 2013 and Tittonell, 2014 defined ecological intensification as a practice that encompasses actively managing farmland towards increasing intensity of ecological processes supporting production, regulating biotic pests, enhancing pollination and nutrient cycling. In other terms, it can be understood as making smart use of services and functions offered by nature at levels of both field and landscape towards increasing agricultural productivity at the same time reducing dependence on agrochemicals (Kovács-Hostyánszki et al., 2017). Essentially ecological intensification aims to enhance crop productivity through improving biodiversity and related ecosystem services (Garibaldi et al., 2019; Kleijn et al., 2019).

Circular Agriculture

An effective means towards administration of soil organic inputs for improvement of soil fertility and sustainability of cropping system is Circular agriculture. Its development is a significant wat to ease starvation, encourage sustainability of global grain supply chain and reduce cost of grain apart from reducing environmental impact from modern agricultural practices (Li et al., 2021). Hence, it is perceived that circular agriculture is strongly suggested towards filling gaps and assist in activities of wider agricultural society (El Janati et al., 2021). Circular agriculture targets trials of farm-based rural budget along with environmental issues suggesting effective models for managing agricultural remains through integrating those in production processes along with improving local food markets. Bekchanov and Mirzabaev et al., 2018 in their study presented that circular agriculture showed that some organic waste products when recycled for use as soil organic inputs have resulted in enhanced soil organic matter and reduced nutrient deficiencies in soil.

Sustainable Intensification

Emerged as a set of guiding principles towards enhancing sustainability and productivity of agriculture, Sustainable intensification is goal-oriented with emphasis on necessity to improve food production along with reducing environmental footprint on farmland (Pretty et al., 2014). In a broader sense sustainable intensification aims to enhance livestock and crop yields and related economic returns. Successful implementation of sustainable intensification can be quantified through metrics which measure system outputs in terms of yield; impact on soil quality, natural resources and ecosystems; and inputs required for achieving yield.

Figure 6. Schematic representation of sustainable intensification

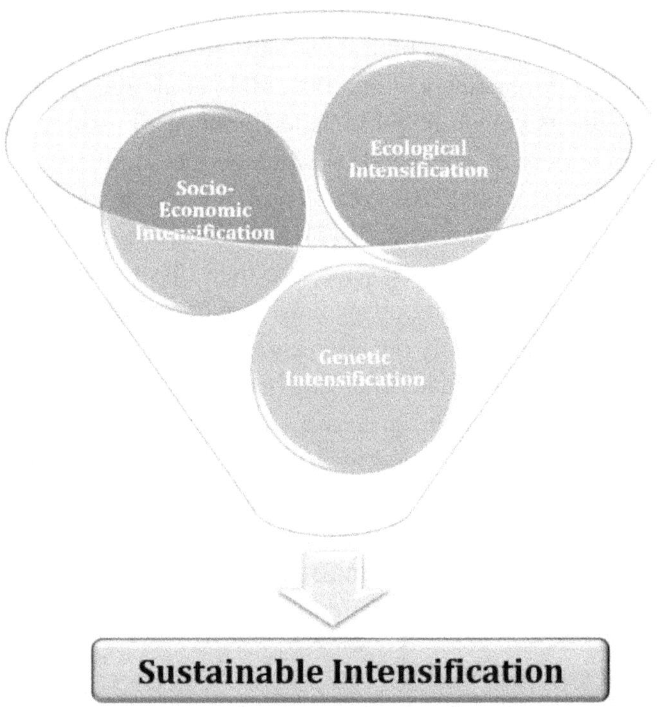

Sustainable intensification was originally focused on need to move beyond seed, pesticide and fertilizer technologies which support high-yield modern agricultural systems to more restorative systems of production, reducing dependence on external inputs, while leveraging internal resources and ecological processes towards nutrient supply and pest control (Cassman and Grassini, 2020). Though Sustainable intensification was proposed with limited-holder farming systems in mind, it acquired global popularity as it amalgamates several technologies, fields and locally-available resources towards developing solutions that are context-specific. Despite its wide spread and promotion by governments its implementation is still restricted to developing countries (Fischer and Connor, 2018; Tseng et al., 2020; Jain et al., 2020).

ORGANIC FARMING

Owing to the fact that organic farming is the most adopted technique of sustainable agricultural practices, this is discussed in detail in the following sections. Organic farming is the land use process which strives to reduce negative environmental impacts including nutrient leakage, biodiversity loss and degradation of soil (Lori et al., 2017). Among several causes, one of the principal motives for extensive loss of biodiversity and habitats is attributed to rigorous land use (Habel et al. 2019) along with higher inputs of pesticides and nutrients (Isbell et al. 2013), narrow crop rotations, elimination of landscape elements (Fabian et al. 2013) etc. Loss of biodiversity induces undesirable effects on successive portions of food chain (Heldbjerg et al. 2016), along with reducing the function as well as quality of agricultural land as a habitat owing to degradation of conditions for reproduction (Quinn et al. 2017).

Ample of meta-analyses and reviews during the last three decades have presented the optimistic consequences of organic farming on biodiversity of both above and below ground in correlation to conservative farming practices (Azeez 2000; Mondelaers et al. 2009; Hole et al. 2005; Rahmann 2011; Tuck et al. 2014). The specific enhancements bought about include abundance in soil microbes and their activity (Lori et al. 2017), enhanced variety of arthropods (Lichtenberg et al. 2017), field birds (Chamberlain et al. 1999), arable flora (Campiglia et al. 2018) etc. In an agrarian landscape, biodiversity is greatly governed by the type of land use (Stoeckli et al. 2017). Further, it has been reported that heterogenous agricultural landscapes boost diversity of species and also their dissimilar ecosystem services (Landis, 2017). Practices like diversified crop rotations, changes in crops of spring and winter along with perennial fodder crops tend to enhance the biodiversity owing to agricultural management taking place at various times of a year resulting in providing alternatives during harvest (Gottwald and Stein- Bachinger 2016).

Organic farming practices are fetching progressively important owing to their resilience; persistent needs for protection of soil, water and air; need for improvement in the socio-economic status of rural communities, farm workers and farmers; for provision of health, safety and nutritious horticulture products to rapidly enhancing population of the world. Accordingly, there is a need for substantial new developments in organic farming for to effective and efficient supply of food for humans. Development of organic farming systems than can have greater productivity over the existing systems require creativity and novelty.

Challenges of Organic Farming

Challenges in organic farming are usually in relation to their evaluation for their feasibility owing to several changes that are only observed over long term; successful organic farming mostly depends on the local conditions. Some factors in relation to transition to organic agriculture comprise of implications for labour and inputs like nutrients, water energy etc, crop variation and yield and farm production in total (FAO, 1998). Although labelling and guidelines pertaining to this method shaped robust market towards organic harvests, certainly organic farming is understood to be the lone legally well-defined approach for sustainable agriculture. Nevertheless, being rigid in nature, a barrier for entry exists for prospective practitioners. While existing stringent requirements aid in the objective of environmental protection, challenges associated with organic farming become higher for farmers to adopt.

Organic farming pursues to reduce use of chemical and synthetic inputs towards achieving sustainability, there are dimensions in sustainability that are not necessarily compatible with organic practices (FAO, 1998). As reported that though organic farming work well in smaller scales, it tends to use more arable land and water for achieving same yield in comparison to conservative agricultural practices (Miller, 2017; Savage, 2015).

Organic Farming and Soil Health

The importance / significance of organic farming lies in the fact that it recognizes that the whole environment where plant grows is much extra over the sum of its discrete parts, and that all living things are co-dependent and consistent. Organic practices of farming encompass giving soil and cultivating environment as a resource to be conserved for future. Further, it focusses on providing the plant with well-adjusted food supply through nourishing soil living creatures surviving in manures, compost and organic materials. It is understood that organic manures like slurry from biogas, agricultural residues, farmyard manures, earthworms help in enhancing soil fertility. Application of such soil amendments change environment of the rhizosphere through affecting aeration, porosity, water holding capacity, temperature and soil microflora as these comprise necessary ingredients like nitrogen, potash, phosphorus, calcium, sulphur and magnesium.

Pest, Disease and Weed Control through Organic Farming

Chief strategy of limiting pests and diseases in organic farming is via providing good plant nutrition and management; using cover crops, crop rotations that alter the field ecology leading to efficient disruption of habitat for weeds, disease organisms and insects. Weed control is bought about by crop rotation, hand-weeding and mechanical tillage along with cover crops, flame weeding, mulches and related management approaches. Pests are kept on check through diverse population of soil organisms, birds and beneficial insects. In situations of uncontrolled pest populations policies such as mating disruption, use of insect predators, barriers and traps are adopted. In conditions where above methods fail, as a last resort, specific non-synthetic, botanical pesticides are applied. Organic foods are characterized by minimal processing in order to maintain integrity of food precisely without addition of any preservatives and artificial ingredients (Singh, 2021).

Nutrition of Organic Foods

As per report in 2002, it is said that organic food contains lower pesticide remains over conservative foods where in 13 percent of organically produced samples were tested against 71 percent of conventionally produced samples. Nevertheless, it is to be noted that a definitive study has not yet been carried out owing to the assembly of parameters intricate in bringing out a fair comparison among organic and conventional foods which include variety of crop, time after harvest, handling methods post-harvest, climate and soil type that significantly govern nutritional quality.

Organic Farming and Rural Development

Contribution of organic farming to vibrant rural economies represent real opportunities at various scales through sustainable development. In fact, some evident growth in organic sector is seen in terms of new employment chances in agricultural processes and connected services, along with advantages to environment that these bring that can be significant both economically and also for social unity of rural areas. Moreover, incentives and economic support for farmers interested in transition to organic making has designed to help rural sector to grow more and also support related businesses throughout food chain (Singh, 2021).

Organic Farming and Biodiversity Enhancement

Biodiversity is both an instrument and aim in organic agriculture, having a natural ecological balance above and below ground is important to the accomplishment of organic farming. Organic farming practices generate favourable conditions for soil abiotic and biotic funds through – green manuring and organic fertilization; crop rotations and strip-cropping; avoiding use of herbicides and pesticides, minimum tillage etc. Scientific research presented that organically managed soils pointedly enhanced total density, biological activity, and diversity of soil microbes, which in turn enhances nutrient cycling and soil structure. Healthy soil is understood to be base for production of food, while plant and animal diversity on land prevents pests and disease outbreaks.

Organic Farming and Conservation of Nature

Ample studies across the globe indicated that organic agricultural practices safeguard non-agricultural biodiversity offering viable alternatives in categories of protected areas allowing human activities. Farmlands bordering and connecting protected areas if adopt organic methods, loss of wildlife, contamination of soil, water and air can be restricted. These act as buffer zones which are critical in successful conservation of protected areas. Further, organic agricultural practices enhance ability of population to live in harmony with nature also deriving economic benefits. Straight impact of organic agriculture on ecosystems is witnessed on various scales like farm margins, on-farm and ecosystem (Singh, 2021).

CONCLUSION

Sustainable agriculture usually encompasses variety of production practices including organic and conventional, integrated system of plant and animal production towards producing long-term results. A truly sustainable agriculture closely intertwines the components of environment, economic and social sustainability. Hence, promoting sustainable agricultural systems need policies which integrate these three components.

In conclusion, it is understood that organic agriculture is a system that favours renewable resources, practicing options like recycling and returning of soil nutrients from waste products, using environment's own systems towards control of diseases and pests in crops, avoiding use of synthetic pesticides, fertilizers, herbicides etc. Future studies should be focussed on developing various systems towards sale of organic produce, targeting organic practices to areas that are environmentally delicate, persuading exchange of technical data among farmers etc.

REFERENCES

Abdelrahman, H., Cocozza, C., Olk, D. C., Ventrella, D., Montemurro, F., & Miano, T. (2020). Changes in labile fractions of soil organic matter during the conversion to organic farming. *Journal of Soil Science and Plant Nutrition*, 1–10.

Ahirwar, N. K., Singh, R., Chaurasia, S., Chandra, R., & Ramana, S. (2020). Effective role of beneficial microbes in achieving the sustainable agriculture and eco-friendly environment development goals: A review. *Frontiers in Microbiology*, *5*, 111–123.

Altieri, M. A., & Nicholls, C. I. (2020). Agroecology and the reconstruction of a post-COVID-19 agriculture. *The Journal of Peasant Studies*, *47*(5), 881–898.

Andrieu, N., Howland, F., Acosta-Alba, I., Le Coq, J. F., Osorio-Garcia, A. M., Martinez-Baron, D., ... Chia, E. (2019). Co-designing climate-smart farming systems with local stakeholders: A methodological framework for achieving large-scale change. *Frontiers in Sustainable Food Systems*, *3*, 37.

Azeez, G. (2000). *The biodiversity benefits from organic faming.* Soil Association/WWFUK.

Bekchanov, M., & Mirzabaev, A. (2018). Circular economy of composting in Sri Lanka: Opportunities and challenges for reducing waste related pollution and improving soil health. *Journal of Cleaner Production*, *202*, 1107–1119.

Bommarco, R., Kleijn, D., & Potts, S. G. (2013). Ecological intensification: Harnessing ecosystem services for food security. *Trends in Ecology & Evolution*, *28*(4), 230–238.

Briones, M. J. I., & Schmidt, O. (2017). Conventional tillage decreases the abundance and biomass of earthworms and alters their community structure in a global meta-analysis. *Global Change Biology*, *23*(10), 4396–4419.

Campiglia, E., Radicetti, E., & Mancinelli, R. (2018). Floristic composition and species diversity of weed community after 10 years of different cropping systems and soil tillage in a Mediterranean environment. *Weed Research*, *58*(4), 273–283.

Cassman, K. G., & Grassini, P. (2020). A global perspective on sustainable intensification research. *Nature Sustainability*, *3*(4), 262–268.

Chamberlain, D. E., Wilson, J. D., & Fuller, R. J. (1999). A comparison of bird populations on organic and conventional farm systems in southern Britain. *Biological Conservation*, *88*(3), 307–320.

De Pinto, A., Cenacchi, N., Kwon, H. Y., Koo, J., & Dunston, S. (2020). Climate smart agriculture and global food-crop production. *PLoS One*, *15*(4), e0231764.

Delbaere, B., Mikos, V., & Pulleman, M. (2014). European Policy Review: Functional agrobiodiversity supporting sustainable agriculture. *Journal for Nature Conservation*, *22*(3), 193–194.

El Janati, M., Akkal-Corfini, N., Bouaziz, A., Oukarroum, A., Robin, P., Sabri, A., ... Thomas, Z. (2021). Benefits of Circular Agriculture for Cropping Systems and Soil Fertility in Oases. *Sustainability*, *13*(9), 4713.

Elevitch, C. R., Mazaroli, D. N., & Ragone, D. (2018). Agroforestry standards for regenerative agriculture. *Sustainability*, *10*(9), 3337.

Erisman, J. W., Eekeren, N. V., Doorn, A. V., Geertsema, W., & Polman, N. (2017). Measures for nature-inclusive agriculture. *Alterra-rapport-Wageningen University and Research Centre*, (2821).

Erisman, J. W., van Eekeren, N., de Wit, J., Koopmans, C., Cuijpers, W., Oerlemans, N., & Koks, B. J. (2016). Agriculture and biodiversity: A better balance benefits both. *AIMS Agriculture and Food*, *1*(2), 157–174.

Fabian, Y., Sandau, N., Bruggisser, O. T., Aebi, A., Kehrli, P., Rohr, R. P., ... Bersier, L. F. (2013). The importance of landscape and spatial structure for hymenopteran-based food webs in an agro-ecosystem. *Journal of Animal Ecology*, *82*(6), 1203–1214.

Firth, A. G., Baker, B. H., Brooks, J. P., Smith, R., Iglay, R. B., & Davis, J. B. (2020). Low external input sustainable agriculture: Winter flooding in rice fields increases bird use, fecal matter and soil health, reducing fertilizer requirements. *Agriculture, Ecosystems & Environment*, *300*, 106962.

Fischer, R. A., & Connor, D. J. (2018). Issues for cropping and agricultural science in the next 20 years. *Field Crops Research*, *222*, 121–142.

Food and Agriculture Organization of the United Nations (FAO). (1988). *Report of the FAO Council, 94th Session*. FAO.

Garibaldi, L. A., Pérez-Méndez, N., Garratt, M. P., Gemmill-Herren, B., Miguez, F. E., & Dicks, L. V. (2019). Policies for ecological intensification of crop production. *Trends in Ecology & Evolution*, *34*(4), 282–286.

Giller, K. E., Hijbeek, R., Andersson, J. A., & Sumberg, J. (2021). Regenerative Agriculture: An agronomic perspective. *Outlook on Agriculture*, *50*(1), 13–25.

Gliessman, S. (2013). Agroecology: Growing the roots of resistance. *Agroecology and Sustainable Food Systems*, *37*(1), 19–31.

Gliessman, S. R. (2018). Defining agroecology. *Agroecology and Sustainable Food Systems*, *42*, 599–600.

Gomiero, T., Pimentel, D., & Paoletti, M. G. (2011). Environmental impact of different agricultural management practices: Conventional vs. organic agriculture. *Critical Reviews in Plant Sciences*, *30*(1-2), 95–124. doi:10.1080/07352689.2011.554355

Gottwald, F., & Stein-Bachinger, K. (2016). *Landwirtschaft für Artenvielfalt-Ein Naturschutzstandard fürkologisch bewirtschaftete Betriebe. 2.* WWF-Deutschland. www. landwirtschaft-artenvielfalt. de

Habel, J. C., Ulrich, W., Biburger, N., Seibold, S., & Schmitt, T. (2019). Agricultural intensification drives butterfly decline. *Insect Conservation and Diversity, 12*(4), 289–295.

Halberg, N. (2012). Assessment of the environmental sustainability of organic farming: Definitions, indicators and the major challenges. *Canadian Journal of Plant Science, 92*(6), 981–996.

Heldbjerg, H., Fox, A. D., Levin, G., & Nyegaard, T. (2016). The decline of the Starling Sturnus vulgaris in Denmark is related to changes in grassland extent and intensity of cattle grazing. *Agriculture, Ecosystems & Environment, 230*, 24–31.

Hellin, J., & Fisher, E. (2019). Climate-smart agriculture and non-agricultural livelihood transformation. *Climate (Basel), 7*(4), 48.

Hirschfeld, S., & Van Acker, R. (2021). ecosystem services in permaculture systems. *Agroecology and Sustainable Food Systems, 45*(6), 794–816.

Hole, D. G., Perkins, A. J., Wilson, J. D., Alexander, I. H., Grice, P. V., & Evans, A. D. (2005). Does organic farming benefit biodiversity? *Biological Conservation, 122*(1), 113–130.

Holmgren, D. (2020). *Essence of permaculture*. Melliodora Publishing.

Hunt, J. R., Celestina, C., & Kirkegaard, J. A. (2020). The realities of climate change, conservation agriculture and soil carbon sequestration. *Global Change Biology, 26*(6), 3188–3189.

Ibeawuchi, I. I., Obiefuna, J. C., & Iwuanyanwu, U. P. (2015). Low External Input Agricultural Farming System for the Increase in Productivity of Resource Poor Farmers. *Soil and Water, 5*(2).

Isbell, F., Reich, P. B., Tilman, D., Hobbie, S. E., Polasky, S., & Binder, S. (2013). Nutrient enrichment, biodiversity loss, and consequent declines in ecosystem productivity. *Proceedings of the National Academy of Sciences of the United States of America, 110*(29), 11911–11916.

Jain, M., Solomon, D., Capnerhurst, H., Arnold, A., Elliott, A., Kinzer, A. T., ... Weinstein, C. (2020). How much can sustainable intensification increase yields across South Asia? A systematic review of the evidence. *Environmental Research Letters, 15*(8), 083004.

Jansson, C., Faiola, C., Wingler, A., Zhu, X. G., Kravchenko, A., de Graaff, M. A., ... Beckles, D. M. (2021). Crops for Carbon Farming. *Frontiers in Plant Science, 12*, 938.

Karadağ, H. (2018). Biodynamic Agriculture in Turkey; Istafil Biodynamic Organic Farmland Sample. In *4th International Conference on Engineering and Natural Science. Book of Abstracts* (*Vol. 116*). Academic Press.

Karasawa, T., Takebe, M., Sato, F., Komada, M., Nagaoka, K., Takenaka, M., ... Kato, N. (2015). Trends of lettuce and carrot yields and soil enzyme activities during transition from conventional to organic farming in an Andosol. *Soil Science and Plant Nutrition, 61*(2), 295–311.

Kassam, A., Friedrich, T., & Derpsch, R. (2019). Global spread of conservation agriculture. *The International Journal of Environmental Studies, 76*(1), 29–51.

Khadse, V. A., Ghatol, P. U., Thakare, P. D., & Bhale, V. M. (2021). *Adoption of biodynamic organic agriculture practices on farmer's field of Vidarbha region*. Academic Press.

Kleijn, D., Bommarco, R., Fijen, T. P., Garibaldi, L. A., Potts, S. G., & van der Putten, W. H. (2019). Ecological intensification: Bridging the gap between science and practice. *Trends in Ecology & Evolution*, *34*(2), 154–166.

Kovács-Hostyánszki, A., Espíndola, A., Vanbergen, A. J., Settele, J., Kremen, C., & Dicks, L. V. (2017). Ecological intensification to mitigate impacts of conventional intensive land use on pollinators and pollination. *Ecology Letters*, *20*(5), 673–689.

Kragt, M. E., Blackmore, L., Capon, T., Robinson, C. J., Torabi, N., & Wilson, K. A. (2014). *What are the barriers to adopting carbon farming practices?* (No. 1784-2016-141870).

Krebs, J., & Bach, S. (2018). Permaculture—Scientific evidence of principles for the agroecological design of farming systems. *Sustainability*, *10*(9), 3218.

Kurgat, B. K., Lamanna, C., Kimaro, A., Namoi, N., Manda, L., & Rosenstock, T. S. (2020). Adoption of climate-smart agriculture technologies in Tanzania. *Front. Sustain. Food Syst*, *4*(3), 55.

Lal, R. (2020). Regenerative agriculture for food and climate. *Journal of Soil and Water Conservation*, *75*(5), 123A–124A.

Landis, D. A. (2017). Designing agricultural landscapes for biodiversity-based ecosystem services. *Basic and Applied Ecology*, *18*, 1–12.

Li, B., Feng, Y., Xia, X., & Feng, M. (2021). Evaluation of China's Circular Agriculture Performance and Analysis of the Driving Factors. *Sustainability*, *13*(4), 1643.

Lichtenberg, E. M., Kennedy, C. M., Kremen, C., Batary, P., Berendse, F., Bommarco, R., ... Crowder, D. W. (2017). A global synthesis of the effects of diversified farming systems on arthropod diversity within fields and across agricultural landscapes. *Global Change Biology*, *23*(11), 4946–4957.

Lori, M., Symnaczik, S., Mäder, P., De Deyn, G., & Gattinger, A. (2017). Organic farming enhances soil microbial abundance and activity—A meta-analysis and meta-regression. *PLoS One*, *12*(7), e0180442.

Mier y Terán Giménez Cacho, M., Giraldo, O. F., Aldasoro, M., Morales, H., Ferguson, B. G., Rosset, P., ... Campos, C. (2018). Bringing agroecology to scale: Key drivers and emblematic cases. *Agroecology and Sustainable Food Systems*, *42*(6), 637–665.

Miller, H. (2017). How College Students Are Being Misled About 'Sustainable' Agriculture. *National Review*.

Mondelaers, K., Aertsens, J., & Van Huylenbroeck, G. (2009). A meta-analysis of the differences in environmental impacts between organic and conventional farming. *British Food Journal*.

Nabi, A., Narayan, S., Afroza, B., Mushtaq, F., Mufti, S., Ummyiah, H. M., & Magray, M. M. (2017). Biodynamic farming in vegetables. *Journal of Pharmacognosy and Phytochemistry*, *6*(6), 212–219.

Nandwani, D., & Nwosisi, S. (2016). Global trends in organic agriculture. In *Organic farming for sustainable agriculture* (pp. 1–35). Springer.

Nicholls, C. I., & Altieri, M. A. (2018). Pathways for the amplification of agroecology. *Agroecology and Sustainable Food Systems*, *42*(10), 1170–1193.

Plieninger, T., Torralba, M., Hartel, T., & Fagerholm, N. (2019). Perceived ecosystem services synergies, trade-offs, and bundles in European high nature value farming landscapes. *Landscape Ecology*, *34*(7), 1565–1581.

Ponzio, C., Gangatharan, R., & Neri, D. (2013). Organic and biodynamic agriculture: A review in relation to sustainability. *International Journal of Plant and Soil Science*, 95–110.

Prestele, R., Hirsch, A. L., Davin, E. L., Seneviratne, S. I., & Verburg, P. H. (2018). A spatially explicit representation of conservation agriculture for application in global change studies. *Global Change Biology*, *24*(9), 4038–4053.

Pretty, J., & Bharucha, Z. P. (2014). Sustainable intensification in agricultural systems. *Annals of Botany*, *114*(8), 1571–1596.

Quinn, J. E., Awada, T., Trindade, F., Fulginiti, L., & Perrin, R. (2017). Combining habitat loss and agricultural intensification improves our understanding of drivers of change in avian abundance in a North American cropland anthrome. *Ecology and Evolution*, *7*(3), 803–814.

Rahmann, G. (2011). Biodiversity and organic farming: what do we know? *vTI Agriculture and Forestry Research, 3*, 189-208.

Rockström, J., & Sukhdev, P. (2016). *How food connects all the SDGs. Keynote Speech: Prof. Johan Rockström & CEO Pavan Sukhdev*. Stockholm Resilience Centre, Stockholm University.

Romanyà, J., & Casals, P. (2019). Biological nitrogen fixation response to soil fertility is species-dependent in annual legumes. *Journal of Soil Science and Plant Nutrition*, 1–11.

Runhaar, H. (2017). Governing the transformation towards 'nature-inclusive' agriculture: Insights from the Netherlands. *International Journal of Agricultural Sustainability*, *15*(4), 340–349.

Savage, S. (2015). *The lower productivity of organic farming: A new analysis and its big implications*. Academic Press.

Seneviratne, S. I., Phipps, S. J., Pitman, A. J., Hirsch, A. L., Davin, E. L., Donat, M. G., ... Kravitz, B. (2018). Land radiative management as contributor to regional-scale climate adaptation and mitigation. *Nature Geoscience*, *11*(2), 88–96.

Singh, M. (2021). Organic farming for sustainable agriculture. *Indian Journal of Organic Farming*, *1*(1).

Stoeckli, S., Birrer, S., Zellweger-Fischer, J., Balmer, O., Jenny, M., & Pfiffner, L. (2017). Quantifying the extent to which farmers can influence biodiversity on their farms. *Agriculture, Ecosystems & Environment*, *237*, 224–233.

Tittonell, P. (2014). Ecological intensification of agriculture—Sustainable by nature. *Current Opinion in Environmental Sustainability*, *8*, 53–61.

Tseng, M. C., Roel, A., Deambrosi, E., Terra, J. A., Zorrilla, G., Riccetto, S., & Pittelkow, C. M. (2020). Towards actionable research frameworks for sustainable intensification in high-yielding rice systems. *Scientific Reports*, *10*(1), 1–13.

Tuck, S. L., Winqvist, C., Mota, F., Ahnström, J., Turnbull, L. A., & Bengtsson, J. (2014). Land-use intensity and the effects of organic farming on biodiversity: A hierarchical meta-analysis. *Journal of Applied Ecology*, *51*(3), 746–755.

Valkama, E., Kunypiyaeva, G., Zhapayev, R., Karabayev, M., Zhusupbekov, E., Perego, A., ... Acutis, M. (2020). Can conservation agriculture increase soil carbon sequestration? A modelling approach. *Geoderma*, *369*, 114298.

Wang, R., van Dorp, L., Shaw, L. P., Bradley, P., Wang, Q., Wang, X., ... Balloux, F. (2018). The global distribution and spread of the mobilized colistin resistance gene mcr-1. *Nature Communications*, *9*(1), 1–9.

Westerink, J., Plomp, M., Ottburg, F., Zanen, M., & Schrijver, R. (2018). Farmers for Nature: the ultimate nature-inclusive agriculture? *Alterra-rapport-Wageningen University and Research Centre*, (2858).

White, C. (2020). Why Regenerative Agriculture? *American Journal of Economics and Sociology*, *79*(3), 799–812.

Wilhelm, M., Davin, E. L., & Seneviratne, S. I. (2015). Climate engineering of vegetated land for hot extremes mitigation: An Earth system model sensitivity study. *Journal of Geophysical Research, D, Atmospheres*, *120*(7), 2612–2623.

Yang, F., Tian, J., Fang, H., Gao, Y., Xu, M., Lou, Y., ... Kuzyakov, Y. (2019). Functional soil organic matter fractions, microbial community, and enzyme activities in a mollisol under 35 years manure and mineral fertilization. *Journal of Soil Science and Plant Nutrition*, *19*(2), 430–439.

Chapter 8
Green Roof Farming and the Urban Environment:
A Way Towards Sustainability

Aakansha Goswami
https://orcid.org/0000-0002-3350-1747
Department of Biotechnology, KVSCOS, Swami Vivekanand Subharti University, Meerut, India

Nisha Malik
Department of Biotechnology, KVSCOS, Swami Vivekanand Subahrti University, Meerut, India

ABSTRACT

Our material lives and comfort have increased as a result of industrialization and urbanization, but changes in land use and land cover pattern in metropolitan areas have diminished vegetative covers. Many problems arise due to decline in vegetation such as urban heat islands, global warming, air pollution, and so on. Several efforts are in vogue to address these issues, but adoption of green roofs is one of the most eco-friendly concepts in India. This natural solution plays a key role in urban transformation towards a sustainable environment by improving the urban environment and lowering the urban heat island effect. In this chapter, prefabricated layers with the growth medium were explained to build the green roof. Both soil (sandy or clayey) and soilless (cocopeat or compost) media were utilized to raise the vegetative layers. The depth of the substrate determines whether the green roof system is intensive or extensive, and based on this, the crop for growing on the green roof is chosen.

INTRODUCTION

Because of economic progress and high population growth, urbanisation is becoming an unavoidable phenomenon. We are losing our agricultural land every hour in the repercussion of ever-growing urbanisation sprawl. Urban growth is leading to many negative impacts such as degradation of air and water quality, loss of farmland and forests, as well as socio-economic effects of economic disparities. Installing green roof system in residential and commercial buildings is a viable option to tackle this situation

DOI: 10.4018/978-1-7998-8434-7.ch008

in the near future. This farming system can offer a multitude of benefits, ranging from environmental and ecological concerns to the aesthetic impact of new and existing construction. This system can be utilised to compensate for increased urban congestion by providing a visual and recreational getaway from metropolitan landscapes' "concrete jungles." They can also help rural structures blend in with their environment, which is a big plus in locations with a lot of natural beauty (Onmura et al., 2001).

Green roof also known as vegetable roofs or inhibit roofs these systems have progressed in recent years from ad hoc case-specific designs to engineered roofing systems, which consist of a combination of inverted roof components overlaid by various components such as a drainage layer, filter cloth, growth media, and specialist plants. For thousands of years, green roofing systems have been effectively used in limited conventional building applications, dating back to ancient Mesopotamia (Peck et al., 1999). The concept of a green roof is not a new one, as its basic functionality has been used for millennia. More recently, increased urbanisation and the encroachment of "urban sprawl" on green belt regions has resulted in a "greying" of the environment and a reduction in usable space. Developed awareness of the benefits that can be gained with green roofs has increased as a result of improved urban density in order to decrease urban sprawl and the resultant loss of geographical identity, as well as increased environmental awareness. There are two types of green roofs: extensive and intensive. Extensive roofs are those with a substrate depth of less than 15cm and are frequently limited to grasses and drought-tolerant plants like Sedums due to their shallow depths. An extended roofing system has the advantage of being lighter due to less growing material and being less expensive than an intensive green roof in terms of both capital and operation and maintenance costs. Intensive roofs require deeper substrate depths of over 15cm and can sustain a wider range of plant life, such as shrubs and small trees. Intensive roofs are built as public spaces and, like landscaping on the ground level, require regular maintenance. Engineered Green roof systems have been shown in studies to reduce the effects of the urban heat island effect (Kohler et al., 2002; Wong et al., 2003(a), Banting et al., 2005). Green roofs increase the number of evapo-transpiring plants in the built environment, reduce the number of surfaces available to re-radiate solar energy, trap airborne particles, and sequester gaseous pollutants like carbon dioxide (Nowak and Crane, 1998). As a result of these combined impacts, the environment is less affected by smog and is generally cooler (Banting et al., 2005).

Green roof installation is commonly considered to be one of the most eco-friendly technique to achieve more comfortable internal building temperatures while using less energy for heating and cooling (Onmura et al., 2001). They help to decrease the internal building temperatures in hot areas by shielding building rooftops from direct sunshine and therefore minimising the thermal variations that roofing system components ordinarily encounter due to cycle heating and cooling. Given the foregoing, the choice of green roof plants has a significant impact on the thermal efficiency of the roof. However, the majority of studies used sedum plants to measure the thermal performance of green roofs. Green roof solutions are becoming increasingly important in urban areas around the world.

A green roof's interacting components produce an ecosystem that mimics some important qualities of ground-level plants that are absent from a conventional roof. Green roofs, which are made up of a drainage layer, a solid matrix "soil" layer, and vegetation, reduce thermal gain directly beneath the roof while also improving the water balance between evaporation and runoff. Green roofs thus serve as a barrier between rainwater and sewage pipes, as well as insulation between solar radiation and interior building spaces. Furthermore, the photosynthesis of the plants atop the green roofs reduces greenhouse gas emissions in the surrounding air. This chapter presents reviews of significant research work carried out by various investigators in relation to the topic under the following sub-divisions:

- Monolithic green roof
- Energy balance models
- Thermal performance
- Environmental benefit
- Economic cost

MONOLITHIC GREEN ROOF

Green roofs, also known as garden roofs, vegetable roofs, and inhibit roofs, are engineering systems made up of prefabricated layers that come together to form a single monolithic system in which vegetation can grow. Green roof culture layer differs from regular soil in that it requires less depth for vegetable growth and is extremely light (Haghighi, 2014). The design and construction of the roof structure and the growing media vary, according to Sailor (2008, 2011), but commonly includes a geo-textile filter, drainage layer, root barrier, and waterproof membrane. The depth of the growing media can range from 20 mm (for vast systems using sedum mats) to 1500 mm, depending on the vegetative layer (for intensive systems containing large shrubs and trees). Figure 1 shows the fundamental components of a green roof.

Green roofs, according to Gedge and Frith (2004), are one of the most common approaches for implementing sustainable development principles in the construction industry, as well as a successful and crucial approach, particularly for densely populated cities with significant climate impacts. Green roofs have been shown in empirical research to retain storm water, preventing sewage treatment centres from being flooded, lowering ambient air temperature, and decreasing the urban heat island effect. They can also extend the life of the roof, filter air pollution, and offer a layer of insulation for the building below, lessening the energy demand of the structure. Finally, green roofs can serve as a community green space, boosting the sense of well-being and visual delight in metropolitan areas that are normally devoid of green space (Cheney, 2004).

Figure 1. Basic components of green roof

Extensive Green Roof

The enormous green roofs are normally inaccessible to the general public, but they provide various environmental benefits, including storm water mitigation and building structure protection. Succulents and moss are common choices for large-scale green roofs. (2006) (Cutlip). According to Newton (2007), a thin layer of substrate (20-200 mm) can be made of recycled materials on large green roofs. They are low-maintenance, low-cost, and require little or no irrigation. Plants such as masses, succulents, wild flowers, and grasses can be grown. They may live on a shallow, low-nutrient base, and sedums, which have dense covering and are drought and wind resilient, can also be planted. A sedum mat is a foundation layer of polyester, Hessian, or porous polythene over which 20 mm of growing medium is spread in an enormous green roof. This is strewn with sedum cuttings, which grow to maturity in the substrate. The Sedum blanket (with the 20mm of growth medium) is rolled out into 50–70 mm of growing medium (regular approach) or directly into a moisture retention blanket (ultra light weight method) when it is placed (Harrison, 2000).

Intensive Green Roof

The intensive green roof, according to Worden et al., (2004), is a traditional-style roof garden with large trees and shrubs, similar to a park and easily accessible to the public. They have a level roof, but they are labor-intensive to maintain and care for, and they require frequent irrigation maintenance and upkeep. Jenrick (2005) found that an intense green roof adds to the weight of the roof because it demands higher soil depth, which necessitates a roof that can withstand the added structural loads. Intensive green roofs also have thicker vegetation above the roof insulation and are better suited to plants of various sizes. This sort of green roof is commonly referred to as a roof garden, and it can only be put to structures that can withstand the additional weight of the additional soil layers, depending on the total weight of the additional soil layers. In terms of both construction and maintenance costs, it is the most expensive green roof kind. They are totally accessible and provide an additional space for building users throughout the year.

Green Roof Layers

Green roofs are most typically used on low-slope or flat roofs, where the lesser roof pitch allows for easier application and architectural inventiveness. The underlying roofing system provides for greater water filtering via a Green roof, but it has the disadvantage of requiring the complete roofing system to be removed for waterproofing membrane repair if the waterproofing membrane ruptures. Vegetation, growing medium (substrate), filter cloth, drainage layer, root repellant system, and waterproofing membrane are all components of a green roof system (Peck and colleagues, 1999). The purpose and materials that can be utilised for the components of a green roof are specified by Denardo, et al., (2003) in Table 1.

Architects and engineers must consider the dead load of moist soil, plants, and other materials, as well as the potential live load of humans or moving machinery, whether constructing a new building with a green roof or retrofitting an existing building to support a green roof. Intensive green roofs with trees may contain more than 600 mm of soil and planters, with a wet weight of >1 tonne/m2, resulting in a total loading of >10 kN/m2 when combined with crowds (Scrivens, 1980a).

Table 1. Basic components of green roof

Components	Functions	Materials used
Water proofing membrane	Prevents water from entering the building and allows for storm water runoff during rainy periods.	Modified bitumen reinforced with fiberglass or non-woven polyester.
Root repellent	Prevents plants from damaging the waterproof membrane in the long term.	Copper linings
Drainage	Maintains proper drainage to the overlying growth media and ensures that plants do not drown from exposure to excess water and that water vapour can be released.	Porous mats, granular media or polystyrene
Filter cloth	Ensures that fine sediments from the substrate do not clog the drainage layer and helps prevent roots from reaching the repellent system.	Non-woven, non biodegradable polyester or polypropylene polyethylene mats
Growing medium	Used as fire resistance, insulation and Water proofing protection and as the growing media for vegetation.	Mixtures of in organics (vermiculite, clay, volcanic rock, coarse sands) and organics
Vegetation	Insulates the system, protects biodiversity, air particle filter, aids in storm water management, CO2 sequester and O_2 producer and serves as a transpiration media.	Perennial, biennials or annuals

Vegetative Layer

A self-sustaining green roof requires minimum maintenance, including irrigation. As a result, green roof plants must be able to withstand a variety of extreme circumstances. Summer water shortages are frequently the most severe stressors, which are worsened by excessive heat and high winds. This is typical of Canterbury's climate, where temperatures can exceed 30 degrees Celsius and wind speeds average 15 kilometres per hour (Snodgrass and Mcintyre, 2010). Due to exposure and the free-draining nature of the terrain, droughts are prevalent. Plant life is further harmed by high winds and intense sun radiation caused by the exposed location. As a result, only plants that can thrive in these conditions should be cultivated. Because necessary watering is frequently mentioned as a potential deterrent to green roof installation, designs that require it should be avoided. An extensive system should be relatively maintenance-free after it has been setup (Macivor and Ranalli, 2011).

ENERGY BALANCE MODELS

To study the role of roofs in the urban heat island and building energy consumption rates, a simple methodology known as "energy balance" modelling is available. It refers to the physical fact that energy cannot be created or destroyed, so the solar and long wave radioactive energy received by a rooftop layer during any time interval must exactly equal, or "balance," the energy gained by that layer (Oke, 1987).

Figure 2. Surface energy balance of green roof

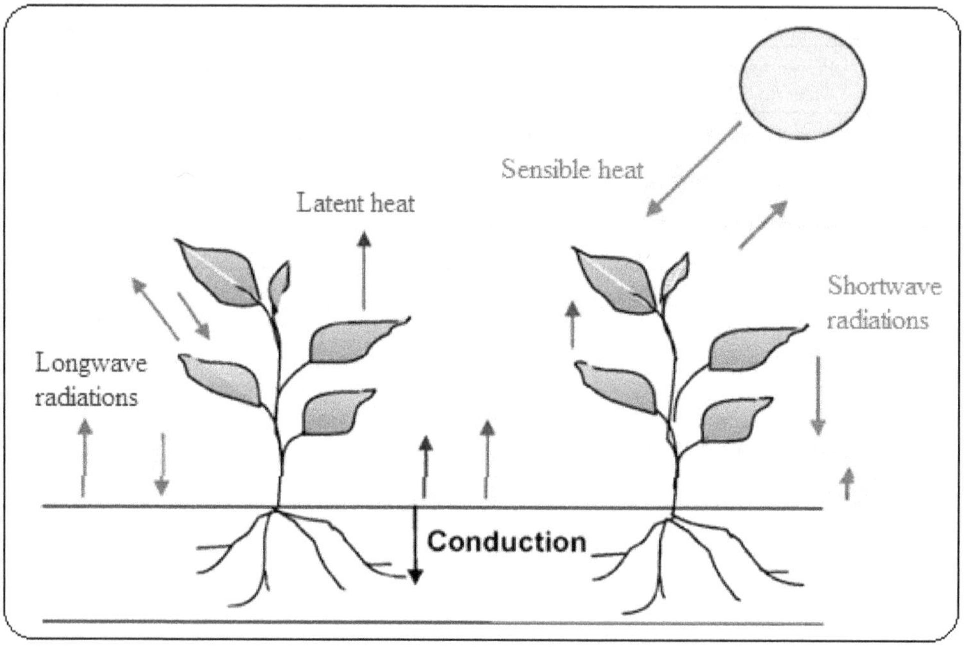

THERMAL PERFORMANCE

Reduce the amount of heat flowing into the interior of buildings in hot regions, and out of them in cold climates, in order to reduce energy use in structures and encourage sustainable development practises. Green roofs are widely regarded as a future instrument for the development of comfortable and energy-efficient buildings in a variety of climates (Onmura *et al.,* 2001). Green roofs reduce the surface temperatures of conventional roofs in hot and humid areas by sheltering the roof from direct sunshine, minimising the thermal fluctuation experienced by the roof (particularly the critical component; the waterproofing membrane). This reduction in heat loading means the roof system will last longer and the roof sealant will be less stressed (Kohler *et al.,* 2002).

ENVIRONMENTAL BENEFIT OF GREEN ROOF

Green roofs have a number of environmental advantages and can help with concerns like as biodiversity, climate change mitigation, flood avoidance, and urban green space expansion. The benefits of green roofs were summarised by Brownlie (1990), Johnston (1995), Johnston and Newton (1993), and Osmundson (1999) as absorption of air pollutants and dust, reduction of the "urban heat island" effect, increased humidity, absorption of noise, absorption of electromagnetic radiation, helping to absorb greenhouse gases (particularly CO2) and giving off oxygen, and use of recycled materials.

ECONOMIC COST OF GREEN ROOF

The initial expense and follow-up maintenance work, according to Peck et al. (1999), are two major impediments to wider use of green roof systems. An large green roof system would cost two to five times more in the initial investment than a normal bare roof structure. Green roofs, on the other hand, could be less expensive in the long term than tiled roofs if correctly planned and designed, because they endure longer and provide many other physical and intangible benefits to building owners. Wong (2003) conducted research and discovered that the long-term economic benefits of green roofs outweigh the upfront costs. However, the most important benefits of green roofs, such as creating a cooler microclimate in metropolitan areas and retaining runoff, are difficult to quantify or evaluate. These advantages, when combined with a green roof's better roof durability and thermal insulation, easily surpass the higher initial expenses for most projects.

Green roofs, according to Peck et al. (1999), provide shade and insulation as roof membrane life increases, leading in energy savings and mitigation of the urban heat island effect. When compared to non-shaded buildings, media depth, shade from plant material, and transpiration can reduce solar energy gain by up to 90%. When external temperatures are between 25°C and 30°C, green roofs reduce inside temperatures by 3 to 4°C. The use of power for air conditioning can be reduced by up to 8% for every 0.5oC decrease in interior building air temperature. When compared to a normal roof, air temperatures above the building have been demonstrated to be 30 degrees cooler when the roof is vegetated Dunnett and Kingsbury (2004).

METHODOLOGY

The main goal of this research is to look into the possibilities and perspectives of green roof systems in subtropical climates. Through comparative field studies in selected places, this pilot project seeks to explore the thermal performance of typical green roof systems of extensive and intense under climate settings. This chapter discusses the technique used, as well as the materials used, to design and build green roof systems as part of urban agriculture in order to assess the urban heat island effect in urban pockets. Within the scope of research, the assessment of energy balance and temperature distributions in roof layers, as well as the definition of the function of green roof systems in terms of sustainability, were explored.

The Material Selection for Fabrication of Green Roof

The performance requirements of a typical green roof in a semi-arid region were taken into consideration, and the essential materials of the green roof system used to build the layers are listed in Table 2.

Framework for Decision Support System

A green roof is a difficult project to design. Due to the numerous aspects involved, selecting an effective green roof system can be a difficult undertaking. To make this decision-making process more efficient, a framework is required. The establishment of a framework to describe the decision process involved in the design of green roof systems is discussed in this green roof. Figure 3 shows a framework for a decision assistance system.

Table 2. Various materials used in fabrication of green roof

S. No	Layers	Materials used
1	Roof deck	Precast reinforced concrete slab with RCC (1 cement:1.5 sand:3 Jally)
2	Water proofing membrane	Grey colour PVC sheet
3	Root barrier	White colour woven poly propylene sheet
4	Drainage	Block colour PVC egg shape mat
5	Filter fabric	Green colour woven PVC mat
6	Growing medium or substrate	(i) Sandy clay loam used for soil substrate (ii) Coco peat used for soil less substrate
7	Vegetation	(i) Tomato (*Lycopersican esculentum*) variety Co.3 (ii) Mari gold (*Tagetes spp*)

Temperature Trend Analysis

The local temperature is one of the most important climatic variables for tracking changes in the atmosphere caused by industry, population growth, and large urbanisation. The current study aims to promote green roof systems in urban areas such as the National Capital Region (NCR), by examining annual and seasonal temperature fluctuations and trends for mean maximum and minimum temperatures over the last three decades, as well as their relationship with urbanisation.

Functional element of green roof

A layered method of roofing is generally used in green roof design to assist plant development and preserve water for plant uptake while reducing ponding on the roof surface. Water drains vertically through the medium and then horizontally down a waterproofing layer towards the exit on the rooftops. The green roof was already well-established, and the layers usually included:

- Vegetation,
- Growing medium or substrate,
- Filter fabric,
- Drainage layer,
- Root barrier,
- Water proofing layer and
- Roof deck and the schematic representation of green roof is given below-

Vegetation

Vegetable and flower crops were raised in this study to evaluate the thermal performance of green roof systems, and the crops were planted in a 1m2 area of green roof in soil less and soil media at the same time. Tomato crop foliage covers under an intensive green roof with no soil and no substrate. In Table-2 lists the crop experiment details as well as the experimental details for intense and extensive green roofs.

Figure 3. A framework on decision support system for green roof systems

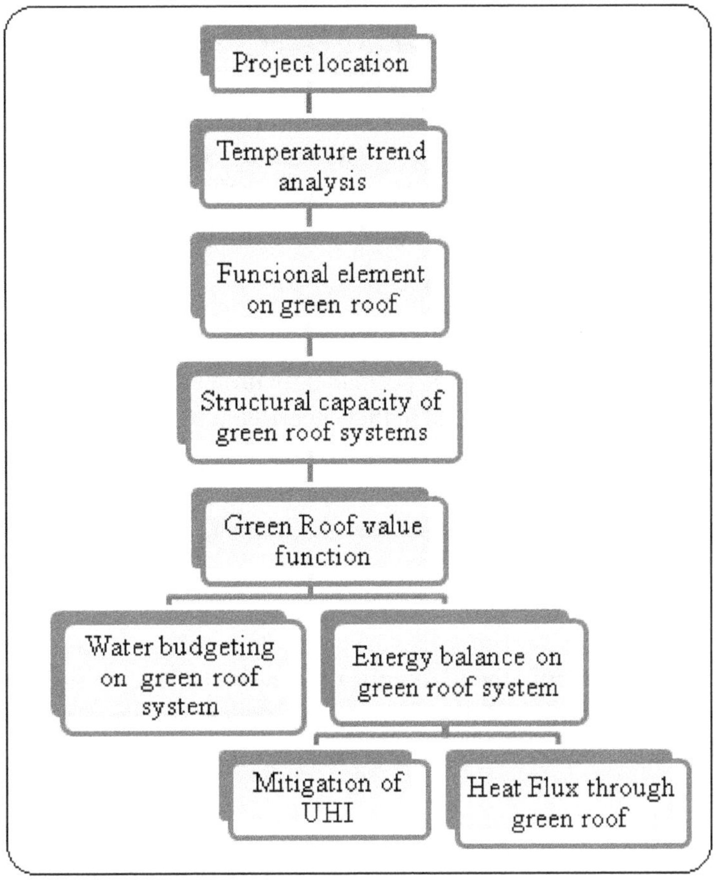

Figure 4. Schematic representation of green roof layer

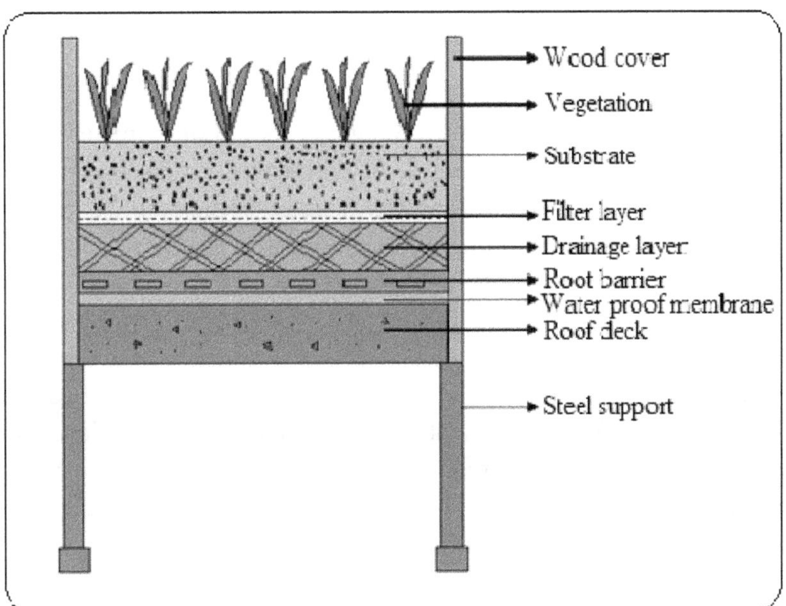

Growing Medium or Substrate

At a 1:1 ratio, vermi-compost was used to make two types of growing media or substrates. It was spread over several layers of filter layer, drainage layer, root barrier, water proofing layer, and precast reinforced concrete slab with area of 1 m^2 that was soil less (coco peat) and soil (sandy clay loam). The soil and soilless substrate depths were determined based on the effective root depth of tomato and marigold crops, which were 30 cm and 15 cm, respectively. The substrate layer, as well as the physical features of soil less and soil substrate, were supported by the wood cover in green roof systems.

Filter Layer

Above the drainage layer is a filter layer that is permanently protected against clogging by fine particles in the growth medium. This was accomplished in this experiment by utilising a woven filter cloth to hold these small particles, with a filter layer size of 1m 1m 0.003. The provision of a filter cloth layer and the growth medium Different parameters were used to characterise the thermo-physical properties of green roof layers such as the water proofing layer, root barrier, drainage layer, filter layer, and substrate layer.

Drainage Layer

To swiftly remove surplus water from the plant root zone, a drainage layer is inserted between the root barrier and the growing media. A vegetation-free zone was created surrounding all drains, with drainage layer dimensions of 1m 1m 0.005m and drainage layer provision.

Root Barrier

Root barrier is a layer of a green roof system that was installed above the water proofing layer, depending on the scheme, to prevent root penetration of the waterproofing membrane. The layer's dimensions were 1m x 1m x 0.003m, with a root barrier included.

Water Proofing Layer

To prevent water damage through the deck layer, the green roof system features an effective and reliable water proofing layer placed above the roof deck, with exact dimensions of 1m 1m 0.005m. The water proofing layer is completely waterproof and has a longer predicted life expectancy than any other component of the green roof system, including the water proofing layer.

Roof Deck

The cornerstone of a green roof is the roof deck layer, which was built to have adequate structural capacity and structural integrity to safely and efficiently resist all loads, effects of loads, and influences that may fairly be expected. Five precast reinforced concrete slabs (1:1.5:3) were produced as roof decks, each with a dimension of 1 m 1 m 0.05 m. One concrete slab was chosen as the reference roof among the five numbers of precast reinforced concrete slabs.

Steel Support

The steel support is designed and built to handle the load of the different layers of the green roof system, such as the roof deck, water proof membrane, root barrier, drainage layer, filter layer, growing media, and vegetation, with dimensions of 1m 1m 0.5m.

Table 3. Crop experiment details

S. No.	Component	Details	
1	Crop	Tomato	Marigold
2	Variety	Co.3	MDU.1
3	Spacing	30 cm × 30 cm	30 cm × 45 cm
4	Duration	110 days	120 days
5	No. of plants	12	9

Intensive Green Roof

- Long and short wave radiation received and emitted by foliage, long and short wave radiation exchange between foliage and substrate, sensible heat flux, and latent heat flux are the four components of the energy balance at foliage. In both extensive green roof treatments, the values of all four aforementioned components increased from the initial stage to harvesting stage for tomato crops.
- The maximum value of long and short wave radiation absorbed and emitted by foliage, as well as radiation exchange between foliage and substrate, in the intense green roof with soilless substrate. Because of the rise in LAI and latent heat of vaporization at foliage observed in this medium, the maximum values of energy balance components were achieved in dirt less substrate when both treatments were compared.
- From the inception stage through the harvesting stage, the net heat flow in intensive green roofs with soil and soilless substrate was studied.
- Long and short wave radiation absorbed and emitted by the substrate, long and short wave radiation exchange between foliage and substrate, sensible heat flux, latent heat flux, and substrate conduction are the five components of the energy balance at the substrate. In both extensive green roof treatments, the values of all components decreased from the beginning stage through the harvesting stage of the tomato crop.
- The sensible heat flux in an intensive green roof with no soil substrate is the maximum value of long and short wave radiation absorbed and emitted by foliage, as well as the radiation exchange between foliage and substrate. Because of increased shade effect and decreased wind speed, latent heat of vaporisation at substrate, and temperature differential between foliage and substrate, the maximum values of energy balance components were obtained in soil less substrate when both treatments were compared.
- From the starting stage to the harvesting stage, the net heat flux at the substrate in an intensive green roof with soil less and soil substrate was investigated. Due to the shadowing impact and

lower temperature difference between foliage and substrate reported in this medium, higher values were recorded.

- At 2 p.m., the extensive green roof system achieved maximal mitigation of the urban heat island effect. At the harvesting stage of the tomato crop, values of 3.5oC and 3.3oC were reported using soil less and soil substrate medium for this green roof. In an intense green roof system, the minimal heat flow through the roof was achieved at 2 p.m. Due to an increase in LAI, latent heat vaporization at foliage, latent heat of vaporization at substrate, and a reduction in the temperature differential between foliage and substrate, a higher value was achieved in soil rather than substrate.
- The highest tomato output of 4.4 kg/ m2 was attained under an intensive green roof with no soil substrate, whereas the fruit yield was 3.8 kg/ m2 in the soil substrate treatment. When compared to the figure of 0.74 kg/mm/m2 recorded in the soil less substrate treatment, the intensive green roof with soil substrate had a higher water efficiency of 0.86 kg/mm/m2.

Figure 5. Tomato crops at harvesting stage in intensive green roofs

Extensive Green Roof

- The marigold crop was grown on a large green roof, and its LAI increased as the crop grew in all treatments. The maximal LAI was obtained at the harvesting stage of treatments (intense green roof with soil less and soil), which were approximately 0.6 and 0.5, respectively.
- The maximum value of long and short wave radiation received and emitted by foliage, as well as radiation exchange between foliage and substrate, under the large green roof with soilless substrate. Because of enhanced LAI and latent heat of vaporization at foliage, the maximum values of energy balance components were obtained in soil less substrate when compared to both treatments.
- From the early stage through the harvesting stage, the net heat flux in an extended green roof with soil and soilless substrate was studied.
- Because of increased shade effect and decreased wind speed, latent heat of vaporization at substrate, and temperature differential between foliage and substrate, maximum values of energy balance components were obtained in soil less substrate when compared to soil substrate.
- From the starting stage through the harvesting stage, the net heat flux at the substrate was studied in an extended green roof with soil and soilless substrate.
- At 2 p.m., the maximum mitigation of the urban heat island effect was achieved, as was the case with the vast green roof system. At the harvesting stage of the marigold crop, values of 2.8°C and 2.6°C were reported using soil less and soil substrate media for this green roof. At 2 p.m., the vast green roof system likewise achieved the lowest heat flow through the roof. At the harvesting stage

of the marigold crop, values of 8.6 W/ m2 and 6.3 W/ m2 were reported with soil less and soil substrate media for this green roof. The causal factors for the higher values achieved in the soilless substrate are the same as those for the intensive green roof system.

Figure 6. Marigolds at harvesting stage in extensive green roofs

Economics

- Environmentally friendly mitigation of UHI, which is required even if it is not economically viable; nevertheless, in this case, the payback period is only 4.6 years, demonstrating that adapting to the NCR climate is economically practical.
- The conclusions that could be drawn from the entire study as follows:
    a. The current study created a straightforward methodology for evaluating the thermal performance of various green roof systems. In order to evaluate the characteristics of green roof thermal behaviour, a parametric study was performed for extensive and intensive green roofs.
- In all seasons, green roofs considerably reduced heat flow in buildings with poor insulation. The thicker the roof substrate, the lower the amount of heat that enters the structure. The amount of heat lost through evapo-transpiration is influenced by the moisture content of the soil. Heat is transferred out of the roof due to enhanced evapo-transpiration effects when the substrate is devoid of soil. The substrate's thermal conductivity increases with moisture level, indicating that dryer soil conditions provide superior thermal insulation. The findings revealed that green roof systems, both intense and extensive, have a good contribution to energy and environmental performance.
- Finally, a multi-criteria comparison was conducted between two green roof systems by assigning a score to various variables that measure energy and environmental performance.

SCOPE OF FUTURE RESEACH WORK

- On a small scale, the thermal performance of wide and intensive green roof systems was investigated in this study. However, on any existing or newly constructed roof, a full green roof research should be conducted in terms of thermal performance and economic viability on a wider scale.
- An examination of several energy balance models can aid in maximizing the thermal value of green roof systems.

- The lowering of heat flux inside the roof is aided by a biometric assessment of the various vegetation layers.

In urban horticulture, the green roof system is quite important. Green roofs provide a number of potential advantages, including improving the urban environment and lowering the urban heat island effect. The design and manufacturing of a green roof system for crop cultivation was completed, and its thermal performance in terms of lowering heat flux and moderating urban heat island effect was assessed, as well as its economic viability. The purpose of the green roof design and manufacturing was to see if it was suitable. The goal of the study was to assess the effectiveness of two types of green roof systems: intensive and extended green roof systems, which were categorised according to substrate depth. Tomatoes with a 30 cm effective root depth were grown on an intensive green roof, whereas marigolds with a 15 cm effective root depth were planted on an extended green roof. The heat balance at the leaf and substrate surfaces was estimated using the FASST energy balance model. Some of the primary elements impacting heat transfer for a green roof that were considered for the study were foliage height, leaf area index (LAI), fractional vegetation coverage, albedo, and stomatal resistance. By treating the leaf as a solid body in which air circulates, the heat and mass fluxes in the foliage were examined.

CONCLUSION

Depending on meteorological conditions, urban thermo-physical, geometrical qualities, and human heat sources present in the area, urban climate differs from that of rural areas, and the magnitudes of the changes can be rather substantial at times. The urban landscape has a considerable impact on heat, moisture, and momentum fluxes, and the contrast between urban and ‘undisturbed’ climates is heightened by the addition of anthropogenic heat, moisture, and pollution to the atmosphere. Green roofs are becoming more widely acknowledged as a cutting-edge, environmentally friendly solution to urban environmental issues. Green roofs have a number of advantages, including reducing solar energy penetration into buildings, lowering cooling loads, reducing the urban heat island effect, and extending the life of roofing materials. Two green roof systems, intense and extensive green roofs, were evaluated in this study under subtropical and semi-arid conditions. Also analysed and explored were the microclimatic and biophysical features of extensive and intense green roofs under various substrates, as well as their heat flux patterns and thermal benefits. The findings may help to guide future green roof design and uptake.

REFERENCES

Banting, D., Doshi, H., Missios, A., Currie, A., & Verrati, M. (2005). Report on the Environmental Benefits and Costs of Green Roof Technology for the City of Toronto. Prepared for City of Toronto and Ontario centres of Excellence – Earth and Environmental Technologies (OCE-Etech).

Brownlie, S. (1990). Roof Gardens A Review. Peterborough Nature Conservancy Council.

Cheney, C. (2004). A busy spring for rooftops in New York City. *Journal of Green Roof Infrastructure Monitor*, *6*(1), 16–20.

Cutlip, J. (2006). Green roofs a sustainable technology. *Docstoc Report*, 23.

Denardo, J. C., Jarret, A. R., Manbeck, H. B., Beattie, D. J., & Berghage, R. D. (2003). Stormwater detention and retention abilities of green roofs. In *World water and environmental resources congress and related symposia*. American Society of Civil Engineers. 10.1061/40685(2003)310

Dunnett, N., & Kingsbury, N. (2004). *Planting green roofs and living walls. Academic Press.*

Gedge, D., & Frith, M. (2004). Green roofs benefits and cost implications. Birmingham City Council.

Harrison, H. W. (2000). *Roofs and Roofing*. Building Research Establishment Report 16.

Jenrick, K. (2005). *Green roofs: a horticultural perspective*. Ritak Vitabuc Gardens.

Johnston, J. (1995). Rooftop Wildlife. *Enact*, 3.

Johnston, J., & Newton, J. (1993). *Building green- A guide to using plants on roofs, walls and pavements*. London Ecology Unit.

Kohler, M., Schmidt, G. F. W., Wilhelm Grimme, F., Laar, M., Lúcia de Assunção Paiva, V., & Tavares, S. (2002). Green roofs in temperate climates and in the hot-humid tropics-far beyond the aesthetics. *Environmental Management and Health*, *13*(4), 382–391. doi:10.1108/09566160210439297

Macivor, J. S., Ranalli, M. A., & Lundholm, J. T. (2011). Performance of dryland and wetland plant species on extensive green roofs. *Annals of Botany*, *107*(4), 671–679. doi:10.1093/aob/mcr007 PMID:21292676

Newton, J. (2007). *Building greener: guidance on the use of green roofs, green walls and complementary features on buildings*. CIRIA Report 16.

Nowak, D. J., & Crane, D. E. (1998). The urban forest effects (UFORE) model Quantifying Urban Forest Structure and Functions. *Integrated Tools Proceedings*, 1-7. https://nrs.fs.fed.us/pubs/gtr/other/gtr_nc212/gtr_nc212co.pdf

Oke, T. R. (1987). *Boundary Layer Climates* (2nd ed.). Methuen.

Onmura, S., Matsumoto, M., & Hokoi, S. (2001). Study on the evaporative cooling effect of roof lawn gardens. *Energy and Building*, *33*(7), 653–666. doi:10.1016/S0378-7788(00)00134-1

Osmundson, T. (1999). *Roof gardens: History, design and construction*. W.W. Norton & Company.

Peck, S., Callaghan, C., Bass, B., & Kuhn, M. (1999). *Greenbacks from green roofs: forging a new industry in Canada*. Status report on benefits, barriers and opportunities for green roof and vertical garden technology diffusion.

Peck, S., & Khun, M. (2001). Design the guidelines of green roof system. Canada Mortgage and Housing Corporation. http://www.cmhc-sghl.gc.ca/en/imquaf/himu/himu_002.cfm

Peck, S., & Kuhn, M. (2001). *Design guidelines for green roofs*. Canada Mortgage and Housing Corporation. https://www.cmhc-schl.gc.ca/

Sailor, D. J. (2008). *A green roof model for building energy simulation programs*. Portland State University. doi:10.1016/j.enbuild.2008.02.001

Sailor, D. J., Elley, T. B., & Gibson, M. (2011). Exploring the building energy impacts of green roof design decisions – a modeling study of buildings in four distinct climates. *Journal of Building Physics*, *35*(4), 372–391. doi:10.1177/1744259111420076

Scrivens, S. (1980a). Roof gardens: Construction. *Architects Journal*, 445-449.

Wong, N. H., Chen, Y., Ong, C. L., & Sia, A. (2003). Investigation of Thermal Benefits of Rooftop Garden in the Tropical Environment. *Building and Environment*, *38*(2), 261–270. doi:10.1016/S0360-1323(02)00066-5

Wong, N. H., Chen, Y., Ong, C. L., & Sia, A. (2003a). Investigation of thermal benefits of rooftop garden in the tropical environment. *Building and Environment*, *38*(2), 261–270. doi:10.1016/S0360-1323(02)00066-5

Wong, N. H., Cheong, D. K. W., Yan, H., Soh, J., Ong, C. L., & Sia, A. (2002). The effects of rooftop garden on energy consumption of a commercial building in Singapore. *Energy and Building*, *35*(4), 353–364. doi:10.1016/S0378-7788(02)00108-1

Wong, N. H., Cheong, D. K. W., Yan, H., Soh, J., Ong, C. L., & Sia, A. (2003b). The effects of rooftop gardens on energy consumption of a commercial building in Singapore. *Energy and Building*, *35*(4), 353–364. doi:10.1016/S0378-7788(02)00108-1

Wong, N. H., Tay, S. F., Wong, R., Ong, C. L., & Sia, A. (2003). Life cycle cost analysis of rooftop gardens in Singapore. *Building and Environment*, *38*(3), 499–509. doi:10.1016/S0360-1323(02)00131-2

Chapter 9
Constructed Wetland-Microbial Fuel Cells (CW-MFCs): Latest Research and Future Perspectives

Maitreyie Narayan
G. B. Pant University of Agriculture and Technology, Pantnagar, India

Praveen Solanki
G. B. Pant University of Agriculture and Technology, Pantnagar, India

R. K. Srivastava
G. B. Pant University of Agriculture and Technology, Pantnagar, India

ABSTRACT

By 2025, two-thirds of the world's population may be facing water shortages, according to the World Wildlife Federation. By 2030, water demand is forecast to increase by 40%. Over 1.2 billion are basically living in areas of physical water scarcity. And almost 1.6 billion face economic water shortage. And as our population continues to grow, there's just going to be more problems. So, we are in extreme need of that creation which can treat used water and can also provide us electricity so that we can use that water for irrigation purposes, toilet flush, and industrial purposes. In recent years the research work reported about an innovative constructed wetland microbial fuel cell (CW-MFCs) for more deduction of pollutants and instantaneous bioelectricity generation. A microbial fuel cell joined with constructed wetland (CW-MFC) is a novel device to treat the wastewater and create power that has more volume for wastewater treatment, is easy to repair, environmentally friendly, requires less space, and is also most cost-efficient than other devices.

DOI: 10.4018/978-1-7998-8434-7.ch009

INTRODUCTION

Energy shortage and environment pollution are the two severe challenges that human beings are facing today. The scarcity of good quality water in the world is a growing problem and the need to purify wastewater for reuse is becoming increasingly important. Organic pollution in wastewater is traditionally treated by aerobic treatment, a very energy intensive and costly process. The present energy sources are mostly high power consuming and structure is also on shaky ground. Constructed wetland (CW) is a man-made wetland which utilizes natural process to treat and improve the quality of water. CW is recognized as an eco-friendly technology and has been used as part of the wastewater treatment process in treating a broad range of wastewater like industrialized wastewater (Vrhovsek et al., 1996), percolated wastewater through landfill (Bulc, 2006), wastewater through agriculture (Minghui et al., 2011), rainstorm and snowstorm water and drainage from acid mines (Nyquist and Greger, 2009), sewage wastewater (Narayan et al., 2018) and domestic wastewater (Solanki et al., 2018a). Constructed wetlands (CW) are "*Constructed wetlands can be defined as engineered water saturated areas in which the natural removal processes for the water pollutants are reproduced and enhanced in order to optimize the purification performances*" (Hammer, 2020). Constructed wetlands are treatment systems that use processes that are evolved by nature itself which involves wetland flora, fauna, soils, water and air and their connected assemblages improves the quality of wastewater (U.S. EPA, 2000). It take benefit of many of the processes like physical, chemical and biological that occurs in natural wetlands, but does so within more controllable surroundings (Narayan et al., 2018).

In recent years Microbial fuel cell (MFC) technology has emerged a lot which not only can recover renewable energy from waste but also helps in the treatment of it. It has drawn great attention of scientists and researchers in recent years because MFC converts chemical energy into electrical energy through biodegradation of waste (Logan, 2008; Puig et al., 2012; Chhimwal et al., 2017; Corbella et al., 2014; Liu et al., 2013). Generally, a MFC consists of an anode and a cathode chamber. Electrochemically active microorganisms are used as biocatalyst to convert the biodegradable compounds into electricity (Corbella et al., 2015). When organic matter at the anodic region was oxidized, electrons (e-) and protons (H^+) were produced and then moved to the cathode. MFC generates electrical current through the transfer of electrons from anode to cathode via the external circuit.

A microbial fuel cell attached with constructed wetland (CW-MFC) system is a latest piece of equipment that embeds the MFC into the constructed wetland (CW) to extravagance the wastewater and produce electricity. CW-MFCs have the advantages of both CWs and MFCs for wastewater treatment (Narayan et al., 2018).

INTRODUCTION OF CW-MFCs

An increasing number of studies have examined the use of microbial fuel cell-coupled constructed wetlands (CW-MFCs) for wastewater treatment. In this system approach, a bacterium oxidizes organic or inorganic both the matter in wetland and electricity is produced with the substances (Oon et al., 2015). Technology combining microbial fuel cells with constructed wetland (CW-MFC) has guaranteed for both wastewater management and electricity generation (Corbella et al., 2015; Doherty et al., 2015; Oon et al., 2015). Constructed wetland–microbial fuel cells (CW–MFCs) are new and latest devices with a insubstantial combination of non-natural ecosystems called wetlands (constructed wetlands) and microbial fuel cell techniques (MFC), in which on account of through rhizosphere effect of wetland plant electricity generation can be enhanced (Villasenor et al., 2013; Liu et al., 2013) and due to the synergistic effect of the 2 units contaminants in the wastewater can be efficiently removed (Yadav et al., 2012; Fang et al., 2013). Both in the constructed wetland and the microbial fuel cell reduction and oxidation processes take place in their anaerobic and aerobic zones and these similarities of both the units are the bases of the combination. However, in the production materials and reactor configurations they have great differences.

Structural Design Of CW-Mfcs

CW-MFCs were normally constructed in PVC pipes and filled with gravels (4–6 mm in diameter) and some other materials like sand, pebbles, clay etc. They are compactly packed forming a layer inside the pipes. The upper end of microcosm was left open, while the lower end was sealed with epoxy materials. In this different sampling ports were provided for collecting samples according to the requirement. Stoppers were used to close the open ends of the sampling ports. For measuring electricity, an arrangement similar to a traditional MFC was made inside the constructed wetland microcosm. A cathode electrode was placed near to the rhizospheric zone. Table 1, has shown the different material and sizes of CW-MFCs. It was introduced with an idea that this zone will be more aerobic as compared to the deeper zones due to the air diffusion from the immediate outer atmosphere and air leakage from the rhizosphere. The anode was placed near to the bottom of microcosm with an idea that this zone will be comparatively anaerobic and suitable for anodic reaction of MFC. Insulated copper wires were used to connect both the electrodes (Yadav et al., 2012). Wetland plant is planted, any wetland plants can be used depends upon the availability.

Table 1. Architecture of CW-MFCs

S.N.	Material used for construction of CW-MFCs	Size	Reference
1.	PVC pipes	62 cm length, 10.5 cm diameter	Yadav et al., 2012a
2.	Polyacrylic plastic cylinders	Diameter 20cm	Fang et al., 2017
3.	polyethylene plastic containers	52 cm length, 16 cm diameter	Wang et al., 2017a
4.	Polyacrylic plastic cylinders	50 cm length, 30 cm diameter	Fang et al., 2015
5.	Acrylic column	75 cm length, 18 cm diameter	Oon et al., 2015
6.	Cylindrical plastic pipes	58.5 cm length, 7 cm diameter	Yadav et al., 2012b
7.	Polyvinyl chloride pipes	10 cm diameter	Wang et al., 2017b
8.	polyacrylic plastic cylinder	52.5 cm length, 30 cm diameter	Fang et al., 2013
9.	polyacrylic plastic column	40 cm length, 10 cm diameter	Liu et al., 2014
10	PVC pipes	60 cm length, 10.5 cm diameter	Srivastava et al., 2015
11.	polymethyl methacrylate cylinder	20 cm length, 13 cm diameter	Lu et al., 2015
12.	acrylic column	75 cm length, 18 cm diameter	Oon et al., 2016
13.	polycarbonate plastic	25 cm length, 15 cm diameter	Cao et al., 2017
14.	glass columns	6 cm length, 3.5 cm diameter	Zhang et al., 2017
15.	polycarbonate plastic cylinder	50 cm length, 30 cm diameter	Liu et al., 2013
16.	organic glass pipe	50 cm length, 30 cm diameter	Song et al., 2017

Maneuver of CW-MFCs

Several designs for CW-MFCs for wastewater treatment have been introduced from the present available researches; it was found that in some CW-MFCs the wastewater is fed from the top of the microcosm. Yadav et al., 2012 fed wastewater from the top of the CW-MFCs, similar pattern was also applied by Wang et al., 2017; Turker et al., 2017. Some researchers pumped wasterwater into the reactor from the bottom with a peristaltic pump Fang et al., 2015; Oon et al., 2015 in order to maximize the redox gradient. Different maneuver of CW-MFCs has been elaborated in table 2. This maneuver reduces the dissolved oxygen (DO) at the anode compartment and at the cathode region there is more availability of oxygen due to rhizospheric zone just above of it. At the start glass wool is used as a partition to provide a 'sharp' redox potential. But the natural redox gradient is afforded by an up flow management of the CW-MFC system which is the result of large electrode separation and contributes to a great extent to the ohmic resistance of the unit (Cheng et al., 2006; Liu et al., 2008; Fan et al.,2012; Ahn and Logan, 2012; Srikanth and Venkata Mohan, 2012).

Table 2. Different maneuver of CW-MFCs

S.N.	Material of electrode	Mode	Fed	Reference
1.	Graphite	Batch	Downflow	Yadav et al., 2012
2.	Graphite	Continous	Upflow	Zao et al., 2013
3.	GAC, SSM	Continous	Upflow	Liu et al., 2013
4.	GAC, SSM	Continous	Upflow	Fang et al., 2013
5.	GAC, SSM	Continous	Upflow	Liu et al., 2014
6.	GAC, SSM	Continous	Upflow	Fang et al., 2015
7.	Granular graphite	Simultaneous upflow- downflow		Doherty et al., 2015
8.	Graphite plate	Horizontal flow		Villasenor et al., 2013

Functioning of CW-MFCs as Wastewater Treatment

As almost zero energy consumption system, CW-MFCs have been fruitfully applied to the treatment of wastewater. In the first round of investigations indicated that CW-MFCs perform in the same way to the traditional CW studies by eliminating 75% (Yadav et al., 2012) and 76.5% (Zhao et al., 2013) of COD. In treating wastewater, wetland flora also plays an important role, the removal efficiencies of CW-MFC planted with *S. validus* was respectively 5.8% for COD, 7.2% for NO_3^--N, and 23.9% for NH_3-N much higher than that of unplanted system (Wang et al., 2017a). According to the (Fang et al., 2015) the highest decolorization rate reached 95.6, when the COD concentration was 300 mg/L while the ABRX3 amount was 30%. The maximum decolorization rate was obtained when HRT was 3 days. Oon et al., 2015 investigate that the oxidation reduction potential (ORP) and dissolved oxygen (DO) report of the aerobic and anaerobic regions were well developed in the upper and lower bed respectively, of UFCW–MFC and biodegradation of organic matter, nitrification and denitrification was also investigated and the removal efficiencies of COD, NO_3-, NH_4 $^+$ were 100%, 40%, and 91%, respectively. Yadav et al., 2012b, in another trial performed 4 experiments first with normal-CW microcosm; second with SRB enriched CW microcosm; third with SRB and IS enriched, CW microcosm; and fourth with SRB and copper doped IS enriched CW microcosm. The microcosms were tested for color removal of methyl orange dye from synthetic wastewater. The SRB enriched CW achieved highest removal (99.5, 98.53, 97.05, 94.62, and 96.52%) and the normal-CW achieved lowest removal (93, 89.18, 48.72, 50.14, and 52.34%) from wastewater containing 500, 1000, 1500, 2000 and 2500 mg l−1 dye respectively. Higher removal was achieved in SRB and IS enriched CW in comparison to other CW microcosms. Turker and yakar, 2017 study deals with performance assessment of a hybrid constructed wetland-microbial fuel cell (HCW-MFC) for boron (B) treatment. The results indicated that 63.4% boron removal was achieved for wastewater containing an average 12.3 mg L^{-1}inflow of boron concentrations. Oon et al., 2016 performed HCW-MFC experiment and which was able to remove 47.5% and 19.1% nitrate and nitrite from synthetic wastewater, respectively. Table 3. has the data related to comparison of COD removal efficiency with different concentrations. The average influent of COD was 624 mg/L, NO_3- was 142 mg/L and NH_4^+ was 40 mg/L, respectively and their removal efficiencies in day 1 HRT were 99%, 46%, and 96%, respectively. Due to lower dissolved oxygen in the system, NO_3- removal was relatively higher in the closed circuit system. Zang et al., 2017 investigates the removal rate and degradation pathway of Sulfamethoxazole (SMX) in bioelectrochemical system (BES) and the removal dynamics of SMX in a BES determined by stacked constructed wetland-coupled microbial fuel cells (CW-MFCs). The outcome bring into being that SMX (30 mg L^{-1}) was quickly degraded in the BES, and the SMX deletion kinetics was pretend well by a first-order kinetic model ($R2 > 0.93$).

Table 3. Comparison of COD removal efficiency with concentration

S.N.	COD concentration (mg/L)	COD removal efficiency (%)	Reference
1.	135	85.7	Fang et al. (2015)
2.	180	85.7	Fang et al. (2013)
3.	250	94.8	Liu et al. (2013)
4.	250	95.0	Liu et al. (2014)
5.	560	90.0	Villasenor et al. (2013)
6.	583	64.0	Doherty et al. (2015)

Functioning of CW-MFCs as Electricity Production

COD Load Effect on Electricity Production

COD loading greatly affects the performance of CW-MFCs. Higher COD removal results in producing more electrons and protons and these electrons are transferred to the cathode via external circuit. Fang et al., 2015 in his work, showed the effects of hydraulic residence time (HRT), reactive brilliant red X-3B (ABRX3) proportion and COD concentration on the electricity production of CW-MFC. The maximum electrical energy was produce when HRT was 3 days. The distribution of electrons in the system may be influence, the ABRX3 proportion (calculated as COD) in the wastewater in electricity production. With an increasing ABRX3 proportion, the power density of CW- MFC decreased concurrently. The COD concentration slightly influenced the CW-MFC performance. When the COD concentration was

300mg/L, the highest power density reached 0.852W/m^3. Yadav et al., 2012 investigated electricity generation along with COD, during treatment of 1000 mg l^{-1} initial dye concentration containing wastewater his results were the maximum power density of 15.73 mW m^{-2} and maximum current density of 69.75 mA m^{-2} achieved. In the Turker and yakar, 2017 research the maximum power and current density were recorded as 78 mWatt/m^2 and 105 mA/m^2 for the last unit of the hybrid system, they further suggest that the power and the current density were decreased with the increasing boron concentration.

Types of Electrodes in Electricity Production

Wang et al., 2017 studied the effects of electrode material and substrate concentration on the bioelectricity generation. He uses our materials including carbon fiber felt (CFF), stain-less steel mesh (SSM), graphite rod (GR), and foamed nickel (FN) as air-cathode and anode for each system. The obtained maximum power densities of Systems 1, 2, 3 and 4 were 4.80, 2.30, 3.35, and 5.11 mW m^{-2}, respectively. Liu et al., 2017 investigated three bio-cathode materials including granular activated carbon (GAC), carbon cloth (CC) and stainless steel mesh (SSM), they were compared and evaluated. The highest maximum power density of 55.05mWm^2 was achieved by GAC–SSM bio-cathode. The CW-MFC with a cathode diameter of 25 cm showed the best electrode performance, and the highest voltage and power density were 560 mV and 0.88 W/m3, respectively according to Fang et al., 2017. Song et al., 2017 notify that 20 cm distance between the anode and cathode produced a most favorable removal of chemical oxygen demand (COD) of 94.90% with a 0.15 W/m3 power density, 0.31% coulombic efficiency and 339.80 W internal resistances.

Wetland Plants Role in Electricity Production

Wang et al., 2017 in his study of, CW-MFCs associate three macrophytes (*J. effuses, T. orientalis* and *S. validus*) for domestic sewage treatment, also unplanted CW-MFC was performed as a control system. Macrophyte *T. orientalis* and *S. validus* can significantly strengthen the bioenergy output in CW-MFC. Highest current (94.27 mA m−2) and power densities (21.53 mW m−2) were obtained in CW-MFC planted with *T. orientalis.* Earlier studies have confirmed that when *Canna indica, Ipomoea aquatic* and *Typha latifolia* were planted in constructed wetland united with MFC systems, contaminant removal was enhanced during wastewater treatment and electricity was generated more efficiently (Fang et al., 2013; Oon et al., 2015; Yadav et al., 2012). Other species including *Arundo donax, Spartina anglica, Lemna minuta, Glyceria maxima* and *Arundinella anomala* have been studied in PMFCs and showed current production associated with plant rhizodeposits (Hubenova and Mitov, 2012; Strik et al., 2011; Timmers et al., 2012; Wetser et al., 2015). Through the optimization of plant selection, reactor configuration, or nutrient media, power production has been improved (Chen et al., 2012; Helder et al., 2012; Wetser et al., 2015).

Prospects for MFC-CWs

MFC is a promising technology for generation of electricity from organic substances, especially from organic waste of different origin. The most important disadvantage of MFC technology is the small power density; this can be fix by either isolation of effective microorganisms that can proficiently transport electrons to anode or by producing engineered strains through recombinant DNA technology that demonstrate better electron relocate rates. Numerous reports have established that rather than pure cultures, consortium of several bacteria illustrate better electron transfer rates to the anode. Various bacterial strains have been exposed to generate mediators which proficiently move electrons to the anode. Immobilized enzymes and cells will also helpful in more electricity production and treatment of wastewater also. Some soil enzymes such as dehydrogenase and urease in wetland matrix strongly correlated with bioelectric production, thus these enzymes might play important roles in bioelectric production with CW-MFCs, and further research had to be carried out in this. Microbes (Cytophagales, Clostridium sp., and Dechloromonas sp., etc.) found in the rhizosphere showed a capability to decompose refractory contaminants, more research in these bacteria is also new area of research. Methane production can be postponed by incorporating a PMFC into a rice microcosm (Arends et al., 2014).

CONCLUSION

Before MFC technology is used in practical applications, substantial efforts are required to optimize its performance. However, the biggest advantage of combining the two units is treatment of wastewater and electricity generation but electricity production is low; more research has to be done in this area to increase its amount. Among the different types of reactor designs, the two-chamber MFCs, suffers from high internal resistance and additional energy consumption. Single chamber MFCs are more advantageous. The implications for CW-MFCs should be investigated further.

REFERENCES

Ahn, Y., & Logan, B. E. (2012). A multi-electrode continuous flow microbial fuel cell with separator electrode assembly design. *Applied Microbiology and Biotechnology*, *93*(5), 2241–2248. doi:10.100700253-012-3916-4 PMID:22314518

Arends, J., Speeckaert, J., Blondeel, E., De Vrieze, J., Boeckx, P., Verstraete, W., Rabaey, K., & Boon, N. (2014). Greenhouse gas emissions from rice microcosms amended with a plant microbial fuel cell. *Applied Microbiology and Biotechnology*, *98*(7), 3205–3217. doi:10.100700253-013-5328-5 PMID:24201892

Bulc, T. G. (2006). Long term performance of a constructed wetland for landfill leachate treatment. *Ecological Engineering*, *26*(4), 365–374. doi:10.1016/j.ecoleng.2006.01.003

Cao, X., Wang, H., Li, X. Q., Fang, Z., & Li, X. N. (2017). Enhanced degradation of azo dye by a stacked microbial fuel cell-biofilm electrode reactor coupled system. *Bioresource Technology*, *227*, 273–278. doi:10.1016/j.biortech.2016.12.043 PMID:28040648

Chen, Z., Huang, Y. C., Liang, J. H., Zhao, F., & Zhu, Y. G. (2012). A novel sediment microbial fuel cell with a biocathode in the rice rhizosphere. *Bioresource Technology*, *108*, 55–59. doi:10.1016/j.biortech.2011.10.040 PMID:22265978

Chhimwal, M., Narayan, M., & Srivastava, R. K. (2017). Electricity generation from microbial fuel cell by using different bio-wastes as substrate. *Environment and Ecology*, *35*(4A), 2958–2964.

Corbella, C., Garfí, M., & Puigagut, J. (2014). Vertical redox profiles in treatment wetlands as function of hydraulic regime and macrophytes presence: Surveying the optimal scenario for microbial fuel cell implementation. *The Science of the Total Environment*, *470*, 754–758. doi:10.1016/j.scitotenv.2013.09.068 PMID:24184552

Corbella, C., Guivernau, M., Viñas, M., & Puigagut, J. (2015). Operational, design and microbial aspects related to power production with microbial fuel cells implemented in constructed wetlands. *Water Research*, *84*, 232–242. doi:10.1016/j.watres.2015.06.005 PMID:26253894

Doherty, L., Zhao, Y., Zhao, X., Hu, Y., Hao, X., Xu, L., & Liu, R. (2015). A review of a recently emerged technology: Constructed wetland–microbial fuel cells. *Water Research*, *85*, 38–45. doi:10.1016/j.watres.2015.08.016 PMID:26295937

Fan, Y., Han, S. K., & Liu, H. (2012). Improved performance of CEA microbial fuel cells with increased reactor size. *Energy & Environmental Science*, *5*(8), 8273–8280. doi:10.1039/c2ee21964f

Fang, Z., Cao, X., Li, X., Wang, H., & Li, X. (2017). Electrode and azo dye decolorization performance in microbial-fuel-cell-coupled constructed wetlands with different electrode size during long-term wastewater treatment. *Bioresource Technology*, *238*, 450–460. doi:10.1016/j.biortech.2017.04.075 PMID:28463809

Fang, Z., Song, H. L., Cang, N., & Li, X. N. (2013). Performance of microbial fuel cell coupled constructed wetland system for decolorization of azo dye and bioelectricity generation. *Bioresource Technology*, *144*, 165–171. doi:10.1016/j.biortech.2013.06.073 PMID:23867535

Fang, Z., Song, H. L., Cang, N., & Li, X. N. (2015). Electricity production from Azo dye wastewater using a microbial fuel cell coupled constructed wetland operating under different operating conditions. *Biosensors & Bioelectronics*, *68*, 135–141. doi:10.1016/j.bios.2014.12.047 PMID:25562740

Hammer, D. A. (Ed.). (2020). *Constructed wetlands for wastewater treatment: municipal, industrial and agricultural*. CRC Press. doi:10.1201/9781003069850

Helder, M., Strik, D. P., Hamelers, H. V., & Buisman, C. J. (2012). The flat-plate plant-microbial fuel cell: The effect of a new design on internal resistances. *Biotechnology for Biofuels*, *5*(1), 1–11. doi:10.1186/1754-6834-5-70 PMID:22998846

Hubenova, Y., & Mitov, M. (2012). Conversion of solar energy into electricity by using duckweed in direct photosynthetic plant fuel cell. *Bioelectrochemistry (Amsterdam, Netherlands)*, *87*, 185–191. doi:10.1016/j.bioelechem.2012.02.008 PMID:22445768

Liu, H., Cheng, S., Huang, L., & Logan, B. E. (2008). Scale-up of membrane-free single-chamber microbial fuel cells. *Journal of Power Sources*, *179*(1), 274–279. doi:10.1016/j.jpowsour.2007.12.120

Liu, S., Song, H., Li, X., & Yang, F. (2013). Power generation enhancement by utilizing plant photosynthate in microbial fuel cell coupled constructed wetland system. *International Journal of Photoenergy*, *2013*, 2013. doi:10.1155/2013/172010

Liu, S., Song, H., Wei, S., Yang, F., & Li, X. (2014). Bio-cathode materials evaluation and configuration optimization for power output of vertical subsurface flow constructed wetland—Microbial fuel cell systems. *Bioresource Technology*, *166*, 575–583. doi:10.1016/j.biortech.2014.05.104 PMID:24956029

Logan, B. E. (2008). *Microbial Fuel Cell*. John Wiley & Sons.

Lu, L., Xing, D., & Ren, Z. J. (2015). Microbial community structure accompanied with electricity production in a constructed wetland plant microbial fuel cell. *Bioresource Technology*, *195*, 115–121. doi:10.1016/j.biortech.2015.05.098 PMID:26066972

Minghui, L., Wen, Z., Yu, X. I. A., & Yongsheng, G. (2011). Study on removal efficiencies of pollutant from constructed wetland in aquiculture waste water around Poyang Lake. *Procedia Environmental Sciences*, *10*, 2444–2448. doi:10.1016/j.proenv.2011.09.380

Narayan, M., Solanki, P., & Srivastava, R. K. (2018). Treatment of sewage (domestic wastewater or municipal wastewater) and electricity production by integrating constructed wetland with microbial fuel cell. In *Sewage*. IntechOpen. doi:10.5772/intechopen.75658

Nyquist, J., & Greger, M. (2009). A field study of constructed wetlands for preventing and treating acid mine drainage. *Ecological Engineering*, *35*(5), 630–642. doi:10.1016/j.ecoleng.2008.10.018

Oon, Y. L., Ong, S. A., Ho, L. N., Wong, Y. S., Dahalan, F. A., Oon, Y. S., Lehl, H. K., & Thung, W. E. (2016). Synergistic effect of up-flow constructed wetland and microbial fuel cell for simultaneous wastewater treatment and energy recovery. *Bioresource Technology*, *203*, 190–197. doi:10.1016/j.biortech.2015.12.011 PMID:26724550

Oon, Y. L., Ong, S. A., Ho, L. N., Wong, Y. S., Oon, Y. S., Lehl, H. K., & Thung, W. E. (2015). Hybrid system up-flow constructed wetland integrated with microbial fuel cell for simultaneous wastewater treatment and electricity generation. *Bioresource Technology*, *186*, 270–275. doi:10.1016/j.biortech.2015.03.014 PMID:25836035

Praveen, S., Narayan, M., Rabha, A. K., & Srivastava, R. K. (2018). Assessment of Cadmium Scavenging Potential of Canna indica L. *Bulletin of Environmental Contamination and Toxicology*, *101*(4), 446–450. doi:10.100700128-018-2416-3 PMID:30116850

Praveen, S., Rabha, A. K., Maitreyie, N., & Srivastava, R. K. (2018). Relative comparison for phytoremediation potential of Canna and Pistia for wastewater recycling. *Environment and Ecology*, *36*(1A), 316–320.

Puig, S., Coma, M., Desloover, J., Boon, N., Colprim, J., & Balaguer, M. D. (2012). Autotrophic denitrification in microbial fuel cells treating low ionic strength waters. *Environmental Science & Technology*, *46*(4), 2309–2315. doi:10.1021/es2030609 PMID:22257136

Song, H., Zhang, S., Long, X., Yang, X., Li, H., & Xiang, W. (2017). Optimization of bioelectricity generation in constructed wetland-coupled microbial fuel cell systems. *Water (Basel)*, *9*(3), 185. doi:10.3390/w9030185

Srikanth, S., & Mohan, S. V. (2012). Influence of terminal electron acceptor availability to the anodic oxidation on the electrogenic activity of microbial fuel cell (MFC). *Bioresource Technology*, *123*, 480–487. doi:10.1016/j.biortech.2012.07.049 PMID:22940358

Srivastava, P., Yadav, A. K., & Mishra, B. K. (2015). The effects of microbial fuel cell integration into constructed wetland on the performance of constructed wetland. *Bioresource Technology*, *195*, 223–230. doi:10.1016/j.biortech.2015.05.072 PMID:26144020

Strik, D. P., Timmers, R. A., Helder, M., Steinbusch, K. J., Hamelers, H. V., & Buisman, C. J. (2011). Microbial solar cells: Applying photosynthetic and electrochemically active organisms. *Trends in Biotechnology*, *29*(1), 41–49. doi:10.1016/j.tibtech.2010.10.001 PMID:21067833

Timmers, R. A., Rothballer, M., Strik, D. P., Engel, M., Schulz, S., Schloter, M., Hartmann, A., Hamelers, B., & Buisman, C. (2012). Microbial community structure elucidates performance of Glyceria maxima plant microbial fuel cell. *Applied Microbiology and Biotechnology*, *94*(2), 537–548. doi:10.100700253-012-3894-6 PMID:22361855

Türker, O. C., & Yakar, A. (2017). A hybrid constructed wetland combined with microbial fuel cell for boron (B) removal and bioelectric production. *Ecological Engineering*, *102*, 411–421. doi:10.1016/j.ecoleng.2017.02.034

U.S. EPA. (2000). Constructed Wetlands Treatment of Municipal Wastewaters. United States Environmental Protection Agency.

Villasenor, J., Capilla, P., Rodrigo, M. A., Canizares, P., & Fernandez, F. J. (2013). Operation of a horizontal subsurface flow constructed wetland–microbial fuel cell treating wastewater under different organic loading rates. *Water Research*, *47*(17), 6731–6738. doi:10.1016/j.watres.2013.09.005 PMID:24074815

Vrhovšek, D., Kukanja, V., & Bulc, T. (1996). Constructed wetland (CW) for industrial waste water treatment. *Water Research*, *30*(10), 2287–2292. doi:10.1016/0043-1354(96)00114-5

Wang, J., Song, X., Wang, Y., Bai, J., Li, M., Dong, G., Lin, F., Lv, Y., & Yan, D. (2017). Bioenergy generation and rhizodegradation as affected by microbial community distribution in a coupled constructed wetland-microbial fuel cell system associated with three macrophytes. *The Science of the Total Environment*, *607*, 53–62. doi:10.1016/j.scitotenv.2017.06.243 PMID:28686895

Wang, J., Song, X., Wang, Y., Zhao, Z., Wang, B., & Yan, D. (2017). Effects of electrode material and substrate concentration on the bioenergy output and wastewater treatment in air-cathode microbial fuel cell integrating with constructed wetland. *Ecological Engineering*, *99*, 191–198. doi:10.1016/j.ecoleng.2016.11.015

Wetser, K., Sudirjo, E., Buisman, C. J., & Strik, D. P. (2015). Electricity generation by a plant microbial fuel cell with an integrated oxygen reducing biocathode. *Applied Energy*, *137*, 151–157. doi:10.1016/j.apenergy.2014.10.006

Yadav, A. K., Dash, P., Mohanty, A., Abbassi, R., & Mishra, B. K. (2012). Performance assessment of innovative constructed wetland-microbial fuel cell for electricity production and dye removal. *Ecological Engineering*, *47*, 126–131. doi:10.1016/j.ecoleng.2012.06.029

Yadav, A. K., Jena, S., Acharya, B. C., & Mishra, B. K. (2012). Removal of azo dye in innovative constructed wetlands: Influence of iron scrap and sulfate reducing bacterial enrichment. *Ecological Engineering*, *49*, 53–58. doi:10.1016/j.ecoleng.2012.08.032

Zhang, S., Yang, X. L., Li, H., Song, H. L., Wang, R. C., & Dai, Z. Q. (2017). Degradation of sulfamethoxazole in bioelectrochemical system with power supplied by constructed wetland-coupled microbial fuel cells. *Bioresource Technology*, *244*, 345–352. doi:10.1016/j.biortech.2017.07.143 PMID:28780269

Chapter 10
Use of Ecofriendly Fertilizers and Crop Residues for Enhancing Crop Productivity and Sustainable Agriculture

Divya Singh
Banaras Hindu University, Varanasi, India

Jay Kumar
Banaras Hindu University, Varanasi, India

Garvita Singh
Gargi College, University of Delhi, New Delhi, India

Ashok Kumar
Banaras Hindu University, Varanasi, India

ABSTRACT

Due to the exploding world population, there is a tremendous increase in demand for agricultural output, which has resulted in the pressure for large-scale crop production. Therefore, chemical fertilizers are used extensively to increase the agricultural output, which has caused hazardous effects on human health, soil quality, and the environment. To combat these problems, beneficial microbes have been exploited and used as biofertilizers. It has been well documented that certain microorganisms including bacteria, fungi, and cyanobacteria have plant growth promoting properties and can be safely used as alternative sources of fertilizer. These microbes when used in combination with crop residues and compost show better results in terms of crop yield and soil fertility. The chapter highlights the potential use of various types of biofertilizers in sustainable agriculture with special emphasis on their beneficial role in nutrient uptake, plant growth promotion, and protection against several plant pathogens in an eco-friendly manner.

DOI: 10.4018/978-1-7998-8434-7.ch010

INTRODUCTION

Agriculture sector is the backbone of the economic growth and food security of any country. During the past few decades, there is tremendous growth in the world population and food demand has become a challenging problem. The production of food is not enough to meet the increasing demand especially in developing countries. To meet the increasing food demand of the human population, modern improved agricultural practices involving extensive use of chemical fertilizers are followed. Chemical fertilizers used in agriculture are usually manufactured by industrial manipulation of substances using predefined quantities of nitrogen, phosphorus and potassium. Although chemical fertilizers increase the yield of crop plants, but cause adverse effects on soil health (Sujanya & Chandra 2011). As such growth and yield of many high yielding varieties of crop plants are dependent on the application of high doses of synthetic fertilizers. Excessive uses of chemical fertilizers have led to adverse effects on the soil texture, structure and fertility together with harmful effects on natural microflora. Moreover, it is costly, toxic in nature and poses threat to nature by causing air, water and soil pollution. These fertilizers are also known to cause eutrophication of water bodies such as ponds, lakes and rivers mainly due to their accumulation in the groundwater (Savci 2012). To overcome and reduce the above problems, there is an urgent need to adopt the strategies of sustainable agriculture to ensure biosafety.

Sustainable agriculture has been rightly defined as "the large scale production of safe, high quality and valuable agricultural products without damaging the natural environment." In recent years, organic farming employing biofertilizers has gained much attention due to its several advantages over chemical fertilizers. Biofertilizers are regarded as one of the chief components of integrated nutrient management in sustainable agriculture because they are eco-friendly, cost-effective and provides renewable input of plant nutrients to substitute chemical fertilizers (Raja 2013; Naik et al., 2019). Generally, biofertilizers are living or latent formulation of efficient strains of microorganisms (fungi, bacteria, and cyanobacteria) that colonizes the rhizosphere or plant's interior and promotes plant growth when applied to seed, plant surfaces, or soil by increasing the availability of nutrients in a form which is readily absorbed by target plants (Vessey 2003). It also enriches the environment of the soil by performing various processes such as nitrogen fixation, mineralization, solubilization of phosphate, production of plant growth promoting substances, humification of organic matter, recycling of nutrients, improvement of soil stability, providing biological control agents, and promotion of mycorrhizal symbiosis. They also play important role in the bioremediation of soils contaminated with non-biodegradable and xenobiotic compounds (Corpoica 2007; Rivera-Cruz et al. 2008). They are known to increase crop yield by producing plant growth promoting substances viz., phytohormones (indole-3-acetic acid), siderophore (for iron sequestration) and enzymes (1-aminocyclopropane-1-carboxylate deaminase). Certain important functions of eco-friendly biofertilizers are illustrated in Figure 1.

Bhardwaj et al. (2014) reported that the use of biofertilizers increases crop production up to 10-40% by improving the amount of proteins, vitamins, and nitrogen fixation in the plants. Several microorganisms namely *Aspergillus fumigatus*, *Aspergillus niger* (Ogbo 2010) *Azotobacter, Azospirillum, Rhizobium* (Aziz et al. 2012; Dogan et al. 2011), *Anabaena oryzae*, *Nostoc calcicola* (Youssef & Ali 1998), *Azolla pinnata* (Mazid & Khan 2015) and several others are capable of forming an association with a variety of host plants and have been exploited as biofertilizers (Nosheen et al., 2021). Besides microorganisms, other natural inputs including dead and decaying remains of organic matter, domestic sewage, crops residues and animal manure are also used as biofertilizers (Chirinos et al. 2006). The annual production of biofertilizers in India is 10,000 mt and an average of around 64g/ha of biofertilizers is utilized per year in our country (Mishra & Dash 2014).

The aforementioned applications of biological fertilizer are bound to bring social, economical as well as environmental benefits to both farmers and consumers. However, the implementation of biofertilizers in agricultural fields requires detection and monitoring of environmental factors involved in physiological and metabolic processes, risk assessment, procurement of biological inoculants, capital investment, and skilled personnel (Plaster 2000; Vanegas 2003). This chapter briefly describes the potential benefits of adopting clean and green technology of biofertilizers usages for maintaining the health of the soil and nourishing the plants with particular emphasis on their role in sustainable agriculture.

Figure 1. Different benefits of biofertilizers for enhanced crop productivity and sustainable agriculture

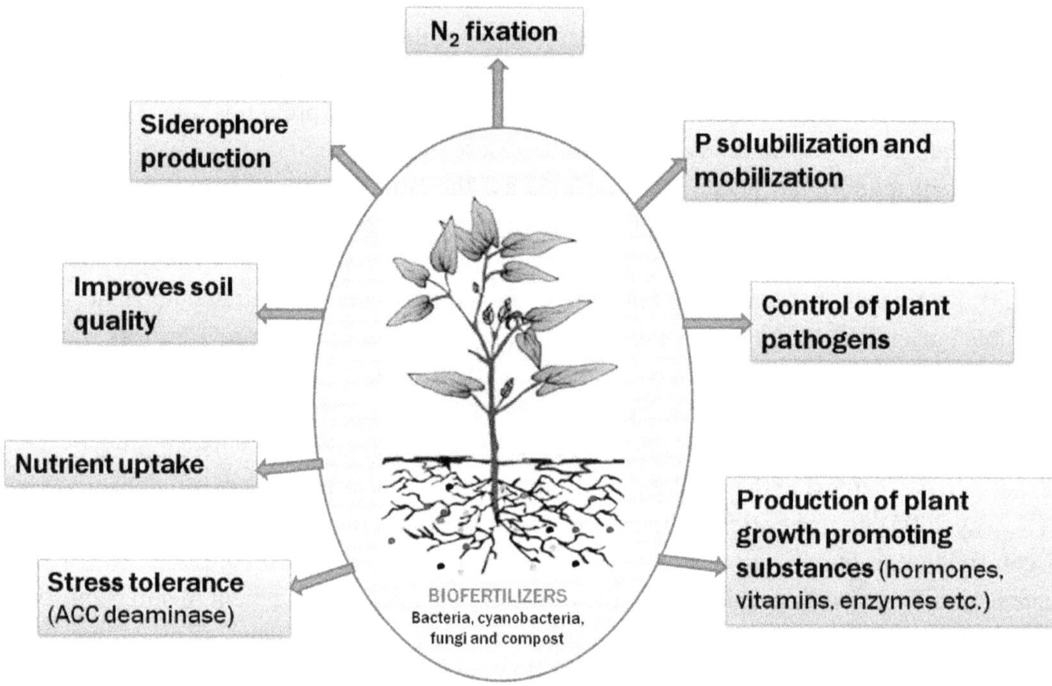

TYPES OF BIOFERTILIZERS

Biofertilizers are classified into different types mostly based on functions they perform, their nature, associated hosts and mode of applications. Table 1 represents the types of biofertilizer and their nature (Table 1).

Table 1. Different types of biofertilizers and the microbes involved

S. No.	Types	Microbes
1	**Nitrogen-fixing**	
(a)	Free-living	Bacteria: *Azotobacter, Azospirillum, Beijerinkia, Clostridium, Klebsiella*, Cyanobacteria: *Anabaena, Nostoc, Aulosira*
(b)	Symbiotic	*Rhizobium, Frankia, Anabaena azollae*
(c)	Associative symbionts/endophytes	*Azospirillum, Enterobacter, Herbaspirrilum, Acetobacter diazotrophicus*
2	**Phosphate-solubilizing**	
(a)	Fungi	*Penicillium sp., Aspergillus awamori*
(b)	Bacteria	*Bacillus megaterium, Bacillus subtilis, Bacillus circulans, Pseudomonas striata*
3	**Phosphate-mobilizing**	
(a)	Arbuscular mycorrhiza	*Glomus* sp., *Gigaspora* sp., *Acaulospora sp., Scutellospora sp., Sclerocystis* sp
(b)	Ectomycorrhiza	*Laccaria* sp., *Pisolithus* sp., *Boletus* sp., *Amanita* sp
(c)	Ericoid mycorrhizae	*Pezizella ericae*
(d)	Orchid mycorrhiza	*Rhizoctonia solani*
4	**Micronutrients solubilizers**	
(a)	Zinc and Silicate	*Bacillus* sp.
(b)	**Plant Growth Promoting**	Majority of rhizobacteria and endophytic bacteria
(c)	**Compost biofertilizers**	Cellulolytic fungi, *Azotobacter*
(d)	**Nanofertilizers**	Mostly bacteria

(Source: http://www.agritech.tnau.ac.in/orgfarm/orgfarm biofertilizer technology)

Nitrogen-Fixing

Rhizobium

Rhizobia are gram-negative, symbiotic, nitrogen-fixing soil bacteria belonging to the Rhizobiaceae family (Alphaproteobacteria) that infect the roots of different leguminous plants, multiply and form nodules (Allito et al. 2015). It has been used as one of the most efficient nitrogen fixers for several decades as it increases crop yield by carrying out the conversion of atmospheric dinitrogen (N_2) into usable forms (Sharma et al. 2011). As high as 50-300 kg N/ha/year is fixed by legumes. Rhizobial inoculums are usually applied in the form of peat granules and allowed to trickle down into the soil where seeds are present (Chen 2006). There are many efficient species of *Rhizobium* such as *R. leguminosarum*, *R. trifolii*, *R. lupini*, *R. phaseoli*, *R. meliloti* etc (Afzal & Bano 2008; Bardin et al. 2004) which are collectively called Rhizobia. They are known to colonize the roots of important leguminous plants namely *Pisum sativum*, *Cicer arietinum*, *Phaseolus vulgaris*, *Glycine max* etc. Additionally, certain non-symbiotic rhizobacteria are also capable of fixing nitrogen in non-leguminous plants like *Parasponia* through non-obligate interaction (Verma et al. 2010).

Leguminous plants in particular require high quantities of nitrogen for the synthesis of various cellular components/biomolecules but they are not able to fix atmospheric N_2 alone and thus depend on certain partner to perform the above function (Mohapatra et al. 2013). Atmospheric N_2 fixation in such plants is fixed by the symbiont *Rhizobia* which establish an active association with the plant in a complex process. During this process, bacterial infection occurs and changes the fate of differentiated cortical cells resulting in the formation of root nodules (Suzaki et al. 2015). The site of bacterial colonization within root nodules and the symbiotic relationship between the host plant and *Rhizobium* are controlled by Nod (nodulation) factors released by *Rhizobia* (Maillet et al. 2011). Different sets of *nod* genes have been reported and their roles in nodulation and nitrogen fixation have been well documented. Typically, the process of N_2 fixation is catalyzed by the enzyme nitrogenase which is made up of two components: i) dinitrogenase (Mo-Fe protein) containing both iron (Fe) and molybdenum (Mo) as its cofactor, and ii) dinitrogenase reductase (Fe-protein) containing iron (Fe) as its cofactor. Both these components function in a coordinated manner to reduce N_2 to NH_3 (Santi et al. 2013). At least three types of nitrogenase complexes based on different cofactors associated with the enzyme have been reported (Ahemad & Kibret 2014; Burns & Hardy 2012). Nitrogenase is synthesized by a *nif* regulon which consists of more than 20 genes in several operons. Both symbiotic and N_2-fixing microbes harbor *nif* genes and share almost identical features. Expression of *nif* gene gets downregulated in the presence of oxygen; however, N_2-fixing microbes have developed efficient protective mechanism(s) against O_2-mediated inactivation of nitrogenase. In legumes-*Rhizobia* symbiosis, leghemoglobin is synthesized in the nodules that have a high affinity for oxygen, binds to oxygen and keeps the enzyme in active form (Shamseldin 2013).

Rhizobium inoculants when used as biofertilizer have been reported to significantly increase the yield and productivity of certain crops like alfalfa, pea, sugar beet (Ramachandran et al. 2011) and ground nut. In the case of groundnut, application of *Rhizobium* led to the increment of various beneficial features such as enhanced nitrogenase and nitrate reductase activity and leghemoglobin content within 50 days after inoculation (Sharma et al. 2011).

Azotobacter

Azotobacter is a genus of gram-negative, aerobic, free-living, nitrogen-fixing, heterotrophic bacteria belonging to the Azotobacteriaceae family (Gammaproteobacteria). They are present in neutral and alkaline soils except for *Azotobacter chroococcum* which occurs commonly in fertile soils (Moraditochaee et al. 2014). Other species of this genus includes *A.vinelandii*, *A. chroococcum*, *A. beijerinckii*, *A. armeniacus*, *A. paspali* and *A. nigricans* which are used as biofertilizer for a variety of crops including oat, wheat, barley, castor, maize, sunflower, sorghum, jute, mustard, rice, sesame, linseeds, sugar beets, rubber, coffee cotton, tobacco, tea, and coconuts (Wani et al. 2013; dos Santos et al., 2020). It has been reported that *Azotobacter* can supplement 15-93 kg N/ha/year in the case of *Paspslum notatum* (tropical grass) (Youssef & Eissa 2014). Besides nitrogen fixation, *Azotobacter* has other important roles such as secretion of vitamins (thiamine and riboflavin) and production of phytohormones (indole acetic acid, cytokinins and gibberellins). Furthermore, it also elevates plant growth by producing substances that promote mineral uptake, seed germination and root development by inhibiting certain root pathogens (Mathivanan et al. 2015). Martin et al. (2011) have reported that one of the strains of *Azotobacter*, *A. indicum* is capable of producing many antifungal compounds that reduce the mortality rate of seedlings by preventing the growth of several pathogenic fungi i.e. it functions as a biocontrol agent. In addition to this, *Azotobacter* has also been known to improve the morphology of roots in rice (Choudhury & Kennedy 2004).

Azospirillum

Azospirillum is another gram-negative, free-living, motile, heterotrophic bacterium belonging to Rhodospirillaceae family (Alphaproteobacteria). They do not form nodules, can thrive in flooded conditions and show favorable effects on crop yields under both greenhouse and field conditions (Mehnaz 2015). The genus *Azospirillum* includes diverse species namely *A. brasilense, A. lipoferum, A. halopraeferens, A. amazonense and A. irakense*. Among these, *A. lipoferum* and *A. brasilense* are the most beneficial species (Mishra et al. 2013). It also forms an associative symbiosis with various crop plants such as sugarcane, maize, pineapple, sorghum, pearl millet etc. All these plants follow the Hatch-Slack pathway (C_4 carbon fixation) of photosynthesis and form organic salts of malic and aspartic acid which are used in N_2 fixation by the bacteria (Mishra & Dash 2014). The nitrogen-fixing ability of *Azospirillum* is about 20-40 kg/ha/year and causes 10-15% increase in the yield of cereal crops (Mohapatra et al. 2013). Moreover, it has been reported that inoculation of the *Azospirillum* improves root morphology by enhancing the number of lateral roots and escalating the formation of root hairs (Mehdipour-Moghaddam et al. 2012). This elevates the water status of plants and also helps in growth and development (Ilyas et al. 2012). Apart from this, inoculation of *A. brasilense* sp. 245 in maize crops resulted in the boosting of plant growth due to the production of phytohormones (IAA, gibberellins, and cytokinin) which in turn, aided in mineral uptake by the plants (Steenhoudt & Vanderleyden 2000). It has been also observed that co-inoculation of *Azospirillum* and *Pseudomonas* alters the culturable bacterial community residing in the rhizosphere of wheat. It also modifies the pattern of carbon source utilization of the naturally occurring soil microflora at both tillering and grain filling stages (Naiman et al. 2009).

Cyanobacteria (Blue-Green Algae)

Cyanobacteria are photosynthetic, nitrogen-fixing prokaryotes living in a variety of habitats. They are considered as one of the best alternatives to increase soil fertility and crop productivity as they contribute to nitrogen fixation, nutrient cycling, decomposition of organic matter, detoxification of heavy metals, inhibition of growth of certain pathogenic microorganisms and produce some bioactive compounds such as hormones, vitamins, and enzymes (Singh et al. 2016). In addition, it has been also reported that cyanobacterial biofertilizers are capable of tolerating pesticide levels recommended for field applications (Kaushik 2013; Singh et al. 2016). Cyanobacteria can perform nitrogen fixation in free-living form as well as in symbiotic association with plants (Fay 1983). It can fix nitrogen up to an extent of 25-30 kg/ ha/year when applied in paddy fields under ideal conditions and are placed next to the legume-*Rhizobium* symbiosis in terms of total nitrogen fixation (Kaushik 2013). Different species of N_2-fixing hetrocystous *Anabaena, Nostoc, Calothrix, Aulosira, Tolypothrix, Plectonema* etc. have been identified as dominant flora of rice fields (Prasad & Prasad 2001; Sahu et al. 2012). Additionally, the role of certain species of *Aphanothece, Gloeocapsa*, and *Aphanocapsa* has been also implicated in active N_2 fixation. In cyanobacteria, heterocysts are the sole site of nitrogen fixation under aerobic condition because these cells create a micro anaerobic environment to protect nitrogenase enzyme from inactivation by oxygen (Kaushik 2013). Some of the important cyanobacteria found in the paddy fields and used as biofertilizers are given in Figure 2.

Studies conducted by Karthikeyan et al. (2007) have demonstrated that cyanobacteria chiefly *Nostoc* and *Anabaena* were able to form associations with wheat roots grown under *in vitro* conditions. Additionally, the nitrogen-fixing capability of *Anabaena* sp. strain PCC7120 has been improved using genetic engineering techniques (Pandey et al. 2013). Apart from fixing nitrogen, cyanobacteria such as *Cylindrospermum musicola* also assist plants by releasing growth promoting substances such as phytohormones and vitamins which have been found to enhance root development and yield of rice plants (Venkataraman & Neelakantan 1967). Furthermore, the application of cyanobacterial biofertilizers to the wheat crop caused remarkable changes in plant morphology along with increment in plant height, weight and produce of wheat crop. These changes are ascribed to the IAA-like compounds secreted by the cyanobacterial strains (Karthikeyan et al. 2007). It has been also reported by Youssef & Ali (1998) that *Anabaena oryzae*, *Nostoc calcicola* and *Spirulina* sp. reduced the rate of infection by root-knot nematode *Meloidogyne incognita* in cowpea plants

Additionally, there are several cyanobacterial strains that are known to form a symbiotic association with ferns, liverworts, fungi and flowering plants (Roychowdhury et al. 2014). Among all these, *Azolla* an aquatic fern has been well exploited as biofertilizers in rice cultivation as it harbors an efficient cyanobiont, *Anabaena azollae*. *Azolla* can be applied as green manure in the fields before rice planting. *A. azollae* lives inside the leaf cavities of *Azolla* where it fixes nitrogen. After death, *Azolla* gets easily decomposed in the soil/water and releases a considerable amount of nitrogen, potassium, phosphorus, zinc, molybdenum, iron and other micronutrients (Ghany et al. 2013). The most common species of *Azolla* that is used on a commercial scale in India is *Azolla pinnata* (Mazid & Khan 2015). Some other species of *Azolla* have also been widely used e.g. *Azolla caroliniana*, *Azolla mexicana*, *Azolla microphylla* and *Azolla filiculoides* (Mishra et al. 2013).

Figure 2. Photographs of cyanobacterial species isolated from rice field and used as biofertilizers for crop improvement

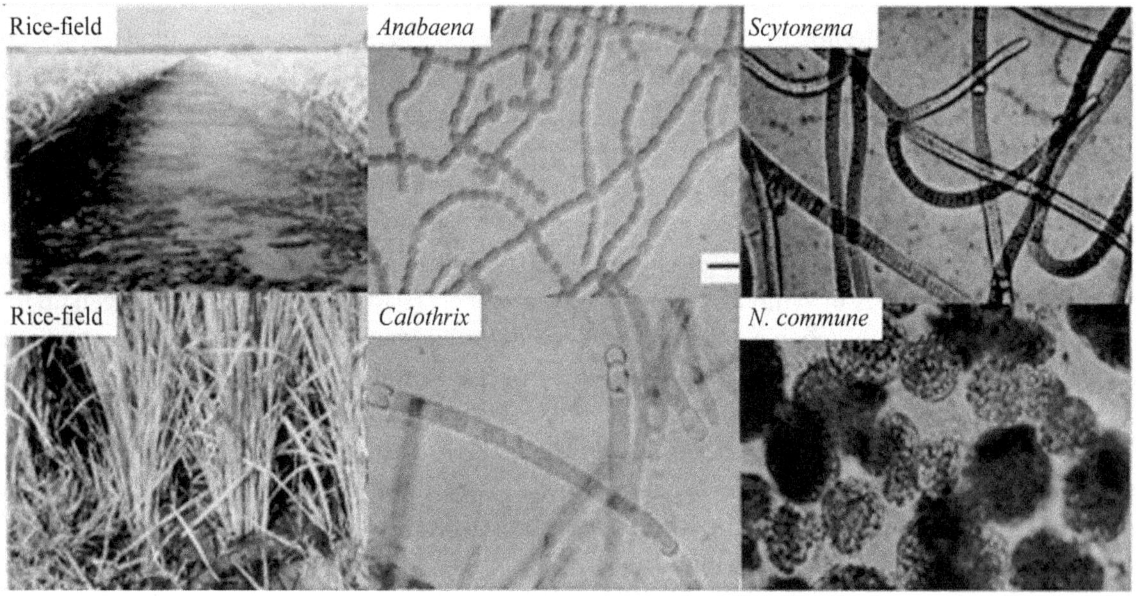

Phosphate-Solubilizing

Fungi

Some fungi have been potentially used as biofertilizers because of their ability to dissolve phosphate such as *Aspergillus fumigatus*, *Aspergillus niger*, *Aspergillus awamori*, *Penicillium digitatum* etc. (Souchie et al. 2006; Wakelin et al. 2004; Whitelaw 1999). Approximately 0.1-0.5% of the total fungal populations show P-solubilizing property in the soil and these fungi retain the solubilizing property even after repeated sub-culturing in the laboratory (Kucey 1988). According to Lebedeva et al. (1999), fungi are more competent than bacteria in mineral P solubilization since they are saprophytic in nature and can reach the patch of soil easily. Additionally, P-solubilizing fungi produce acids in large amount than bacteria and thus show higher P-solubilizing activity (Venkateswarlu et al. 1984). As such most of the phosphorus present in the soil is insoluble and in complex form and cannot be absorbed by the plants because plants can take up phosphate in only monobasic and dibasic soluble forms (Mahdi et al. 2012). In this scenario, these fungi play a pivotal role in making phosphate available to the plants by solubilizing insoluble phosphate through the secretion of organic acids particularly, gluconic and citric acids which in turn, lowers the pH of the soil resulting in the dissolution of bound phosphates (Khan et al. 2010). Fungi not only perform mineral solubilization but also exhibit other properties such as biological control, production of secondary metabolites and showing tolerance to heavy metals (Khan et al. 2010; Lebedeva et al. 1999) which make them a suitable candidate for use as biofertilizers.

Bacteria

Several soil and plant rhizospheric bacteria notably the species of *Pseudomonas*, *Bacillus*, *Micrococcus*, *Flavobacterium*, *Rhizobium*, *Burkholderia*, *Achromobacter*, *Agrobacterium*, *Acetobacter* and *Erwinia* have been documented to be involved in the solubilization of phosphate. Nevertheless, it has been reported that the majority of phosphate-solubilizing bacteria (PSB) generally resides in the rhizosphere as compared to non-rhizospheric soil (Youssef & Eissa 2014). These bacteria can carry out both solubilization as well as mineralization of phosphate. Solubilization of insoluble phosphorus into soluble form is achieved by the secretion of organic acids which chelate the cations attached to phosphate, while, mineralization of organic phosphorus is accomplished by the catalytic action of diverse phosphatases (McComb et al. 2013). PSB biofertilizers have been marketed as 'phosphobacterin' which are usually cells of *Bacillus megaterium* var. *phosphaticum* and increases crop yield by 10-20% (Mohapatra et al. 2013). Han et al. (2006) reported that mixed application of *Bacillus megaterium* var. *phosphaticum* and *Bacillus mucilaginosus* increased the uptake of phosphorus and potassium by pepper and cucumber plants. Besides making phosphorus available to the plants in soluble form, PSB also stimulate plant growth by increasing the rate of biological nitrogen fixation (Mohammadi & Sohrabi 2012). Recently, a bacterial strain NII-0909 of *Micrococcus* sp. has been isolated which exhibited the property of siderophore production in addition to phosphate solubilization (Dastager et al. 2010).

Phosphate-Mobilizing

There is another category of symbiotic fungi which mobilizes phosphate and is considered to be a promising biofertilizer. These symbiotic fungi form relationship with roots of the higher plant either in an intracellular or extracellular mode which is known as mycorrhizae. Mycorrhizal association presents an example of mutualism whereby the host plant provides carbohydrates to the fungi and in return, fungi contribute additional water and micronutrients to the plants. As such mycorrhizal association turns out to be advantageous for both the fungi and the host plant (Frey-Klett et al. 2011). Depending upon the colonization of fungus, mycorrhizae are further categorized into two types: ectomycorrhiza and endomycorrhiza. In ectomycorrhizae, fungal hyphae (specifically basidiomycetes) occupy the intercellular spaces on the root surface of trees such as oak, eucalyptus, pine, peach etc. Whereas in endomycorrhiza, fungal hyphae (endosymbionts belonging to the genera *Glomus*, *Rhizophagus* etc.) enter inside the intracellular spaces of the root cortical cells of orchids, crop plants, grasses, and some woody plants (Aggarwal et al. 2011; Tisserant et al. 2013). These endosymbionts often contain vesicles (arbuscles) for storing nutrients and channeling the stored nutrients into the roots of the host plant and thus known as vesicular-arbuscular mycorrhizal (VAM) fungi (Aggarwal et al. 2011). Apart from phosphorus, VAM also provides other nutrients like manganese, sulphur, zinc and iron to the plant leading to enhanced growth and yield (Kogel et al. 2006). Pellegrino & Bedini (2014) showed that inoculation of VAM fungi in the field resulted in increased productivity of chickpea. Additionally, VAM interactions benefit plants with an improved ability to resist drought and salinity stress (Lehmann et al. 2014). Furthermore, it functions as a biocontrol agent to protect the plant from the attack of pathogens (Anandraj et al. 2001) and has been also used as a bioremediation agent to decontaminate polluted soils (Chibuike & Obiora 2014).

Micronutrients Providing Microbes

Undoubtedly, nitrogen-fixing and phosphate-solubilizing microorganisms benefit plants in many ways; there is an utmost requirement of micronutrients such as zinc, copper, iron, etc. for better growth, development and yield of crop plants. Zinc plays an essential role in plant growth and development because its deficiency causes a decrease in the nutritional quality of crops along with a decline in yield (Mishra & Dash 2014). Plants can not readily absorb zinc since it is present in the form of insoluble complexes in the soil (Alloway 2008). Several bacterial species perform the transformation of unusable forms of soil zinc to usable zinc which becomes readily available to the plants. These bacteria are typically known as zinc solubilizers and include species of *Bacillus*, *Pseudomonas*, *Acinetobacter*, *Gluconacetobacter* etc. (Di Simine et al. 1998; Fasim et al. 2002; Saravanan et al. 2007). These bacteria also show other valuable traits such as the production of antibiotics, phytohormones, siderophores, antifungal substances, vitamins, and hydrogen cyanide (Rodr´ıguez & Fraga 1999). Bacteria-mediated zinc solubilization is performed mainly by the chelation of ligands, presence of oxidoreductive systems on cell surface and production of siderophores (Saravanan et al. 2011; Wakatsuki 1995). Khande et al. (2017) have examined the role of zinc solubilizing and biofortification property of *Bacillus* strain in improving the productivity of wheat and soybean plants. Furthermore, zinc solubilizing capacity of *Gluconacetobacter diazotrophicus* was characterized using different Zn compounds. It was observed that *G. diazotrophicus* effectively solubilized ZnO by producing gluconic acid which assisted in the solubilization process (Saravanan et al. 2007).

In addition to Zn-solubilizing bacteria, silicate-solubilizing bacteria (SSB) also play significant role because they not only solubilize insoluble forms of silicates, but can also solubilize potassium and phosphates present in the soil (Vasanthi et al. 2012). These bacteria are generally distributed in small population in water, soil and in silicate minerals and include *Bacillus mucilaginosus*, *B. flexus*, *B. megaterium* and *Pseudomonas fluorescens* (Vasanthi et al. 2016). They carry out the conversion of polymeric silica into monomeric form which is imperative for the biogeochemical cycling of silicate in the biosphere.

Additionally, they are also involved in the release of nutrients viz., calcium and magnesium from the silicates (Lauwers 1974). Solubilized silica gets readily absorbed by the plant roots in the form of ortho-silicic acid (H_4SiO_4) and stored as silica gel leading to its further deposition in sclerenchyma, epidermal cells, inflorescence brackets and vascular bundles (Lanning et al. 1958; Rodrignes & Datnoff 2005). Few researchers have evaluated the role of SSB in enhancing the rice crop yield growing under a lowland ecosystem (Peera et al. 2016). Similarly, some researchers have demonstrated antagonistic activity of SSB against certain crop pathogens namely *Rhizoctonia solani*, *Magnaporthae grisae*, *Macrophomina pheasolina* and *Alternaria alternate* (Naureen et al. 2015). Maleva et al. (2017) have illustrated the positive impact of SSB-based biofertilizers on the structural and functional aspects of photosynthetic machinery of *Brassica juncea*. Altogether SSB enriched biofertilizers can be considered as a cheap and eco-friendly fertilizer especially for the reclamation of barren soil for increasing the agricultural output.

Plant Growth Promoting Rhizobacteria (PGPR)

PGPR include a group of soil bacteria inhabiting the rhizosphere of many plant species. They establish a loose symbiotic relationship with plants and show beneficial effects to the plants (Ahmad et al. 2008). Interactions of PGPR with plants have been exploited commercially to achieve sustainable agriculture since PGPR exert diverse advantageous effects on the plants through direct and indirect mechanisms. Direct mechanisms involve enabling nutrient acquisition via processes of N_2 fixation, phosphate solubilization, potassium solubilization, siderophore production and modulating levels of phytohormones, whereas, indirect mechanisms comprise of inhibiting the attack of several pathogens on plants using antibiosis, induced systemic resistance, production of lytic enzymes and exopolysaccharides (Egamberdieva & Lugtenberg 2014; Mahanty et al. 2017; Singh et al., 2019). Additionally, some PGPR also confers other valuable traits to plants such as tolerance to abiotic stresses, detoxification of heavy metals, biological control of pests and insects (Egamberdieva & Lugtenberg 2014). *Pseudomonas*, *Bacillus*, *Rhizobium*, *Sinorhizobium*, *Bradyrhizobium*, *Enterobacter*, *Klebsiella*, *Azobacter*, *Azosprillum*, *Variovorax*, *Serratia* etc. are the most common examples of plant growth-promoting rhizobacteria that help in encouraging overall growth and development of crop plants (Glick 2012; Mahanty et al. 2017). Inoculation of these PGPR were found to have profound impact on the productivity of canola, maize, wheat, potato, barley, oat, soy beans, tomato, peas, lentil, cucumber and radicchio (Gray & Smith 2005). PGPR are also useful in the revitalization of soil quality by transforming barren, infertile, poor quality soil into fertile and cultivable soil (Gabriela et al. 2015). PGPR has now become an attractive option for biofertilizers.

Compost Biofertilizers

Composting is a biological process of decomposition and recycling of complex organic biodegradable wastes including animal wastes, crop residues and agro-industrial wastes into simple, hygienic, humus rich product. This humus rich product is known as compost and has been successfully used for a long time as a biofertilizer and soil conditioner (Popkin 1995). The process of biological transformation of these organic wastes into compost is mainly performed by microorganisms such as cellulolytic fungi and *Azotobacter*. These microbes produce nutrients by breaking down organic matter which ultimately improves the physical and chemical properties of soil (Yadav 2002). In addition to this, compost also increases the water holding capacity of the soil. It also offers other advantages such as its application in any type of soil, stimulation of agricultural output, augmentation of plant growth promotion, reduction of runoff, and is economically beneficial for farmers (Mills 2006). Composts also function as a biocontrol agent against several plant pathogens and have been used as an alternative to replace chemical fungicides (Zhang et al. 1998). Tanu et al. (2004) noted that the use of poultry manure drastically increased the dry weight and essential oil content of *Java citronella* plants. Furthermore, the efficiency of compost as biofertilizers for the growth of paddy has been documented by Jelin et al. (2013).

Nano-Fertilizers

Soil infertility and soil pollution problems are dominating due to the imbalanced use of NPK, chemical pesticides and the presence of high bicarbonate content in the irrigation water (Narimani et al. 2010). With the advent of nanotechnology, researchers are making serious efforts to explore the possible use of this technology in the formulation and application of biofertilizers in the field of sustainable agriculture. A nano-based fertilizer refers to a component in the range of nanometer scale that delivers nutrients to crop plants. This novel nano-fertilizer embraces a wide range of applications in agriculture such as plant disease control, enhanced nutrient uptake, minimal waste production, improved plant growth, and reduced toxicity (Fasusi et al., 2021). Additionally, nano fertilizers also offer controlled release and targeted delivery of nutrients, reduced loss of nutrients and seem the most potential alternative for sustainable agriculture (Sekhon 2014). In nano-fertilizers, nutrients molecules can be encapsulated by nanomaterials, coated with a thin protective layer, or delivered as emulsions or as nanoparticles. In certain cases, nanofertilizers may have nanodevices to synchronize the release of nutrients for the uptake by crop plants to prevent undesirable loss of nutrients from the soil, water and air (DeRosa et al. 2010).

Impacts of several organic, inorganic, and composite nano-fertilizers have been tested on various plants to assess their effects on growth, development, and crop productivity. Nano-fertilizers are applied to the crops through the spray on soil, root placement, foliar spray, soaked roots or seeds for targeted delivery. N, P, K, ZnO, Fe_2O_3, silicon dioxide ($Si/SiO/SiO_2$), TiO_2 and carbon nanotube are successfully used as nanofertilizers for different crops (Dimkpa & Bindraban 2017; Raliya et al. 2017). Nano-based fertilizers are more advantageous as compared to chemical and traditional biofertilizers because they are needed in very low quantity and the mode of application is simple and cheaper.

FORMULATION AND APPLICATION OF BIOFERTILIZERS

Certain parameters are followed while formulating biofertilizers such as viability and number of inoculums, capable of improving soil fertility, potential to stimulate plant growth and productivity and should be non-toxic. An ideal formulation should have the following desirable features: i) it should allow supplementation of nutrients, ii) should be cost-effective and have adaptable pH (Catroux et al. 2001; Herrmann & Lesueur 2013), iii) must be eco-friendly, non-toxic and biodegradable in nature, iv) should permit fast and controlled release of microbes into the soil (Malusa et al. 2012), v) should have long shelf-life and must be able to retain its viability under extreme environmental conditions (Malusa et al. 2012). The most widely used formulations so far are liquid, peat, granules and freeze-dried powders.

Liquid Formulation

Due to the ease of handling and inoculating, liquid formulations have been used extensively on a large scale. Liquid biofertilizers are generally maintained in broth cultures, organic oils, water or polymer-based suspensions (Herrmann & Lesueur 2013).

Peat Formulations

It comprises partly decayed materials which impart nutrition and a protective platform for the growth of different microbes both on the surface of the material and inside the material (Bashan et al. 2014).

Granules

Granules composed of tiny marbles, silica or calcite grains of varied sizes and coated with the desired microorganisms with the help of an adhesive are mostly prepared in the form of powder-type inoculum. These are preferred over peat formulations because they are easy to handle and are dust-free (Bashan et al. 2014).

Freeze-Dried Powders

This type of formulation is made by using soil and carriers which can be organic, inorganic, synthetic but inert in nature. Carrier plays a crucial role as a vehicle for carrying viable biofertilizers from the factory to the field in healthy condition (Bashan et al. 2014).

Biofertilizers in any formulation (with or without carrier) are commonly applied to seeds or seedlings and soil. In general, irrespective of varying inoculating methods, the number of live microbes reaching the soil from various biofertilizers formulation is always much lesser than the number of microflora present in natural soil or rhizosphere. It is thus necessary for the exogenously added microbes to undergo multiplication to express useful characters for the benefits of plants. Among the numerous methods of application of biofertilizers, seed inoculation has gained worldwide popularity (Khan & Zaidi 2007). During seed inoculation, seeds are soaked with liquid culture to allow the coating of microbes on the surface. This coating process has been reported to be improved in the presence of adhesive substances like gum arabic (Khan et al. 2010). However, there is an alternative method for the application of biofertilizers known as soil application in which the microbial inoculums are introduced directly into the soil. This method offers some merits over seed inoculation such as; i) there is a possibility of larger microbial population per unit area, ii) chances of contact with chemically treated seeds are greatly reduced, iii) it is a fast process as soaking of seeds with liquid cultures is no longer required, and iv) inoculants can better tolerate low moisture surroundings as compared to carrier-based inoculants (Khan et al. 2010). Seedling root dip is another method of biofertilizers application where the microbial inoculants are suspended in water and the seedling roots are dipped in the suspension for some time (20-30 min) before transplantation in the field (Mohapatra et al. 2013).

BENEFITS

Increase in Crop Yields

Biofertilizers increase crop yield by enhancing the growth of plants in sustainable agriculture. Biofertilizers contain natural components which are not harmful to the plants. Microbial biofertilizers provide nutrients to the plants in two different ways; first by secreting plant growth promoting substances and secondly by converting the complex substances into usable forms. Biofertilizers release nutrients slowly in the soil and contribute to the residual pool of organic N, P and K, thereby reducing the leaching of nutrients. These compounds increase crop yields by increasing shoot development and root proliferation. Generally, biofertilizers are known to increase crop yield by 15-35%. For example, algal biofertilizers enhanced the rice crop yields at rates ranging from 10-45% (Bhardwaj et al. 2014; Carvajal-Muñoz & Carmona-Garcia 2012). Microbial biofertilizers increase the crop yield by nitrogen fixation, phosphate solubilization, mineralization of soil and production of siderophores, phytohormones, and specific enzymes (Kumar et al. 2017).

Soil Structure

The use of microbial biofertilizers improves the soil structure by influencing the aggregation of the soil particles. As a result of the good structure provided to the soil, the root growth of the plant is promoted. Biofertilizers strengthen the soil profile, leave water sources untainted and improve plant growth without detrimental side effects. Biofertilizers also increase the organic matter content of the soil (Gabriela et al. 2015).

Improve Water Holding Capacity

Microbial colonization induces tolerance to the drought in plants by improving leaf water and turgor pressure, maintaining stomatal functioning and transpiration, and increasing the root length and development. Biofertilizers increase the soil organic matter, therefore improving the water retention and buffering capacity of the soil against acidity, alkalinity, salinity, pesticides and toxic chemicals (Mills 2006).

Cost-Effective

Biofertilizers are cost-effective and easy to use. Due to low cost, even low-income group farmers can also use in the crop fields. Biofertilizers can be prepared from easily available organic materials such as cow dung, rice husks, soil, bamboo and vegetables etc. They can reduce the input expenses by replacing the cost of chemical fertilizers. Farmers can prepare the inoculums themselves (Raja 2013).

Eco-Friendly

Biofertilizers are eco-friendly because the components of biofertilizers are not harmful to the plants and environment. Biofertilizers can replace chemical fertilizers, as the latter is beneficial for plants but cause various hazardous to plants. Additionally, in many cases, chemical fertilizers decrease plant growth and pollute the environment by releasing harmful chemicals. Microbes-based biofertilizers also play important roles in the cycling of nutrients. Plants and microbes are important components of the ecosystem and their interactions help in regulating the biogeochemical cycle in the environment (Delong 2009; Kour et al., 2020). If the soil is free of undesirable chemicals, it will retain its fertility, which will be beneficial for the plants as well as the environment.

Biocontrol Agent

Certain biofertilizers also function as biocontrol agent suppressing soil-borne diseases. They destroy the harmful disease causing agents from the soil. By the use of biofertilizers, plants can also be protected against drought and other abiotic stresses. Biofertilizers contain various microbes (bacteria, fungus or protozoan) which can be utilized to target a specific type of pest. *Azotobacter* enrich the soil with antibiotic pesticide and inhibit the spread of soil-borne pathogens like *Pythium* and *Phytophthora*. *Bacillus thuringiensis* is used as an important biocontrol agent because it synthesizes different proteins which target specific insect larvae (Kumar et al., 1996; Lemaux, 2008; Aloo et al., 2019).

Tolerance to Abiotic Stresses

Abiotic stresses such as heavy metal contamination in soils, nutrient deficiency, drought, salinity and temperature are considered to be the major constraints that are reducing crop yield. *Pseudomonas fluorescens* MSP-393 has been shown to produce osmolytes and salt-stress induced proteins that combat the harmful effects of salt (Paul & Nair 2008). Egamberdiyeva (2007) has demonstrated that calcisol produced by PGPRs viz., *Bacillus polymyxa*, *P. alcaligenes* and *Mycobacterium phlei* confers protection against high temperatures and salinity stress. Moreover, the application of arbuscular mycorrhiza to rice plants exposed to drought stress resulted in increased photosynthetic efficacy and antioxidative response (Ruiz-Sanchez et al. 2010). Several strains of *Pseudomonas* have been reported to overcome the negative effects of cadmium in canola and barley plants through the production of indole-3-acetic acid (IAA), siderophore and 1-aminocyclopropane-1- carboxylate (ACC) deaminase (Baharlouei et al. 2011).

LIMITATIONS

Undoubtedly, biofertilizers are widely used and seem promising alternatives for synthetic fertilizers but few limitations need the attention of agricultural scientists and other researchers. Some of the important limitations are briefly mentioned below:

- The main limitation of the use of biofertilizers is quality assurance. Nutrient compositions, poor quality and the efficacy of most of the biofertilizers are still poorly explored. It is essential to monitor the quality of biofertilizers in terms of microbial inoculum number, activity of potential characters, viability, survival under field conditions etc to assure proper functioning in sustainable agriculture.
- There is a need to introduce certain markers in the potential biofertilizers to track their presence and growth in fields where they are applied. This is essential to assure the farmers/users about the validity of biofertilizers.

- Before the release of biofertilizers it must be ensured that these are capable to grow and retain the desired characters in all types of agro-climatic conditions. Alternatively, region specific species/ strains of biofertilizers should be developed.
- There is inadequate awareness about the use and benefits of biofertilizers among the users. Extensive awareness programs are needed for skill development in the production and application of biofertilizers. Users should be well informed about the benefits of biofertilizers.
- Simple and cost-effective methods for storage and application should be developed.
- Government should provide sufficient subsidiary for the production of biofertilizers so that it is available at low cost. NGO should be associated with promoting the uses and sale of biofertilizers.

CONCLUSION AND FUTURE REMARKS

Biofertilizers represent an indispensable strategy in modern agricultural practices as it emphasizes on efficient use of resources with the minimum generation of undesirable substances which are not only detrimental to human health but also adversely disrupt the ecological balance. The use of biofertilizers has substantially reduced our dependency on chemical fertilizers and thus playing important role in protecting the health of the soil. In recent years, biofertilizers have become an important constituent of integrated nutrient management and organic farming because these accomplish several mechanisms to promote plant growth and productivity. Moreover, they render land more suitable for cultivation by carrying out nutrient acquisition, regulating nitrogen and phosphorus cycles in the soil ecosystem and production of plant growth promoting compounds. Biofertilizers are also important as they release nutrients in a slow and controlled manner in the soil which reduces the loss of nutrients due to agricultural run-off. Besides this, recent advancements in molecular biotechnology have led to the production of improved strains of microbes for use as biofertilizers by overexpressing the gene involved in the biosynthesis of phytohormones/siderophores/antagonistic compounds etc. Although, an indefinite number of unexplored plant growth promoting microbes exists in the nature, there is a need to identify and characterize these beneficial strains to raise more potent strains. It would be worthwhile to apply the metagenomics approach to identify and exploit beneficial genes/enzymes which have remained unexplored to date. Furthermore, detailed knowledge of plant-microbe interactions and their genomics would be useful in developing crop-specific strains of biofertilizers. Altogether, the success of biofertilizers in achieving environment-friendly agro-ecosystem relies upon the proper selection of microbes, strain modification using genetic engineering, retaining their viability during storage and transportation and skillful formulation as well as application of these microbes in the field for future crop improvement.

ACKNOWLEDGMENT

DS is grateful to DST-INSPIRE, Department of Science and Technology, New Delhi, for the award of Junior Research Fellowship (DST/INSPIRE Fellowship/2014/296, IF140707). JK is thankful to University Grants Commission (UGC), New Delhi, for providing financial assistance (23/06/2013 (I) EU-V). AK is thankful to Department of Biotechnology and Indian Council of Agricultural Research, New Delhi, for continuously supporting works on plant growth promoting bacteria by providing financial supports in the form of research projects.

REFERENCES

Afzal, A., & Bano, A. (2008). *Rhizobium* and phosphate solubilizing bacteria improve the yield and phosphorus uptake in wheat (*Triticum aestivum* L.). *International Journal of Agriculture and Biology, 10*, 85–88.

Aggarwal, A., Kadian, N., Tanwar, A., Yadav, A., & Gupta, K. K. (2011). Role of arbuscular mycorrhizal fungi (AMF) in global sustainable development. *Journal of Applied and Natural Science, 3*(2), 340–351. doi:10.31018/jans.v3i2.211

Ahemad, M., & Kibret, M. (2014). Mechanisms and applications of plant growth promoting rhizobacteria: Current perspective. *Journal of King Saud University. Science, 26*(1), 1–20. doi:10.1016/j.jksus.2013.05.001

Ahmad, F., Ahmad, I., & Khan, M. S. (2008). Screening of free-living rhizospheric bacteria for their multiple plant growth promoting activities. *Microbiological Research, 163*(2), 173–181. doi:10.1016/j.micres.2006.04.001 PMID:16735107

Allito, B. B., Nana, E. M., & Alemneh, A. A. (2015). *Rhizobia* strain and host-legume interaction effects on nitrogen fixation and yield of grain legume: A review. *Molecular Soil Biology, 6*, 1–6.

Alloway, B. J. (2008). Zinc in soils and plant nutrition. Academic Press.

Aloo, B. N., Makumba, B. A., & Mbega, E. R. (2019). The potential of Bacilli rhizobacteria for sustainable crop production and environmental sustainability. *Microbiological Research, 219*, 26–39. doi:10.1016/j.micres.2018.10.011 PMID:30642464

Anandraj, M., Venugopal, M.N., Veena, S.S., Kumar, A., & Sarma, Y.R. (2001). Ecofriendly management of disease of species. *Indian Species, 38*, 28-31.

Aziz, G., Bajsa, N., & Haghjou, T. (2012). Abundance, diversity and prospecting of culturable phosphate solubilizing bacteria on soils under crop-pasture rotations in a no-tillage regime in Uruguay. *Applied Soil Ecology, 61*, 320–326. doi:10.1016/j.apsoil.2011.10.004

Baharlouei, K., Pazira, E., & Solhi, M. (2011). Evaluation of inoculation of plant growth-promoting rhizobacteria on cadmium. In *International Conference on Environmental Science & Technology*. IACSIT Press.

Bardin, S. D., Huang, H. C., Pinto, J., Amundsen, E. J., & Erickson, R. S. (2004). Biological control of *Pythium* damping-off of pea and sugar beet by *Rhizobium leguminosarum* bv. *viceae. Canadian Journal of Botany, 82*(3), 291–296. doi:10.1139/b04-003

Bashan, Y., de Bashan, L. E., Prabhu, S. R., & Hernandez, J. P. (2014). Advances in plant growth-promoting bacterial inoculant technology: Formulations and practical perspectives. *Plant and Soil, 378*(1-2), 1–33. doi:10.100711104-013-1956-x

Bhardwaj, D., Ansari, M. W., Sahoo, R. K., & Tuteja, N. (2014). Biofertilizers function as key player in sustainable agriculture by improving soil fertility, plant tolerance and crop productivity. *Microbial Cell Factories, 13*(1), 66. doi:10.1186/1475-2859-13-66 PMID:24885352

Burns, R. C., & Hardy, R. W. (2012). Nitrogen fixation in bacteria and higher plants. Springer Science & Business Media, 21, 192.

Carvajal-Muñoz, J. S., & Carmona-Garcia, C. E. (2012). Benefits and limitations of biofertilization in agricultural practices. *Livestock Research for Rural Development*, *24*, 43.

Catroux, G., Hartmann, A., & Revellin, C. (2001). Trends in rhizobial inoculant production and use. *Plant and Soil*, *230*(1), 21–30. doi:10.1023/A:1004777115628

Chen, J.-H. (2006). The combined use of chemical and organic fertilizers and/or biofertilizer for crop growth and soil fertility. In *Proceeding of International Workshop on Sustained Management of the Soil-Rhizosphere System for Efficient Crop Production and Fertilizer Use*. Land Development Department.

Chibuike, G. U., & Obiora, S. C. (2014). Heavy metal polluted soils: Effect on plants and bioremediation methods. *Applied and Environmental Soil Science*, *2014*, 752708. doi:10.1155/2014/752708

Chirinos, J., Leal, A., & Montilla, J. (2006). Uso de insumos biológicos como alternativa para la agricultura sostenible en la zona sur del Estado Anzoátegui. *Revista Digital Ceniap Hoy*, *11*, 1–7.

Choudhury, M. A., & Kennedy, I. R. (2004). Prospects and potentials for system of biological nitrogen fixation in sustainable rice production. *Biology and Fertility of Soils*, *39*(4), 219–227. doi:10.100700374-003-0706-2

Corpoica, Gobernación de Antioquia, & FAO. (2007). *Manual de buenas prácticas agrícolas en la producción de fríjol voluble*. Medellín: FAO.

Dastager, S. G., Deepa, C. K., & Pandey, A. (2010). Isolation and characterization of novel plant growth promoting *Micrococcus* sp NII-0909 and its interaction with cowpea. *Plant Physiology and Biochemistry*, *48*(12), 987–992. doi:10.1016/j.plaphy.2010.09.006 PMID:20951599

Delong, E. F. (2009). The microbial ocean from genomes to biomes. *Nature*, *459*(7244), 200–206. doi:10.1038/nature08059 PMID:19444206

DeRosa, M. R., Monreal, C., Schnitzer, M., Walsh, R., & Sultan, Y. (2010). Nanotechnology in fertilizers. *National Nanotechnology Journal*, *5*(2), 91. doi:10.1038/nnano.2010.2 PMID:20130583

Di Simine, C. D., Sayed, J. A., & Gadd, G. M. (1998). Solubilization of zinc phosphate by a strain of *Pseudomonas fluorescens* isolated from a forest soil. *Biology and Fertility of Soils*, *28*(1), 87–94. doi:10.1007003740050467

Dimkpa, C. O., & Bindraban, P. S. (2017). Nanofertilizers: New products for the industry? *Journal of Agricultural and Food Chemistry*. Advance online publication. doi:10.1021/acs.jafc.7b02150 PMID:28535672

Dogan, K., Kamail, C. I., Mustafa, G. M., & Ali, C. (2011). Effect of different soil tillage methods on rhizobial nodulation, biomass and nitrogen content of second crop soybean. *African Journal of Microbiological Research*, *5*(20), 3186–3194. doi:10.5897/AJMR11.165

dos Santos, R.M., Diaz, P.A.E, Lobo, L.L.B., & Rigobelo, E.C. (2020). Use of plant growth-promoting rhizobacteria in maize and sugarcane: characteristics and applications. *Frontiers in Sustainable Food Systems*, *4*(136), 1-15.

Egamberdieva, D., & Lugtenberg, B. (2014). Use of Plant Growth-Promoting Rhizobacteria to Alleviate Salinity Stress in Plants. In *Use of Microbes for the Alleviation of Soil Stresses* (Vol. 1, pp. 73–96). Springer. doi:10.1007/978-1-4614-9466-9_4

Egamberdiyeva, D. (2007). The effect of plant growth promoting bacteria on growth and nutrient uptake of maize in two different soils. *Applied Soil Ecology*, *36*(2-3), 184–189. doi:10.1016/j.apsoil.2007.02.005

Fasim, F., Ahmed, N., Parsons, R., & Gadd, G. M. (2002). Solubilization of zinc salts by a bacterium isolated from the air environment of a tannery. *FEMS Microbiology Letters*, *213*(1), 1–6. doi:10.1111/j.1574-6968.2002.tb11277.x PMID:12127480

Fasusi, O.A., Cruz, C., & Babalola, O.O. (2021). Agricultural sustainability: microbial biofertilizers in rhizosphere management. *Agriculture, 11*(163), 1-19.

Fay, P. (1983). *The blue-greens*. Edward Arnold.

Frey-Klett, P., Burlinson, P., Deveau, A., Barret, M., Tarkka, M., & Sarniguet, A. (2011). Bacterial-fungal interactions: Hyphens between agricultural, clinical, environmental, and food microbiologists. *Microbiology and Molecular Biology Reviews*, *75*(4), 583–609. doi:10.1128/MMBR.00020-11 PMID:22126995

Gabriela, F., Casanovas, E. M., & Quillehauquy, V. (2015). *Azospirillum* inoculation effects on growth, product quality and storage life of lettuce plants grown under salt stress. *Scientia Horticulturae*, *195*, 154–162. doi:10.1016/j.scienta.2015.09.015

Ghany, T. M. A., Alawlaqi, M. M., & Al Abboud, M. A. (2013). Role of biofertilizers in agriculture: A brief review. *Mycopathologia*, *11*, 95–101.

Glick, B.R. (2012). Plant growth-promoting bacteria: mechanisms and applications. *Scientifica*, 15.

Gray, E. J., & Smith, D. L. (2005). Intracellular and extracellular PGPR: Commonalities and distinctions in the plant-bacterium signaling processes. *Soil Biology & Biochemistry*, *37*(3), 395–412. doi:10.1016/j.soilbio.2004.08.030

Han, H. S., Supanjani, E., & Lee, K. D. (2006). Effect of co-inoculation with phosphate and potassium solubilizing bacteria on mineral uptake and growth of pepper and cucumber. *Plant, Soil and Environment*, *52*(3), 130–136. doi:10.17221/3356-PSE

Herrmann, L., & Lesueur, D. (2013). Challenges of formulation and quality of biofertilizers for successful inoculation. *Applied Microbiology and Biotechnology*, *97*(20), 8859–8873. doi:10.100700253-013-5228-8 PMID:24037408

Ilyas, N., Bano, A., Iqbal, S., & Raja, N. I. (2012). Physiological, biochemical and molecular characterization of *Azospirillum* sp. isolated from maize under water stress. *Pakistan Journal of Botany*, *44*, 71–80.

Jelin, J., Dhanrajan, M. S., & Mariappan, V. (2013). Assessment of compost as a bio-fertilizer for the growth of paddy. *Journal of Environmental Biology*, *34*, 975–979. PMID:24555324

Karthikeyan, N., Prasanna, R., Lata, N., & Kaushik, B. D. (2007). Evaluating the potential of plant growth promoting cyanobacteria as inoculants for wheat. *European Journal of Soil Biology*, *43*(1), 23–30. doi:10.1016/j.ejsobi.2006.11.001

Kaushik, B. D. (2013). Developments in cyanobacterial biofertilizer. *Proceeding of Indian Natural Science Academy*, *80*(2), 379–388. doi:10.16943/ptinsa/2014/v80i2/55115

Khan, M. S., & Zaidi, A. (2007). Synergistic effects of the inoculation with plant growth promoting rhizobacteria and arbuscular mycorrhizal fungus on the performance of wheat. *Turkish Journal of Agriculture and Forestry*, *31*, 355–362.

Khan, M. S., Zaidi, A., Ahemad, M., Oves, M., & Wani, P. A. (2010). Plant growth promotion by phosphate solubilizing fungi- current perspective. *Archives of Agronomy and Soil Science*, *56*(1), 73–98. doi:10.1080/03650340902806469

Khande, R., Sushil, K. S., Ramesh, A., & Mahaveer, P. S. (2017). Zinc solubilizing *Bacillus* strains that modulate growth, yield and zinc biofortification of soybean and wheat. *Rhizosphere*, *4*, 126–138. doi:10.1016/j.rhisph.2017.09.002

Kogel, K. H., Franken, P., & Huckelhoven, R. (2006). Endophyte or parasite- what decides? *Current Opinion in Plant Biology*, *9*(4), 358–363. doi:10.1016/j.pbi.2006.05.001 PMID:16713330

Kour, D., Rana, K.L., & Yadav, A.N. (2020). Microbial biofertilizers: bioresources and eco-friendly technologies for agricultural and environmental sustainability. *Biocatalysis and Agricultural Biotechnology, 23*, 1-11.

Kucey, R. M. N. (1988). Effect of *Penicillium billai* on the solubility and uptake of P and micronutrients from soil by wheat. *Canadian Journal of Soil Science*, *68*, 261–270. doi:10.4141/cjss88-026

Kumar, J., Singh, D., Ghosh, P., & Kumar, A. (2017). Endophytic and epiphytic modes of microbial interactions and benefits. In *Plant-Microbe Interactions in Agro-ecological Perspectives* (vol. 2, pp. 227–253). doi:10.1007/978-981-10-5813-4_12

Lanning, F. C., Ponnaiya, B. W. X., & Crumpton, C. F. (1958). The chemical nature of silica in plants. *Plant Physiology*, *33*(5), 339–343. doi:10.1104/pp.33.5.339 PMID:16655143

Lauwers, A. M., & Heinen, W. (1974). Biodegradation and utilization of silica and quartz. *Archives of Microbiology*, *95*(1), 67–78. doi:10.1007/BF02451749 PMID:4365644

Lebedeva, E., Nazarenko, A. V., & Vedenyapina, E. G. (1999). Soil micromycetes in the area of "Severstal" metallurgical works (Cherepovets, Russia). *Botanica Lithuanica*, *3*, 107–110.

Lehmann, A., Veresoglou, S. D., Leifheit, E. F., & Rillig, M. C. (2014). Arbuscular mycorrhizal influence on zinc nutrition in crop plants- a meta-analysis. *Soil Biology & Biochemistry*, *69*, 123–131. doi:10.1016/j.soilbio.2013.11.001

Lemaux, P. G. (2008). Genetically engineered plants and foods: A scientist's analysis of the issues. *Annual Review of Plant Biology*, *59*(1), 771–812. doi:10.1146/annurev.arplant.58.032806.103840 PMID:18284373

Mahanty, T., Bhattacharjee, S., Goswami, M., Bhattacharyya, P., Das, B., Ghosh, A., & Tribedi, P. (2017). Biofertilizers : A potential approach for sustainable agriculture development. *Environmental Science and Pollution Research International*, *24*(4), 3315–3335. doi:10.100711356-016-8104-0 PMID:27888482

Mahdi, S. S., Talat, M. A., Hussain, D. M., Hamid, A., & Ahmad, L. (2012). Soil phosphorus fixation chemistry and role of phosphate solubilizing bacteria in enhancing its efficiency for sustainable cropping- a review. *Journal of Pure & Applied Microbiology*, *6*, 1–7.

Maillet, F., Poinsot, V., André, O., Puech-Pagès, V., Haouy, A., Gueunier, M., Cromer, L., Giraudet, D., Formey, D., Niebel, A., Martinez, E. A., Driguez, H., Bécard, G., & Dénarié, J. (2011). Fungal lipochitooligosaccharide symbiotic signals in arbuscular mycorrhiza. *Nature*, *469*(7328), 58–63. doi:10.1038/nature09622 PMID:21209659

Maleva, M., Borisova, G., Koshcheeva, O., & Sinenko, O. (2017). Biofertilizer based on silicate solubilizing bacteria improves photosynthetic function of *Brassica juncea*. *AGROFOR International Journal*, *2*, 13–19.

Malusa, E., Sas-Paszt, L., & Ciesielska, J. (2012). Technologies for beneficial microorganisms inocula used as biofertilizers. *TheScientificWorldJournal*, *2012*, 491206. doi:10.1100/2012/491206 PMID:22547984

Martin, X. M., Sumathi, C. S., & Kannan, V. R. (2011). Influence of agrochemicals and *Azotobacter* sp. application on soil fertility in relation to maize growth under nursery conditions. *Eurasian Journal of Biosciences*, *5*, 19–28. doi:10.5053/ejobios.2011.5.0.3

Mathivanan, R., Umavathi, S., Ramasamy, P. K., & Thangam, Y. (2015). Influence of vermicompost on the activity of the plant growth regulators in the leaves of the Indian butter bean plant, *Dolichos lab lab* L. *International Journal of Advanced Research in Biological Sciences*, *2*, 84–89.

Mazid, M., & Khan, T. A. (2015). Future of bio-fertilizers in Indian agriculture: An overview. *Irish Journal of Agricultural and Food Research*, *3*, 10–23.

McComb, R. B., Bowers, G. N. Jr, & Posen, S. (2013). *Alkaline phosphatase*. Springer Science and Business Media.

Mehdipour-Moghaddam, M. J., Emtiazi, G., & Salehi, Z. (2012). Enhanced auxin production by *Azospirillum* pure cultures from plant root exudates. *Journal of Agricultural Science and Technology*, *14*, 985–994.

Mehnaz, S. (2015). *Azospirillum*: a biofertilizer for every crop. In *Plant microbes symbiosis: applied facets* (pp. 297–314). Springer.

Mills, T. (2006). Composting cafeteria residuals with earthworm. *BioCycle*, *47*, 54–55.

Mishra, D. J., Singh, R., Mishra, U. K., & Kumar, S. S. (2013). Role of bio-fertilizer in organic agriculture: A review. *Research Journal of Recent Sciences*, *2*, 39–41.

Mishra, P., & Dash, D. (2014). Rejuvenation of biofertilizer for sustainable agriculture and economic development. *Consilience: The Journal of Sustainable Development*, *11*, 41–61.

Mohammadi, K., & Sohrabi, Y. (2012). Bacterial biofertilizers for sustainable crop production: A review. *Journal of Agricultural and Biological Science*, *7*, 307–316.

Mohapatra, B., Verma, D. K., Sen, A., Panda, B. B., & Asthir, B. (2013). Bio-fertilizers- a gateway to sustainable agriculture. *Popular Kheti*, *1*, 97–106.

Moraditochaee, M., Azarpour, E., & Bozorgi, H. R. (2014). Study effects of bio- fertilizers, nitrogen fertilizer and farmyard manure on yield and physiochemical properties of soil in lentil farming. *International Journal of Biosciences*, *4*, 41–48.

Naik, K., Mishra, S., Srichandan, H., Singh, P.K., & Sarangi, P.K. (2019). Plant growth promoting microbes: Potential link to sustainable agriculture and environment. *Biocatalysis and Agricultural Biotechnology, 21*, 1-12.

Naiman, A. D., Latrónico, A., & de Salamone, I. E. (2009). Inoculation of wheat with *Azospirillum brasilense* and *Pseudomonas fluorescens*: Impact on the production and culturable rhizosphere microflora. *European Journal of Soil Biology*, *45*(1), 44–51. doi:10.1016/j.ejsobi.2008.11.001

Narimani, H., Rahimi, M. M., Ahmadikhan, A., & Vaezi, B. (2010). Study on the effects of foliar spray of micronutrient on yield and yield components of durum wheat. *Applied Scientific Research*, *2*, 168–176.

Naureen, Z., Aqeel, M., & Hassan, M. N. (2015). Isolation and screening of silicate bacteria from various habitats for biological control of phytopathogenic fungi. *American Journal of Plant Sciences*, *6*, 2850–2859. doi:10.4236/ajps.2015.618282

Nosheen, S., Ajmal, I., & Song, Y. (2021). Microbes as biofertilizers, a potential approach for sustainable crop production. *Sustainability, 13*, 1-20.

Ogbo, F. C. (2010). Conversion of cassava wastes for biofertilizer production using phosphate solubilizing fungi. *Bioresource Technology*, *101*(11), 4120–4124. doi:10.1016/j.biortech.2009.12.057 PMID:20138509

Pandey, S., Shrivastava, A. K., Rai, R., & Rai, L. C. (2013). Molecular characterization of Alr1105 a novel arsenate reductase of the diazotrophic cyanobacterium *Anabaena* sp. PCC7120 and decoding its role in abiotic stress management in *Escherichia coli*. *Plant Molecular Biology*, *83*(4-5), 417–432. doi:10.100711103-013-0100-x PMID:23836391

Paul, D., & Nair, S. (2008). Stress adaptations in a plant growth promoting Rhizobacterium (PGPR) with increasing salinity in the coastal agricultural soils. *Journal of Basic Microbiology*, *48*(5), 1–7. doi:10.1002/jobm.200700365 PMID:18702073

Peera, S. K. P. G., Balasubramaniam, P., & Mahendran, P. P. (2016). Effect of silicate solubilizing bacteria and fly ash on silicon uptake and yield of rice under lowland ecosystem. *Journal of Applied and Natural Science*, *8*(1), 55–59. doi:10.31018/jans.v8i1.746

Pellegrino, E., & Bedini, S. (2014). Enhancing ecosystem services in sustainable agriculture: Biofertilization and biofortification of chickpea (*Cicer arietinum* L.) by arbuscular mycorrhizal fungi. *Soil Biology & Biochemistry*, *68*, 429–439. doi:10.1016/j.soilbio.2013.09.030

Plaster, E. (2000). *La ciencia del suelo y su manejo*. Thompson Editores.

Popkin, R. (1995). Good news for waste watchers: Recycling, composting show results for the future. *Environmental Protection Agency Journal*, *21*, 188–190.

Prasad, R. C., & Prasad, B. N. (2001). Cyanobacteria as a source biofertilizer for sustainable agriculture in Nepal. *Botanica Orientalis*. *Journal of Plant Sciences*, 127–133.

Raja, N. (2013). Biopesticides and biofertilizers : Ecofriendly sources for sustainable agriculture. *Journal of Biofertilizers & Biopesticides*, *4*(1), 1–2. doi:10.4172/2155-6202.1000e112

Raliya, R., Saharan, V., Dimkpa, C., & Biswas, P. (2017). Nanofertilizer for precision and sustainable agriculture: Current state and future perspectives. *Journal of Agricultural and Food Chemistry*. Advance online publication. doi:10.1021/acs.jafc.7b02178 PMID:28835103

Ramachandran, V. K., East, A. K., Karunakaran, R., Downie, J. A., & Poole, S. P. (2011). Adaptation of *Rhizobium leguminosarum* to pea, alfalfa and sugar beet rhizosphere investigated by comparative transcriptomics. *Genome Biology*, *12*(10), 106–109. doi:10.1186/gb-2011-12-10-r106 PMID:22018401

Rivera-Cruz, M., Trujillo, A., Córdova, G., Kohler, J., Caravaca, F., & Roldán, A. (2008). Poultry manure and banana waste are effective bio-fertilizer carriers for promoting plant growth and soil sustainability in banana crops. *Soil Biology & Biochemistry*, *40*(12), 3092–3095. doi:10.1016/j.soilbio.2008.09.003

Rodr'ıguez, H., & Fraga, R. (1999). Phosphate solubilizing bacteria and their role in plant growth promotion. *Biotechnology Advances*, *17*(4-5), 319–339. doi:10.1016/S0734-9750(99)00014-2 PMID:14538133

Rodrignes, F. A., & Datnoff, L. E. (2005). Silicon and rice disease management. *Fitopatologia Brasileira*, *30*(5), 457–469. doi:10.1590/S0100-41582005000500001

Roychowdhury, D. E., Paul, M. A., & Banerjee, S. K. (2014). A review on the effects of biofertilizers and biopesticides on rice and tea cultivation and productivity. *International Journal of Science. Engineering & Technology*, *2*, 96–106.

Ruiz-Sanchez, M., Aroca, R., Munoz, Y., Polon, R., & Ruiz-Lozano, J. M. (2010). The arbuscular mycorrhizal symbiosis enhances the photosynthetic efficiency and the antioxidative response of rice plants subjected to drought stress. *Journal of Plant Physiology*, *167*(11), 862–869. doi:10.1016/j.jplph.2010.01.018 PMID:20227134

Sahu, D., Priyadarshani, I., & Rath, B. (2012). Cyanobacteria- as potential biofertilizer. *CIBTech Journal of Microbiology*, *1*, 20–26.

Santi, C., Bogusz, D., & Franche, C. (2013). Biological nitrogen fixation in non-legume plants. *Annals of Botany*, *111*(5), 743–767. doi:10.1093/aob/mct048 PMID:23478942

Saravanan, V. S., Kumar, M. R., & Sa, T. M. (2011). Microbial zinc solubilization and their role on plants. In D. K. Maheshwari (Ed.), *Bacteria in Agrobiology, Plant Nutrient Management* (pp. 47–63). Springer. doi:10.1007/978-3-642-21061-7_3

Saravanan, V. S., Madhaiyan, M., & Thangaraju, M. (2007). Solubilization of zinc compounds by the diazotrophic, plant growth promoting bacterium *Gluconacetobacter diazotrophicus*. *Chemosphere*, *66*(9), 1794–1798. doi:10.1016/j.chemosphere.2006.07.067 PMID:16956644

Savci, S. (2012). An agricultural pollutant: Chemical fertilizer. *International Journal of Environmental Sciences and Development*, *3*, 73–80. doi:10.7763/IJESD.2012.V3.191

Sekhon, S. (2014). Nanotechnology in agri-food production: An overview. *Nanotechnology, Science and Applications*, *7*, 31–53. doi:10.2147/NSA.S39406 PMID:24966671

Shamseldin, A. (2013). The role of different genes involved in symbiotic nitrogen fixation-review. *Global Journal of Biotechnology & Biochemistry*, *8*, 84–94.

Sharma, P., Sardana, V., & Kandola, S. S. (2011). Response of groundnut (*Arachis hypogaea* L.) to *Rhizobium* inoculation. *Libyan Agriculture Research Centre Journal Interaction*, *2*, 101–104.

Singh, D., Ghosh, P., Kumar, J., & Kumar, A. (2019). Plant growth-promoting rhizobacteria (PGPRs): functions and benefits. In D. P. Singh, V. K. Gupta, & R. Prabha (Eds.), *Microbial Interventions in Agriculture and Environment* (pp. 205–227). doi:10.1007/978-981-13-8383-0_7

Singh, J. S., Kumar, A., Rai, A. N., & Singh, D. P. (2016). Cyanobacteria: A precious bio-resource in agriculture, ecosystem, and environmental sustainability. *Frontiers in Microbiology*, *7*, 1–19. doi:10.3389/fmicb.2016.00529 PMID:27148218

Smith, B. E., Richards, R. L., & Newton, W. E. (2013). Catalysts for nitrogen fixation: nitrogenases, relevant chemical models and commercial processes. Springer Science & Business Media.

Souchie, E. L., Saggin-Ju'nior, O. J., Silva, E. M. R., Campello, E. F. C., Azco'n, R., & Barea, J. M. (2006). Communities of P-solubilizing bacteria, fungi and arbuscular mycorrhizal fungi in grass pasture and secondary forest of Paraty, RJ- Brazil. *Anais da Academia Brasileira de Ciências*, *78*(1), 183–193. doi:10.1590/S0001-37652006000100016 PMID:16532216

Steenhoudt, O., & Vanderleyden, J. (2000). *Azospirillum*, a free living nitrogen-fixing bacterium closely associated with grasses: Genetic, biochemical and ecological aspects. *FEMS Microbiology Reviews*, *24*(4), 487–506. doi:10.1111/j.1574-6976.2000.tb00552.x PMID:10978548

Sujanya, S., & Chandra, S. (2011). Effect of part replacement of chemical fertilizers with organic and bio-organic agents in ground nut, *Arachis hypogea*. *Journal of Algal Biomass Utilization*, *2*, 38–41.

Suzaki, T., Yoro, E., & Kawaguchi, M. (2015). Leguminous plants: Inventors of root nodules to accommodate symbiotic bacteria. *International Review of Cell and Molecular Biology*, *316*, 111–158. doi:10.1016/bs.ircmb.2015.01.004 PMID:25805123

Tanu, P. A., & Adholeya, A. (2004). Effect of different organic manures/composts on the herbage and essential oil yield of *Cymbopogon winterianus* and their influence on the native AM population in a marginal alfisol. *Bioresource Technology, 92*, 311-319.

Tisserant, E., Malbreil, M., Kuo, A., Kohler, A., Symeonidi, A., Balestrini, R., Charron, P., Duensing, N., Frei dit Frey, N., Gianinazzi-Pearson, V., Gilbert, L. B., Handa, Y., Herr, J. R., Hijri, M., Koul, R., Kawaguchi, M., Krajinski, F., Lammers, P. J., Masclaux, F. G., ... Martin, F. (2013). Genome of an arbuscular mycorrhizal fungus provides insight into the oldest plant symbiosis. *Proceedings of the National Academy of Sciences of the United States of America*, *110*(50), 20117–20122. doi:10.1073/pnas.1313452110 PMID:24277808

Vanegas, R. (2003). *La transición: la búsqueda del cambio hacia sistemas sustentables de producción agropecuaria. Foro Agrario «Sistemas Agrícolas Sustentables.* Quito Universidad Central del Ecuador.

Vasanthi, N., Saleena, L. M., & Raj, S. A. (2012). Silicon in day today life. *World Applied Sciences Journal*, *17*, 1425–1440.

Vasanthi, N., Saleena, L. M., & Raj, S. A. (2016). Silica solubilization potential of certain bacterial species in the presence of different silicate minerals. *Silicon*, *10*(2), 267–275. doi:10.100712633-016-9438-4

Venkataraman, G. S., & Neelakantan, S. (1967). Effect of cellular constituents of the nitrogen fixing blue-green alga *Cylindrospermum musciola* on the root growth of rice seedlings. *The Journal of General and Applied Microbiology*, *13*(1), 53–61. doi:10.2323/jgam.13.53

Venkateswarlu, B., Rao, A. V., Raina, P., & Ahmad, N. (1984). Evaluation of phosphorus solubilization by microorganisms isolated from arid soil. *Journal of the Indian Society of Soil Science*, *32*, 273–277.

Verma, J. P., Yadav, J., Tiwari, K. N., Lavakush, S. V., & Singh, V. (2010). Impact of plant growth promoting rhizobacteria on crop production. *International Journal of Agricultural Research*, *5*(11), 954–983. doi:10.3923/ijar.2010.954.983

Vessey, J. K. (2003). Plant growth promoting rhizobacteria as biofertilizers. *Plant and Soil*, *255*(2), 571–586. doi:10.1023/A:1026037216893

Wakatsuki, T. (1995). Metal oxidoreduction by microbial cells. *Journal of Industrial Microbiology*, *14*(2), 169–177. doi:10.1007/BF01569900 PMID:7766210

Wakelin, S. A., Warren, R. A., Harvey, P. R., & Ryder, M. H. (2004). Phosphate solubilization by *Penicillium* sp. closely associated with wheat roots. *Biology and Fertility of Soils*, *40*(1), 36–43. doi:10.100700374-004-0750-6

Wani, S. A., Chand, S., & Ali, T. (2013). Potential use of *Azotobacter chroococcum* in crop production: An overview. *Current Agriculture Research Journal*, *1*(1), 35–38. doi:10.12944/CARJ.1.1.04

Whitelaw, M. A. (1999). Growth promotion of plants inoculated with phosphate-solubilizing fungi. *Advances in Agronomy*, *69*, 99–151. doi:10.1016/S0065-2113(08)60948-7

Yadav, S. P. (2002). Performance of effective microorganisms (EM) on growth and yields of selected vegetables. *Nature Farming & Environment*, *3*, 35–38.

Youssef, M. M., & Ali, M. S. (1998). Management of *Meloidogyne incognita* infecting cowpea by using some native blue green algae. *Anzeiger für Schädlingskunde, Pflanzenschutz. Umweltschutz*, *71*, 15–16.

Youssef, M. M., & Eissa, M. F. (2014). Biofertilizers and their role in management of plant parasitic nematodes. A review. *E3 Journal of Biotechnology & Pharmaceutical Research*, *5*, 1–6.

Zhang, W., Han, D. Y., Dick, W. A., Davis, K. R., & Hoitink, H. A. J. (1998). Compost and compost water extract-induced systemic acquired resistance in cucumber and *Arabidopsis*. *Phytopathology*, *88*(5), 450–454. doi:10.1094/PHYTO.1998.88.5.450 PMID:18944926

Chapter 11
Green Synthesis Technology (GST):
A Potential Remedy for Nanotoxicity

Manjul Mungali
https://orcid.org/0000-0002-6620-9005
Swami Vivekanand Subharti University, Meerut, India

Amit Kumar
Swami Vivekanand Subharti University, Meerut, India

Nisha Malik
Swami Vivekanand Subharti University, Meerut, India

ABSTRACT

Nanotechnology offers new materials and applications that will benefit society, yet there is growing concern about the potential health and environmental impacts of the production and use of nanoscale products. Although hundreds of studies of nanomaterial hazards have been reported, in green nanotechnology, nanomaterials or nanoparticles are synthesized or developed by biological approaches, such as biogenesis (biosynthesis) that has vast applications in the biomedical, food, and agricultural fields. Additionally, green nanotechnology offers many potential benefits, such as the enhancement of biomedical diagnostics, improved food quality and safety, reduction of agricultural inputs, and enrichment of nanoscale nutrient absorption from the soil. The green synthesis method, which is an alternative to chemical and physical methods, provides an eco-friendly way of synthesizing nanoparticles as this method does not require costly, harmful, and toxic chemicals.

DOI: 10.4018/978-1-7998-8434-7.ch011

INTRODUCTION

Nanotechnology

Nanotechnology is an enabling technology that is expected to result in major changes across many industry sectors and to contribute to novel materials, devices and products. Depending on the area of application there are different timelines for the beginning of industrial prototyping and nanotechnology commercialization. First generation products are already in the market such as paints, coatings and cosmetics. More products such as pharmaceuticals, diagnostics and applications in energy storage and production are in development. Many studies have tried to estimate the prospect of the nanotechnology market with different data.

Nanotechnologies have become a significant priority worldwide. Several manufactured nanoparticles - particles with one dimension less than 100 nm – are increasingly used in consumer products. At nanosize range, the properties of materials differ substantially from bulk materials of the same composition, mostly due to the increased specific surface area and reactivity, which may lead to increased bioavailability and toxicity.

Some natural particles are of nano-scale such as colloidal humus and ultrafine particles in atmospheric emissions. Environmental nanoparticles are commonly formed as either weathering byproducts of minerals, as biogenic products of microbial activity, or as growth nuclei in super-saturated fluids. Manufactured nanoparticles can be inorganic like nanopowders of metal oxides and metal salts like CdS (quantum dots) or organic chemicals and polymers like dendrimers (Ryan, J.N. and Elimelech, M, 1996, Lead, J.R. and Wilkinson, K.J. 2006, Nel, A. et al., 2006).

Among potential environmental applications of nanoparticles, remediation of contaminated groundwater with nanoscale iron is one of the most prominent example. Regarding personal-care products, nanoparticles of titanium dioxide and zinc oxide are included in toothpaste, beauty products, sunscreens and textiles (Yuranova, T et al., 2007). Metal oxide-based nanomaterials and/or nanoparticles are also increasingly used in fillers, opacifiers, ceramics, coatings, catalysts, semiconductors, microelectronics, prosthetic implants and drug carriers. Photocatalytic properties of TiO2 may be used for solar-driven self-cleaning coatings and for biocidal/antiproliferative applications (Cai, R et al., 2006, Blake, D.M. et al., 1999). In the form of manufacturing and household waste the metal oxide nanoparticles are likely to end up in natural water bodies. For perspective on potential "nanopollution", one may consider that 2 g of 100 nm size nanoparticles contains enough material to provide every human worldwide with 300,000 particles each (Chang, H., et al., 2005, Hardman. R. 2006). Nanoparticles can cross even the strongest biological barrier such as blood-brain barrier (Oberdörster, E. 2004, Lockman, P. et al., 2003). The chemical synthesis method of nanoparticles is depicted in figure 1.

As an example, in a literature the researcher showed that exposure to fullerenes (C60) caused oxidative damage in the brain of fish. Despite that, nanosized materials were till recently treated as variations of the technical material or existing formulation and thus not requiring separate registration (Oberdörster, E. 2004).

Various studies have shown that, because of their small size, unbound nanoparticles can be inhaled and enter the bloodstream via the lungs, disperse throughout the body and penetrate other organs, however it should be noted that many of these studies use instillation rather than inhalation as a way to administer nanoparticles, and generally in a greater quantity dose than reflects actual practical circumstances. It has also been shown that, depending on their properties, some manufactured nanoparticles can be damaging to cells. Little data is available on the toxicology, release, environmental behavior and safety of nanomaterials.

Nanotechnologies and manufactured nanomaterials, as with any new technology, may bring many advances to society and benefits for the environment, but also pose new challenges in health, environment safety and possible impacts on society. Because of the very broad range of potential applications using nanotechnology and the wide variety of characteristics displayed by manufactured nanomaterials, detailed discussion of both benefits and of health and environmental risks should take place at the level of individual nanotechnology applications.

Figure 1. Different approach of nanoparticles synthesis

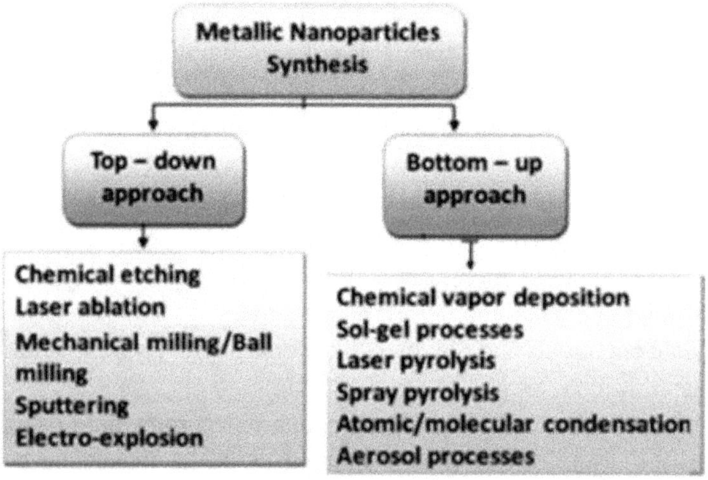

Figure 2. Chemically prepared gold nanoparticles (AuNPs) suspended in liquid

Nanoparticles toxicity (Nanotoxicity)

The toxicity of Nanoparticles can be attributed to the features below:

1. Particle size and surface area are important material characteristics from a toxicological perspective. As the size of a particle decreases, its surface area increases and also allows a greater proportion of its atoms or molecules to be displayed on the surface rather than the interior of the material. The change in the physicochemical and structural properties of engineered nano material with a decrease in size could be responsible for a number of material interactions that could lead to toxicological effects.
2. Surface charge of the particle is responsible for electrostatic interactions.
3. Hydrophobicity and sometimes lipophilic groups which may allow the particle to interact with proteins and membranes respectively.
4. Nanostructure could cause inhibition of enzyme activity either competitive or noncompetitive.
5. Accumulation of an inert particle in the body could also trigger tissue formation around the foreign entity and thus leading to formation of a scar tissue (K. P. Singh and Manjul Mungali 2011)

Kinetics of Nanoparticles

Active cell uptake includes endocytosis or phagocytosis by specialised cells while NPs can also penetrate nonphagocytic mammalian cell membranes like that of red blood cells and become internalized at vesicles and mitochondria causing serious cell damage and cell death, figure 3 represented the pathway for nanoparticle entry into the cell (Moore, 2006). One characteristic common to both organic and inorganic NPs is that finer the NP, the stronger the adverse health response seems to be. NP size (surface area), surface charge and chemical nature influence the severity of adverse health reactions (Reijnders, 2006). Although some of the effects of NP pollution on human health have similar consequences whether caused by organic or inorganic NPs. Due to the different physicochemical surface characteristics of organic and inorganic NPs, regulations and control mechanisms for the protection of the environment may differ substantially in some respects depending on the type of NP.

The fate of nanomaterials in aqueous environment is controlled by many biotic/abiotic processes such as solubility/dispersability, interactions between the nanomaterials and natural/anthropogenic chemicals in the ecosystem. In addition, ecological risk assessment is essential to understand environmental implications of nanomaterials. Before unknowingly dumping a huge amount of dangerous nanomaterials into the environment, we need to investigate the solubility and degradability of engineered NPs in soils and waters, to establish baseline information on their safety, toxicity and adaptation of soil and aquatic life (K. P. Singh and Manjul Mungali 2011).

Green Synthesis Technology (GST)

The Green synthesis method is an option to chemical and physical methods, which provides an eco friendly way of synthesizing nanoparticles. This method does not require costly, harmful and toxic chemicals. Metallic nanoparticles with diverse shapes, sizes, contents and physicochemical properties can be synthesized by this alternative method. In green synthesis method we use biological organisms such as bacteria, actinobacteria, yeasts, molds, algae and plants, or their products. Nanoparticle synthesis is achieved by molecules in biological organisms, such as proteins, enzymes, phenolic compounds, amines, alkaloids and pigments by reduction of metallic compounds. Concerning the environmental protection, green synthesis of nanoparticle has gained friendly and growing demand. It has vast applications in the biomedical, food, and agricultural fields. Additionally, green nanotechnology offers many potential benefits, such as the enhancement of biomedical diagnostics, improved food quality and safety, reduction of agricultural inputs, and enrichment of nanoscale nutrient absorption from the soil. The green synthesis method of nanoparticles is shown in figure 5.

Figure 3. Potential endocytotic pathways for nanoparticle entry into cells

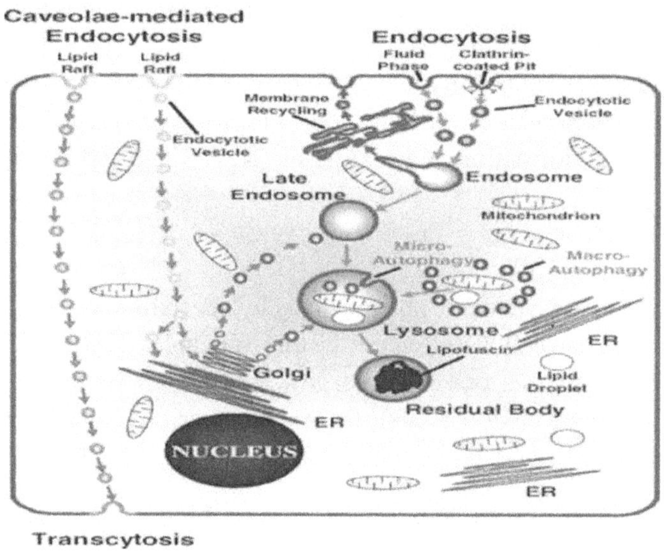

Figure 4. Kinetics of nanoparticles

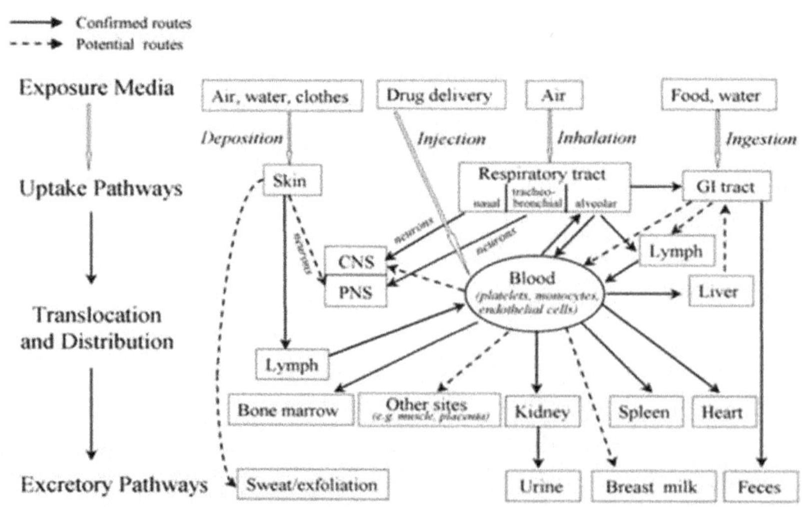

Figure 5. Methods used in green synthesis of nanoparticles

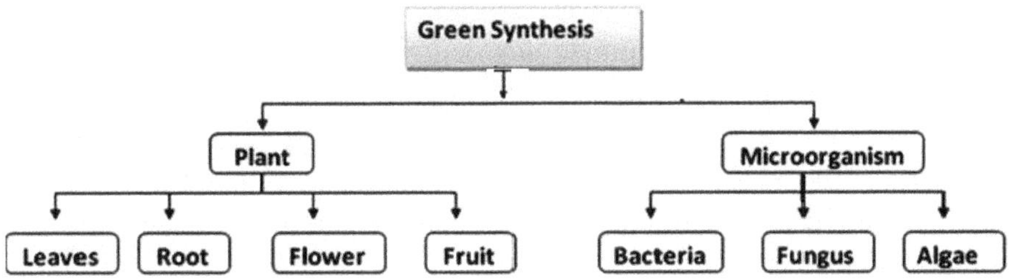

Figure 6. Green synthesis of silver nanoparticles from aloe vera plant extract

The enzymatic synthesis of nanoparticles can be done as extra-cellular or intra-cellular and it is achieved by simply-cultured yeasts and molds. The size of nanoparticles in enzymatic synthesis depends upon the incubation conditions and the concentration of metallic ion solutions. However, some molds are pathogenic in nature so we are frequently using them.

Medicinal plants have been used as a home remedy from ancient time due to its variety of metabolites. These metabolites can reduce the metal ions and assist synthesize of metal nanoparticles from plant extracts. These nanoparticles are having strong binding affinity with many functional groups of the plant extracts. Green synthesis method is the remedy for nanotoxicity. Medicinal Plants and specially those plants that have huge potential for detoxification, reduction and accumulation of metals, are good candidates for easy, fast and low cost materials for making herbal products with antimicrobial properties and removing metal-borne pollutants. The preparation of nanoparticles is begin in extracts obtained from plant parts such as leaves, roots and fruits into the aqueous solution of metal ions.

CONCLUSION

Nanotechnologies and manufactured nanomaterials, as with any new technology, may bring many advances to society and benefits for the environment, but also pose new challenges in health, environment safety and possible impacts on society. Because of the very broad range of potential applications using nanotechnology and the wide variety of characteristics displayed by manufactured nanomaterials, detailed discussion of both benefits and of health and environmental risks should take place at the level of individual nanotechnology applications.

Particle size and surface area are important material characteristics from a toxicological perspective. As the size of a particle decreases, its surface area increases and also allows a greater proportion of its atoms or molecules to be displayed on the surface rather than the interior of the material. The change in the physicochemical and structural properties of engineered nano material with a decrease in size could be responsible for a number of material interactions that could lead to toxicological effects. Chemical composition of the particles is responsible for their reactivity. Active cell uptake includes endocytosis or phagocytosis by specialised cells while NPs can also penetrate nonphagocytic mammalian cell membranes like that of red blood cells and become internalized at vesicles and mitochondria causing serious cell damage and cell death.

Concerning the environmental protection, green synthesis of nanoparticle has gained friendly and growing demand. It has vast applications in the biomedical, food, and agricultural fields.

REFERENCES

Ryan, J. N., & Elimelech, M. (1996). Colloid mobilization and transport in groundwater. *Colloids and Surfaces. A, Physicochemical and Engineering Aspects*, *107*, 1–56. doi:10.1016/0927-7757(95)03384-X

Lead, J. R., & Wilkinson, K. J. (2006). Natural aquatic colloids: Current knowledge and future trends. *Environmental Chemistry*, *3*, 159–171. doi:10.1071/EN06025

Nel, A., Xia, T., Mädler, L., & Li, N. (2006). Toxic potential of materials at nanolevel. *Science*, *311*(5761), 622–627. doi:10.1126cience.1114397 PMID:16456071

Yuranova, T., Laub, D., & Kiwi, J. (2007). Synthesis, activity and characterization of textiles showing self-cleaning activity under daylight irradiation. *Catalysis Today*, *122*(1-2), 109–117. doi:10.1016/j.cattod.2007.01.040

Cai, R., Van, G. M., Aw, P. K., & Itoh, K. (2006). Itoh, K. Solar-driven self-cleaning coating for a painted surface. *Comptes Rendus. Chimie*, *9*(5-6), 829–835. doi:10.1016/j.crci.2005.04.007

Blake, D. M., Maness, P.-C., Huang, Z., Wolfrum, E. J., Jacoby, W. A., & Huang, J. (1999). Application of the photocatalytic chemistry of titanium dioxide to disinfection and the killing of cancer cells. *Separation and Purification Methods*, *28*(1), 1–50. doi:10.1080/03602549909351643

Chang, H., Jwo, C. S., Lo, C. H., Tsung, T. T., Kao, M. J., & Lin, H. M. (2005). Rheology of CuO nanoparticle suspension prepared by ASNSS. *Reviews on Advanced Materials Science*, *10*, 128–132.

Hardman, R. (2006). A toxicologic review of quantum dots: Toxicity depends on physicochemical and environmental factors. *Environmental Health Perspectives*, *114*(2), 165–172. doi:10.1289/ehp.8284 PMID:16451849

Oberdörster, E. (2004). Manufactured nanomaterials (fullerens, C60) induce oxidative stress in the brain of juvenile largemouth bass. *Environmental Health Perspectives*, *112*(10), 1058–1062. doi:10.1289/ehp.7021 PMID:15238277

Lockman, P., Oyewumi, M., Koziara, J., Roder, K. E., Mumper, R. J., & Allen, D. D. (2003). Brain uptake of thiamine-coated nanoparticles. *Journal of Controlled Release*, *93*(3), 271–282. doi:10.1016/j.jconrel.2003.08.006 PMID:14644577

Singh & Mungali. (2011). *Water Pollution and Management*. Academic Press.

Moore, M. N. (2006). Do nanoparticles present ecotoxicological risks for the health of the aquatic environment? *Environment International*, *32*, 967–976.

Reijnders, L. (2006). Cleaner nanotechnology and hazard reduction of manufactured nanoparticles. *Journal of Cleaner Production*, *14*, 124–133.

ADDITIONAL READING

Chang, M. R., Lee, D. J., & Lai, J. Y. (2007). Nanoparticles in wastewater from a science-based industrial park – Coagulation using polyaluminium chloride. *Journal of Environmental Management*, *85*(4), 1009–1014. doi:10.1016/j.jenvman.2006.11.013 PMID:17202026

Gozalvez-Zafrilla, J. M., Sanz-Escribano, D., Lora-Garcia, J., & Hidalgo, M. C. L. (2008). Nanofiltration of secondary effluent for wastewater reuse in the textile industry. *Desalination*, *222*(1-3), 272–279. doi:10.1016/j.desal.2007.01.173

Huang, L., Wang, D., Tang, H., & Wang, S. (2006). Separation and purification of nano-Al13 by UF method. *Colloids and Surfaces. A, Physicochemical and Engineering Aspects*, *275*(1-3), 200–208. doi:10.1016/j.colsurfa.2005.09.040

Nowack, B., & Bucheli, T. D. (2007). Occurrence, behaviour and effects of nanoparticles in the environment. *Environmental Pollution*, *150*(1), 5–22. doi:10.1016/j.envpol.2007.06.006 PMID:17658673

Park, E.-J., Yi, J., Chung, K.-H., Ryu, D.-Y., Choi, J., & Park, K. (2008). Oxidative stress and apoptosis induced by titanium dioxide nanoparticles in cultured BEAS-2B cells. *Toxicology Letters*, *180*(3), 222–229. Advance online publication. doi:10.1016/j.toxlet.2008.06.869 PMID:18662754

Young, S., Balluz, L., & Malilay, J. (2004). Natural and technological hazardous material releases during and after natural disasters: A review. *The Science of the Total Environment*, *322*(1-3), 3–20. doi:10.1016/S0048-9697(03)00446-7 PMID:15081734

Chapter 12
Sustainable Watershed Management:
A Novel Approach for Green Technology

Sonal Sareen Pathak
Mata Gujri Mahila Mahavidyalaya (Autonomous), India

Sandhya Batalia
Mata Gujri Mahila Mahavidyalaya (Autonomous), India

ABSTRACT

In the present scenario, adequate water supply for agriculture, inland transportation, public health services, power production, and other industries is a subject of concern. The proper watershed management structure is essential to achieve sustainable land and water resource management and for eradication of rural poverty. Thus, a watershed management plan will achieve both socio-economic and ecological goals for sustainable water management. The integrated watershed management plan includes application of cost-effective soil water nutrient management practices. The chapter focuses on the watershed management system, its types and multidisciplinary approaches, and the impact of watershed management programs in improving economy and crop productivity. Thus, watershed serves as an important unit for the management of interdependent natural resources like water, land, and vegetation. However, in the future, the inevitable integration of information and computer technology into environmental technologies is promising solutions for the advanced management of water resources.

INTRODUCTION

The aim of green technology is to make people life better without compromising the quality of life. The power to heal our future is the necessity of green technology. It is therefore friendly way to protect the earth's natural resources and the environment. According to Mahatma Gandhi, *"Earth provides enough to satisfy every man's need but not every man's greed"*. India is endowed with many natural resources but 50% of the resources are degraded due to deforestation, soil erosion and indiscriminate use of water.

DOI: 10.4018/978-1-7998-8434-7.ch012

This leads to constraints in productivity like water scarcity, inherent low soil fertility and poor water management practices. Water is a basic natural resource which nurtures life and 70% of the population relies on agriculture and 2/3 part of the cropped area is dependent on rainfall without any protective irrigation. There is an urgent need to develop natural resource management techniques to sustain livelihood. The main component of integrated water resource management is the sustainable water supply (Raju *et al* 2008). The water as natural resource has many uses in domestic purpose (drinking, cleaning, sanitization, cooking), industrial operations (energy production, agriculture, manufacture of goods). Therefore, the three goals of accessing sustainability in water supply include environmental integrity, economic stability and social responsibility. The basic elements of Sustainable watershed management are interconnected as:

Figure 1. Basic elements of sustainable watershed management

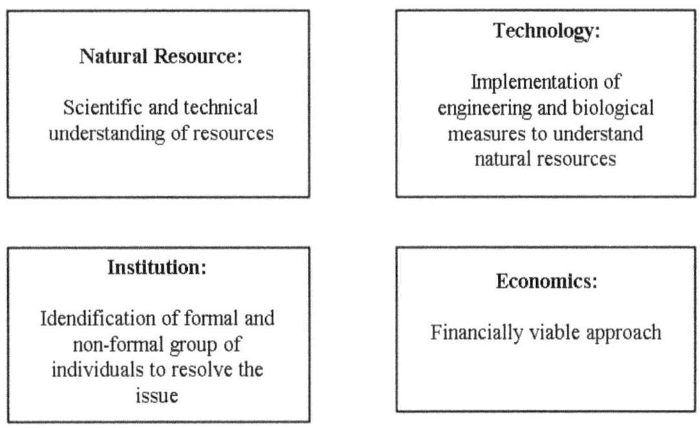

Watershed management is a geo-hydrological unit draining to a common point by system of drains. It is thus the land and water area bounded by a drainage area on the earth's surface from which runoff resulting from precipitation flows past a single point into a large stream, river or ocean (Shukla 2015). The water which comes from rainfall and storm water is collected in watershed which is separated from other watershed by naturally elevated areas. Thus, "a watershed is an area of land and water bounded by a drainage divide within which the surface runoff collects and flows out of a watershed through a single outlet onto a river lake or pond". The main aim of watershed management is to conserve the soil, plant and water resources of a catchment while benefiting humanity. It combines environmental, social, and economic concerns with approaches to treat forest, range, agricultural and urban parcels in an integrated manner (Naiman, 1994). The problem of degraded land, water and its management is complex and multi-dimensional and its development aims to develop human resource in watershed development and management and generate awareness about the importance of sustainable development and maintenance of existing work force working in the watershed development and develop skill in the rural youth to work in the watershed development based on watershed management approach and developing natural resources on sustainable basis (Suryawanshi *et al* 2012).

The factors affecting watershed management are Watershed characters, climatic factors, land use pattern and water resources. Depending upon the size, drainage, shape and pattern, the watersheds may be classified as follows:

- Macrowatershed (>50,000Hectare)
- Subwater shed (10,000-50,000 Hectare)
- Milliwater shed (1000-10,000 Hectare)
- Microwater shed (100-1000 Hectare)
- Mini water shed (1-100 Hectare)

WATERSHED MANAGEMENT- A MULTIDISCIPLINARY APPROACH

The approach for watershed invites integrated inputs from various disciplines for the development of watersheds in accordance with their characteristics. The objectives of watershed management include:

- To increase the agricultural production
- Conservation of soil and water that is to control damaging runoff and degradation
- Rational utilization of natural resources like water and soil
- To minimize risk of flood, drought and landslides
- To protect and enhance the water resource originated in watershed
- To manage and utilize runoff water
- To increase infiltration of rain water
- To enhance ground water recharge

Figure 2. Components of Sustainable watershed management

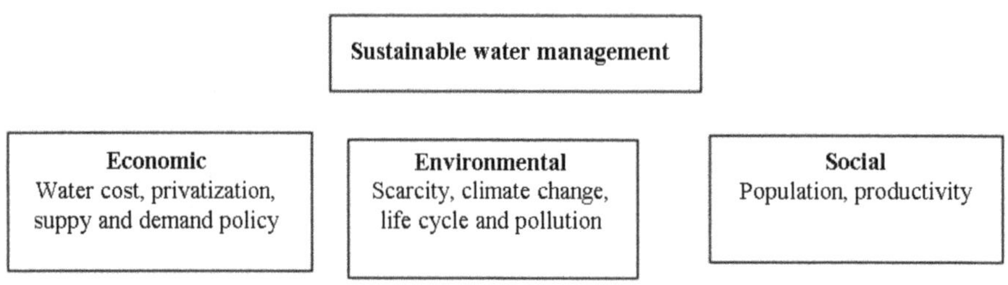

Basically, the components of Watershed development program includes various sectors like:

1. Soil and land management
2. Water management
3. Crop management
4. Afforestation
5. Fodder development
6. Livestock management

7. Rural energy management
8. Farm activities
9. Development of community skills and resources

Objectives of Watershed Management

The objectives of Watershed Development Programme (WDP) are diverse. The watershed Management and development aims to balance the conservation and use by humans of land and water resources within a watershed. The overall attributes of watershed developmental approach are promoting economic development of rural area, employment generation and restoring ecological balance. The other objectives are:

1. Environmental: For protecting vegetative cover for the whole year, protecting fertile topsoil, utilizing the land and increasing ground water recharge.
2. Economic: It draws attention for increasing cropping, intensity through inter and sequence cropping, maximizing farm income through agricultural related activities such as dairy, poultry etc. and sustained livelihood status of the watershed community
3. Institutional: Formation of watershed committees and self-help groups
4. Social: Alleviation of poverty awareness generation, improving skills of local community, women participation etc.
5. Equity: To develop equitable distribution of benefits of land and water resource development and biomass production, involvement of village communities in participating, planning etc.

Watershed Management in India

The program for water shed management in India mainly adopts a common strategy of multiple resource management to develop plans which are environmentally, socially and economically sustainable. To accelerate the pace of development of wastelands/degraded lands the Government had set up the National Wastelands Development Board in 1985 under the Ministry of Environment and Forests. Later a separate Department of Wastelands Development in the Ministry of Rural Development and Poverty Alleviation was created in 1992 and the National Wastelands Development Board was transferred to it. In April 1999, Department of Wastelands Development was renamed as the Department of Land Resources to act as the nodal agency for land resource management. (https://vikaspedia.in/social-welfare/rural-poverty-alleviation-1/schemes-department-of-land-resources/watershed-management-programmes)

In India, WDP has been sponsored and implemented by Government of India from early 1970's. Various Programmes like Drought Prone Area Programme (DPAP, 1973-74), Desert Development Programme (DDP, 1977-78), River Valley Project (RVP, 1948), National Watershed Development Project for Rain fed Area (NWDPRA, 1990-91) and Integrated Waste Land Development Programme (IWDP, 1989-90) were launched in various hydro-ecological areas which were affected by water stress and drought like situations. Entire water shed development Programme was initially based on structural driven compartmental approach of soil conservation and rain water harvesting during 1980's. The integrated WDP with participating approach was emphasized during mid 1980's and early 1990's. This approach was focused on raising crop productivity and livelihood improvement in watersheds (Wani *et al* 2005) along with soil and water conservation measures. Further under the chairmanship of Prof. C.H. Hanumantha Rao, it was strongly felt that there was a need to move away from the conventional approach which lack optimum use of resources, sustainable outcome and integrated planning. Thus, the new guideline was recommended in 1995, which gave emphasis to collective action and community participation. Later on, DPAP, DDP and IWDP were consolidates as the Watershed Development Component of Prime Minister Krishi Sinchayee Yojna (WDC-PMKSY).

The watershed guidelines were revised in 2001 as Haryali Guidelines which involve the Panchayat Raj Institutions (PRIs) more meaningfully in planning, implementation and management of economic development activities in rural areas.The following are the objectives of projects under Hariyali (https://megsoil.gov.in/docs/HARAYALI_Guidelines.pdf):

1. Harvesting every drop of rainwater for purposes of irrigation, plantations including horticulture and floriculture, pasture development, fisheries etc. to create sustainable sources of income for the village community as well as for drinking water supplies.
2. Ensuring overall development of rural areas by creating regular sources of income for the Panchayat from rainwater harvesting and management.
3. Employment generation, poverty alleviation, community empowerment and development of human and other economic resources of the rural areas.
4. Mitigating the adverse effects of extreme climatic conditions such as drought and desertification on crops, human and livestock population for the overall improvement of rural areas.
5. Restoring ecological balance by harnessing conserving and developing natural resources i.e. land, water, vegetative cover especially plantations.
6. Encouraging village community towards sustained community action for the operation and maintenance of assets created and further development of the potential of the natural resources in the watershed.
7. Promoting use of simple easy and affordable technological solutions and institutional arrangements that make use of, and build upon, local technical knowledge and available materials.

Neeranchal Committee in 2005 suggested a shift to a deeper concern with livelihood issues. It is World Bank assisted National Watershed Management Project to strengthen and provide technical assistance to Watershed component of PMKSY. It is mainly implemented in nine participating states - Andhra Pradesh, Chhattisgarh, Gujarat, Jharkhand, Madhya Pradesh, Maharashtra, Odisha, Rajasthan and Telangana. It is primarily designed to bring about institutional changes in watershed and rainfed agricultural management practices in India. It also devises strategies for the sustainability of improved watershed management and support equity livelihood and incomes through forward linkages, on a platform of inclusiveness and local participation.

Types of Watershed Management Programs

The Watershed management structure comprise of different types:

1. Broad beds and furrows are laid within the field boundaries. It mainly conserves soil moisture during rainy days and is therefore helpful to control soil erosion. It is suitable when the slope of the land is less than 3% and is laid within the field boundaries. Broad bed is a raised land and it enables the soil to preserve water level for a longer time. A furrow is a narrow trench made in ground by

a plough. It mainly controls the soil erosion and conserves soil moisture during rainy days. It also stimulates plant growth.

Figure 3. Broad bed and furrows

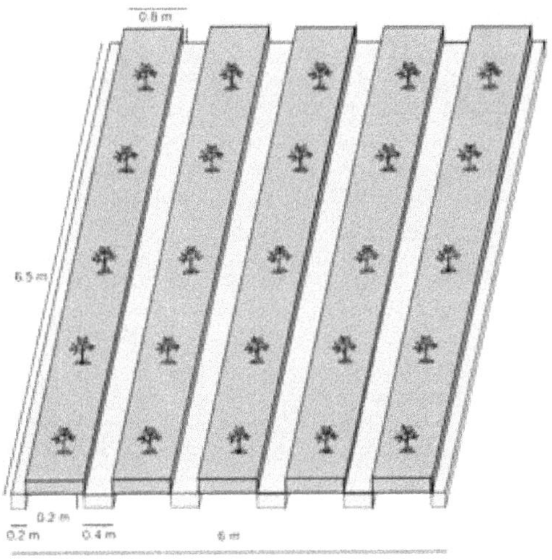

Figure 4. Contour bund

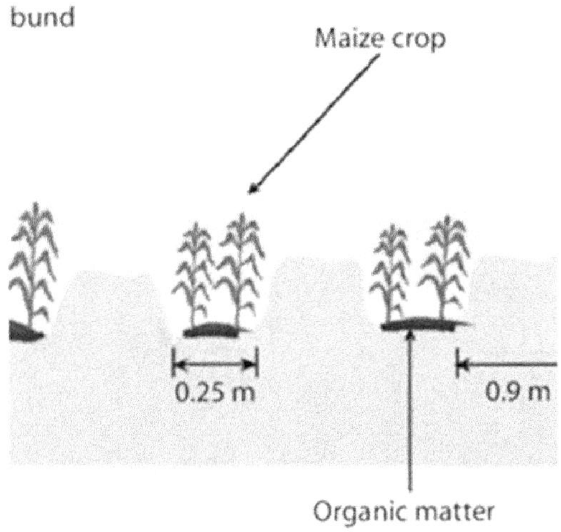

2. Contour Bund, which functions to intercept the runoff flowing down the slope by a closed or open embankment. It can be adopted on all types of soil and helps to control the runoff velocity and retain moisture in the field. It consist of constructing narrow based trapezoidal earthen embankment at intervals along the contour to impound runoff water behind them so that all that stored water is absorbed gradually into the soil profile for crop use. A series of such bunds divide the area into strips and act as barrier to the flow of water. It is construction of small embankment to reduce the length of the slope, to reduce the velocity of the runoff water, to hold the water for a longer period so that more water infiltrates into the ground and less water is runoff. Bunds are generally employed to control soil erosion and moisture conservation. Bunds are constructed along the contours with some minor alterations to meet the practical situations.
3. Bench terracing is the basic type of farmland in mountains. Bench terrace are a soil and water conservation measures used on sloping land with relatively deep soils to retain moisture and to control soil erosion. When the land slope is more, bunding is not effective for preventing soil erosion. It enables cultivation by bringing sloping land into different level strips. It mainly has step like fields along contours by half cutting and half filling. The level slopes depend on vertical and horizontal intervals. It is practiced on steep hill slopes and involves converting the original ground into level step like fields constructed by half cutting and half filling. Terracing is of 3 kinds: bench terraced farmland, sloping terraced farmland and combination level terraced farmland and natural slope.

Figure 5. Bench terracing

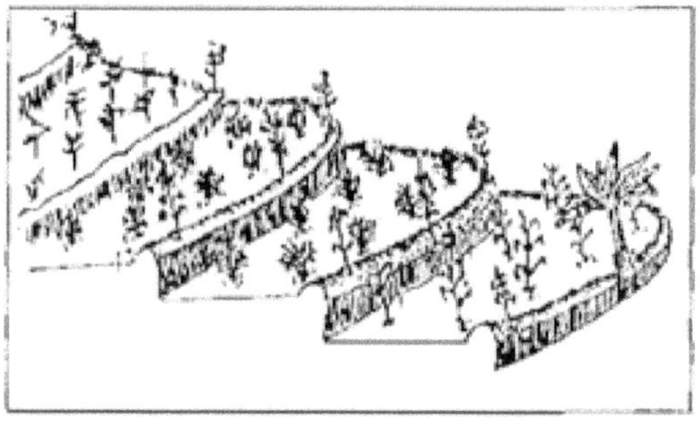

Typical bench terrace

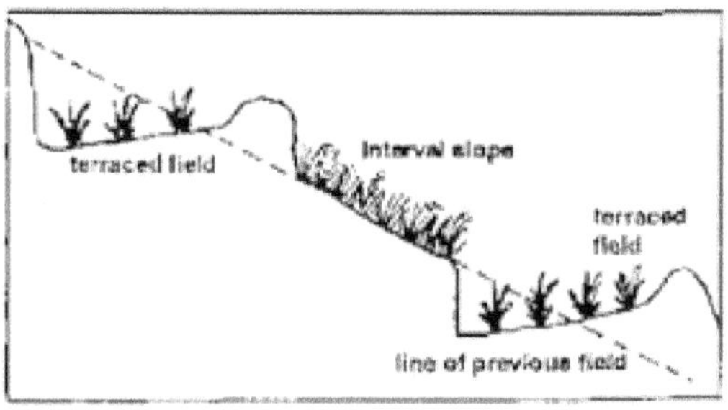

Modified bench terrace with Interval slope (China)

4. Check dam consist of low weir constructed along the gullies. The stored water improves soil moisture of the adjoining area. It is also known as gully plug and are actually dams built across channels or gillies to reduce the intensive flow to monitor and to trap the sediments. It can also control the erosive power and maintain the river channel morphology. Check dams also improve the water availability. These are constructed of variety of materials like rock, sand bags filled with peagravel or rocks. They are small barriers built along the direction of water flow on shallow rivers and streams for the purpose of water harvesting. The small dams retain excess water flow during monsoon rains in a small catchment area. They are relatively small and temporary structures and consist of barriers constructed on ditches small streams and channels. A checkdam interferes with the flow in upstream and downstream channels and dissipates the energyof flowing water. In practice, it is observed that a larger drainage area produces increased runoff volumes and flow rates that must be compatible with the check dam design criteria. They may be in form like loose stone, woody vegetation, masonry and log.
5. Percolation pond has a shallow depression created at lower portions in natural or diverted stream course. It augments the ground water recharge and is preferable under gentle sloping stream. It is a small water harvesting structure, constructed across the natural stream to harvest and impound the runoff from the catchments for a longer time to facilitate storage.
6. Micro-catchments of sloping land which is useful for in-situ moisture conservation with staggered planting and erosion control for tree crops suitable for dry land horticulture and agro-forestry

Figure 6. Check dam
(https://megamanual.geosyntec.com/npsmanual/image/checkdams.jpg)

Figure 7. Percolation pond

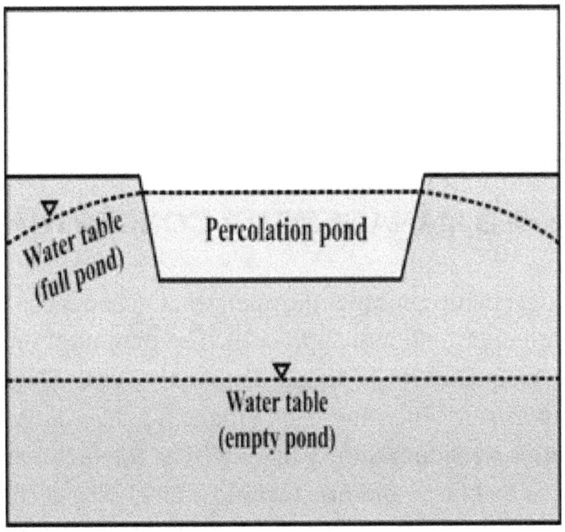

Approaches for Watershed Management

The appropriate management of watershed can be achieved by two approaches:

1. Integrated Watershed Management Approach: It involves integration of technologies in sustainable manner to meet the basic need of people and animals for the optimum development of land, water and plant resources within the natural boundaries of drainage area. This approach aims to improve the standard of living of common people by increasing his earning capacity by offering all facilities required for optimum production (Singh, 2000). In India large water resource management

projects have adverse socio-economic consequences. The indiscriminate expansion of marginal lands and over utilization of existing irrigation water resources result in improper water management. Therefore, integrated watershed management is followed in which water is managed in an appropriate way at lowest appropriate level. The process of planning and implementing management of water resources by integrating biophysical, socio-economic and institutional aspects is defined as integrated management of watershed. Earlier the practice of soil conservation was only performed wherein the welfare of land users was neglected. So, it was felt that for proper sustainable watershed management, there is need to integrate water and soil conservation with social and economic development. The use of appropriate technology, model information system and resource mapping through geographical system is model approach for integrated management for sustainable watershed. The improvement of living conditions of poorer by proper income generation and restoration of ecological balance by cost affective technologies and community participation are the key factors promoting sustainable water shed management practices.

2. Consortium/ Conglomerate Approach: It involves collective action and community participation including of primary stakeholders, government and non-government organizations, and other institutions. The cooperative action and public participation enhance the sustainable management of natural resources. It is multidisciplinary approaches which require proficiency in field of engineering agronomy, forestry, horticulture, animal husbandry, entomology, social science, economics and marketing. The expertise of different area to expand the effectiveness of watershed interventions an initiatives id the primary goal for this approach. The participation of local authorities, NGO's municipalities and private companies with the farmers is essential for the success of watershed management program

INTEGRATED WATERSHED MANAGEMENT: CONCEPTUAL FRAMEWORK

The sustainable watershed management combines the micro-watershed management (bottom up approach) with macro watershed management (top down approach). The planning, implementation, operation and maintenance stages are the prerequisite for sustainable management (Sriyanaa *et al* 2020). The micro watershed is a village level approach which involves village heads, village development agencies, and environmental care communities with community leaders. The top down approach or macro watershed involves different institutions at technical and non technical level. Supporting institutions consist of offices in Central Java Province (Central Office of Pemali Juana River Region, Central office Watershed Management, Central Java Environment and Forestry Service, Central Java Province Water Resources, and Spatial Planning), higher education institutions, non-governmental organizations communities, Central Java Watershed Management Coordination Forum, and Watershed Forum.

Soil bioengineering in watershed management is new approach for integrated management. Soil bioengineering is one important component in meeting specific watershed objectives, such as flood control, storm water management, stabilization, environmental. It is a stabilization method that uses living plants as primary structural component for restoring urban streams (Gholami *et al* 2017). It involves ecological engineering solutions providing benefit to both human and nature. It often incorporates woody vegetation for the restoration of ecological systems. The combination with the conventional technologies is beneficial and successful approach. It is based on integrated ecological principles and sound engineering practices.

Figure 8. Integrated watershed management: conceptual framework
(https://ars.els-cdn.com/content/image/1-s2.0-S2095633920300034-gr3.jpg)

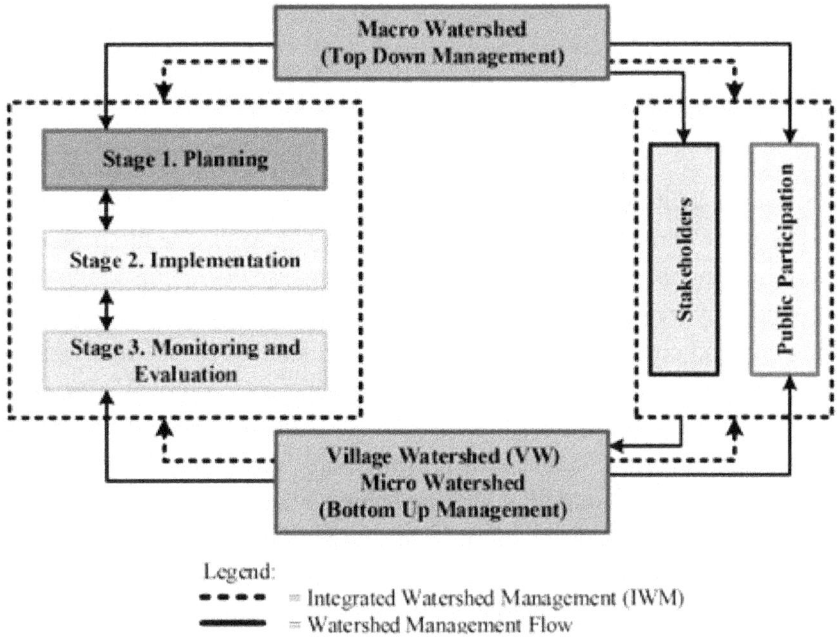

Jhabua Watershed Case Study: Bringing Back the Water

Jhabua is located in the central Indian state of Madhya Pradesh was a heavily forested area. Over the last four to five decades, more than 80 per cent of its forests were severely degraded. The causes included intensive cultivation on uplands and clearing of forests by timber contractors. Loss of forests led to loss of land productivity, diminishing employment opportunities, and subsequently, the outward migration of local people. The impact was felt more deeply by the tribals who comprise 83 per cent of the district's population, and whose survival is closely linked to forests. Over this period, Jhabua witnessed some severe droughts and famines. It had highly undulated and sparsely distributed forest cover wherein 7% arable land including cultivable and 18% notified as forest land. The average rainfall was only 50 mm per annum and therefore classified as drought prone region

In October 1994, the state government launched an ambitious programme in response to the problems. The Rajiv Gandhi Mission for Watershed Development (RGMWD) was an effort to involve the people in a large-scale land and water management programme (Varade *et al* 2011). The target was to improve 1.2 million hectares of land spread over several watersheds by the year 2000. It was envisaged that more than 2.9 million hectares of land area—spanning 6,691 villages through 5,024 watersheds, would be greened. Jhabua watershed is located at altitude of 380mm to 540 m with an area of 5400sq km. The use of watershed management helped the farmers to increase the productivity of crops like maize, cotton, peanut soyabean, gram blackbeans oilseeds etc. The groundwater table was found to be increased by 0.64 m on an average in 19 watersheds due to which consequently the cropped area and area under Rabi (winter crops) was also increased (Mahapatra 2002). The irrigated areas were increased to 1,115 ha in 18 micro-watersheds, which were nearly doubled the irrigated area of 1994-1995. As 92% of the population (predominantly tribals) was engaged in farming, WDP has significantly reduced poverty and increased employment opportunities in the area which has become the site of learning to the farmers, researchers and policy makers.

CONCLUSION

In India, the productive agriculture land is in constant process of degradation thereby converting to the wasteland. The wasteland development is of great significance to harness the full potential of the available land and resources. Rain water harvesting and watershed management are important tools for sustainable water management and play major role in soil erosion and water conservation. Rain water harvesting is a system of collecting and concentrating the rain water which can be used for irrigation. Successful sustainable watershed management (SWM) combine's scientific and technical information with cultural and societal values Watershed is an essential in everyday life and is dependent on activities like agriculture and industry, power production, inland transportation etc. Therefore, construction of watershed is essential to achieve sustainable water and land management (Sharma 2002). A good watershed planner should have detailed information about the components and process of sustainable watershed management (Joshi *et al* 2008). This is a prerequisite for the sustainable management of agriculture resources. Thus, in a particular area the emerging watershed framework builds on existing management programs and resources but has its goal for sustainable watershed system as integrated system. The integrated approach of SWM provides a context for integration using practical, tangible management units that people understand, better understanding and appreciation of nature, generating ecologically-based, innovative, cost-effective solutions for stronger working relationships. The successful development of watershed management plan depends on the collaborative and comprehensive processes. Many researchers suggest that the use of knowledge from society with plans and projects are more accurate and technically sound for proper development program. This study illustrates the planning of water management with the use of watershed technology considering climate topography soil vegetation land use types etc for the conservation of water resources.

REFERENCES

Gholami, L., Karimi, N., & Kavian, A. (2017). Soil bioengineering Methods used in water management and stabilization slopes steep. *Ecohydrology*, *4*(1), 149–162.

Guidelines for Hariyali. (2003). https://megsoil.gov.in/docs/HARAYALI_ Guidelines.pdf

Joshi, P. K., Jha, A. K., Wani, S. P., Sreedevi, T. K., & Shaheen, F. A. (2008). *Impact of Watershed Program and Conditions for Success: A Meta-Analysis Approach*. Global Theme on Agroecosystems, International Crops Research Institute for the Semi-Arid Tropics and National Centre for Agricultural Economics and Policy Research.

Mahapatra & Richard. (2002). Jhabua: A Green Miracle. In Making Water Everybody's Business - Practice and Policy of Water Harvesting Centre for Science and Environment. Academic Press.

Naiman, R (1994). Watershed Management: Balancing Sustainability and Environmental Change. In *Technology & Engineering*. Springer.

Raju, K. V., Aziz, A., Sundaram, M. S. S., Sekher, M., Wani, S. P., & Sreedevi, T. K. (2008). *Guidelines for Planning and Implementation of Watershed Development Program in India: A Review*. Global Theme on Agroecosystems Report no. 48. Monograph. International Crops Research Institute for the Semi-Arid Tropics.

Sharma, R. (2002). *Watershed Development Adaptation Strategy for Climate Change. Paper presented in South Asia expert workshop on Adaptation to Climate Change for Agricultural Productivity, organized by the Government of India*, UNEP and CGIAR.

Shukla, H. (2015). Watershed Management: Its Role In Environmental Planning And Management. *IOSR Journal of Environmental Science, Toxicology, and Food Technology, 1*(5), 8-11.

Singh, R. V. (Ed.). (2000). Watershed planning and management. Yash Publishing House.

Sriyanaa, I. (2020). Watershed management index based on the village watershed model (VWM) approach towards sustainability. *International Soil and Water Conservation Research, 8*(1), 35–46. doi:10.1016/j.iswcr.2020.01.003

Suryawanshi, S. P., & Kamble, A. (2012) Watershed management in India. *Proceeding of International Conference SWRDM-2012*.

Varade, Wankhade, Yadav, & Khandare. (2011). Watershed management as a tool for changing the Kaleidoscope of central India: A case study from Jhabua district of Madhya Pradesh, India. *Nature Environment and Pollution Technology, 10*(5), 589–594.

Wani, S. P., & Garg, K. K. (2009). Watershed Management Concept and Principles. In *Best-bet Options for Integrated Watershed Management Proceedings of the Comprehensive Assessment of Watershed Programs in India*. ICRISAT.

Wani, S. P., & Ramakrishna, Y. S. (2005). Sustainable management of rainwater through integrated watershed approach for improved rural livelihoods. Watershed management challenges: Improved productivity, resources and livelihoods.

Watershed management programmes. (n.d.). https://vikaspedia.in/social-welfare/rural-poverty-alleviation-1/schemes-department-of-land-resources/watershed-management-programmes

Chapter 13
Management of Insects/Nematodes for Improving the Crop Productivity

Anju Rani
Swami Vivekanand Subharti University, Meerut, India

Permod Kumar
Swami Vivekanand Subharti University, Meerut, India

Amit Kumar
Swami Vivekanand Subharti University, Meerut, India

Gyanika Shukla
CCS University, Meerut, India

Raj Singh
Maharishi Markandeshwar University (Deemed), India

ABSTRACT

More than five million tons of chemical pesticides are applied annually throughout the world. Out of this, about 65% are used for agriculture, and the remainder by public health agencies and government agencies. The primary source of pesticides in ecosystems is agriculture and forestry. Pesticide poisoning is a major public health problem. These chemical pesticides affect nonpathogenic organisms, and due to that, environmental problems arise. To avoid the use of these hazardous chemicals, integrated pest management strategies are beneficial in which cultural, physical and mechanical, biological and chemical control measures are included. IPM strongly focuses on supporting natural mortality factors like pests and natural enemies. IPM requires a great deal of sampling, monitoring, and decision making to be effective.

DOI: 10.4018/978-1-7998-8434-7.ch013

INTRODUCTION

Plant-parasitic nematodes are root feeders and completing their life cycles in the root zone and with many plants. Some are endoparasitic and feed within the tissue of roots, tubers, buds, seeds, etc. while others are ectoparasitic, feeding externally through plant walls. Even a single endoparasitic nematode can kill a plant or reduce its productivity. A few species of nematodes are highly host-specific like *Heterodera-glycines* on soybeans and *Globoderarostochiensis*on potatoes. But in general, nematodes have a wide host range. Endoparasitic root feeders are root-knot nematodes (*Meloidogyne* species), cyst nematodes (*Heterodera* species), and the root-lesion nematodes (*Pratylenchus* species). Nematodes feed directlyand cause drastically decrease a plants uptake of nutrients and water. They have the greatest impact on the crop productivity when they attack the roots of seedlings immediately after seed germination. Not only nematodes but there are several species of insects and mites pests have been reported to attack agricultural as well as horticultural crops and cause severe damage and affect farmers economy. There are some chewing, sucking, lapping insects belongs to ordes Orthoptera, Hymenoptera, Lepidoptera, Coleoptera, Hemiptera etc. Some insects are host specific while others have wide host range. Insect and nematode control is essentially prevention, because once a plant is parasitized it is not possible to kill the pest. The most sustainable approach to insect and nematode control integrates several tools and strategies. A balanced soil ecosystem will support a wide variety of biological control.

Integrated practices are very useful and significantly control the pest without harming environment. Such integrated program is called Integrated Pest Management (IPM). IPM is effective to all types of pests from insects and mites, to fungi, bacteria, weeds and nematodes. There are number of control methods but cultural and biological methods preferred first to save environmental pollution, crops and beneficial pests because pesticides (fungicides, insecticides, nematicides, weedicides etc.) kill about 98% of the pest and 99% of the enemies.

Integrated Pest Management was first introduced by RF Smith and R Van de Bosch in 1967 and formalized by US National Academy of sciences in 1969.Integrated Pest Management is known as crop protection or pest management system which comes under farming systems. It is a sustainable agricultural approach with a sound ecological foundation.Integrated pest management used in commercial agriculture since the 1970s. It is a systematic approach to pest management which focuses on preventing plant pathogens. It involves monitoring pest populations, identifying pests and choosing a combination of tactics to keep pest populations at an acceptable level. Tactics include cultural, mechanical, biological and chemical methods of the pest management. IPM stresses trying the least toxic methods first. IPM has a traditional low input approach. According to Rabb (1972) integrated pest management is the "intelligent selection and use of pest control actions that will ensure favorable economic, ecological, and sociological consequences". This is underlined in the economic threshold concept where as pesticides are not applied unless pest levels are high enough to potentially reduce profits. Economic thresholds ensure that there is a sound economic foundation for the use of pesticides.

There are other several definitions given by different scientists-

Integrated control is describe as:

- Applied pest control which blends and coordinates natural (biological) and synthetic (Chemical) control. Chemical management is used as essential and in a way that is least disruptive to biological management. Integrated manipulation may also make use of certainly occurring natural control just as natural control affected with the guide of controlled or welcomed on biotic vendors. (Stern et al., 1959).

- The application of pests by means of integrating host resistance, botic and chemical controls in a way that reduces economic, wellbeing and ecological risks. (CPM Crop Protection Manager 1997).
- IPM is a decision aid machine for the choice and uses of pest manage tactics, singly or harmoniously coordinated into a management approach, based totally on value/advantage analyses that keep in mind the pursuits of and influences on manufacturers, society, and the surroundings. (Marcos Kogan. 1998).
- 'A way to deal with the management of pests in open offices that joins biological, social, physical, and chemical devices in a way that limits financial, wellbeing, and environmental risk (Bajwa & Kogan 2002).
- Integrated Pest Management (IPM) is a methodology to forestall and smother pests with least effect on human health, the environment and non-target life forms (Dreistadt and Steve, 2004).

Agriculture Sustainability is the replacement of chemical compounds and gives information on natural data sources. Sustainability embrace complex methodology which keeps up high efficiency and farm benefits without hurting nature and assets.

Reasons to Apply IPM

- It assists with keeping a fair biological system and advances a healthy environment: Chemicals can decimate beneficial parasites (fungi) and insects and influence soil and soil microflora and effect the environment.
- Pests can become protection from pesticides thus pesticides can be insufficient.
- Applications of IPM keep up a decent open picture: IPM ensure home and business structure, economy of farmers, animals wellbeing, human health. (Stern et al., 1959; Dreistadt and Steve, 2004).

Objectives of IPM

- To recognize and checking pest issues.
- To diminish the underlying inoculums and adequacy.
- To increment the obstruction of the host and postpone the beginning of the disease.
- To moderate the secondary cycle.
- To keep record and assess the program.

Components of IPM

There are three fundamental segments – monitoring, prevention, and intervention.

Monitoring

Monitoring is the establishment of IPM which gives information on current irritation and yield circumstance which will assist with choosing the most ideal mixes of the pest management procedures. Pheromone traps are most successful monitoring gadget and extremely modest to buy and give benefits including recognizable proof of pest issues (hot spots), (Dreistadt & Steve 2004).

Prevention

It incorporates social and physical control measures.

Cultural Techniques

Cultural strategies remember area of the crop for the field and Selection of the resistant varieties (Insect, pests, or diseases), disinfection (end of the pest), crop rotation (intercropping, blended trimming etc.), culturing, soil and water the management, plant nourishment, collecting and capacity and safeguarding biodiversity are utilized for anticipation from contamination just as infestation. (Gupta Suman, 2010).

Area of the Crop in the Field

Location of the developing yields in the field is most appropriate to atmosphere, soil, pH and geology gives them amicable conditions fitting field preparation.

Soil Solarization

Soil solarization in summer in the intercropping period helps to kill heat sensible microbes at profundities up to 20 cm, similar to root-hitch nematodes. Soil amendments with fertilizers or composts can stimulate characteristic microbial soil vegetation and have are pressive impact on phytoparasitic nematode populations, (Goswami and Mittal, 2004).

Determination of Resistant Varieties

Plants varieties which are tolerance to insect, pest and microbes can be acquired through Plant breeding, plant tissue culture and recombinant DNA technology by resistant genes to the affected varieties and then that varieties become impervious to explicit plant microorganisms. Joining of hereditary material from Bacillus thuringiensis (Bt) in cotton, corn and potatoes makes the plant tissues poisonous to the growths, creepy crawlies and nematodes.

Intercropping System

Crop rotation is the most established and least expensive technique embraced in agriculture to destruction of specific kinds of microbes from infested soil. Intercropping system gives great anticipation from microorganisms and weeds. It is significant for soil richness the executives, disease management, nematode and weed management. Mixed trimming, intercropping and utilization of barrier crop essential for the accomplishment of economical agriculture.

Management of Soil and Water

Tillage turns the dirt, crop trash and weeds before the planning of seed bed. Culturing gives ideal air circulation to advantageous microorganisms and upgrading supplement cycle and natural issue decay. In water management, rain water management lessens evapotranspiration and Irrigation water management incorporates water system plan at fitting time with legitimate amount of water without causing alkalinity, acidity and water logging. (Marcos Kogan, 1998).

Crop Nutrition Management

Plants require a legitimate given amount and blend of nutrient to deliver great yield, lack of nutrients can restrain the plant development and plants become more susceptible to microbes and insect because of nutrient insufficiency. Distinctive sort of soil contain various kinds of Nurient improve soil fertility. The nonstop reusing of nutrients into and out of the soil includes complex biological and chemical interaction.

Harvesting and Storage

Different crops have different time of sowing, different time of harvesting and collection. After collection of seed, good quality of seed collect for storage and before storage that room should be cleaned from microbes, insect and any harmful biological organisms. If these is no management for harvesting of crop, storage of crop then the quantitatively reduce the food.

Preserving Biodiversity

Biodiversity can be saved by ensuring natural habitats close to farmland. Cautious administration of trees, plants, edges and fences are significant for wildlife habitat, gainful and ruthless animal and insects.

Physical Control Measures

Mulching

Mulching is the methods wherein plant deposits are utilized to cover the dirt surface. Straw, wood and bark chips are utilized for this methods. Mulching check the spread of microbial infection and insect infection in the dirt and plants become healthy on the grounds that numerous microorganisms stay over wintering and over summering in the dirt which may slaughter in the wake of mulching.

Pruning

Cutting and removal of old, infested and infected twig of the plants

Hand Picking

Removal of insects and their eggs by hands provides positive control of insects. Hand picking is least expensive practice.

Eradication

Collection and destruction of infected crop debris from the field.

Trapping

Using traps to avoid the pests like *Helicoverpa, Spodoptera* and beetles etc. is beneficial method.

Regulatory Method

Plant Quarantine

The term 'Quarantine' means 40 days period. The purpose of the health authorities was to establish adequate detention period. This term adopted to cover protective methods for the exclusion of pests and diseases of agricultural and horticultural crops. The first legal restrictions to hinder the spread of disease were enacted against human disease. Plant quarantine may be defined as the restriction imposed by duly constituted authorities on the crop yield, movement and existence of plants or their material to prevent introduction or spread of a pest. When the plant pathogens introduced into new area, they are likely to cause much more catastrophic epidemics than do the existing pathogens. There are some plant disease epidemics that occurred through out the world, for example the downy mildew of grapes, bacterial canker of citrus, banana bunchy top, wart of potato, it is very difficult to predict accurately whether an exotic organism will become established, and once established, become economically important.Out of these diseases bunchy top of banana and potato wart introduced in India from Sri Lanka and Netherland respectively.

Objectives and Major Activities of Plant Quarantine

1. Inspection of imported agricultural commodities for preventing the introduction of exotic pests and diseases inimical to Indian Fauna and Flora and provide certificates.
2. Undertaking Post Entry Quarantine Inspection (PEQI) in respect of identified planting materials.
3. Conducting the Pest Risk Analysis to finalize phytosanitary requirements for import of plant/plant material.
4. Prevention of the introduction and spread of exotic pests those are destructive to crops by regulating/restricting the import of plants or their products.
5. Inspection of agricultural commodities meant for export as per the requirements of importing countries under International Plant Protection Convention (IPPC)
6. Removal of exotic pests and diseases those are already introduced for containing/ controlling them by adopting domestic quarantine regulations.

Intervention

1. Biological control measures
2. Chemical control measures

Biological Control Measures

Definition of bio-control: It is a component of an integrated pest management and defined as the reduction of pest population by using natural enemies. Bioagents includes predators, parasites, fungi, bacteria, viruses etc. It is a beneficial action to control phytopathogenic organisms to protect agricultural as well as horticultural crops.

Types of Natural Enemies

Parasites

Insects parasitize other phytopathogenic insects by laying their eggs in or on the body of host and kill that host. Ichneumonid wasps (e.g., *Megarhyssamacrurus* female) prey mainly on caterpillars of butterflies and moths. Braconid wasps: Tiny wasps attack caterpillars and greenfly. A common parasite of the cabbage white caterpillar seen as clusters of sulphur yellow cocoons bursting from collapsed skin of caterpillar. Chalcid wasps: (smallest of insects) parasitize eggs/larvae of greenfly, whitefly, cabbage caterpillars, scale insects and Strawberry. Tachinid flies parasitize a wide range of insects including caterpillars, adult and larva, beetles, true bugs, and others (Stern et al., 1959). *Encarsia formosa* is a parasitoid of whitefly, a sap-feeding insect and cause wilting and black sooty moulds. The wasp lays its eggs in young whitefly 'scales', turning them black as the parasite larvae pupates. *Eretmocerus* spp. parasitoid upon white flies while *Aphidiuscolemani* kill aphids. *Gonatocerusashmeadi* control the glassy-winged sharpshooter. *Homalodiscavitripennis* controlled ~95% of the pest density. *Trichogramma* species attack eggs of moths and butterflies. (Smith 1996; Rabb and Guthrie.1970).

Some fungi trap the nematodes which differ by their trapping mode: network traps (*Arthrobrotrysoligospora*, *A. superba*), constrictive rings (*A. anchonia*, *A.dactyloides*, *Dactylariabrochopaga*, etc.), or adhesive knobs (*Monacrosporiumcionopagum*, *Dactylellalobata*). Some fungi, like *Dactylaria candida* present two types of trapping mechanism like adhesive knobs, and constrictive but non-adhesive rings.

Although fungi are saprophytes, but they can trap the nematodes in a larval or free adult stage in the soil. It has been proven that the production of peptides by extracellular proteases hydrolyzing the nematodes cuticles may induce trap formation. Some fungi are able to infect and destroy the eggs of nematode. Most of the fungi that have this mode of action are saprophytes, and cansecondarily invade already-dead eggs. Egg parasitic fungi are *Aecilomyces*, *Pochonia* and *Verticilium* genera. *Paecilomyceslilacinus* and *Pochoniachlamydosporia* are probably the most effective egg-parasites. *P. lilacinus*has been proven to successfully control root-knot nematodes, *M. javanica* and *M. incognita* on tomato, egg-plant and other vegetable crops.

Predators

Birds, fish, amphibians, reptiles, small mammals, and arthropods. Arthropods (insects, mites and spiders) are the most important predators in pest management including lady beetles, ground beetles, green lacewings, assassin bugs, predaceous bugs, minute pirate bugs, predatory mites, and spiders. Predators usually deposit their eggs near their prey so the immatures can immediately find their host and begin feeding. Nymph and adult stages of flower bug (*Oriustristicolor, O. insidiosus*) feed on aphids, insect eggs, thrips and young caterpillars. The two spotted stink bug, *Perillusbioculatus* feed on larvae of the Colorado potato beetle and sunflower beetle and numerous species of caterpillars. Numerous species of lady beetles are common predators of aphids and other small insects and their eggs. Lacewing adults and larvae feed on aphids, small caterpillars, insects eggs and mites. Spiders consume all types of insects.

Figure 1. Braconid wasp larvae

Figure 2. Ichneumonid wasps

Figure 3. Adult lady beetle

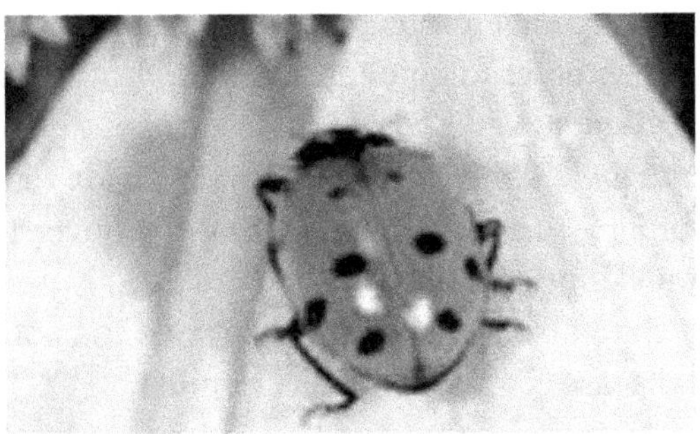

Figure 4. Lacewing larva

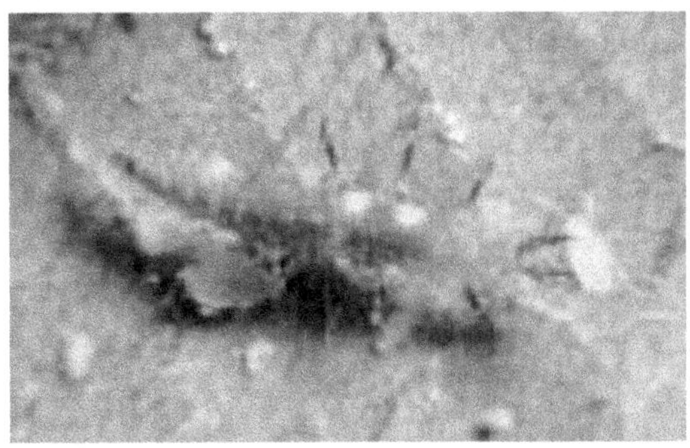

Fungi, viruses, bacteria, protozoans, and other microorganisms cause infection in insects Insect-parasitic nematodes are also included in this group of natural enemies. These are small worms that attack and kill insects that live in moist habitats. A few species are currently being sold commercially for insect control. Insect pathogens, Some insect pathogens and nematodes are commercially available and can be to achieve biological control of specific pests. Both diseases and nematodes, like parasites are specific to certain species or groups of pests; they do not harm nontarget organisms, like beneficial insects, animals, humans, or plants. They can quickly spread through an insect population causing rapid mortality in a short period of time, and can be important in the natural control of pest populations.

Table 1. List of natural enemies

Predators	Pathogenic Pests
Big-eyed bugs	Aphids, leafhoppers, lygus bugs (nymphs), spider mites
Green lacewings	Aphids, leafhoppers, plant bugs (immature), spider mites
Syrphid flies	Aphids
Lady beetles	Aphids, scales, spider mites
Minute pirate bugs	Aphids, spider mites, thrips
Parasitoids	**Pathogenic pests**
Ichneumonids	Aphids (larvae), cutworms, loopers
Braconids	Aphids, cutworms
Chalcids	cutworms, loopers
Trichogrammasps	caterpillars, maggots
Tachinids	Codling moths, cutworms, grasshoppers

Bioagents

Bacillus thuringiensis (Bt), a naturally occurring bacterium which releases a toxin in the gut of insect and degrades the lining of stomach and death of insect takes place. Bt effective against larval stages of insect pests like, Colorado potato beetles, and cabbage loopers. *Bacillus papillae* and *B. lentimorbus* cause "milky disease" of white grub. A beneficial soil bacterium, *Saccharopolysporaspinosa*, produces a natural metabolite known as 'spinosad'. It has been formulated as a microbial insecticide and effective against a wide variety of insect pests including European corn borer, diamondback moth, leafminers, Colorado potato beetle etc. *Metarhizium, Beauveria, Entomophthora* and *Zoopthora. Metarhiziumaniso-pliae, B. bassiana, Entomophthoramuscae, ZoophthoraRadicans, Z. phytonomi*attacks on a wide range of insects, such as grasshoppers, true bugs, aphids, caterpillars, beetles, adult flies and the hover fly, potato leafhopper, many aphids, alfalfa weevil and subterranean insects like ants and termites. *Nosema locustae* a protozoan used to control grasshoppers. Viruses are also play an important role to control insect pests. Nucleopolyhedrosis (entomopathogenic viral disease) control alfalfa caterpillar.

Biopesticides

According to European Commission, biopesticides have been defined as "form of pesticide based on micro-organisms or natural products". Hance pesticide which derived from natural material like animals, plants, microbes are called biopesticides. The total world production of biopesticides is over 3,000 tons/yr, which is increasing at a rapid rate. India has a vast potential for biopesticides. Plant extracts or biopesticides having antimicrobial activity due to presence of terpenoids, essential oils, phenolics and polyphenols, alkaloids and polypeptides and called natural products. Most of the biopesticides find use in public health, except a few that are used in agriculture. Besides, transgenic plants and beneficial organisms known as bio-agents which are used for insect pest management in India (Suman Gupta,2010; Palmer, Bromley, and Brandenburg, 2007; Nesci, Bluma and Etcheverry, 2005).

Table 2. List of biopesticides registered as insecticides. Act, 1968

Name of the Biopesticide
Bacillus thuringiensis (Bt) var. israelensis
Bacillus thuringiensis(Bt) var. kurstaki
Bacillus thuringiensis (Bt) var. galleriae
Bacillus sphaericus (Bs)
Trichodermaviride (Tv)
Trichodermaharzianum (Th)
Cymbopogan
Pseudomonas fluoresens
Beauveriabassiana
NPV of Helicoverpaarmigera
NPV of Spodopteralitura
Neem based pesticides

Microbial Biopesticides

Biopesticides like *Bacillus thuringiensis var. israelensis, kurstaki, galleriae, Trichodermaviride, Trichodermaharzianum, Pseudomonas fluoresens, Beauveriabassiana* are microbial pesticides and used against plant pathogenic fungi, bacteria, insect pests and nematodes. Plants contain antimicrobial compounds which can be toxic to pathogens and these natural plant products can be used to control pest population. Plant growth promoting rhizobacteria are bacteria that colonize plant roots, and in doing so, they promote plant growth and reduce infection as well as infestation. Oomycetes (*Catenari-aanguillulae, Myzocytiumlenticulare, M. anomalum*with biflagellated zoospores able to encyst in the nematode's cuticle), Zygomycetes (*Meristracumasterospermum*, with spherical conidia, which produce a germinative filament after adhesion that germinates and creates new conidiospores), Deuteromycetes (*Meriaconiospora*with club-like spores, which fix on the host via their anterior part), and Basidiomycetes (*Nematoctonusleiosporus*, with adhesive spores) and Hyphomycetes (*Hirsutellasp*) fungi parasitize nematodes. Exclusively endoparasitic fungi must live their entire cycle in the host nematodes. Along with fungi some insects can also be parasites to nematodes like *Entomobryoidesdissimilis*or nematodes predatory mites (*Alliphis* spp., *Alicorhagia* spp.)

Biopesticides Obtained from Plant Extracts

Transgenic crops that act as insecticides began in 1996 with a genetically modified potato that produced the *cry* protein, derived from the bacterium Bacillus thuringiensis, which is toxic to beetle larvae such as the Colorado potato beetle. The technique is expected to replace many other insecticides, which are losing effectiveness due to the spread of pesticide resistance.

Plant extracts obtained from *Calocedrusmacrolepis* var. *formosana, O. acutidens,*cymbopogan (*Cymbopogon citrates*), *Ocimumgratissimum, Thymus vulgaris,* Bergamot (*Citrus hystrix*), neemproducts, tea tree (*Melaleucaalternifolia*), *Asarumheterotropoides* var. *mandshuricum,* Pudhina, Thriphala, Garlic, show antimicrobial activity against Leaf folder, late blight of potato, bacterial leaf blight, *Helminthosporium* leaf spot, aphids, Jassids, fruit borer, stem borer. Active ingredients of neemlikeAzadirachtin, Nimbin, Nimbidin, Nimbinene, Nimbridic acidand Azadirone have antifungal and insecticidal properties and are effective against more than 300 species of arthropods, 11 species of nematodes, 12species of fungi and three viruses (Rabb and Guthrie, 1970).

Table 3. List of botanical biopesticides effective against insect pests

Plant product used as biopesticide	Target pests
Neem	A variety of sucking and chewing insect.
Ryania	Caterpillars (corn earworm, and others) and thrips
Rotenone	Leaf-feeding insects, such as aphids, certain beetles (bean leaf beetle, Colorado potato beetle, cucumber beetle, flea beetle, strawberry leaf beetle) and caterpillars, fleas and lice on animals.
Limonene and Linalool	Fleas, aphids and mites, fire ants, wasps and house crickets.
Sabadilla	Squash and stink bugs, thrips, caterpillars leaf hoppers.
Pyrethrum / Pyrethrins	Ants, aphids, roaches, flies, and ticks

Chemical Control Measures

Chemicals used to control plant pathogenic fungi, insects, weeds, nematodes are called pesticides. Insecticide, weedicides, Acaricide, Miticide, Fungicide, Herbicide, nematicides are jointly called pesticides (Stern et al., 1959).

Chemical pesticides increase metabolic detoxification and decreased target sensitivity. Along with these pesticides are toxic to other beneficial microorganisms as well as insects. If the non-chemical practices are not sufficient to control pest population, chemical substances are allowed for use to control plant pathogens. (Aspelin and Grube, 1999).

INSECTICIDES AND NEMATICIDES

The chemicals which kill insects and nematodes known as insecticides and nematicides respectively. On the basis of mode of action pesticides are of two types: Systemic and contact. Systemic pesticides become incorporated and distributed systemically throughout the whole plant. When insects and feed on the plant, they ingest the chemicals while Contact pesticides are toxic to insects upon direct contact. These may be inorganic or organic, these are metals include arsenates, copper and fluorine and sulfur compounds, last one is commonly used. Contact pesticides usually have no residual activity. Insecticides and Nematicides are divided into the following four groups:

Fumigants

1. D-D: 1,2-dichloropropane,1,3-dichloropropene are broad spectrum effective nematicide until but 1,2-dichloropropane alone is inactive nematicide and 1,3-Dichloropropene is sixth most abundantly used pesticide in the United States. It also has both fungicidal and insecticidal activity.
2. Methyl bromide**:** It is a broad-spectrum fumigant toxic to nematodes. Now a days it is banned by Government agencies due to its carcinogenic properties.
3. Chloropicrin**:** One of the oldest soil fumigants, chloropicrin's primary agricultural use in soils is as a fungicide, although it does have herbicidal and nematocidal activity.
4. Metam Sodiumis: A soil fumigant used to control nematodes, fungi, insects, and weeds.
5. Sodium Tetrathiocarbonate: It is more recently registered soil fumigant active against fungi, insects and nematodes.

Seed Treatment

Treated seeds with specific pesticides (fungicide, insecticide or nematicide) prevent pre-inoculation of the pathogen. Seed treated with emisan protect from fungal infection, chlorpyriphos reduce white grub population, higher dose of chlorpyriphose kill cotton surface weevil. Avicta Complete bean, Avicta 500 FS and Apron Maxx provide immediate and reliable complete protection to soybean plant against insects and nematodes like Root knot, lesion, cyst, stunt etc. nematodes and thrips, leafhopper, white grub, beetles, maggots etc. and fungi like *Phytophthora, Fusarium, Pyhtium, Rhizoctonia, Phomopsis* etc. Seed treatment with *Trichodermaviride* 2-4g/100 gm and per kg, Carbandazim @ 1g/100 gm seed, *Pseudomonas fluorescens* @ 10gm/kg, Chlorpyriphos @ 4 ml/kg seed, Endosulfan @ 7ml / kg seeds, Thiram 75% WP, Carboxin 75% WP, *Bacillus subtilis* and *Pseudomonas flourescens*2.5 – 5 kg in 100kg FYM, *Pseudomonas flourescens,* @ 10gm/kg. Captan 75 WS @ 1.5 to 2.5 gma.i./litre for soil drenching. Carbofuran/Carbosulfan 3% (w/w) etc. effective against soil and seed borne infection caused by fungi, insects and nematodes of the crops like agricultural, vegetable and horticultural crops.

Carbamates

Carbamate have a much shorter duration of action and are somewhat less toxic. Aldicarb (granular formulation and systemic) and active against a wide variety of nematodes as well as insects and mites and is useful in a variety of soil types throughout the world, e.g., Carbofuran (granular and liquid formulation and systemic insecticide). Oxamyl is the only nematicide with downward-moving systemic activity.

Organochlorides

DDT, the best known organochloride, was created by Swiss scientist Paul Müller and got Nobel Prize for Physiology or Medicine in 1948. DDT was introduced in 1944. DDT open sodium channels in the insect's nerve cells. But DDT reduce the thickness of egg shells of predatory birds and reducing bird populations. Most of organochlorides are banned now a day due to toxic effects on beneficial insects and other organisms.

Organophosphates

Organophosphates show cumulative toxic effect to wildlife, so multiple exposures to the chemicals amplifies the toxicity. Ethoprop (G orEC), a non-systemic insecticide/nematicide show the mobility in the soil. Fenamiphos show systemic insecticidal activity. It is widely used as an ematicide. Fosthiazateis developed in1992 as system icorganophosphorus nematicide having broad-spectrum activity and active against *Pratylenchuspenetrans, Meloidogyne* spp and *M. arenaria* on peanut but not for *M. javanica* and *Rotylenchulusreniformis* (Van Damme et al., 2005; Verdejo-Lucas et al., 2003).

Biochemicals

The combination of extracts of *Opuntialindheimeri, Quercusfalcate, Rhusaromatica, Rhizophoramangle* (sincocin). It showed effectivity against citrus nematode, reniform and cyst nematode.

Pyrethroid

Pyrethroid pesticides mimic the insecticidal activity of the natural compound pyrethrum, which is found in pyrethrins. Pyrethroids are non-persistent sodium channel modulators and are less toxic than organophosphates and carbamates and applied to control household pests (Thomas and Kintrup 1991).

Neonicotinoids (Broad Spectrum)

Neonicotinoids are synthetic analogues of the natural insecticide nicotine. These chemicals are acetylcholine receptor agonists (a substance which initiates a physiological response when combined with a receptor). They applied as sprays, drenches, seed and soil treatments. It has recently comes under scrutiny (critical observation) for allegedly pernicious (causing great damage) effects on honeybees and its potential to increase the susceptibility of rice to planthopper attacks.

Ryanoids

These are synthetic analogues with the same mode of action as ryanodine, a naturally occurring insecticide extracted from *Ryaniaspeciosa.* They bind to calcium channels in cardiac and skeletal muscle, blocking nerve transmission.

CONCLUSION

Pesticides are widely used to protect agricultural and horticultural crops from plant pathogenic organisms like insects, fungi, nematodes etc. But chemical pesticides produce hazardous effects on the crops, other beneficial plant, insects and microorganisms like bioagents, predators and parasitic insects. Hence many chemical pesticides are banned now a days due to their lethal effects. Instead of chemical applied pest control is introduced known as integrated pest management. IPM has several tactics prevent and suppress pests without harmful impact on human health, the environment and nontarget organisms.

Farmers purchase fungicides at how any cost to protect their crops but fungicides affect adversely on soil fertility, soil microorganisms and other organisms like earthworms and even biological nitrogen fixing bacteria and algae also affected by the use of fungicides. Fungicides also cause pollution. Instead of using fungicides farmers should use biocontrol agents and plant extracts but these sources should be provided by government authorizing institutes commercially at very cheap rates comparison to fungicides to the farmers. So that our environment and living world will be safe.

REFERENCES

Aspelin, L., & Grube, A. H. (1999). Pesticides Industry Sales and Usage, 1996 and 1997, U.S. *Environmental Protection Agency.*, *733*, 99–001.

Crop Protection Manager. (1997). Insect Management- Insecticides Do Have a Role in IPM. Author.

Dreistadt, S. H. (2004). *Pests of Landscape Trees and Shrubs: An Integrated Pest Management Guide, Publication 3359*. University of California Division of Agriculture and Natural Resources.

Goswami, B. K., & Mittal, A. (2004). Management of root-knot nematode infecting tomato by *Trichodermaviride* and *Paecilomyceslilacinus. Indian Phytopathology*, *57*, 235–236.

Gupta, S. (2010). Biopesticides: An eco-friendly approach for pest control. *Journal of Biopesticides*, *3*(1), 186–188.

Kogan, M. (1998). Integrated Pest Management: Historical Perspectives and Contemporary Developments. *Annual Review of Entomology*, *43*(1), 243–270. doi:10.1146/annurev.ento.43.1.243 PMID:9444752

Nesci, A., Bluma, R., & Etcheverry, M. (2005). *In vitro* selection of maize rhizobacteria to study potential biological control of *Aspergillus*section *Flavi. European Journal of Plant Pathology*, *113*, 1–13. doi:10.100710658-005-5548-3

Palmer, W. E., Bromley, P. T., & Brandenburg, R. L. (2007). *Wildlife & pesticides - Peanuts*. North Carolina Cooperative Extension Service.

Rabb, R.L., & Guthrie, F.E. (1970). *Concepts of pest management*. Academic Press.

Smith, S. M. (1996). Biological control with Trichogramma: Advances, successes, and potential of their use. Annual Review of Entomology, 375–406. doi:10.1146/annurev.en.41.010196.002111

Stern, V. M., Smith, R. F., van den Bosch, R., & Hagen, K. S. (1959). The integrated control concept. *Hilgardia*, *29*(2), 81–101. doi:10.3733/hilg.v29n02p081

Suman, G. (2010). Biopesticides: An eco-friendly approach for pest control. *Journal of Biopesticides*, *3*(1), 186–188.

The Children's Health Act of 2000, PL 106-310, Title V, Section 511, enacted October 17, 2000. Codified in: Title 42, U.S.Code, Section 300w- 3(a)(1)(E).

Thomas, J., & Kintrup, J. (1991). Pyrethroids as household insecticides: Analysis, indoor exposure and persistence. *Fresenius' Journal of Analytical Chemistry*, *340*(7), 446–453. doi:10.1007/BF00322420

Van Damme, V., Hoedekie, A., & Viaene, N. (2005). Long-term efficacy of *Pochoniachlamydosporia* for management of *Meloidogynejavanica* in glasshouse crops. *Nematology*, *7*, 741–745.

Verdejo-Lucas, S., Sorribas, F. J., Ornat, C., & Galeano, M. (2003). Evaluating *Pochoniachlamydosporia* in a double cropping system of lettuce and tomato in plastic houses infested with *Meloidogynejavanica. Plant Pathology*, *52*(4), 521–528. doi:10.1046/j.1365-3059.2003.00873.x

Chapter 14
Green Biotechnology:
Potential and Prospects to Tackle Future Challenges

Alok Kumar Singh
Acharya Narendra Deva University of Agriculture and Technology, India

Ashish Pandey
Saint Mark's College, USA

Devendra Kumar
https://orcid.org/0000-0001-9516-4830
Swami Vivekanand Subharti University, Meerut, India

Reshu Chaudhary
https://orcid.org/0000-0002-9566-5368
Swami Vivekanand Subharthi University, Meerut, India

Rakesh Singh Sengar
https://orcid.org/0000-0002-7282-4135
Sardar Vallabhbhai Patel University of Agriculture and Technology, Meerut, India

Ashutosh Singh
Rani Lakshami Bai Central Agricultural University, Jhansi, India

Nisha Malik
Swami Vivekanand Subharti University, Meerut, India

Narendra Singh
BUA&T, Banda, India

Aditi Chatterjee
University of California, USA

D. K. Dwivedi
Acharya Narendra Deva University of Agriculture and Technology, Ayodhya, India

ABSTRACT

In this era, agriculture is one of the central issue. In the world where scarcity and ailments associated with malnourishment is high and anticipated to extend within the future, new approaches have to be necessitated. Traditional techniques alone cannot meet the obligatory measures hence alternative measures need to be created. Thus, green biotechnology is the crucial requirement. Therefore, this paper narrates the significance of green biotechnology to cope with future challenges, emphasizing the applications of the techniques and the necessity for its implementation.

DOI: 10.4018/978-1-7998-8434-7.ch014

INTRODUCTION

Green Biotechnology is the combination of practices that utilize natural resources and biodiversity wealth to sustain agriculture and quality production. Green Biotechnology is the rapidly growing field within modern biotechnology by deploying natural resources as the source of renewable energy to produces foods but it is in that it is also in the form of biofuels and, as a novel means to generate novel pharmaceutical products. It produces more efficient crop plants with better quality foods and nutrition, and economically beneficial *Green biotechnology* focused on agriculture as working field with the conserving natural resources. Green biotechnological approaches and applications involve development of new crop genotypes of agricultural interest, production of biofertilizers and biopesticides, using in vitro cultivation and cloning plants. It uses environmental- friendly solutions alternative to traditional agriculture, horticulture, and animal breeding processes for selection, breeding, and management of crops for more economical production which are accounted as follows (Vaghasiya and Shiroya, 2015):

- Utilization of bacteria for increased plant growth Development of grains having pest resistance
- Exploiting microbes for better plant growth, tolerance to abiotic stresses and ripen earlier
- Genetic alterations in plant genotypes to develop tolerance to fungal pathogens
- Genetic modifications to resist frost stress in plants
- Using phytoremediation to get rid of heavy metals like lead, nickel, silver, chromium etc.
- Plant genetic engineering to express pesticides

Using bacterial cultures to assure better crop yield instead of herbicides and pesticides. The world population is anticipated to achieve over ten billion within the year 2050, whereas agricultural production is growing at the slower rate of regarding 1.8 percent annually (Altman, 1999). All human beings depend on agriculture to provide the quality of food they need. However, traditional agricultural experiences face some serious constraints, including market restrictions, limited natural resources, and scarcity of natural and natural resources and inherent biological genetic limitations.

Green biotechnological approaches is the need of the day because only traditional agriculture practices cannot meet the demands. The Green Revolution helped a lot by increasing wheat production by 10 folds in India and many other countries, thus producing three times that amount. However, this has a boundary and needs to back up a solution to provide better plants. Reproduction is necessary with green technical knowledge because traditional methods alone cannot meet needs. Yes, we need a different approach to increase yields. Therefore, using green biotechnology-based technology can turn into a “Evergreen revolution”. To reduce the demand for fertilizers and pesticides under climate change, to balance world population growth and its impact on future generations, adapt crops to competition and climate change, and maintain the new technologies needed to increase profits and increase the productivity of many ways. More sustainable and sustainable work to increase food security and at the same time reduce land Negative impacts of strong agriculture. Therefore, there is an urgent need to adopt green biotechnology engineering approach.

GREEN BIOTECHNOLOGY

Major Challenges of Agriculture

1. *Population growth:* More food requirements to meet the needs of the ongoing population
2. *Less acreage*: Shortage of arable land is a major concern in some regions,
3. *Less water*: Limited water resource for agricultural consumption.
4. *Environmental protection*: Agriculture must contribute towards reduction of greenhouse gas emissions, by using biomass from plants for future energy production.
5. *Abiotic stresses:* Agriculture will have to adjust to extreme weather conditions and enhance tolerance to abiotic stresses such as drought, flood, salinity, etc.
6. *Biotic stresses:* such is weeds, diseases, insects, etc. can have a huge impact on agriculture, so it should find measures to enhance tolerance to these problems.

Therefore, new technological solutions need to be found to overcome the above challenges.

The Need for Green Biotechnology Adoption

It is found that in US, organic fanning is the fastest growing food trend. In the United States, organic crops are well known for being the fastest growing food. According to the USDA chemical research study, farmland is said to be smaller than traditional agriculture. In developing countries, deforestation of forest areas and burning, slashing them on agricultural land, release tremendous amount of carbon dioxide into the atmosphere and ultimately detrimental to the water cycle and wildlife. If all the agricultural land in the world were to move to organic farming to reduce pesticide use and nitrogen production, you would add 50% arable land. Therefore, half of the forest is cultivated needs to be chopped out from other forests and raises livestock to meet the fertilizers needed to fertilize these resources. Deforestation produces large amounts of carbon dioxide. This means that most natural gas is not a viable alternative, so the problem now is to reduce pesticides and nitrogen and export more food on land without cutting down some of the world's forests. Are there any options? Yes, the answer is green biotechnology. This technology is the combination of cultivation approaches that can nourish people with more nutrition, and can sustain agriculture with less fertilizer, less watering, and less pesticide use.

Approaches and Techniques used in Green Biotechnology

Tissue Culture and Micropropagation

Tissue culture is the cultivation of plant cells, tissue or specially formulated nutrient media under aseptic condition. Using the right condition, an entire plant can be regenerated from a single cell. There are different tissue culture methods which are as follows:

Another culture

One of the tissue culture method to develop improved varieties in a short time period has been employed in the successful development of double haploid line of rice, wheat, sorghum, barley and other field crops. Anther culture also helps genetic improvement of scented *Indica rice* by the use of androclonal variation (Roy and Mandal, 2005).

Embryo Rescue

It involves the culture of immature embryos of plants in a special medium to prevent abortion of the young embryo and to support its germination. This is used routinely in breeding parental lines having different or incompatible genome such as in introducing traits of wild relatives into cultivated crops. Wild rice is a rich source of traits for resistance to pests and abiotic stresses.

The development of a new rice plant type for West Africa had combined yield traits of the Asian *Oryza sativa* parent with local adaptation traits from African rice *Oryza glaberrima* (Anonymous, 2014). So embryo rescue technique is utilized for the transfer of resistant gene from the wild type to the cultivated ones. This technique has significant advantages over traditional clonal propagation techniques.

Micropropagation

A tissue culture method developed for the production of disease-free, high quality planting material and for the rapid production of many uniform plants is now possible to provide clean and uniform planting materials, in plantations like banana, pine, date, rubber tree, etc., field crops like sugarcane, pineapple, tomato, potato etc., root crops like cassava, yam, sweet potato etc. (Ogero et al. 2012, Kumar et al., 2020) and agronomical crops like sugarcane (Kumar et al., 2020), ornamental plants like orchids such as *Coelogyne stricta* (through pseudobulb segment culture) (Basker and Bai, 2006), *Dendrobiwn crysanthum* (Rao and Barman, 2014), anthuriums, gerbera (*Gerbera jamesonii*) (Bhargava et al. 2013), and *Begonia rubrovenia* var. meisneri which is a rare and endemic plant of Meghalaya (Mittal and Decker, 2013). Many medicinal plants can be regenerated through this technique such as *Drymaria cordata* (Tejavathi and Indira, 2013), *Tylophora indica* (Kaur et al. 2011), *Centella asiatica* (Singh et al. 2010), various *Decalepsis* species (Sharma and Shahzad, 2014), *Saraca asoca* (Subbu et al. 2008), *Tagetes erecta* (Deka and Arjuna, 2014), and *Gaultheria fragrantissima* (Ranyaphia et al. 2011), the valuable oil yield of which is known to be higher in the north-eastern part of India. Conservation of biodiversity of these important medicinal plants of India through this technique can be seen in a review paper (Sharma *et al.* 2010). Many important spice crops like turmeric (*Curcuma longa*) (Mittal and Decker, 2013), ginger, chilli (*Capsicum annuum* L.) (Reddy et al. 2014), garlic (Taskin et al. 2013) also use this technique for improvement of the crops. This technique also offers effective strategy for rapid propagation and mass multiplication of economically important species of various *Dendrocalamus species* (Singh et al. 2014) (Kapruwan et al. 2014) keeping in view of their sustainable development and utilization. Plant which is critically endangered to Meghalaya, *Ilex khasiana* (Dang et al. 2011) and the only pitcher plant in India which is available only in the state of Meghalaya, *Nepenthes khasiana* (Anonymous, 2001) also use this technique for its mass multiplication including a giant fern, *Cyathea gigantean* (Das *et al.*, 2013) one of the endangered plant available in the North-Eastern states. Most of the plants mentioned above are endangered and important, so by the adoption of this technique it can be of much achievement. Micropropagated plants are found to establish more quickly, grow more vigorously and taller, have a shorter and more uniform production cycle and produce higher yields than conventional propagules (Kumar et al., 2019; Kumar et al., 2020).

Molecular Breeding and Marker-Assisted Selection

For identification of specific genes, a short segment of nucleotide sequence known as molecular markers are used. The markers are located near the DNA sequence of the desired gene. Therefore, this map of the markers and genes on specific chromosomes can be created by identifying the location of the gene with the help of the marker. With complete genetic maps and advanced knowledge, the location of markers and genes as well as their distance from other known genes can be detected. Using a very detailed genetic maps and better knowledge of the genetic structure of a plant's DNA researchers can analyze a tiny bit of tissue from a newly germinated seedling and hence can concentrate only on a specific trait. Currently this technique is already in use for crops where gene and the markers for a specific trait are known (Malik et al., 2020). It is being used in the efficient introgression of important genes into rice such as bacterial blight resistance, increased beta carotene content, submergence tolerance, etc. There are already more than 2200 simple microsatellite repeats identified for rice. Molecular markers are also used to determine the genetic profile of a line of variety. The information on genetic diversity of the lines is utilized in selecting for extremely unrelated parents useful for hybrid seed technology and also provides details on the parentage of the line, the possible traits and the unique identity of the plant useful for germplasm collection database (Malik et al., 2020). This technique has been used for genetic diversity analysis in many plants like maize (Malik et al., 2020), soybean (Ghosh *et al.* 2014), cucurbit species and many more using different markers like SSRs and RAPDs and thus can be utilized for widening the genetic base. Differences between coconut plants can also be investigated with SCoT signaling analysis (Rajesh et al. 2015).

Genetic Engineering and Genetically Modified (GM) Crops

Genetic engineering is the process in which recombinant DNA technology can be used to change the genetic structure of cells. The ability to manipulate individual genes and to transfer genes between species that would not readily interbreed is what distinguishes genetic engineering from traditional plant breeding. Unlike traditional breeding practices, genetic engineering allows the direct access and transfer of one or just few genes between related or distant species.

Use of Genetic Engineering

This strategy is used when all other strategies have been exhausted and when: (Anonymous, 2014)

1. the trait to be introduced is not present in the germplasm of the crop;
2. the trait is very difficult to improve by conventional breeding methods; and
3. it will take very long time to introduce or improve such trait in the crop by conventional breeding methods.

The length of time in developing transgenic plant depends upon the gene, crop species, available resources and regulatory approval (Choudhary *et al.* 2014). It varies from 6 to 15 years before a new transgenic plant or hybrid is ready for commercial release. There has been a consistent increase in the global area planted to transgenic or GM crops or biotech crops from 1996 up to the present. ISAAA's Annual Global Status Report downloadable at the ISAAA website: http://www.isaaa.org presents an up-to-date record of the number of countries growing GM crops, the number of hectares planted, the benefits of biotech resources, information on farmers growing biotech resources and current information on expected future events and the direction the technology will takes. Transgenic crops which are planted commercially are herbicide tolerant soybean, maize, canola, cotton; insect resistant maize and cotton; and virus resistant squash and papaya: in India BT cotton being the only commercially available crop.

With the help of genetic engineering, many of the properties that can be incorporated into plants are known as aggregate properties, such as insect resistance and herbicide tolerance in combination in corn and cotton crops that are available commercially. The processing of many substances like biofertilizers, biopesticides and many biocontrol agents have been introduced by genetic engineering. *Burkholderia cepacia* (Devi et al. 2012), a naturally occurring rhizobacter found in North-East India can be used as a biocontrol agent for the control of many diseases such as wilt and damping off diseases in French beans. B. *cepacia* has also been reported to increase growth of corn, increase crop yield and degrade a wide variety of pesticides. *Withania somnifera* (Thilip et al. 2015), an important medicinal plant which yield pharmaceutically active compounds called with anolides can be genetically modified by agrobacterium mediated transformation through sonication and heat treatment for hairy root induction as the hairy root has the ability to synthesize with a ferine A and with anolide A, both steroidal lactones of medicinal value. It is found that it is possible to produce transgenic crops which are substantially equivalent to the controlled non-transformed lines (Baudo *et al.* 2006). There are various future initiatives in genetic engineering which will make more direct contributions to food quality, clean environment, pharmaceutical production and livestock feeds. GM seed tend to be more expensive, but in return it reduces expenses in several areas, such as the cost of pesticides, machines and labour. But above all: yields generally increase considerably because plants' own mechanism protect them from harmful insects and more effective weed management reduces harvest losses which used to be considered inevitable. Farmers have widely adopted GM technology. Between 1996 and 2011, the total surface area of land cultivated with GM crops had increased by a factor of 94, from 1700 sq. kms. (4200000 acres) to 1600000 km^2 (395 million acres (160 million hectares) in 29 countries including India. For more researches on GM crops one can also refer to the book of 'A decade of EU-funded GMO research (2001-2010) (Anonymous, 2010).

The above techniques can also be used in combination, as in case of genetic improvement of ginger (Suma *et al.* 2008), *Dichanthittm annitlatum* (Vaghasiya and Shiroya, 2015), and many other crops as well.

Production of Green Energy

Consumption of raw materials such as bioethanol from edible crops, biomass and lignocellulosic products and biodiesel from used vegetable oils and animal fats, and biogas which is produced by fermentation of plant and animal wastes is very good source for energy production to combat green house gas emission and environmental pollution which otherwise with other traditional energy products like fossil fuels is damaging to the environment and destined to unavoidable environmental pollution. The first generation biofuels were made with edible sugars and starches which had restrictions such as competition for land and water for food and fibre with increasing demand for animal feeds while the next generation or second and third generation biofuels are being developed from inedible lignocellulosic compounds (Mittal and Decker, 2013) (Asgher et al., 2014) that include woody biomass and wood wastes, crop residues, switch grass, municipal wastes and algae which do not compete directly with food production and can often be produced on marginal or unused croplands. Plant microbe interactions can also be a promising strategy for bioenergy crop production as these clops are not utilized for human consumption, it is ideal for developing and evaluating noval technologies and applications. Lands unfit for food production can be made into use and planted with low input perennial crops capable of producing high biomass yield and these energy crops inoculated with beneficial endophytes can also be employed to phytoremediate land for future crop production. Although these bioenergies can not be a replacement but it can surely supplement our use. The Indian Railways has started led to use the oil from the Jatropha plant to power its diesel engines with great success, currently diesel locomotives running from Thanjavur to Nagore section and Titruchirapalli to Lalgudi, Dendigul and Karur sections (Anonymous, 2012).

Numerous researches have been conducted on the field of Biotechnology which can also be referred in Annual Report of Indian Agricultural Research Institute, New Delhi, Biotechnology and Management of Bioresources Division, TERI, New Delhi and many more.

Role of Green Biotechnology to Fight and Adapt to Climate Change

Changing meteorological conditions associated with climate change will definitely have an impact on agriculture yields. Green Biotechnology offers a toolbox which can help farmers produce food sustainably through green house gas emission reduction, crop adaptation, crop protection and increased yield from less available arable land.

Greenhouse Gas Reduction

Agricultural practices such as deforestation, cattle feedlots and fertilizer use contribute to green house gas emissions. So, green biotechnology can help farmers to produce food sustainably through less fuel consumption on farms through a reduced need to spray crops, carbon sequestration and reduced fertilizer use and nitrous oxide emissions. Crops developed with green biotechnology like GM herbicide tolerant and GM insect resistant crops reduce the need for tillage or ploughing allowing farmers to adopt conservation or "no-till" farming practices and less pesticide spraying. This results in less fuel use and less CO_2 emissions. The development of crops that use nitrogen efficiently which requires less nitrogen fertilizer also helps in reduction of NO_2 emissions.

Crop Adaptation

Green Biotechnology can play a vital role in improving yields by using water more sustainably, thus helping to cope with water scarcity which works in two ways, i.e. by reducing water loss and by improving drought tolerance. Water loss can be reduced by developing practices to reduce the amount of ploughing before planting the-crops which means the soil surface is not inverted and hence traps the soil moisture. Drought tolerance can be improved by using drought tolerant crops.

Protected and Increased Yield with Less Surfaces

Crops adopted by Green Biotechnology helps 'farmers to increase yield while also using fewer precious natural resources and being more resistant to pests and disease that are likely to spread in today's changing climate.

CONCLUSION

Green Biotechnology as we can see is changing the condition of plants in three main areas i.e., controlled and sustainable growth (vegetative, nutritive, production and fertility). Protect plants from the threats of biotic and abiotic stress. and Increasing the awareness by producing specialty foods, bio chemicals and pharmaceuticals (Altman 1999). To cope up with the problems that we are facing today and tend to be faced in upcoming days, it is the need of an hour to adopt Green Biotechnology and find measures to fight with it. The challenges are related to the fight against climate change, food safety and security, scarcity of petroleum products due to limited fossil fuel resources, increasing population, fight against diseases, poverty and social exclusion. Crop production needs to meet the requirements of rapid growth in demand while ensuring environmental sustainability. Preservation of natural resources and to support livelihood of the fans and followers around the world are the major concerns. Food that are tolerant to different climate changes, food with higher yield, disease resistant, and development of alternative renewable resources that are environmental friendly, (some of which are just a waiste and regarded as unwanted materials can therefore be a useful one) should be the main focus today's research activities which I must say can be achieved through Green biotechnology. In India biotech applications to agriculture is still in the nascent early stage and its benefits have not reached to the majority of the population. Why is it so? The main reason behind this is the lack of awareness among the people.

Therefore, more awareness programmes must be found in this field. For an agricultural country like India, the adoption of an agriculture technology needs to be done carefully, keeping in mind the interest of farmers. Regarding the for' or 'against' the technique, one should be fully aware of the pros and eons and should look into both sides of it scientifically. First of all, one should be able to differentiate between a fact and a myth appropriately. Evaluation tests, approval decision and scientific, technical and other information should be provided to the public promptly and appropriately to accept science. People need to consider the scale of risks and different benefits. We should make use of the application of all available technologies without discrimination and prejudice with respect the order and ethical principles. However only a structured dialogue with policy makers, stakeholders and the public based on sound science and empirical evidence will clear the way for a balanced assessment of the benefits and risks of biotechnology. Thus, it is necessary to ensure the local people in the importance of technologies developed for betterment of crops and promote awareness and educate about the important indigenous crops available in the region and importance towards the conservation strategy and the techniques need to be adopted. So that they revealed to deal with the crisis that they faced to be in near future. The agricultural practices adopted in the North-Last region is still not been promising. The area still uses old age shifting (Jhum) cultivation which has many disadvantages and the region, once opulently endowed with rich genetic diversity of plants, has been stripped due to human interference by implementation of irrational land use system. With rapid surge in human and livestock population and scientifically seized land, farmers are growing rapidly in large numbers of people and animals and in growing demand for food, feed, oil, food, textiles, timber and other development projects. The farmers have been forced to exploit forestland and groundwater resources at sub-optimal level by ignoring it's this potential in complete defiance. This causes slow-moving losses, loss of biodiversity, severe erosion that will lead to the depletion of living things, land production levels and the ability to transport their contents. This has resulted in progressive decrease in forest cover, serious soil erosion leading to depletion of plant nutrients, gradual degradation and decline in land productivity and its carrying capacity, silting of major river basin causing recurrent floods in the plains, and drying up of perennial steams as well as ecological imbalances. Destroying and receding, large features of river expansion cause frequent flooding of fields, causing permanent steam damage and environmental instability. What are the particularly heavy resources that require specific location measures to store, use, and manage these resources to maximize productivity without sacrificing quality. The scientific goal of biotechnology and world agriculture is is the improvement of the genetics of our crops. Since, the country's economy depends on agriculture and a large proportion of its population depends on agriculture, so there is an alarming situation we need to deal with "green biotechnology".

REFERENCES

A decade of EU-funded GMO research 2001-2010. (2010). European Commission Directorate-General for Research and Innovation Biotechnologies, Agriculture, Food. Publications Office of the European Union.

Agricultural Biotechnology (A Lot More than Just GM Crops). (2014). International Service for the Acquisition of Agri-Biotech Applications (ISAAA) Resources Publications.

Altman, A. (1999). Plant biotechnology in the 21st century: the challenges ahead: Review, EJB. *Electronic Journal of Biotechnology*, *2*(2), 51–55. doi:10.2225/vol2-issue2-fulltext-1

Asgher, M., Bashir, F., & Hafiz, M. N. I. (2014). A comprehensive ligninolytic pre-treatment approach from lignocellulose green biotechnology to produce bio-ethanol. *Chemical Engineering Research & Design*, *92*(8), 1571–1578. doi:10.1016/j.cherd.2013.09.003

Basker, S., & Narmatha Bai, V. (2006). Micropropagation of Coelogyne stricta (D.Don) Schltr. via Pseudobulb Segment Cultures. *Tropical and Subtropical Agroecosystems*, *6*, 31–35.

Baudo, M. M., Lyons, R., Powers, S., Pastori, G. M., Edwards, K. J., Floldsworth, M. J., & Shewry, P. R. (2006). Transgenesis has lessimpact on the transcriptome of wheat grain than conventional breeding. *Plant Biotechnology Journal*, *4*(4), 369–380. doi:10.1111/j.1467-7652.2006.00193.x PMID:17177803

Bhargava, Dilta, Gupta, Dhiman, & Modgil. (2013). Studies on micropropagation of gerbera (Gerbera jamesonii Bolus). *Indian Journal of Applied Research*, *3*, 8–11. doi:10.15373/2249555X/NOV2013/4

Biotechnology and Management of Bioresources Division (BMBD) Brochure. (n.d.). The Energy and Resources Institute (TERI).

Chandrasekaran, Raju, Varutharaju, Asiarn, & Shajahan. (2015). Improved Agrobacterium rhizogenes-mediated hairy root culture system of Withania somnifera (L.) Dunal using sonication and heat treatment. *Biotech, 3*, 297.

Choudhary, B., Gheysen, G., Buysse, J., van der Meer, P., & Burssens, S. (2014). Regulatory options for genetically modified crops in India [Review]. *Plant Biotechnology Journal*, *12*(2), 135–146. doi:10.1111/pbi.12155 PMID:24460889

Dang, J. C., Kumaria, S., Kumar, S., & Tandon, P. (2011). Micropropagation of Ilex- khasiana, a critically endangered and endemic holly of Northeast India. *AoB Plants*, *2011*, 1–7. doi:10.1093/aobpla/plr012 PMID:22476482

Deka, B., & Arjuna, M. (2014). Effect of Plant Growth Regulators on in vitro propagation of Tagetes erecta Linn [Review]. *Indian Journal of Basic and Applied Medical Research*, *3*, 15–23.

Devi, S. I., Somkuwar, B., Potshangbam, M., & Talukdar, N. C. (2012). Genetic characterization of Burkholderia cepacia strain from Northeast India: A potential bio-control agent. *Advances in Bioscience and Biotechnology*, *3*(8), 1179–1188. doi:10.4236/abb.2012.38144

Green Biotechnology and Climate Change. (2009). Factshect produced by EuropaBio. The European Association for Bioindustries.

Jayceta Ghosh, R. D. (2014). An Assessment of Genetic Relatedness between Soybean [Glycine max (L.) Merrill] Cultivars Using SSR Markers. *American Journal of Plant Sciences*, *5*(20), 3089–3096. doi:10.4236/ajps.2014.520325

Kapruwan, S., Bakshi, M., & Kaur, M. (2014). Effect of growth regulators on the in vitro multiplication of Dendrocalamus Hamiltonii. *International Journal of Engineering Research and Applications*, *4*, 83–86.

Kaur, H., Anand, M., & Goyal, D. (2011). Establishment of an efficient protocol for micropropagation of stem explants of Tylophora indica, an important medicinal plant. *African Journal of Biotechnology*, *10*, 6928–6932.

Kumar, D., Sengar, R. S., Malik, N., Yadav, M. K., Gupta, S., Chand, P., Singh, G., & Kumar, P. (2020). In vitro evaluation to intensify the differential morphogenetic response through plant growth regulators and antibiotic supplementation in sugarcane. *Plant Physiol. Rep.*, *25*(2), 335–346. doi:10.100740502-020-00516-6

Kumar, D., Sengar, R. S., Yadav, M. K., anchand, P., Singh, G., & Gupta, S. (2019). Evaluation of Sterilant Effect on In-vitro Culture Establishment in Sugarcane Variety Co 0118. *International Journal of Current Microbiology and Applied Sciences*, *8*(07), 1226–1233. doi:10.20546/ijcmas.2019.807.146

Kumar, J., Shukla, S. M., Bhat, V., & Gupta, S. (2005). In Vitro Plant Regeneration and Genetic Transformation of Dichanthium annulatum. DNA and Cell Biology, 24, 670-679.

Malik, N., Kumar, D., & Babu, B. K. (2020). Analysis of genetic divergence and population structure through microsatellite markers in normal and quality protein maize genotypes from NW Himalayan region of India. *Vegetos*, *33*(1), 194–202. doi:10.100742535-020-00100-1

Mittal & Decker. (2013). Application of biotechnology for biofuels: Transforming biomass to biofuels. *Biotech, 3*, 341-343.

Ogero, Mburugu, Mwangi, Ngugi, & Ombori. (2012). Low Cost Tissue Culture Technology in the Regeneration of Sweet Potato (Ipomoea batatas (L) Lam). *Research Journal of Biology, 2*, 51-58.

Pursuit and Promotion of Science: The Indian Experience. (2001). Indian National Science Academy (INSA).

Raiesh, Sabana, Rachana, Rahman, Jerard, & Karun. (n.d.). Genetic relationship and diversity among coconut (*Cocos nucifera* L.) accessions revealed through SCoT analysis. *Biotech, 3*, 304.

Rama Subbu, R., Chandraprabha, A., & Sevugaperumal, R. (2008). Invitro Clonal propagation of vulnerable medicinal plant, Saraca asoca (Roxb.) De Wilde. *Natural Product Radiance, 7*, 338–341.

Ramez, N. (2013). *Greener than green: Biotech and the future of agriculture (Part I)*. http://www.geneticliteracyprojectorg/2013/04/221greener-than-green-biotech-andthefuture-of-agriculture/

Ranyaphia, R. A., Maoa, A. A., & Borthakurb, S. K. (2011). Direct organogencsis from leaf and internode explants of in vitro raised wintergreen plant (Gaultheria fragrantissima). *Science Asia, 37*(3), 186–194. doi:10.2306cienceasia1513-1874.2011.37.186

Reddy, M. K., Srivastava, A., Kumar, S., Kumar, R., Chawda, N., Ebert, A. W., & Vishwakanna, M. (2014). Chilli (Capsicum Annum L.) Breeding In India: An Overview. *SABRAO Journal of Breeding and Genetics, 46*, 160–173.

Roy & Mandal. (2005). Anther culture response in indica rice and variations in major agronomic characters among the androclones of a scented cultivar, Karnal local. *African Journal of Biotechnology, 4*, 235–240.

Sharma, S., & Shahzad, A. (2014). An overview on Decalepsis: A Genus of Woody Medicinal Climbers. *Journal of Plant Science Research, 1*, 104.

Singh, Jaiswal, & Jaiswal. (2014). Thidiazuron-Induced Anatomical Changes and Direct Shoot Morphogenesis In Dendrocalainus strictus nees. *Canadian Journal of Pure and Applied Sciences, 8*, 2901–2904.

Singh, S., Gautam, A., Sharma, A., & Batra, A. (2010). *Centella asiatica* (L.): A Plant with Immense Medicinal Potential but Threatened. *International Journal of Pharmaceutical Sciences Review and Research, 4*, 9–17.

Someswar & Barman. (2014). In Vitro Micropropagation of Dendrobium chtysanthum Wall. ex Lindl. -A Threatened Orchid. Scholars Academic Journal of BIosciences, 2, 39-42.

Strategies for Agricultural Research in the North-East. (2001). National Academy of Agricultural Sciences.

Study on The Indian Green Biotech Market and EU Collaboration Opportunities. (2012). *European Business and Technology Centre (EBTC)/Euro India Research Centre.*

Suma, Keshavachandran, & Nybe. (2008). Agrobacterium tumefaciens mediated transformation and regeneration of ginger (*Zingiber officinale Rosc.*). *Journal of Tropical Agriculture*, *46*, 38–44.

Taskin, H., Baktemur, G., Kurul, M., & Buyiikalaca, S. (2013). Use of Tissue Culture Techniques for Producing Virus-Free Plant in Garlic and Their Identification through Real-Time PCR. *TheScientific-WorldJournal*, *2013*, 1–5. doi:10.1155/2013/781282 PMID:23935432

Tejavathi, D. H., & Indira, M. N. (2013). Regeneration of Shoots from Leaf Callus Cultures of Dryntaria cordata (1) willd ex Roem and Schult. *Indian Journal of Fundamental and Applied Life Sciences*, *3*, 111–115.

Vaghasiya & Shiroya. (n.d.). *A Review Green biotechnology - a help to the environment*. http://www pharmatutor.org/articles/review-green-biotechnology-helpenvironment

Chapter 15
Impact, Mitigation, Adaptation, and Food Security for Promotion of Sustainable Agriculture Through Green Technology

Ashu Singh
Sardar Vallabhbhai Patel University, India

Pragati Misra
Sam Higginbottom University of Agriculture Technology and Sciences, India

Pradeep Kumar Shukla
Sam Higginbottom University of Agriculture Technology and Sciences, India

ABSTRACT

Over the past few decades, the man-induced changes in the climate of the earth have become the focus of scientific and social attention. The most imminent of these is the increased levels of greenhouse gases (GHGs) such as carbon dioxide, methane, and nitrous oxide in the atmosphere leading to global warming. The recent report of the Intergovernmental Panel on Climate Change (IPCC) has reconfirmed the increasingly strong evidence of global climate change and projected that the global average temperature of the air above the earth's surface would rise by 1.8-6.4°C by the end of the century depending upon the developmental pathways of the countries. For the Indian region (South Asia), the IPCC has projected 0.5-1.2°C rise in temperature by 2020, 0.88-3.16°C by 2050, and 1.56-5.4 °C by 2080, depending on the scenario of future development. Overall, the temperature increase is likely to be much higher in winter (rabi) than in the rainy (kharif) season. Precipitation is likely to increase by 15-40% by the end of the century.

DOI: 10.4018/978-1-7998-8434-7.ch015

INTRODUCTION

The inevitable changes in climate resulting from the 'greenhouse effect' are expected to have a major impact on Indian agriculture. The economic viability and environmental amenity of agriculture depend heavily on soil health. In order to better predict the likely impact of climate change, a proper understanding of how soil health will be affected, is required. The objective of this article is to assess the impact of climate change on soil health, and to analyze the mitigation and adaptation strategies in the context of Indian agriculture (Figure 1).

Soil as a Source and Sink at of Greenhouse Gases

Soil is one of the important and sink of GHGs thus directly influencing climate change and indirectly through the production and consumption of ammonia nitrogen oxide and carbon monoxide. It contributes about 20% to the total emission of carbon dioxide through soil respiration and root respiration, 12% of methane and 60% of anthropogenic nitrous oxide emissions (IPCC 1996). As per India's Initial National Communication (NATCOM, 2004) to the UNFCCC, the emission fronts the anthropogenic activities in India amounted to 817 Tg of CO_2. 18.1 Tg of CH_4 and 0.18 Tg of N_2O for the base year 1994. The per capita CO_2emission was 0.87 t CO_2 which was only 4% of the US and 23% of the global average per capita CO_2 emissions.

Agriculture sector in India contributes 28% the total GHG emissions. The emissions from agriculture are primarily due to methane emission from rice fields (23%), enteric fermentation in in ruminant animals (59%), and nitrous oxides from application of manures and fertilizer soils. Rice is generally grown in waterlogged condition, which creates an anoxic environment and is conducive for methane production by the anaerobic methanogenic bacteria, Methane is also produced as a result of microbial activity in anaerobic, conditions in the rumen of cattle, buffalo, and other livestock for the degradation of cellulosic feeds and roughages eaten up by the animals. Anaerobic decomposition of dung also contributes to methane emissions. Agricultural soils particularly those fertilized with inorganic or organic N, are major sources of nitrous oxide. Nitrification and Denitrification are the processes that contribute towards nitrous oxide emission from soil.

Impact of Climate Change on Soil Health

The soil system responds to the short-term events such as rainfall, and also undergoes long-term changes such as physical and chemical weathering due to climatic change. The potential impacts on soil health resulting due to climate change would be in the organic matter supply, temperature regimes, hydrology and salinity.

The following are the major consequences of global climate change on soil health.

- Soil carbon levels are expected to decrease due to decreased net primary production. Any gains by increased plant water use efficiency, due to elevated CO_2 are likely to be outweighed by increased carbon mineralization after episodic rainfall, and reduced annual and growing season rainfall. The quality of soil organic matter may also shift where the more inert components of the carbon pool prevail. The residues of crop under elevated CO_2concentration will have higher C:N ratio and this may reduce their rate of decomposition and nutrient supply. Increase of soil temperature will in-

Figure 1. Overview of mitigation strategy for climate change

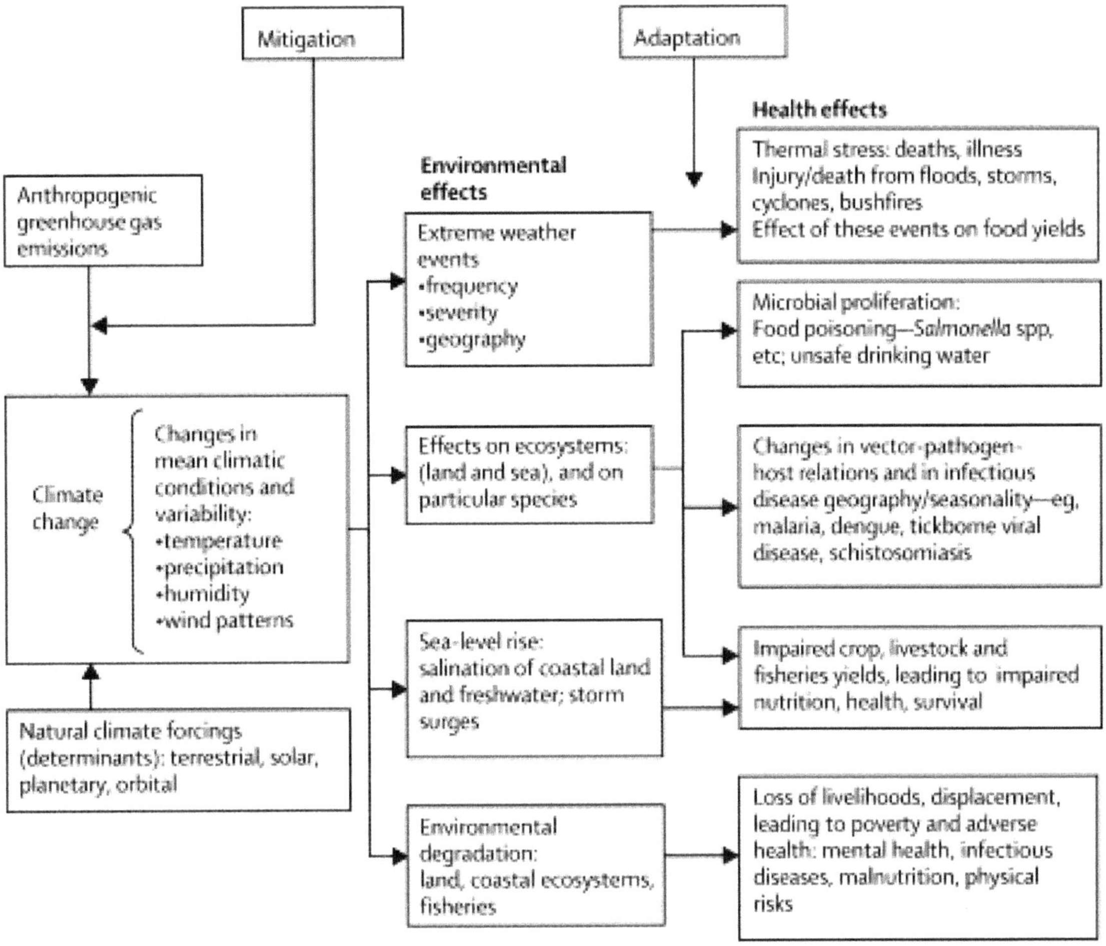

crease N mineralization, but its availability may decrease due to increased gaseous losses through processes such as volatilization and denitrification.

- With increased temperature and higher rate of evapo-transpiration, soil will be drier. This may also result in lowering of groundwater table at some places. The melting of glaciers in the Himalayas will increase water availability in the Ganges and their tributaries in the but in the long run the availability of water will decrease considerably. The soil water balanced in different parts of India will be distributed and the quality of groundwater along the coastal tract will be more affected due to intrusion of sea water.
- Change in rainfall volume and frequency, and increasing events of intense wind may alter the severity, frequency and extent of soil erosion. Therefore, an increased risk of soil erosion and nutrient loss due to reduced vegetation cover in combination with torrential rainfall and greater wind intensities is expected.
- Rise in sea level may lead to salt-water ingression in the coastal lands turning them less suitable for conventional agriculture. Transient salinity may increase as capillary rise dominates, bringing salts into the root zone in salt affected soils. Leaching during torrential rainfall events may

be limited due to surface sealing. Increased subsoil drying increases concentration of salts its the soil solution. Conversely, the severity of saline scalds due to secondary salinization may abate, as groundwater levels fall in line with reduced rainfall.

- Soil biodiversity and its functions are expected to change under conditions of elevated CO_2 and changed moisture and temperature regimes. As soil biodiversity regulates nutrient dynamics and many disease risks, nutrient availability to crops and pastures could change as per the exposure to soil-borne diseases.

Mitigation of GHG Emission from Soil

The strategies for mitigating methane emission from rice cultivation could be altering water management, particularly promoting mid-season aeration by short-term drainage; improving organic matter management by promoting aerobic degradation through composting or incorporating it into soil during off-season drained period; use of rice cultivars with few unproductive tillers, high root oxidative activity and high harvest index; and application of fermented manure like biogas slurry in place of unfermented farmyard manure. Methane emission from ruminants can be reduced by altering the feed composition, either to reduce the percentage which is converted into methane or to improve the milk and meat yield.

The most efficient management practice to reduce nitrous oxide emission is site-specific nutrient management. The emission could also be reduced by nitrification inhibitors such as nitropyrene and dicyandiamide. There are some plant-derived organics such as neem oil, neem cake and Karana seed extract which can also act as nitrification inhibitors.

Mitigation of CO_2 emission from agriculture can be achieved by increasing carbon sequestration in soil through manipulation of soil moisture and temperature, setting aside surplus agriculture land, and restoration of soil carbon in degraded land. Soil management practices such as reduced tillage, manuring, residue incorporation, improving soil biodiversity, micro aggregation, and mulching can play important roles in sequestering carbon in soil. Adaptation Strategies to Climate Change with Respect to Soil Health Potential adaptation strategies to deal with the impact of climate change are developing cultivars tolerant to heat and salinity stress and resistant to flood and drought, modifying crop management practices, improving water management, adopting new farm techniques such as resource conserving technologies (RCTs), crop diversification, improving pest management, better weather forecasts and crop insurance and harnessing the indigenous technical knowledge of farmers. As far as soil health is concerned the impact of climate change needs to be addresses through soil organic matter.

Efficient use of Resources

Resource conservation technologies involving zero-or minimum tillage with direct seeding permanent or semi-permanent residue cover, and crop rotation have potential to improve the efficiency of use of natural resources, including water air, fossil fuel and soil. Among other things the efficiencies gained include less land and time needed to produce the required staple cereals, allowing farmers to diversify crops and cropping patterns or pursue other gainful activities. The technologies improve the sustainability of the cropping system by conserving the resource base and higher input use efficiency and also mitigating GHG emission (Aggarwal and Pathak. 2009).

Developing Climate-ready Crops

One important adaptive option includes developing cultivars resistant to climate change. Crop varieties that are resistant to lodging (e.g., short rice cultivars) may withstand strong winds during the sensitive stage of crop growth. The yield ceiling must be raised and the yield gap narrowed while maintaining sustainable production and a friendly environment. Development of new varieties with higher yield potential and stability is complementary to bridging the yield gap. A combination of farm-level adaptation and economic adjustments such as increased investment in agriculture infrastructure, and reallocation of land and water would be desired in the agricultural sector (Pathak and Wassmann 2017).

Crop Diversification

Diversification of crop and livestock varieties including the replacement of plant types cultivars hybrids, and animal breeds with new varieties intended for higher drought or heat tolerance, have been advocated as having the potential to increase productivity in the face of temperature and moisture stresses. Diversity in seed genetic structure and composition has been recognized as an effective defense against disease and pest outbreak and climate hazards. Moreover, the demand for high value foods such as fruits, Vegetables, dairy, meat, eggs and fish is increasing because of growing the income and urbanization. This is reducing the demand for traditional rice and wheat. Diversification from rice-wheat to high value commodities will increase income and result in reduced water and fertilizer use. However, there is a need to quantity the impacts of crop diversification on income, employment, soil health, water use and greenhouse gas emission. The most significant problem to overcome is that diversification is costly in terms of the income opportunities that farmers forego i.e., switching crop varieties can be expensive, making crop diversification typically less profitable than specialization. Moreover, traditions can often be difficult to overcome and will dictate local practices.

Adjusting Cropping Season

Adjustment of planting dates to minimize the effect of temperature increase-induced spikelet sterility can be used to reduce yield instability, by avoiding the flowering period to coincide with the hottest period. Adaptation measures to reduce the negative effects of increased climatic variability, as normally experienced in arid and semi-arid tropics, may include changing the cropping calendar to take advantage of the yet period and to avoid extreme weather events during the growing season. Cropping systems may have to change by including suitable cultivars (to counteract compression of crops development), increasing crop intensities (i.e., the number of successive crop produced per unit area per year) or planting different type of crops. Farmers will have to adapt to changing hydrological regimes by changing crops.

Water Management

In-situ soil-water management, particularly in mid and semi-arid regions where crop growth is severely limited by water deficit even if nutrient availability is adequate, is important for enhancing productivity and soil organic C content of soil. Water harvesting techniques and micro catchments are extremely beneficial in production in arid climates (WWEC, 2010).

Harnessing the Indigenous Technical Knowledge of Farmers

Farmers in south Asia, often poor and marginal, are experimenting with the climatic variability for centuries. There is a wealth of knowledge for a range of measures that can help in developing technologies to overcome climate vulnerabilities. There is a need to harness that knowledge and fine-tune them to suit the modern needs.

At the global scale, the key greenhouse gases emitted by human activities are:

Figure 2. Global greenhouse gas emissions by gas

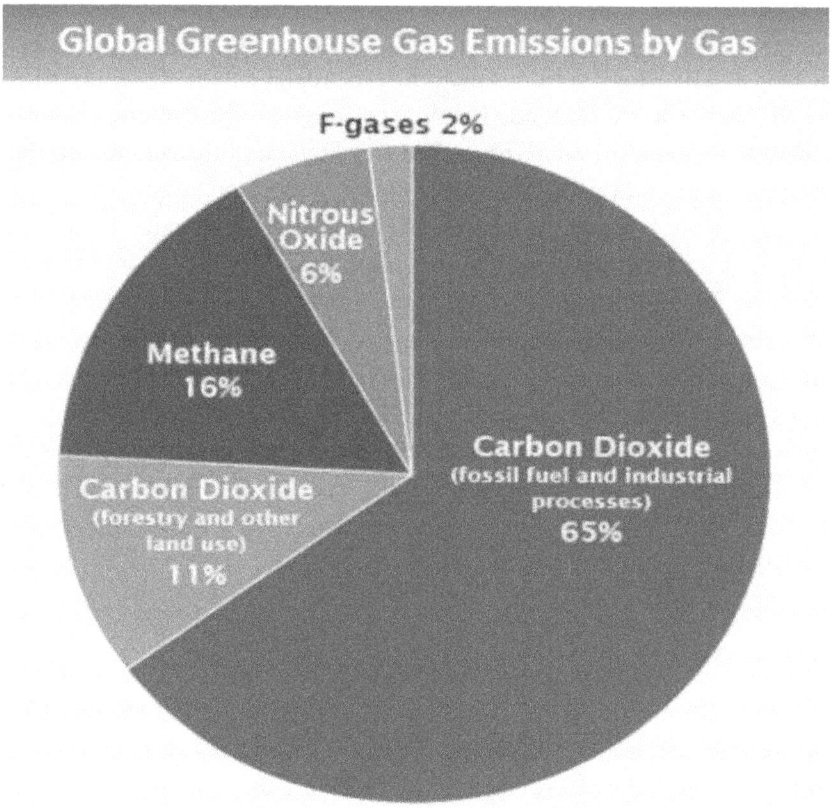

- **Carbon dioxide (CO2):** Fossil fuel use is the primary source of CO_2. CO_2 can also be emitted from direct human-induced impacts on forestry and other land use, such as through deforestation, land clearing for agriculture, and degradation of soils. Likewise, land can also remove CO_2 from the atmosphere through reforestation, improvement of soils, and other activities.
- **Methane (CH4):** Agricultural activities, waste management, energy use, and biomass burning all contribute to CH_4 emissions.
- **Nitrous oxide (N2O):** Agricultural activities, such as fertilizer use, are the primary source of N_2O emissions. Fossil fuel combustion also generates N_2O.

- **Fluorinated gases (F-gases):** Industrial processes, refrigeration, and the use of a variety of consumer products contribute to emissions of F-gases, which include hydro fluorocarbons (HFCs), per fluorocarbons (PFCs), and sulfur hexafluoride (SF_6).
    a. Black carbon is a solid particle or aerosol, not a gas, but it also contributes to warming of the atmosphere.

Global Emissions by Economic Sector

Global greenhouse gas emissions can also be broken down by the economic activities that lead to their production.

Figure 3. Global greenhouse gas emissions by economic sector

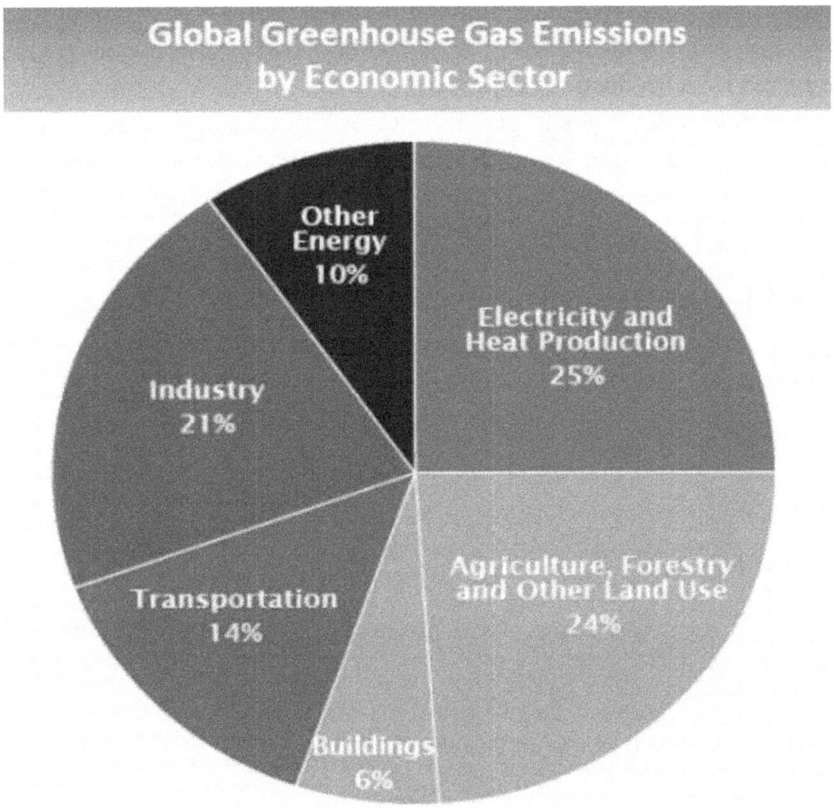

- **Electricity and Heat Production** (25% of 2010 global greenhouse gas emissions): The burning of coal, natural gas, and oil for electricity and heat is the largest single source of global greenhouse gas emissions.
- **Industry** (21% of 2010 global greenhouse gas emissions): Greenhouse gas emissions from industry primarily involve fossil fuels burned on site at facilities for energy. This sector also includes emissions from chemical, metallurgical, and mineral transformation processes not associated with

energy consumption and emissions from waste management activities. (Note: Emissions from industrial electricity use are excluded and are instead covered in the Electricity and Heat Production sector.)

- **Agriculture, Forestry, and Other Land Use** (24% of 2010 global greenhouse gas emissions): Greenhouse gas emissions from this sector come mostly from agriculture (cultivation of crops and livestock) and deforestation. This estimate does not include the CO_2 that ecosystems remove from the atmosphere by sequestering carbon in biomass, dead organic matter, and soils, which offset approximately 20% of emissions from this sector.
- **Transportation** (14% of 2010 global greenhouse gas emissions): Greenhouse gas emissions from this sector primarily involve fossil fuels burned for road, rail, air, and marine transportation. Almost all (95%) of the world's transportation energy comes from petroleum-based fuels, largely gasoline and diesel.
- **Buildings** (6% of 2010 global greenhouse gas emissions): Greenhouse gas emissions from this sector arise from onsite energy generation and burning fuels for heat in buildings or cooking in homes. (Note: Emissions from electricity use in buildings are excluded and are instead covered in the Electricity and Heat Production sector.)
- **Other Energy** (10% of 2010 global greenhouse gas emissions): This source of greenhouse gas emissions refers to all emissions from the Energy sector which are not directly associated with electricity or heat production, such as fuel extraction, refining, processing, and transportation.

Figure 4. Trends in global emissions

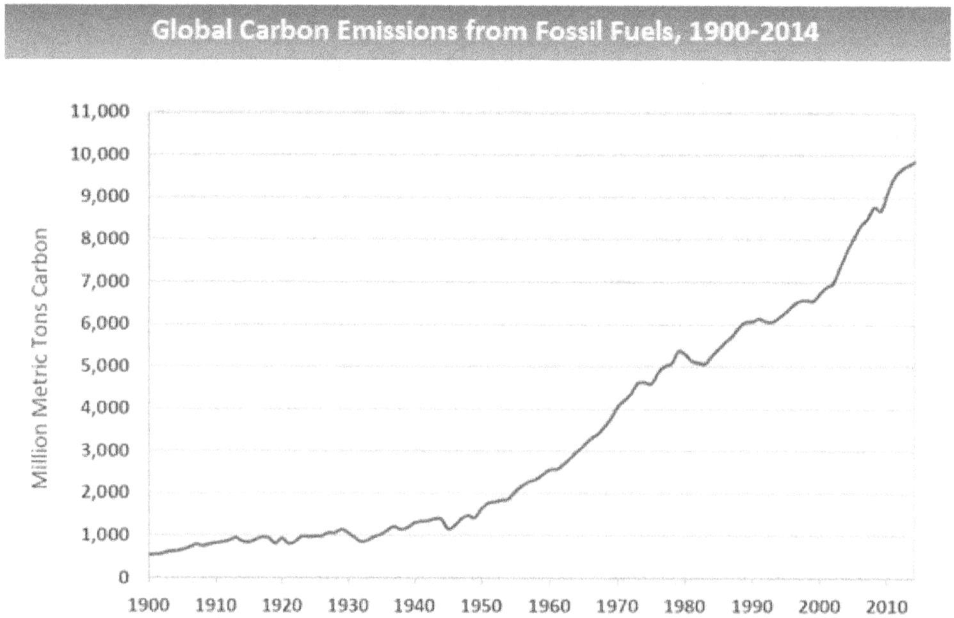

Figure 5. Emissions by country

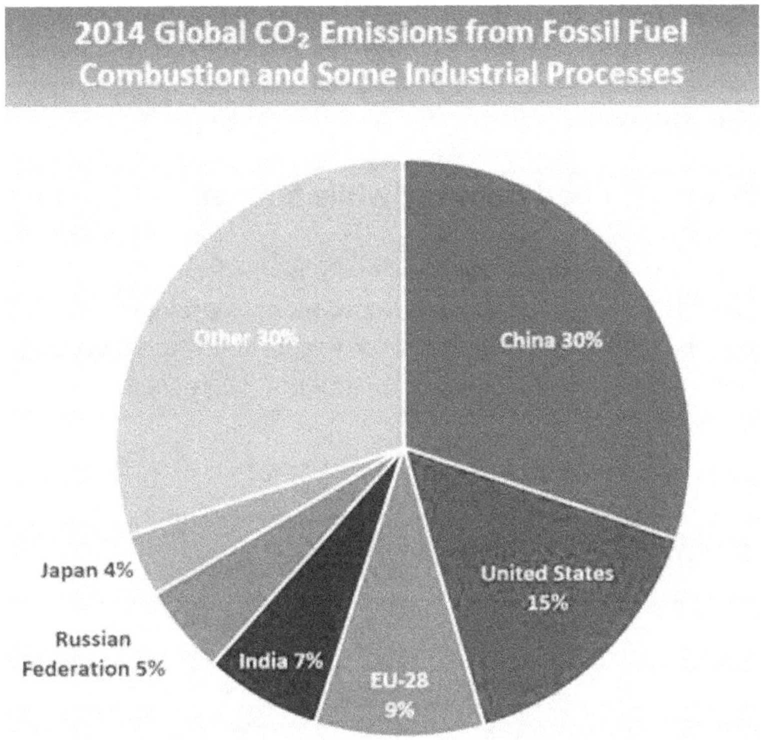

Global Hot Spots of Food Insecurity

Number of food-insecure people in the world has increased to 1020million (FAO, 2009), and the U.N. Millennium Development Goals of cutting hunger to half by 2015 will not be met. Increase in prices of food staples in 2007-08 has reduced the access to food and exacerbated food insecurity for poor people around the world. Over and above the problem of access to food, the supply at food is also limited by soil degradation (IPCC, 2007), especially that caused by accelerated erosion, nutrient deficiency, and the depletion of soil organic carbon (SOC) pool. Regions with a severe problem of food insecurity include South Asia (or SA comprising of Afghanistan, Pakistan, India, Nepal, Bhutan, Bangladesh and Sri Lanka) (FAO, 1996; Banik, 2007; Agarwal et al., 2009), And Sub-Saharan Africa (SSA) (Anonymous, 2009). Farmers in SA and SSA are small landholders, resource-poor and highly vulnerable to climate-related and other natural and anthropogenic disruptions. The subsistence level threshold of per capita availability of cereals is 190 kg ha^{-1}(Funk and Brown, 2009). With this criterion, those regions which will face a serious challenge of food-insecurity through 2030 and beyond are SA and the entire continent of SSA. Whereas the per capita food production will decline in all regions of SA, the production is below the threshold optimum in Western Asia (760 kg/person) and SA (193 kg/person). Assuming that access to food through better income can make a major impact in reducing regional and local food production gaps, food insecurity is not an issue in Western Asia because of the oil-based revenue. However, food-insecurity will remain a serious issue in SA. By 2050, the number of food-insecure people in SA may increase to 373 million, and the per capita production levels may reach similar to those of the 1960s and

1970s (Funk and Brown, 2009). The problem of food-insecurity is even more daunting in SSA where the per capita cereal production is below the threshold level (190 kg/person) in 2007 and is projected to decrease more drastically by 2030, The problem may be exacerbated by the projected climate: change and the increase in drought stress (Verhagen *et al.*, 2004), and the attendant increase in risks of soil degradation and land desertification.

Strategies of Advancing Food Security while Mitigation Climate Change

Improving agricultural production is the basic strategy for both advancing food security and mitigation climate change. Indeed, agriculture must be engine of economic development SA, SSA, and elsewhere in the developing world. It is difficult to achieve poverty reduction without prior investment in agriculture (Lipton, 2005), especially in countries where majority of the population is rural and is dependent on agriculture.

Table 1. Per capita cereal Production in 2007 and 2030 in different regions of Asia and the world (Modified from Funk and Brown, 2009)

	Region	Per capita Cereal Production (kg ha-1)	Per Capita Cropland Area(ha)
	2007	**2030**	**2007**
World	354	306	0.11
North America 1374	1236	0.23	
Asia			
Eastern Asia	314	276	0.06
Southern Asia	231	193	0.09
South-East Asia	368	356	0.10
Western Asia	204	160	0.10
Central Asia	541	-	0.32
Sub-Saharan Africa			
Eastern Africa	131	84	0.09
Middle Africa	62	38	0.06
Northern Africa	190	180	0.11
Southern Africa	182	189	0.07
Western Africa	189	166	0.16

Elsewhere may not be able to advance economically and industrially without replacing extractive and subsistence farming (Funk and Brown, 2009) by science-based agriculture comprising of proven technologies and recommended management practices (RMPs). In addition to advancing food security and alleviating poverty, raising low crop yields, and breaking the agronomic barriers in SA and SSA (especially in rainfed /dry farming agro ecosystems), SCS may also be the judicious strategy to adapt to climate change. The fact is that even the modest increase (20% to 30%) in production in rain fed agriculture in SA and SSA through SCS can have a drastic positive impact on the standard of living and wellbeing of hundreds of millions of people now living below the poverty line.

CLIMATE CHANGE AND FOOD SECURITY

Agriculture, climate and food security Agriculture is important for food security in two ways: it produces the food people eat; and (perhaps even more important) it provides the primary source of livelihood for 36 percent of the world's total workforce. In the heavily populated countries of Asia and the Pacific, this share ranges from 40 to 50 percent, and in sub-Saharan Africa, two-thirds of the working population still make their living from agriculture (ICAR, 2012). If agricultural production in the low-income developing countries of Asia and Africa is adversely affected by climate change, the livelihoods of large numbers of the rural poor will be put at risk and their vulnerability to food insecurity increased.

PROTECTING FOOD SECURITY THROUGH MITIGATION OF CLIMATE CHANGE

In the long term, mitigating climate change will be critical to avoiding future breakdowns in food and livelihood systems and sharp increases in the number of food-insecure people worldwide. Historically, land conversion from forest to pasture- or cropland, and intensive crop and livestock production practices have been important sources of greenhouse gas emissions. Food systems also have enormous potential to mitigate climate change, however, particularly at the production end of the food chain. Moreover, many of the most effective mitigation measures also represent highly effective adaptation strategies, especially for commercial agriculture. Investing in wider adoption of best practices for mitigation in the food and agriculture sector could therefore have multiple payoffs for food security, including contributing to the stability of global food markets and providing new employment opportunities in the commercial agriculture sector, as well enhancing the sustainability of vulnerable livelihood systems. Such practices include: reducing emissions of CO_2, such as through reduction in the rate of land conversion and deforestation, better control of wildfires, adoption of alternatives to the burning of crop residues after harvest, reduction of emissions from commercial fishing operations, and more efficient energy use by forest dwellers, commercial agriculture and agro-industries; reducing emissions of methane and nitrous oxide, such as through improved nutrition for ruminant livestock, more efficient management of livestock waste and of irrigation water on rice paddies, more efficient applications of nitrogen fertilizer on cultivated fields, and reclamation of treated municipal wastewater for aquifer recharge and irrigation; sequestering carbon, such as through improved management of soil organic matter, with conservation agriculture involving permanent organic soil cover, minimum mechanical soil disturbance and crop rotation (which also saves on fossil fuel usage); improved management of pastures and grazing practices on natural grasslands, including by optimizing stock numbers and rotational grazing; introduction of integrated agroforestry systems that combine crops, grazing lands and trees in ecologically sustainable ways: use of degraded, marginal lands for productive planted forests or other cellulose biomass for alternative fuels; and carbon sink tree plantings. According to the most recent data released by IPCC, clearing of forested area for agriculture accounted for 17.4 percent of total greenhouse gas emissions in 2000, with emissions from intensive crop and livestock production contributing another 13.5 percent. By contrast, studies carried out by the World Resources Institute (WRI) indicate that energy sector emissions attributable to agricultural and food processing use of fossil fuels account for only 2.4 percent of greenhouse gas emissions (WRI, 2006). The share of total transportation emissions attributable to food system activities is not identified, but as total Climate change and food security: a framework document 60 emissions for all forms of transport for all purposes came to just 13.1 percent, the part attributable to transport of food commodities and products is likely to be low. In the United Kingdom, the Carbon Trust, established in 2001 with government funding, has promoted the concept of the "carbon footprint". By undertaking a carbon investigation of their supply chains, all businesses can minimize the carbon emitted at every stage of a product's life cycle, from source to shelf, consumption and disposal. The total amount of carbon emitted to arrive at a final product is that product's carbon footprint (Carbon Trust). Contributions of agriculture and forestry to greenhouse gas emissions Source: Adapted from IPCC, 2007b. Application of this concept to the food system has led some to argue that food products imported from developing to developed countries, particularly horticultural products transported by air, should not be traded because of their high carbon footprints. However, the principal suppliers of these foods are small-scale farmers who are just emerging into commercial markets, whose livelihood systems are still precarious and whose household food security would be seriously jeopardized if new overseas market opportunities were suddenly denied them. Moreover, as already described, the carbon footprint of food processing and transport is negligible compared with the emissions generated by production processes in the food system. Therefore, although there are opportunities for reducing the carbon footprint of food at all stages of the food chain, the focus of mitigation efforts in the food system should be on introducing agricultural production practices that reduce emissions or increase carbon sequestration.

REDUCING EMISSIONS

Good options exist for reducing the current level of agriculture-related emissions and, in the process, introducing more sustainable farming practices that strengthen ecosystem resilience and provide more security for agriculture-based livelihoods in the face of increased climatic variability. These are discussed in the following sections. Reducing agricultural and forestry emissions of carbon dioxide. The primary source of carbon emissions in the food and agriculture sector is land conversion from forested area to cultivated or grazing land. Carbon emissions can be reduced through more efficient energy use by mechanized agriculture and agro-industries, and through adoption of alternatives to the common practice of burning crop residues after harvest. However, the amounts involved are minor compared with the potential contribution that reducing the rate of deforestation could make. As already noted, intentional land conversion and deforestation, also referred to as anthropogenic land-use change, currently accounts for an important share of greenhouse gas Energy 25.9 Greenhouse gas emissions by sector in 2004 Industry 19.4 Forestry 17.4 Agriculture 13.5 Transport 13.1 Buildings 7.9 Waste and wastewater 2.8 Protecting food security through mitigation of climate change 61 emissions. Moreover, the reduction in global forested area caused by land clearing and unsustainable logging (in which cut trees are not replaced with new plantings) has reduced the capacity of the world's forests to store carbon. Evidence shows that Amazon deforestation, related to agricultural expansion for livestock grazing and the production of livestock feed and biofuel crops, already contributes substantially to global anthropogenic CO_2 emissions (Carvalho *et al.*, 2004). Continued intensification of the global livestock industry and growing demand for liquid biofuel crops will create additional pressure to clear tropical forests worldwide unless policies are put in place to manage the process sustainably. UNFCCC and the Kyoto Protocol recognize the potential role of forests in providing a variety of adaptive ecosystem services in addition to mitigating climate change through carbon sequestration. These services include biodiversity preservation, watershed protection on mountain slopes, control of desertification, and maintenance of the environmental integrity of fragile coastal zones. However, current rates of forest degradation and deforestation are threatening the capacity of the world's forests to perform these multiple roles. Cyclical loss and regrowth of trees and forests is a natural process. Forests are regularly ravaged by the spread of plant pests and fire. Natural fires maintain forest health by clearing away dense brush and dead wood and allowing new growth to emerge; they are also part of the life cycles of some species. The natural burning of trees and other organic matter releases CO2 into the atmosphere, while the decay of dead plants produces methane. These emissions of greenhouse gases are normally compensated for by the process of photosynthesis in living plants, especially the new vegetation that springs up on cleared land and needs CO2 in order to grow. In recent times, however, a still largely uncontrolled process of deforestation resulting from human activity has been altering this natural balance. Changes in temperature ranges and precipitation, attributable to climate change, can harm forests further. Droughts and forest fires are expected to increase, with devastating effects on forests that are already stressed by human activity. There are indications that the Amazon is drying out, which could lead to dangerous fires and desertification. Invasive insect species may also damage forest health. Insects' role in boreal ecology is to decompose litter, supply food for birds and small animals and eliminate diseased trees, but insect attacks are likely to increase in frequency and intensity as established forests succumb to the physiological stress associated with warmer weather (Greenpeace Online). In Canada, for example, more than 12 million ha of forests have been lost in recent years, owing to mountain pine beetle attacks, which are more common when winters are mild. Forests' capacity to play their natural role in maintaining climatic stability is closely linked to food systems' response to the challenge of climate change (Pathak, 2013). To slow down and eventually reverse the still largely uncontrolled deforestation process, forest clearing, grazing in forested areas, cutting of trees for fuel wood and commercial logging must all become planned activities, based on trade-offs between benefits and costs at different spatial and temporal scales. Action is needed on several fronts, especially through an integrated approach that simultaneously addresses the global demand for additional land to produce food and fuel, the dependence on forests as a source of livelihood for many rural people in developing countries, and the economic value of ecosystem services provided by forests. The actions required include creating economic alternatives to reduce the incentive for clearing forests or using forest resources unsustainably, promoting second-generation biofuels to avoid land clearing for biofuel crops, and enforcing more strictly the regulations that discourage potential investors from setting wildfires to clear land for commercial development. Controlling frontier expansion in tropical rain forests can make an important contribution to climate change mitigation, but often the sole option for preserving forested area is through intensifying agricultural production on the better land. It has been demonstrated that when intensification involves increased fertilizer inputs, the related emission increases are far less than the avoided emissions of organic carbon from the forests that have been preserved (Mullen 2005). Use of carbon offset schemes to pay rural households for sustainable management of the forested areas that they rely on for fuel and other forest products can provide the incentive to Climate change and food security: a framework document 62 stop them cutting wood to sell as timber, fuel wood or charcoal. To be effective, however, this approach needs to be accompanied by public or private sector investment in alternative sources of timber and cooking fuel to meet the growing demand. Reducing agricultural emissions of methane and nitrous oxide Digestive processes and wastes from ruminant livestock that eat a great deal of fibrous material are an important source of methane, especially in intensive production units, where large numbers of animals are concentrated in relatively small spaces. Worldwide, ruminant livestock are the largest source of methane from human-related activities (EPA Online). Through the process of enteric fermentation, which is unique to ruminant animals such as cattle, sheep and goats, unused carbon is released in the form of methane during the digestion of fibrous materials in the diet. Methane emissions from animal manure are also considerable, and increasing rapidly. These two sources account for 60 percent of agricultural emissions of methane and about 30 percent of total anthropogenic methane emissions. The other main source of agricultural methane is rice, accounting for almost 40 percent of agricultural methane emissions and about 20 percent of all human-caused methane emissions (GHG Online a). Although nitrous oxide is a relatively less important greenhouse gas in terms of share, it is highly potent, and derives almost entirely from manure, cultivated soils that have been fertilized with organic matter or inorganic compounds containing nitrogen, and nitrogen fixing legumes. The following sections discuss methods for minimizing emissions of methane and nitrous oxide from agricultural activities. Reducing methane emissions from ruminant livestock: Methane emissions per animal and per unit of livestock product are high when the animals' diet is poor (EPA Online). Range-fed beef cows are the most important source of methane from enteric fermentation because they are very large animals, even compared with dairy cows; their diets, consisting mainly of forages of varying quality, are generally poorer than those in the dairy or feedlot sectors; the level of management is usually not as good; and the beef cow population is very large. Better grazing management and dietary supplementation have been identified as the most effective ways of reducing emissions from this sector because they improve animal nutrition and reproductive efficiency. There are several technologies for reducing methane release from enteric fermentation. The basic principle is to increase the digestibility of feedstuffs, by either modifying feed or manipulating the digestive process. Most ruminants in developing countries, particularly in Africa and south Asia, have a very fibrous diet. Technically, these diets are relatively easy to improve through the use of feed additives or supplements. However, such techniques are often beyond the reach of smallholder livestock producers, who lack the capital, and sometimes the knowledge, to implement changes. Often, technical improvements may not be economical, such as where there is lack of demand or insufficient infrastructure. Even in Australia, for example, many opportunities to reduce emissions, such as through dietary fat supplementation or increased grain feeding, are not part of the low-cost range-fed dairy system, which focuses on per hectare rather than per cow production (Eckard, Dalley and Crawford, 2000) Another approach is to increase the level of starch or rapidly fermentable carbohydrates in the diet, thereby reducing excess hydrogen and the subsequent formation of methane. These too are measures that extensive range-fed production systems may not be able to adopt without external support, but national or regional planning strategies in areas where such systems are important could promote change. More advanced technologies that are being studied but are not currently operational would be applicable to free-ranging ruminants. Livestock fed on improved diets produce more milk and meat per animal. This increased production efficiency reduces the amount of methane emitted per unit of production and the size of the herd required to produce a given level of product. Because many developing countries are striving to increase production from ruminant animals (primarily milk and Protecting food security through mitigation of climate change 63 meat), improvements in production efficiency are urgently needed to meet goals while avoiding increase methane emissions. In the United States, the Environmental Protection Agency (EPA) reports that more efficient livestock production has already led to increased milk production, and methane emissions have decreased over the last several decades (EPA Online). Technically speaking, the potential for efficiency gains – and therefore for methane reductions – is even larger for beef and other ruminant meat production, which is typically based on poorer management, including inferior diets. Better grazing management and dietary supplementation have been identified as effective ways of improving efficiency and reducing emissions from this sector. In evaluating the emission reductions obtained from dietary modifications, it is important to consider that the feed and feed supplements used to enhance productivity and reduce methane emissions may have considerable embodied greenhouse gas emissions that have a negative affection the balance. Increased reliance on mechanized production of feed grains for both ruminants and non-ruminants has made the livestock food chain more fossil fuel intensive. Relying more on non-ruminant sources of animal protein (pigs, poultry, fish) in the diet can mitigate emissions from enteric fermentation and contribute to food security by improving the livelihoods of livestock-dependent households and adding diversity to the diet. Most of the increase in demand for animal protein to 2030 and beyond is projected to occur in emerging developing countries in Asia, where pig and poultry meat is preferred, so the relative share of beef in total animal protein consumption is likely to decline over time. If rising costs for water, feed and fuel trigger significant increases in the market price for ruminant livestock products, the result could be a shift in consumer behavior among those who currently prefer beef. Reducing methane emissions from rice: At between 50 and 100 million tonnes of methane a year, rice agriculture is a large source of atmospheric methane, possibly the greatest of the human-incurred methane sources. The warm, waterlogged soil of rice paddies provides ideal conditions for methanogens, and although some of the methane produced is usually oxidized by metamorphose in the shallow overlying water, the vast majority is released into the atmosphere (GHG Online). As the world population increases, reducing rice agriculture remains largely untenable as a strategy for reducing methane emissions from paddy rice fields. However, substantial reductions are possible through a more integrated approach to rice paddy irrigation and varietal selection. Many rice varieties can be grown under much drier conditions than those traditionally employed, with large reductions in methane emission without any loss in yield. Intermittent and/or alternating dry-wet irrigation of rice fields can be employed with these varieties. Applying the principles of conservation agriculture to crops such as irrigated rice would provide chances for reducing the water consumption of this cropping system and, by changing the soil environment from mostly anaerobic to aerobic, could also make it easier to fine-tune the irrigation pattern to reduce the emission of methane. There is also great potential for improved varieties of rice that can produce much larger crops per area of rice paddy, thereby allowing for reduced areas of rice paddies without reducing production. The addition of compounds that favor the activity of other microbial groups over that of the methanogens, such as ammonium sulphate, has proved successful under some conditions. Reducing methane emissions from manure: Although manure is the residue from animals' digestive processes – so is a waste product – it contains important amounts of nitrogen, phosphates and potassium that provide valuable soil nutrients when applied to farmers' fields. Poor manure management can increase the loss of pollutants to the environment, however. Nitrogen in manures can be lost as nitrate, nitrous oxide (a greenhouse gas) or ammonia (a constituent of acid rain and a cause of terrestrial eutrophication). Phosphorus-rich manure particles can be washed into watercourses, and can raise soil phosphorus contents to levels where phosphorus leaching begins. Climate change and food security: a framework document 64 If manure is managed as a liquid substance, it decays and forms methane (University of Hertfordshire Online). In the wild, animal manure is spread over a wide area and decomposes aerobically in the oxygen in the natural environment. Intensive livestock rearing methods cause high concentrations of manure to build up in relatively small areas, however, leading to a predominance of anaerobic (oxygen-free) decomposition of the manure, which produces methane (GHG Online) (Darwin 1999). There are options for managing manure in ways that do not contribute to greenhouse gas accumulation. Methane is not released when manure is managed as a solid substance through composting and drying, or is applied and worked into the fields without being left to stand. Moving away from intensive rearing methods to increased grazing time for animals, so greater dispersal of their manure, also increases aerobic rather than anaerobic decomposition and reduces the rate of methane production (Denman *et al.* 2007). The temperature at which manure is stored can have a significant effect on methane production. In farming systems where manure is stored in stables, such as in pig farms where effluents are stored in a pit in the cellar of the stable, emissions can be higher than when manure is stored outside at lower ambient temperatures. Greenhouse gas production can also be reduced through deep cooling of manure. Cooling of pig slurry can reduce indoor methane and nitrous oxide emissions by 21 percent (Sommer, Peterson and Møller, 2004). Trapping the methane released by livestock manure, for example in slurry tanks, has already proved very successful in reducing methane emissions to the atmosphere. The recovered methane, often called "biogas", can be flared off as CO2 or used as a fuel. The capture and burning of methane released from animal wastes is an increasingly applied form of energy generation and forms the basis for several carbon reduction and trading projects. Biogas is typically made up of 65 percent methane and 35 percent CO2, so the combustion of methane releases CO_2, but this is 23 times less noxious in terms of global warming impact than methane is. A further mitigation dividend is obtained when combustion provides an energy source to replace the use of fossil fuels. There are various storage systems for exploiting this huge potential, including covered lagoons and other structures for liquid storage, such as pits and tanks. These are suitable for large- or small-scale systems and cover a wide range of technological options and degrees of sophistication. Covered lagoons and biogas systems produce a slurry that reduces methane emission when applied to rice fields, instead of untreated dung (Sanchez *et al.,* 2005). The wider use of biogas systems (either for generating energy for on-farm use or for delivering electricity to the public net) depends on the relative prices of other energy sources. Until recently, biogas systems have not usually been competitive without subsidies, apart from in remote locations where electricity and other forms of energy are unavailable or unreliable. However, interest in this source of energy is growing. It is assumed that manure emissions in cool climates could be reduced by 50 percent through adoption of an alternative management option to replace the storage of manure as liquid slurry in open pits. In warmer climates, where methane emissions from liquid slurry are estimated to be more than three times as high (IPCC, 2007), a reduction potential of 75 percent is considered reasonable. Reducing nitrous oxide emissions from agricultural soils: A major direct source of nitrous oxide from agricultural soils is the widespread increase in the use of synthetic nitrate-based fertilizers, driven by the need for greater crop yields and by more intensive farming practices. Where large applications of these fertilizers are combined with irrigation practices that saturate soils, the resulting lack of oxygen in the soil produces conditions that are favorable to anaerobic conversion of solid nitrates and nitrites into nitrogen-containing gases (denitrification) and release of large amounts of nitrous oxide into the atmosphere. The widespread and often poorly controlled use of animal waste as fertilizer can also lead to substantial emissions of nitrous oxide from agricultural soils. The ammonia in urea-based fertilizers and manures vaporizes when exposed to the air. Ammonia, a compound containing nitrogen and hydrogen, can also be a source of nitrous oxide through volatilization following fertilizer application or during storage of manure. Protecting food security through mitigation of climate change 65 Some additional nitrous oxide is thought to arise from agricultural soils through the planting of leguminous crops that fix nitrogen, but the importance of this source is not yet clear. Nitrogen leaching and runoff from agricultural soils is another source of nitrous oxide emissions. After fertilizer application or heavy rain, large amounts of nitrogen may leach from the soil into drainage ditches, streams, rivers and eventually estuaries. Part of the nitrous oxide produced in agricultural soils is emitted to the atmosphere as soon as drainage water is exposed to the air, and another part is deposited in aquatic and estuarine sediments and emitted from there after undergoing denitrification (GHG Online d). Net nitrogen use in farming affects climate change, because it is linked to nitrous oxide emissions, and water pollution, because nitrates pollute soil, fresh and marine waters. Net nitrogen use can be measured relatively easily by recording the amounts of nitrogenous fertilizers and manures that are used on the farm, adding the nitrogen estimated to be fixed by legumes, and subtracting the nitrogen harvested in the crop and by-products. The net climate change impact is calculated by deducting the sequestration of greenhouse gases absorbed by the additional plant growth caused by fertilizer use from the temperature-forcing impacts of nitrogen fertilizers. The best way to manage human interference in the nitrogen cycle is to maximize the efficiency of nitrogen uses (Subramanian, 2017). Better targeting of fertilizer applications, in both space and time, can significantly reduce releases of nitrous oxides from agricultural soils. Land management strategies that consider the optimum amounts of fertilizer necessary for maximum crop yield and minimum waste are crucial, both environmentally and economically. The exact form of nitrogen-based fertilizer and the best time of year to use it are other key factors on which to base fertilization campaigns. Rapid incorporation and shallow injection of livestock wastes reduce nitrogen loss to the atmosphere by at least 50 percent, and deep injection into the soil essentially eliminates the loss (Sikka and Sulochana 1980). Crop rotations that efficiently recycles these nutrients, and fertilizer applications near to when they are needed by crops reduce the potential for further loss. RICMS uses a variety of these methods to increase the efficiency of nitrogen fertilizer in rice production. Options for reducing emissions from grazing systems are also important. Adding nitrification inhibitors to urea or ammonium fertilizer compounds before application can substantially reduce emissions of nitrous oxide (Boschetti *et al.*, 2007). On pastures, this technology inhibits the production of nitrous oxide from animal urine (McIntyre *et al.*, 2009). Balanced feeding is also important; for example, feed that is high in nitrogen will produce manure with high nitrogen content, which emits greater levels of nitrous oxide than manure with low nitrogen content does. Land drainage is another option for reducing nitrous oxide emissions before nitrogen enters the next phase of the nitrogen cascade. The compacting of soil by traffic, tillage and grazing livestock can reduce its oxygen content and enhance conditions for denitrification. Reducing soil wetness through better drainage can increase oxygen content and may reduce nitrous oxide emission significantly, especially in more humid environments.

CONCLUSION

For India's agricultural production systems to be viable into the future, there is a need to identify soil management systems that are climate change compatible, where soil organic C is maintained or enhanced and soil water is used efficiently to compensate for less available water. The importance of maintaining groundcover and residue retention to protect the soil from erosion and water loss will be greater. We need to urgently take steps to increase adaptive capacity to mitigate climate change impact. This would require increased adaptation research, capacity building, development activities and changes in land-use management. A win-win solution is to start with such adaptation strategies that are needed for sustainable development. Policies and incentives should be evolved that would encourage farmers to sequester carbon in the soil and improve soil health, and use water and energy more efficiently.

REFERENCES

Agarwal, S., Satyavada, A., Patra, P., & Kumar, R. (2008). Strengthening functional community–provider linkages: Lessons from the Indore urban health programme. *Global Public Health: An International Journal for Research, Policy and Practice*, *3*(3), 308–325. doi:10.1080/17441690701592957

Denman, K. L., Astuk, S. N., & Zimmerman, A. S. (2007). Couplings between Changes in the Climate System and Biogeochemistry. In *Climate Change 2007: The Physical Science Basis, Contribution of Working Group I to the Fourth Assessment Report of the Intergovernmental Panel on Climate Change*. Cambridge University Press.

Darwin, R. F. (1999). A Farmer's View of the Ricardian Approach to Measuring Agricultural Effects of Climatic Change. *Climatic Change*, *41*(3-4).

Darwin, R., Tsigas, M., Lewandrowski, J., & Raneses, A. (1995). World agriculture and climate change: economic adaptations. Agricultural Economic Report 703, US Department of Agriculture.

Sommer, S. G., Petersen, S. O., & Møller, H. B. (2004). Algorithms for calculating methane and nitrous oxide emission from manure management. *Nutrient Cycling in Agroecosystems*, *69*(2), 143–154. doi:10.1023/B:FRES.0000029678.25083.fa

Pathak, H., Sankhyan, S., Dubey, D. S., Bhatia, A., & Jain, N. (2013). Dry direct-seeding of rice for mitigating greenhouse gas emission: Field experimentation and simulation. *Paddy and Water Environment*, *11*(1-4), 593–601. doi:10.100710333-012-0352-0

Pathak, H., & Wassmann, R. (2007). Introducing greenhouse gas mitigation as a development objective in rice-based agriculture: I Generation of technical coefficients. *Agricultural Systems*, *94*, 807–825.

FAO/LEG. (1996). Regional compendium on aquaculture and inland fisheries legislation (Asian Region). FAO Development Law Service. FAO.

IPCC. (2007). *Climate Change 2007: The Physical Science Basis. Contribution of Working Group I to the Fourth Assessment Report of the Intergovernmental Panel on Climate Change* (S. Solomon, D. Qin, M. Manning, Z. Chen, M. Marquis, K. B. Averyt, M. Tignor, & H. L. Miller, Eds.). Cambridge University Press.

Verhagen, A., Dietz, A. J., Ruben, R., Van Dijk, H., De Jong, A., Zaal, F., De Bruin, M., & Van Keulen, H. (2004). Climate change and food security in the drylands of West Africa. In A. J. Dietz, R. Ruben, & A. Verhagen (Eds.), *The Impact of Climate Change on Drylands with a focus on West Africa* (pp. 365–384). Kluwer Academic Publishers.

ICAR. (2012) Degraded and Wastelands of India-Status and Special Distribution. Indian Council of Agricultural Research & National Academy of Agricultural Sciences.

Funk, C. C., & Brown, M. E. (2009). Declining per capita agricultural production and warming oceans threaten food security? *Food Security*, *1*(3), 271–289. doi:10.100712571-009-0026-y

Carvalho, J. A., Costa, F. S., Veras, C. A. G., Sandberg, D. V., Alvarado, E. C., Gielow, R., Serra, A. M., & Santos, J. C. (2004). Biomass fire consumption and carbon release rates of rainforest-clearing experiments conducted in northern Mato Grosso, Brazil. *Journal of Geophysical Research*, *106*(D16), 17877–17887. doi:10.1029/2000JD900791

Boschetti, L., Roy, D., Barbosa, P., Boca, R., & Justice, C. O. (2007). A MODIS assessment of the summer 2007 extent burned in Greece. *International Journal of Remote Sensing*, *29*(8), 2433–2436. doi:10.1080/01431160701874561

McIntyre, D., Herren, H. R., Wakhungu, J., & Watson, R. T. (2009). *International Assessment of Agricultural Knowledge, Science and Technology for Development (IAASTD): Global Report*. Island Press.

Subramanian, A. (2017). *Transforming Indian Agriculture: By Loving Some Agriculture Less and the Rest More*. Lecture at National Academy of Agricultural Sciences.

Sikka, D. R., & Gadgil, S. (1980). On the Maximum Cloud Zone and the ITCZ over Indian, Longitudes during the Southwest Monsoon. *Monthly Weather Review*, *108*(11), 1840–1853. doi:10.1175/1520-0493(1980)108<1840:OTMCZA>2.0.CO;2

Mullen, Orden, & Gulati. (2005). *Agricultural policies in India*. International Food Policy Research Institute (IFPRI).

WWEC. (2010). *National Statistics on Water and Wastewater Treatment*. WWEC.

Chapter 16
Integrative Agronomic Paradigm for Efficient Phytoremediation of Metal-Contaminated Soil:
Agronomic Practices for Efficient Phytoremediation

Shweta Saraswat
https://orcid.org/0000-0003-0504-7812
Department of Environmental Sciences, G. B. Pant University of Agriculture and Technology, Pantnagar, India

J. P. N. Rai
Department of Environmental Sciences, G. B. Pant University of Agriculture and Technology, Pantnagar, India

ABSTRACT

Soil pollution by heavy metals and metalloids has led to significant environmental and human health hazards. Conventional remediation technologies for trace elements are cost-intensive and degrade soil quality. Phytoremediation offers an in-situ, ecological, and economic technology for remediation of metals from soil employing hyperaccumulators. However, selection of phytoremediation technology for a given site depends on soil conditions, contaminant nature and concentration, coverage and depth of polluted area, nature of plant used, etc. As phytoremediation is a slower process, it alone becomes inefficient and thus requires integration of various agronomic practices such as plant genotypes, soil properties, irrigation, fertilizers and chelators addition, crop rotation, planting methods, weed and pest control, and inoculation of plant growth promoting rhizobacteria and mycorrhizae to up-scaling phytoremediation technology for reclaiming and restoring contaminated sites effectively in a short time span.

DOI: 10.4018/978-1-7998-8434-7.ch016

INTRODUCTION

In recent years, terrestrial environments have been contaminated by various inorganic and organic wastes due to extensive mining operations, industrial emissions and effluent discharge, application of sewage sludge and waste disposals. Soil pollution by trace elements, both metals and metalloids, along with organic pollutants has led to significant environmental effects e.g. barren soil, water pollution, loss of biodiversity, and health effects such as biomagnification of heavy metals in food chain (Suman et al., 2018; Ashraf et al., 2019). The frequent application of pesticide and chemical fertilizers, sewage irrigation and discharge of untreated and partially treated effluent of various industries has exposed the soil with various toxic elements and compounds. The inorganic pollutants viz. heavy metals are of much concern owing to have recalcitrant nature, which adversely affects soil structure, reduce organic matter, decrease microbial diversity, replace essential elements, and ultimately reduce crop productivity. Such degraded lands get converted into wasteland due to losing its capacity to support plants' growth and lead to great economic loss. Therefore, remediation of such metal contaminants are urgently required by adoption of ecofriendly technologies for effective wasteland management (Gerhardt et al., 2017; Hasan et al., 2019) as well as gaining economic benefits from harmless agriculture producing food, fibre, fodder and biomass for energy generation.

Generally, soil remediation measures are mainly based on mechanical or physio-chemical techniques, such as soil incineration, excavation, washing, solidification, and electric field application (DalCorso et al., 2019). However, these conventional approaches are costly, inefficient in case of low contaminant concentrations' energy intensive, generate secondary waste and negatively affect soil quality (Ali et al., 2013; DalCorso et al., 2019). However, phytoremediation, a gentle remediation process, employs metal accumulator plants and associated microbes to remove inorganic and organic contaminants including heavy metals, petroleum products, solvents, explosives, polycyclic aromatic hydrocarbons, pesticides and other agrochemicals from contaminated soil and water in an eco-friendly, non-invasive and cost-effective manner (Salt et al., 1998). Plants extend their root system into the soil matrix and establish robust rhizospheric ecosystem to accumulate heavy metals and modulate their bioavailability, thereby reclaiming the polluted soil and stabilizing soil fertility (Jacob et al., 2018). Usually, the metal contaminants accumulated in roots later transported to the shoots, which are further harvested and thus metal is recovered from the plant biomass. Movement of metals across the root membrane, loading and translocation of metals through the xylem and sequestration and detoxification of metals at cellular and whole plant levels are important mechanisms adopted by metal accumulator plants. A wide range of metals accumulating plants, whose number exceeding 400, capable of accumulating high concentrations (approx. 1% dry wt. of plant) have been identified so far. Plants capable of accumulating more than 1,000 mg kg^{-1} dry weight of metals like Ni, Co, Cu and Pb and 100 and 10,000 mg kg−1 for Cd and Zn, respectively, in their shoots are termed hyperaccumulators (Baker et al., 2000), whilst those that grow fast, produce high biomass and tend to accumulate moderate concentrations of metals are called accumulators (Ginneken et al. 2007). The hyperaccumulator plant species tolerate and accumulate extraordinary metal levels in their aerial parts without showing any significant toxicity symptoms and possess the ability to tolerate, grow and reproduce themselves in highly polluted soil. Phytoremediator plants use different adaptive mechanisms to accumulate or exclude metals, thus maintaining their growth. Phytoremediation involves various specified mechanisms, such as:

- phytoextraction – absorption of metals from the soil by plant roots and translocation to other tissues;
- phytovolatilisation – volatilisation of metals by plants to the atmosphere;
- phytostabilisation – stabilisation of metals into a less soluble form; and
- rhizofiltration – absorption, concentration or precipitation of metals by plant roots (Salt et al., 1998).

Thlaspi caerulescens or *Alyssum bertolonii* accumulate high amounts of one or more metals in their aboveground tissue (Ebbs et al., 1997). Likewise, wild plant species such as *Cynodon dactylon*, *Hirsfeldia incana*, *Malva nicaeensis*, *Sylibum marianum* and *Conyza discoridies* have also been reported the most promising species for Pb and/or Zn phytoextraction (Celestino et al., 2006; Abou-Shanab et al., 2007). About 80 *Brassica* spp. are known as metal hyperaccumulators including *Thalspi caerulescens*, *Alyssum murale*, *T. rotundifolium*, *Dicoma niccolifera* for Ni, Zn, Cr, As and Pb uptake (Rathore et al., 2017). As such, metal hyperaccumulators are of great potential but their use in the field is limited because of their slow growth, low biomass production, metal toxicity, climate specificity, and other abiotic and biotic stress factors e.g. water stress, soil acidity or salinity, nutrient deficiency, frost, soil erosion or compaction, herbivory, pests, etc. which severely limit the performance and establishment of hyperaccumulator plant species used for metal remediation and site restoration.

Selection of Plant Species

The careful selection of plant species and optimization of growth are key elements in successful phytoremediation of metal-contaminated soils under different pedo-climatic conditions. Effective remediation of contaminated site depends on proper selection of plant species as a given plant species with various ecotypes, cultivars, varieties or clones can differ greatly in their response against metal contaminants (Vyslouzilova et al., 2003; Marmiroli et al., 2011). The main criteria for plant selection to be used for phytoremediation is high tolerance to the metal contaminant(s) and other additional abiotic and biotic factors, such as water stress, soil acidity or salinity, high biomass and metal accumulation potential, deep root system, fast growth and easy harvest and economics etc. Besides, the potency of metal remediator plant is also based on site-specific conditions: nature and concentration of contaminant(s), depth of contamination, climate, altitude and soil characteristics. In addition, phytoremediator plant must be native and particularly of non-invasive nature. Non-native species have the advantage of aggressively degrading contaminants, and at the same time supporting large biomass. However, these species must be carefully managed, need frequent irrigation, fertilizer application, and pesticide/herbicide treatments. These may well be targeted species for phytoremediation; however, by themselves, they cannot restore the natural ecosystem, rather, may also disturb natural plant succession in the region. While taking into account the cost of restoration and management along with remediation, allowing native species to naturally revegetate is an attractive option. Both native and non-native species have shown promise in degrading contaminants. Therefore, our rationale has been that the success with either approach depends not only on the plant's ability to survive, adapt, and succeed in degrading contaminants at polluted sites, but also on how one can optimize conditions best suited to plant growth using agronomic practices. Growing metal hyperaccumulator plants with these agronomic interventions can substantially increase plant biomass, augment the ability to absorb, uptake and compartmentalize heavy metals and eventually land reclamation.

SELECTION OF AGRONOMIC PRACTICES FOR PHYTOREMEDIATION

The process of phytoremediation is mainly based on plant-soil interactions that involve the mass transfer of a pollutant from the bulk soil to the plant biomass. Therefore, the management of plant and soil system would positively affect phytoremediation efficiency. As, phytoremediation strategies are often implemented in nutrient-poor, imbalanced pH and water stressed soils, therefore, to enhance phytoremediation efficiency, appropriate agronomic practices including plant species selection, soil management practices, fertilization, irrigation, crop rotation, intercropping/row cropping, plant densities, planting and harvesting methods, weed and pest control (Claus et al., 2007; Vangronsveld et al., 2009; Kidd et al., 2015) along with biostimulation and bioaugmentation must be adopted to improve plant biomass and heavy metals uptake from contaminated environment (Vamerali et al., 2012). In foregoing paragraphs, various major and allied agronomic practices which may be adopted for effective phytoremediation of contaminated sites have been discussed in detail.

Major Agronomic Practices

Soil pH

Metal solubility in soils is strongly influenced by soil pH (Cappuyns et al., 2014). Most metals are relatively more mobile under acid and oxidizing conditions and are strongly retained under alkaline and reducing conditions. Relatively small changes in pH value may cause both an increase and a decrease in the concentration of metals by several orders of magnitude. The solubility of heavy metal compounds, especially in the form of oxides, hydroxides, carbonates or mineral forms, depends on the pH value (van der Sloot et al., 2007). The regulation of soil pH using elemental sulphur and calcium hydroxide can increase trace metal absorption in plant. However, addition of citric acid, malic acid and other low molecular weight organic acids in metal contaminated soil can solubilise and promote the accumulation of heavy metal in plants (Hinsinger et al., 2006; Haoliang et al., 2007). Heavy metal uptake from the soil increases with decreasing soil pH, because low pH increases metal availability (Wang et al., 2006). Protons have a higher affinity for negative charges on the colloids, thus they compete with metal ions of these sites and release metals. The soluble metal forms of both Cadmium and Zinc were greatly increased with decreasing soil pH, which significantly enhanced metal uptake in *Thlaspi caerulescens* (Wang et al., 2006). Such behaviour be attributed to the fact that plant roots can decrease the soil/sediment pH by releasing protons, low molecular weight organic acids, or by redox reactions (including sulphide and iron oxidation) resulting in greater metal bioavailability. In another study, Eriksson (1989) reported decreased Cadmium content in ryegrass at decreased soil pH from 5 to 4 by adding sulphur. This can be due to either interactive effect between Cd^{2+} and H^{+} at the root surface or root damage at low pH. Thus, metal uptake by plant roots can be low as well as high under low pH conditions because of interaction between metal ions and protons. As such, different plant species behaved differently when exposed to a range of different pH values. Different soils had varied responses to acidification treatment and a different optimum pH may exist, which need to be identified to avoid unnecessarily extreme acidification of soils. Therefore, to optimize the use of plant-based remediation, there is dire need to understand the mechanisms behind plant induced pH-changes.

Soil Fertilization

Appropriate fertilization regimes in contaminated and nutrient limiting soils could enhance growth and survival of plants, rhizospheric fungal and bacterial communities (Susarla et al., 2002) along heavy metal uptake in plant tissues through rapid biomass production (Smical et al., 2008). Fertilizer application and the root exudates led to acidification and ion exchange are responsible for higher metal uptake in plants. It was also observed that decrease in soil pH from 7.2 to 6.9 increased $NaNO_3$-extractable Cd and Zn by 35 fold and 8 fold respectively (Kashem & Singh 2002; Kayser et al., 2003). Zhuang et al. (2005) reported that soil acidification by $(NH_4)_2SO_4$ and NH_4NO_3 enhanced Zn and Cd uptake in shoots of mustard plant. Souri & Karimi (2017) also observed an increase in phytoextraction ability of As hyperaccumulator, *Isatis cappadocica* while exogenous addition of sodium nitroprusside (SNP) as nitric oxide supplement. The highest level of arsenate (1500 μM) reduced the plant growth and chlorophyll content, while SNP alleviated these inhibitory effects. The application of SNP significantly increased As concentration in the root and shoots compared with As-stressed plants.

Nutrient fertilization in contaminated soil can also promote plant growth and biomass, and thereby increase the total amount of heavy metals accumulated in plants (Bani et al., 2007). It can change the form and activity of heavy metals, and then affect absorption and accumulation of heavy metals in plants. It is further added that the interactions among fertilizer, trace metals and major nutrients (N, P and S) are key-players in metal uptake by plants (Mench 1998). Fertilizers can influence metal speciation and complexation in contaminated soils and therefore should be selected according to the remediation technique to be implemented. In the case of phytoextraction, ammonium, sulfate or chloride based fertilizers may be applied so as to enhance phytoavailability and hence uptake of metallic cations, while in the case of phytostabilization of trace metals, certain fertilizers can minimize plant metal uptake. Fertilization effects may also vary largely depending on the plant physiological requirements. The application of elemental sulphur may increase the metal solubility for plant uptake (Wenger et al., 2002; Wang et al., 2008; Iqbal et al., 2012). Microbial oxidization of elemental sulphur leads to the production of sulphuric acid and consequent decrease of soil pH (depending on the soil pH buffering capacity). Sulphur oxidation along with microbial and root respiration might also lead to partly anoxic conditions in the rhizosphere, inducing reductive dissolution of MnO_2, which further enhanced Zn and Cd solubility (Iqbal et al., 2012). Chloride can increase crop Cd uptake due to the formation of chloride complexes particularly in alkaline soils (Hattori et al., 2006). Therefore, the nutrient needs of selected crops during plant growth cycle should be assessed by monitoring toxicity symptoms of the plants. Further, to maintain vigorous plant growth, development, and health, optimum soil nutrient level need to be also maintained and monitored periodically throughout remediation cycles.

Organic Matter Addition

In general, soil organic amendments e.g. composts, manure, sewage sludge or biosolids increase soil microbial biomass and enzyme activity, and enhance N transformation rates (i.e. ammonification, nitrification). Organic amendments result in higher plant biomass production, which improve the metal phytoextraction efficiency. Organic matter contains essential micro- and macronutrients, decrease bulk density, improve water infiltration and water holding capacity. Decomposition of organic amendments leads to changes in soil chemical properties, which affects the soil metal mineralization, fractionation, its solubility and distribution (Premi et al., 2013). Organic matter influence the behavior of heavy metals in the soil by:

- releasing heavy metals associated with the organic matter,
- extracting or mobilizing heavy metals from the complexes and;
- improving soil microbial populations which affect heavy metal mobility and availability to the plant through release of chelating agents, acidification, phosphate solubilisation and redox changes (Rathore et al., 2017).

However, the metal(loid) bioavailability depends on the nature of the organic matter, and on the particular soil type and elements concerned (Clemente et al., 2005). A co-solubilization of heavy metals such as Cu, Ni, and Pb, and the likely formation of organo-metal complexes, increases the potential for their migration and bioavailability. Some forms of unstabilized organic matter such as manure or biosolids may increase metal mobility due to their high dissolved organic matter content that can complex metals and facilitate their movement through soil (Almas et al., 1999; Tandy et al., 2009). Humic acid–metal interactions provoke flocculation that increases solubilisation and extractable fraction of heavy metals. Thus, humic acid-rich materials can be a useful amendment for phytoextraction of metals. Farm yard manure and vermicompost were also effective in improving root and shoots dry matter of mustard in metal contaminated soil through solubilisation and thus increase metal uptake (Singh et al., 2007).

Soil Tillage

There is scarce information on the effect of tillage practices (i.e. conventional versus reduced or no tillage), or ploughing methods, on the mobility or uptake of trace elements for to maximize phytoremediation potential. The other agronomic practices such as tillage, co-cropping, crop rotation and coppicing can also be employed. Pioneer studies suggested that tillage helps in turning of heavy metals into the root zone where remediation takes place, which promotes the contact between plants and heavy metals, and improves the efficiency of phytoremediation. Vamerali et al. (2012) compared distinct tillage and ploughing methods on crop, trace metals (As, Cd, Co, Cu, Pb, Zn) concentration and growth on metal-polluted pyrite wastes. Tillage mode, either surface or sub-soiling tillage, was not found to be critical, although sub-soiling enhanced aboveground productivity. Overall, this mechanical activity can have variable effects on the performance of soil re-vegetation, and primarily depends on the properties of the parent materials. Therefore, mechanical decompression of soil is of crucial importance prior to initiating remediation of contaminated sites.

Crop Rotation

Crop rotation or alternative cropping patterns can significantly influence the extraction or immobilization of contaminants, as well as soil protection and quality. Moreover, they can be of high importance in terms of plant productivity and nutrition, enhancing biodiversity, and reduce the influence of weeds, diseases, insects and pests. The use of a single species (even an hyperaccumulator) cannot ensure continuous soil covering and metal uptake. On the other hand, phytoremediation should be continuous and efficient for the longest possible time as such, the soils should not be left bare after harvesting the crop. Banuelos (2000) recommended rotation of metals-accumulating plant species with agronomic crops to maintain adequate growth and yield of phytoremediators. Incorporating crops with deep rooted plants into rotations favour a complete and deeper use of soil resources, or target deeper contaminants. Puschenreiter et al. (2013) observed an increase in shoot uptake of Cd and Zn in barley crop after two years of phytoextraction with *Salix smithiana*. This suggests that during the early phytoextraction period root exudates might have solubilized more metals than the amount removed from the labile pool. In such case, longer periods for crop rotation would be expected. Earlier studies demonstrated good results with three year crop rotations of sunflower – tobacco – corn, or a two year rotation of sunflower – tobacco followed by winter fodder pea in Zn-contaminated agricultural soils at the Bettwiesen site, Switzerland (Nehnevajova et al., 2009; Herzig et al., 2014). After five years of the sunflower– tobacco rotation, bioavailable Zn concentrations were reduced up to 70% in top soils (Herzig et al., 2014). At Cu-contaminated Biogeco site, sunflower–tobacco and tobacco–sunflower–vetiver rotations were also successfully implemented for 6-years (Kolbas 2012). Accordingly, such crop rotation could be used to generate profitable crops, including the production of safe (low Cd) stock fodder fortified with Zn and green manure for micronutrient-deficient soils, or bioenergy.

The incorporation of cover crops can also bring additional benefits such as contributing to the maintenance of soil organic material in upper soil layers and an improved soil aggregation, or a promotion in biological soil tillage through root development, or weed and pest control. Legume crops can fix nitrogen (N), thereby improving soil N status for subsequent crops planted in the following season. Saraswat and Rai (2011a) also recommended the application of leguminous plant i.e. *Leucaena leucocephala* for phytoextraction of Cd and Zn along with nitrogen fixation in metal polluted soils.

Intercropping

The use of a variety of vegetation leads to a greater chance of success and is preferred over monocultures due to the following advantages:

- mixed stands may lose only one or two species to a disease, while monocultures may be entirely susceptible so that one event can destroy the entire phytotechnology system;
- intercropping support more diverse microbial communities that possibly promote an enhancement of the process;
- synergistic effects such as nutrient cycling can occur in mixed stands;
- the biodiversity and potential habitat restoration qualities are promoted;
- a higher volume of soil is explored by roots, then root metal uptake can contemporary occur at different soil depths.

Therefore, the application of intercropping for more than two hyperaccumulators can improve metal remediation efficiency. The relevant plant species selected for phytoremediation have been studied in monocultures, which exhibited decline in biomass yield due to the depletion of nutrients, occurrence of diseases, pests, and weeds, and have a negative effect on soil fertility (Facknath & Lalljee 2000; Lasat 2000; Mench et al., 2010). However, intercropping systems can promote plant-soil and plant-plant interactions resulting in improved nutrient and metal bioavailability, soil protection and crop yield (Shen et al., 2013). Intercropping with leguminous plants reduce nutrient deficiency by supplying additional nitrogen and phosphorus through biological N_2 fixation and mobilization of phosphorus due to rhizosphere acidification respectively (Stroo & Jencks 1982). Both leguminous and non-leguminous species can reduce reliance on inorganic fertilizers. Alder (*Alnus* sp.), has also been recognized as intercropping candidate, through its symbiosis with N-fixing microorganisms (e.g. *Frankia*), ectomycorrhizae and arbuscular mycorrhizae (Taleshi et al., 2009). It can effectively colonize at nutrient deficient, metal contaminated sites and mine spoils characterized by low soil fertility as pure stands or in crop rotations with other fast growing trees (Roy et al., 2007). In recent studies, Cd concentration in the leaves of mustard with legume intercrops was observed higher (30.5 mg kg^{-1}) than the control (26.8 mg kg^{-1}). Further, Cd removal by mustard was accelerated when it was intercropped with leguminous plants (Yang et al., 2016). Another intercropping effect was observed by Wieshammer et al (2007), where two metal-accumulating species, *Salix caprea* and *Arabidopsis halleri*, were combined in a pot experiment to support the metal accumulation in. *S. caprea*. Unlike field conditions, this intercropping reduced shoot biomass and shoot metal concentrations in both species, perhaps due to competition for metal uptake and possibly for nutrients and water. A similar metal competition effect has been reported by Hu et al (2013), where Zn and Cd hyperaccumulator *Sedum alfredii* tended to decrease the Cd accumulation in the co-cropped *Ipomoea aquatica*. Wu et al (2012) performed field trials on a Cd, Cu and polychlorinated biphenyls (PCBs) contaminated agricultural field in the Zhejiang province (China). Monocultures and intercroppings of three species were established: the Cd/Zn-hyperaccumulator *Sedum plumbizincicola*, the Cu-tolerant and accumulating *Elsholtzia splendens,* and the legume *Medicago sativa*. Intercropping pattern had no effect on the plant uptake of trace metals or PCBs. However, greater reductions in soil Cd and Zn were observed in some combinations. Also, *S. plumbizincicola* intercropped with *M. sativa* together with liming was found suitable strategy for remediation of multi-metal contaminated soil.

Planting Methods and Density

The establishment, development and maintenance of dense plant cover in metal contaminated sites are a critical phase for phytoremediation, which mainly depends on the metal concentration, soil fertility and climate. Some crops e.g. grass species, forage crops and sunflower can be directly sown in contaminated sites, while others e.g. Tobacco, Subabool require pre-cultivation of seedlings and planting for increasing plant survival. In some cases, it is recommended to use quite double seed density, given the unfavourable conditions for plant germination in contaminated sites, and thinning of supplementary plants (Vamerali et al., 2012). Seeds should be sown at approx. 3-4 cm depth (depending on the soil type and water holding capacity) for better stand development, and shortly after sowing pest control is important (e.g. snails, rabbits and deer). After 10-12 days of germination, extra seedlings should be uprooted to provide appropriate spacing of upto 20 cm between plants within rows. It is recommended to evaluate the survival and growth of phytoremediators in contaminated soil prior to planting in the field so as to insure their tolerance (Kidd et al., 2015). Further, dense planting of phytoremediator is required to achieve high biomass production and metal accumulation. Li et al. (2009) compared the effects of two planting methods of high-biomass tree, *Averrhoa carambola*, on a Cd-contaminated site. Plots were planted with seedlings at high planting density and compared with grafted seedlings planted at a low density. It was concluded that Cd phytoextraction using high-density seed-seedling method was a feasible option to clean up agricultural soils slightly contaminated by Cd. In case of sunflower plants, where maximum shoot metal removal is required for metal phytoextraction, row distance of 30-50 cm and plant distance within rows of 20 or 10 cm, respectively, resulting in a plant density of 85,000 - 170,000 plants ha^{-1} can be adopted (Kidd et al., 2015). However, Tobacco plants used for phytoremediation require pre-growth of seedlings in pots prior to transplantation into the contaminated field. Tobacco clones, developed by *in vitro* micropropagation, have also shown sufficient growth on three contaminated sites in Rafz and Bettwiesen (Switzerland) and Lommel (Belgium) in field experiments. Prior to planting on site, the *in vitro* propagated tobacco seedlings (4-5 weeks old) need to be transplanted for 2 weeks into soil in covered plates to permit rooting (under 14 h photoperiod; at 25/22 °C in a growth chamber), followed by 2-3 weeks outdoor conditioning. Thus, earlier studies demonstrated that higher planting densities may also lead to a plateau of biomass production, but a higher risk of disease due to easier spreading and infection should not be underestimated. Additional research and development for planting methods are therefore required to maximize metal removal efficiency of phytoremediator plants from contaminated sites.

Irrigation Management

As soil moisture affects both plant growth and element transport in soil, proper water management especially in sub-arctic, arid and semi-arid areas that undergo relatively long periods of drought is essential to accomplish phytoremediation successfully. Prolonged drought induced stress reduces plant growth and enhances sensitivity to pathogens. Considering the potential risks of pollutant mobilization, this practice should be managed carefully. A site water balance should be studied considering the natural rainfall, the crop transpiration and the water holding capacity. Onder et al. (2016) also reported that the level of irrigation and deficit irrigation practices positively affected the leaf area, the chlorophyll content of the young leaves, the fresh and dry weights of shoot, the root dry weight, and the level of Ni removed through the action of plant organs. Claus et al. (2007) suggested the positive role of additional irrigation to enhance Zn uptake and removal by *Phragmites australis*. Increasing soil humidity by Tobacco mutant lines showed outstanding performance in enhancing shoot dry weight, drought resistance, and Cd, Zn and Cu removal from contaminated soil (Kolbas 2012). However, in arid areas, irrigation of contaminated sites may increase the treatment costs. Irrigation with wastewater may be a sustainable approach to meet both the plant water requirements and an alternative treatment of wastewater, especially since phytomanaged sites are already under continuous monitoring. However, such options have yet to be practiced or tested at a field scale.

Weed and Pest Control

Weeds normally dominate plant communities in the early stages of natural re-vegetation of contaminated sites as pioneer plants. Biomass crops used for phytoremediation of polluted soils may have slower growth rates under high weed populations than those recorded in optimal conditions. Therefore, weed control should be an integrative part of phytomanagement, especially in early stages of the site management. In addition, weed control at appropriate stage, helps to provide high soil moisture content resulting in greater biomass of hyperaccumulators.

Chemical (e.g. herbicide) and mechanical (e.g. tilling, disking, mulching) weed control methods can be adopted particularly in early stages. Herbicides can be applied before or after the emergence of the phytoremediating species. Application of pre-emergence herbicides ensures good weed control, quick emergence, and establishment of selected plants. Post-emergence herbicides control weeds that emerge later in the growing season (Lasat 2002). At a contaminated field site (Leipzig, Germany), application of herbicides before plant harvesting improved the phytoextraction of Cr, Ni and Zn by a factor of 2-3 in comparison to control plants (Claus et al., 2007). Biological pest control by inclusion of plant species antagonistic of crop pests may be an option for reducing the use of agrochemicals in contaminated sites. Rotational crop cultivars can also act as natural pesticides. Following a particular crop with a taxonomically different crop (one not susceptible to the same pests) often eliminates pest problems and can increase growth and yield (Stapleton & Banuelos 2009). Regarding the pest control, not enough information is available for the hyperaccumulators.

Biomass Harvest Management

Since phytoremediation objectives are often ensure high plant productivity, and in the case of phytoextraction to maximize shoot metal removal, the optimum harvesting time may differ from harvesting times that are considered appropriate for conventional agricultural purposes. For example, in a phytoextraction trial on a former sewage sludge deposit site (Leipzig, Germany) harvesting of *Phragmites australis* in October and November instead of December (which is more common practice) resulted in an increase in plant dry biomass and higher metal extraction. On the other hand, multiple harvesting of this common reed over the whole growing season did not result in any beneficial increase in dry biomass production (Claus et al., 2007). In phytoextraction methods based on the annual cultivation of tobacco, the topping of plants before flowering, and then shoot harvest can enhance the development of bottom suckers and allow a second harvest, especially in southern Europe and depending on water supply especially from August to October (Kolbas 2012). Harvesting in sequential manner may also reduce the metal content in soil up to the extent which will be far less than the threshold for toxic effects on remaining crops.

The plant material generated may need to be collected and treated as hazardous waste. Even if the plant waste is not classified as a hazardous waste, it may be sufficiently contaminated to require special handling as per relevant national laws. Several options exist, including use of biomass as energy source (direct-fired or as a biofuel feedstock), recovering inorganic constituents i.e. precious metals) from the plant tissues, or supplementing animal feed supplies. If the plant species selected is efficient, the higher the biomass removed from the site, the greater the proportion of contaminant removed. Assuming that harvesting operations are managed under safe conditions not exposing the workers to occupational health hazards, the strategy by which the plant biomass is harvested is not a technical problem. Currently the problem that remain unsolved is related to the available post-harvesting biomass disposal options: this is one of the problems inhibiting the widespread use of this remediation technique. To decrease handling, processing, and potential landfilling costs, waste volume can be reduced by thermal, microbial, physical, or chemical means. Only for radionuclide phytoremediation, safe disposal of harvested biomass is required.

Allied Agronomic Practices

Use of Bio-inoculants (plant growth promoting rhizobacteria (PGPR), mycorrhizal fungi and earthworms)

Phytoremediation efficiency of plants can be enhanced using bioinoculants such as bacteria, fungi, or algae which are characterized for their plant growth promoting traits and resistance to different heavy metals (Table 1). Inoculation of PGPR in combination with plant enhances plant growth and thereby phytoremediation efficiency (Whiting et al. 2001, Abou-Shanab et al., 2003; Ma et al. 2009; Ullah et al., 2015) by adopting various mechanisms such as oxidation-reduction of metals, secretion of biosurfactants, release of chelating agents i.e. organic acids, amino acids, siderophores and enzymes, acidification, phosphate solubilisation, (Abou-Shanab et al., 2003; Montiel-Rozas et al., 2016), synthesis of phytohormones precursors (Perveen et al., 2002; Ahmad et al., 2008) etc.. *Pseudomanas* sp. are well reported to solubilize Zn, Ni and Cu in contaminated soil and promote growth of *Ricinus communis* (Rajkumar & Freitas 2008) and increase plant biomass and Cd accumulation in *Brassica napus* (Amico et al., 2008) in metal-stressed soil. Janssen et al. (2015) also demonstrated the positive role of rhizobacteria to improve plant biomass and metal uptake in metal remediator Willow species. Nayak et al. (2018) also observed the enhancement of toxic Cr (VI), Fe, and other heavy metals phytoremediation by the synergistic combination of native *Bacillus cereus* strain and Vetiveria zizanioides L. Other practices such as inoculation with mycorrhizal fungi has also been reported to improve plant growth and/or modify soil metal mobility and their uptake/translocation by woody crops (Sell et al., 2005; Baum et al., 2006; Langer et al., 2012). The influence of bacterial and fungal activity on trace element mobility and its application in phytoremediation has also been emphasized (Lebeau et al., 2008; Sessitsch et al., 2013). Besides, earthworms which play important role in agricultural production, can also be used for expedite the process of metal remediation by two means: first, they can improve the physico-chemical characteristics of soil, develop soil fertility, and enhance plant biomass; second, they may have direct or indirect effects on the bioavailability of heavy metals in soil while feeding the soil, and excretion. Nannoni et al. (2014) also demonstrated that earthworm activity in red soil not only reduce soil pH, but also significantly increase DTPA-extractable Zn content, thus promotes the absorption and accumulation of metals in plants.

Chelating Agents

In order to increase metal bioavailability, uptake and translocation of metals, natural and/or synthetic chelators are also amended in contaminated soil. Chelators are used to increase metal bioavailability in contaminated soil. Since plant uptake requires metals in mobile form, the negative charge of various soil particles tends to attract and bind heavy metals that are cations and prevent them from becoming soluble and diffuse to root surface. Using chelating agents such problems have been overcome. However, the effects of chelating agents on growth performance and heavy metal uptake may vary and depend on the interaction among chelating agents-metals-soils. A wide range of natural and synthetic chelators such as ethylenediamine-N, N'-disuccinic acid (EDDS), ethylenediaminetetraacetic acid (EDTA), nitrilotriacetic acid (NTA), diethylenetriaminepentaacetic acid (DTPA), $CaCl_2$, citrate, oxalate, malate, succinate, tartrate, phthalate, salicylate and acetate etc. have been used for "chelate-induced hyperaccumulation". Han et al. (2018) also suggested that citric acid and EDTA positively affected the growth, photosynthetic properties and heavy metal accumulation of *Iris halophila Pall.* cultivated in Pb mine tailings. Further, microorganisms and plants are known to produce natural chelating agents such as Oxalic acids, malic acids, and Citric acids that complex with metals and solubilise in water be absorbed by plants. Chen et al. (2006) studied the potential effects of chelate (glucose and citric acid) amendment on phytoextraction of copper and microbial community composition in soil under laboratory condition using a Cu accumulator, *Elsholtzia splendens*, and a non-accumulator,*Trifolium repens* and observed chelate-facilitated phytoremediation of Cu with no negative effect on microbial community. Chigbo & Batty (2015) also compared the efficiency of a synthetic chelate i.e. EDTA, citric acid, and their combination for phytoremediation in Cu-pyrene co-contaminated soil and reported that test plant *Zea mays* can be effective only with the help of both kind of chelators applied together.

Although, chelate-assisted phytoremediation has been proposed as an effective tool for the extraction of heavy metals from soil by plants, its side effects on soil, plant and environment are usually neglected.

The more efficient chelators pose higher risks of plant toxicity, as synthetic chelators are themselves very phytotoxic, therefore a caution must be practiced while applying the chelators for metal phytoremediation.

Transgenic Approaches for Enhancing Phytoremediation

To overcome the challenges adhered with phytoremediation such as slow growth, difficult germination and extraction of only one or two metals by plants, transgenic technologies can be adopted to develop genetically modified strains of metal hyperaccumulators able to tolerate and resist heavy metals, produce more biomass, synthesize stress proteins/heat shock proteins, metal-binding proteins and peptides such as glutathiones, phytochelatins and metallothioneins that act as metal chelator, scavenger of free radicals, and repair damage caused by the metal induced oxidative stress intracellularly (Saraswat & Rai 2011b). Besides, the incorporation of metal transporter genes into hyperaccumulator plants might further augment phytoremediation capabilities. Many reports of transgenic plants are available with increased metal tolerance and accumulation (Table 2). So far, most of the transgenic plants rely on overexpressing genes involved in the biosynthesis pathways of metal-binding proteins and peptides (Zhu et al., 1999; Lee et al., 2003), genes that can convert a toxic ion into a less toxic form (Dhanker et al., 2003). Although efforts have been made in strengthening phytoremediation technology by incorporating agronomic inputs for accelerating remediation of contaminated soils, however, these technologies needs further fine tuning by considering the adoption of following aspects in future:

1. The uptake, translocation, accumulation and detoxification of specific metal in hyperaccumulator plants and their molecular mechanisms be addressed. The use of genetic engineering to modify plants for metal uptake, transport and sequestration may open new avenues for enhancing phytoremediation potential. Moreover, genes related to metal metabolism, biochemical and molecular mechanisms and regulation related to metal uptake like exclusion, apoplastic barriers, xylem loading, root-to-shoot transfer, concentration, storage and detoxification for plants are new arena of discovery that will pave way for development of efficient transgenic phytoremediator plants.
2. An in-depth understanding of the mechanisms of rhizospheric interaction, uptake, transport and sequestration of metals in hyper-accumulators will redesign novel transgenic plants with improved remediation traits.
3. Use of chelators and chemical surfactants, which often become toxic and leach into ground water or soil, may result undesirable environmental consequences. To avoid such conditions, it is pertinent to develop natural or artificial chelating agent possessing low toxicity, easy degradablility, high selectivity and adequate efficiency.
4. Region specific systematic large-scale field studies should be conducted to explore a feasible plant cultivation management device. Optimizing fertilization technology to improve soil environment, plant growth and heavy metal absorption, and thereby improving the phytoremediation efficiency are the need of hour.
5. Apart from this, the metal recovery process especially from accumulator plants still remain challenge and calls for thorough investigation, so that the harvested biomass of phytoremediators can be utilized safely, metals be utilized profitably and as such the environmental risks will be minimized.

Table 1. Role of PGPRs in enhancing phytoremediation of heavy metals in contaminated soils. (Adapted from Rathore et al., 2017)

PGPR and strain	Species	Metal	Effect(s)
Achromobacter xylosoxidans	*B. juncea*	Pb, Cu	Increased root and shoot length, fresh and dry weight and metal uptake (Whitfield et al., 2003)
Pseudomonas aeruginosa	*B. juncea*		Increased root and shoot biomass (Ma et al., 2009)
Pseudomonas tolaasii ACC23, Pseudomonas fluorescens ACC9	*B. napus*	Cd	Increased root and shoot growth and Cd uptake (Sinha & Mukherjee 2008)
Pseudomonas fluorescens G10, Microbacterium sp. G16	*B. napus*	Pb	Increased root elongation of rape seedlings and Pb uptake (Dellamico et al., 2008)
Pseudomonas sp. RJ10, Bacillus sp. RJ16	*B. napus*	Cd	Enhanced root and shoot biomass and Cd uptake (Sheng et al., 2008)
Azotobacter chroococcum HKN-5 Bacillus megaterium HKP-1 Bacillus mucilaginosus HKK-1	*B. juncea*	Pb, Zn, Cu	Increased Pb, Zn, Cu removal (Sheng & Xia 2006)
Bacillus subtilis SJ-101	*B. juncea*	Ni	Increased plant biomass and Ni accumulation (Zaidi et al., 2006)
Streptomyces tendaeF4	*Helianthus annuus*	Fe and Cd	promoted plant growth and uptake of Fe and Cd (Dimpka et al., 2009)
Rahnella sp. JN6	*Polygonum pubescens*	Cd, Pb, Zn	Increased dry weights, concentrations and uptake of Cd, Pb and Zn in both aerial and root tissues (Ha et al., 2013)
Pseudomonas brassicacearum strain Zy-2-1	*Medicago lupulina*	Cu	copper resistance, increased plant growth, nodulation, produced IAA, siderophores and ACC deaminase (Kong et al., 2017)

Table 2. Genes introduced into plants and the effects of their expression on heavy metal tolerance, accumulation, or volatilization. (Adapted from Yang et al., 2005)

Gene	Product	Source	Target	Maximum observed effect[a]
merA	Hg(II) reductase	Gram-negative bacteria	*Liriodendron tulipifera, Nicotiana tabacum*	50 mmol l_1 HgCl2; 500 mg HgCl2 kg–1
merA	Hg(II) reductase	Gram-negative bacteria	*Arabidopsis thaliana*	V: Hg-volatilization rate increase 10-fold T: 10mmol l_1 CH3HgCl (440-fold)
merB	Organomercurial lyase	Gram-negative bacteria	*A. thaliana*	V: Up to 59 pg Hg (0) mg–1 fresh biomass min–1
APS1	ATP sulfurylase	*A. thaliana*	*B. juncea*	A: Two-fold increase in Se concentration
MT-I	MT	Mouse	*N. tabacum*	T: 200 mmol l_1 CdCl2 (20-fold)
CUP1	MT	*Saccharomyces cerevisiae*	*B. oleracea*	T: 400 mmol l_1 CdCl2 (approximately 16-fold)
gsh2	GSH synthase	*E. coli*	*B. juncea*	A: Cd concentrations 125%
gsh1	γ-Glu-Cys synthase	*E. coli*	*B. juncea*	A: Cd concentrations 190%
NtCBP4	Cation channel	*N. tabacum*	*N. tabacum*	T: 250 mmol l_1 NiCl2 (2.5-fold), Pb-sensitive A: Pb concentrations 200%
ZAT1	Zn transporter	*A. thaliana*	*A. thaliana*	T: Slight increase
TaPCS1	PC	*Triticum aestivum*	*Nicotiana glauca R. Graham*	A: Pb concentrations 200%

[a]Relative values refer to control plants not expressing the transgene. References are given in the text. A: accumulation in the shoot; GSH: glutathione, MT: metallothionein; T: tolerance; V: volatilization.

REFERENCES

Abou-Shanab, R., Ghanem, N., Ghanem, K., & Al-Kolaibe, A. (2007). Phytoremediation potential of crop and wild plants for multimetal contaminated soils. *Research Journal of Agriculture and Biological Sciences*, *3*, 370–376.

Abou-Shanab, R. A., Angle, J. S., Delorme, T. A., Chaney, R. L., Berkum, P. V., Moawad, H., Ghanem, K., & Ghozlan, H. A. (2003). Rhizobacterial effects on nickel extraction from soil and uptake by *Alyssum murale*. *The New Phytologist*, *158*(1), 219–224. doi:10.1046/j.1469-8137.2003.00721.x

Ahmad, F., Ahmad, I., & Khan, M. S. (2008). Screening of free-living rhizospheric bacteria for their multiple plant growth promoting activities. *Microbiological Research*, *163*(2), 173–218. doi:10.1016/j.micres.2006.04.001 PMID:16735107

Almas, A., Singh, B. R., & Salbu, B. (1999). Mobility of Cadmium-109 and Zinc-65 in soil influenced by equilibration time, temperature, and organic matter. *Journal of Environmental Quality*, *28*(6), 1742–1750. doi:10.2134/jeq1999.00472425002800060008x

Amico, E. D., Cavalca, L., & Andreoni, V. (2008). Improvement of *Brassica napus* growth under cadmium stress by cadmium-resistant rhizobacteria. *Soil Biology & Biochemistry*, *40*(1), 74–84. doi:10.1016/j.soilbio.2007.06.024

Ali, H., Khan, E., & Sajad, M. A. (2013). Phytoremediation of heavy metals-concepts and applications. *Chemosphere*, *91*(7), 869–881. doi:10.1016/j.chemosphere.2013.01.075 PMID:23466085

Ashraf, S., Ali, Q., Zahir, Z. A., Ashraf, S., & Asghar, H. N. (2019). Phytoremediation: Environmentally sustainable way for reclamation of heavy metal polluted soils. *Ecotoxicology and Environmental Safety*, *174*, 714–727. doi:10.1016/j.ecoenv.2019.02.068 PMID:30878808

Baker, A. J. M., McGrath, S. P., Reeves, R. D., & Smith, J. A. C. 2000. Metal hyperaccumulator plants: A review of the ecology and physiology of a biological resource for phytoremediation of metal-polluted soils. In Phytoremediation of Contaminated Soil and Water. Lewis Publishers.

Bani, A., Echevarria, G., Sulçe, S., Morel, J. L., & Mullai, A. (2007). In-situ phytoextraction of Ni by a native population of *Alyssum murale* on a ultramafic site (Albania). *Plant and Soil*, *293*(1-2), 79–89. doi:10.100711104-007-9245-1

Banuelos, G. S. (2000). Factors influencing field phytoremediation of selenium-laden soils. In N. Terry & G. Banuelos (Eds.), *Lewis Publishers* (pp. 41–60). CRC.

Baum, C., Hrynkiewicz, K., Leinweber, P., & Meissner, R. (2006). Heavy metal mobilization and uptake by mycorrhizal and nonmycorrhizal willows (*Salix x dasyclados*). *Journal of Plant Nutrition and Soil Science*, *169*(4), 516–522. doi:10.1002/jpln.200521925

Cappuyns, V., Alian, V., Vassilieva, E., & Swennen, R. (2014). pH dependent leaching behaviour of Zn, Cd, Pb, Cu and As from mining wastes and slags: Kinetics and mineralogical control. *Waste and Biomass Valorization*, *5*(3), 355–368. doi:10.100712649-013-9274-3

Celestino, M. D. R., Font, R., Rojas, R. M., & Bailon, A. D. H. (2006). Uptake of Lead and Zinc by wild plants growing on contaminated soils. *Industrial Crops and Products*, *24*(3), 230–237. doi:10.1016/j.indcrop.2006.06.013

Chen, Y., Wang, Y., Wu, W., Lin, Q., & Xue, S. (2006). Impacts of chelate-assisted phytoremediation on microbial community composition in the rhizosphere of a copper accumulator and non-accumulator. *The Science of the Total Environment*, *356*(1-3), 247–255. doi:10.1016/j.scitotenv.2005.04.028 PMID:15935447

Chigbo, C., & Batty, L. (2015). Chelate-assisted phytoremediation of cu-pyrene-contaminated soil using *Z. mays*. *Water, Air, and Soil Pollution*, *226*(3), 1–10. doi:10.100711270-014-2277-2

Claus, D., Dietze, H., Gerth, A., Grosser, W., & Hebner, A. (2007). Application of agronomic practice improves phytoextraction on a multipolluted site. *Journal of Environmental Engineering and Landscape Management*, *15*(4), 208–212. doi:10.3846/16486897.2007.9636932

Clemente, R., Walker, D. J., & Bernal, M. P. (2005). Uptake of heavy metals and As by *Brassica juncea* grown in a contaminated soil in Aznalcollar (Spain): The effect of soil amendments. *Environmental Pollution*, *138*(1), 46–58. doi:10.1016/j.envpol.2005.02.019 PMID:15894412

DalCorso, G., Fasani, E., Manara, A., Visioli, G., & Furini, A. (2019). Heavy metal pollutions: State of the art and innovation in phytoremediation. *International Journal of Molecular Sciences*, *20*(14), 3412. doi:10.3390/ijms20143412 PMID:31336773

Dellamico, E., Cavalca, L., & Vincenza, A. (2008). Improvement of *Brassica napus* growth under cadmium stress by cadmium-resistant rhizobacteria. *Soil Biology & Biochemistry*, *40*(1), 74–84. doi:10.1016/j.soilbio.2007.06.024

Dhankher, O. P., Shasti, N. A., Rosen, B. P., Fuhrmann, M., & Meagher, R. B. (2003). Increased cadmium tolerance and accumulation by plants expressing bacterial arsenate reductase. *The New Phytologist*, *159*, 431–441.

Ebbs, S. D., Lasat, M., Brady, D. J. M., Cornish, J., Gordon, R., & Kochian, V. (1997). Phytoextraction of cadmium and zinc from a contaminated soil. *Journal of Environmental Quality*, *26*, 1424–1430.

Eriksson, J. E. (1989). The influence of pH, soil type and time on adsorption and uptake by plants of Cd added to the soil. *Water, Air, and Soil Pollution*, *48*(3-4), 317–335. doi:10.1007/BF00283334

Facknath, S., & Lalljee, B. (2000). *Allelopathic strategies for eco-friendly crop protection*. Springer. doi:10.1007/978-94-011-4173-4_3

Gerhardt, K. E., Gerwing, P. D., & Greenberg, B. M. (2017). Opinion: Taking phytoremediation from proven technology to accepted practice. *Plant Science*, *256*, 170–185. doi:10.1016/j.plantsci.2016.11.016 PMID:28167031

Ginneken, L. V., Meers, E., Guisson, R., Ruttens, A., Elst, K., Tack, F. M. G., Vangronsveld, J., Diels, L., & Dejonghe, W. (2007). Phytoremediation for heavy metal-contaminated soils combined with bioenergy production. *Journal of Environmental Engineering and Landscape Management*, *15*(4), 227–236. doi:10.3846/16486897.2007.9636935

Han, Y., Zhang, L., Gu, J., Zhao, J., & Fu, J. (2018). Citric acid and EDTA on the growth, photosynthetic properties and heavy metal accumulation of *Iris halophila Pall.* cultivated in Pb mine tailings. *International Biodeterioration & Biodegradation, 128*, 15–21. doi:10.1016/j.ibiod.2016.05.011

Haoliang, L., Chongling, Y., & Jingchun, L. (2007). Low-molecular-weight organic acids exuded by Mangrove (*Kandelia candel* (L.) Druce) roots and their effect on cadmium species change in the rhizosphere. *Environmental and Experimental Botany, 61*(2), 159–549. doi:10.1016/j.envexpbot.2007.05.007

Hasan, M. M., Uddin, M. N., Ara-Sharmeen, F. I., Alharby, H., Alzahrani, Y., & Hakeem, K. R. (2019). Assisting phytoremediation of heavy metals using chemical amendments. *Plants, 8*, 295. doi:10.3390/plants8090295 PMID:31438507

Hattori, H., Kuniyasu, K., Chiba, K., & Chino, M. (2006). Effect of chloride application and low soil pH on cadmium uptake from soil by plants. *Soil Science and Plant Nutrition, 52*(1), 89–94. doi:10.1111/j.1747-0765.2006.00007.x

Herzig, R., Nehnevajova, E., Pfistner, C., Schwitzguebel, J. P., Ricci, A., & Keller, C. (2014). Feasibility of labile Zn phytoextraction using enhanced tobacco and sunflower: Results of five- and one-year field scale experiments in Switzerland. *International Journal of Phytoremediation, 16*(7-8), 735–754. doi:10.1080/15226514.2013.856846 PMID:24933882

Hinsinger, P., Plassard, C., & Jaillard, B. (2006). Rhizosphere: A new frontier for soil biogeochemistry. *Journal of Geochemical Exploration, 88*(1-3), 210–213. doi:10.1016/j.gexplo.2005.08.041

Hu, J. L., Chan, P. T., Wu, F. Y., Wu, S. C., Zhang, J. H., Lin, X. G., & Wong, M. H. (2013). Arbuscular mycorrhizal fungi induce differential Cd and P acquisition by Alfred stonecrop (*Sedum alfredii* Hance) and upland kangkong (*Ipomoea aquatica* Forsk.) in an intercropping system. *Applied Soil Ecology, 63*, 29–35. doi:10.1016/j.apsoil.2012.09.002

Iqbal, M., Puschenreiter, M., Oburger, E., Santner, J., & Wenzel, W. W. (2012). Sulfur-aided phytoextraction of Cd and Zn by *Salix smithiana* combined with *in situ* metal immobilization by gravel sludge and red mud. *Environmental Pollution, 170*, 222–231. doi:10.1016/j.envpol.2012.07.008 PMID:22842051

ITRC (Interstate Technology & Regulatory Council). (2009). *Phytotechnology Technical and Regulatory Guidance and Decision Trees*. Author.

Jacob, J. M., Karthik, C., Saratale, R. G., Kumar, S. S., Prabakar, D., Kadirvelu, K., & Pugazhendhi, A. (2018). Biological approaches to tackle heavy metal pollution: A survey of literature. *Journal of Environmental Management, 217*, 56–70. doi:10.1016/j.jenvman.2018.03.077 PMID:29597108

Janssen, J., Weyens, N., Croes, S., Beckers, B., Meiresonne, L., Peteghem, P. V., Carleer, R., & Vangronsveld, J. (2015). Phytoremediation of metal contaminated soil using Willow: Exploiting plant-associated bacteria to improve biomass production and metal uptake. *International Journal of Phytoremediation, 17*(11), 1123–1136. doi:10.1080/15226514.2015.1045129 PMID:25942689

Kashem, M. A., & Singh, B. R. (2002). The effect of fertilizer additions on the solubility and plant-availability of Cd, Ni and Zn in soil. *Nutrient Cycling in Agroecosystems, 62*(3), 287–296. doi:10.1023/A:1021226201136

Kayser, A., Wenger, K., Keller, A., Attinger, W., Felix, H. R., Gupta, S. K., & Schulin, R. (2003). Enhancement of phytoextraction of Zn, Cd and Cu from calcareous soil: The use of NTA and sulfur amendments. *Environmental Science & Technology*, *34*(9), 1778–1783. doi:10.1021/es990697s

Kidd, P., Mench, M., Alvarez-Lopez, V., Bert, V., Dimitriou, I., Friesl-Hanl, W., Herzig, R., Olga Janssen, J., Kolbas, A., Müller, I., Neu, S., Renella, G., Ruttens, A., Vangronsveld, J., & Puschenreiter, M. (2015). Agronomic practices for improving gentle remediation of trace element-contaminated soils. *International Journal of Phytoremediation*, *17*(11), 1005–1037. doi:10.1080/15226514.2014.1003788 PMID:25581041

Kolbas, A. (2012). *Phenotypic traits and development of plants exposed to trace elements; use for phytoremediation and biomonitoring* (PhD thesis). University of Bordeaux.

Langer, I., Santner, J., Krpata, D., Fitz, W. J., Wenzel, W. W., & Schweiger, P. F. (2012). Ectomycorrhizal impact on Zn accumulation of *Populus tremula* L. grown in metalliferous soil with increasing levels of Zn concentration. *Plant and Soil*, *355*(1-2), 283–297. doi:10.100711104-011-1098-y

Lasat, M. M. (2000). Phytoextraction of metals from contaminated soil: A review of plant/soil/metal interaction and assessment of pertinent agronomic issues. *Journal of Hazardous Substance Research*, *2*, 1–25.

Lasat, M. M. (2002). Phytoremediation of toxic metals: A review of biological mechanisms. *Journal of Environmental Quality*, *31*, 109–120. PMID:11837415

Lebeau, T., Braud, A., & Jezequel, K. (2008). Performance of bioaugmentation-assisted phytoextraction applied to metal contaminated soils: A review. *Environmental Pollution*, *153*(3), 497–522. doi:10.1016/j.envpol.2007.09.015 PMID:17981382

Lee, S., Moon, J. S., Ko, T. S., Petros, D., Goldsbrough, P. B., & Korban, S. S. (2003). Overexpression of *Arabidopsis* phytochelatin synthase paradoxically leads to hypersensitivity to cadmium stress. *Plant Physiology*, *131*(2), 656–663. doi:10.1104/pp.014118 PMID:12586889

Li, J. T., Liao, B., Dai, Z. Y., Zhu, R., & Shu, W. S. (2009). Phytoextraction of Cd-contaminated soil by carambola (*Averrhoa carambola*) in field trials. *Chemosphere*, *76*(9), 1233–1239. doi:10.1016/j.chemosphere.2009.05.042 PMID:19541343

Ma, Y., Rajkumar, M., & Freitas, H. (2009). Inoculation of plant growth promoting bacterium *Achromobacter xylosoxidans* strain Ax10 for the improvement of copper phytoextraction by *Brassica juncea*. *Journal of Environmental Management*, *90*(2), 831–837. doi:10.1016/j.jenvman.2008.01.014 PMID:18329785

Marmiroli, M., Visioli, G., Maestri, E., & Marmiroli, N. (2011). Correlating SNP genotype with the phenotypic response to exposure to cadmium in populus spp. *Environmental Science & Technology*, *45*(10), 4497–4505. doi:10.1021/es103708k PMID:21491876

Mench, M. J. (1998). Cadmium availability to plants in relation to major long-term changes in agronomy systems. *Agriculture, Ecosystems & Environment*, *67*(2-3), 175–187. doi:10.1016/S0167-8809(97)00117-5

Mench, M., Lepp, N., Bert, V., Schwitzguébel, J.P., Gawronski, S., Schröder, P., & Vangronsveld, J. (2010). Successes and limitations of phytotechnologies at field scale: outcomes, assessment and outlook from COST Action 859. *Journal of Soil and Sediment Contamination, 10*, 1039-1070.

Mohsin, M., Kuittinen, S., Salam, M. M. A., Peräniemi, S., Laine, S., Pulkkinen, P., Kaipiainen, E., Vepsäläinen, J., & Pappinen, A. (2019). Chelate-assisted phytoextraction: Growth and ecophysiological responses by *Salix schwerinii ELWolf* grown in artificially polluted soils. *Journal of Geochemical Exploration, 205*, 205. doi:10.1016/j.gexplo.2019.106335

Montiel-Rozas, M. M., Madejon, E., & Madejon, P. (2016). Effect of heavy metals and organic matter on root exudates (low molecular weight organic acids) of herbaceous species: An assessment in sand and soil conditions under different levels of contamination. *Environmental Pollution, 216*, 273–281. doi:10.1016/j.envpol.2016.05.080 PMID:27267743

Nannoni, F., Rossi, S., & Protano, G. (2014). Soil properties and metal accumulation by earthworms in the Siena urban area. *Applied Soil Ecology, 77*, 9–17. doi:10.1016/j.apsoil.2014.01.004

Nayak, A., Panda, S., Basu, A., & Dhal, N. (2018). Enhancement of toxic Cr (VI), Fe, and other heavy metals phytoremediation by the synergistic combination of native *Bacillus cereus* strain and *Vetiveria zizanioides* L. *International Journal of Phytoremediation, 20*(7), 682–691. doi:10.1080/15226514.2017.1413332 PMID:29723050

Nehnevajova, E., Herzig, R., Bourigault, C., Bangerter, S., & Schwitzguebel, J. P. (2009). Stability of enhanced yield and metal uptake by sunflowers mutants for improved phytoremediation. *International Journal of Phytoremediation, 11*(4), 329–346. doi:10.1080/15226510802565394

Onder, D., Onder, S., Daghan, S., & Uygu, V. (2016). Influence of different irrigation level and different nickel (ni) doses on phytoremediative capacity of *Tagetes erecta* l. *Fresenius Environmental Bulletin, 25*, 191–199.

Perveen, S., Khan, M. S., & Zaidi, A. (2002). Effect of rhizospheric microorganisms on growth and yield of greengram *(Phaseolus radiatus)*. *Indian Journal of Agricultural Sciences, 72*, 421–423.

Premi, O. P., Kandpal, B. K., Rathore, S. S., Shekhawat, K., & Chauhan, J. S. (2013). Green manuring, mustard residues recycling and fertilizer application affects productivity and sustainability of Indian mustard (*Brassica juncea* L.) in Indian semi arid tropics. *Industrial Crops and Products, 41*, 423–429. doi:10.1016/j.indcrop.2012.04.030

Puschenreiter, M., Wittstock, F., Friesl-Hanl, W., & Wenzel, W. W. (2013). Predictability of the Zn and Cd phytoextraction efficiency of a *Salix smithiana* clone by DGT and conventional bioavailability assays. *Plant and Soil, 369*(1-2), 531–541. doi:10.100711104-013-1597-0

Rajkumar, M., & Freitas, H. (2008). Influence of metal resistant-plant growth-promoting bacteria on the growth of *Ricinus communis* in soil contaminated with heavy metals. *Chemosphere, 71*(5), 834–842. doi:10.1016/j.chemosphere.2007.11.038 PMID:18164365

Rathore, S. S., Shekhawat, K., Dass, A., Kandpal, B. K., & Singh, V. K. (2017). Phytoremediation mechanism in Indian mustard (*Brassica juncea*) and its enhancement through agronomic interventions. *Proceedings of the National Academy of Sciences, 2017. Section B: Biological Sciences.*

Roy, S., Khasa, D. P., & Greer, C. W. (2007). Combining alders, frankiae, and mycorrhizae for the re-vegetation and remediation of contaminated ecosystems. *Canadian Journal of Botany*, *85*(3), 237–251. doi:10.1139/B07-017

Salt, D. E., Smith, R. D., & Raskin, I. (1998). Phytoremediation. *Annual Review of Plant Physiology and Plant Molecular Biology*, *49*(1), 643–668. doi:10.1146/annurev.arplant.49.1.643 PMID:15012249

Saraswat, S., & Rai, J. P. N. (2011a). Prospective application of *Leucaena leucocephala* for phytoextraction of Cd and Zn and nitrogen fixation in metal polluted soils. *International Journal of Phytoremediation*, *13*(3), 1–18. doi:10.1080/15226514.2010.483259 PMID:21598792

Saraswat, S., & Rai, J. P. N. (2011b). Complexation and detoxification of Zn and Cd in metal accumulating plants. *Reviews in Environmental Science and Biotechnology*, *10*(4), 327–339. doi:10.100711157-011-9250-y

Sell, J., Kayser, A., Schulin, R., & Brunner, I. (2005). Contribution of ectomycorrhizal fungi to cadmium uptake of poplars and willows from a heavily polluted soil. *Plant and Soil*, *277*(1-2), 245–253. doi:10.100711104-005-7084-5

Sessitsch, A., Kuffner, M., Kidd, P., Vangronsveld, J., Wenzel, W., Fallmann, K., & Puschenreiter, M. (2013). The role of plant-associated bacteria in the mobilization and phytoextraction of trace elements in contaminated soils. *Soil Biology & Biochemistry*, *60*, 182–194. doi:10.1016/j.soilbio.2013.01.012 PMID:23645938

Shen, J., Li, C., Mi, G., Li, L., Yuan, L., Jiang, R., & Zhang, F. (2013). Maximizing root/rhizosphere efficiency to improve crop productivity and nutrient use efficiency in intensive agriculture of China. *Journal of Experimental Botany*, *64*(5), 1181–1192. doi:10.1093/jxb/ers342 PMID:23255279

Sheng, X. F., & Xia, J. J. (2006). Improvement of rape (*Brassica napus*) plant growth and cadmium uptake by cadmium-resistant bacteria. *Chemosphere*, *64*(6), 1036–1042. doi:10.1016/j.chemosphere.2006.01.051 PMID:16516946

Sheng, X. F., Xia, J. J., Jiang, C. Y., He, L. Y., & Qian, M. (2008). Characterization of heavy metal-resistant endophytic bacteria from rape (*Brassica napus*) roots and their potential in promoting the growth and lead accumulation of rape. *Environmental Pollution*, *156*(3), 1164–1170. doi:10.1016/j.envpol.2008.04.007 PMID:18490091

Singh, G., Brar, M. S., & Malhi, S. S. (2007). Decontamination of chromium by farm yard manure application in spinach grown in two texturally different Cr-contaminated soils. *Journal of Plant Nutrition*, *30*(2), 289–308. doi:10.1080/01904160601118125

Sinha, S., & Mukherjee, S. K. (2008). Cadmium-induced siderophore production by a high Cd-resistant bacterial strain relieved Cd toxicity in plants through root colonization. *Current Microbiology*, *56*(1), 55–60. doi:10.100700284-007-9038-z PMID:17899260

Smical, A. I., Hotea, V., Oros, V., Juhasz, J., & Pop, E. (2008). Studies on transfer and bioaccumulation of heavy metals from soil into lettuce. *Environmental Engineering and Management Journal*, *7*(5), 609–615. doi:10.30638/eemj.2008.085

Souri, Z., & Karimi, N. (2017). Enhanced Phytoextraction by As Hyperaccumulator *Isatis cappadocica* Spiked with Sodium Nitroprusside. *Soil & Sediment Contamination*, *26*(4), 457–468. doi:10.1080/15320383.2017.1326457

Stapleton, J. J., & Banuelos, G. S. (2009). Biomass crops can be used for biological disinfestation and remediation of soils and water. *California Agriculture*, *63*(1), 41–46. doi:10.3733/ca.v063n01p41

Stroo, H. F., & Jencks, E. M. (1982). Enzyme activity and respiration in minesoils. *Soil Science Society of America Journal*, *46*(3), 548–553. doi:10.2136ssaj1982.03615995004600030021x

Suman, J., Uhlik, O., Viktorova, J., & Macek, T. (2018). Phytoextraction of heavy metals: A promising tool for clean-up of polluted environment? *Frontiers in Plant Science*, *9*, 1476. doi:10.3389/fpls.2018.01476 PMID:30459775

Susarla, S., Medina, V. F., & Mc Cutcheon, S. C. (2002). Phytoremediation: An ecological solution to organic chemical contamination. *Ecological Engineering*, *18*(5), 647–658. doi:10.1016/S0925-8574(02)00026-5

Taleshi, S., Dhumal, K., Alipour, A., Espahbodi, K., & Ghasemi, O. (2009). Impact of alder (*Alnus subcordata*) in fertility of forest soil. *Research Journal of Environmental Sciences*, *3*(6), 640–644. doi:10.3923/rjes.2009.640.644

Tandy, S., Healey, J. R., Nason, M. A., Williamson, J. C., & Jones, D. L. (2009). Remediation of metal polluted mine soil with compost: Co-composting versus incorporation. *Environmental Pollution*, *157*(2), 690–697. doi:10.1016/j.envpol.2008.08.006 PMID:18819736

Ullah, A., Heng, S., Munis, M. F. H., Fahad, S., & Yang, X. (2015). Phytoremediation of heavy metals assisted by plant growth promoting (PGP) bacteria: A review. *Environmental and Experimental Botany*, *117*, 28–40. doi:10.1016/j.envexpbot.2015.05.001

Vamerali, T., Marchiol, L., Bandiera, M., Fellet, G., Dickinson, N. M., Lucchini, P., Mosca, G., & Zerbi, G. (2012). Advances in agronomic management of phytoremediation: Methods and results from a 10-year study of metal-polluted soils. *Italian Journal of Agronomy*, *7*(4), 323–330. doi:10.4081/ija.2012.e42

van der Sloot, H. A., van Zomeren, A., Meeussen, J. C. L., Seignette, P., & Bleijerveld, R. (2007). Test method selection, validation against field data, and predictive modelling for impact evaluation of stabilised waste disposal. *Journal of Hazardous Materials*, *141*(2), 354–369. doi:10.1016/j.jhazmat.2006.05.106 PMID:16889893

Vangronsveld, J., Herzig, R., Weyens, N., Boulet, J., Adriaensen, K., Ruttens, A., Thewys, T., Vassilev, A., Meers, E., Nehnevajova, E., van der Lelie, D., & Mench, M. (2009). Phytoremediation of contaminated soils and groundwater: Lessons from the field. *Environmental Science and Pollution Research International*, *16*(7), 765–794. doi:10.100711356-009-0213-6 PMID:19557448

Vyslouzilova, M., Tlustos, P., & Szakova, J. (2003). Cadmium and zinc phytoextraction potential of seven clones of *Salix* spp. planted on heavy metal contaminated soils. *Plant, Soil and Environment*, *49*(12), 542–547. doi:10.17221/4191-PSE

Wang, A. S., Angle, J. S., Chaney, R. L., Delorme, T. A., & Reeves, R. D. (2006). Soil pH effects on uptake of Cd and Zn by *Thlaspi caerulescens. Plant and Soil, 281*(1-2), 325–337. doi:10.100711104-005-4642-9

Wang, Y. P., Li, Q. B., Hui, W., Shi, J. Y., Lin, Q., Chen, X. C., & Chen, Y. X. (2008). Effect of sulphur on soil Cu/Zn availability and microbial community composition. *Journal of Hazardous Materials, 159*(2-3), 385–389. doi:10.1016/j.jhazmat.2008.02.029 PMID:18394797

Wenger, K., Kayser, A., Gupta, S. K., Furrer, G., & Schulin, R. (2002). Comparison of NTA and elemental sulfur as potential soil amendments in phytoremediation. *Soil & Sediment Contamination, 11*(5), 655–672. doi:10.1080/20025891107023

Whitfield, L., Richards, A. J., & Rimmer, D. L. (2003). Effects of mycorrhizal colonization on *Thymus polytrichus* from heavy metalcontaminated sites in north England. *Mycorrhiza, 14*, 47–54. PMID:14564560

Whiting, S. N., De Souza, M., & Terry, N. (2001). Rhizosphere bacteria mobilize Zn for hyperaccumulator by *Thlaspi caerulescens. Environmental Science & Technology, 35*(15), 3144–3150. doi:10.1021/es001938v PMID:11505990

Wieshammer, G., Unterbrunner, R., Garcia, T. B., Zivkovic, M. F., Puschenreiter, M., & Wenzel, W. W. (2007). Phytoextraction of Cd and Zn from agricultural soils by *Salix* ssp and intercropping of *Salix caprea* and *Arabidopsis halleri. Plant and Soil, 298*(1-2), 255–264. doi:10.100711104-007-9363-9

Wu, L. H., Li, Z., Han, C. L., Liu, L., Teng, Y., Sun, X. H., Pan, C., Huang, Y. J., Luo, Y. M., & Christie, P. (2012). Phytoremediation of soil contaminated with cadmium, copper and polychlorinated biphenyls. *International Journal of Phytoremediation, 14*(6), 570–584. doi:10.1080/15226514.2011.619227 PMID:22908627

Yang, L., Kai, L., Yong, L., Yang, W. W., Fuzhong, Z. P., Jian, Z., Chen, L., Gao, S., & Zhang, L. (2016). Cadmium contamination of soil and crops is affected by intercropping and rotation systems in the lower reaches of the Minjiang River in south-western China. *Environmental Geochemistry and Health, 38*(3), 811–820. doi:10.100710653-015-9762-4 PMID:26323960

Zaidi, S., Usmani, S., Singh, B. R., & Musarrat, J. (2006). Significance of *Bacillus subtilis* strain SJ-101 as a bioinoculant for concurrent plant growth promotion and nickel accumulation in *Brassica juncea. Chemosphere, 64*(6), 991–997. doi:10.1016/j.chemosphere.2005.12.057 PMID:16487570

Zhu, Y. L., Pilon-Smits, E. A. H., Tarun, A. S., Weber, S. U., Jouanin, L., & Terry, N. (1999). Cadmium tolerance and accumulation in Indian mustard is enhanced by overexpressing g-glutamylcysteine synthetase. *Plant Physiology, 121*(4), 1169–1177. doi:10.1104/pp.121.4.1169 PMID:10594104

Zhuang, P., Ye, Z. H., Lan, C. Y., Xie, Z. W., & Shu, W. S. (2005). Chemically assisted phytoextraction of heavy metal contaminated soils using three plant species. *Plant and Soil, 276*(1-2), 153–162. doi:10.100711104-005-3901-0

Chapter 17
Polyhydroxyalkanoates Production in Transgenic Plants:
Green Plastics for Better Future and Environmental Sustainability

R. S. Sengar
https://orcid.org/0000-0002-7282-4135
Sardar Vallabhbhai Patel University of Agricultue and Technology, India

Reshu Chaudhary
https://orcid.org/0000-0002-9566-5368
Keral Verma Subharti College of Science, Subharti University, Meerut, India

Ashish Pandey
Saint Mark's College, USA

Aditi Chatterjee
University of California, San Francisco, USA

ABSTRACT

In the present time, polyhydroxyalkanoates have established themselve as the alternatives of petroleum-based synthetic polymers due to their biodegradability and eco-friendly nature. In the last two decades, several scientists have engaged in search of cost-effective alternatives of producing PHA at larger scale. Therefore, many plant species have been genetically engineered for this purpose. The major obstacles in producing PHA polymers in transgenic plants are the regulation of the appropriate monomer's composition and ratio synthesized in their cells. Efforts are on the way to encounter these difficulties as soon as possible. The research is also going on to enhance the production and accumulation of these biopolymers in transgenic plants. Among the targeted cell organelles, plastids have been considered as the best sites for higher production of polyhydroxyalkanoates at cheaper costs because of its maternal inheritance and its being unaffected by gene silencing. Polyhydroxyalkanoates production technologies are still costly, but these could be cost effective in the near future.

DOI: 10.4018/978-1-7998-8434-7.ch017

INTRODUCTION

Over the past 60 years, the development of synthetic polymers has reduced the men's dependence upon the utilization of plant product based polymers such as rubber, cotton, wood etc. But the excessive and ever growing demands of utilizing synthetic polymers raise the global environmental concerns. Therefore, the scientific community has engaged to discover alternatives of synthetic polymers. As an alternative of synthetic polymers, polyhydroxyalkanoates (PHAs) came into existence. These are the natural products of certain bacterial cells. Therefore, scientists decided to produce biological polymers by using bacterial cells as bioreactors. This was first time exploited in 1980 by Imperial Chemical Industries (Anderson and Dawes, 1990). This industry generated a large set up for the bulk production of polyhydroxyalkanoates using polymer accumulating bacterial strains such as *Ralstonia eutropha*. But the cost of producing PHAs through this way was very high. So that again there were also the requirements of cost effective alternatives for producing biopolymers. The development of advanced biotechnological tools attracted the concern of global scientists towards the utilization of plant bioreactors for producing renewable biological polymers. Therefore, the scientific community moved towards the utilization of plant bioreactors for generating polyhydroxyalkanoates (PHAs) in a cost effective and eco-friendly manner. For this purpose, a variety of transgenic plant species including *Arabidopsis*, tobacco, rapeseed, cotton, alfalfa, flax, sugarcane, *Camellina* and oil palm were tested at larger scales and generated new dimensions of producing biopolymers.

Polyhydroxyalkanoates (PHAs) are biopolyesters that are synthesized naturally in a broad range of bacterial cells such as *Alcaligenes eutrophus* and many other species as an inert carbon and energy reserves accumulated in cytoplasm upto about 80% of the total dry weight in the form of round shape granules with a diameter of 0.2-1.0 µm (Sabbagh and Muhamad, 2017). These polymers are made up of about 600-35000 identical monomer units. Polyhydroxyalkanaotes act as water insoluble storage compounds which are synthesized under environmental stress conditions in the excess of carbon and the limiting quantities of important growth nutrients such as nitrogen, phosphorus, iron, magnesium, potassium, sulfur, zinc or oxygen (Masood et al., 2014). These biopolymers are depolymerized during the exhausted carbon source conditions. Thus, the degraded products could be used by microbes as an energy and carbon source (Anderson and Dawes, 1990).

Polyhydroxyalkanoates (PHAs) are considered similar to the conventional plastics in reference of its properties such as thermoplastic and polypropylene nature (Anjum et al., 2016). Instead of petrochemical plastics, PHAs are natural, non-toxic, biodegradable and renewable (Sharma et al., 2016). These properties make PHA an attractive alternative of petrochemical plastic. In near future, it is hopefully projected that the production of synthetic plastic polymers could possibly be reached up to about eight hundred ten million tons (810 tons) by the end of 2050 (Gumel et al., 2013). As it is well known that the plastic pollution has been an unbeatable burning issue across the globe. Therefore, it has been urgent necessity to find out the eco-friendly alternatives of synthetic plastics. Thatswhy, the scientific community and industries are engaging to produce synthetic polymers through natural means. Generally microbial bioreactors are utilized in producing PHAs at larger scales but the whole process of bioplastic polymer production is still highly expensive than the process of producing petrochemicals based synthetic polymers due to the cost of the nutrition for microbial cultures (Baiker, 2017). So that the scientific community has engaged in optimizing transgenic plants as novel bioreactors for production of PHAs at cheaper costs. The present chapter summarizes the informations about PHA, its structure, biosynthesis and the current status of transgenic plants which were engineered for producing PHAs at cheaper costs.

PHA Structure and Biosynthesis

Polyhydroxyalkanoates are generally linear polyesters consisting of several 3-(R)-hydroxy fatty acid monomers (HA) linked together by ester bonds. These ester bonds are produced by the linkage of carboxylic group of one monomer unit to the hydroxyl group of another monomer unit (Sudesh *et al.*, 2000; Lenz and Marchessault, 2005). On the basis of the presence of carbon atoms in the monomers, polyhydroxyalkanoates are generally categorized into two major groups. The first group is called short chain length polyhydroxyalkanoates (scl-PHAs) and the second group is medium chain length polyhydroxyalkanoates (mcl-PHAs). Short chain length polyhydroxyalkanoates (scl-PHAs) generally consist of 3-5 carbon atoms whereas; the medium chain length polyhydroxyalkanoates (scl-PHAs) have 6-14 carbons. Under natural conditions, the short chain length polyhydroxyalkanoates (scl-PHAs) are synthesized in *Cupriavidus necator* while the medium chain length polyhydroxyalkanoates (scl-PHAs) are accumulated in *Pseudomonas* species. The examples of short-chain length PHAs are P(3HB) [Poly-3-hydroxybutyrate)], P(4HB) [Poly-4-hydroxybutyrate] and P(3HV) [Poly(3-hydroxyvalerate)] or the copolymer P(3HB-co-3HV) whereas, the P(3HHx) [Poly-3-hydroxyhexanoate), P(3HO) [Poly-3-hydroxyoctanoate] and copolymer P(3HHx-co-3HO) are considered as medium-chain length PHAs (Kim and Lenz, 2001). Each polyhydroxyalkanoate (PHA) polymer generally consists of about 1000 to 10000 monomers but most of them are synthesized by short chain length (SCL) monomer units (Van der walle *et al.*, 2001). The chemical structure of polyhydroxyalkanoates (PHAs) is depicted in fig. 1.

Figure 1. The chemical structure of polyhydroxyalkanoates (PHAs). The pendant R groups (shaded boxes) vary in chain length from one carbon (C1) to over 14 carbons (C14). Structures shown here are poly-3-hydroxybutyrate (PHB) [R=methyl], poly-3-hydroxyvalerate (PHV) [R=ethyl], and poly-3-hydroxyhexanoate (PHH) or poly-4-hydroxybutyrate (P4HB) [R=propyl]. Adapted from (Suriyamongkol et al., 2007)

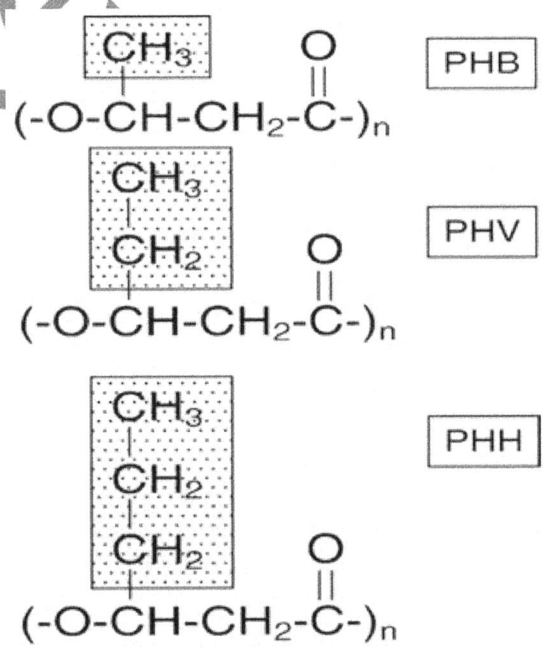

The most of our knowledge about biosynthesis of polyhydroxyalkanoate is mainly based upon the studies on the production of polyhydroxybutyrate (PHB) in cytoplasm of a gram negative soil bacterium *Ralstonia eutrophus* or *Alcaligens eutrophus* bacteria. This bacterium has the capability of producing polyhydroxyalkanoate in a natural way and could accumulate the polymer up to 85% of its total dry body weight when grown on culture media with excess of glucose. In this way, it acts as an energy source but the production of this polymer is limited when there is growth limiting conditions such as lacking of macro-elements such as nitrogen, phosphorus, trace elements or the lack of oxygen in culture media (De Koning, 1995). Previous studies reported that the polyhydroxybutyrate (PHB) could be depolymerized into acetoacetate and further into acetyl coenzyme A (CoA) by applying growth limiting conditions (Steinbuchel and Valentin, 1995). Therefore, it is clearly demonstrated that the acetyl coenzyme A acts as a precursor of polyhydroxybutyrate biosynthesis in bacterial cell. Polyhydroxybutyrate (PHB) decomposes into 3-hydrobutyrate (3-HB) monomers that can be used by fungi and bacteria as carbon sources. The biosynthesis of PHB was first time described in 1973 in a bacterium *Ralstonia eutrophus* by Gottingen and Hull (Senior and Dawes, 1973). There are three key enzymes namely acetoacetyl-CoA reductase, 3-ketothiolase and PHA synthase which leads to the production of plolyhydroxybutyrate by using acetyl-CoA. PHA synthase uses Co-A thioester of (r)-hydroxy fatty acids as substrate. The enzyme popularly known as 3-ketothiolase encoded by gene phaA or phbA is mainly responsible for catalyzing the reversible condensation of two molecules of acetyl-CoA into acetoacetyl-CoA molecule. The acetoacetyl-CoA reductase encoded by *phaB or phbB* gene reduces acetoacetyl-CoA into R-(-)-3-hydroxybutyryl-CoA. After that the R-(-)-3-hydroxybutyryl-CoA finally polymerizes into polyhydroxybutyrate (PHB) by the action of PHA synthase enzyme encoded by a gene called *phaC or phbC* (Yunus *et al.,* 2008; Kosseva and Rusbandi, 2018). The polyhydroxyalkanoates (PHAs) biosynthesis pathway is schematically depicted in fig. 2. Polyhydroxyalkanoates are generally biosynthesized through two possible routes. The first route is based upon β-oxidation pathway intermediates and also on alkanoic acids. In this process, the levorotary S-3-hydroxyacyl-CoA is converted into R-3-hydroxyacyl-CoA, a dextrorotatory enantiomer by the action of an enzyme epimerase. In the second route, the fatty acid biosynthesis intermediates such as R-3-hydroxyacyl-ACP are used. In this process, the acyl carrier protein (ACP) is replaced by coenzyme A using an important enzyme 3-hydroxyacyl-CoA-ACP transacylase. Both these processes are completed by a gene called phaC [Kosseva and Rusbandi, 2018].

PHA Production in Transgenic Plants

The production of PHAs by using microorganisms is costly because of various factors such as variety and amount of nutrition supplied for microbes, optimized growth environment and sterilized conditions (Din *et al.,* 2012). Therefore in comparison of microbes, transgenic plants are considered as cheaper eco-friendly alternatives. The biosynthesis of PHAs in transgenic plants mainly depends upon mineral salts, water, light and carbon dioxide (CO_2). The PHAs production in transgenic plants is generally based upon the availability of acetyl CoA, a primary substrate for PHA biosynthesis, because the plant cells do not have the abilities to degrade PHA as the microbes do. Acetyl-CoA is the main metabolite of plant's catabolic and anabolic processes. The plant cellular compartments such as cytoplasm, mitochondria, peroxisomes and plastids are rich in acetyl-CoA. Therefore, the scientists targeted these compartments as the major sites for producing and accumulating various PHAs in transgenic plants. The literature showed that the first experimental research attempt for producing PHA was successfully achieved in the cytoplasm of transgenic *Arabidopsis thaliana*. After this work several another research experiments

Figure 2. The medium chain length polyhydroxyalkanoates (mcl-PHA) biosynthesis pathways occur in the peroxisomes of transgenic plants. Here, enzyme ACD = Acyl-CoA dehydrogenase; ECH = Enoyl-CoA hydratase; HCD = L-3-hydroxyacyl-CoA dehydrogenase and KT = β-ketothiolase. Reproduced from Dobrogojski et al. (2016)

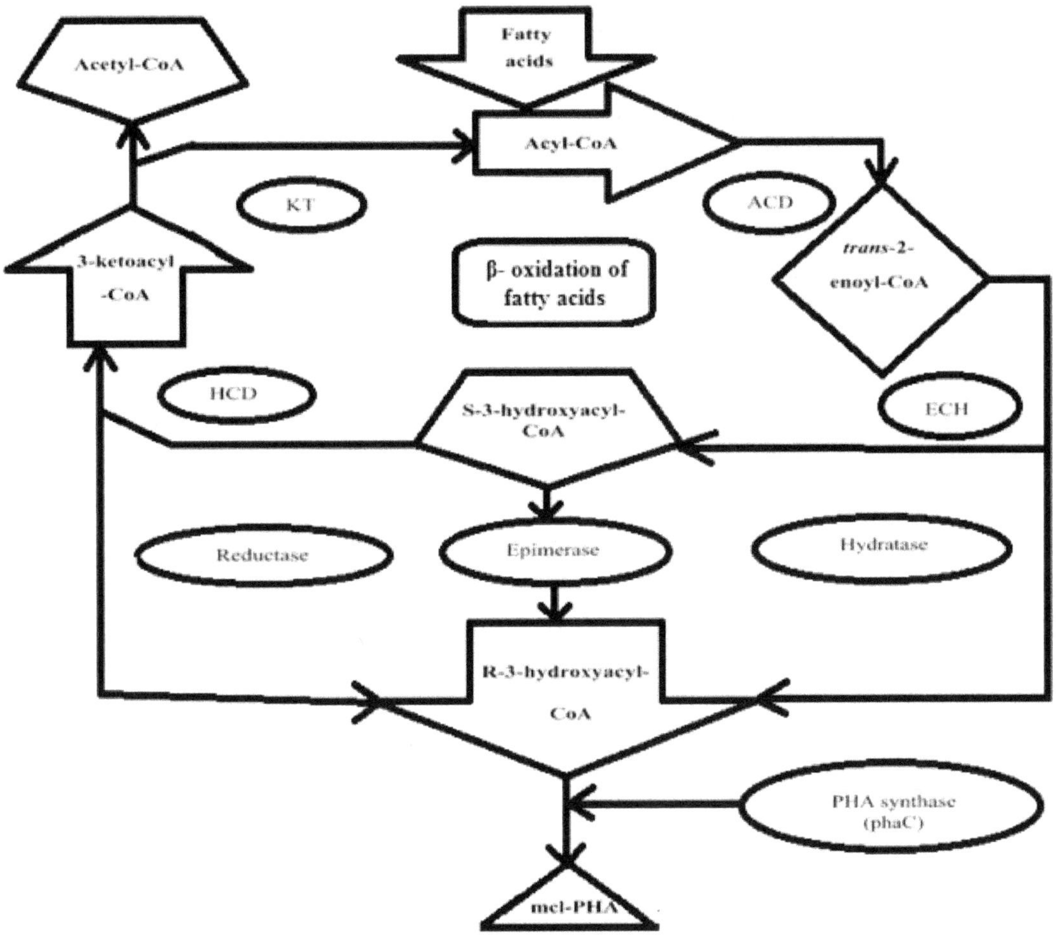

were also conducted using various plant species. But the deficiency of acetyl CoA and acetoacetyl-CoA because of their utilization in plant hormone and steroid biosynthesis pathways limited the production of PHAs inside plant cell cytoplasm. Like as cytoplasm, mitochondria also have the limitations of the deficiency of acetyl-CoA because of its utilization during cellular respiration. Plastids appear to the best site for PHA biosynthesis in plants because there acetyl-CoA is present in higher concentrations and mainly utilized for the biosynthesis of fatty acids. The plastids are the organelle which work properly despite the structural changes and have the ability to store larger starch granules. But the plastids do not have the stocks of beta ketothiolases. The beta ketothiolases is located in cell's cytoplasm. This problem could be overcome by applying specific DNA encoding plastid targeted sequences inserting in the vectors. The peroxisomes also considered as high potential sites for the production of PHAs in transgenic plant cells because of having high reductive strength of NADH and their beta oxidation of fatty acids. Peroxisomes are important cell organelles because of synthesizing medium chain length polyhydroxyalkanoates (mcl-PHAs). Since the last two decades several scientists are doing research on producing PHAs in transgenic plants. The detailed informations regarding the current status of transgenic plants developed for producing polyhydroxyalkanoates are given in table 1.

Table 1. The current status of transgenic plants engineered for production of polyhydroxyalkanoates (PHAs)

Plant species	Gene used	Cell Organelle	PHA produced	References
Arabidopsis thaliana	----	Cytoplasm	P3HB	Poirier *et al.*, 1992
Arabidopsis thaliana	*phbA, phbB, phaC*	Plastids	P3HB	Nawrath *et al.*, 1994
Arabidopsis thaliana	*phbB, phbC*	Cytoplasm	P3HB	Poirier *et al.*, 1995
Arabidopsis thaliana	*phaC1*	Peroxisomes	mclPHA	Mittendorf *et al.*, 1998
Arabidopsis thaliana	*phbA, phbB, phaC*	Plastids	P3HB-3HV	Slater *et al.*, 1999
Arabidopsis thaliana	*phbA, phbB, phaC*	Plastids	P3HB-3HV	Valentin *et al.*, 1999
Arabidopsis thaliana	*phbA, phbB, phaC*	Plastids	P3HB	Bohmert *et al.*, 2000
Arabidopsis thaliana	*phbA, phbB, phbC*	Plastids	P3HB	Bohmert *et al.*, 2002
Arabidopsis thaliana	*phbA, phbB, phaC*	Peroxisomes	scl-mclPHA	Arai *et al.*, 2002
Arabidopsis thaliana	*phaA, phaB, phaC*	Plastids	P3HB	Kourtz *et al.*, 2005
Arabidopsis thaliana	*phbA, phbB, phaC*	Cytoplasm	P3HB-co-3HV	Matsumoto *et al.*, 2005
Arabidopsis thaliana	*phbA, phbB, phaC*	Peroxisomes	scl-mclPHA	Matsumoto *et al.*, 2006b
Arabidopsis thaliana	*phaA, phaB, phaC*	Plastids	P3HB	Kourtz *et al.*, 2007
Arabidopsis thaliana	*phaC1, phaA, phaB*	Plastids	scl-mclPHA	Matsumoto *et al.*, 2009
Arabidopsis thaliana	*phaA, phaB, phaC*	Peroxisomes	scl-PHA	Tillbrook *et al.*, 2011
Beta vulgaris	*phaA, phaB, phaC*	Plastids	P3HB	Menzel *et al.*, 2003
Brassica napus	*phbA, phbB, phaC*	Plastids	P3HB	Houmiel *et al.*, 1999
Brassica napus	*phbB, phaC*	Plastids	P3HB-co-3HV	Slater *et al.*, 1999
Brassica napus	*phaA, phaB, phaC*	Plastids	P3HB, P3HB-co-3HV	Valentin *et al.*, 1999
Brassica napus	*phbA, phbB, phaC*	Cytoplasm	P3HB	Poirier and Gruys, 2001
Camellina sativa	*phaA, phaB, phaC*	Plastids	P3HB	Patterson *et al.*, 2011
Camellina sativa	*phaA, phaB, phaC*	Plastids	P3HB	Malik *et al.*, 2015
Elaeis guineensis	*bktB, phaB, phaC*	Plastids	P3HB	Parveez *et al.*, 2015
Glycine max	*phbA, phbB, phaC*	Vacuoles	P3HB	Schnell *et al.*, 2012
Gossypium hirsutum	*phaB, phaC*	Cytoplasm, Plastids	P3HB	John and Keller, 1996
Linum usitatissimum	*phbA, phbB, phbC*	Plastids	P3HB	Wrobel-Kwiatkowska *et al.*, 2004
Linum usitatissimum	*phaA, phaB, phaC*	Plastids	P3HB	Wrobel-Kwiatkowska *et al.*, 2007
Linum usitatissimum	*phbA, phbB, phbC*	Plastids	P3HB	Wrobel-Kwiatkowska *et al.*, 2009
Linum usitatissimum	*phbA, phbB, phbC*	Fibers	P3HB	Kulma *et al.*, 2015

continues on following page

Table 1. Continued

Plant species	Gene used	Cell Organelle	PHA produced	References
Linum usitatissimum	*phaC1*	Peroxisomes	mclPHA	Wrobel-Kwiatkowska *et al.*, 2019
Medicago sativa	*phbA, phbB, phaC*	Plastids	P3HB	Saruul *et al.*, 2002
Nicotiana tabacum	*phbB, phaC*	Cytoplasm	P3HB	Nakashita *et al.*, 1999
Nicotiana tabacum	*phbA, phbB, phbC*	Cytoplasm	P3HB	Nakashita *et al.*, 2001a
Nicotiana tabacum	*phbA, phbB, phbC*	Plastids	sclPHA	Nakashita *et al.*, 2001b
Nicotiana tabacum	*phbA, phbB, phaC*	Plastids	P3HB	Arai *et al.*, 2001
Nicotiana tabacum	*phbA, phbB, phbC*	Plastids	P3HB	Bohmert *et al.*, 2002
Nicotiana tabacum	*phaA, phaB, phaC*	Plastids	P3HB	Suzuki *et al.*, 2002
Nicotiana tabacum	*phbA, phbB, phbC*	Plastids	P3HB	Lossl *et al.*, 2003
Nicotiana tabacum	*phaA, phaB, phaC*	Plastids	P3HB	Arai *et al.*, 2004
Nicotiana tabacum	*phaC2, aadA*	Plastids	mclPHA	Wang *et al.*, 2005
Nicotiana tabacum	*phaB, phaC*	Cytoplasm	P3HB	Matsumoto *et al.*, 2011
Nicotiana tabacum	*phaA, phaB*	Plastids	P3HB	Bohmert-Tatarev *et al.*, 2011
Nicotiana tabacum	*phaB, phaC*	BY-2 Cells	P3HB	Yokoo *et al.*, 2015
Oryza sp.	*phbB, phbC*	Cytoplasm	P3HB	Endo *et al.*, 2006
Panicum virgatum	*phaA, phaB, phaC*	Plastids	P3HB	Somleva *et al.*, 2008
Panicum virgatum	*phaA, phaB, phaC*	Plastids	P3HB	Somleva and Ali, 2010
Panicum virgatum	*phaA, phaB, phaC, FBPase, SBPase*	Plastids	P3HB	Somleva *et al.*, 2012
Populus sp.	*phbA, phbB, phaC*	Plastids	P3HB	Dalton *et al.*, 2011
Saccharum officinarum	*phaA, phaB, phaC*	Plastids	P3HB	Brumbley *et al.*, 2003
Saccharum officinarum	*cTP-CPL, HCHL*	Plastids	pHBA	McQualter *et al.*, 2005
Saccharum officinarum	*phaA, phaB, phaC*	Cytoplasm, Plastids	P3HB	Petrasovits *et al.*, 2007
Saccharum officinarum	*phaA, phaB, phaC*	Plastids	P3HB	Purnell *et al.*, 2007
Saccharum officinarum	*phaA, phaB, phaC2, phaJ2, FatB2, KasA1*	Peroxisomes	scl-mclPHA	Anderson *et al.*, 2011
Saccharum officinarum	*phaA, phaB, phaC*	Peroxisomes, Vacuoles	sclPHA	Tilbrook *et al.*, 2011
Saccharum officinarum	*phaA, phaB, phaC*	Plastids	P3HB	Petrasovits *et al.*, 2012
Saccharum officinarum	*NphT7, phaA, phaB, phaC*	Plastids	mclPHA	McQualter *et al.*, 2015
Solanum tuberosum	*phbA, phbB, phaC*	Plastids	P3HB	Bohmert *et al.*, 2002
Solanum tuberosum	----	Cytoplasm	mclPHA	Romano *et al.*, 2003
Tamarix sp.	*phbB, phbC*	Cytoplasm	P3HB	Endo *et al.*, 2006
Solanum tuberosum	*phaC1, phaG*	Plastids	mclPHA	Romano *et al.*, 2005
Zea mays	----	Peroxisomes	P3HB	Hahn *et al.*, 1999
Zea mays	*phbA, phbB, phaC*	Plastids	P3HB	Poirier and Gruys, 2001
Zea mays	*IlvA, phaA, phbB, phbC*	Plastids	P3HB	Mitsky *et al.*, 2003

PHA, polyhydroxyalkanoate; **P(3HB)**, poly-3-hydroxybutyrate; **P(3HB-co-HV)**, poly(3-hydroxybutyrate-co-3-hydroxyvalerate); **Scl-mclPHA**, short-chain-length to medium-chain-length PHA; **mclPHA**, medium chain-length PHA.

CONCLUSION

The environmental pollution generated through petroleum based synthetic polymers has become a very big global challenge. The production of synthetic polymers is increasing day by day and now it has appeared in an unbeatable form of pollutants. The management of the plastic and its products is not an easy task; it takes several hundreds of years to be decomposed. Therefore, it is a need of present time to find out eco-friendly biodegradable alternatives. Polyhydroxyalkanoates (PHAs) has appeared as a smart choice of scientific community as well as industry in the form of plastic alternatives. PHAs are the major class of biodegradable biopolymers which are biosynthesized by microorganisms in a natural way. The production of PHAs by using microorganisms is costly because of various factors such as variety and amount of nutrition supplied for microbes, optimized growth environment and sterilized conditions. Therefore in comparison of microbes, transgenic plants are considered as cheaper eco-friendly alternatives. The biosynthesis of PHAs in transgenic plants mainly depends upon mineral salts, water, light and carbon dioxide (CO_2). Some cellular compartments such as cytoplasm, mitochondria, peroxisomes and plastids have been targeted as important sites for producing and accumulating PHAs in transgenic plants. Since the last two decades several scientists have engaged in research for optimizing transgenic plants as bioreactors for producing PHAs. Several plant species including *Arabidopsis thaliana*, *Camellina*, tobacco, sugarcane, maize, rapeseed, flax, cotton, oil palm have been genetically engineered for producing PHAs. But till date, no one plant species is released for commercial production of biopolymers. The research is on the way, things are optimizing and we hopefully expect that the transgenic plants would be available in near future for producing PHAs at larger scales.

REFERENCES

Anderson, A. J., & Dawes, E. A. (1990). Occurrence, metabolism, metabolic role, and industrial uses of bacterial polyhyroxyalkanoates. *Microbiological Reviews*, *54*(4), 450–472. doi:10.1128/mr.54.4.450-472.1990 PMID:2087222

Anderson, D., Gnanasambandam, A., Mills, E., O'Shea, M., Nielsen, L., & Brumbley, S. (2011). Synthesis of short-chain length/medium chain length polyhydroxyalkanoate (PHA) copolymers in peroxisomes of transgenic sugarcane plants. *Tropical Plant Biology*, *4*(3), 170–184. doi:10.100712042-011-9080-7

Anjum, A., Zuber, M., Zia, K. M., Noreen, A., Anjum, M. N., & Tabasum, S. (2016). Microbial production of polyhydroxyalkanoates (PHAs) and its copolymers: A review of recent advancements. *International Journal of Biological Macromolecules*, *89*, 161–174. doi:10.1016/j.ijbiomac.2016.04.069 PMID:27126172

Arai, Y., Nakashita, H., Doi, Y., & Yamaguchi, I. (2001). Plastid targeting of polyhydroxybutyrate biosynthetic pathway in tobacco. *Plant Biotechnology (Sheffield, England)*, *18*(4), 289–293. doi:10.5511/plantbiotechnology.18.289

Arai, Y., Nakashita, H., Suzuki, Y., Kobayashi, Y., Shimizu, T., Yasuda, M., Doi, Y., & Yamaguchi, I. (2002). Synthesis of a novel class of polyhydroxyalkanoates in *Arabidopsis* peroxisomes and their use in monitoring short-chain-length intermediates of β-oxidation. *Plant & Cell Physiology*, *43*(5), 555–562. doi:10.1093/pcp/pcf068 PMID:12040103

Arai, Y., Shikanai, T., Doi, Y., Yoshida, S., Yamaguchi, I., & Nakashita, H. (2004). Production of polyhydroxybutyrate by polycistronic expression of bacterial genes in tobacco plastid. *Plant & Cell Physiology, 45*(9), 1176–1184. doi:10.1093/pcp/pch139 PMID:15509840

Bohmert, K., Balbo, I., Kopka, J., Mittendorf, V., Nawrath, C., Poirier, Y., Tischendorf, G., Trethewey, R. N., & Willmitzer, L. (2000). Transgenic *Arabidopsis* plants can accumulate polyhydroxybutyrate to up to 4% of their fresh weight. *Planta, 211*(6), 841–845. doi:10.1007004250000350 PMID:11144269

Bohmert, K., Balbo, I., Steinbuchel, A., Tischendorf, G., & Willmitzer, L. (2002). Constitutive expression of the beta-ketothiolase gene in transgenic plants: A major obstacle for obtaining polyhydroxybutyrate-producing plants. *Plant Physiology, 128*(4), 1282–1290. doi:10.1104/pp.010615 PMID:11950977

Bohmert-Tatarev, K., McAvoy, S., Daughtry, S., Peoples, O. P., & Snell, K. D. (2011). High levels of bioplastic are produced in fertile transplastomic tobacco plants engineered with a synthetic operon for the production of polyhydroxybutyrate. *Plant Physiology, 155*(4), 1690–1708. doi:10.1104/pp.110.169581 PMID:21325565

Brumbley, S. M., Petrasovits, L. A., Bonaventura, P. A., O'Shea, M. J., Purnell, M. P., & Nielsen, L. K. (2003). Production of polyhydroxyalkanoates in sugarcane. *Proc. of Int. Soc. of Sugarcane Tech., Mol. Biol. Workshop, 4*, 31.

Dalton, D. A., Ma, C., Shrestha, S., Kitin, P., & Strauss, S. H. (2011). Trade-offs between biomass growth and inducible biosynthesis of polyhydroxybutyrate in transgenic poplar. *Plant Biotechnology Journal, 9*(7), 759–767. doi:10.1111/j.1467-7652.2010.00585.x PMID:21265995

De Koning, G. (1995). Physical properties of bacterial poly (R)-3-hydroxyalkanoates). *Canadian Journal of Microbiology, 41*(13), 303–309. doi:10.1139/m95-201

Dobrogojski, J., Spychalski, M., Lucinski, R., & Borek, S. (2018). Transgenic plants as a source of polyhydroxyalkanoates. *Acta Physiologiae Plantarum, 40*(9), 162–178. doi:10.100711738-018-2742-4

Endo, N., Yoshida, K., Akiyoshi, M., & Manji, S. (2006). Hybrid fiber production: A wood and plastic combination in transgenic rice and *Tamarix* made by accumulating poly-3-hydroxybutyrate. *Plant Biotechnology (Sheffield, England), 23*(1), 99–109. doi:10.5511/plantbiotechnology.23.99

Gumel, A. M., Annuar, M. S. M., & Chisti, Y. (2013). Recent advances in the production, recovery and applications of polyhydroxyalkanoates. *Journal of Polymers and the Environment, 21*(2), 580–605. doi:10.100710924-012-0527-1

Hahn, J. J., Eschenlauer, A. C., Sleytr, U. B., Somers, D. A., & Srienc, F. (1999). Peroxisomes as sites for synthesis of polyhydroxyalkanoates in transgenic plants. *Biotechnology Progress, 15*(6), 1053–1057. doi:10.1021/bp990118n PMID:10585189

Houmiel, K.L., Slater, S., Broyles, D. P. (1999). (β-hydroxybutyrate) production in oilseed leukoplasts of *Brassica napus. Planta, 209*, 547-550.

John, M. E., & Keller, G. (1996). Metabolic pathway engineering in cotton: Biosynthesis of polyhydroxybutyrate in fiber cells. *Proceedings of the National Academy of Sciences of the United States of America, 93*(23), 12768–12773. doi:10.1073/pnas.93.23.12768 PMID:11038522

Kim, Y. B., & Lenz, R. W. (2001). Polysters from microorganisms. *Advances in Biochemical Engineering/Biotechnology*, *71*, 51–79. doi:10.1007/3-540-40021-4_2 PMID:11217417

Kosseva, M. R., & Rusbandi, R. (2018). Trends in the biomanufacture of polyhydroxyalkanoates with focus on downstream processing. *International Journal of Biological Macromolecules*, *107*, 762–778. doi:10.1016/j.ijbiomac.2017.09.054 PMID:28928063

Kourtz, L., Dillon, K., Daughtry, S., Madison, L. L., Peoples, O., & Snell, K. D. (2005). A novel thiolase-reductase gene fusion promotes the production of polyhydroxybutyrate in *Arabidopsis. Plant Biotechnology Journal*, *3*(4), 435–447. doi:10.1111/j.1467-7652.2005.00136.x PMID:17173631

Kourtz, L., Dillon, K., Daughtry, S., Madison, L. L., Peoples, O., & Snell, K. D. (2007). Chemically inducible expression of the PHB biosynthetic pathway in *Arabidopsis. Transgenic Research*, *16*(6), 759–769. doi:10.100711248-007-9067-1 PMID:17279436

Kulma, A., Skorkowska-Telichowska, K., Kostyna, K., Szatkowskia, M., Skała, J., Drulis-Kawa, Z., Preisnera, M., Zuk, M., Szperlika, J., Wang, Y. F., & Szopaa, J. (2015). New flax producing bioplastic fibers for medical purposes. *Industrial Crops and Products*, *68*, 80–90. doi:10.1016/j.indcrop.2014.09.013

Lenz, R. W., & Marchessault, R. H. (2005). Bacterial Polyesters: Biosynthesis, Biodegradable Plastics and Biotechnology. *Biomacromol*, *6*(1), 1–8. doi:10.1021/bm049700c

Lossl, A., Eibl, C., Harloff, H. J., Jung, C., & Koop, H. U. (2003). Polyester synthesis in transplastomic tobacco (*Nicotiana tabacum* L.): Significant contents of polyhydroxybutyrate are associated with growth reduction. *Plant Cell Reports*, *21*(9), 891–899. doi:10.100700299-003-0610-0 PMID:12789507

Malik, M. R., Yang, W., Patterson, N., Tang, J., Wellinghoff, R. L., Preuss, M. L., Burkitt, C., Sharma, N., Ji, Y., Jez, J. M., Peoples, O. P., Jaworski, J. G., Cahoon, E. B., & Snell, K. D. (2015). Production of high levels of poly-3-hydroxybutyrate in plastids of *Camelina sativa* seeds. *Plant Biotechnology Journal*, *13*(5), 675–688. doi:10.1111/pbi.12290 PMID:25418911

Masood, F., Yasin, T., & Hameed, A. (2014). Polyhydroxyalkanoates—What are the uses? Current challenges and perspectives. *Critical Reviews in Biotechnology*, *35*(4), 514–521. doi:10.3109/07388551.2014.913548 PMID:24963700

Matsumoto, K., Aoki, E., Takase, K., Doi, Y., & Taguchi, S. (2006a). *In vivo* and *in vitro* characterization of Ser477X mutations in polyhydroxyalkanoate (PHA) synthase 1 from *Pseudomonas* sp 61-3: Effects of beneficial mutations on enzymatic activity, substrate specificity, and molecular weight of PHA. *Biomacromolec*, *7*(8), 2436–2442. doi:10.1021/bm0602029 PMID:16903693

Matsumoto, K., Arai, Y., Nagao, R., Murata, T., Takase, K., Nakashita, H., Taguchi, S., Shimada, H., & Doi, Y. (2006b). Synthesis of short-chain-length/medium-chain-length polyhydroxyalkanoate (PHA) copolymers in peroxisome of the transgenic *Arabidopsis thaliana* harboring the PHA synthase gene from *Pseudomonas* sp. 61-3. *Journal of Polymers and the Environment*, *14*(4), 369–374. doi:10.100710924-006-0035-2

Matsumoto, K., Morimoto, K., Gohda, A., Shimada, H., & Taguchi, S. (2011). Improved polyhydroxybutyrate (PHB) production in transgenic tobacco by enhancing translation efficiency of bacterial PHB biosynthetic genes. *Journal of Bioscience and Bioengineering*, *111*(4), 485–488. doi:10.1016/j.jbiosc.2010.11.020 PMID:21185778

Matsumoto, K., Murata, T., Nagao, R., Nomura, C. T., Arai, S., Arai, Y., Takase, K., Nakashita, H., Taguchi, S., & Shimada, H. (2009). Production of short-chain-length/medium-chain-length polyhydroxyalkanoate (PHA) copolymer in the plastid of *Arabidopsis thaliana* using an engineered 3-ketoacyl-acyl carrier protein synthase III. *Biomacromolec*, *10*(4), 686–690. doi:10.1021/bm8013878 PMID:19265441

Matsumoto, K., Nagao, R., Murata, T., Arai, Y., Kichise, T., Nakashita, H., Taguchi, S., Shimada, H., & Doi, Y. (2005). Enhancement of poly(3-hydroxybutyrate-co-3-hydroxyvalerate) production in the transgenic *Arabidopsis thaliana* by the *in vitro* evolved highly active mutants of polyhydroxyalkanoate (PHA) synthase from *Aeromonas caviae*. *Biomacromolecules*, *69*(4), 2126–2130. doi:10.1021/bm050113g PMID:16004454

McQualter, R. B., Fong Chong, B., O'Shea, M., Meyer, K., van Dyk, D. E., Viitanen, P. V., & Brumbley, S. M. (2005). Initial evaluation of sugarcane as a production platform for a *p*-hydroxybenzoic acid. *Plant Biotechnology Journal*, *3*(1), 29–41. doi:10.1111/j.1467-7652.2004.00095.x PMID:17168897

McQualter, R. B., Petrasovits, L. A., Gebbie, L. K., Schweitzer, D., Blackman, D. M., Chrysanthopoulos, P., Hodson, M. P., Plan, M. R., Riches, J. D., Snell, K. D., Brumbley, S. M., & Nielsen, L. K. (2015). The use of an acetoacetyl-CoA synthase in place of a β-ketothiolase enhances poly-3-hydroxybutyrate production in sugarcane mesophyll cells. *Plant Biotechnology Journal*, *13*(5), 700–707. doi:10.1111/pbi.12298 PMID:25532451

Menzel, G., Harloff, H. J., & Jung, C. (2003). Expression of bacterial poly (3-hydroxybutyrate) synthesis genes in hairy roots of sugar beet (*Beta vulgaris* L.). *Applied Microbiology and Biotechnology*, *60*(5), 571–576. doi:10.100700253-002-1152-z PMID:12536258

Mitsky, T. A., Slater, S., Reiser, S. E., Hao, M., & Houmiel, K. L. (2003). *Multigene expression vectors for the biosynthesis of products via multienzyme biological pathways.* US Patent Application 2003/0182678.

Mittendorf, V., Robertson, E. J., Leech, R. M., Kruger, N., Steinbuchel, A., & Poirier, Y. (1998). Synthesis of medium-chain-length polyhydroxyalkanoates in *Arabidopsis thaliana* using intermediates of peroxisomal fatty acid β-oxidation. *Proceedings of the National Academy of Sciences of the United States of America*, *95*(23), 13397–13402. doi:10.1073/pnas.95.23.13397 PMID:9811811

Mozejko-Ciesielska, J., & Kiewisz, R. (2016). Bacterial polyhydroxyalkanoates: Still fabulous? *Microbiological Research*, *192*, 271–282. doi:10.1016/j.micres.2016.07.010 PMID:27664746

Nakashita, H., Arai, Y., Shikanai, T., Doi, Y., & Yamaguchi, I. (2001a). Introduction of bacterial metabolism into higher plants by polycistronic transgene expression. *Bioscience, Biotechnology, and Biochemistry*, *65*(7), 1688–1691. doi:10.1271/bbb.65.1688 PMID:11515561

Nakashita, H., Arai, Y., Suzuki, Y., & Yamaguchi, I. (2001b). Molecular breeding of transgenic tobacco plants which accumulate polyhydroxyalkanoates. *RIKEN Rev*, *42*, 67–70.

Nakashita, H., Arai, Y., Yoshioka, K., Fukui, T., Doi, Y., Usami, R., Horikoshi, K., & Yamaguchi, I. (1999). Production of biodegradable polyester by a transgenic tobacco. *Bioscience, Biotechnology, and Biochemistry*, *63*(5), 870–874. doi:10.1271/bbb.63.870 PMID:27385570

Nawrath, C., Poirier, Y., & Somerville, C. (1994). Targeting of the polyhydroxybutyrate biosynthetic-pathway to the plastids of *Arabidopsis thaliana* results in high-levels of polymer accumulation. *Proceedings of the National Academy of Sciences of the United States of America*, *91*(26), 12760–12764. doi:10.1073/pnas.91.26.12760 PMID:7809117

Parveez, G. K. A., Bahariah, B., Ayub, N. H., Masani, M. Y. A., Rasid, O. A., Tarmizi, A. H., & Ishak, Z. (2015). Production of polyhydroxybutyrate in oil palm (*Elaeis guineensis* Jacq.) mediated by microprojectile bombardment of PHB biosynthesis genes into embryogenic calli. *Front Plant Sci*, *6*, 598. doi:10.3389/fpls.2015.00598 PMID:26322053

Patterson, N., Tang, J., Cahoon, E. B., Jaworski, J. G., Wang, W., Peoples, O. P., & Snell, K. D. (2011). *Generation of high polyhydroxybutryate producing oilseeds*. International Patent Application WO/2011/034945.

Petrasovits, L. A., Purnell, M. P., Nielsen, L. K., & Brumbley, S. M. (2007). Production of polyhydroxybutyrate in sugarcane. *Plant Biotechnology Journal*, *5*(1), 162–172. doi:10.1111/j.1467-7652.2006.00229.x PMID:17207265

Petrasovits, L. A., Zhao, L., McQualter, R. B., Snell, K. D., Somleva, M. N., Patterson, N. A., Nielsen, L. K., & Brumbley, S. M. (2012). Enhanced polyhydroxybutyrate production in transgenic sugarcane. *Plant Biotechnology Journal*, *10*(5), 569–578. doi:10.1111/j.1467-7652.2012.00686.x PMID:22369516

Poirier, Y., Dennis, D. E., Klomparens, K., & Somerville, C. (1992). Polyhydroxybutyrate a biodegradable thermoplastic, produced in transgenic plants. *Science*, *256*(5056), 520–523. doi:10.1126cience.256.5056.520 PMID:17787950

Poirier, Y., & Gruys, K.J. (2001). Production of PHAs in transgenic plants. *Biopolyesters,* 401-435.

Poirier, Y., Somerville, C., Schechtman, L. A., Satkowski, M. M., & Noda, I. (1995). Synthesis of high-molecular-weight poly ([R]-(-)-3-hydroxybutyrate) in transgenic *Arabidopsis thaliana* plant cells. *International Journal of Biological Macromolecules*, *17*(1), 7–12. doi:10.1016/0141-8130(95)93511-U PMID:7772565

Purnell, M. P., Petrasovits, L. A., Nielsen, L. K., & Brumbley, D. W. (2007). Spatiotemporal characterisation of polyhydroxybutyrate accumulation in sugarcane. *Plant Biotechnology Journal*, *5*(1), 173–184. doi:10.1111/j.1467-7652.2006.00230.x PMID:17207266

Romano, A., Raemakers, K., Bernardi, J., Visser, R., & Mooibroek, H. (2003). Transgene organization in potato after particle bombardment-mediated co-transformation using plasmids and gene cassettes. *Transgenic Research*, *12*(4), 461–473. doi:10.1023/A:1024267906219 PMID:12885167

Romano, A., Van der Plas, L. H. W., Witholt, B., Eggink, G., & Mooibroek, H. (2005). Expression of poly-3-(R)-hydroxyalkanoate (PHA) polymerase and acyl-CoA-transacylase in plastids of transgenic potato leads to the synthesis of a hydrophobic polymer, presumably medium-chain-length PHAs. *Planta*, *3*(220), 455–464. doi:10.100700425-004-1349-8 PMID:15351883

Sabbagh, F., & Muhamad, I. (2017). Production of polyhydroxyalkanoate as secondary metabolite with main focus on sustainable energy. *Renewable & Sustainable Energy Reviews*, *72*, 95–104. doi:10.1016/j.rser.2016.11.012

Saruul, P., Srienc, F., Somers, D. A., & Samac, D. A. (2002). Production of a biodegradable plastic polymer, poly-beta-hydroxybutyrate in transgenic alfalfa. *Crop Science*, *42*, 919–927.

Schnell, J. A., Treyvaud-Amiguet, V., Arnason, J. T., & Johnson, D. A. (2012). Expression of polyhydroxybutyric acid as a model for metabolic engineering of soybean seed coats. *Transgenic Research*, *4*(21), 895–899. doi:10.100711248-011-9575-x PMID:22183687

Senior, P. J., & Dawes, E. A. (1973). The regulation of poly ß-hydroxybutyrate metabolism in *Azotobacter beijerinckii. The Biochemical Journal*, *134*(1), 225–238. doi:10.1042/bj1340225 PMID:4723225

Sharma, L., Srivastava, J. K., & Singh, A. K. (2016). Biodegradable polyhydroxyalkanoate thermoplastics substituting xenobiotic plastics: a way forward for sustainable environment. In A. Singh, S. Prasad, & R. Singh (Eds.), *Plant responses to xenobiotics* (pp. 317–346). Springer. doi:10.1007/978-981-10-2860-1_14

Slater, S., Mitsky, T. A., Houmiel, K. L., Hao, M., Reiser, S. E., Taylor, N. B., Tran, M., Valentin, H. E., Rodriguez, D. J., Stone, D. A., Padgette, S. R., Kishore, G., & Gruys, K. J. (1999). Metabolic engineering of *Arabidopsis* and *Brassica* for poly (3-hydroxybutyrateco-3-hydroxyvalerate) copolymer production. *Nature Biotechnology*, *17*(10), 1011–1016. doi:10.1038/13711 PMID:10504704

Somleva, M., & Ali, A. (2010). *Propagation of transgenic plants.* International Patent Application WO/2010/102220.

Somleva, M., Chinnapen, H., Ali, A., Snell, K. D., Peoples, O. P., Patterson, N., Tang, J., & Bohmert-Tatarev, K. (2012). *Increasing carbon flow for polyhydroxybutyrate production in biomass crops.* US Patent Application 2012/0060413.

Somleva, M. N., Snell, K. D., Beaulieu, J. J., Peoples, O. P., Garrison, B. R., & Patterson Nii, A. (2008). Production of polyhydroxybutyrate in switchgrass, a value-added co-product in an important lignocellulosic biomass crop. *Plant Biotechnology Journal*, *6*(7), 663–678. doi:10.1111/j.1467-7652.2008.00350.x PMID:18498309

Steinbuchel, A., & Valentin, H.E. (1995). Diversity of bacterial polyhydroxyalkanoic acids. *FEMS Microbiol Lett, 128*, 219-228.

Sudesh, K., Abe, H., & Doi, Y. (2000). Synthesis, structure and properties of polyhydroxyalkanoates: Biological polyesters. *Progress in Polymer Science*, *25*(10), 1503–1555. doi:10.1016/S0079-6700(00)00035-6

Suriyamongkol, P., Weselake, R., Narine, S., Moloney, M., & Shah, S. (2007). Biotechnological approaches for the production of polyhydroxyalkanoates in microorganisms and plants—A review. *Biotechnology Advances*, *25*(2), 148–175. doi:10.1016/j.biotechadv.2006.11.007 PMID:17222526

Suzuki, Y., Kurano, M., Arai, Y., Nakashita, H., Doi, Y., Usami, R., Horikoshi, K., & Yamaguchi, I. (2002). Enzyme inhibitors to increase poly-3-hydroxybutyrate production by transgenic tobacco. *Bioscience, Biotechnology, and Biochemistry*, *66*(12), 2537–2542. doi:10.1271/bbb.66.2537 PMID:12596845

Tilbrook, K., Gebbie, L., Schenk, P. M., Poirier, Y., & Brumbley, S. M. (2011). Peroxisomal polyhydroxyalkanoate biosynthesis is a promising strategy for bioplastic production in high biomass crops. *Plant Biotechnology Journal*, *9*(9), 958–969. doi:10.1111/j.1467-7652.2011.00600.x PMID:21447054

Valentin, H. E., Broyles, D. L., Casagrande, L. A., Colburn, S. M., Creely, W. L., DeLaquil, P. A., Felton, H. M., Gonzalez, K. A., Houmiel, K. L., Lutke, K., Mahadeo, D. A., Mitsky, T. A., Padgette, S. R., Reiser, S. E., Slater, S., Stark, D. M., Stock, R. T., Stone, D. A., Taylor, N. B., ... Gruys, K. J. (1999). PHA production, from bacteria to plants. *International Journal of Biological Macromolecules*, *25*(1-3), 303–306. doi:10.1016/S0141-8130(99)00045-8 PMID:10416678

Van der walle, G.A.M., De Koning, G.J.M., Weusthuis, R.A., & Eggink, G. (2001). Properties, modifications and applications of biopolyesters. *Adv Bioch Eng/Biotech, 71*, 263-291.

Wang, Y. H., Wu, Z. Y., Zhang, X. H., Chen, G. Q., Wu, Q., Huang, C. L., & Yang, Q. (2005). Synthesis of medium-chain-length-polyhydroxyalkanoates in tobacco via chloroplast genetic engineering. *Chinese Science Bulletin*, *50*(11), 1113–1120. doi:10.1360/982005-130

Wrobel-Kwiatkowska, M., Kropiwnicki, M., Zebrowski, J., Beopoulos, A., Dyminska, L., Hanuza, J., & Rymowicz, W. (2019). Effect of mcl-PHA synthesis in flax on plant mechanical properties and cell wall composition. *Transgenic Research*, *28*(1), 77–90. doi:10.1007/s11248-018-0105-y PMID:30484148

Wrobel-Kwiatkowska, M., Skorkowska-Telichowska, K., Dyminska, L., Mączka, M., Hanuza, J., & Szopa, J. (2009). Biochemical, mechanical, and spectroscopic analyses of genetically engineered flax fibers producing bioplastic (poly-β-hydroxybutyrate). *Biotechnology Progress*, *25*(5), 1489–1498. doi:10.1002/btpr.194 PMID:19572280

Wrobel-Kwiatkowska, M., Zebrowski, J., Starzycki, M., Oszmianski, J., & Szopa, J. (2007). Engineering of PHB synthesis causes improved elastic properties of flax fibers. *Biotechnology Progress*, *23*(1), 269–277. doi:10.1021/bp0601948 PMID:17269698

Wrobel-Kwiatkowska, M., Zebrowski, J., & Szopa, J. (2004). Polyhydroxybutyrate synthesis in transgenic flax. *Journal of Biotechnology*, *107*(1), 41–54. doi:10.1016/j.jbiotec.2003.10.005 PMID:14687970

Yokoo, T., Matsumoto, K., Ooba, T., Morimoto, K., & Taguchi, S. (2015). Enhanced poly (3-hydroxybutyrate) production in transgenic tobacco BY-2 cells using engineered acetoacetyl-CoA reductase. *Bioscience, Biotechnology, and Biochemistry*, *79*(6), 986–988. doi:10.1080/09168451.2014.1002448 PMID:25647430

Yunus, A. M. M., Parveez, G. K. A., & Ho, C. L. (2008). Transgenic plants producing polyhydroxyalkanoates. *Asia-Pacific Journal of Molecular Biology and Biotechnology*, *16*(1), 1–10.

Chapter 18
Green Chemistry and Its Role for Environmental Sustainability

Aakansha Goswami
https://orcid.org/0000-0002-3350-1747
Swami Vivekanand Subahrti University, India

Manjul Mungali
https://orcid.org/0000-0002-6620-9005
Swami Vivekanand Subharti University, India

ABSTRACT

The development of synthetic chemical products has increased in the last few decades. Some synthetic chemicals are very vital (e.g., pharmaceutical products, fertilizers, textiles, etc.), and the dependency of humans on these chemicals has led to an increase in their production. Green chemistry covers the entire life cycle of a chemical product and aims to reduce hazardous substances from the design process, manufacturing, use, and final disposal. Green chemistry is also used for the removal of waste or harmful metals from water sources or for the treatment of wastewater. In this context, the elevated level of Arsenic concentrations in natural waters now becomes a global problem. It has a lethal effect on human health. The threshold concentration of Arsenic has been re-established at 0.01mg/L by WHO. The chapter focuses on green chemistry, and it describes the preparation of economic bio-sorbent for the removal of Arsenic using wood pulp and MGA-NCC bio-nanosorbent synthesis process.

INTRODUCTION

Green chemistry is an important concept to the waste management industry because it focuses on reducing the amount of waste created instead of simply focusing on how to handle waste that has already been created. Due to this focus, green chemistry helps with source reduction, which is when products are designed, manufactured, packaged, and used in a way that limits the amount or toxicity of waste created (Nguyen et al 2013).

DOI: 10.4018/978-1-7998-8434-7.ch018

The first goal of source reduction is to reduce the overall amount of waste that is produced. The second goal is to conserve resources by not using raw, virgin materials. In other words, by following source reduction, fewer raw materials will have to be used to produce products. Green chemistry supports source reduction because it reduces the overall amount of hazardous substances used, which leads to a decrease in the hazardous waste produced and the need for raw materials to make hazardous substances (Berlioz et al., 2018).

As a concept, it emerged in the 1990s, and in order to further focus the efforts of chemists towards it, the 12 principles detailed here were published. They were created by Paul Anastas and John Warner, (Anastas and Warner, 1998) and are essentially a checklist of ways to reduce both the environmental impact and the potential negative health effects of chemicals and chemical synthesis. The 12 principles are-

Waste Prevention

This tenet simply states that chemical processes should be optimized to produce the minimum amount of waste possible. A metric, known as the environmental factor (or E factor for short), was developed to gauge the amount of waste a process created, and is calculated by simply dividing the mass of waste the production process produces by the mass of product obtained, with a lower E factor being better.

Atom Economy

Atom economy is a measure of the amount of atoms from the starting material that are present in the useful products at the end of a chemical process. Side products from reactions that aren't useful can lead to a lower atom economy, and more waste. In many ways, atom economy is a better measure of reaction efficiency than the yield of the reaction; the yield compares the amount of useful product obtained compared to the amount you'd theoretically expect from calculations. Therefore, processes that maximise atom economy are preferred.

Less Hazardous Chemical Synthesis

Ideally, we want chemicals we create for whatever purpose to not pose a health hazard to humans. We also want to make the synthesis of chemicals as safe as possible, so the aim is to avoid using hazardous chemicals as starting points if safer alternatives are available.

Designing Safer Chemicals

This principle links closely to the previous one. Chemists must aim to produce chemical products that fulfill their role, be that medical, industrial, or otherwise, but which also have minimal toxicity to humans. The design of safer chemical targets requires knowledge of how chemicals act in our bodies and in the environment. In some cases, a degree of toxicity to animals or humans may be unavoidable, but alternatives should be sought.

Safer Solvents and Auxiliaries

Many chemical reactions require the use of solvents or other agents in order to facilitate the reaction. They can also have a number of hazards associated with them, such as flammability and volatility. Solvents might be unavoidable in most processes, but they should be chosen to reduce the energy needed for the reaction, should have minimal toxicity, and should be recycled if possible.

Design for Energy Efficiency

Energy-intensive processes are frowned upon in green chemistry. Where it is possible, it is better to minimise the energy used to create a chemical product, by carrying out reactions at room temperature and pressure. Considerations of reaction design also have to be made; removal of solvents, or processes to remove impurities, can increase the energy required, and by association increase the process's environmental impacts.

Use of Renewable Feedstocks

The perspective of this principle is largely towards petrochemicals: chemical products derived from crude oil. These are used as starting materials in a range of chemical processes, but are non-renewable, and can be depleted. Processes can be made more sustainable by the use of renewable feedstocks, such as chemicals derived from biological sources.

Reduce Derivatives

Protecting groups are often used in chemical synthesis, as they can prevent alteration of certain parts of a molecule's structure during a chemical reaction, whilst allowing transformations to be carried out on other parts of the structure. However, these steps require extra reagents, and also increase the amount of waste a process produces. An alternative that has been explored in some processes is the use of enzymes. As enzymes are highly specific, they can be used to target particular parts of a molecule's structure without the need for the use of protecting groups or other derivatives.

Catalysis

The use of catalysts can enable reactions with higher atom economies. Catalysts themselves aren't used up by chemical processes, and as such can be recycled many times over, and don't contribute to waste. They can allow for the utilisation of reactions which would not proceed under normal conditions, but which also produce less waste.

Design for Degradation

Ideally, chemical products should be designed so that, once they have fulfilled their purpose, they break down into harmless products and don't have negative impacts on the environment. Persistent organic pollutants are products which don't break down and can accumulate and persist in the environment; they are typically halogenated compounds, with DDT being the most famous example. Where possible, these chemicals should be replaced in their uses with chemicals that are more easily broken down by water, UV light, or biodegradation.

Real Time Pollution Prevention

Monitoring a chemical reaction as it is occurring can help prevent release of hazardous and polluting substances due to accidents or unexpected reactions. With real time monitoring, warning signs can be spotted, and the reaction can be stopped or managed before such an event occurs.

Safer Chemistry for Accident Prevention

Working with chemicals always carries a degree of risk. However, if hazards are managed well, the risk can be minimised. This principle clearly links with a number of the other principles that discuss hazardous products or reagents. Where possible, exposure to hazards should be eliminated from processes, and should be designed to minimise the risks where elimination is not possible.

GREEN CHEMISTRY FROM THEORY TO PRACTICE

As part of the economic structure and the rapid extraction of natural resources, environmental problems have been considered in the past. The developed ways of using materials (feedstocks), the initial design of chemical processes, the toxic properties of products, energy consumption and other criteria involved in the manufacture of products have taken several years to consider (life cycle, recycling, etc).

The term 'use and generation of hazardous substances' is another part of the concept of Green Chemistry. Green Chemistry often covers all substances that are part of the process, rather than concentrating solely on those harmful substances that may be accidentally generated in a system. Green Chemistry also acknowledges that there are major ramifications for the use of toxic chemicals, ranging from regulatory concerns, storage and transport, waste processing and liability issues (Bhattacharya et al., 2018). The rapid growth of new chemical technology and the large number of new chemical products in recent decades have turned environmental activists' attention to remedial steps for adverse effects (monitoring environmental pollution, reduction of pollutants, recycling, etc). But the truth is that design and creativity in production systems, taking resources into account, is the most efficient way to minimize the negative impacts. Materials, atomic economy, the use and processing of toxic secondary materials and eventually the product life cycle and their realistic recycling into new materials.

Treatment of Wastewater by Absorbents

Water pollutants especially from effluents of industrial and domestic wastewater have become concern nowadays. The toxic content in water pollutants can affect the health of the people and aquatic environment due to non-biodegradable characteristic and accumulate in living tissue (Memon, J. R. et al 2008). At present, the industrial and agricultural activities increased from year to year thus generate large amount of wastewater contain toxic pollutant in water body. Besides, the availability of clean water to sustain in future becomes most challenging for researcher worldwide. Therefore, in order to minimize the pollution problem in the effluents, many treatments are introduced in physical, chemical and biological such as membrane filtration, adsorption, chemical precipitation, and microbial degradation. Among this adsorption process that involves rapid separation of adsorbate from liquid to a surface is most economical with less contamination and effective for dye removal treatment technique due to the energy efficient, low treatment cost, suitability, simple of design and easy to operate [28]. However, in adsorption process choosing an efficient adsorbent is important during adsorption process because to ensure easy separation from wastewater after treatment, Heavy metal in aquatic stream may release into wastewater from various industries such as batteries, mining, fertilizer industries, refining ores, electroplating etc. that possess toxicity and harm the environment. The heavy metal that hazardous to human life due to their toxicity and bio-accumulation tendency in nature are Cr, Fe, Se, V, Cu, Co, Ni, Cd, Hg, As, Pb, Zn etc (Chandra V et al 2017).

CURRENT POSITION OF AWARENESS

The complex survey of the related literature highlights the following points:

- In order to boost sorption performance and environmental steadiness of biosorbents, extraordinary prominence is mandatory, making them feasible for commercial use.
- Arsenic reduction has mainly been attempted using inorganic coagulants. Arsenic decontamination using organic biosorbents has not received much attention.
- Inorganic nano materials have been advanced in order to reduce arsenic species. Recently, toxicity concerns of these health compounds have been raised.
- In order to increase their affinity towards target organisms, organic nanoparticles can easily be functionalized with different chemical groups. The complexity of organic materials reflects the achievement of structural order with the full structure built from the nested levels of structural hierarchy in which models can be shaped by self-assembled organic materials over several length scales.
- Cellulose for the sorption of cationic species is the most abundant and low-cost renewable polymer resource and requires modifications for the sorption of anionic species.
- No attention was paid to the physico-chemical relationship between arsenic and cellulose, which is strongly needed to understand the mechanistic aspects of various arsenic sorption organisms.

ACHIEVEMENTS OF GREEN NANOTECHNOLOGY

Green nanotechnologies have been the key technical justification for solving numerous problems in water treatment. Many of them have increased the trust that green nano-techniques will help achieve the UN Millennium Development Target of halving the lack of access to hygienic water by the population by 2015. Supporters of this approach state that green nanotechnologies can defeat unanswered confrontations linked to the removal of water contaminants, while being very powerful, capable, long-lasting and rational at the same time. A key field for green nanotechnology research involving treatment, desalination and reuse has been the state of clean drinking water. Green nanotechnology's ambitious assertion is that improvements in productivity made by the use of nanotechnology would result in less and more sustainable consumption with environmental savings. Economical materials and end markets have been affected by improved efficiency (Dedicate K. et al 2017). Economic goods tend to contribute to consumption rising forever. Fortunately, this technical advancement is often necessary to achieve results that are ecologically positive and socially just.

Research findings have clearly demonstrated the green nanotechnology achievements mentioned below:

- Green nano-sorbents can remove polluted water particles.
- In the background of nano-catalysts and redox dynamic nanoparticles, toxic organic fluids can be converted into nontoxic materials.
- Bacteria can be neutralised by the application of nano-biocides without producing products.
- For water treatment, carbon nano tube filters and cartridges are used.
- For water treatment and decontamination, reverse osmosis membranes by nano-composites, carbon nano tube membranes and polymeric nano fibrous membranes have been used.

NANOCRYSTALLINE CELLULOSE: SUITABLE GREEN BIONANOSORBENT

Nanocrystalline cellulose (NCC) is a new type of material composed of high aspect ratio (length to width ratio) nano scale cellulose fibrils that are commonly used in various technological areas. NCC is prepared by chemical, enzymatic, or mechanical handling or permutation of cellulose from cellulose. Multiphase materials where at least one of the particle phases is less than 100 nm in size are denoted by bio nanosorbents. A massive skylight of potential has emerged at the back of nanosorbents to cross the limits of traditional micro-scale sorbents (Adriano DC. 2020). For water purification purposes, recent studies have used cellulosic nano fibres in membrane formation.

It is clear that these hydroxyl groups and their ability to form hydrogen bonding play a major role in guiding nano crystallite crystalline packing. To add stable negative or positive electrostatic charges, the surface of nano cellulose can be chemically functionalized. In terms of enhanced properties than MCC, NCC provides interesting potential. These fibrils have excellent mechanical properties and numerous free surface reactive-OH groups that offer great functionalization possibilities (G. Owens 2017). One advantage of using nano cellulose is that, assuming the nanoparticles remain well distributed; a small amount may have an incredibly wide surface area. NCC's unique properties, such as high rigidity and strength, high specific surface area, low thermal expansion coefficient, optical clarity and self-assembly efficiency, have attracted great interest in using these properties in different fields through the development of bio nano composites (Berlioz S et al 2018). Many fundamental or technical studies have been aimed at NCC functionalization in order to improve the properties of the material. Recently, the functionalization of NCC has received a lot of attention and is one of the key areas of research in functional materials production.

METHODOLOGY

In order to encourage biosorption by nanotechnology, the concepts of Green Nanotechnology can be extended in order to produce additional environmentally friendly biosorbents. Biosorbents are chemically engineered to be energy efficient, safe, green and bio-waste-friendly. Via design and engineering at the nanoscale, green nanotechnology looks for such targets, even linking donations from other legislation, counting physics, chemistry, biology, materials and environmental sciences utilizing friendly nanotechnology purposes that are endorsed as green for bio sorbent development.

Bio sorbent Preparation Protocol

MCC was isolated with minor modifications for the optimization of experimental conditions.

Grafting

Grafting of diethylene triamine was carried out with minor modifications for the optimization of experimental conditions.

Grafting of L-cysteine was carried out with minor modifications for the optimization of experimental conditions.

Silylation

Silylation of 3-Aminopropyltrimethoxy silane (APTMS) was performed on NCC with minor modifications to optimize experimental conditions.

The 3-Mercaptopropyltrimethoxy silane (MPTMS) silylation was performed on NCC with minor modifications for optimization of experimental conditions.

Cationization

The cationization of diethyl amine was performed on NCC with minor modifications to optimise experimental conditions.

The cationization of mercapto glycolic acid on NCC was performed with minor modifications to optimise experimental conditions.

SUMMARY

The toxicity of the water environments of the earth is so high that all life forms are detrimentally affected. By using advances in water purification technology, wealthy people can be relieved of this risk to a degree. The presence of several xenobiotics in the aqueous environment, however, severely affects weak humans and other species. Heavy metals/metalloids are the ubiquitous trace contaminants of the marine environment in the midst of complex pollutants present in waste water, since they are persistent, non-biodegradable. They grow in living tissues, causing humans and animals to have different diseases and disorders.

Many waste water treatment technologies are available with their own benefits and drawbacks. However, high running costs, the use of large quantities of chemicals, technological limitations, tedious nature. Recent research into heavy metal biosorption has built a strong knowledge base and highlighted the enormous potential of this successful green phenomenon. Price, sorption efficacy and environmental stability of the raw material are priorities for the commercialization of this process. Research interests have centered on enhancing the sorption efficacy and environmental stability of the biosorbent in order to achieve these requirements. By adding desired molecules that have functional groups responsible for metal binding, the efficacy of biosorbents is correlated with the newly synthesised surface. In improving the sorption efficacy and stability of biomaterials, surface chemistry has played an important role. Therefore, eco-friendly charging of the surface area of pristine biomaterials is strategically significant and is the current area of study.

FUTURE PERSPECTIVES

Green Nanotechnology is software that pays attention to consequences. It is focused on reducing the risk of using nanomaterials, minimising the risk of using nanomaterials, and reducing the risk of producing unnecessary chemical intermediates and end-products by using nanomaterials. Right now, there is an open window of opportunity, and we need to take advantage of it by using Green Nanotechnology thoughtfully and productively (Bodek I. et al 2018).

A satisfactory notification of the greener intent of nanotechnology is planned by the 12 principles of Green Chemistry. Green nanotechnologies are used to boost pressing environmental challenges such as over-exploitation, climate change, the reduction of natural assets from the perception of sustainable invention and investment, particularly in the fields of water decontamination, innovation in renewable energy, waste management/environmental remediation, and novel materials (Chen YN.et al 2017).

In addition, an extensive area of research could be opened up by synthesizing large structures such as nanorods, nanosphere, nanoflowers, nanoflakes and nanocones with improved functional and structural properties. By means of calculated synthesis approaches, several structures of nanomaterials with potential properties could be generated. Future research, however, should be explored through the development of new methods for wastewater treatment.

REFERENCES

Abraham, E., Deepa, B., Pothan, L. A., Jacob, M., Thomas, S., & Anandjiwala, R. (2020). Extraction of nanocellulose fibrils from lignocellulosic filaments: A novel methodology. *Starch Polymers*, *86*, 1468–1475.

Adriano, S. K., & Shah, B. A. (2020). Danger of arsenic pollution in groundwater influencing the Ganga alluvial plain. *Ecological Health Perspectives*, *112*, 19–20.

Anastas, P. T., & Warner, J. C. (1998). *Green Chemistry: Theory and Practice*. Oxford University Press.

Berlioz, S., Molina-Boisseau, S., Nishiyama, Y., & Heux, L. (2018). Gas-Phase Surface Esterification of Cellulose Microfibrils and Whiskers. *Biomacromolecules*, *10*(8), 2144–2151. doi:10.1021/bm900319k PMID:19572699

BGS (British Geological Survey)/DPHE. (2018). *Arsenic pollution of groundwater in Bangladesh 2: Final report*. BGS Technical Report WC/00/19.

Bhattacharya, P., Ahmad, K. M., Hasan, M. A., Broms, S., Fogelstrom, J., Jacks, G., Sracek, O., & Routh, J. (2018). Versatility of arsenic in groundwater in a piece of Brahmanbaria area, NE Bangladesh. *Acta Hydrochimica et Hydrobiologica*, *31*, 97–107.

Bodek, I., Lyman, W. J., Reehl, W. F., & Rosenblatt, D. H. (2018). *Natural Inorganic Chemistry: Properties, Processes and Estimation Methods*. Pergamon Press.

Bondeson, D. (2018). Mathew An and Oksman K. Advancement of the Isolation of Nanocrystals From Microcrystalline Cellulose by Acid Hydrolysis. *Cellulose (London, England)*, *13*, 171–180. doi:10.100710570-006-9061-4

Chakraborti, D., Rahman, M. M., Paul, K., Chowdhury, U. K., Sengupta, M. K., Lodh, D., Chanda, C. R., Saha, K. C., & Mukherjee, S. C. (2017). Arsenic catastrophe in the Indian subcontinent: What exercises have been realized? *Talanta*, *58*, 3–22. doi:10.1016/S0039-9140(02)00270-9 PMID:18968730

Chen, J., Lin, N., Huang, J., & Dufresne, A. (2017). Profoundly alkynyl-functionalization of cellulose nanocrystals and progressed nanocomposites there of by means of snap science. *Polymer Chemistry*, *6*, 4385–4395. doi:10.1039/C5PY00367A

Chen, Y. N., Chai, L. Y., & Shu, Y. D. (2017). Investigation of arsenic (V) adsorption on bone roast from watery arrangement. *Diary of Hazardous Materials*, *160*, 168–172. doi:10.1016/j.jhazmat.2008.02.120

Cheng, G., Varanasi, P., Li, C., Liu, H., Melnichenko, Y.B., Simmons, B.A., Kent, M.S., & Singh, S. (2017). *Progress of Cellulose Crystalline Structure and Surface Morphology of Biomass as a Function of Ionic Liquid Pretreatment and Its Relation to Enzymatic Hydrolysis*. doi:10.1021/bm101240z

Cui, H., Qi, L., Gao, S., & Shang, J. (2017). Strong adsorption of arsenic species by indistinct zirconium oxide nanoparticles. *Journal of Industrial and Engineering Chemistry*, *18*(4), 1418–1427. doi:10.1016/j.jiec.2012.01.045

Cullen, W. R., & Reimer, K. J. (2017). *Arsenic speciation in the climate. Substance Reviews, 89, 713-764.*

Darn, V. B. H., Doan, H. D., Dang-Vu, T., & Lohi, A. (2017). Harmony and energy of biosorption of cadmium(II) and copper(II) particles by wheat straw. *Bioresource Technology*, *100*, 211–219.

Memon, J. R., Memon, S. Q., Bhangera, M. I., Memon, G. Z., Turki, A. E., & Allen, G. C. (2008). Characterization of banana peel by scanning electron microscopy and FT-IR spectroscopy and its use for cadmium removal. *Colloids and Surfaces. B, Biointerfaces*, *66*(2), 260–265. doi:10.1016/j.colsurfb.2008.07.001 PMID:18760572

Nguyen, T. A. H., Ngo, H. H., Guo, W. S., Zhang, J., Liang, S., Yue, Q. Y., Li, Q., & Nguyen, T. V. (2013). Applicability of agricultural waste and byproducts for adsorptive removal of heavy metals from wastewater. *Bioresource Technology*, *148*, 574–585. doi:10.1016/j.biortech.2013.08.124

Owens, G., Bhattacharya, P., & Nadebaum, P. (Eds.). (2018). *Managing arsenic in the climate: From soil to human wellbeing*. CSIRO Publishing.

Chapter 19
Crop Residue Recycling for Improving Crop Productivity and Soil Health

S. S. Rathore
Indian Agricultural Research Institute, India

Kapila Shekhawat
Indian Agricultural Research Institute, India

ABSTRACT

Crop residues are nutrient-rich precious bio-resources having huge potential of enhancing crop productivity and maintaining soil health. The potential benefits of on-farm crop residue utilization have been well-established and well-realized, especially after the popularization of conservation agriculture. The soil health in terms of physico-chemical and biological properties significantly improves both under crop residues incorporation or retention. Yet, huge quantities of crop residues are burnt on-farm to clear the fields. Efficient management of crop residues as a component of conservation agriculture could be a non-intrusive approach for providing economically and environmentally sustainable solutions for resource-scarce and fragile agro-ecologies. This chapter is a compilation of the concerted efforts made by several researchers in understanding the role of crop residues for soil health improvement, standardizing the on-farm crop residue recycling under varied farm situations, and appropriate mechanization required for their successful use.

INTRODUCTION

India produces a huge amount of crop residues biomass (685 MT annually). The crop residues include arable crop residues such as straw, stover/stalk, stubbles and seeds/pods. The processed crop residues include husk, seeds (broken/ shriveled), bagasse, molasses and roots. The crop residues are mainly used as animal feeding and their bedding, thatching of homes and as a source of fuel for domestic and industrial purposes. After putting aside the residues for agricultural and non-agricultural uses, 34% of

DOI: 10.4018/978-1-7998-8434-7.ch019

it remains as surplus, which is either burnt or is being wasted. This surplus crop residue biomass has the potential to generate 4.15 EJ bio-energy, which can meet out 17% of India's total primary energy consumption (Hiloidhari et al., 2014). The stems, leaves, chaff and husks that remain in the fields after crops are harvested for grain, seed or fiber play a critical role in soil quality and environmental issues since they are primary inputs of elemental carbon (C) into soil systems (Fig 1). Crop residues have been referred to as 'wastes' but as a natural and valuable resource are also considered to be 'potential black gold.' The negative connotation of 'residues' may refer to the remains after a part is taken or something leftover or useless. On the contrary, crop residues offer a large, but finite potential mechanism for C sequestration and nutrient cycling (Reicosky and Wilts, 2005).

Figure 1. Schematic representing the role of crop biomass in the agricultural ecosystems carbon cycle (Source: Reicosky and Wilts, 2005)

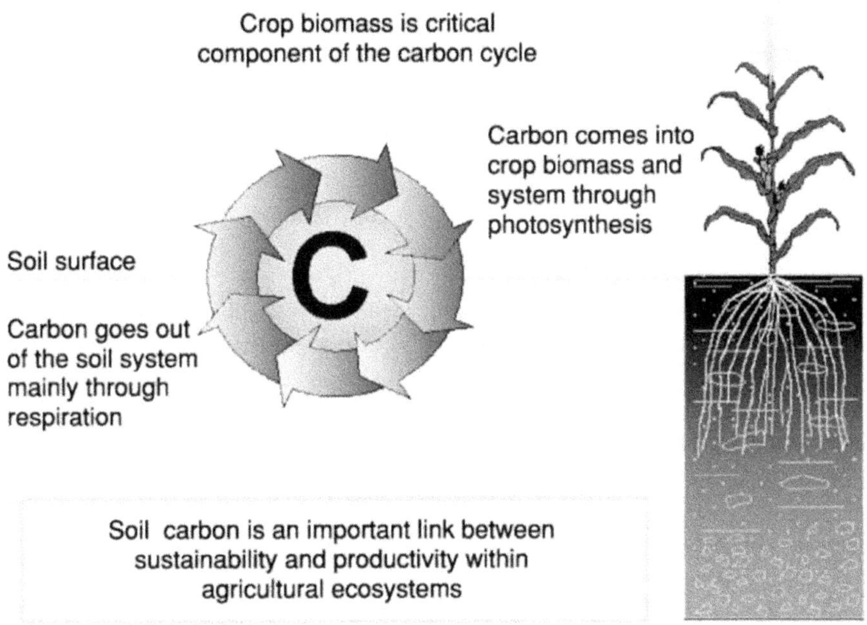

CROP RESIDUE GENERATION IN INDIA

The amount of crop residues produced in India is estimated to be approximately 620 million tons (Mt) and an increasing trend in crop residues generation has been observed with increasing food grain production in the country (Fig. 2). Half the quantity of agro-residues thus produced finds use as roofing material, animal feed, fuel and packing material, while the other half is disposed-off by burning in the field. About 50% of the total residues generated are produced by three crops viz. rice, wheat and oilseed. Rice residues account for 135 Mt annually. The rice-wheat (RW) system accounts for nearly 1/4th of the total crop residues produced in India. The surplus crop residues (i.e., total residues produced minus the quantity used for various purposes) are typically burned on-farm. The amount of surplus crop residues available in India is estimated to range from 84–141 Mt year^{-1} where the cereal crops contribute about

58%. Of the 82 Mt of surplus crop residues from the cereal crops, 44 Mt is from rice followed by 24.5 Mt from wheat, which are mostly burned on-farm.

Figure 2. Annual crop residue generated in India since 1950-51

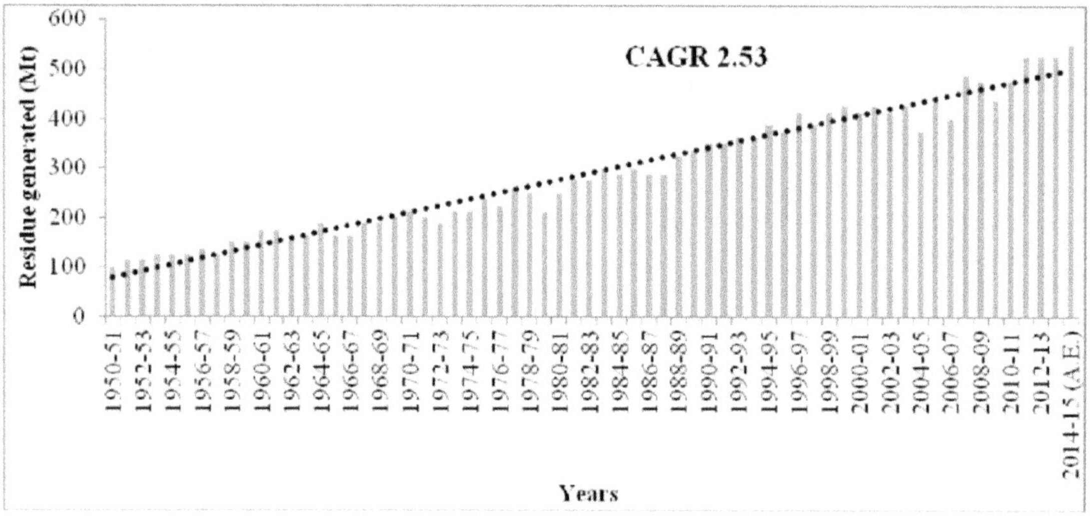

The cereal crops contribute the most in total crop residues generation, followed by oilseed and fibre crops. The share of various crop groups in residue generation and the surplus has been shown in Fig. 3.

Figure 3. Residue generation (A) and the share of unutilized residues in total residues generated (B) by different crops in India (MNRE, 2009)

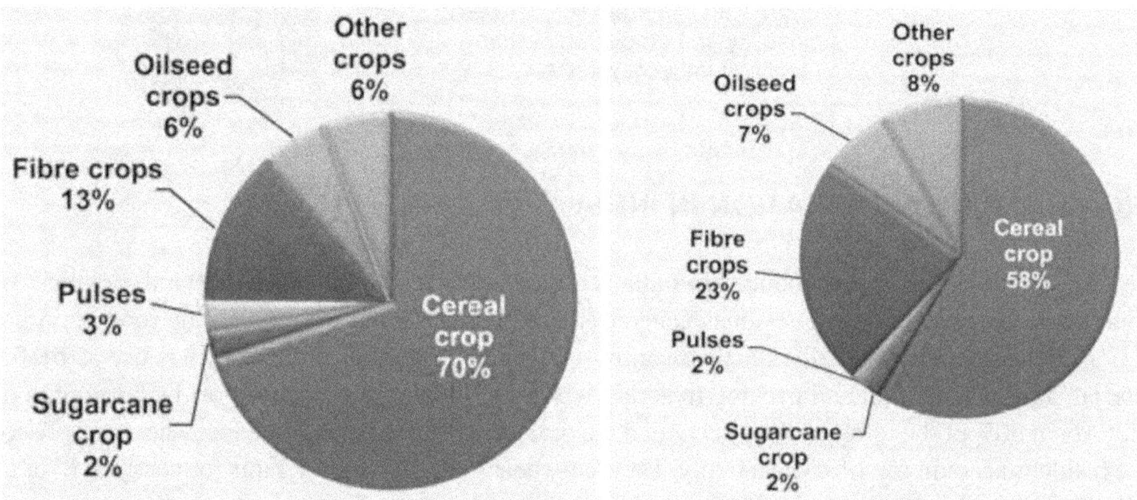

The major states of India which contribute towards the crop residues generation in different crops has been given in table 1, which indicates that the states producing the highest crop residues are also responsible for huge residue burning. A huge area under cultivation in these states has great potential of the residue use for maintaining higher soil fertility.

Table 1. Annual crop residues generation from 5 major states of India (Mt/yr)

State	Crop residues generation	Crop residues surplus	Crop residues Burnt*
Uttar Pradesh	59.9	13.5	13.3
Punjab	50.8	24.8	9.8
Maharashtra	46.5	14.7	7.8
United Andhra Pradesh	43.9	6.96	6.46
West Bengal	35.9	4.2	10.8
India (Total)	501.7	140.8	91.3

Source: MNRE, 2009
*based on IPCC coefficient

Crop Residue: A Precious Natural Resource

Crop residues are rich sources of nutrients which is vital for the sustainability of agricultural ecosystems. If some quantity of fertilizers is substituted through crop residues recycling in soil, a significant reduction in the energy inputs and green house gases emitted through the synthesis of fertilizers and their emissions from the field after fertilizer use can be curtailed. Besides accumulating in the seed or grain part, a majority of the nutrients, especially macro-nutrients remain in the straw. About 25% of nitrogen (N) and phosphorus (P), 50% of sulphur (S) and 75% of potassium (K) uptake by cereal crops is retained in crop residues, making them valuable nutrient sources (Singh, 2003).

The nitrogen, phosphorus and potassium content in the straw of major cereal crops is shown in table 2.

Table 2. Nutrient concentration (%) in straw/stalk in major cereal crops

Straw/stalk	Nitrogen	Phosphorus	Potassium
Paddy	0.36	0.8	0.71
Wheat	0.53	0.10	1.10
Sorghum	0.40	0.23	2.17
Maize	0.42	1.57	1.65
Pearl millet	0.65	0.75	2.50

Source: ICAR, 2006

Straw contains around 15-25% moisture content and has a net calorific value of around 13 MJ/kg at this moisture content. The straw of oilseed rape has higher levels of calcium, thus besides for soil nutrient recycling, it is also important for problem soil improvement purposes. Potassium content in plant residues could be 59 to 88% of the total K removal by the crops. The content of secondary and micronutrients in crop residues may also be 80 to 95% of the total removal by the crops. The rice straw contains 89% Mg and 93% Ca of the total removal by the crop. Similarly, 85% of Zn, 83% of Fe, 93% of Mn and 95% of Cu removed by the rice crop was in the straw (Lakshmanan et al., 2005). Recycling the crop residues can thus return back most of the nutrients that are taken up as cations (K, Ca, Mg, Fe, Mn, Zn, Cu) removed by the crop. In most field crops, residues (stem and leaves) generally exceed the grain or any other economic produce not only in biomass but also in the content of plant nutrients other than N, P and S.

But, most crop residues have a C: N ratio wider than 50 (Table 3), whereas, soil microorganisms, in general, require 30 parts of C and 1 part of N (C: N ratio of 30:1) for their growth. Thus when crop residues or other materials of a C:N ratio wider than 30 are incorporated in to soil, soil microorganisms utilize native soil N for their growth and this creates a temporary N hunger for the growing crop plants, which become pale and show stunted growth. The solution to the problem is that crop residues should be composted for reducing their C:N ratio where immediate effect in the current season is expected. Also, appropriate crop residues should be recycled in the soil keeping in view the availability of fertilizers and the crop requirement. In the tropics and sub-tropics, the cereal-based systems usually dominates where most crop residues are rich in lignin and have high C/N ratios that are generally greater than 60 (Cadisch and Giller 1997; Handayanto et al. 1997). Although crop residues are often on the soil surface, they are more likely to partially incorporate and decompose as the season progresses adding to SOC (Parker 1962). However, the wide C/N ratio leads to prolonged N immobilization by microorganisms, rendering N unavailable for crop growth in the short term (Giller et al. 1997). Thus, high N inputs are required when crop residues with high C:N ratio are used as mulch cover.

Table 3. Carbon (%), nitrogen (%) and C:N ratio of crop residues produced by major cereal crops

Residues	Carbon (% of dry matter)	Nitrogen (% of dry matter)	C:N ratio
Rice straw	40.8	0.46	89
Wheat straw	40.0	0.33	124
Oat straw	40.3	0.52	83
Barley straw	40.9	0.32	128
Ragi straw	39.5	0.64	62
Sorghum stalk and leaves	45.7	0.49	63
Maize stalk and leaves	44.2	0.84	53
Cotton stick	44.6	0.62	72
Sugarcane leaves	39.4	0.35	113

Source: Tiwari, 1995

The decomposition of crop residues also depends up on the lignin, hemi-cellulose and poly-phenol content. Lignin is a recalcitrant fraction which is highly resistance to microbial decomposition and the decomposition rates of crop residues are often inversely related to their lignin content. Poly-phenol concentration in plant tissue also reduces its rate of decomposition by binding to protein and forming complexes resistant to decomposition. The bio-chemical composition of various crop residues has been shown in table 4.

Table 4. Some biochemical parameters (%) of crop residues of major cereals

Residues	Cellulose	Hemi-cellulose	Lignin	Protein
Rice-straw	35.8	22.9	11.6	3.1
Wheat-straw	37.8	26.9	18.8	2.4
Barley straw	40.4	25.3	17.8	3.9
Oat-straw	36.4	26.6	18.0	2.7
Groundnut shells	16.8	42.2	15.7	14.8

Source: Gaur, 2001

Crop Residue Burning: Environmental Concerns

The on-farm burning of crop residues is done primarily to clear-off the fields. Cropping system intensification has aggravated the problem of on-farm residue burning due to very less turn-around period. Also, the harvesting through combines, shortage of human labour and high cost of removing the crop residues has left farmers with no option but to burn. Presently, 22 Mt (around 80% of total) rice straw produced annually in Punjab is subjected to burning to clear the fields for timely sowing of wheat (Yadvinder-Singh et al., 2010). During burning of crop residues around 80% of carbon is lost as carbon-di-oxide and a small fraction is evolved as carbon mono-oxide also (Ponnamperuma, 1984). The impact of residue burning on soil health and environment is given below:

- It release soot particles and smoke causing human/ animal health problems leading to various air-borne/lung diseases
- It results in loss of plant nutrients like N, P, K and S. Up to 80% loss of N and S, 25% of P and 21% of K occurs during burning of crop residues Yadvinder-Singh et al., 2005).
- Burning is a waste of valuable resources (Source of C, bio-active compounds, feed and energy for rural households and small industries)
- Heat generated from the burning of crop residues elevates soil temperature causing death of active beneficial microbial population
- The immediately increase in the exchangeable NH_4 nitrogen and bicarbonate-extractable P content, without any build up of nutrients
- It leads to emission of GHGs namely CO_2, CH_4 and NO causing global warming
- It also leads to increase in the levels of aerosols, acid deposition, depletion of the O_3 layer

Improvement in Soil Health using Crop Residues

On-farm residue recycling is a knowledge-intensive farming approach which can manage agro-ecosystems for improved and sustained productivity and increased profits while preserving and enhancing the resource base and the environment (Stigter et al., 2018). Crop residues as an important source of organic matter improve soil physical, chemical and biological properties (Kumar and Goh 2000). The improvement in soil health with crop residues is a complex process, where various chemical transformations takes place within soils as shown in fig.4. The incorporation of crop residues into the soil has a large effect on the soil parameters, while the effect may remain small on crop yield during the short term of 1-3 years but the effect becomes apparent with time (Gupta et al., 2007). Crop residues when applied to soil have a significant effects on soil organic matter, and physical, chemical and biological properties of soil (Kumar and Goh, 2000; Bijay- Singh et al., 2008; Chauhan et al., 2012) which are discussed as under:

Figure 4. Effects of crop residues and their management on physical, chemical and biological parameters of soil quality and overall soil fertility
(Source: Singh and Rengel, 2007)

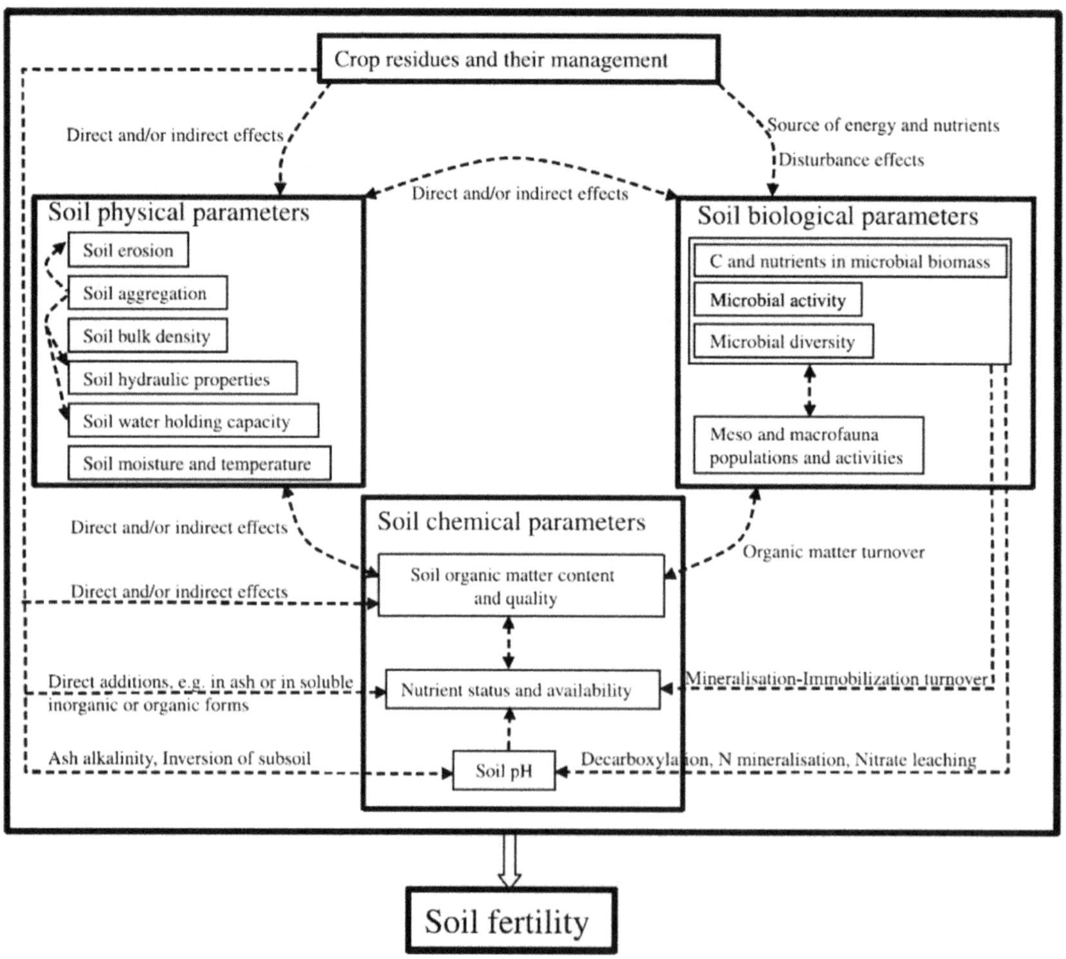

Soil Physical Properties

The crop residue when retained on the soil intercepts rainfall and prolongs its availability through slow release. They act as barriers, reducing the runoff velocity and giving the water more time to infiltrate. The higher water infiltration and reduced evaporation losses also reduce the nutrient losses through erosion. They buffers the soil against forces of raindrop impact and wind shear Surface residues acts as mulch which moderate soil temperatures primarily by insulating the soil surface from the sun's radiant energy, reduce evaporation during summers and lower soil temperature during winters. In Indian context, rice residue retention in wheat accompanied with zero-tillage in rice-wheat system has been reported beneficial in reducing terminal heat effect in wheat which is an emerging concern globally in view of climate change (Jat et al., 2009). Since organic matter (OM) is known to maintain soil aggregate stability (Shekhawat et al., 2016), the addition of crop residues often improves soil structure and aggregation. Residue retention with zero tillage caused more stable macro aggregates and increased the protection of C and N in the micro aggregates within the macro aggregates compared to conventionally tilled raised beds (Lichter et al., 2008). Crop residues and tillage management can also affect the leaching of the nutrients, which may pollute the groundwater or surface waters.

Soil Chemical Properties

Soil organic matter is one of the primary contributors/indicator to soil health and soil quality. Crop residues are precursors to soil organic matter (SOM). Surface retention of sufficient crop residues under CA increases SOM, biological activity and carbon sequestration which enhances long term sustainability. Incorporated crop residues decompose 1.5 times faster than surface placed residues and greater losses of N through leaching or denitrification. The crop residues are rich source of potash and crop residues retention with zero-tillage have been reported beneficial for enriching the soil with available potash (K). Crop residues contain large quantities of K, and their recycling can markedly increase K availability in soils (Chatterjee and Mondal, 1996). Recycling of crop residues can improve crop yields at low rates of K application and can decrease crop response to applied K. Buresh et al., (2010) reported that for rice-maize with 5 t/ha rice and 12 t/ha maize yield, the retention of maize residues can markedly reduce the net K export but does not eliminate the deficit in K balance when rice residues are not retained. Retention of all maize and rice residues is required to achieve near-neutral K balances.

Cooler soil temperatures will slow nutrient release from soil organic matter, reduce diffusion of nutrients to the plant roots, and can affect root growth. Initially, when no-till is first adopted the increased carbon (C) from the crop residues causes immobilization of soil N as decomposing microorganisms use soil N to maintain their C:N ratios during the decomposition process. With time the turn-over, or breakdown, of soil organic matter reaches a new equilibrium and the pool of potentially mineralizable N increases resulting in more plant-available nitrate (NO_3)-N and ammonium (NH_4)-N. Soil P and K tend to be immobile in the soil and without tillage and soil mixing, immobile nutrients may accumulate at the soil's surface (0-5 cm). An understanding of how nutrients move and react in the soil is necessary for proper fertilizer management in reduced tillage systems. In summary, research results do emphasize that return of crop residues do increase soil organic matter, and additionally, the increase is greater as more fertilizer N is used. This represents a build up of a potentially larger labile pool of organic N in no-till systems. How much of that labile pool will be utilized by the subsequent crops will largely be influenced by the amount of such organic N that is mineralized during the growing season. This labile pool of organic matter is redistributed to the top 5 cm of soil in continuous no-till production systems while it is somewhat uniformly distributed throughout the profile in conventional tillage systems. Such distribution of organic matter in contrasting tillage systems influences the dynamics and efficiency of N as the rate of microbial activity increases at the soil-residue interface. There is potentially a greater likelihood of more immobilization, denitrification or leaching of applied nitrogen in no-till systems because of the increased microbial activity at the residue-soil interface in no-till systems. For this reason, poorer yield response to suboptimal N application in no-till corn (Bandel et al., 1975) is possibly not due as much to losses of soil N from the rooting zone as it is to a shift in N content of soil N components resulting in more total N being immobilized with no-till.

Soil Biological Properties

Crop residue as an organic nutrient source improves biological activity by providing habitat and an energy source for soil organisms including earthworms and microbes. If adequate levels of crop residues are retained on the soil surface and combined with reduced or ZT, the C sequestered will correspond in reduction in CO_2 emissions. Conservation of soil resources requires proper management of crop residues. The primary limiting factor for microbial growth in most soils is the C energy source. An abundance of C and other nutrients are returned to the soil through decomposition of crop residues and biological nutrient cycling.

Environmental Benefits

CA involving zero-till and surface managed crop residue systems are excellent opportunity to eliminate burning of crop residues which contribute to large amount of green house gases like CO_2, CO, NO_2, SO_2 and large amount of particulate matter. Crop residue utilization in soils also reduces non-point pollution. Burning of crop residues, also contributes to considerable loss of plant nutrients, which could be recycled when properly managed. Large scale burning of crop residues is also a serious health hazard. Crop residue retention also mitigates flooding by holding water and avoiding run off into streams and rivers by reducing surface runoff and decreases sedimentation. Further, an improvement has been also recorded in the ground water quality by denaturing and filtering of pollutants.

Crop Residue Management under Rainfed Ecologies

The positive effects of crop residue management under rainfed conditions are even more beneficial as rainfall distribution is often erratic and seasonal dry spells are common. Mulch retention under low rainfall (1000 mm per season) increases infiltration of rainfall and reduces evaporative losses and sustains crop production by substantially increasing soil organic carbon (Govaerts et al. 2009). Carbon increases due to crop residue retention when SOC is maintained and the dissipation through the oxidation process is less, however it may not contribute to the absolute increase. Crop residues provide soil cover which decreases run-off and soil loss especially on low slopes (Adekalu et al. 2007). If organic matter is low and sand content is high, as under arid and semi-arid ecologies, up to 90% cover may be necessary to improve infiltration and reduce run-off and soil erosion. Crop residues have low thermal conductivity such that mulching can reduce soil temperature for optimal germination and root development in hot environments (Lal 1978; Riddle et al. 1996). They insulate the soil surface and increase resistance to heat and vapour transfer leading to increased available soil water (Hatfield and Prueger 1996; Dexter 1997; Cook et al. 2006).

Residue Retention *vis-à-vis* Residue Incorporation

Although, the amount and rate of mineralizable N under zero tillage is more than the conventional systems due to microbial activity in no-till soils, the residue retention as mulch has less surface area exposed to microbial action since they remain undisturbed over residues incorporation and mixing in the plow layer. Thus, as basal the N recommendations should be higher in no-till systems than conventional tillage systems, till a new steady state equilibrium between immobilization and mineralization is reached at a later phase and supply of N from the labile organic pool increases. Studies have also shown that soils high in soil organic matter retain more moisture, especially when residues are retained on the soil surface as compared to when they are incorporated.

The time of incorporation is also very important. Incorporation of wheat straw 10 days before rice transplanting and rice straw 3 weeks before wheat sowing gave 0.25 and 0.42 t ha^{-1} higher yields of rice and wheat, respectively, over incorporation of wheat straw at 3 days before rice transplanting and rice straw at 2 weeks before wheat sowing Kavinandan et al. (1987). The incorporation of rice straw into the soil produced significantly higher yields of wheat (3.51 t ha^{-1}) compared to when rice straw was removed (2.91 t ha^{-1}) (Salim, 1995). Singh et al. (1996) reported that in Pantnagar (India), incorporation of rice straw 3 weeks before wheat sowing significantly increased wheat yields on a clay loam but not on a sandy loam soil. The lower yields with straw incorporation, might be due to immobilization of N and slow decomposition of rice straw at low temperatures during wheat growth. However, with the advancement of time with decomposition in rice straw higher wheat yield and N uptake was recorded due to favourable soil moisture regime, regulation of soil temperature, control of weeds (Shekhawat et al., 2017) and an increase in the microbiological activity.

Farm Machinery for Efficient Residue Management

Among the crop residues of various crops, the rice straw is of poor quality for forage, bioconversion, and engineering applications. The limited alternative uses compel farmers to burn the rice straw. The in-field retention of crop residues can play an important role in replenishing soil quality and reducing environmental pollution from stubble burning, but the availability of appropriate machinery for managing *in-situ* crop residue is a major concern among farmers. Loose straw as well as anchored stubbles are left on the surface of the field after harvesting of crops through combines. The characteristics of machines required for crop residue management includes; low fuel consumption, economically beneficial, enable the timely/early sowing of succeeding crop at the given moisture and reduce the weed seed bank and total water requirement of the crop. The seeding through advanced zero-till machines are capable of cutting through loose straw, penetration into soil and placing seed at proper depth. Generally, four types of furrow openers i.e. single disc opener, double disc opener, triple disc opener i.e. double disc opener equipped with either powered or unpowered rotary disc coulter and star wheel punch planter are being introduced, which are successful in rice-wheat cropping systems.

Single Disc Opener

The single disc type furrow opener cuts a furrow slice in the soil and pushes it to the side, thereby causing disturbance to the top layer soil. The boot for seed tube is placed at backside of disc. Generally, a single disc of 34 cm diameter sharpened at an angle of 9-100 and with concavity of about 2-2.5 cm is used. The openers are mounted at disc angle of 60 (with horizontal) and tilt angle of 3° (with vertical) to move the soil laterally. These furrow openers are recommended for tilled but trashy field conditions.

Double Disc Opener

These openers are provided with two flat and sharpened discs opposed to each other and set at a small angle to the direction of travel as well as to vertical with included angle of about 100. Discs are positioned in such a way that they form a V-groove in the soil by pushing the soil downward and sideways. The penetration of discs is obtained by applying downward force. The seed boot is located between the two discs. The openers are used in various soil conditions, especially tilled and trashy fields.

Happy and Turbo Seeders

The Happy and Turbo Seeder technology provides an alternative to burning for managing rice residues and allows direct drilling of wheat in standing as well as loose residues. The Happy Seeders were especially designed for retention and incorporation of rice residue in the field. But the use of the variant depends on residue condition, its amount and the time left for wheat sowing. A 'Turbo Happy Seeder' can be used successfully *in situ* by retaining on soil surface for sowing of the wheat (Sidhu et al., 2015).

Bed Planter-Cum-Zero Till Drill

Raised bed planter-cum-ZT drill is a two-in-one version, where the loose straw after combining could be collected with field balers. The drill can be used directly without any surface manipulation of residue and a system for combining, field baling and zero-tillage could be a viable option. To manage the straw from combine harvested wheat fields, the straw combine is also used extensively. The straw combine harvests the uncut straw as well as pick up the combine ejected loose straw from the field, chops the straw into fine pieces and blows it into an enclosed trolley trailed behind the tractor. The field could be drilled directly with rigid tines mounted inverted-T openers.

Crop Residue Retention in Major Cropping Systems

Rice-Wheat System

Rice-wheat (RW) system of the Indo-Gangetic Plains has played a significant role in the food security of India. However, sustainability of this important cropping system is at risk due to deterioration of soil health, mounting pressure on natural resources and emerging challenges of climate change. Zero tillage and crop residue management could assist in achieving sustainable productivity and allow farmers to reduce nutrient and water inputs, and reduce risk due to climate change.

About 40% of N, 30–35% of P, 80–85% of K, and 40–50% of S absorbed by rice crop remain in the vegetative parts at maturity. Similarly, about 25-30% of N and P, 35-40% of S, and 70-75% of K uptake are retained in wheat residues. Typical amounts of nutrients in rice straw at harvest are 5–8 kg N, 0.7–1.2 kg P, 12–17 kg K, 0.5–1 kg S, 3–4 kg Ca, 1–3 kg Mg, and 40–70 kg Si per ton of straw on a dry weight basis (Dobermann and Witt, 2000). The mean N, P and K amounts in rice straw was 6.2 kg N, 1.1 kg P and 18.9 K kg per ton of straw van Duivenbooden (1992). Besides NPK, one ton of rice and wheat residues contain about 9-11 kg S, 100 g Zn, 777 g Fe and 745g Mn. Rice and subsequent wheat remove large amounts of K annually i.e. from 210 to 360 kg K_2O $ha^{-1}yr^{-1}$. The K in rice straw left in the field is a readily available nutrient source that is released quickly for plant use.

Rice–Rice Cropping System

In contrast to rice–wheat cropping systems, the majority of the studies on rice–rice cropping systems show that incorporation of rice residues in more beneficial for enhancing yield and N use efficiency. Ismunadji (1978) reported that incorporation of 10 t rice straw ha^{-1} increased grain yield by 18.2% and over untreated control in Indonesia. In another experiment where residue incorporation for more than 6 years, the rice yield increased by 0.4 to 0.7 t ha^{-1} compared with fields that used to receive chemical fertilizers and where rice straw was either removed or burned in Philippines (Ponnamperuma, 1984). The beneficial effects of rice straw incorporation were, however, small during the initial 2–6 years of the study, depending up on the soil and climate the beneficial effect may come even after 15 years of the study also, where an increase of 0.70 t ha^{-1} in rice grain yield was observed with rice straw incorporation over removal (Beaton et al., 1992). When crop residues were allowed to decompose for 2 or more weeks before rice transplanting, the adverse effects of phytotoxicity and N immobilization on crop growth was reduced (Ali et al., 1995; Wu et al., 1997). Thus, the length of the period allowed for decomposition of crop residues before the sowing/planting of the next crop also affects the agronomic response to applied residues.

Rice–Legume Cropping System

Green gram (*Vignaradiata*), black gram (*Vigna mungo*), cowpea (*V. unguicuata*), soybean (*Glycine max*), and groundnut (*Arachishypogae*) are the important summer food legumes that fit into the rice-based multiple cropping systems; they leave substantial amount of residues. The quantity of N in above-ground residues remaining after grain harvest ranges from 17 kg to 101 kg N ha^{-1}, and N fertilizer equivalent ranges from 37 to 100 kg N ha^{-1}, with a mean value of 40–45 kg N ha^{-1} (Buresh and De Datta, 1991). In addition to above-ground residues, roots of food legumes contain up to 40 kg N ha^{-1} at final harvest, which is progressively released during crop growth. Residues of grain legumes after harvesting pods have a lower N content than those of green manure but rapidly release N in tropical flooded soils. The potential benefits of legume residues and legume green manures in rice other than fixed N are described by Buresh and De Datta (1991). Studies with 15N-labeled legume residues indicate that N recovered by subsequent rice ranges from 25 to 45% of the N originally contained within the legume residues (Yadvinder-Singh et al., 2004).

In a rice–wheat-mung bean system, the incorporated straw of mungbean grown after wheat just before rice planting and observed an increase in rice yield equal to the application of 60 kg fertilizer N ha^{-1} (Rekhi and Meelu, 1983; Bhandari et al., 1992). Prasad and Palaniappan (1987) reported than incorporation of mungbean and soybean residues along with recommended fertilizers produced the highest rice yield. Incorporation of lentil residues, resulted in recycling of about 50–60% of the total N, P, and K removed by the lentil crop (Prasad et al., 1990). Sangakkau (1987) reported that rice yields after ploughing residues of mungbean were 3.2 and 3.4 t ha^{-1}, respectively, as compared to the 3.0 t ha^{-1} after fallow.

In the rice-soybean cropping system, Adisarwanto et al. (1996) reported that application of rice straw increased soybean yield by 103% at two sites in Indonesia and lentil yields by 34% over control. In field trials on groundnut-rice, groundnut-maize-rice, and maize-soybean-rice in Peru, Loli and Chuguizuta (1993) reported that incorporation of legume residues in rice significantly increased rice yields, while incorporation of maize residue inhibited germination and reduced rice yields.

Other Rice-Based Cropping Systems

Sugar beet produces large quantities of crop residues. In most cases, residues from sugar beet resulted in higher yields of the following crop (Watanabe, 1989). In a rice-sugar beet cropping sequence in Punjab (India), incorporation of sugar beet tops containing 90 kg N ha^{-1} increased rice grain yield by 52% over untreated control. Rice yield increased significantly up to 120 kg N ha^{-1} without and up to 80 kg N ha^{-1} with sugar beet tops (Kapur and Kanwar, 1994). The estimated urea N equivalent of sugar beet tops was 37 kg N ha^{-1} in rice and 19 kg N ha^{-1} in the following crop of sugar beet.

In rice-potato-sesame and rice-potato-mung bean cropping sequences, incorporation of crop residues along with 75% of the recommended N, P, and K fertilizers consistently increased the productivity of constituent crops in the two sequences (Jayaram et al., 1990). In a 2-year field experiment on a red yellow podzolic soil on a rice-maize-cowpea sequence in Indonesia, crop residues supplied a significant amount of N to the following rice crop (Sisworo et al., 1990). In a rice-potato-groundnut rotation, Sanyal et al. (1993) recorded similar yields of rice with 100% recommended NPK fertilizer applied alone and incorporation of crop residue along with 75% of recommended NPK fertilizers. Crop residue incorporation increased nutrient uptake by 14%. Suyamto (1993) observed that in a rice-maize cropping system, the application of 5 t rice straw ha^{-1} significantly increased the rice yield by 26% over control. The beneficial effect of rice straw on crop yield was equivalent to 73 kg K ha^{-1}. In a rice-cotton-rice cropping system in Tamil Nadu (South India), incorporating 22.6 t cotton sticks ha^{-1} (adding 270 kg N hal) on a clay loam soil significantly increased the rice yield by 22.2% over no stick incorporation (Budhar and Palaniappan, 1999). These data suggest that cotton residues can be incorporated in the following rice for higher yields and fertilizer N dose can be reduced.

Constraints in Crop Residue Utilization

- The major constraint in crop residue use for soil health improvement is its competitive uses from alternative economic options like livestock feed and bedding and as a source of fuel.
- The availability and beneficial impact of crop residues use largely depends on the local biophysical and socioeconomic environment.
- Retention of crop residues may depress yields in the short term due to immobilization of N which contrasts sharply with farmers' needs. Thus, the apprehension of initial lower crop yields and/or economic returns discourage farmers to use crop residues.

- Optimum fertiliser application becomes necessary even after crop residue use and hence a net balance reflecting the combination of fertilizers and crop residue needs to be worked out at all situations.
- A very high cost is involved in removing or transportation of these bulk residues to the place of application by conventional methods.
- The loose residue left after crop harvesting through combines and also anchored residues, offer difficulty in the field operations related to land preparation, sowing, pesticide and fertilizer application. These routine farm operations may require specialized equipments
- Weed control under a residue covered field is very difficult and it also increases the dependence on chemicals for weed management.
- The collection, transportation and management of crop residues is a costly matter, especially under the shortage of human labour.
- The unavailability of alternative economically viable solutions and long period required for composting force the farmers to burn the residues for clearing-off the field.
- The less turn-around period between the harvest of rice and sowing of succeeding wheat under irrigated ecologies leaves very limited options for residue management.
- Burning of crop residues on field is also a layman's way to control weeds and insect-pest-diseases.

FUTURE LINE OF RESEARCH

A comprehensive appraisal of the benefits and constraints related to crop residue management would encourage the farmers for adoption of residue recycling and their ecological management. The following research efforts if carried out could make the crop residue use successful in a sustainable manner:

- Understanding the interactions between the type and rate of crop residue application, the contribution to nutrient enhancement in soil and the potential for crop yield improvement are needed.
- Future research needs should be devoted to identifying appropriate nutrient management strategies in CA systems together with crop residue retention and crop rotations to boost crop productivity.
- Technologies for efficient use of crop residues thorough basic, strategic and applied research needs to be developed, especially under conservation agriculture perspective.
- The appropriate management strategies for residue utilization needs to be formulated through precise quantification of the generation of crop residues including total production from different crops, their quality, utilization and amount burnt on-farm, for evolving management strategies.
- A blanket approach of residue management is not going to be beneficial; rather development of region-specific crop residue inventories should be made under variable cropping schemes.
- With substantial alterations and improvement in machines could promote a large scale adoption by resource poor and low-skilled farmers also. Eg. Happy Seeder
- A clear guideline of recommendation for optimum irrigation, fertilizer management, pest management and long-term effects on soil health would facilitate easy adoption crop residue management techniques.
- The attempts should be made to quantify the economic, social and environmental benefits under different situations for practical understanding and easy adoption by farmers.

- Form a basis for policy level issues in relation to C sequestration, erosion control, fertilizer-use efficiency and incentives to retain crop residues. Developing crop varieties to produce more root biomass to improve the natural soil resource base.
- Standardize the use of microbe-mediated faster decomposition of residues for *in-situ* incorporation.

CONCLUSION

With cropping system intensification we have overstretched the resource use and signs of productivity and profitability fatigue have been quite visible. Crop residues are the principal element of permanent soil cover can be a green technological panacea for the low fertility and problem soils. When retained on the soil surface, they offer various benefits in terms of crop productivity and soil health. The surface residue protects the erosion and weathering of topsoil which is enriched with organic matter. It also promotes soil infiltration, maintains soil temperature and maintains soil bio-diversity. Designing long-term experiments to study the impact of CA on soil health, water and nutrient use efficiency, C sequestration, GHGs emission and ecosystem services through simulation models for prediction of impact of crop residue use on crop growth, soil properties, crop yield and ultimately farm income.

REFERENCES

Adekalu, K. O., Olorunfeli, I. A., & Osunbitan, J. A. (2007). Grass mulching effect on infiltration, surface runoff and soil loss of three agricultural soils in Nigeria. *Bioresource Technology*, *98*, 912–917. PMID:16678407

Adisarwanto, T., Utomo, W. H., Kirchhof, G., & So, H. B. (1996). Composition of food legume crops to different soil management practices after rainfed lowland rice East Jawa. *Proc. Management of Clay Soils for Rainfed Lowland Rice-Based Cropping Systems*, 142–147.

Ali, A., Medhi, D. N., Dekamedhi, B., & Baroova, R. (1995). Effect of rice straw in combination with diVerent levels of nitrogen, phosphorus and potash on transplanted rice. *Journal of Agricultural Soil Science Society of North East India*, *8*, 248–250.

Bandel, V. A., Dziernia, G., Stanford, G., & Legg, J. O. (1975). N behavious under no-tillage and conventional corn culture. First year results using unlabeled N fertilizer. *Agronomy Journal*, *67*, 782–786.

Beaton, J. D., Hasegawa, M., Jiung-Chang, X., Keng, J. C. W., & Halstead, E. H. (1992, September). Influence of intensive long-term fertilization on properties of paddy soils and sustainable yields. In *Proceedings of International Symposium on Paddy Soils* (pp. 246-251). Academic Press.

Bhandari, A. L., Sood, A., Sharma, K. N., & Rana, D. S. (1992). Integrated nutrient management in a rice-wheat system. *Journal of the Indian Society of Soil Science*, *40*(4), 742–747.

Budhar, M. N., & Palaniappan, S. (1994). Effect of incorporation of preceding cotton (Gossypium species) on succeeding rice (Oryza sativa). *Indian Journal of Agronomy*, *39*(3), 461–462.

Buresh, R. J., & De Datta, S. K. (1991). Nitrogen dynamics and management in rice-legume cropping systems. *Advances in Agronomy*, *45*, 1–59.

Buresh, R. J., Pampolino, M. F., & Witt, C. (2010). Field-specific potassium and phosphorus balances and fertilizer requirements for irrigated rice-based cropping systems. *Plant and Soil*, *335*(1), 35–64.

Cadisch, G., & Giller, K. E. (1997). *Driven by nature: plant residue quality and decomposition*. CAB International.

Chatterjee, B. N., & Mondal, S. S. (1996). Potassium nutrition under intensive cropping. *Journal of Potassium Research*, *12*(4), 358–364.

Chauhan, B. S., Singh, R. G., & Mahajan, G. (2012). Ecology and management of weeds under conservation agriculture: A review. *Crop Protection (Guildford, Surrey)*, *38*, 57–65.

Cook, H. F., Valdes, G. S., & Lee, H. C. (2006). Mulch effects on rainfall interception, soil physical characteristics and temperature under Zea mays L. *Soil & Tillage Research*, *91*(1-2), 227–235.

De, R., Rao, Y. Y., & Ali, W. (1983). Grain and fodder legumes as preceding crops affecting the yield and N economy of rice. *The Journal of Agricultural Science*, *101*(2), 463–466.

Dexter, A. R. (1997). Physical properties of tilled soils. *Soil & Tillage Research*, *43*(1-2), 41–63.

Dobermann, A., & Witt, C. (2000). The potential impact of crop intensification on carbon and nitrogen cycling in intensive rice systems. *Carbon and nitrogen dynamics in flooded soils*, 1-25.

Gaur, A. C. (2001). Organic manure: A basic input in organic farming. *Indian Farming*, *51*(3), 3–11.

Giller, K. E., Cadisch, G., Ehaliotis, C., Adams, E., Sakala, W. D., & Mafongoya, P. L. (1997). Building soil nitrogen capital in Africa. *Replenishing Soil Fertility in Africa, 51*, 151-192.

Govaerts, B., Verhulst, N., Castellanos-Navarrete, A., Sayre, K. D., Dixon, J., & Dendooven, L. (2009). Conservation agriculture and soil carbon sequestration: Between myth and farmer reality. *Critical Reviews in Plant Sciences*, *28*(3), 97–122.

Gupta Raj, K., & Seth, A. (2007). A review of resource conservation technologies for sustainable management of the rice–wheat cropping systems of the Indo-Gangetic Plains. *Crop Protection (Guildford, Surrey)*, *26*, 436–447.

Handayanto, E., Giller, K. E., & Cadisch, G. (1997). Regulating N release from legume tree prunings by mixing residues of different quality. *Soil Biology & Biochemistry*, *29*(9-10), 1417–1426.

Hatfield, J. L., & Prueger, J. H. (1996). Microclimate effects of crop residues on biological processes. *Theoretical and Applied Climatology*, *54*(1), 47–59.

Hiloidhari, M., Das, D., & Baruah, D. C. (2014). Bioenergy potential from crop residue biomass in India. *Renewable & Sustainable Energy Reviews*, *32*, 504–512.

Hossain, M. D. (2009) Nutrient and Residue Management for Improving Productivity and N Use Efficiency of RiceWheat-Mungbean Systems in Bangladesh. *Proceedings of the International Plant Nutrition Colloquium XVI*. https://escholarship.org/uc/item/2r29j9t0

Houng, K. H., & Hwa, C. (1975). The effect of crop resdiues on the growth of following crops II. EVect of decomposing rice straw and bagasse on the nutrient uptake of rice seedlings. *Journal of Chinese Agricultural Chemistry Society*, *13*, 412–415.

ICAR. (2006). Handbook of Agriculture. Indian Council of Agricultural Research.

Ismunadji, M. (1978). Utilization of cereal crop residues and its agricultural significance in Indonesia. Central Research Institute for Agriculture.

Jat, M. L., Singh, R. G., & Saharawat, Y. S. (2009). Innovations through conservation agriculture: progress and prospects of participatory approach in the Indo-Gangetic plains. *4th World Congress on Conservation Agriculture*, 60-64.

Jayaram, D., Chatterjee, B. N., & Mondal, S. S. (1990). Effect of FYM, crop residues and fertilizers management in sustaining productivity under intensive cropping. *Journal of Potash Research*, *6*, 172–179.

John, P. S., Prasad, R., Pandey, R. K., Buresh, R. J., & Prasad, R. (1989). Lowland rice response to urea following three cowpea cropping systems. *Agronomy Journal*, *81*, 853–857.

Kapur, M. L., & Kanwar, R. S. (1994). Sugarbeet top as a source of N for maize and rice grown in rotation with sugarbeet in subtropical north-west India. *Tropical Agriculturist*, *71*, 12–16.

Kavinandan, S. K., Gupta, J. P., & Mahapatra, I. C. (1987). Studies on residue management and biofertilizers in rice-wheat sequence. *Indian Journal of Agronomy*, *32*, 278–279.

Kulkarni, K. R., & Pandey, R. K. (1988). Annual legumes for food and as green manure in a rice-based cropping system. In *Sustainable Agriculture-Green Manure in Rice Farming* (pp. 280–289). Int. Rice Res. Inst.

Kumar, K., & Goh, K. M. (2000). Crop residues and management practices: Effects on soil quality, soil nitrogen dynamics, crop yields, and nitrogen recovery. *Advances in Agronomy*, *68*, 197–319.

Lakshmanan, R., Prasad, R., & Jain, M. C. (2005). Yield and uptake of micronutrients by rice as influenced by variety and nitrogen fertilization. *Archives of Agronomy and Soil Science*, *51*, 1–14.

Lal, R. (2005). World crop residues production and implications of its use as a biofuel. *Environment International*, *31*, 575–584. PMID:15788197

Lehmann, J., Gaunt, J., & Rondo, M. (2006). Bio-char sequestration in terrestrial ecosystems—A review. *Mitigation and Adaptation Strategies for Global Change*, *11*, 403–427.

Lichter, K., Govaerts, B., & Six, J. (2008). Aggregation and C and N contents of soil organic matter fractions in the permanent raised-bed planting system in the Highlands of Central Mexico. *Plant and Soil*, *305*, 237–252.

Loli, O. O., & Chuguizuta, F. (1993). Effect of crop rotations and incorporation of residues on yield of dry season rice (*Oryza sativa* L.). *Proceedings of the 11th Latin-American Congress and the 2nd Cuban Congress of Soil Science*, 1418–1424.

MNRE (Ministry of New and Renewable Energy Resources). (2009). *Govt. of India*. www.mnre.gov.in/biomassrsources

Parker, D. T. (1962). Decomposition in the field of buried and surface-applied cornstalk residue. *Soil Science Society of America Journal, 26*, 559–562.

Ponnamperuma, F. N. (1984). Straw as a Source of Nutrients for Wet-Land Rice. In Organic Matter and Rice. IRRI.

Prasad, R., John, P. S., George, M., Singh, S., & Sharma, S. N. (1990). Effect of lentil residue management on the productivity and NPK removal by lentil-rice double cropping. *Lens Newsletter, 17*(1), 5–9.

Prasad, R., & Palaniappan, S. P. (1987). Pulse crop residue as N source in rice-based cropping system. *International Rice Research Newsletter, 12*(1), 31.

Rathore, A. L., Pal, A. R., & Sahu, K. K. (1998). Tillage and mulching eVects on water use, root growth and yield of rainfed mustard and chickpea grown after lowland rice. *Journal of the Science of Food and Agriculture, 78*, 149–161.

Reicosky, D. C., & Wilts, A. R. (2005). Crop-residue management. Encyclopedia of Soils in the Environment, 334–338. doi:10.1016/b0-12-348530-4/00254-x

Rekhi, R. S., & Meelu, O. P. (1983). Effect of complementary use of moong straw and inorganic fertilizer N on the nitrogen availability and yield of rice. *Oryza, 20*, 125–129.

Riddle, W. C., Gillespie, T. J., & Swanton, C. J. (1996). Rye mulch characterization for the purpose of microclimatic modelling. *Agricultural Meteorology, 78*, 67–81.

Salim, M. (1995). Rice crop residue use for crop production. In Organic Recycling in Asia and the Pacific. United Nations.

Sangakkau, U. K. (1987). Legume for intensification of rice-based farming systems of mid country Sri Lanka. In *Food legumes and Improvement for Asian Farming System pp* (pp. 222–223). Int. Rice Res. Inst.

Sanyal, B., Mondal, S. S., & Chatterjee, B. N. (1993). Fertilizer management with bulky organic manure in rice-potato-groundnut sequence for sustaining productivity. *Journal of Potash Research, 9*, 218–227.

Sharma, A. R., & Mitra, B. N. (1992). Integrated nitrogen management in rice (*Oryza sativa*)- wheat (*Triticum aestivum*) cropping system. *Indian Journal of Agricultural Sciences, 62*, 70–72.

Sharma, P. K., De Datta, S. K., & Redulla, C. A. (1989). Effect of percolation rate on nutrient kinetics and rice yield in tropical rice soils. II. Role of straw amendments. *Plant and Soil, 119*, 121–126.

Shekhawat, K., Rathore, S. S., & Dass, A. (2017). Weed menace and management strategies for enhancing oilseed brassicas production in the Indian sub-continent: A review. *Crop Protection (Guildford, Surrey)*. Advance online publication. doi:10.1016/j.cropro.2017.02.017

Shekhawat, K., Rathore, S. S., & Kandpal, B. K. (2016). Crop establishment techniques affect productivity, sustainability, and soil health under mustard-based cropping systems of Indian semi-arid regions. *Soil & Tillage Research, 158*, 137–146.

Sidhu, H.S., Singh, M., & Humphreys, E. (2007). The Happy Seeder enables direct drilling of wheat into rice stubble. *Australian Journal of Experimental Agriculture, 47*, 844–854.

Sidhu, H.S., & Singh, M., Singh, Y., & Blackwell, J. (2015). Development and evaluation of the Turbo Happy Seeder for sowing wheat into heavy rice residues in NW India. *Field Crops Research*, *184*, 201–212.

Singh. (2004). Long-term effects of organic inputs on yield and soil fertility in rice-wheat rotation. *Soil Society of American Journal.*, *68*(3), 845–853.

Singh. (2005). Crop residue management for nutrient cycling and improving soil productivity in rice-based cropping systems in the tropics. *Advances in Agronomy*, *85*, 269–407.

Singh. (2010). *Options for Effective Utilization of Crop Residues. Research Bulletin 3/2010*. Punjab Agricultural University.

Singh, B., & Rengel, Z. (2007). The role of crop residues in improving soil fertility. In *Nutrient cycling in terrestrial ecosystems* (pp. 183–214). Springer.

Singh, Y. (2003). *Crop residue management in rice-wheat system. Addressing resource conservation issues in rice-wheat Consortium for the Indo-Gangetic Plains*. CIMMYT.

Singh, Y., Singh, D., & Tripathi, R. P. (1996). Crop residue management in rice-wheat cropping system. In *Abstracts of Poster Sessions, 2nd International Crop Science Congress*. National Academy of Agricultural Sciences.

Sisworo, W. H., Mitrosuhardjo, M. M., & Rasjid, H. (1990). The relative roles of N fixation, crop residues and soil in supplying N in multiple cropping systems in a humid, tropical upland cropping system. *Plant and Soil*, *121*, 73–82.

Stigter, K., Ramesh, K., & Upadhyay, P. K. (2018). Mulching for microclimate modifications in farming–an overview. *Indian Journal of Agronomy*, *63*(3), 89–97.

Suyamto, S. (1993). Direct and residual eVect of potassium fertilizer on rice-maize cropping rotation on Vertisols. *Indonesian Journal of Crop Science*, *8*(2), 29–38.

Tamak, J. C., Narwal, S. S., & Singh, S. (1993). Effect of rice residues incorporated in soil on seedling emergence, growth and yield of lentil (*Lens esculanta*). *Legume Research*, *16*, 77–78.

Tiwari, K. N. (1995). Urvarak aur Khad. Indian Council of Agricultural Research.

Trenbath, B. R. (1993). Intercropping for the management of pests and diseases. *Field Crops Research*, *34*, 381–405.

Van Duivenbooden, N. (1992). *Sustainability in terms of nutrient elements with special reference to West Africa. Agrobiological Research Centre*. CABODLO.

Watanabe, J. (1989). Increasing fertility with crop residues and green manures. *Noko To Engei*, *64*, 223–228.

Wu, J. G., Ren, C., & Li, H. Z. (1997). The effects of undecomposed organic materials on rice yield in paddy soils. *Jilin Nongye Daxue Xuebao*, *19*, 68–73.

Chapter 20
Green Technology for Weed Management to Promote Agro-Eco System Health and Sustainable Development

Devendra Kumar
https://orcid.org/0000-0001-9516-4830
Swami Vivekanand Subharti University, India

Nisha Malik
Swami Vivekanand Subharti University, India

Vivek
Sardar Vallabhbhai Patel University of Agriculture and Technology, India

S. K. Sinha
Charles R. Drew University of Medicine and Science, USA

ABSTRACT

In the agriculture system, weed emergence is going to a major constraint to cropped as well as non-cropped areas. They compete with crop plants and hinder their uptake of nutrients, water, and space by suppressing their growth, which ultimately reduces crop yield as well as quality of produce. Due to the dynamic nature of weeds, adoption attains a presence of serious magnitude in different parts of the country. Thus, the prior need is to adopt practices that lead to prevention of weed establishment within the cultivated crop and also protect the environment from these hazards. In this chapter, the authors discuss the management strategies to mitigate the adverse effect of weeds on agricultural productivity and environmental health using green technology. Green technology is a combination of management techniques to keep the soil alive and in good health by the adoption of eco-friendly management practices such as soil solarization, plant arrangement, crop rotation, cultural practices, etc. Some future prospects of weeds management are also discussed here.

DOI: 10.4018/978-1-7998-8434-7.ch020

INTRODUCTION

In India, weeds are going to become major biotic constraints in cropped and non-cropped land due to its dynamic nature, adoption to high-input and intensive cropping system. They cause economic losses by competing or suppressing crop plants for nutrients, water and space resulting in significant crop yield reduction (Auskalniene *et al.,* 2010). Weed species associated with an ecosystem depends upon the environmental characteristics, grown crop, management practices and other biotic and abiotic factors (Yirefu *et al.,* 2012). Due to the weeds, about of 10% part in crop yield damaged in cereals, legumes, pulses, vegetables, horticultural and other commercial crops. In India, the losses caused by weeds (33%) are more than the losses caused by any other disease (20%) or insect pests (26%). Weeds also create problems in the forestry and water bodies. They reduce the quality of far produce and serve as alternate hosts for diseases and pastes (Rao 2000). Due to the climate change, it is expected that growth of C_3 plants would be enhanced more as compared to C_4 plants because of CO_2 enrichment. Since majority of crops belong to C_3 whereas large numbers of weeds belong to C_4 category therefore, it may have significant implications for crop-weed interaction. Many crops such as rice and maize face the dual problem of increased competition from weeds and reduced yield in a harsh environment due to global warming. Climate change is expected to promote proliferation of new weed species and cause shifts in the composition of weed flora. Despite the adoption of technologies, some of the weeds have attained a presence of serious magnitude in different parts of country.

Weed Problems in India

Weed management is a major issue in agriculture production system because their diversification with crops is not fully understood. In India, a major portion of weed management is being usually carried out using chemical herbicides. Currently, wheat, rice, soybean and sugarcane account for major share of the total herbicides consumption. It has been estimated that herbicides are currently being used on more than 20 million ha, which constitute about 10% of the total cropped area in the country. Herbicides and any other pesticide are poisonous in nature and results in toxicity to the user as well as the environment, if used injudiciously. Different herbicides may be required in different crops and recommended dose of application also differ often to control the same weeds. Some herbicides also affect soil microbes and carry-over on the succeeding crops in cold climates. Continuous use of herbicide results in weed flora shifts i.e. major weed from a minor weed. To fulfill the near future demand for food by 2050, there is urgent requirement to refining management strategies for alleviating their adverse effect on agricultural productivity and environmental health. Green technology leads to cultivation of land and raising crops in eco-friendly environment to keep the soil alive and good health. It refers the use of eco-friendly techniques and biological materials (biofertilizers, crop residue, animal, farm wastes and aquatic wastes) for releasing nutrients to crops and organic wastes. No single strategy of weed management provides the desired level of weed control. Therefore, weed management can be carried out in combination of weed management strategies.

WEED MANAGEMENT STRATEGIES

Preventative Management Strategies to Control Weed

Preventive management refers to any method used to prevent weed establishment within the cultivated crop. In these approaches, understanding of weed biology (weed life cycle, reproduction and spread) is an important factor to develop management strategy. The dispersal of weeds from infested field to new areas can be avoided by using some of the strategies *viz.* obtaining certified weed free seeds, farm equipments cleaning before moving one location to another, irrigation water screening to prevent movement of weed seeds from irrigation ditches, use of well rotten decomposed farm yard manure (FYM), distortion of parasite weeds prior to flowering, deep tillage during hot summer months etc. (Table and Fig 1). Movement or existence of invasive weeds is brought under regulation so that introduction or spread of a weed pest may be prevented or restricted.

Mechanical and Physical Management to Weed Control

Manual and mechanical control of weeds primarily consists of removing the weeds physically. This management includes the use of preplant tillage or pre-cultivation like ploughing or disking which are helpful in reducing the rate and spread of certain perennial weeds or also kill seedling and bury seeds below germination zone. The use of new generation machines for tillage, inter-cultivation, soil solarization, stale seed bed, spraying, sowing and harvesting and residue management provides cost effective means of weed management (Fig 2).

Table 1. Preventive weed management strategies

Practice	Effect upon weeds
Cleaning of machinery	Prevents movement of weed seeds from one location to another
Certified seeds	Minimizes risk of weed mixed within seed
Irrigation water screening	Prevent weed seeds movement through irrigation channels
Deep tillage and preplant tillage	Damages perennial roots and rhizomes buries seeds to deeply emerge, bring weed seeds to surface leads to killing of weeds
Cultivation after planting	Removes emerged weeds from the crop
Stale seed bed	Flushing of weeds from the soil before planting
Hand weeding	Physically weed removal
Organic fertility sources	Favor crops over faster- growing weeds due to slow release of nutrients
Drip irrigation	Direct water to the crop rather than to weeds
Mulch	Smoothes weeds: delays emergence of weeds
Using transplants	Competitive advantage to crop
Competitive cultivars	Improve competitive ability to crop against weeds
Increase plant density	Suppresses weeds by shedding
Clcaning after harvest	Prevents seed set by residue weeds
Cover crops	Suppress weeds, improves soil health

Figure 1. Preventive Management Strategies

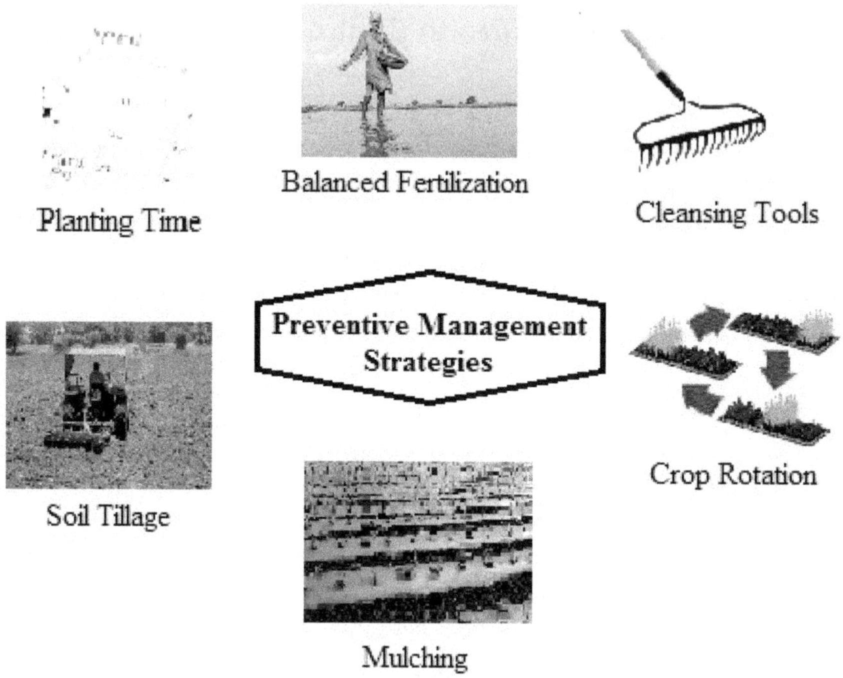

Hand Weeding/Hand Pulling

Hand hoeing or manual weeding with hand tools such as sickle and *khurpi or* animal-drawn mechanical implements such as blade harrows are still the most common practices performed for weed control in millets. Hand weeding before flowering followed by burning can be an effective and practicable method of checking seed production. Profuse emergence of new inflorescence from below ground plant parts has also been observed within a short span of 7-10 days of hand weeding or hoeing, this warrant of frequent repetitive measures. Mechanical weeding provides excellent physical conditions to the crop growth by way of soil aeration through stirring of soil.

Tillage/Inter-cultivation

Deep tillage during summer months causes seed desiccation and places then below the root zone preventing seed germination (Malhi and Lemke 2007; Chauhan and Johnson 2009).

Stale Seed Bed

It comprises of allowing weed to germinate (either after rainfall or through light irrigation) on a prepared fine seed bed, then removing weed seedling via tillage or flame weeding. It provides an opportunity for germination and growth of crop before the next flush of weeds. The use of false seed bed instead of standard seed bed may reduce weed reduce weed emergence upto 80%.

Soil Solarization

It is generally conducted over 4-6 weeks during the warmer time of year with high solar irradiation and minimal cloud and precipitation. Solarization refers to the covering moist soil with white or black polyethylene sheet for 6-7 weeks or longer, so can increase the soil temperature almost 10ºC which results in killing of pests, fungi, bacteria and weed seeds. Weed seeds and young seedlings come under the heat, moisture and through direct contact by with the plastic are killed due to scorching. In general soil solarization is more effective against annual weed species and less effective against perennial weeds. It has been proven to be the most effective method in controlling broomrape in open crops.

Figure 2. Mechanical and Physical Weed Management Strategies

Cultural Management Strategies to Control Weed

Cultural management refers to any technique/strategy (Table 2) that make the crop more competitive against weeds (Azad et al., 1988) such as growing competitive crops in rotation, timely cultivation, mulches, alteration in planting dates, proper fertilization and irrigation, mixed and intercropping with fast canopy forming crops.

Crop Rotation

Crop rotation denotes the change into practice of crop cultivation sequence for a specific period of time in the same area/ field. It helps in constructing healthy soil by fulfillment of micronutrients as well as reduces the reliance of one kind of insect, pest and weed emergence in the field. Crop rotation has a great impact on weed species and density suppression in comparison to tillage (Barberi and Lo Cascio 2001).

Using different cultural strategies, sequence of growing crops with different vegetation and variations in the lifestyle cycles lessen the weeds emergence because of the complexity in crop cultivation sequence (Bhan and Kumar 1997; Jatt et al., 2009). Change to the crop sequence, such as planting of chickpea, wheat, barley after mustard is the handiest and generally used approach for decreasing weeds infestation. Crop rotation of legume crops with non-legumes, narrow leaf grasses with broad leaf forage crops, and the changes in hot/ humid season vegetation with cool season crops are some of the examples of reducing weed species emergence. As per the survey data, 84% of the farmers indicated the crop rotation with forage crops notably decreased the weeds problem. Fifty percent of data revealed the reduction in annual grasses, whereas 35% suggested the decrease in broad leaf weeds emergence. Thus, crop rotation is much effective with aid of following crop planting practice:

- Crop rotation of two crops, if one of the crops is perennial or any cover crop is used during cultivation.
- Crop rotation of three crops, if two of them are excessive residue producing crops.
- If weeds emerged late, than an early maturing crop should be planted before an annual crop.
- Crop rotation of non-cereal crops with wheat or rice, if field is affected with the crop associated weeds.
- Crop rotation of commercial crops with perennial fodder crops.

Mulches, Cover Crop and Surface Residue

Mulches offer a barricade to the weeds grows under it, as they no longer get proper sun light to subsist (Gill et al., 1992). The most commonly used mulches are obtained from natural sources like dead crop residues, straw, scrap papers, or the various types of organic solids which can efficiently chock the sun light. The mulch crops that can develop faster canopy with higher growth as compared to weeds and completely fulfill/ cover the gap area of crop to suppress the weed growth are considered as ideal mulches (Brainard *et al.,* 2008).

Living Mulches

The plants grow within the cultivated crops are comes under living mulches. The plants come under this category are not directly gives any benefit to farmer or not harvested with the cultivated crop. Basically, short duration legume, grasses and forage crops having higher growth rate are planted for this motive due to their faster coverage of gap within the cash crop. Living mulches offer ecological advantages to the cultivated cash crop via suppression of weed emergence and controlling pest population (Mohler and Calloway 1992). Soil erosion also be controlled, as these provides the barrier to the irrigation water flow which sustain the soil fertility level (Sharma *et al.,* 2010). For example, *Winterrye* can be used as living mulch with zucchini squash for controlling weeds emergence (Walter and Young 2008).

Tillage/Inter-cultivation

Deep tillage during summer months causes seed desiccation and places then below the root zone preventing seed germination (Malhi and Lemke 2007; Chauhan and Johnson 2009).

Figure 3. Cultural Strategies

Cover Crops

Cover crops are the plants used to control weeds emergence and growth by altering the environmental temperature, light intensity, humidity as well as maintaining soil moisture. It may be either live or dead mulches, sometimes the intercrop. The residue of these crops helps to augment the soil fertility as well as restrict the soil erosion. Cover crops may be categorized into two different classes. The crops come under the first category are planted during the hostile ecological situations or the period of off-season to destroy or harvest before the sowing of cash crop. Whereas the crops used as living mulches or intercrop within the cash crop considered among second category. The first category of cover crops restricts the weeds emergence by their remain residue after harvesting or destroying (Price *et al.,* 2008), while the cover crops of second category controls the weed growth via its rapid growth and faster canopy developing ability. The Parasitic weeds can be controlled by intercropping of soybean, cowpea, groundnut with Pearl millet and Sorghum (Iqbal *et al.,* 2007). Mulching of green manuring crops like sunhem*p* and *Sesbania* also be used to reduce the weed emergence or weed management strategy in dry-seeded rice (Singh *et al.,* 2007).

Surface Residue

The residue remains after the destroying or harvesting the cover crops is called surface residue. If this residue left within the field without tillage, it suppresses the weeds emergence (Gangwar *et al.,* 2006). It directly restricts the weeds emergence by limiting the availability of sunlight, aeration (Mahapatra *et al.,* 2004; Khaliq *et al.,* 2010). Due to the no tillage, seeds of weeds are not capable to get mild touch of light and air to germinate (Cromar *et al.,* 1999).

Table 2. Preventive and cultural management strategies to control weeds

Cultural practices	Category	Prevailing effect	Examples
Crop rotation	Preventive strategy	Stoppage of weed establishment	Alternation between winter-spring-summer crops
Primary tillage	Preventive strategy	Stoppage of weed establishment	Deep ploughing in summer season, changes in tillage and deep ploughing
Seed bed preparation	Preventive strategy	Stoppage of weed establishment	False or stale seed bed preparation
Soil solarization	Preventive strategy	Stoppage of weed establishment	Black or transparent polyethylene covering
Irrigation and drainage	Preventive strategy	Stoppage of weed establishment	Changes in irrigation method, use of sprinkler or drip irrigation
Surface residue	Cultural strategy	Stoppage of weed establishment	Stubble cultivation
Alteration in plantation date row arrangement	Cultural strategy	Enhancement in crop competitiveness	High seed rate, reduction in inter row distance, late sowing or changes in transplanting date
Crop genotype	Cultural strategy	Enhancement in crop competitiveness	Use of early canopy developing varieties with high emergence ability
Cover crops	Cultural strategy	Enhancement in crop competitiveness and higher growth	Sowing or planting any cover crop between the gap or area
Plantation of Intercrop	Cultural strategy	Stoppage of weed establishment, Enhancement in crop competitiveness	Any crop that can be sown or planted as intercrop
Fertilizer application	Cultural strategy	Stoppage of weed establishment, Enhancement in crop competitiveness	Use of nano-fertilizers or bio-fertilizers with slow nutrient release, use bio-control system

Alteration in Planting Date and Spatial Arrangement

It had seen that modification in crop planting time, density and arrangement pattern can reduce weed emergence and increase crop competitiveness (Malik *et al.,* 1988). The reduced parasitism of *Orabanche* had been observed if late planting of mustard (in last week of October), and *Phalaris minor* in wheat as a result of specific weed and host plant differential response to low temperatures in Indian conditions (Mahajan *et al.,* 2002).

Planting Time

Planting time for a crop may vary from year to year depending on weather and soil conditions. In some cases, it is wise to seed or transplant a cash crop early to get canopy closure as soon as possible. Later, planting allows one or two pre-cultivations of weeds, and also can give the cash crop a jump start because of warmer soils (Lenssen 2008).

Arrangement

Narrow row (<30cm) are beneficial in reducing weed competition and increasing crop yield (Chauhan and Johnson 2010). If narrow row spacing is possible with available planting and cultivation equipment and if it can be done without negatively affecting yield, it can be tried as a weed management tactic.

Crop Density

Many researchers have demonstrated that by increasing crop density weed competition decreases, though this strategy poses several risks. This is best suited to seed crops (such as corn and wheat) and not well-suited to most fruiting crops for which increased plant density results in reduced fruit size. Higher plant density for row crops also helps to buffer against losses caused by mechanical injury from cultivation (Ghuman *et al.,* 2008).

Crop Genotype Choice

Use of different crop genotype which have rapid canopy developing ability and cover up the area or suppress the weed emergence (Mishra and Singh 2009). These types of genotypes include the early maturing and higher growth rate varieties with short life cycle (Chauhan *et al.,* 2001; Gealy *et al.,* 2014). Plantation of such genotypes will directly reduce the weed emergence and requirement of weed management strategies.

Irrigation Management

It has been observed that less infestation of the parasitic weed in raya/mustard grown under flooded irrigation compared to sprinkler irrigation or on conserved moisture (Gealy *et al.,* 2014). Weeds do not survive under an extended period of inundation availability of water and undulating topography (Baraibar *et al.,* 2009).

FUTURE OPPORTUNITIES FOR WEED MANAGEMENT

For weed management in Indian agriculture, innovative approaches are being attempted to check the weed immergence in an effective and eco-friendly manner. Following opportunities are available for weed management.

Crops Having Herbicide Tolerance

The growing of herbicide tolerant crop may change the spectrum of weed management in India. Herbicide tolerant crops are grown in many parts of the world, and expected that such crops will be introduced into India in near future after the legitimate concerns are adequately addressed. Crop-wise herbicide used in India is shown in Fig 4.

Figure 4. Crop-Wise Herbicide used in India

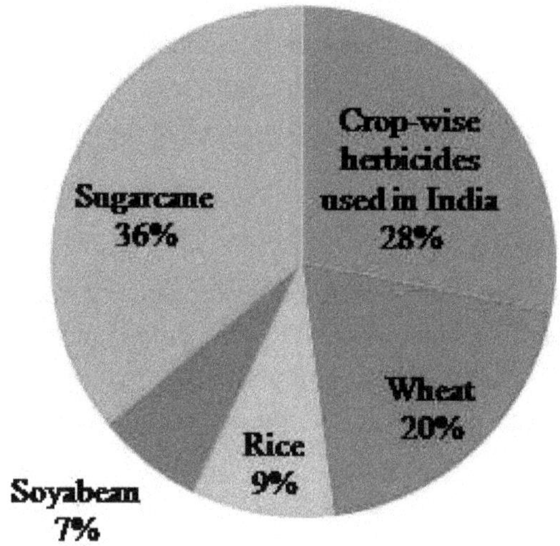

Bio-Herbicides

Biotechnology research can provide breakthrough for development of bio-herbicides, which are host-specific and environment friendly and new strains of bio-agents effective against problematic weeds can be identified. Research on bioactive compounds that may act as herbicide establishment is essential to develop weed management strategies.

Nano-Herbicides

Research on nano-herbicides and other slow release chemicals is continuing, and is expected to yield results in the coming years. The long-term control of weeds avoiding repeated application of herbicides in application zones can be provided by controlled/slow release of herbicide molecules. These nano-herbicide formulations minimize the spread of herbicide residues in the environment and longevity of the herbicide by protecting it from environmental degradation.

Use of Sensor-Based Farming

Remote sensing techniques-based farming is being developed for site-specific weed monitoring. It will be helpful to avoid wastage of herbicides and minimize residue hazards.

CONCLUSION

- Effective decontamination techniques for active and transformation products will provide opportunity for mitigation of residue hazards.

- Research on crop-weed interaction during climate will be helpful to further insights for developing more effective weed management strategies.
- Utilization of weeds biomass as compost, in medicinal and bioremediation programmes.

Green technology in Indian economy has a great role in mitigating the increasing food security problem in India. This leads to cultivation and raising of crops in eco-friendly environment which promotes and enhances agro-ecosystem health, including biodiversity, which is accomplished using biological, mechanical and on-farm agronomic practices. Nowadays, weeds represent a major challenge by prohibiting the use of synthetic herbicides and desiccants. Despite research and advances in weed management practices, these remains as a constant threat to productivity, profitability and sustainability in agriculture. Therefore, green technology focuses upon management strategies designed to prevent weeds as well as the productivity of crop vigorous enough to out compete weeds.

REFERENCES

Auskalniene, O., Psibisauskiene, G., Auskalnis, A., & Kadzys, A. (2010). Cultivar and plant density influence on weediness in spring barley crops. *Zemdirb Agric*, *97*, 53–60.

Azad, B. S., Singh, H., & Dubey, A. N. (1988). Control of Phalaris minor in wheat by cultural and chemical methods. *Pesticides*, *22*, 41–42.

Baraibar, B., Westerman, P. R., Carrion, E., & Recasens, J. (2009). Effects of tillage and irrigation in cereal fields on weed seed removal by seed predators. *Journal of Applied Ecology*, *46*(2), 380–387. doi:10.1111/j.1365-2664.2009.01614.x

Barberi, P., & Lo Cascio, B. (2001). Long-term tillage and crop rotation effects on weed seed bank size and composition. *Weed Research*, *41*(4), 325–340. doi:10.1046/j.1365-3180.2001.00241.x

Bhan, V. M., & Kumar, S. (1997). Integrated management of Phalaris minor in rice–wheat ecosystems in India. *Proceedings of international conference on ecological agriculture: towards sustainable development,* 400-415.

Brainard, D. C., Bellinde, R. R., Hahn, R. R., & Shah, D. A. (2008). Crop rotation, cover crop, and weed management effects on weed seed banks and yields in snap bean, sweet corn, and cabbage. *Weed Science*, *56*(3), 434–441. doi:10.1614/WS-07-107.1

Chauhan, B. S., & Johnson, D. E. (2009). Influence of tillage systems on weed seedling emergence pattern in rainfed rice. *Soil & Tillage Research*, *106*(1), 15–21. doi:10.1016/j.still.2009.10.004

Chauhan, B. S., & Johnson, D. E. (2010). Implications of narrow crop row spacing and delayed Echinochloa colona and Echinochloa crusgalli emergence for weed growth and crop yield loss in aerobic rice. *Field Crops Research*, *117*(2-3), 177–182. doi:10.1016/j.fcr.2010.02.014

Chauhan, B. S., Yadav, A., & Malik, R. K. (2001). Competitive wheat genotypes under zero tillage-an important tool to manage resistant *Phalaris minor. Indian Journal of Weed Science*, *33*, 75–76.

Cromar, H. E., Murphy, S. D., & Swanton, C. J. (1999). Influence of tillage and crop residue on post dispersal predation of weed seeds. *Weed Science*, *47*(2), 184–194. doi:10.1017/S0043174500091608

Gangwar, K. S., Singh, K. K., Sharma, S. K., & Tomar, O. K. (2006). Alternative tillage and crop residue management in wheat after rice in sandy loam soils of Indo-Gangetic Plains. *Soil & Tillage Research*, *88*(1-2), 242–252. doi:10.1016/j.still.2005.06.015

Gealy, D. R., Anders, M., Watkins, B., & Duke, S. (2014). Crop performance and weed suppression by weed-suppressive rice cultivars in furrow- and flood-irrigated systems under reduced herbicide inputs. *Weed Science*, *62*(2), 303–320. doi:10.1614/WS-D-13-00104.1

Ghuman, R. S., Brar, L. S., & Walia, U. S. (2008). Role of variety and plant geometry on weed management in transplanted rice (Oryza sativa L.). *Indian Journal of Weed Science*, *40*, 137–141.

Gill, K. S., Arshad, M. A., Chivundu, B. K., Phiri, B., & Gumbo, M. (1992). Influence of residue mulch, tillage and cultural practices on weed mass and corn yield from three field experiments. *Soil & Tillage Research*, *24*(3), 211–223. doi:10.1016/0167-1987(92)90088-S

Iqbal, J., Cheema, Z. A., & An, M. (2007). Intercropping of field crops in cotton for the management of purple nutsedge (*Cyperus rotundus* L.). *Plant and Soil*, *300*(1-2), 163–171. doi:10.100711104-007-9400-8

Jat, M. L., Singh, S., Rai, H. K., Chhokar, R. S., Sharma, S. K., & Gupta, R. K. (2005). Furrow irrigated raised bed planting technique for diversification of rice–wheat system of Indo-Gangetic plains. *J Jpn Assoc Int Coop Agric For*, *28*, 25–42.

Khaliq, A., Matloob, A., Irshad, M. S., Tanveer, A., & Zamir, M. S. I. (2010). Organic weed management in maize through integration of allelopathic crop residues. *Pakistan Journal of Weed Science Research*, *16*, 409–420.

Lenssen, A. W. (2008). Planting date and preplant weed management influence yield, water use, and weed seed production in herbicide-free forage barley. *Weed Technology*, *22*(3), 486–492. doi:10.1614/WT-08-009.1

Mahajan, G., Brar, L. S., & Walia, U. S. (2002). Minor response in wheat in relation to planting dates, tillage and herbicides. *Indian Journal of Weed Science*, *34*, 213–215.

Mahapatra, B. S., Mishra, A., & Kumar, A. (2004). *Green manuring: a basic concept and its role in weed control. Training manual: advances in weed management*. G.B. Pant University of Agriculture and Technology.

Malhi, S. S., & Lemke, R. (2007). Tillage, crop residue and N fertilizer effects on crop yield, nutrient uptake, soil quality and nitrous oxide gas emissions in a second 4-year rotation cycle. *Soil & Tillage Research*, *96*(1-2), 269–283. doi:10.1016/j.still.2007.06.011

Malik, R. K., Balyan, R. S., & Bhan, V. M. (1988). Effect of herbicides and planting dates of weeds in chickpea. *Indian Journal of Weed Science*, *20*, 75–81.

Mishra, J. S., & Singh, V. P. (2008). Competitive ability of wheat (Triticum aestivum) cultivars against wild oat (Avena ludoviciana) infestation under zero tillage. *Indian Journal of Agricultural Sciences*, *78*, 838–841.

Mohler, C. L., & Calloway, M. B. (1992). Effects of tillage and mulch on the emergence and survival of weeds in sweet corn. *Journal of Applied Ecology*, *29*(1), 21–34. doi:10.2307/2404343

Price, A. J., Stoll, M. E., Bergtold, J. S., Arriaga, F. J., Balkcom, K. S., Kornecki, T. S., & Raper, R. L. (2008). Effect of cover crop extracts on cotton and radish radicle elongation. *Communications in Biometry and Crop Science*, *3*, 60–66.

Rao, V. S. (2000). *Principles of weed science* (2nd ed.). Science Publishers Inc Enfield. doi:10.1201/9781482279603

Sharma, A. R., Singh, R., Dhyani, S. K., & Dube, R. K. (2010). Effect of live mulching with annual legumes on performance of maize (Zea mays) and residual effect on following wheat (Triticum aestivum). *Indian Journal of Agronomy*, *55*, 177–184.

Singh, S., Ladha, J. K., Gupta, R. K., Bhushan, L., Rao, A. N., Sivaprasad, B., & Singh, P. P. (2007). Evaluation of mulching, intercropping with Sesbania and herbicide use for weed management in dry-seeded rice (Oryza sativa L.). *Crop Protection (Guildford, Surrey)*, *26*(4), 518–524. doi:10.1016/j.cropro.2006.04.024

Walters, S. A., & Young, B. G. (2008). Utility of winter rye living mulch for weed management in zucchini squash production. *Weed Technology*, *22*(4), 724–728. doi:10.1614/WT-08-035.1

Yirefu, F., Tana, T., Tafesse, A., & Zekarias, Y. (2012). Competitive ability of sugarcane (*Saccharum officinarum* L.) cultivars to weed interference in sugarcane plantations of Ethiopia. *Crop Protection (Guildford, Surrey)*, *32*, 138–143. doi:10.1016/j.cropro.2011.10.016

Chapter 21
Sustainable Agriculture Through Organic Farming

Permod Kumar
Swami Vivekanand Subharti University, India

Raj Singh
Maharishi Markandeshwar University (Deemed), India

Anju Rani
Keral Verma Subharti College of Science, Swami Vivekanand Subharti University, India

Amit Kumar
Subharti College of Science, Swami Vivekanand Subharti University, India

Gyanika Shukla
Ch. Charan Singh University, Meerut, India

ABSTRACT

The chapter deals with the concept of organic farming as an effective way to promote sustainable agriculture worldwide. The mordent approaches of conventional farming uses chemical pesticides and fertilizers that affect the environment adversely by increasing population and diminishing soil fertility and quality. In such a situation, organic farming may be proved as a significant approach to restore the environmental condition and nutrition security for the purpose of survival. There is a need of raking awareness and research in the area of organic farming to develop a cheap and eco-friendly approach for sustainable agriculture.

INTRODUCTION

Every day, farmers and ranchers around the world develop new, innovative strategies/technique to produce and distribute food, fuel and fiber sustainably. While these strategies vary greatly, they all embrace three pillars or goals of Sustainability: Profit (over the long term); Stewardship (of our nation's land,

DOI: 10.4018/978-1-7998-8434-7.ch021

air and water) and quality of life (for farmers, ranchers and their communities) (Gaur 2006; Soumya 2015). Sustainable agriculture with low technological input is limiting dependence on organic fertilizers, pesticides, other agriculture chemical etc. Disease and plants/ crops pests are managed through crop rotation, organic insecticides, bio- herbicidal, combined with a good crop management. Cattle, fisheries, livestock, poultry must be returned by the integrated so that is part of the organic farming (Kratochvil 2005). Through the management of good soil plant nutrient requirements can be found as well as environmental conditions and biodiversity can be improved and protected without depending on chemical fertilizers, pesticides etc.

Table 1. World distribution of organic agriculture land by region 2014 (FiBL survey 2016)

Continent	Organic agriculture land (mio. ha.)	Non-agriculture land (mio. ha.)
Africa	1.3	11.8
Asia	3.6	6.3
North America	3.1	0.06
Latin America	6.8	3.0
Oceania	17.3	-
Europe	11.6	16.3

Why Organic Farming is needed

1. to produce nutritionally of high quality food (no chemical residues);
2. to keep our healthy environment;
3. to maintain long term fertility of the soil;
4. to recycle agriculture product or by product within the farm unit;
5. to reduce the problem of water logging, acidity and salinity *viz.* deterioration of soil quality with existing practices;
6. to maintain genetic diversity of life (plant/animal).
7. to maintain the healthy and sustainable production;
8. to increase the farm income (Vaarst 2010).

The idea, organic agriculture began in the early part of the twentieth century (1920-organic 1.0), primarily in Europe, but also in the United States. The pioneers of the early organic movement were motivated by a desire to reverse the perennial problems of agriculture – soil erosion, soil depletion, decline of crop varieties, low quality food, livestock feed, and rural poverty etc. The soil's health and vitality were believed to be embodied in its biology and in the organic soil fraction called humus. The strategy of soil management is known as ''humus farming" (Alroe et al., 2006; Kuepper, 2010).

J. I. Rodale (father of the modern organic farming), developed the main source of information about "non-chemical" farming methods and was heavily influential in the development of organic production methods. The term 'organic' was first used in relation to farming by Northbourne (1940) in his book *Look to the Lad*. Organic farming was also determined in EU regulation EWG No. 2092/91. Considering the actual farming practices in Central Europe, organic farming is regarded as the production system with the highest degree of ecological compatibility (Eichenberger and Vogtmann, 1981; Lindenthal et al., 1996; Bmlf, 1999). This statement is based on the principles of organic farming, which are to a great extent corresponding with the criteria for a (ecologically strong) sustainable development. The concept of organic farming can be tested and applied for the safety of flora and fauna. Organic farming considers the important aspects like sustainable natural resources and environment. It is a production system, which favours maximum use of organic materials like crop residues, farmyard manure (FYM), compost, green manure, oil cakes, bioforming, bio-fertilizers, bio-pesticides, bio-gas slurry, microbial products like *Azotobacter, Trichoderma, Pseudomonas, Beauveria,* Blue green algae (BGA), *Azolla, Bacillus spp.* etc., have increased the yield and also played important role for minimizing the harmful effect of pesticides and herbicides (Silwana 2007). There is urgent need to involve more research for the development of eco-friendly production technology.

Moreover, in its basic standards International Federation of Organic Agriculture movements (IFOAM 2000) also formulated some principle social, environmental and economic aims for organic farming; "As such, the objective of sustainability lies at the heart of organic farming" (Lampkin, 1994; Bhattacharya and Chakraborty 2005).

Sustainable agriculture as given out by the Food and Agriculture Organization that "The successful management of resources for agriculture to satisfy changing human needs while maintaining or enhancing the quality of environment and conserving natural resources".

IFOAM, an international organization was established in 1972 for organic farming organizations and defines the goal of organic farming as:

Organic agriculture is a production system that sustains the health of soils, ecosystems and people. It relies on ecological processes, biodiversity and cycles adapted to local conditions, rather than the use of inputs with adverse effects. Organic agriculture combines tradition, innovation and science to benefit the shared environment and promote fair relationships and a good quality of life.

Principle of Organic Farming

According to united states agency for international development (USAID) in 1943 "organic farming is a system which avoids the use of synthetic inputs (fertilizers, pesticides, hormones, feed additives etc) and the maximum extent feasible rely upon animals/plants manures, crop rotations, crop residues, off-farm organic waste, mineral grade rock additives and biological system of nutrient mobilization and plant protection". According to FAO 2008, on the basis of agriculture sustainability, organic farming is depending on four basic principles are as follow:

- **Health:** Organic agriculture must contribute to the health of soil, plants, animals etc. It is the sustenance of mental, physical, ecological and social well being. For instance, it provides pollution and chemical free, nutritious food items for living being.
- **Fairness:** Organic farming should build on relationships that ensure fairness regarding the common environment and life opportunities
- **Ecological Balance:** Organic farming should be based on living ecological systems and cycles; work with them, emulate them and help sustain them
- **Care:** Organic agriculture should be practiced in a careful and responsible manner to benefit the present and future generations and the environment.

These basic principles provide organic farming with a platform for ensuring the health of environment for sustainable development.

Organic Source of Nutrients

The organic sources mainly depends on macro and micro nutrients besides supplying N, P, and K also make unavailable sources of elemental nitrogen, bound phosphates, and decomposed plant residues into an available form to facilitate the plants to absorb the nutrients. Its encouraged the growth, activity of mycorrhizae, beneficial microorganisms in the soil and also helpful in alleviating the increasing incidence or deficiency of secondary and micronutrients, so finally, capable of sustaining high crop productivity and soil health (Nambiar *et al.,* 1992). The farmers can in turn, get good remuneration from organically produced crops/ plant and if included in high value crop rotations, that is, aromatic rice (*Oryza sativa* L.), table pea (*Pisum sativum* L.), onion (*Allium cepa* L.) and dhaincha (*Sesbania aculeate* L.) (Singh *et al.,* 2001; Kalyan 2005; Yadav *et al.,* 2013), Breeding maize with hairy stems againts *Ostrinia nubilalis* (corn borer), Producing protein and fat from food waste with larvae of the Black Soldier Fly (*Hermetia illuscens*), Novel bio control agents in plant and livestock health management *viz.* expl. *Duddingtonia flagrans* by Niggli (2014) of highly demand at national and international levels.

Currently, Frick (2016) describes the differences between organic and non-organic milk and meat, especially in terms of fatty acid composition and the concentrations of certain essential minerals and antioxidants. Some important findings are as following:

1. both organic milk and meat contain around 50% more beneficial omega-3 fatty acids than conventionally produced products;
2. organic milk contains 40% more conjugated linoleic acid, which is beneficial to human health;
3. organic milk contains slightly higher concentrations of iron, vita. E and some carotenoids and;
4. conventional milk contained on average 74% more of the essential mineral iodine.

The differences mainly stem from the different feed composition for conventionally and organically farmed animals. Organic standards heavily restrict the content of concentrated feed (maize, grains, soy) in fodder for ruminants (beef, sheep, goats). The standards also require poultry and pigs to have access to open grazing areas, which lead them to eat more grass and herbs. Niggli (2016) concludes "That there are composition differences between organic and conventional food. Taken together, the three studies on crops, meat and milk suggest that a switch to organic fruit, vegetables, meat and dairy products would provide significantly higher amounts of dietary antioxidants and omega-3 fatty acids."

PROS AND CONS OF ORGANIC FARMING

Pros

1. Organic farming is more economical to compare the other farming system. It rises of benefits *via* reduced soil erosion and carbon erosion, less use of water, nutrient- contamination by pesticides and increased biodiversity.

2. Organic farming is capable of producing the same crop variants that are produced *via* conventional farming methods, even as it brings 50% down the expenditure on fertilizers and energy and retains 40% more top soil.
3. Soil management is effectively addressed by organic farming. It involves techniques like crop rotation and inter-cropping and makes extensive use of green manure, which helps even damaged soil *i.e.* prone to erosion and salinity are able to feed on micro nutrients.
4. Its farming helps the farmers clear the weeds, the use of mechanical and chemical applications. Organic way of farming relies on practices such as hand weeding and enhancement of soil with mulch, garlic and clove oil, corn gluten meal, table salt and borax, to get rid of insects and weeds, while ensuring the crop quality.
5. Farming in the organic way is environment-friendly and non-toxic, as it uses green pesticides such as neem, composed tea and spinosad. These pesticides help in identifying and removing diseased and dying plants in time.

Figure 1. Organic farming cycle
(Vats and Sharma, 2011)

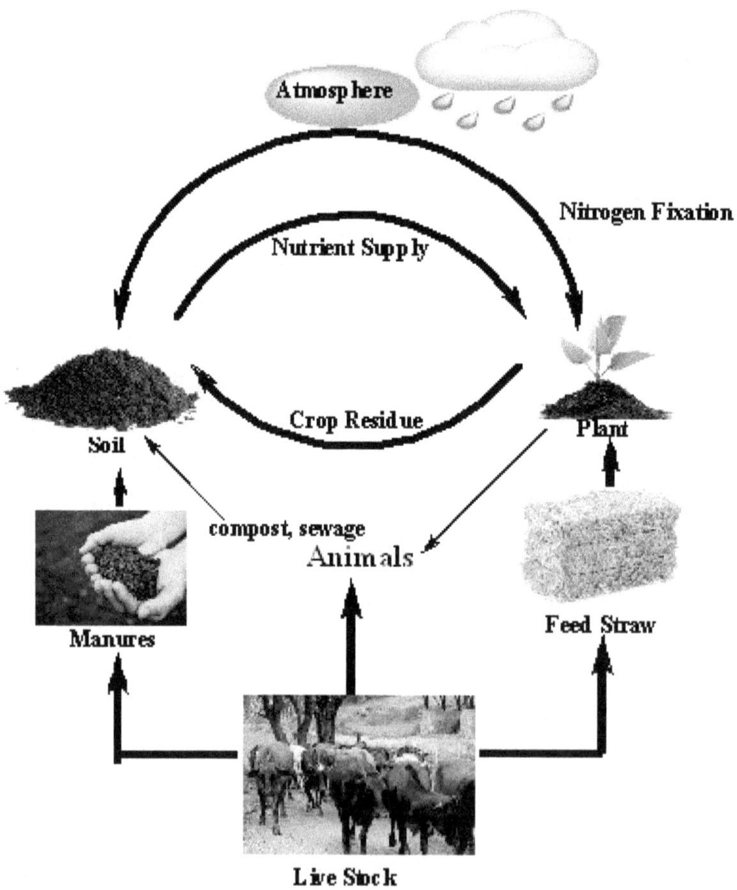

Cons

1. Organic methods of farming; they concluded small yields when compared to conventional farming methods (UN Environmental program 2008).
2. It is capable of catering to a very small consumer group, the expanding cropland is dramatically destroying world ecosystem.
3. Organic farms which produce potatoes, seed grass and sugar beet are barely producing half of the total output produced by conventional farming, in the same area (Danish Environmental Protection Agency).
4. Organic agriculture is hardly contributing to addressing the issue of global climate change. It does reduce carbon dioxide emissions to a certain extent, but there is no dramatic contribution.
5. The increased risk of *E. coli.* infection by the consumption of organic food.

Eco-Friendly Technology of Organic Farming

Organic farming is essentially an agriculture employing a knowledge/understanding of naturally occurring processes. According to Albert Howard (father of modern organic agriculture) to point out the organic farming techniques in his book ''An Agriculture Testament thousands of years of agriculture''Organic farming relies heavily on the natural break down of organic matter, using techniques (Dubey and Shukla, 2014) like green manure and composting, to maintain nutrients taken from the soil by the previous crops. Some important techniques are following:

- **Cultivation:** it includes mon or polyculture, its increases crop resistance against pests and diseases.
- **Crop rotation:** its control the build up of pest and diseases. Soil remains higher quality and pests of a particular plant will not be allowed more than one year to breed and thrive. Some area of poor soil, time to time recover by leaving it empty *i.e.* part of a crop rotation. So finally result in reduction of a pest and/or improvements of the soil.
- **Intercropping and companion planting:** Intercropping allows to "layer" the crops so that they compete less for sun, water and nutrients etc, so plants grow stronger and makes them more resistant to pests. Companion planting (CP) places similar crops together which also actually benefit from each other.
- **Mulching:** By mulching a soil it naturally represses weeds and the soil does not dry, allowing healthy growth of plants. Cover crops have a same weed-controlling function. Plants which overhang close to the ground can be easily attacked by ground dwelling animals such as caterpillars, or decompose through contract with the soil. So it can be used for number of ways to reduce pests.
- **Tillage:** The breaking up of the soil *i.e.* to improve soil profile as well as for primary weeding or mechanical control. It needs to be done at the right time. If the loosening of soil needs to be done in any situation where the soil has become consolidated (after rainy and winter session) deep tillage can cause eutrophication, loss of nutrients etc.

In contrast, organic farming tends to tolerate some pest populations while taking a longer-term approach. The organic pest control involves techniques like encouraging predatory beneficial insects and microorganisms, careful crop selection and crop rotation. Each of these techniques also provides other benefits- soil protection, fertilization, pollination, water conservation etc (Vats and Sharma 2011). The

Green Revolution and its chemical based technology are losing its appeal as dividends are falling and returns are unsustainable. Pollution and climate change are other negative externalities caused by use of fossil fuel based chemicals. As of 2012, the market for organic farming and other products has reached $63 billion worldwide. Finally,Organic farming to improve soil fertility, environmentally safe, economically viable and socially replicable.

Figure 2. Organic Certification Process

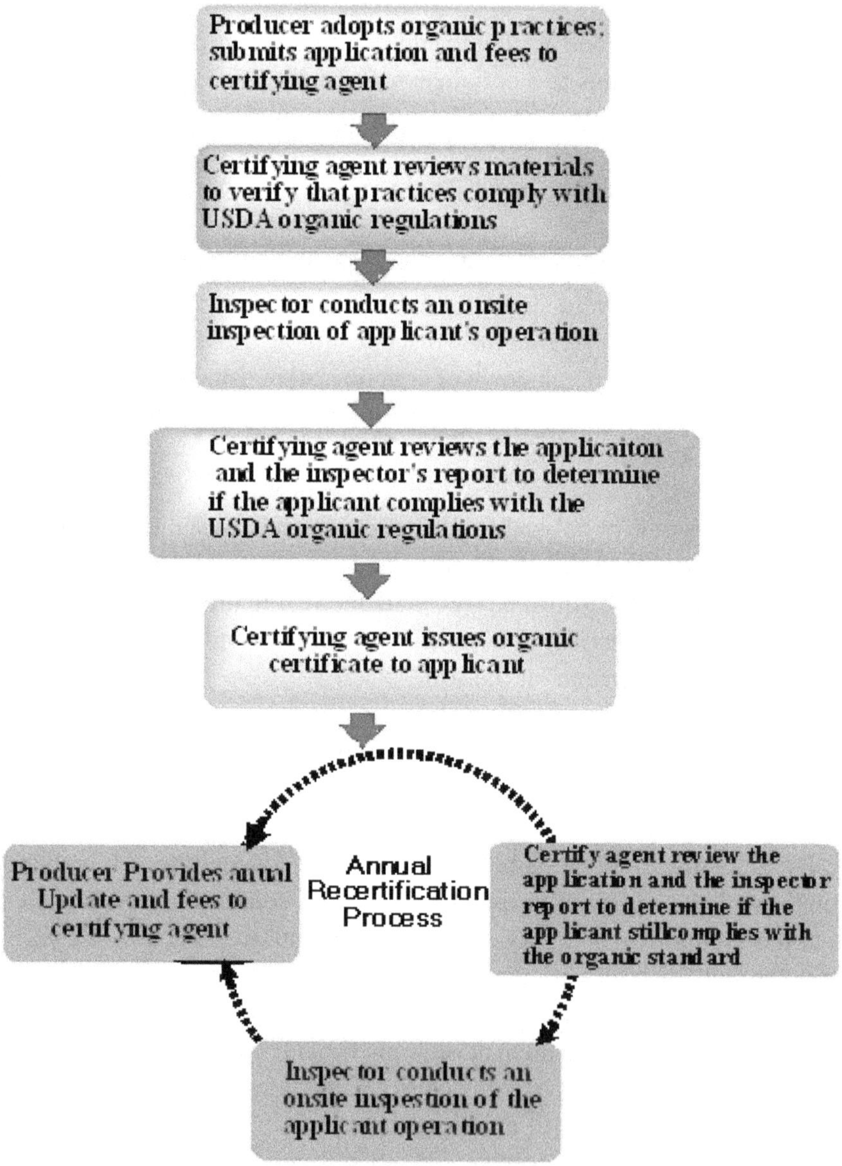

ORGANIC CERTIFICATE

Organic certification verifies that your farm or handling facility located anywhere in the world complies with the agreement U.S. Department of Agriculture (USDA), organic regulations and allows you to sell, label, and represent your products as organic. The USDA National Organic Program (NOP) administers these regulations, with substantial input from its citizen advisory board and the public. The owner's farm or handling facility would be certified by a private, foreign, or State entity. Certifying agents are responsible for ensuring that USDA organic products meet or exceed all organic standards. Certification provides the consumer, whether end-user or intermediate processor, assurance of the organic product's integrity (Narayanan, 2005; Huber, 2007; Baier, 2012).

Who needs to be Certified?

If the owner's farm or business receives more than $5,000 in gross annual organic sales, it must be certified.

Resources

www.attra; ncat.org; www.ioia.net; www.ifoam.org; www.mosesorganic.org; www.ncat.org; www.omri.org; www.ota.com; www.ams.usda.gov/nop.

Benefits of Organic Farming in Support for Environmental Health

According to the American Public Health Association (AHA) which is in savoir of environment, environment health organic farming mainly depends on "The Basic Six" of public health efforts on soil, on air, on water, on foods, and on waste management (Slamet, 2011). According to Rockets (2007) organic farming a lot of benefit to the healthy environment, some are following:

1. **Quality of Soil** keeping the nature of the physical, chemical and biological good soil is important in organic farming.
2. **Quality of Air** Organic farming proves to be capable of minimizing global climate change due to green house gases emissions in organic farming this emission is lower than conventional agriculture. Organic farming doesn't use synthetic nitrogen fertilizer so there is no nitrogen oxides emission from artificial fertilizers. The use of petroleum is low thus lowering emissions of carbon dioxide. More importantly, organic farming provides shelter (sink) for carbon dioxide through increased content of organic matter in soil and the closure of the ground surface with plants covering the ground.
3. **Waste Management** it reduce the amount of waste through recycling waste into organic fertilizer. Livestock manure, straw and other agricultural waste as long as it is considered waste, thus being a material that has a value as a source of nutrients and organic substances for organic farming (Departemen, 2004).

CURRENT DYNAMIC STRATEGIES OF ORGANIC FARMING

Organic agriculture is practiced in 172 countries, and 43.7 million hectares of agricultural land are managed organically by approximately 2.3 million farmers. The global sales of organic food and drink reached 80 billion US dollars in 2014. Undoubtedly, innovation in organic farming will increase yields on organic farms. But we need to produce 70% more food than we do with today's conventional yields to feed the world in 2050.

Table 2. Organic Agriculture 2016: statistics Key indicators and Countries (FiBL survey 2016, based on national data sources and data from certifiers global market: Organic Monitor 2016)

Statistics key indicators	World	Top countries
Countries with organic activities	2014: 172 countries	New countries: Kiribati, Puerto Rico, Suriname, United States Virgin Islands
Organic agricultural land	2014: 43.7 million hectares (1999: 11 million hectares)	Australia (17.2 million hectares; 2013) Argentina (3.1 million hectares) US (2.2 million. hectares, 2011)
Organic share of total agricultural land	2014: 0.99%'	Falkland Islands (Malvinas) (36.3%) Liechtenstein (30.9%) Austria (19.4%)
Wild collection and further, non-agricultural areas	2014: 37.6 million hectares (1999: 4.1 million hectares)	Finland (9.1 million hectares) Zambia (6.8 million hectares) India (4 million hectares)
Producers	2014: 2.3 million producers (1999: 200000 producers)	India (650000; 2013) Uganda (190552) Mexico (169703; 2013)
Organic market size	2014: 80 billion US dollars (1999: 2.5.2 billion US dollars)	US (35.9 billion USD; 27.1 billion earns) Germany (10.5 billion USD; 7.9 billion euros) France (6.8 billion USD; 4.8 billion euros)
Per capita consumption	2014: 11 US dollars (14 euros)3	Switzerland (221 euros) Luxemburg (164 euros) Denmark (162 euros)
Number of countries with organic regulations	2015: 87 countries	
Number of (FOAM affiliates	2015: 784 affiliates from 117 countries	Germany - 91 affiliates China- 57 affiliates India 44 affiliates USA-4o affiliates

Starting with conventional agriculture we'll need to boost yields by 70%. Starting with organic agriculture, we'd need to nearly triple yields in that time – a much larger challenge.

Consumer benefits include the following:

- **Nutrition**: The nutritional value of food is dramatically superior in mineral content to that grown by modern conventional methods. Healthy plants mean healthy people, and such better nourished plants provide better nourishment to people and animals alike.
- **Poison-free**: A major benefit to consumers of organic food is that it is free of contamination with health harming chemicals such as pesticides, fungicides and herbicides, example: In 1972 to2004 the incidence of new cancer cases per year (average for both sexes) has risen from 323 to 488 per 100,000 people. This is an increase of over 50% in just 32 years (New South Wales, Australia).

- **Food tastes better:** organically grown food tastes better than that conventionally grown. The tastiness of fruit and vegetables is directly related to its sugar content, which in turn is a function of the quality of nutrition that the plant itself has enjoyed. The quality of fruit and vegetable are measured by Brix analysis (specific gravity or density).
- **Food keeps longer:** Organically grown plants are nourished naturally, rendering the structural and metabolic integrity of their cellular structure superior to those conventionally grown. As a result, organically grown foods can be stored longer and do not show the latter's susceptibility to rapid mold and rotting.

Figure 3. Different Condition of Sustainable Agriculture at Different Level
(FiBL survey 2016)

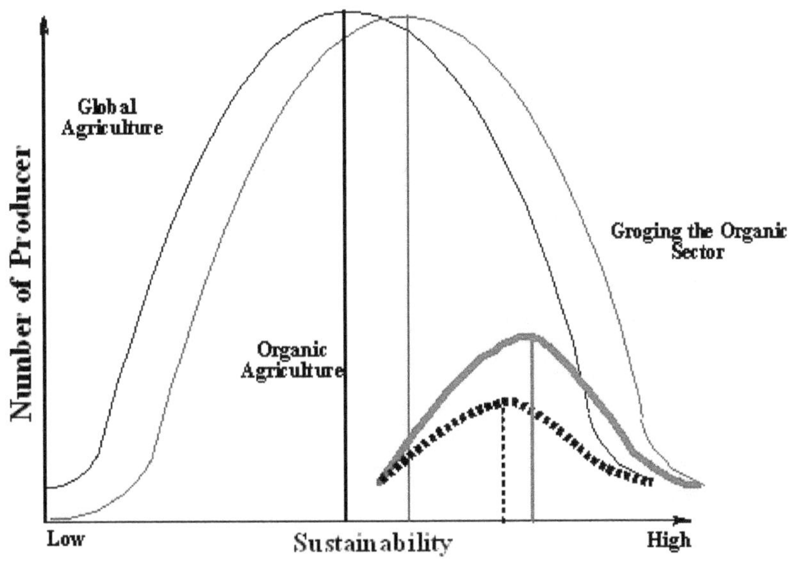

Grower Benefits

Organically healthy plant grown in properly balanced soil resists most diseases and insect pests. These tests are following in different way:

- **Disease and pest résistance:** For instance, in an orange grove infested with scale, he restored the mineral balance to part of the soil and the trees growing in that part became clean while the rest remained diseased. By the same means he grew healthy rosebushes between rows that were riddled by insects, and tomato and cucumber plants, both healthy and diseased, where the vines intertwined. Northern observed that the bugs ate up the diseased and refused to touch the healthy plants.
- **Weed competitiveness:** Weeds are nature's band-aids, placed by the wisdom of creation to heal and restore damaged soils. The crops are healthier and also better able to compete with those weeds that are present.

- **Lower input costs:** Fertilizers are either created *in situ* by green manuring and leguminous crop rotation or on-farm *via* composting and worm farming. Biodynamic farmers use a low cost microbial solution sprayed onto their crops. The creation of living, fertile soil conditions through early corrective soil re-mineralization and other strategic are significant establishment costs that, however, reap ongoing benefits to production at minimal maintenance.
- **Drought resistance:** Organically grown plants are more droughts tolerant (Sangakkara *et al.*, 2008).
- **Added Value:** There is a discerning market of consumers who recognize the greater food value of organic produce and are willing to pay premium prices for it. So the development of sustainable agriculture in which both farmers and consumers takes the initiative in small and large scale (courteney *et al.*, 2006; Lamine and Bellon, 2009).

Significance of Organic Farming

1. Organic Farmers are expected to provide additional information and knowledge relevant to the farmers for developing organic farming activity.
2. The government should resolve various obstacles faced by organic farmers in developing organic agriculture.
3. Business may provide additional information for the corporate world who expound on the results of organic farming, for example eating healthy home-specific business results organic farming or any other business related to organic farming, *i.e.* certification agencies, the sale of organic fertilizer. It could increase people's income so that it can increase the purchasing power of the community (Jimenez, 2010).
4. The Civil Society (Non Government Organization) is encouraging NGO's to do the mentoring activities at the farmers in sustainable development in organic farming.
5. Global Contribution give input on all over the world who are interested in the development of organic farming with a social development approach that could push the growth of organic agriculture movements.

Challenge of Organic Farming in Sustainable Agriculture

The current problems in the 21st century, namely the impact of modern agribusiness technologies (farms depend on pesticides, chemical fertilizers), One of the negative effects is the occurrence of the damage in the quality of the environment globally caused by industrial waste that pollutes the environment (Arsenio and Arsenio, 2010). According to report of FAO in year 2000, the green revolution contributed to more than 20% of global greenhouse emissions. The quality of environmental health has been declining due to excessive use of pesticides in the activities of the green revolution. At present current challenge are organic 3.0.

Organic 1.0, 2.0 and 3.0

The Organic 1.0 (1920-1969) is laid out how people can healthily nourish them whilst protecting the ecological and biodiversity. Organic 2.0 (1970-2014) are enable the market for certified organic products to develop and gain a significant foothold in many parts of the world. The Organic 3.0 (2015…) strategy for Organic 3.0 includes six main features: (i) A culture of innovation; (ii) Continuous improvement toward best practice; (iii) Diverse ways to ensure transparent integrity; (iv) Inclusive of wider sustainability interests; (v) Holistic empowerment from farm to the final consumer; (vi) True value and fair pricing.

Figure 4. Different Ratio of Normal Crops and Organic Crops

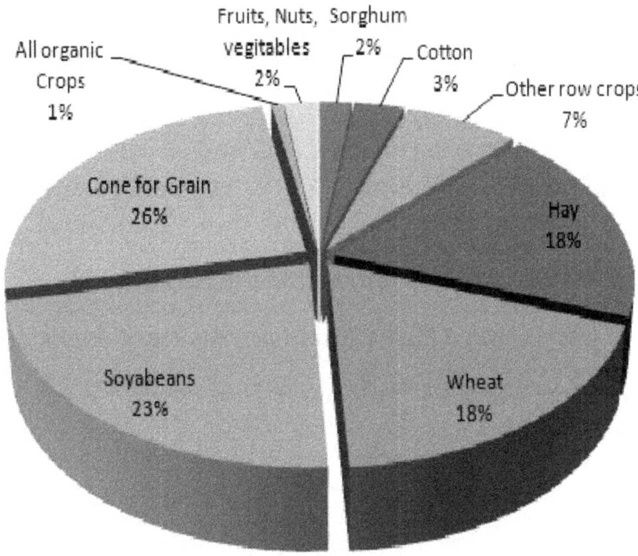

CONCLUSION

Organic agriculture in developing countries is fast growing but is still a small part of the agricultural scene. Their current situation is poor and a conventional pathway to improve yields is costly and difficult. At the same time, adoption of organic agriculture, with support from the government or NGOs can make a big difference, so much so that organic farmers' poverty levels fall and their well-being in the widest sense (including income, health, and education) rises, even above those of conventional farmers. Finally, organic agriculture can much better contribute to the environmental goals of carbon sequestration and maintenance of biodiversity than conventional agriculture.

REFERENCES

Alroe, H. F., Byrne, J., & Glover, L. (2006). *Organic agriculture and ecological justice: Ethics and practice. In Global Development of Organic Agriculture. Challenges and Prospects.* CABI Publishing.

Arsenio, S. K., & Arsenio, V.M.L. (2010). *Ecological Crisis Our Responsibility let us amend our lives.* C & E Publishing Inc.

Baier, A. H. (2012). Organic certification of farms and Business producing Agricultural Products. *USDA Organic*, 1-8. www.attra.ncat.org

Bengtsson, J., Ahnstrom, J., & Weibull, A.-C. (2005). The effects of organic agriculture on biodiversity and abundance: A meta-analysis. *Journal of Applied Ecology*, *42*(2), 261–269. doi:10.1111/j.1365-2664.2005.01005.x

Bhattacharyva, P., & Chakraborty, G. (2005). Current Status of Organic Farming in India and other Countries. *Indian Journal of Fertilisers*, *1*(9), 111-123.

BMLF. (1995). *Organic Farming in Austria*. Wien: BMLF

Departemen, A. R. I. Syaamil Al-Qur'an. (2004). *Environmental Benefits of Organic Farming*. www.organicabout.com

Dubey, R. K., & Nidhi, S. (2014). Organic Farming: An Eco-Friendly Technology and Its Importance And Opportunities in the Sustainable Development. *International Journal of Innovative Research in Science, Engineering and Technology*, *3*, 3.

Frick & Bonn. (2016). *The world of organic agriculture*. FiBL & IFOAM.

Gaur, A. C. (2006). *Handbook of organic Farming and Biofertilizers*. Ambica Book Publication.

Huber, B. (2007). Regulation and Organic Certification. *Disampaikan dalam workshop Biofach*.

IFCAM. (2009). *The World of Organic Agriculture – Statistics & Emerging Trends 2008*. https://www.soel.de/fachtheraaii

IFOAM. (2000). *Basic Standards for Organic Production and Processing*. IFOAM General Assembly. http://www.ifoam.org/

Jimenez, J. (2010). *Social policy and Social change*. Sage.

Kalyan, S. (2005). *Development of sustainable farming system model for the irrigated agro-ecosystem of eastern UP, ICAR, adhoc project. Final annual report, department of agronomy, Institute of agriculture science*. BHU.

Kratochvil, R. (2005). Organic Farming: An approach to make agriculture more sustainable. *Jahrbuch der Österreichischen Gesellschaft für Agrarökonomie*, *12*, 245–260.

Kuepper, G. (2010). *A brief overview of the history and philosophy of organic agriculture*. Kerr Center of Sustainable Agriculture. www.kerrcenter.com

Lampkin, N. H. (1994). *Organic Farming: Sustainable Agriculture in Practice. An International Perspective*. Cab International.

Lindenthal, T., Vogl, C. und Heb, J. (1996). Integrale Schwerpun ktthemen und Methodikkriterien der Forschung im okologischen Landbau [Integral focus topics and methodological criteria of research in oological agriculture]. In *Erstellung eines Strategiepapiers furs die Forschungsforderung. Forderungsdienst*. Wien: BMWFK und BMLF.

Nambiar, K. K. M., Soni, P. N., Vats, M. R., Sehgal, D. K., & Mehta, D. K. (1992). *AICRP on long term fertilizer experiments, Annual Reports 1987-88 and 1988-89*. IARI.

Narayanan, S. (2005). Organic Farming. In *India: Relevance, Problems and Constraints, Occasional, Paper No. 38*. Department of Economic Analysis and Research National Bank for Agriculture and Rural Development.

Niggli, U. (2014). *How will organic farming look like in 20 years and to which challenges will it have to respond to? In TIPI Workshop*. BIOFACH.

Sangakkara, P., Bandaranayake, U., & Gyjanayake, J. N. (2008). Organic matter addition in organic farming over dry season. IFOAM and the Consorzio Modena Bio.

Seabridge, K. (2015). *Organic farming statistics 2014*. Department of Environment, Food and Rular Affairs.

Silwana, T. T., Lucas, E. O., & Olaniyan, A. B. (2007). The effect of inorganic and organic fertilizers on the growth and development of component crop in maize/ bean intercrop in eastern cape of south Africa. *Journal of Food Agriculture and Environment*, *5*(1), 262–272.

Singh, K. N, Sharma, I. P., & Srivastava, V. C., (2001). Effect on FYM fertilizer and plant density on productivity of rice-what sequence. *Journal of Research, Birsa Agriculture University*, *13*(2), 159-162.

Soumya, K. M. (2015). Organic Farming: An effective way to promote sustainable Agriculture Development in India. *Journal of the Humanities and Social Sciences*, *20*(6), 31–36.

Susanto. (2015). Sustainable organic farming for Environmental Health: A social development model. *International Journal of Scientific & Technology Research, 4*(5), 196-211.

Vaarst, M. (2010). Organic Farming as a development strategy: Who are interested and who are not. *Journal of Sustainable Development*, *3*(1), 38–50. doi:10.5539/jsd.v3n1p38

Vats, S., & Sharma, V. (2011). *Organic farming history and techniques*. eVaidya Jiwellness Pvt. Ltd.

Yadav, S. K., Babu, S., Yadav, M. K., Singh, K., Yadav, G. S., & Pal, S. (2013). A Review of Organic Farming for Sustainable Agriculture in Northern India. *International Journal of Agronomy*, 1–8.

Chapter 22
Application of Biofertilizers for Sustainable Agriculture

Raj Singh
Maharishi Markandeshwar University (Deemed), India

Sushil Kumar Upadhyay
Maharishi Markandeshwar University (Deemed), India

Amit Kumar
Keral Verma Subharti College of Science, Swami Vivekanand Subharti University, India

Anju Rani
Subharti College of Science, Swami Vivekanand Subharti University, India

Permod Kumar
Subharti College of Science, Swami Vivekanand Subharti University, India

Gyanika Shukla
CCS University, India

ABSTRACT

The use of microbial inoculants as a source of bio-fertilizers has become a hope in the present time. It is an eco-friendly way to fertilize the crop plants; it can solve the problem of high coat of fertilizers and help in saving the economy of the world, resulting in sustainable agricultural development. These biofertilizers use living microorganisms that establish symbiotic and asymbiotic relationships with the plants or are an inoculation of microorganisms which promotes the plant growth by increasing the primary nutrient supply to the host plant and also retains the soil fertility. This chapter describes the facts of experimental results of the use of biofertilizers on different crops in India and as well as in other countries. Biofertilizers are classified on the basis of their association with plants. Generally nitrogenous and phosphatic fertilizers are used in crops plant for higher yield. The Rhizobium, Azotobacter, BGA, Frankia, Bacillus, and Mycorrhiza fulfil the need of fertilizers as microbial inoculants of N and P.

DOI: 10.4018/978-1-7998-8434-7.ch022

INTRODUCTION

The fertilizers are integral part of agricultural and play a very significant role in agricultural productivity from ancient time. During 20th century, significant development in crop yield was achieved in the world through maximum use of chemical fertilizers (Rajasekaran *et al.*, 2012). But we are already familiar with the controversial issues related with the use of chemical fertilizers in agricultural. The non –judicious use of chemical fertilizers pave way for the pollution of soil (Swapana, 2013). Apart from killing living organisms, they reach even the deeper layers through filling and irrigation of the land (Rajasekaran *et al.,* 2012). On finding entry into water supplies, they disrupt the aquatic ecosystem also (Rao and Puttana 2000).

The total consumption of fertilizers in our country is about 9.2 M. tons and is expected to increase to about 20 M. Tons by turn of the 20th century at present the fertilizer consumption varies form 130,125,60 and 70 kg/ha (NPK) for North, south, East and West respectively making for a national average of approximately 90kg/ha. Some states like Punjab are using more than 160 kg nutrients/ha as against some using less than 10 kg nutrients/ha (Roychowdhury et al., 2014) Nitrogenous fertilizers are expensive and produced in industries by Haber - Bosch process utilizing dwindling resources of energy such as petroleum and coal, causing air pollution also. To overcome this problem, it is necessary to develop an alternative method of supplying nutrients to plants. The use of beneficial microorganism as biofertilizers has shown a huge improvement in crop yield in recent years (Roychowdhury et al., 2014).

Fertilizers of Biological Origin

Indian soils are usually poor in organic matter as well as in nitrogen, therefore economic and environmental point of view, in recent years, use of microbial inoculants as a source of bio-fertilizers has become a hope (Kumar, 2014). It is eco-friendly way to fertilize the crop plants; it can solve the problem of high coat of fertilizers and help in saving the economy of the country (Rao, 1993). According to Subba Rao (1982), the bio-fertilizers are 'nutrient inputs of biological origin for plant growth". The biological origin should be referred as microbiological process synthesizing complex compounds and their further release into the soil around the plant roots which are taken up by plant roots (Nelson, 2004). Generally plant are required two types of fertilizers to fulfils their needs: (i) Nitrogenous fertilizers; (ii) Phosphate fertilizers. The material of biological origin which is commonly used to maintain improved soil fertility may be classified as shown in Fig. 1.

Nitrogenous Biofertilizers

Some prokaryotes, converts the gaseous form of atmospheric nitrogen in to organic through biological nitrogen fixation. The prokaryotes which bring about biological nitrogen fixation are known as diazotoroph and this phenomenon is diazotorophy. These are divided into two groups on the basis of their association as symbiotic and asymbiotic nitrogen fixer:

1. **Symbiotic Nitrogen Fixers:** The nodule forming bacteria e.g. *Rhizobium* of Leguminosae family is the best nitrogen fixing symbionts. These are host specific and establish symbiotic relationship with different plants and may or may not develop special structures at the site of nitrogen fixation

(Mishra *et al.*2012,). The following symbiotic microorganisms play significant role in nitrogen fixation.

2. **Rhizobium spp:** Rhizobia are Gram negative soil bacteria capable of forming root nodules in most of the leguminous plants and some non-leguminous plant. In some cases, however, stem nodules are also produced. *Azorhizobium caulinodans* is one such *Rhizobial spp* that nodulates the stems of *Sesbania rostrata*. The root nodulating rhizobia are of common occurrence. These are divided into several species depending upon legume species; they are able to nodulate most of leguminous plants (Table 1).

Figure 1. Classification of Bio-fertilizers

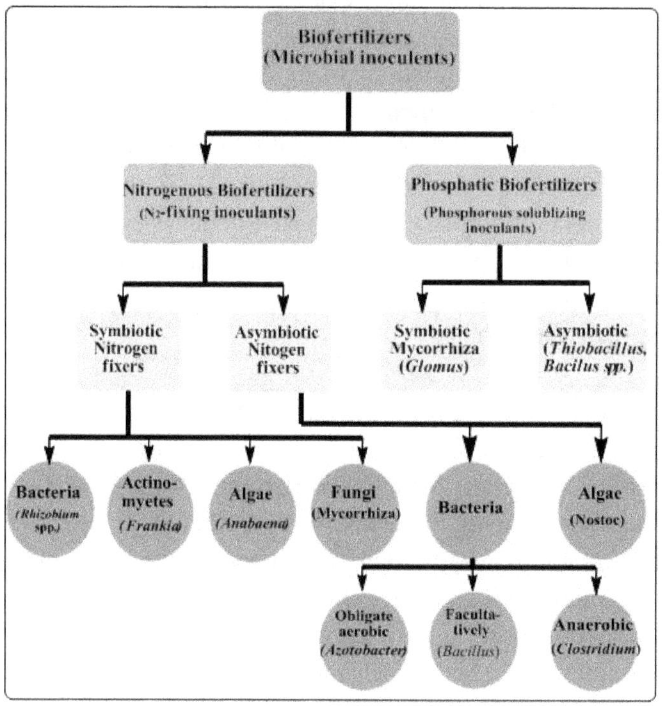

Table 1. List of microorganisms nodulating leguminous plants

Microorganisms able to nodulate	Legume species
R. leguminosarum	Pea (*Pisum sativum*), Khesari (*Lathyrus sativus*) and Kidney bean (*Phaseolus vulgaris*)
R. meliloti	Lucerne (*Medicago sativa*) and Fenugreek (*Trifolium foenumgraecum*)
R. trifoli	Clover (*Trifolium sp.*) and Egyptian clover (*Trifolium alexandrinum*)
R. lupine	Lupines (*Lupinus alba*)

R. lupine is relatively slow growing as compared to the other species of *Rhizobium*. However, the slowest growing rhizobial species is *R. japonicum.*

Rhizobium cells contain genes for nitrogen fixation (*nif* genes) on a megaplasmid. The bacteria enter the roots through root hair, the interaction being highly specific and progressing through several steps, it ultimately results in nodule formation. Many genes of *Rhizobium* as well the host legume are involved in the process. Inside the nodule many bacterial cells change into non-dividing bacteroids, which produce nitrogenase, the enzyme which reduces atmospheric nitrogen into ammonia. Nitrogenase is highly sensitive to oxygen; it is protected from the oxygen by pink pigment leg-haemoglobin which binds to oxygen, produced by the legume and present in the nodules.

The nitrogen fixing ability of legumes inoculated with these Rhizobia ranges from 50 kg to 150 kg per hectare. Following efforts are also being made to increase this ability to fix nitrogen:

1. Efficient rhizobia for various different crops and locations, which are tolerant to various stresses (drought, temperature and pH) are being isolated, maintained and used.
2. A *hup*$^+$ gene (uptake hydrogenase) is used, which helps in recycling of H2 produced during nitrogen fixation. This saves energy required for reduction. Rhizobia strains with *hup*$^+$ gene have been shown to fix more nitrogen.
3. Strains deficient in nitrate fixing ability. Nitrate reductase catalyses the reaction of conversion of NO_3^- to NO_2^-, which inhibit nitrogen fixation.

PREPARATION OF RHIZOBIUM INOCULANTS

In 1895 Nobble and Hiltner prepared a laboratory culture of rhizobia and named "nitragin" after that knowledge of *Rhizobium* –legume symbiosis expanded rapidly devoted for isolation, identification and industrial development. The *Rhizobium* inoculants can be prepared as:

1. **Isolation of Rhizobium:** The healthy root nodules are prepared for isolation of *Rhizobium* inoculums as to obtained contamination free *Rhizobia*. The following steps (Fig. 2) are involved in isolation of *Rhizobium*.
2. **Identification:** Generally *Rhizobium* and *Agrobacterium* grow on YEMA medium. They can be distinguishing with the help of following tests:
    a. **CRYEMA test:** CRYEMA medium is prepared by adding 2.5 ml congo red dye with one liter of YEMA medium. *Rhizobium* utilizers utilize Congo red dye and form white circular and elevated colonies, but *Agrobacterium* display Congo red color.
    b. **Coarbo fuschin test:** The β-polyhydroxy butyrate (PHB) of *Rhizobium* is stain with carbol fuschin dye lactose test.
    c. **Lactose test:** lactose utilizes by *Agrobacterium* by secreting enzyme ketolactose whereas *Rhizobium* can't use lactose.
    d. **Nodulation test:** the surface sterilized legume seeds crown on sand with *Rhizobium* suspension. After 3-4 weeks seedling are taken out carefully. The presence of root nodules prove the selection of right *Rhizobium* strain.
3. **Production of Rhizobial bio-fertilizer:** The pure culture of *Rhizobium* obtained according to fig. 2 served as source of culture for mass production of Rhizobial biofertilizers (Fig. 3)

4. **Inoculation of legume seeds with *Rhizobium*-** The seed inoculation with aqueous suspension of carrier *Rhizobium* for sowing revealed the luxuriant nodulation. The yield of pulse crops can be substantially increased by *Rhizobium* inoculation. The steps of legume seed inoculation with rhizobial culture are given in Fig. No. 4. To protect the rhizobial cell from adverse condition such as dryness, acidity, alkali, excess fertilizers and pecticides, aspecial method of inoculation is adopting called as pelleting.

Figure 2. Isolation of Rhizobium

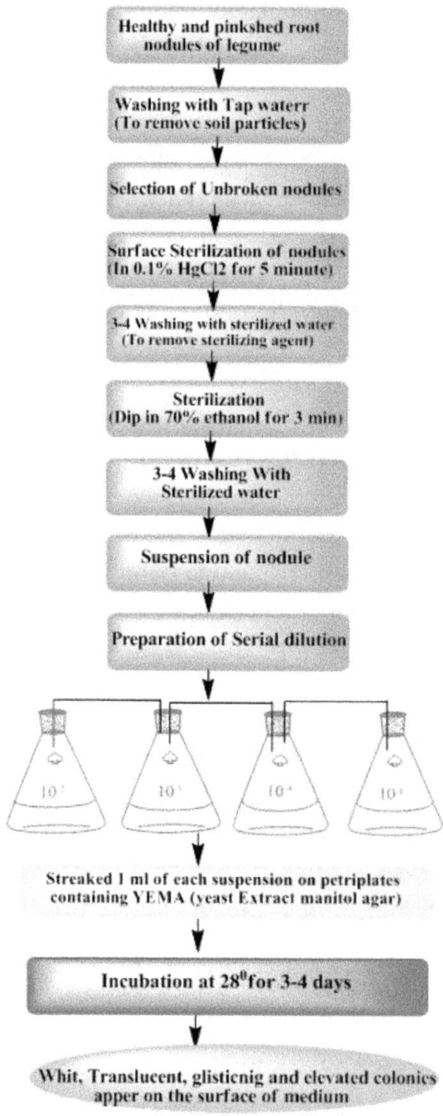

Figure 3. Production of Rhizobial bio-fertilizer

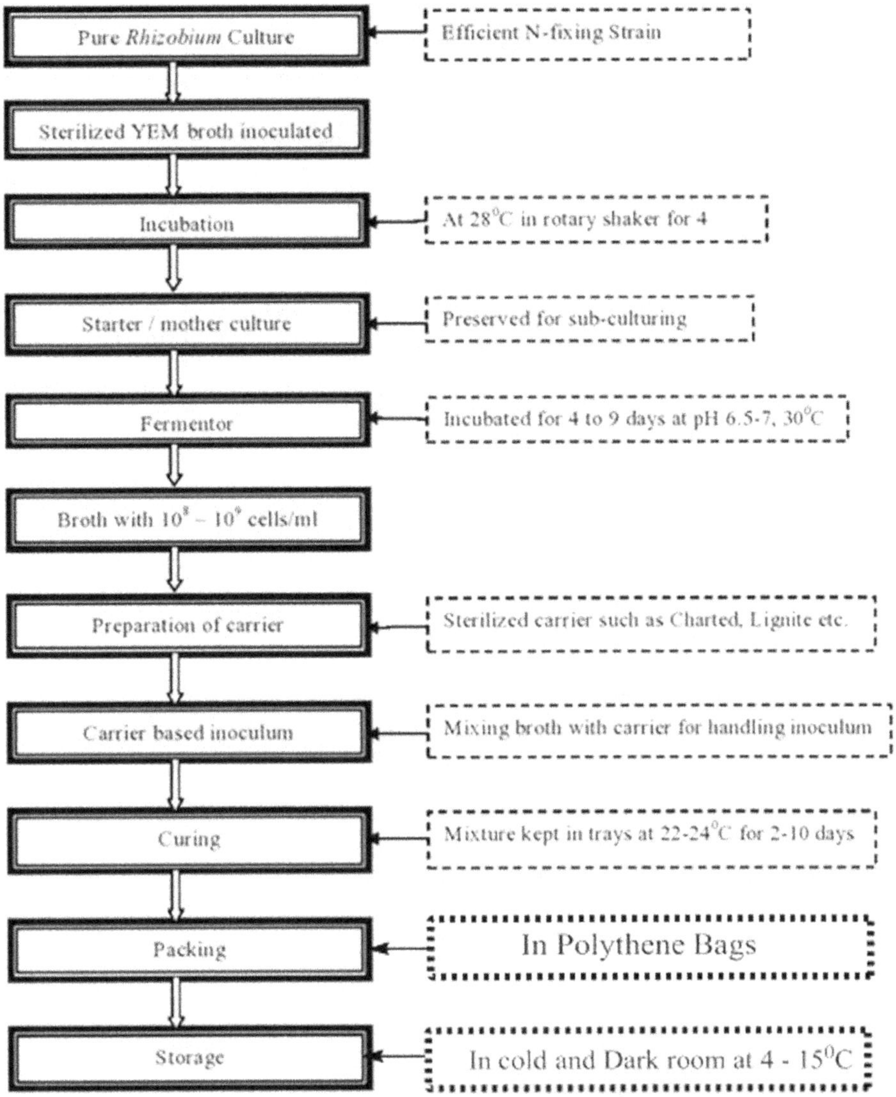

FRANKIA

Frankia discovered by Frank in 1880s so named in honour of Frank. *Frankia* belong to the class actinomycetes and form a symbiotic structure with the roots of non-leguminous plants and called **actinorhiza**. About 24 genera of angiosperm bear actinorhiza root nodules e.g: A*lnus, Hippophae, Gymnostoma, Casurrania, Coriaria etc*. *Frankia* is slow growing microorganism required pH range of 6.7-7.0, BAP medium used to isolate the *Frankia,* Hyphae are thin, cylindrical, hyaline and poorly branched. Spores are produced in sporangia; the spherical vesicles are formed in the tip of hyphae due to swelling of tips. These are the sites of nitrogen fixation like heterocytes in cyanobacteria (Rao Prakasa and Puttana, 2001).

Figure 4. Legume seed inoculation with Rhizobial culture

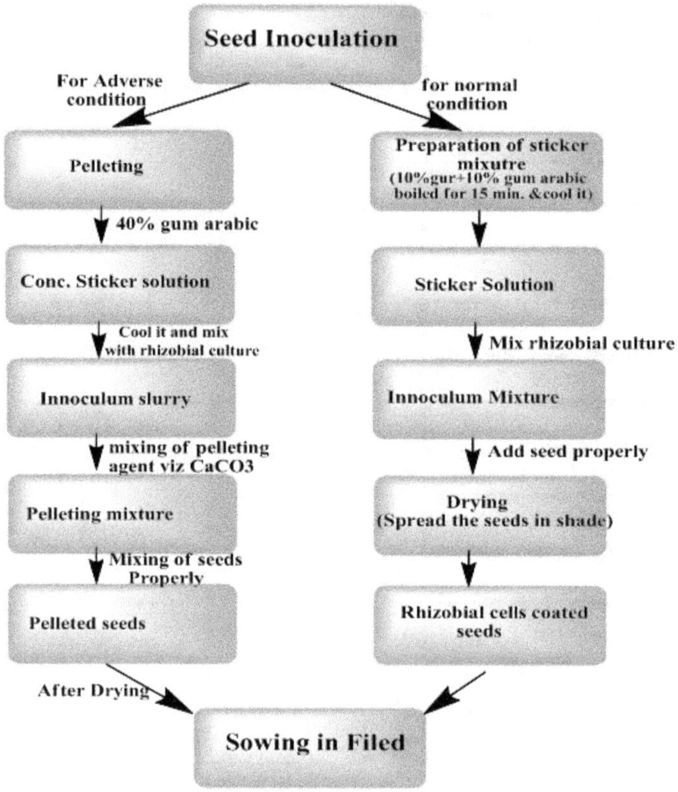

Figure 5. BAP medium

BAP medium

K_2HPO_4	-	0.591 gm
KH_2PO_4	-	0.952 gm
NH_4Cl	-	0.267 gm
$Mg_2SO_4.7H_2o$	-	0.095 gm
$CaCl_2.2H_2O$	-	0.01 gm
FeNa- EDTA	-	0.01 gm
Na-propionate	-	0.48 gm
Trace element soiution	-	1 ml
Vit. Solution	-	1 ml
Distiled water	-	1 litre
Agar	-	16 gm
pH	-	6.7

Isolation of Frankia

Callaham*et al* (1978) for the first time in USA isolated *Frankia* from *Camptonia peregrina* plant. The actinorhizal nodules are large sized; the young nodules are performed to isolate *Frankia*. It can be understand by Fig. 6.

Figure 6. Isolation of Frankia

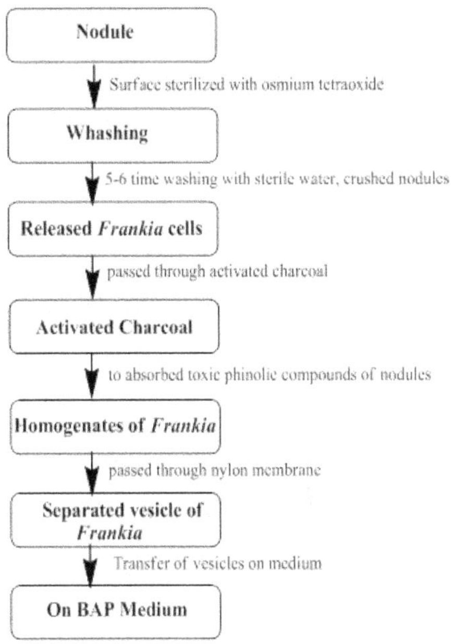

Significance of Frankia

The spores infect the root hair of plant and hyphae penetrate the cortex. The mitotic activities of plant root cell arise increased due to the secretion of growth hormones which results in the development of nodules. The non-leguminous plants also benefited with N- fixation by *Frankia*. It has been observed that the A*lnus* tree increases the N-content of soil about 61.5 to 157 Kg/Ha/annum and *Casuarnia* about 60 Kg/ha/annum.

ALGAE AND AZOLLA

Blue green algae (Cynobacteria) are photosynthetic prokaryotes which fix N_2 symbiotically (e.g: *Anabaena, azollae* with *Azolla* a pteridophytic plant) in BGA N_2 fixation occurs in specialized structures called heterocysts with the help of nitrogenase. The heterocystsact as O_2 proof compartments which protect nitrogenase from O_2 inactivation.

Azolla as Biofertilizers

The BGA is supplied with *Azolla*, an aquatic pteridophytic plant which contains an endophytic BGA, *Anabaena azollae* in its leafcavities, symbyically. The importance of *Azolla* as biofertilizer in rice field was observed byVietman. Now it is very much popular in China, India, Indonesia, Philippines and Bangladesh. Total six species are known *viz, Azolla coroliniana, A. filiculoides, A. maxicana, A. microphylla, A. nilotica, A. rubra* and *A. pinnata*. Out of the *A. pinnata* is more popular as biofertilizer.

Cultivation of Azolla

The mass cultivation of *Azolla* shown in Fig 7.

Figure 7. Cultivation of Azolla

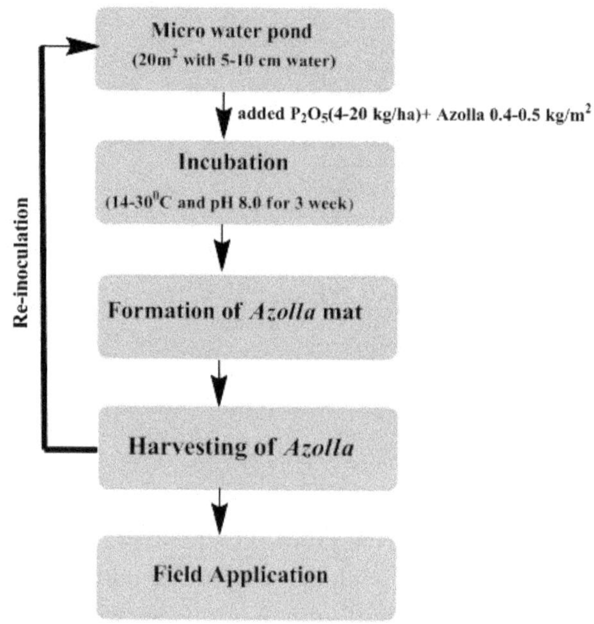

Application of Azolla

The harvested *Azolla* plants dried and used as green manure. Generally, two following methods are used for its application (Tuan, 1979; Singh, 1989).

- Incorporation of *Azolla* in the field prior to rice plantation.
- Transplantation followed by incorporation of *Azolla*.

The report from the IRRI, Manila (Philippines) revealed that incorporation of *Azolla* increased the yield similar to that obtained from 30 kg/ha Urea or ammonium phosphate. The *Azolla* is tolerant against heavy metals *viz.* As, Hg, Pb, Cu, Cd, Cr etc. it absorbs heavy metals into cell walls through generation of specific metal resistant enzymes. Therefore, it can be incorporated in rice field near polluted areas to absorb the harmful metals. Hence, it will be beneficial for organic farming.

ASYMBIOTIC NITROGEN-FIXER

The bacteria and algae (cyanobacteria) play a significant role in nitrogen fixation asymbiotically when they applied to rhizosphere and make it available to crop plant (Kennedy *et al.,* 2004). They also produce growth promoting substance e.g.: antibiotics, hormones and vit. etc; helpful to the plant growth and development. The following microorganisms fix nitrogen asymbiotically:

Bacteria

Generally, *Azotobacter* and *Azospirillum* used as biofertilizers:

1. **Azotobacter:** *Azotobacter* is gram-nigative rod shaped, aerobic free-living bacterium found in the soil and rhizosphere. The most efficient strain of *Azotobacter* fix 30 Kg of N_2 from 100 Kg degrading organic matter. It is recommended as biofertilizer for rice, wheat, millets and cotton *etc. A. chroococcum* is efficient soil and rhizosphere inhabitant (Kumar, 1994). Other species such as *A. beijerincki. A. insignis, A. paspali* and *A.vinelandii* also used as biofertilizer. The following steps are involve in the properties of biofertilizers-
    a. **Isolation of *Azotobacter*:** The A*zotobacter* can be isolated by the soil dilution plating technique on the nitrogen- free medium *viz.* Ashby's medium, Jensen's medium and Burk's medium etc. See Fig 8.

Figure 8. Ashby's medium

Ashby's medium		
Manitols	-	20.0 gm
K_2HPO_4	-	0.2 gm
$Mg_2SO_4.7H_2o$	-	0.2 gm
NaCl	-	0.2 gm
K_2SO_4	-	0.1 gm
$CaCO_3$	-	50 gm
Agar	-	16 g
Distiled water	-	1 litre

Dissolved $MgSO_4.7H_2O$, NaCl, K_2SO_4 and $CaCO_3$ in 200 ml water in 250 ml conical flask. In another flask dissolved K_2HPO_4 in 100 ml of water and mixed all the constituents and make the final volume 1000 ml by addition of remaining distilled water.

b. **Production of Azotobacter biofertilizer:** The mass production of *azotobacter* inoculums can be explain by Fig 9.

Figure 9. Production of Azotobacter biofertilizer

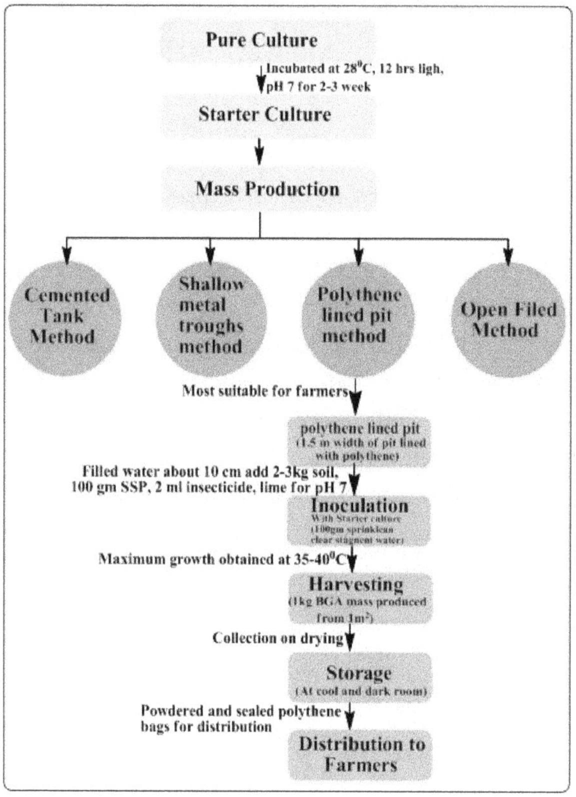

Azospirillum Biofertilizer

For the first time in 1925 Beijerinek discovered the N-fixating bacterium in the roots of digit grass in Brazil and named *Spirillum lipoferum*. Tarrand *et. al.,* (1978) renamed *Spirillum* as *Azospirillum*. It is a gram negative, vibroid bacterium. It grows well and promtes nitrogenase activity at pH 5.6 to 7.2 *Azospirillum* species occurs in association with the roots of many plants of the grass family. e.g. sugarcane, jowar, bajara, and wheat etc (Tien, 1979). It is an associative symbiont because it effectively colonises the roots and infects cortex also with developing any apparent structure on root. *Azospirillum* with fram yard manure (FYM), capable of fixing about 30 Kg N/ha/yr. The field inoculation of crop saving of 15 to 25 Kg N/ha/yr under micro-aerophilic conditions. The *Azospirillum* and *Azotobacter*, besides supplying N to soil, also secrete antibiotics which act as pesticides. There are five species of *Azospirillum viz*

A. lipoferum, A. brasilense, A. amazonens, A. halopraeferns and A. irakense. The first two species are commonly used as biofertilizers. The following steps are involved in the preparation of biofertilizers.

Isolation of Azospirillum

It can be isolated from plant root as well as rhizosphere on Jensen's medium. It can isolate from soil, the steps involved in *Azotobacter* isolation will remain same. If is isolates from root the Fig 10 will follow. The following steps are involved:

Figure 10. Isolation of Azospirillum

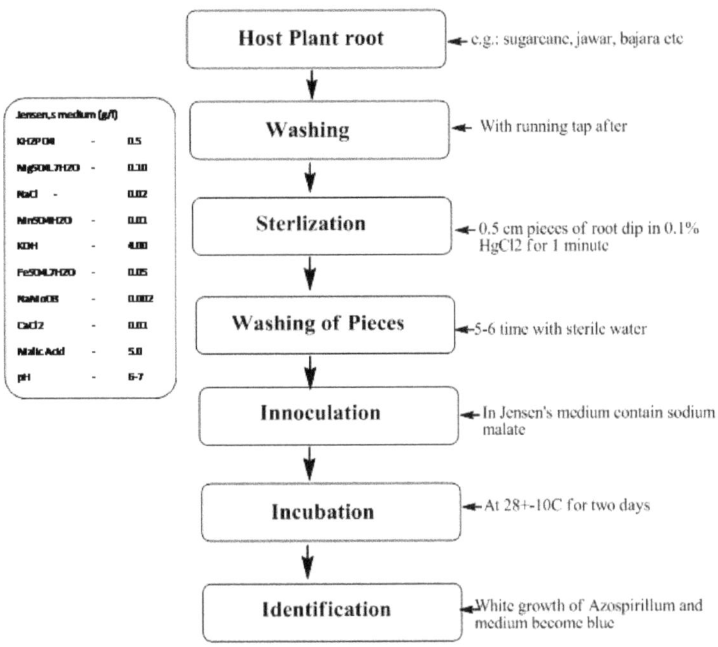

Production of Azospirillum Biofertilizer

The steps involved in Azotobacter isolation will also follow in mass production of *Azospirillum* biofertilizer.

ALGAE AS CYANOBACTERIAL FERTILIZERS

The blue green algae (BGA) also fix N_2 asymbiotically, some cyanobacterial inoculants such as *Nostic, Plectonema, Ocillatoria* and *Tolgpothrix etc*. have been found to be more effective incomposite culture than single culture along with N_2 fixation BGA also provides following advantages:

- Increase soil biomass as humus (Saikia *et al.,* 2011).
- Reclamation of saline soil.

- Provide growth promoting substance eg; plant hormones (IAA, IBA, NAA), vit. Amino acid, antibiotics *etc.*
- Tolerance to pesticides.

Venkataraman in 1961 proposed the term"algalization" to the process of application of blue green algae culture in field as biofertilizer. Venkataraman (1972) developed "algalization technology" in India transferd it to farmer level. The following steps are following to develop BGA biofertilizer-

Isolation of BGA

The BGA are isolated on pringsheims broth or Fogg medium. The BGA colonies collected from flooded rice field or top layer of paddy field is collected from flooded rice field or top layer of paddy field is collected and powered. Generally 12 hrs light is required to culture the BGA, where all steps are more or less similar as mentioned in *Azotobacters* isolation.

Production of BGA Biofertilizers

The pure BGA culture grown in the prescribed medium in conical flask under optimum condition of light and temperature. The culture can be used as starter culture for mass culture of BGA inocula. It can be explained by following Fig 11.

Phosphatic Biofertilizer

Phosphorous is also a macro element next to the nitrogen, play a vital role in the growth and development of microorganism and plants. The most of phosphorous present in soil is not available to plant; it is found in insoluble from as rock phosphate. Therefore plant generally deficit of phosphorus and to full fil the need of phosphorous to plant, chemical fertilizer such as single or triple super phosphate applied to plants. There are so many microorganisms present in soil, convert non available inorganic phosphate present in soil into on (soluble) available from utilizable by crop plants (Oberson *et al.,* 1993). These microorganisms convert insolable phosphorous into soluble phosprausthence called as phosphate solublizing microorganisms (PSM). These includes both bacteria and fungi. On the basis of their association these are also classify into two groups:

1. **Symbiotic phosphate solublizing microorganism:** In this group the fungi occupied a dominant position and play a very significant role in phosphorus nourishment for plants e.g.: mycorrhiza.
2. **Mycorrhiza:** It is a symbiotic relationship between the roots of plants and fungi. The word mycorrhiza derived from Greak word *Mykes* means fungus and *Rhiza* means root *i.e.* fungal root. This word was first used by A. B. Frank in 1885 for fungus-tree association. Mycorrhiza is a specific morphological structure which develops as a result of mutualistic symbiosis between some specific plant roots and fungi, so that the nutrients absorbed from soil by the fungi are released to the roots cells and in turn, the fungus takes its food requirements from the root plant. Plants which suffer from nutrient deficiency specially P or N develop mycorrhiza. Mycorrhizas are of two types on the basis of their position with plant:

a. **Ectomycorrhiza:** It is a form of mycorrhizal association in which there is a well- developed mycelium forming a mental on the outside of the root of the forest trees (Ex: pine, oak, brich and eucalyptus *etc.*). This functionally increases the surface of interface between plant roots and soil. Ectomycorrhizae enhance water and nutrient *viz.* N, P, K and Ca uptake, resulting in greater plant vigor, growth and yield. These absorb and store nitrogen, phosphorus, potassium and calcium in their mantle. These also break down complex organic molecules into simpler, easily available forms; protect the roots from pathogen and produce plant growth promoting substance such as cytokinins.

b. **Endomycorrhiza:** In this type of mycorrhizal association the fungus lives between and within the cells of the cortex of roots of most fruits and horticulture plants (e.g.: coffee, pepper, cardamom and betel *etc.*). Endomycorrhiza are produced both by septate and non-septate fungi. There is a form of endotropic mycorrhizae in which the fungus lives between the cell of the cortex and forms temporary hypal projection that penetrate the cortical cells. These are called Vesicular Arbuscular Mycorrhiza or VAM.

Figure 11. Production of BGA biofertilizers

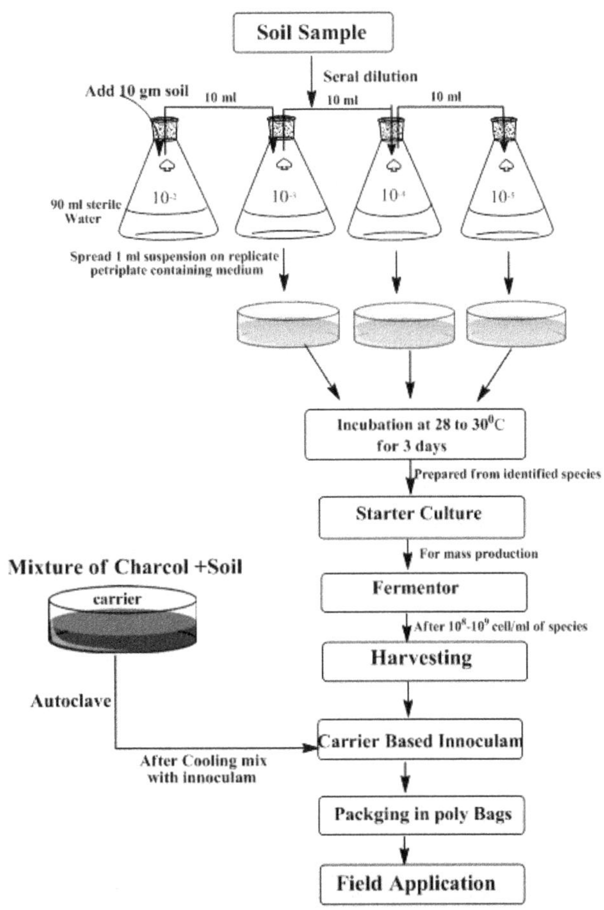

The projections of the VAM hyphae penetrate into cortical cells and may from swollen vesicles or finely branched masses called arbuscales. VAM fungi are two type- sporocarpic (*e.g.* G*lomus* and S*cleocystis*) and non spocarpic (e.g.: *Gigaspora, Scutellospora, Acaulospora* and*Entrophospora*) They particularly help in phosphrous nutrition also produce growth promoting substance and offer resistance against pathogen.

PRODUCTION OF MYCORRHIZA BIOFERTILIZERS

The production of biofertilizers can be explain in the following step:

1. **Isolation of inoculum-** the spore of VAM fungi from soil are isolated by various techniques- floatation method, floatation bubbling methods, sieving method, sucrose centrifugation, airstream fraction etc. out of these sieving and floatation methods are the most commonly used (see in fig 12).
    a. **Sieving method –** in this method soil suspension pass through a series of sieves of 710µm, 250 µm, 75 µm and 45 µm.

Figure 12. Isolation of VAM inoculum

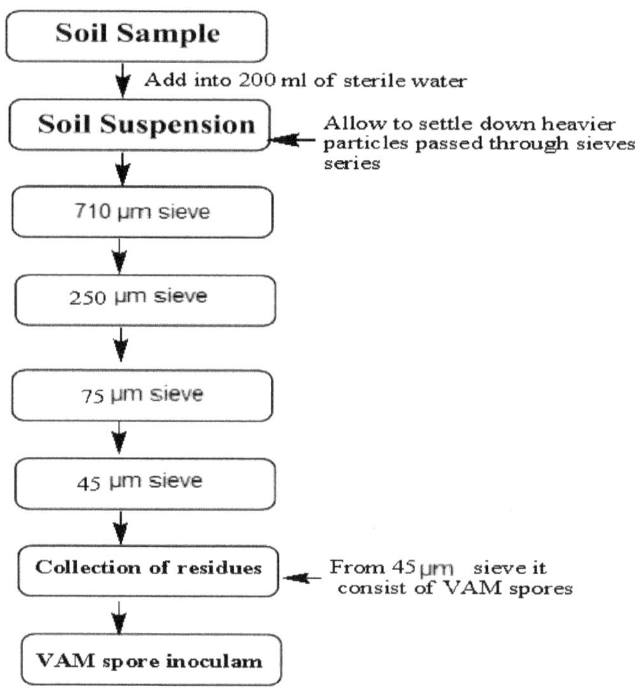

2. **Mass production of mycorrhizal biofertilizers-** Mycorrhizal biofertilizers can be produced by pot culture technique. In this technique host plant, mycorrhizal spore and soil are required. Generally, citrus, sorghum, onion, strawberry and maize host plant are used for large scale production of

biofertilizers. The VAM spore isolated by wet sieving and decantation technique serve as starter inoculums. The production can be explained with help of following fig 13.

Figure 13. Mass production of mycorrhizal biofertilizers

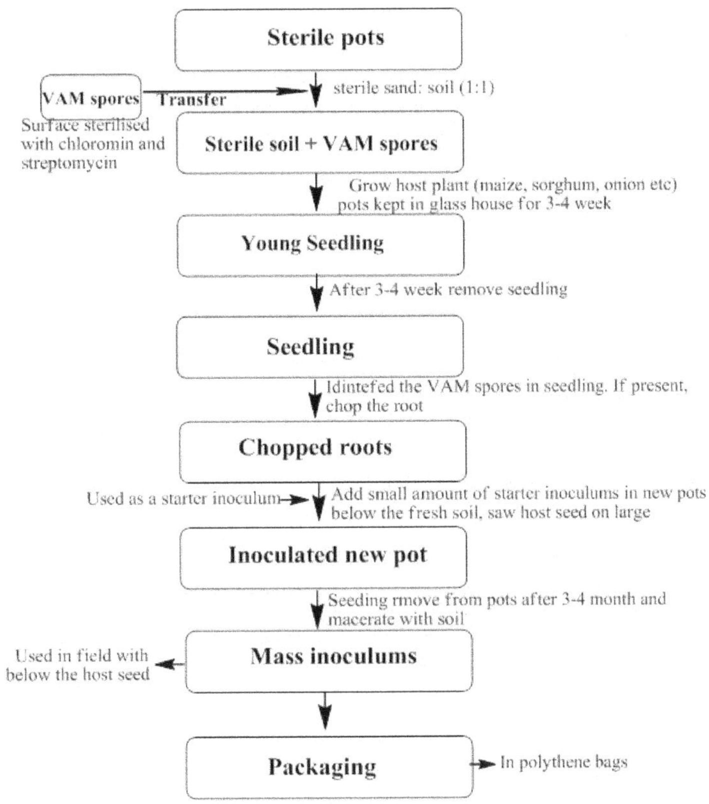

Significance of Mycorrhiza

Following are the same important benefits:

- It increases surface area of feeder root by spreading their mycelia in soil, hence rate of absorption of nutrients and water also increased resulting in enhanced plant growth.
- Their presence reduces the effects of pathogen on plants, hence plant become disease resistance.
- VAM increase the root development in woody plant cutting during vegetative propagation
- VAM increase the selective absorption of N, P, K etc. The available to plants in less amount (Tinker, 1984)
- Pine tree inoculated with VAM grow well because limited roof hairs are present in plant
- VAM reduced plant response to soil stress such as – high salts, heavy metal toxicity, drought, and frost (Cruz *et al.*, 2008).
- VAM produced organic glues which bind soil particles; hence they play a significant role the in-soil conservation and soil fertility.

BACTERIA

There are several species of bacteria serve as phosphate solubilising microorganism (PSM) present in soil *e.g.* species of *Pseudomonas, Bacillus, Flavobacterium* and *Micrococcus etc.* (Oberson et al.,1993). They solubilized rock phosphate, $FePO_4$, $AlPO_4$, apatite and tricalcium phosphate. These secrete organic acids such as acetic acid *viz.* acetic acid, formic acid, lactic acid, propionic acids and succinic acids etc. hence bounded from of phosphate are solubilize and phosphorus molecules (PO_4^{-3}) are available to plants. Production of PSM biofertilizers can be described in following steps.

1. **Isolation:** The Pikovskaya's medium ($MgSO_4.7H_2O$ - 0.10gm/l, KCl – 0.20 gm/l, $(NH_4)_2.SO_4$ – 0.50 gm/l, $MnSO_4$ – 0.002 gm/l, $FeSO_4$ – 0.002 gm/l, $FeSO_4.7H_2O$ – 5.0 gm/l Na_2MoO_4 – 5.0gm/l, Glucose – 10.0 gm/l, yeast extract – 0.5 gm/l, Distilled Water 1000 ml)is prepared with 0.5% gum Arabic, autoclaved and pours in to Petri dishes. Soil sample collected from field from different locations of same field, mixed together. About 1 gm soil sample used to isolate PSM by serial dilution methods.
2. **Production of PSM biofertilizers:** The isolated PSM inoculums is used to produce PSM biofertilizers at large scale.
3. **Significance of PSM:** the seed inoculation or bacterization of PSM is similar as mentioned in *Rhizobium.* The new efficient bacteria (*Pseudomonas striata* and *Bacillus polymyxa*) and fungi (*Aspergillus awamori*) have investigated at IARI, new Delhi. It can be used as PSM this mixture show significant results. These preparations know as "IARI micorphas culture" (Tilak 2016). The PSM solubilise 20 – 30% phosphate which is then absorbed by plant. Present yield increase in crop due to PSM fertilizers varies from 9 to 23% (Motsara and Bisoyi, 2001).

CONCLUSION

For sustainable agriculture development to increase in yield of food crops the use of beneficial microorganisms as biofertilizers has shown a significant improvement in modern era. Therefore the use of chemical fertilizers must be minimize. In this chapter the isolation, production and proper use of biofertilizers has been described. Now the time has come for the farmers to accept this microbial technology. So that farmers would have the opportunity to understand the effects of biofertilizers and are willing to use them. The application of inorganic chemical fertilizers was thus significantly reduced to 30-50%. This helps in the production of organic food as well as reduction of soil and water pollution. Therefore economic and environmental point of view, in recent years, use of microbial inoculants as a source of bio-fertilizers has become a hope for sustainable agriculture.

REFERENCES

Cruz, R., & Husain, T. (2008). Effect of vesicular arbuscular mycorrhiza (VAM) fungi inoculation on coppicing ability and drought resistance of Senna spectabilis. *Pakistan Journal of Botany, 40*(5), 2217–2224.

Kennedy, I. R., Choudhury, A. T. M. A., & Kecskés, M. L. (2004). Non-symbiotic bacterial diazotrophs in crop-farming systems: Can their potential for plant growth promotion be better exploited? *Soil Biology & Biochemistry*, *36*(8), 1229–1244. doi:10.1016/j.soilbio.2004.04.006

Kumar, S. P. (1989). Use of *Azolla* in Asian Agriculture. *Applied Agricultural Research*, *4*, 149–161.

Kumar, T. S. (n.d.). Review on biofertilizers. *IJCRCPS, 1*, 17-23.

Kumar, V. (1994). Nitrogen economy in Indian mustard through use of *Azotobacterchroococcum*. *Crop Research*, *8*, 449–452.

Mishra, D., Rajvir, S., Mishra, U., & Kumar, S. S. (2013). Role of bio-fertilizer in organic agriculture: A review. *Research Journal of Recent Sciences ISSN*, *2277*, 2502.

Motsara & Bisoyi. (2001). Crop demonstration on biofertilizer. National Biofertilizer Development Centre.

Nelson, M. L. (2004). Plant growth promoting rhizobacteria (PGPR): Prospects for new inoculants. *Crop Management*, *2*(1), 301–305. doi:10.1094/CM-2004-0301-05-RV

Oberson, A., Fardeau, J. C., Besson, J. M., & Sticher, H. (1993). Soil phosphorus dynamics in cropping systems managed according to conventional and biological agricultural methods. *Biology and Fertility of Soils*, *16*(2), 111–117. doi:10.1007/BF00369411

Rajasekaran, S., Ganesh Shankar, K., Jayakumar, K., Rajesh, M., Bhaaskaran, C., & Sundaramoorthy, P. (2012). Biofertilizers current status of Indian agriculture. *J. Environ. Bioenergy*, *4*(3), 176.

Rao, E. P., & Puttana, K. (2001). Nitrate, agriculture and environment. *Current Science*, (9), 1163–1168.

Rao, S. (1982). Biofertilizer. In N. S. Subba Rao (Ed.), *Advances in Agricultural Microbiology* (pp. 219–242). Oxford & IBH Pub. Co.

Rao, S. (1993). *Biofertilizers in Agriculture*. Oxford and IBH Publishing Co.

Roychowdhury, D., Paul, M., & Banerjee, S. K. (2014). A review on the effects of biofertilizers and biopesticides on rice and tea cultivation and productivity. *International Journal of Science, Engineering and Technology*, *2*(8), 96–105.

Saikia, D. N., Sarma, J., Dutta, P. K., & Baruah, D. K. (2011). Sustainable productivity of tea through conservation of bio-mass, addition of bio-fertilizers and reduction of inorganic fertilizer. *Two and a Bud, 58*, 109-117.

Swapana, L. A. (2013). Development of biofertilizers and its Future Perspective. *Journal of Pharmaceutics*, (4), 327–332.

Tien, T., Gaskins, M., & Hubbel, D. (1979). Plant growth substances produced by *Azospirillum brasilese* and their effect on growth of pearl millet (*Pennisetum americanum* L.). *Applied and Environmental Microbiology*, *37*(5), 1016–1024. doi:10.1128/aem.37.5.1016-1024.1979 PMID:16345372

Tilak, K. V. B. R., & Manoharachary, C. (2016). Eco-friendly Plant Growth Promoting Rhizobacteria for Crop Improvement. *Microbial Inoculants in Sustainable Agricultural Productivity*, 297-309.

Tuan, T., & Thuiet, R. (1979). *Use of Azolla in rice production in Vietnam. Nitrogen and Rice*. IRRI.

Chapter 23
Eco-Toxicology:
Duckweed Growth Inhibition Toxicity Test

Vinay Panwar
Swami Vivekanand Subharti University, India

Laxman Singh Gautam
SSV PG College, Hapur, India

ABSTRACT

The duckweed test measures the inhibition of growth of the freshwater macrophyte, Lemma sp. This test is based on the OECD Test Guideline 221 and involves exposing Lemna sp. to contaminated water for 2-d. Two toxicity test endpoints are generally measured in this toxicity test: inhibition in growth rate and changes in biomass. The aquatic plant Lemna sp. belongs to the family Lemnacae, which has a number of worldwide species and is commonly referred to as duckweed. Lemna glbba and Lemna minor are species representative of temperature areas and are the most commonly used species in toxicity tests. Both species have a floating or submerged disc old stem (frond), and a very thin root emanates from the centre of the lower surface of each frond. Lemna sp reproduces by vegetative producing new fronds. Young plants have shorter roots and consist of two to three fronds of different sizes. The small size of Lemna, its simple structure, asexual reproduction, and generation time makes plants of this genus very suitable for laboratory testing.

INTRODUCTION

The toxicity test measure the inhibition in growth (and biomass) of *Lemna* sp. after exposure to a industrial water (Table 1). Huebert, D.B. and Shay J.M. (1993) five concentrations of the contaminant or five dilutions of contaminated water are prepared and dispensed into test vessels each concentration prepared in quadruplicate. *Lemna* sp. (7-10 days old) is then added to each test vessel and the number of fronds counted at least three times over 7 days. After 7 days, the specific growth rate is calculated in each replicate. The pH, conductivity and oxygen are also measured in each treatment at the start and end of the toxicity test. (USEPA., 1996).

DOI: 10.4018/978-1-7998-8434-7.ch023

For quality control and quality assurance purposes, a negative control and positive control is included in each toxicity test. The growth of *Lemna* sp, in all other treatment is compared to this control. A positive control is also included in the toxicity test to ensure *Lemna* sp. is responding to a known contaminant in a reproducible way.

For the practical session in this workshop, the toxicity of a sample (e.g. river water) and the reference toxicant KCL will be measured. Due to the limited time available in the workshop, the exposure duration will be 4 days (Monday to Thursday) and only the growth rate will be measured using frond counts at the start and end of the toxicity test. Also, each control and treatment will be tested in triplicate, not quadruplicate as recommended in the standard protocol, information on how to culture and carry out toxicity tests with *Lemna* sp. is provided in the Organization for Economic Co-operation and Development (OECD) 2006.

Table 1. Summary of the test protocol for the Lemna sp. growth inhibition bioassay. (Environment Canada., 1993, 1999). (Lockhart W. L., Billeck B. N., and Baron C. L., 1989)

Test type	Static or semi-static renewal
Temperature	24±2°C
Light quality	Cool White Fluorescent Lighting
Light intensity	2000 **µmol** m^{-2} **s**$^{-1}$
Photoperiod	Continuous light
Test chamber size	20 mL
Test solution volume	10 mL
Renewal of test solution	If required, twice over the 7-day period
Age of test organisms	7-10days
Initial frond density in test	One colony of three fronds each or two colonies chambers of two fronds each, but all treatments must have the same number of fronds and colonies
No. of replicates	5 chambers/concentration
Dilution water	SIS media
Sample concentration	Minimum of 5 and a control
Dilution factor	2 or 3
Test duration	7 days
Endpoint	Growth rate (frond division rate and doublings over 7 days); biomass (dry weight)
Test Acceptability	Number of fronds in controls ≥7 doublings over7 days; Reference toxicant KCI within Cusum Limits (5±2 g/L)

PRACTICAL SESSION OUTLINE

- Ground A B C Sample 1
- Grounds C D E Sample 2

Monday: 1 Set up toxicity text
Thursday: 1 Complete toxicity test and data analysis

Health and Safety

- Wear Laboratory coats
- Wear closed footwear
- Wear appropriate disposable gloves
- Wear safety glasses where appropriate
- Read relevant Stately Data Sheets [SDS]

Monday, Day 0: Set up Toxicity Test (ASTM., 2000)

Equipment Provided Per Group

- Flask of *Lemna* sp.
- Swedish Standard (SIS) *Lemna* growth medium: media (1L)
- SIS media stock solutions (7 stock solution: 3 mL each)
- Reference toxicant, KCI 500 g/L (10 mL)
- Sample (250 mL)
- Test vessels 20 mL glass vlals with lids (55)
- Inoculums loops (6-12)
- Petri dish, large (2)
- Pipettes 10 mL, 1 mL, 200µt and pipette tips
- Beakers for preparing sample dilutions (200 or 250 mL×5)
- Containers for preparing KCI test solutions and control (100 mL×6)
- Glass stirring rod (or extra Inoculums loops)
- Measuring cylinders (2×100mL)
- Containers for measuring physio-chemical of test solutions and original sample (1L)
- Markers
- Waste beaker
- Wash bottles with Milli-Q water
- Toxicity test data sheet
- Random number sheet
- SDS, potassium chloride

Access to

- pH meter (and calibration solution)
- conductivity meter
- Oxy meter

Procedure (Walbridge C. T.,1977).

1. Prepare test solution, Control test solution (SIS media)
    a. Label one measuring cylinder 'media' and label one 100-mL, container 'Control'.

b. Using the ‘media’ measuring cylinder, dispense 100ml of SIS media into the ‘Control’ container.

Reference Toxicant Solutions (KCL). (SSI., 1995)

- Label five 100 mL containers: 1.3 g/L, 5 g/L, 10 g/L and 20 g/L.
- Prepare KCL concentrations using SIS media and the 500g/L KCL stock, using SIS media/

A – Dispense 10 mL of SIS media into each of the five labeled containers then,
A – 1.3 g/L = add 0.25 mL of KCL stock solution of container with media.
B – 2.5 g/L= add 0.5 mL of KCL stock solution to container with media
C – 5.0 g/L = add 1.0 mL KCL stock solution to container with media
D – 10 g/L = add 2.0 mL KCL stock solution to container with media
E – 20 g/L = add 4.0 mL KCL stock solution to container with media

Preparation of Sample Test Solution

1. Label the other measuring cylinder ‘sample’.
2. Dispense 200 mL, of sample into a 200-250 mL container using the ‘sample’ measuring cylinder.
3. Add the seven SIS stock solutions to the sample and mix thoroughly.
    a. Stock 1:2 mL
    b. Stock 2:1 mL
    c. Stock 3:1 mL
    d. Stock 4:1 mL
    e. Stock 5:200µL
    f. Stock 6:1 mL
    g. Stock 7:20 µL
4. Label five 200-250 mL containers 6.3%, 13%, 25%, 50% and 100%
5. 100% Sample = using the ‘sample measuring cylinder, dispense 100 mL of sample (+nutrients) into the ‘100%’ container.
6. 50% Sample= using the ‘sample’ measuring cylinder, dispense 100 mL of the sample (+nutrients) into the ‘50%’ container. Then add 100mL of SIS media using the ‘media’ labeled measuring cylinder. Mix thoroughly.
7. 25% Sample = Using the ‘sample’ measuring cylinder, dispense 100 mL of the 50% sample (+nutrients) into the ‘25%’ container. Then add 100mL of SIS media using the ‘media’ labeled measuring cylinder. Mix thoroughly.
8. 13% Sample = Using the ‘sample’ measuring cylinder, dispense 100 mL of the 25% sample (+nutrients) into the ‘13%’ container. Then add 100mL of SIS media using the ‘media’ labeled measuring cylinder. Mix thoroughly.
9. 6.3% Sample = Using the ‘sample’ measuring cylinder, dispense 100 mL of the 13% sample (+nutrients) into the ‘63%’ container. Then add 100mL of SIS media using the ‘media’ labeled measuring cylinder. Mix thoroughly.

Measure Physio-Chemical Characteristics of Test Solutions. (ISO CD 2009).

Transfer 20-30 mL of each test solution (control, reference toxicant and sample) into a labeled container that is suitable for physio-chemical measurements. Also measure the physio-chemical parameters of the original sample (without nutrients). Measure the pH, conductivity and dissolved oxygen in each treatment and record the values on the data sheet. Follow the instructions for each meter and rinse the probe between each sample type.

Label Test Vials

Label each test vessel (e.g. 20-mL glass vials) with your Letter (A,B, C,D,E or F) sequential number (1 to 55), treatment abbreviation and replicate number (1,2,3,6) An example for Group A is given below .(AFNOR., 1996).

Three replicates for control and each treatment are prepared. Replicates 4 and 5 will be combined on the final day of the toxicity test for measurement of pH, conductivity and dissolved oxygen. (Barnes M, Correll R, Stevens DP. 2003, Dunnett, C.W. 1955,1964).

Table 2. Five replicates for control and each treatment prepared

Treatment	Treatment Abbreviation	Label				
		Replicate 1	Replicate 2	Replicate 3	Replicate 4	Replicate 5
Control	C	A1-C-1	A2-C-2	A3-C-3	A4-C-4	A5-C-5
Reference Toxicant, KCL						
1.3 g/L	1.3	A6-1.3-1	A7.1.3-2	A8-1.3-3	A9-.3-4	A10-1.3-5
2.5 g/L	2.5	A11-2.5-1	A12-2.5-2	A13-2.5-3	A14-2.5-4	A15-2.5-5
5.0 g/L	5	A16-5-1	A17-5-2	A18-5-3	A19-5-4	A20-5-5
10 g/L	10	A21-10-1	A22.10-2	A23-10-3	A24-10-4	A25-10-5
20 g/L	20	A26-20-1	A27-20-2	A28-20-3	A29-20-4	A30-20-5
Sample						
6.3%	S6.3	A31-S6.3-1	A32-S6.3-2	A33-S6.3-3	A34-S6.3-4	A35-S6.3-5
13%	S13	A36-S13-1	A37.S13-2	A38-S13-3	A39-S13-4	A40-S13-5
25%	S25	A41-S25-1	A42=S26-2	A43-S26-3	A44-S27-4	A45-S26-5
50%	S50	A46-S50-1	A47-S50-2	A48-S50-3	A49-S50-4	A50-S50-5
100%	S100	A51-S100-1	A5-S100-2	A52-S100-3	A53-S100-4	A54-S100-5

Dispense Test Solutions into Test Vessels

- Pipette 10 mL of the Control test solution into each of the five Control test vessels
- Pipette 10 mL of the 1.3 g/L KCL solution into each of the five 1.25 g/L test vessels
- Pipette 10 mL of the 2.5 g/L KCL solution into each of the five 2.5 g/L test vessels

- Pipette 10 mL of the 5 g/L KCL solution into each of the five 5 g/L test vessels
- Pipette 10 mL of the 10 g/L KCL solution into each of the five 10 g/L test vessels
- Pipette 10 mL of the 20 g/L KCL solution into each of the five 20 g/L test vessels
- Pipette 10 mL of the 6.3% Sample solution into each of the five 6.3% Sample test vessels
- Pipette 10 mL of the 13% Sample solution into each of the five 13% Sample test vessels
- Pipette 10 mL of the 25% Sample solution into each of the five 25% Sample test vessels
- Pipette 10 mL of the 50% Sample solution into each of the five 50% Sample test vessels
- Pipette 10 mL of the 100% Sample solution into each of the five 100% Sample test vessels

Inoculate Lemna sp. into Test Vessels

Gently pour the duckweed into a large Petri dish
Pour a small amount of SIS media into another Petri dish (approximately half fill the Petri dish)

Select healthy looking duckweed colonies and transfer them into the second Petri dish using an inoculation loop. Healthy colonies should have three fronds and have a consistent green colour. Your will need to collect 55 colonies for use in the toxicity test. (Note: two plants of 2-3 frond colonies may be added to improve the growth rate over the short test duration; check with your demonstrator.)

Transfer one colony from the second Petri dish into each control vial. Place the lid on the vial ensuring that the lid is loose to allow gas exchange, Repeat this process for each treatment. To avoid cross contamination of the inoculums loop and test solution follow the order; (Williams, D.A. 1971, 1972)

Control, 1.3 g/L, 2.5 g/L, 5 g/L, 10 g/L, 20 g/L. With a new inoculums loop; 6.3%, 13%, 25%, 50%, 100%.

Incubate Test

The toxicity test is incubated in a growth chamber at 24±2^{0}C and a light intensity of 60-80μmol photons s^{-1} m^{-2} with continuous illumination [24h light].

Place the test vessels in the incubation cabinet in the order listed on the random number sheet.

Thursday, Day 3: Completion to Toxicity Test and Data Analysis

For a 7-d duckweed toxicity test, (ANOVA is used to calculate the mean average specific growth rate). For the practical session, you will be counting the number of fronds on the final day of the four-day test only. (Norberg-King T.J. 1988)

Equipment Provided per Group

- Containers for measuring physio-chemical of test solutions (11)
- Waste beaker
- Wash bottles with Milli-Q water

Access to:

- pH meter (and calibration solutions)
- conductivity meter
- dissolved oxygen meter

Also:

- Your group's toxicity test vessels
- Your data sheets

Procedure

Count fronds

Count the number of fronds in each test vessel and record the number on the test data sheet. All fronds, regardless size and shape are counted, including and visible protruding new buds. Record the health observation of the fronds using the abbreviations below.

C = Chlorosis (loss of pigment, loss of green colour)
N = Niecrosis (dead tissue seen as small white on brown spots)
Y = Yellow coloration
A = Abnormal sized fronds
G = Gibbosity (humped or swollen appearance of fronds)
D = Destruction of coloby (single fronds separated from colonies)
R = Root destruction
L = Less of buoyancy (unable to float on the surface of the solution)

Measure Physio-Chemical Characteristics of Test Solutions

Combine the test solutions in replicates 4 and 5 for each treatment into a vial for physic-chemical measurements.

Measure the pH, conductivity and dissolved oxygen in each treatment

Data Analysis

Toxicity test endpoint-Growth rate: Growth rate is calculated as specific growth rate (μ), that is, the measure of the rate of change in biomass relative to the initial biomass.

The average specific growth rate (μ) of duckweed is calculated for each control and treatment replicate for each time-point using the equation:

$$\mu i - j = \frac{In\left(Nj\right) - In(Ni)}{t}$$

Where:

μi-j: average specific growth rate from time i to j
Ni: number of fronds on the replicate at time i
Nj: number of fronds in the replicate at time j
t: time period from i to j (e.g. 0-48h, 48-96h, 96-144h etc)

For each replicate, the μ, values are averaged to obtain a mean specific growth rate, μ, for each replicate.

In this experiment, the number of fronds at the beginning (0h) and end (72h) of the test was counted, therefore, only the specific growth rate from t=0-48 h will be calculated.

A spreadsheet has been provided with the formulas already entered. The specific growth rate (μ) and yield will automatically be calculated.

The inhibition in specific growth rate as a percentage is calculated using the following formula;

$$\%Ir = \frac{\left(\mu c - \mu T\right)}{\mu c} \times 100$$

Where:

%Ir = percent inhibition in average specific growth rate
μc = mean value for μ in the control
μT = mean value in the treatment group

In addition to growth rate, the yield (change in biomass) can also be calculated. The yield is sim the number of fronds on the initial day subtracted from the number of fronds day, the total growth of duckweed as doublings over X days is the final number of fronds divided by the initial number of fronds. (Van Ewijk, P.H. and J.A. Hoekstra. 1993).

Extrapolating Toxicity Endpoints

The mean specific growth rate for each treatment is used to calculated toxicity endpoints such as values.

For both the reference toxicant and sample, the toxicity values can be extrapolated using a graphical or statistical method.

1. Graphical method: In the graphical method the mean growth rate of duckweed in each treatment is plotted on the y-axis and the nominal (i.e. theoretical) concentration is plotted on the x-axis. The concentration is theoretical since the concentration has not been measured.
    a. Using either graph paper or a graphical software package such as Excel or Sigma plot mean μ value for each treatment on the y axis and the nominal concentration on the x axis.
    b. Draw a line joining up all the data points in the graph – this line is called the concentration response line. Then draw a horizontal line from the 50% point on the y-ais across until it reaches the concentration response line and at the point draw a line vertically down until reaches the x-axis. This is the concentration of KCI or sample that causes a 50% inhibition

duckweed growth – the EC50. The same process can be done to calculate alternative ECx values. (Brain P. and Cousens R. 1989).

2. Statistical method. (Sokal, R.R. and F.J. Rohlf. 1981), (Miller, R.G., Jr. 1986), (Winer, B.J. 1971). The statistic method fits a particular statistical distribution (i.e. the log-logistic distribution) to the toxicity data using a best fit method and then calculates the concentration that causes the selected percentage effect. Statistical program has been included on the workshop CD. (Bruce R.D. and Versteeg D.J. 1992).

Table 3. Random number sheet

Group X	Group B	Group C	Group D	Group E	Group F
49	11	14	48	13	21
19	50	10	8	9	8
30	53	38	50	51	31
29	33	27			
3	9	12			
2	2	55			
23	28	26			
6	19	47			
9	13	5			
55	16	42			
15	1	43			
42	39	15			
1	20	28			
37	49	48			
45	37	45			
24	29	38			
31	25	4			
47	31	34			
22	52	40			
35	44	2			
34	41	14			
48	18	24			
46	7	20			
17	30	16			
26	53	4			
32	43	19			
5	34	37			
7	26	41			
14	54	30			

continues on following page

Table 3. Continued

Group X	Group B	Group C	Group D	Group E	Group F
11	15	11			
40	45	52			
39	3	39			
25	32	46			
4	4	18			
38	6	3			
10	5	33			
12	40	17			
52	12	29			
8	42	53			
51	55	54			
20	21	50			
27	51	23			
43	11	25			
21	36	10			
18	47	49			
16	27	22			
28	23	7			
54	46	1			
36	24	36			
41	22	32			
13	35	35			
44	17	6			
32	55	14			
44	24	42			
53	20	45			
36	19	10			
50	27	6			
31	6	5			
35	12	39			
49	9	51			
3	34	15			
30	4	54			
37	1	38			
25	47	35			
15	18	55			
55	11	37			
42	51	49			

continues on following page

Table 3. Continued

Group X	Group B	Group C	Group D	Group E	Group F
18	17	21			
23	36	46			
28	2	47			
24	31	13			
12	41	29			
2	26	8			
1	53	43			
38	3	11			
41	21	19			
5	22	4			
10	33	41			
21	44	32			
9	45	1			
22	46	44			
27	29	33			
17	25	25			
47	33	22			
6	8	36			
34	42	16			
11	37	12			
4	30	24			
51	43	17			
45	7	26			
54	28	30			
43	16	3			
16	40	53			
26	50	40			
39	13	50			
7	35	28			
40	15	31			
33	10	48			
29	32	34			
19	23	7			
13	54	20			
8	5	23			
14	48	9			
20	52	52			
46	49	27			
52	14	18			

Table 4a. Lemna sp. Growth rate test

Group Name	-------------------------
Sample	-------------------------
Test start date and time	-------------------------
Test end date and time	-------------------------

Table 4b.

Test Treatment	Day				Test end day	
	1^{0}C	pH	Cond	Do	PC/HA	PH/HA
Original sample						
Control (SIS media)						
13 g/L KCI						
25 g/L KCI						
5 g/L KCI						
10 g/L KCI						
20 g/L KCI						
53% Sample						
13% Sample						
25% Sample						
50% Sample						
100% Sample						

Table 4c.

Sample		No. of Fronds Day 0 Day 3			Health observations (C,N,A,C) Day 0 Day 3	
Vial #	Treatment					
1						
2						
3						
4						
5						
6						
7						
8						
9						
10						
11						
12						
13						

continues on following page

Table 4c. Continued

Sample		No. of Fronds Day 0 Day 3			Health observations (C,N,A,C) Day 0 Day 3	
Vial #	Treatment					
14						
15						
16						
17						
19						
20						
21						
22						
23						
24						
25						
26						
27						
28						
29						
30						

Table 4 (d).

Sample		No. of Fronds Day 0 Day 3		Health observation (C,N,A,C) Day 0 Day 3	
Vial #	Treatment				

continues on following page

Table 4d. Continued

Sample		No. of Fronds Day 0 Day 3		Health observation (C,N,A,C) Day 0 Day 3	
Vial #	Treatment				

C= Chlorosis (loss of pigment, loss of green colour)
N = Necrosis (dead tissue seen as small white or brown spots)
Y = Yellow colouration
A = Abnormal size fronds
G = Gibbosity (humped or swollen appearance of fronds)
D = Destruction of coloby (single fronds separated from colonies)
R= Root destruction
L= Loss of buoyancy (unable to float on the surface of the solution)

REFERENCES

Barnes, M., Correll, R., & Stevens, D. P. 2003. A Simple spreadsheet estimating low-effect concentrations and associated logistic dose response curves. *Abstracts SETAC Asia/pacific Australasian Society for Ecotoxicology Meeting: Solutions to Pollution, Christchurch, New Zealand, September 28-October 1*, 156.

ASTM - American Society for Testing and Materials. (2000). Standard Guide for Conducting Static Toxicity Test With Lemna gibba G3. E 1415-91 (Reapproved 1998). In Annual Book of ASTM Standards, Vol. 11.05 Biological Effects and Environmental Fate; Biotechnology; Pesticides. ASTM.

USEPA - United States Environmental Protection Agency. (1996). OPPTS 850.4400 Aquatic Plant Toxicity Test Using Lemna spp., "Public draft". EPA 712-C-96-156.

AFNOR - Association Française de Normalisation. (1996). XP T 90-337: Détermination de l'inhibition de la croissance de Lemna minor. Author.

SSI - Swedish Standards Institute. (1995). Water quality - Determination of growth inhibition (7- d) Lemna minor, duckweed. SS 02 82 13. Author.

Environment Canada. (1999). Biological Test Method: Test for Measuring the Inhibition of Growth Using the Freshwater Macrophyte, Lemna minor. EPS 1/RM/37. Author.

Environment Canada. (1993). *Proposed Guidelines for Registration of Chemical Pesticides: NonTarget Plant Testing and Evaluation.* Canadian Wildlife Service, Technical Report Series No. 145.

Sims, Whitehouse, & Lacey. (1999). *The OECD Lemna Growth Inhibition Test. Development and Ring-testing of draft OECD Test Guideline*. R&D Technical Report EMA 003. WRc plc - Environment Agency.

OECD. (2000). *Guidance Document on Aquatic Toxicity Testing of Difficult Substances and Mixtures.* OECD Environmental Health and Safety Publications, Series on Testing and Assessment No.23.

ISO CD 2009. Water Quality - Determination of the Toxic Effect of Water Constituents and Waste Water to Duckweed (Lemna minor) - Duckweed Growth Inhibition Test.

Walbridge, C. T. (1977). *A flow-through testing procedure with duckweed (Lemna minor L.).* US EPA Report No. EPA600/3-77 108.

Lockhart, W. L., Billeck, B. N., & Baron, C. L. (1989). Bioassays with a floating plant (Lemna minor) for effects of sprayed and dissolved glyphosate. *Hydrobiologia, 118/119*(1), 353–359. doi:10.1007/BF00027800

Huebert, D. B., & Shay, J. M. (1993). Considerations in the assessment of toxicity using duckweeds. *Environmental Toxicology and Chemistry, 12*(3), 481–483. doi:10.1002/etc.5620120309

Dunnett, C. W. (1955). A multiple comparisons procedure for comparing several treatments with a control. *Journal of the American Statistical Association, 50*(272), 1096–1121. doi:10.1080/01621459.1955.10501294

Dunnett, C. W. (1964). New tables for multiple comparisons with a control. *Biometrics, 20*(3), 482–491. doi:10.2307/2528490

Williams, D. A. (1971). A test for differences between treatment means when several dose levels are compared with a zero dose control. *Biometrics, 27*(1), 103–117. doi:10.2307/2528930 PMID:5547548

Williams, D. A. (1972). The comparison of several dose levels with a zero dose control. *Biometrics, 28*(2), 510–531. doi:10.2307/2556164 PMID:5037867

Sokal & Rohlf. (1981). *Biometry. The principles and practice of statistics in biological research* (2nd ed.). W.H. Freeman and Company.

Miller, R. G. Jr. (1986). *Beyond ANOVA, basics of applied statistics*. John Wiley & Sons.

Winer, B. J. (1971). *Statistical principles in experimental design* (2nd ed.). McGraw Hill.

Van Ewijk & Hoekstra. (1993). Alternatives for the no-observed-effect-level. *Environmental Toxicology and Chemistry, 12*, 187–194.

Bruce, R. D., & Versteeg, D. J. (1992). A statistical procedure for modelling continuous toxicity data. *Environmental Toxicology and Chemistry, 11*, 1485–1494.

Norberg-King, T. J. (1988). *An interpolation estimate for chronic toxicity: The ICp approach.* National Effluent Toxicity Assessment Center Technical Report 05-88.

Brain, P., & Cousens, R. (1989). An equation to describe dose-responses where there is stimulation of growth at low doses. *Weed Research*, *29*, 93–96.

KEY TERMS AND DEFINITIONS

LC_{50} (Median Lethal Concentration): The effective concentration of test material where 50% of the test organisms are killed/immobilized over a specified period of time.

LOEC (Lowest Observed Effect Concentration): The lowest concentration or dilution of a test material that is statistically different from the control group.

NOEC (No Observed Effect Concentration): The highest concentration or dilution of a test material that is not statistically different from the control group.

Chapter 24
Organic Farming:
A Key to Sustainable Development

Nisha Malik
Swami Vivekanand Subharti University, India

Devendra Kumar
https://orcid.org/0000-0001-9516-4830
Swami Vivekanand Subharti University, Meerut, India

Manjul Mungali
Swami Vivekanand Subharti University, India

Prateek Tripathi
https://orcid.org/0000-0002-0958-2428
The Scripps Research Institute, La Jolla, USA

ABSTRACT

The first aim in agriculture has always been to maximize the production and minimize the cost of cultivation without disturbing the environment as well as services it provides to the society. Different types of hazardous chemicals are being used in order to increase farm yield due to which agricultural challenges are also increasing day by day, and they ultimately result in health and environmental problems. Therefore, we need a better and holistic approach towards sustainable development (i.e. organic farming). It is a farming system that avoids the use of synthetic fertilizers and pesticides, thereby minimizing their deleterious effect on human and animal health and natural resources. According to the global organic organization (IFOAM), four main principles, namely the principles of health, ecology, fairness, and care, are the roots from which organic agriculture grows and develops. Thus, this chapter will be an attempt to explain the significance of organic farming that will provide not only health benefits but also environmental benefits.

DOI: 10.4018/978-1-7998-8434-7.ch024

INTRODUCTION

The term organic farming was first used by Lord Northbourne and describes it as "the farm as organism" in his book "Look to the Land" which was published in 1940. Influenced by Sir Albert Howard's work, Lady Eve Balfour did first scientific i.e. comparison of organic and conventional farming. The USDA study defined organic farming focus upon changes in crop type plantation at regular interval (crop rotation), use of surface residues after crop harvesting, post harvest operations, FYM (farm yard manure) in place of chemical fertilizers and biological control systems for effective insect pest management (Tripathi, 2016). In the term of FAO, organic farming refers the distinctive production system which can sustain agro-ecological system health, enrich soil with micro-minerals and prohibit the use of chemical fertilizers by replacing with the bio-fertilizers and bio-control system for plant protection (Helga and Julia 2017).

Organic agricultural based on agro-ecological principles focuses on soil fertility and plant health and no use of agro-chemicals whereas, conventional agriculture uses synthetic pesticides and water-soluble synthetically produced fertilizers (Finamore *et al.,* 2004). Organic farming supports the use of organic fertilizers like compost and farm yard manure and provides plant protection through biological control system (Edwards, 2007). The products or strategies used in organic farming ultimately helps in promoting clean agro-ecosystem to enhance the agricultural productivity such as legume-based organic system which had greater net balances of both C and N than did the conventional one (Drinkwater *et al.,* 1998). Natural insect predators are encouraged, crops are rotated to reduce the insects and pests and to improve soil fertility (Clark *et al.,* 1999; Njoroge 1994). Natural bio-products, waste materials and crop residue such as mulches are also used to control disease and weeds.

In organic farming, the farmers should have to be careful in plant strain or variety selection through plant breeding rather genetic engineering. Use of the crop strains that have ability to fix more carbon and which may decrease the green house gases (GHGs) emission with aid in reversing climate change (Halberg *et al.,* 2005; 2006; Kustermann *et al.,* 2008; Niggli *et al.,* 2009). In chemical farming, a specific insecticide may be applied to kill a particular insect quickly. Farming carried out using chemicals controls can reduce pest populations in a short period of time but causes severe damage to natural predator which finally increase the population of pest (IFOAM 2003a; Weidmann and Kilcher 2011).

Organic farming has grown out to be the best possible relationship between the earth and men. Nowadays, the major problem to apply organic farming is certainly its entry into the policy making arena, global market and the transformation of organic products into commodities (Halberg at al. 2006). In last two decades, a significant sensitization of the global community towards environmental protection and of food quality assurances has been established (Stoll 1986; Heaton 2002; Niggli *et al.,* 2008). Department of Agriculture's National Organic Standards Board (NOSB) established the rules for organic farming in the United States. Organic farming may be considered that it can fulfill both these demands and become the mean for complete development of rural areas (Helga and Julia, 2017). Organic farming is now being popular in mainstream and shows great increment commercially, socially and environmentally. Organic agriculture has both general and legal definitions. Generally, organic agriculture refers to farming systems that prohibit or minimize the use of synthetic pesticides and fertilizers or replace these by bio-products to protect the environment for sustainable development (Mendoza, 2002; IFOAM, 2003b).

PRINCIPLES AND LEGISLATION OF ORGANIC FARMING

To ensure the success of organic farming many principles are applied. The farm is viewed as an ecosystem, realizing that a change in one part of the system may disrupt the interrelationship on the farm. Due to use of fertilizers, high nitrogen levels found to present in the soil which is likely to stimulate weed growth. So, organic farmers will be able to address this problem by planting a crop which utilizes more nitrogen, thus creating a normal balance of soil nutrients (Njoroge 1994; Niggli *et al.,* 2008). According to organic farming, crop rotation is used to maintain habitat as natural as possible. Another principle of organic farming is recycling of organic matters. It results in replenishing the nutrients depleted from the soil during the crop cultivation, encouraging the growth of microorganisms which could regulate to store nutrients in the soil leading to crop growth in most effective manner. In organic farming, plant and animal residues are recycled to ensure self sufficiency of nutrients (Niggli *et al.,* 2009). The International Federation for Organic Agriculture Movement's (IFOAM) states that the definition of organic agriculture is based on four basic principles (Fig 1) as described below:

Health

It should have capability to promote soil health, and can sustain plant growth and productivity without having any impact on human health and environment by avoiding harmful chemicals.

Ecology

It involves utilization of environmentally friendly techniques that relies on the sustaining methods where there is no need to manipulate or harm any ecosystem and can in any condition.

Fairness

It should have ability to build up relationships with environment that provide opportunities with nature fairness.

Care

The farming which can deal in defensive and conscientious manner with nature by protecting human health and environment.

All these principles include the basic objectives that are considered essential for producing high quality food, fiber and other goods in an environmentally friendly manner (Yadav, 2015). The procedure of organic farming legislation is described in the form of flowchart (Fig 2).

Figure 1. Principles of organic farming

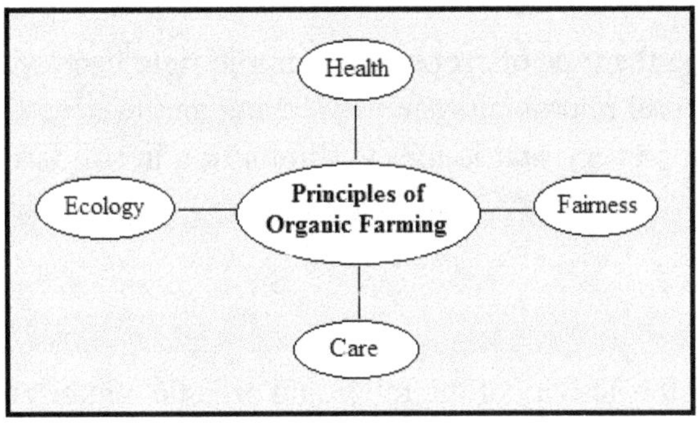

Figure 2. Procedure of organic farming legislation

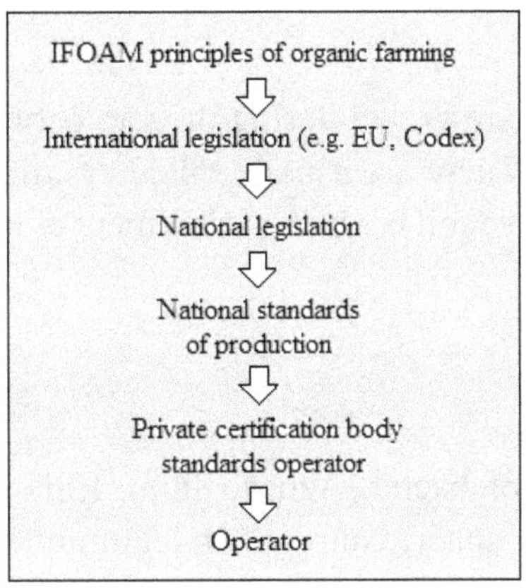

BASIC STEPS AND COMPONENTS OF ORGANIC FARMING

Basic Steps

The basic steps involved in organic farming (Meena *et al.,* 2013) include:

1. Transformation of conventional method into organic management system
2. Protection of human health and ecological biodiversity
3. Use of organic management for plant protection and soil fertilization like FYM and bio-control system (IFOAM, 2003b)
4. Use of green technological strategies for weed management like preventive and cultural management.
5. Animal husbandry incorporation in farming system as an integral part.

Organic Farming Components

Some of the components of organic farming are shown in Fig 3 and discussed in this section:

Crop Rotation

Crop rotation means changing the type of crops grown in the field each season or each year (IIRR and ACT 2005). To allow the natural microbial systems working and to keep the soil healthy crop rotation is must. If we repeat same crop every year it goes to diminution in soil fertility and also can encourage to weeds emergence.

Compost

Compost is the by-product of the activity of microbes and organic matter viz., plant and animal residue. It is cheap, easy to make and is a very effective to improve soil and crop quality. It reforms soil water holding capacity and helps in maintenance of soil moisture.

Green Manures

Green manures are the crops which are grown during the gap between in plantation of next cash crop, not for direct benefit or harvesting. These are usually killed or cultivated and mixed within soil before next crop sowing. These crops improve soil health by fulfillment of nutrient deficiency of soil and reduce use of chemical fertilizers.

Bio-Fertilizers

They are bacterial cultures *viz.,* Azotobacter, Azospirillum, Rhizobium, PSB and Pseudomonas etc., which have the ability of fixing atmospheric nitrogen in leguminous and non- leguminous crops. They have been found to be very effective tools in biological nutrient mobilization and fertility management.

Vermi-Composting

Vermi-composting is a method of composting using earthworms for effective recycling of non-toxic organic wastes to the soil.

Animal Husbandry

Animals' welfare is considered very important in an organic system. However, animal husbandry is also a part of organic agriculture because of their incorporation in agriculture. They provide organic manure and FYM for agriculture which reduces the demand of fertilizers but there should be taken of care by management of animal yard in systematic way so that they didn't able to harm the crop. Food for animals should be grown in an organic manner.

Figure 3. Components of organic farming

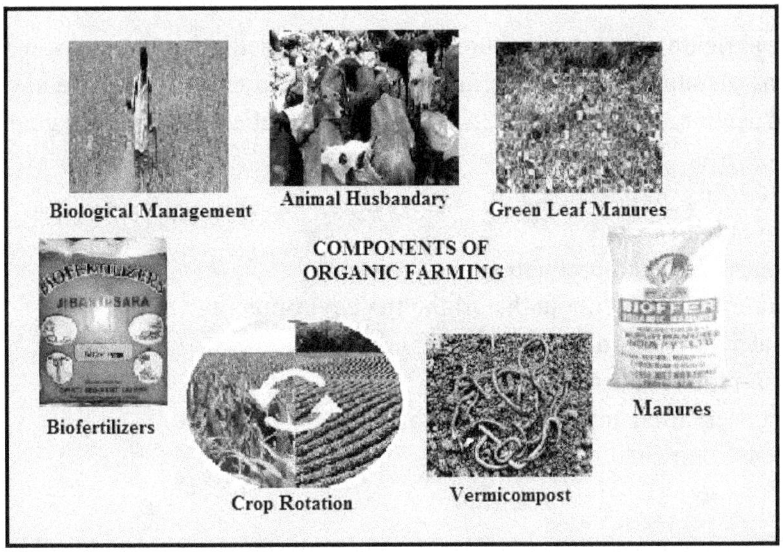

CONCEPT AND SYSTEM

Organic farming is one of the broad-spectrum production systems which support the environment. Organic production systems aimed at achieving agro-ecosystem which is socially and ecologically sustainable (Mendoza, 2002; IFOAM, 2003b, 2005). Among the most stringent definitions is that of US Department of Agriculture, which may define Organic Farming as, "A system that is designed to produce agricultural products using methods and substances that maintain the integrity of organic agricultural products until they reach the consumer" (Ummyiah *et al.,* 2017). Organic farming uses cultural, biological and mechanical methods which provides long term protection to environment and recycling of wastes to return nutrients to the land and handle rule agricultural products without using synthetic herbicide/ pesticides or additives processing (Huber and Van, 2009; Meena *et al.,* 2013). Some plans are required to apply organic farming:

- The composition and commercial availability of each substance, to be used in the operation and where it will be used.
- Monitoring techniques description including frequencies to be used.
- Record maintenance, including sales. The records must be maintained for a minimum of five years and made available during business hours for inspection.
- Physical barrier management or buffer zone on operations with organic and non-organic components.

ORGANIC FARMING GOALS

The goal of organic agriculture is to contribute to the enhancement of eco-system sustainability (Fig 4). Sustainability means maintaining or enhancing the quality of the environment and conserving natural resources for future generations (Paul *et al.,* 2006; Helga and Julia, 2017). These goals are listed below:

- Focus on soil fertility and plant health.
- No use of agro-chemicals.
- Healthy and nutritious food production.
- Pests and diseases control without disturbing the environment.
- Ensure that water stays clean and safe (water conservation).
- Low use of non-renewable energy.
- Biodiversity conservation and maintenance of ecological balance.
- To minimize environmental risks.

Figure 4. Goals of organic farming

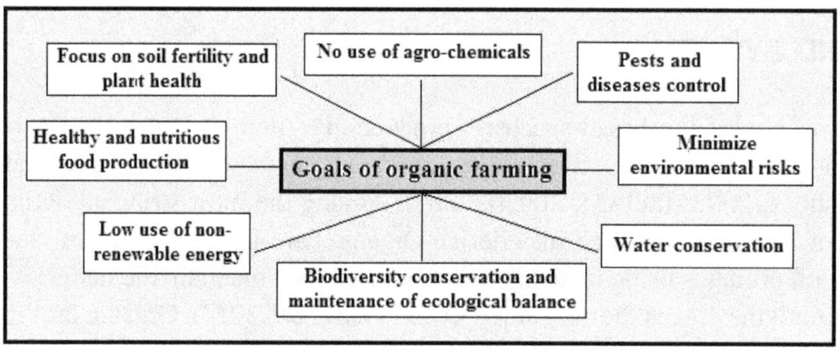

SOIL FERTILITY AND INSECT-PEST MANAGEMENT

A thorough understanding of pest life cycles and interactions is required for effective organic pest control. Pest control in organic farming is similar to integrated pest management at some points. Organic farming tolerates some pest populations while taking a longer- term approach. Pest control by organic farming for cumulative effect uses many techniques like allowing for an acceptable level of pest damage which encourage to predators that are beneficial insects to control pests and encouraging growth of beneficial microorganisms and insects. Second is the selection of disease-resistant varieties, plantation of companion crops that minimize the growth of pests and others is the use of plants that regulates pests life cycle and mitigate the use of biological pesticides and herbicides. Last method is year to year rotating crops to different locations to interrupt pest reproduction cycles; using insect trapping strategies to monitor and control insect populations (Stoll, 1986; IFOAM 2003a).

The base of organic farming are the biological matters that are used in many forms such as compost, manure (organic or green), crop rotation and biological plant protection system. These biological processes are driven by microorganisms such as mycorrhiza, which allows the natural production of nutrients in the soil throughout the growing season, and has been referred to as feeding the soil to feed the plant (Huber and Van, 2009). In chemical farming, individual nutrients like nitrogen are synthesized in a more or less pure form that plants can use immediately, and applied on a manmade schedule. These techniques also provide other benefits *viz.,* soil protection and improvement, fertilization, pollination, water conversation, season extension etc., and these benefits are both complementary and cumulative in overall effect on farm health (Butler et al., 2008).

NUTRIENT MANAGEMENT IN ORGANIC FARMING

Organic farming is often understood as a form of agriculture with use of only organic inputs find the supply of nutrients and management of pests and diseases (IFOAM, 2002, 2003a). Often, organic agriculture has been criticized on the grounds that farm productivity and profitability might not be improved with organic inputs alone, because the availability of organic sources is highly restricted. Organic farming systems defined as the management of soil organic matter for enhancing chemical, biological and physical properties of the soil (Helga and Julia, 2017).

One of the basic principles of soil fertility management in organic system is that plant nutrition depends on biologically-derived nutrients' instead of using readily soluble forms of nutrients; less available forms of nutrients such as those in bulky organic materials are used (Butler *et al.,* 2008). The potential sources of nutrients used in organic farming are animal dung, residues after harvesting the crop, green manure, bio-fertilizers from agro industries and wastes produced after food processing. In this animal dung can be uses as fuel and extensively used in the form of farmyard manure. Development of several compost production technologies like vermi-compositing, phosphor-compositing, N-enriched phosphor-compositing, etc. (IFOAM, 2003b; Edwards, 2007), improve the quality of compost through the enrichment with nutrient- and other biological additives. The manures are capable to fulfill the soil nutrient deficiency by promoting growth of micro flora in soil. Some organic farming strategies tool combined with the high input techniques have been developed to create Integrated Farming Systems such as integrated nutrient and pest management and biological control methods which reduce the chemical need (Paul *et al.,* 2006).

NEED AND REQUIREMENT OF ORGANIC FARMING

Organic farming is an age-old concept in India, which changed during the First Green Revolution. In past time, some strategies and outlook has seen a quantum change worldwide towards agriculture and marketing of food. The focus is now more on quantity and outer appearance rather than intrinsic or nutritional value (Brandt and Molgaard, 2001). Excessive applications of pesticides and fertilizers have caused damage to the soil and environment besides affecting crop production (IFOAM, 2005). Use of pesticides in the world has registered tenfold increase from 1945 to till date and in India, more than 80,000 tonnes of pesticides are used. Pesticides residue is the second largest cancer causing agent, next to cigarette. Besides, the pesticides and fertilizers that persist in the soil are harmful to the beneficial soil microorganism and earthworms, thereby resulting in degradation of soil fertility. Organic farming can mitigate each of these problems. It also greatly helps a farmer to become self-sufficient in his requirements for agro-inputs and reduce the cost. In organic farming system, certain minimum requirements are to be considered so that farm is certified as an organic farm.

Conservation

When a farmer switches over to the system of organic farming from the conventional system of farming, it is known as conservation. The farmers should have a conservation plan prepared if the entire field is not converted into organic at a time. In that case it is necessary to maintain organic and non organic field separately.

Mixed Farming

Animal husbandry, poultry, fisheries, etc., should be practiced in addition to agriculture farming. Shifting cultivation is not allowed.

Cropping Pattern

Crop rotation should be followed if annual crops are grown. Crop rotation should cover green manure as well folder crops. Mono-cropping should be avoided.

Planting

Species and varieties cultivated should be adapted to soil and climatic condition and resistant to pest and diseases. Seeds planting materials should be procured from organic source.

Manuring

Soil fertility should he maintained/enhanced through raising green manure crop, leguminous crops etc. The residues of plants after harvest should be incorporated into the soil far as possible. Biodegradable materials of microbial, plant or animal origin shall be applied as manures.

OPTIONS IN ORGANIC FARMING

There are three on-going options in organic farming to alleviate the problems caused by conventional inorganic farming systems (Dobbs and Smolik, 1996; Finamore *et al.,* 2004) which are shown in Fig 5.

Pure Organic Farming

This accounts use of organic manures and biological methods to control pest problems and complete exclusion of inorganic fertilizers and pesticides.

Integrated Green Revolution Farming

In this type of management system, some types of strategies are used such as intensive use of external inputs, increased irrigation, development of high yielding and hybrid varieties as well as mechanizations of labour with much greater efficiency and limited damage to the environment and human health (Clark *et al.*, 1999).

Integrated Farming

In this type of farming system farmers have to depend on local resources and ecological processes, recycling of agricultural wastes and crop residues which require low input. These farming systems improve the quality of life and reduce the risk of depletion in natural resources, resulting in viable and sustainable agriculture production (Mendoza, 2002).

Figure 5. Types of organic farming

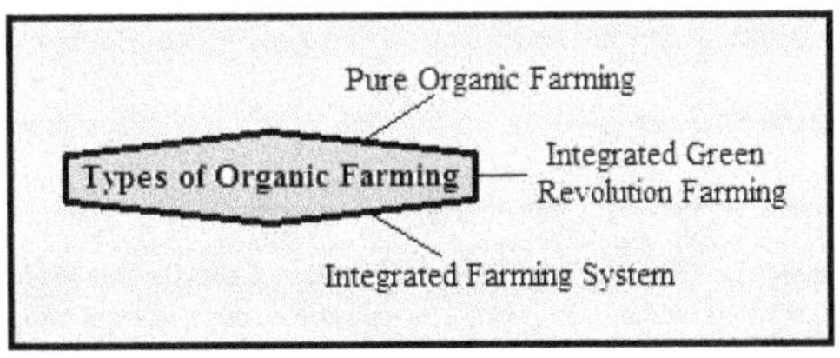

IMPACT OF LOCKDOWN ON ORGANIC FARMING

COVID-19 pandemic has brought threaten to livelihoods as well as food securities (FAO, 2020). During the period of lockdowns, when all the economic activities were almost choked off, home confinement has encouraged Indians to move towards their organic sector for nutrient and delicious food at home for better food safety, quality and security (Singh, 2021). All the Indians were stressed under uncertainties, restriction for export, import, unemployment, and declining supplies. Therefore, it was the primary need of the time to grow food spices and vegetables at the nearest agricultural land. Under clear sky, non-polluted water stations, less contaminated ground water, non-availability of fertilizers (Lamichhane and Reay-Jones, 2021) and more workforces at home has boosted and triggered many Indian farmers to go for organic undergone a fundamental shift to the marginal and labor class to rebuild their livelihood through organic farming in India for lively hood with crop and bovine residue as bio-fertilizers and vermi-compost for their fields. The lockdown came as a blessing as it has changed lives for the better. Organic farming has not only become the source of income but has also helped to keep the families healthy. This type of farming system has given a new lease of life.

ADVANTAGES AND OPPORTUNITIES FOR ORGANIC FARMING

Organic farming offers various advantages (Kristiansen, 2006) over conventional farming. Some of them are shown in Fig 6 and listed below:

- Ecological services provisions such as stability in yield and crop protection *i.e.,* maintenance of ecological balance.
- Minimum chemical residues in food and the environment.
- Equivalent economic performance like conventional farming.
- Animal welfare standards.
- Standard and reliable strategies and certification schemes.
- Production of nutritious and healthy foods.
- Sustainable development and poverty reduction.
- Dynamic review of policies and standards.
- Strong consumer demand and brand recognition. Indigenous knowledge is valued.
- Potential for cooperative rural and regional development.

Figure 6. Advantages of organic farming

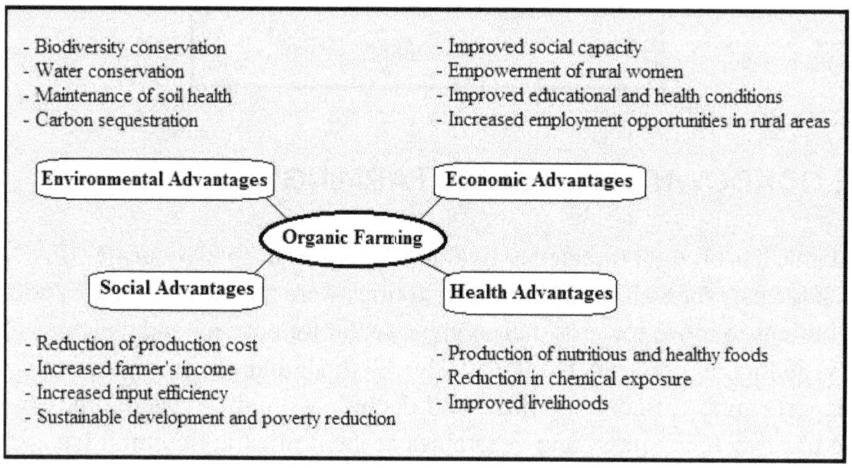

BOTTLENECKS OF ORGANIC FARMING

Besides the various advantages of organic farming, this system is accorded with several disadvantages also (Meena *et al.,* 2013) which are shown in Fig 7 and they are as follows:

- It is a time taking process and low yields.
- Organic inputs are required largely.
- Quality inputs availability is low.
- Less marketing and research facilities.
- Frequent pest and weed control management is required instead of one time chemical applications.

Figure 7. Bottlenecks of organic farming

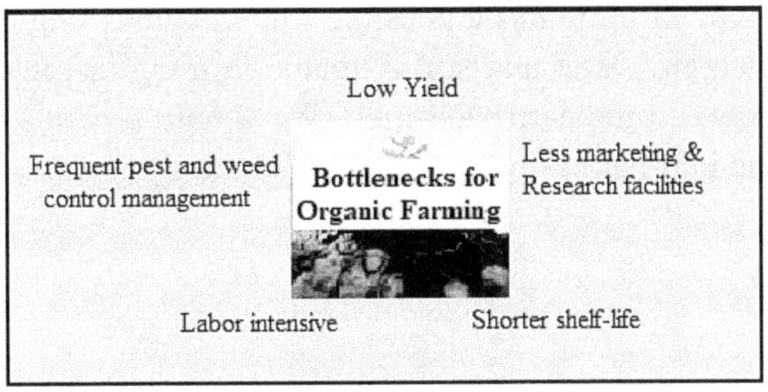

CHALLENGES FOR ORGANIC FARMING

There are many major constraints or problems of organic farming in India (Elayaraja and Vijai, 2020). A range of challenges are highlighted in this section:

1. Maintaining sustainability in the global economy.
2. Maintaining flexible organic standards and certification strategies to overcome issues such as:
    a. Nature conservation and regeneration;
    b. Equitable, affordable and flexible access to certification services;
    c. Responsible labour relations and land tenure arrangements;
    d. Animal welfare;
    e. New inputs such as 'natural' biocides, soil amendments and GMOs; and
    f. Incomplete basis from organic standards including/excluding materials.
3. Developing locally applicable agronomic solutions to production constraints, such as weeds, animal health and soil fertility.
4. Expanding research activities in many disciplines and fosters the integration of knowledge.
5. Preserving food quality while trying to increase productivity.
6. Educating and training at all levels to build capacity, infrastructure and networks.
7. Inadequacies in regulatory and marketing structures.
8. Excessive consumer prices.
9. Inconsistent quality and availability.

Besides these, several other challenges of organic farming includes ecological justice; animal welfare; fair trade; supply chain development; productivity limitations; and regional adaptation and global harmonisation for standards (Halberg *et al.*, 2005).

CONCLUSION

Organic farming applies the strategies to achieve sustainable development of agriculture system without disturbing the natural environment. Organic farming not only provides health benefits but also environmental benefits by reducing the CO2 emissions. This farming system works in harmony with nature which prohibits the use of chemical fertilizers, pesticides, or herbicides and other synthetic additives. In another way organic farming is a kind of agricultural practices which provide its consumers, with fresh, tasty and reliable food. Organic farming has a possible role in conserving our biodiversity and reduce environmental pollutions such air, water and soil. Organic farming has inspired peoples to create the best possible relationship between the earth and men. We should practice organic farming instead of intensive or conventional one in order to protect our natural eco-system.

REFERENCES

Brandt, K., & Mølgaard, J. P. (2001). Organic agriculture: Does it enhance or reduce the nutritional value of plant foods. *Journal of the Science of Food and Agriculture*, *81*(9), 924–931. doi:10.1002/jsfa.903

Butler, G., Nielsen, J. H., Slots, T., Seal, C., Eyre, M. D., Sanderson, R., & Leifert, C. (2008). Fatty acid and fat-soluble antioxidant concentrations in milk from high- and low-input conventional and organic systems: Seasonal variation. *Journal of the Science of Food and Agriculture*, *88*(8), 1431–1441. doi:10.1002/jsfa.3235

Clark, M. S., Horwath, W. R., Shennan, C., Scow, K. M., Lantini, W. T., & Ferris, H. (1999). Nitrogen, weeds and water as yield-limiting factors in conventional, low-input, and organic tomato systems. *Agriculture, Ecosystems & Environment*, *73*(3), 257–270. doi:10.1016/S0167-8809(99)00057-2

Dobbs, T. L., & Smolik, J. D. (1996). Productivity and profitability of conventional and alternative farming systems: A long-term on-farm paired comparison. *Journal of Sustainable Agriculture*, *91*(1), 63–79. doi:10.1300/J064v09n01_06

Drinkwater, L. E., Wagoner, P., & Sarrantonio, M. (1998). Legume-based cropping systems have reduced carbon and nitrogen losses. *Nature*, *396*(6708), 262–265. doi:10.1038/24376

Edwards, S. (2007). The impact of compost use on crop yields in Tigray, Ethiopia. Institute for Sustainable Development (ISD). In *Proceedings of the International Conference: Organic Agriculture and Food Security*. FAO.

Elayaraja, M., & Vijai, C. (2020). Organic farming in India: Benefits and Challenges. *Eur. J. Mol. Clin. Med.*, *7*(11), 3021–3029.

FAO. (2020). *Responding to the impact of the COVID-19 outbreak on food value chains through efficient logistics*. Report of activity No.166 from 04.06.2020 to 06.07.2020. https://www.fao.org/fsnforum/activities/discussions/covid19_logistics

Finamore, A., Britti, M. S., Roselli, M., Bellovino, D., Gaetani, S., & Mengheri, E. (2004). Novel Approach for food safety evaluation. *Journal of Agricultural and Food Chemistry*, *52*(24), 7425–7431. doi:10.1021/jf049097p PMID:15563230

Halberg, N., Alrøe, H. F., & Kristensen, E. S. (2005). Synthesis: Perspectives for organic agriculture in a global context. *Global Development of Organic Agriculture: Challenges and Prospects,CAB Int.*, *12*, 343–367.

Heaton, S. (2002). Assessing organic food quality: Is it better for you? *Proceedings of the UK Organic Research Conference*, 55-60.

Helga, W., & Julia, L. (2017). BIOFACH, "Organic agriculture worldwide." In *Current Statistics*. FiBL.

Huber, M., & Van, de V.L. (2009). *Overview of research linking organic production methods and health effects in the lab, in animals and in humans*. Department of Health Care and Nutrition, Louis Bolk Institute.

IFOAM Norms, . (2002). *IFOAM Basic Standards - IFOAM Accreditation Criteria*. IFOAM.

IFOAM. (2003a). Pest and disease management in organic management. A Latin American perspective. IFOAM.

IFOAM. (2003b). *Training manual for organic agriculture in the tropics*. IFOAM.

IFOAM. (2005). *The IFOAM Norms for organic production and processing including IFOAM Basic Standards and IFOAM Accreditation Criteria*. IFOAM.

IIRR & ACT. (2005). *Conservation agriculture: A manual for farmers and extension workers in Africa*. International Institute of Rural Reconstruction and African Conservation Tillage Network.

Kristiansen, P. (2006). Organic agriculture: opportunities and challenges. *Organic Agriculture: A Global Perspective*, *18*, 421-441.

Küstermann, B., Kainz, M., & Hülsbergen, K. J. (2008). Modeling carbon cycles and estimation of greenhouse gas emissions from organic and conventional farming systems. *Renewable Agriculture and Food Systems*, *23*(1), 38–52. doi:10.1017/S1742170507002062

Lamichhane, J. R., & Reay-Jones, F. P. F. (2021). Editorial: Impacts of COVID-19 on global plant health and crop protection and the resulting effect on global food security and safety. *Crop Protection (Guildford, Surrey)*, *139*, 105383. doi:10.1016/j.cropro.2020.105383 PMID:32921863

Meena, R. P., Meena, H. P., & Meena, R. S. (2013). Organic farming: Concept and components. *Popular Kheti*, *1*(4), 5–14.

Mendoza, T. C. (2002). *An Evaluation study of plan international sustainable agriculture program in Calapan*. Oriental Mindoro.

Niggli, U., Fließbach, A., Hepperly, P., & Scialabba, N. (2009). *Low greenhouse gas agriculture: Mitigation and adaptation potential of sustainable farming systems*. FAO.

Niggli, U., Slabe, A., Schmid, O., Halberg, N., & Schlueter, M. (2008). Vision for an organic food and farming research agenda to 2025. *Organic Knowledge for the Future*.

Njoroge, J. W. (1994). *Field notes on organic farming*. Kenya Institute of Organic Farming.

Paul, K., Acram, T., & John, R. (2006). *Organic agriculture a global perspective*. Csiro Publishing.

Regenerative Agricultural Technologies. (1990 *Trainors Kit*. International Institute of Rural Reconstruction.

Singh, G. (2021). Lockdown cash crunch effects organic farming. *VillageSquare*. https://www.villagesquare.in

Stoll, G. (1986). Natural crop protection based on local farm resources in the tropics and subtropics. Intermediate Technology Publications.

Sustainable agriculture practices and technologies: Guidelines for farmers Africa 2000 network. (1997). UNDP.

Tripathi, V. (2016). Organic farming: A new avenue for rural entrepreneurs for sustainable business. *Int J Educ Management Stud.*, *6*(1), 8–10.

Ummyiah, H. M., Jabeen, N., Afroza, B., Mushtaq, F., Mufti, S., & Nabi, A. (2017). Certification of organic products. *International Journal of Current Microbiology and Applied Sciences*, *6*(11), 2300–2308. doi:10.20546/ijcmas.2017.611.273

Weidmann, G., & Kilcher, L. (2011). *African organic agriculture training*. FiBL.

Yadav, A.K. (2015). *Organic agriculture (Concept, Scenario, Principles and Practices)*. National Project on Organic Farming, Dept. of Agriculture and Cooperation, Govt of India.

Chapter 25
Induced Resistance for Sustainable Management of Wheat Diseases

Prem Lal Kashyap
ICAR-Indian Institute of Wheat and Barley Research (IIWBR), India

Anju Sharma
ICAR-Indian Institute of Wheat and Barley Research (IIWBR), India

Sudheer Kumar
ICAR-Indian Institute of Wheat and Barley Research (IIWBR), India

Rahul Tripathi
ICAR-Indian Institute of Wheat and Barley Research (IIWBR), India

Poonam Jasrotia
ICAR-Indian Institute of Wheat and Barley Research (IIWBR), India

Devendra Pal Singh
ICAR-Indian Institute of Wheat and Barley Research (IIWBR), India

Ravi Shekhar Kumar
ICAR-Indian Institute of Wheat and Barley Research (IIWBR), India

Gyanendra Pratap Singh
ICAR-Indian Institute of Wheat and Barley Research (IIWBR), India

ABSTRACT

The concept of induced resistance (IR) in crop protection has bloomed in the past few decades as it offers improved plant protection in an environmentally safe and economically sound manner. A large number of resistance-inducing compounds of different origins have been reported and tested in different plant-pathogen systems. Several reports indicated that the mechanism of action of these compounds differs from fungicides because they restrict plant pathogen development indirectly via triggering plant defense. Additionally, resistance inducers can be combined with biological control agents and even fungicides, which could result in reduced agrochemical usage in agriculture sector. A number of biological control agents are available for field applications, but further development and effective implementation will require a better understanding of the complex interactions among plants, people, and the environment. To that end, the chapter presents an advanced survey of the nature and practice of biological control based on IR as it is applied to the suppression of wheat diseases.

DOI: 10.4018/978-1-7998-8434-7.ch025

INTRODUCTION

Wheat (*Triticum aestivum* L.) is one of the most important cereal crops which provided daily nourishment to the large section of global population. Several reports indicated that global wheat production would need to enhance by 60% to match the food needs by 2050, which is a difficult task due to dwindling land and water resources and changing climatic conditions (Savadi et al., 2018; Kashyap et al., 2017). Additionally, its yield and grain quality are significantly impacted by different kinds of diseases. Among various diseases, rusts (yellow, brown and black), Karnal bunt, powdery mildew, foliar blights and loose smut are of major economic significance (Kashyap et al. 2019; Jasrotia et al., 2018; Kashyap et al., 2018a; Singh, 2017; Kashyap et al., 2011; Kashyap et al. 2020). Other region specific diseases of minor importance are head scab, flag smut, foot rot, hill bunt, wheat blast and viral diseases. Previous studies reported that diseases usually leads to a 15-20% loss of wheat yield per year, but losses may be as high as 50% when the flag leaf suffers from severe disease during the heading and filling stages (Figueroa et al., 2018; Griffey et al., 1993). According to Park (2007), leaf rust alone can cause up to 60% yield losses while stem rust can cause 100% yield losses in severe conditions. Similarly, the average yield losses due to leaf blight is about 20% in India (Joshi et al., 2004), however, it may be up to 80% under heavy infection (Joshi et al., 2007). To date, ensuring a satisfactory yield and quality of the harvest requires an extensive use of systemic and protectant fungicides for their management. But the continued application of fungicides has several negative consequences including the development of fungicide-resistant races of the pathogen, increase of residue levels, and serious risks for human health (Kashyap et al. 2019; Kashyap et al., 2018b; Bruce et al., 2010). Moreover, wheat breeders have been unable to locate complete resistance against fungal diseases in any of the wheat cultivars, although, some of them shown varying degree of resistance against the diseases. Thus, induced resistance (IR) could be an alternative approach for improving wheat productivity under biotic stress.

Advances in research efforts involving induced resistance (IR) in plants have been accompanied by the emergence of new chemical inducers with higher efficiency, stability and with less impact on the environment than existing ones. These new products are able to enhance agricultural productivity by reducing the yield losses by improving the resistance levels (Oliveira et al., 2016). S-methylbenzo [1,2,3] thiadiazole-7-carbothiate (acibenzolar-S-methyl) (ASM) was the first SAR inductor released for commercial use (Lyon and Newton, 2007). Since then, several other products derived from elicitor molecules released during the early stages of the plant–pathogen interactions have been reported (Wiesel et al., 2014; Angelova et al., 2006). The elicitors developed as products include carbohydrate polymers, lipids, and glycoproteins that are either secreted by microorganisms or derived from the cell walls of fungi, bacteria, or even plants (Thakur and Sohal, 2013). Besides this, compounds that mimic the action of salicylic acid (SA) include 2, 6-Dichloroisonicotinic acid (INA) and ASM have also become available in the market and are currently being used in agriculture. The major merit of these compounds over conventional wheat disease management practices is that by using these, the number of sprays of conventional fungicides may be reduced. All the resistance inducers have a preventive action, enhancing the natural defenses of the plants and reducing the preventive pesticide application treatments or frequency (Llorens et al., 2017). Usually the induced resistance is strong enough to protect plants against moderate level of stresses. However, in cases of high disease pressure that overcome the natural defenses of the plant, the combination of resistance inducers with classical treatments can achieve complete protection. Moreover, a correct schedule of treatments with alternation between these compounds and classical ones could be useful for both reductions of disease severity and chemical residues. Moreover, the use of these inducer agents in combination with fertilizers or the colonization with microbes could be an interesting alternative to preventive and curative treatments. The incorporation of natural resistance inducer in disease management programs, alone or in combination with classical methods, could be a reliable method for reducing the yield losses as well as amount of chemical residues in the environment. Briefly, the chapter provides a current summary of resistance inducers that have been successfully used in the wheat crop to protect against pathogens by activating the plant's own defense.

INDUCED RESISTANCE (IR) AND ITS CLASSIFICATION

Induced resistance (IR) is a physiological state of enhanced defensive capacity of the plants against the attack of pathogens such as fungi, bacteria, viruses, and nematodes, elicited by abiotic and/or biotic stimuli, whereby the plant's basal defenses are induced against subsequent biotic challenges (van Loon et al., 2006). IR can be classified on the basis of the nature of the elicitor and the regulatory pathways into systemic acquired resistance (SAR) and induced systemic resistance (ISR) (Durrant and Dong, 2004; van Loon et al., 1998). SAR can be triggered by exposing the plant to virulent, avirulent, and nonpathogenic microbes (Pieterse et al. 2014). Depending on the host plant and elicitors, a specific time period is required for the activation and establishment of SAR wherein accumulation of pathogenesis-related proteins (chitinase and glucanase), and salicylic acid takes place (Choudhary and Johri, 2009; Walter et al., 2005). Principally, SAR is an enhancement of salicylic acid (SA)-dependent basal resistance, while ISR is an enhancement of jasmonate/ethylene (JA/ET)-dependent basal resistance (Cortes-Barco et al., 2010). Besides this, the mode of action of ISR is based on the defense mechanisms that are activated by inducing agents. Moreover, ISR once expressed, it activates multiple defense mechanisms that include increased activity of chitinases, β-1, 3-glucanases and peroxidases (Anand et al., 2007; Maurhofer et al., 1994). Another important feature of ISR is the wide spectrum of pathogens that can be controlled by a single inducing agent (Hoffland et al., 1996). Thus, ISR appears to be the result of several mechanisms which together are effective against wide range plant pathogens. Induced responses and induced resistance have been well-documented and have been reported in several plant species (Kashyap et al., 2018b; Srivastava et al., 2012; Kashyap and Dhiman 2009; Heil and Bostock, 2002; Karban and Baldwin, 1997). A major advantage of plant resistance inducers, compared to agrochemicals, is that they are compatible with biocontrol agents. Another advantage is that induced resistance via priming is based on an augmentation of basal defense resistance (Bruce et al., 2017; Ahmad et al., 2010) and is controlled by a large number of defense related plant genes. Consequently, induced resistance is a durable form of resistance, since the augmentation of multigenic resistance is difficult to break by pathogens (Gardner et al., 1999; Ahmad et al., 2010). Thus, the prospect of broad-spectrum disease control using the plant's own resistance mechanisms has led to increasing interest in the development of agents which can mimic natural inducers of resistance.

RESISTANCE INDUCING COMPOUNDS AND THEIR APPLICATIONS

Wheat crop can be induced to enhance level of resistance against pathogen infection by providing exogenous application of a variety of abiotic and biotic inducers. Biotic inducers include infection by hemibiotropic pathogens and plant-growth-promoting bacteria, and treatment with non pathogens or cell wall fragments (Walter et al., 2013). Abiotic inducers include chemical compounds which act at various points in the signaling pathways involved in disease resistance, as well as water stress, heat shock, and pH stress etc. (Walters et al., 2005). Resistance induced by these agents is broad spectrum and long lasting, but rarely provides complete control of infection (Věchet and Šerá et al., 2015).

Chemical and Non-Biological Inducers

Chitosan

Chitosan, a β-1,4-D-glucosamine polymer found in the walls of many fungi, is a polycationic polymer that may readily interfere with negatively charged molecules exposed at the cell surface. Kashyap et al. (2015) comprehensively described the application of chitosan in plant growth promotion and protection. Specifically, in wheat, chitosan has been reported to protect wheat from *F. graminearum* infection (Reddy et al., 1999). Chitosan also induced resistance in winter wheat to the snow mould pathogen *Microdochium nivale*, although the effect was variable (Hofgaard et al., 2005). Chitosan treatment of wheat led to increased chitinase gene expression, but chitosan was also found to exert a direct inhibitory effect on *in vitro* growth of *M. nivale* (Hofgaard et al., 2005). Therefore, it is possible that both induced resistance and direct antifungal activity were responsible for the control of *M. nivale* on wheat. The authors suggested that the variability in disease control obtained may have been the result of differences in disease pressure, with higher disease pressures leading to poorer disease control (Hofgaard et al., 2005). The potential ability of chitosan seed treatment to induce resistance against *Fusarium graminearum* in durum wheat was evaluated by Orzali et al. (2014). Both field and greenhouse trials revealed that the chitosan (0.5%) seed treatment provided effective protection against the root and foot rot disease, with results comparable to the copper treatment. Thus, the prior induction of resistance in plants caused by chitosan can be considered as important and eco-friendly strategic tools for controlling seed and soil pathogens and reducing the need of chemicals for plant disease control.

Salicylic Acid (SA) and its Derivatives

It is well known that salicylic acid (SA) participates in regulating defenses in plants and is predominantly associated with resistance against biotrophic and hemibiotrophic pathogens (Grant and Lamb, 2006). SA and its functional analogs [e.g. 2, 6-dichloroisonicotinic acid (INA), benzothidiazoles (BTH)] and β-amino-n-butyric acid (BABA)] are among the most thoroughly studied compounds for inducing disease resistance (Kashyap et al., 2018b; Kim et al., 2013; Conrath, 2006). In addition to the activation of defense gene expression, it was first time reported that SA and its analogues (BTH) protect wheat against powdery mildew in growth cabinets when sprayed with 0.3mM final concentration and when treatments were applied between four and seven days before challenge inoculation (Görlach et al., 1996). Further, they demonstrated that induced resistance to both *Puccinia recondita* and *Septoria* spp. in wheat by BTH treatment and achieved protection against powdery mildew for an entire season with a single application of BTH (30g ha^{-1}). Since then, the molecule has been shown to be effective against a wide range of pathogens on a range of crops and was released commercially under the trade names Bion® (in Europe) and Actigard® (in the USA). It has been noted that under real environmental/field conditions plants respond differently at different times to chemical elicitor treatments such as benzothiadiazoles (BTH). For instance, in a field experiment under natural conditions on the effect of nitrogen fertilization, fungicides and resistance induction on *Fusarium* head blight (FHB) and related mycotoxin accumulation in wheat (Heier et al., 2005), BTH and biomass of the cyanobacterium *Spirulina platensis* were used. The results indicated that these plant defense activators are as effective as conventional fungicide. Immoderate N-fertilization, however, can enhance mycotoxin levels significantly, even under conditions unfavourable for *Fusarium* spp. Indeed, under field conditions, there may be a general high level expression of defence genes compared with glasshouse grown plants (Pasquer et al., 2005). Zhang et al. (2007) evaluated SA, sodium salt of salicylic acid (NaSA), INA, and BABA for the management of *Fusarium* head blight (FHB) of wheat when sprayed on wheat heads at three days prior to pathogen challenge with *Gibberella zeae*, NaSA and INA at 10 mM significantly reduced FHB severity compared to the non-treated disease control. Moreover, NaSA, INA, and BABA at 1 mM significantly reduced FHB severity, when sprayed on wheat heads just beginning to emerge from boot at ten days prior to pathogen inoculation. Interestingly, induced FHB resistance was achieved by treating wheat with INA at concentrations as low as 0.1 mM. These findings indicated that induced systemic resistance was at least partially responsible for the reduction of FHB disease. Thabit (2008) experimentally shown that BTH and SA reduced wheat leaf rust infection through the activation of pathogenesis related protein, including peroxidase and phenolic compounds. Further, the effect of BTH (1mM) and SA (1000 μg ml^{-1}) on ISR was investigated by Al-Maaroof et al. (2014) in moderately susceptible and susceptible wheat genotypes (cv. Tamuz-2 and AL-8/70) against *Puccinia striiformis* f. sp. *tritici*. Resistance was characterized by reduced infection of yellow rust disease. Changes in peroxidase, phenylalanine ammonia-lyase activities and in total phenolic compound content demonstrated that the resistance to *Puccinia striiformis* has been induced by these chemical inductors in wheat genotypes. BTH, BABA and SA at 1 mM dosage were evaluated to manage Karnal bunt (*T. indica*) under glass house conditions by Kashyap et al. (2018b). Results indicated that 1 mM concentration of these compounds is sufficient to control KB infection in greenhouse grown wheat. Moreover, maximal accumulation of PDF1.2 transcripts followed by PR1 and PR5 in independently primed wheat spikes with all tested compounds at three days post-inoculation was observed. Results suggest that these compounds primed the plant for a rapid and intense response to pathogen infection involving augmented activation of SA and JA/ ET mediated defense response and enhanced level of wheat protection.

Heptanoyl salicylic acid (HSA) obtained by esterification of 2-OH benzoic acid with heptanoic acid have been shown to induce protection on wheat against powdery mildew (Muchembled et al., 2006). HSA was reported to lead to 70% protection against *Bgt* only a single spray of a 1 gl^{-1} solution (Muchembled et al., 2006). It has been shown to be twice as efficient as SA in preventive treatments: the protection levels against *Bgt* increased from 50% in plantlets after SA spraying to 95% after HSA spraying (Renard-Merlier et al., 2007), without any direct effect on conidia germination *in vitro*. The heptanoic acid did not reduce *per se* the infection and indeed probably improved the effect of SA by increasing its penetration through the hydrophobic plant cuticle (Muchembled et al., 2006).

Jasmonic Acid

Jasmonic acid (JA) plays an important role in plant resistance to different plant pathogens. The effect of application of JA on the induction of resistance in wheat against *Stagonospora nodorum* was studied by Jayaraj et al. (2004). Pre-treatment of wheat plants with JA significantly reduced (up to 56%) the incidence of leaf blotch disease compared with untreated control plants under greenhouse conditions. Similarly, Dutt et al. (2011) pretreated resistant (HD-29) and susceptible (WH-542) wheat varieties with JA (1 μg μl^{-1}) and then artificially inoculated with sporidial suspension of *Tilletia indica* to study its influence in reducing KB infection. JA was found to improve the plant defense against KB as its exogenous application resulted in decrease in coefficient of infection (CI) in both susceptible and resistant varieties following pathogen inoculation. Further, they noticed that JA induced cystatin gene expression more in resistant variety than susceptible one. In case of powdery mildew of wheat, exogenous MeJA (1 mM L^{-1}) significantly enhanced the powdery mildew resistance of the susceptible varieties (cv. Chinese Spring and Pumai 9) (Duan et al., 2014). It was observed that after treatment, MeJA pass into plant tissue via stoma and then hydrolysis to JA by esterase in the cytoplasm to achieve long-distance signal transduction and communication between plants to induce defense responses. As a result, after inoculation with *Bgt*, endogenous JAs accumulated rapidly, reached the maxima at 2- 5 h post-inoculation (hpi), then decreased rapidly, and the concentration was almost the same as that of un-inoculated control at 96 hpi. The results showed that JA induce wheat plants to elevate expression level of *PR1*, *PR2*, *PR3*, *PR4*, *PR5*, *PR9*, *PR10* and *Ta-JA2* by 10 folds in different varieties after MeJA application. Motallebi et al. (2015) studied the effect of methyl jasmonate on enzyme activities in wheat genotypes infected by the crown and root rot pathogen *Fusarium culmorum.* The activities of pathogen responsive defense-related enzymes (SOD, CAT, POX, PPO, LOX and PAL), total phenols and callose contents were higher in Sumai3, while treatment with MeJA (1 mM) significantly increased these enzymes activities and total phenols content in Falat, signifying the most sensitive cultivar which had a weak reaction to the pathogen but a strong response to MeJA treatment.

Lipopolysaccharides

Lipopolysaccharides (LPS) and lipooligosaccharides from the outer surface of Gram negative bacteria are known to induce a number of disease resistance components (Ranf, 2016; Erbs and Newman, 2003). For instance, the spent growth medium and purified exopolysaccharides (EPS) from *Pantoea agglomerans* have been shown to prime suspension-cultured wheat cells (Ortmann and Moerschbacher, 2006). Culture filtrate of *P. agglomerans* sprayed on wheat leaves suppressed infection of *Puccinia recondita* f. sp. *tritici* (Kempf and Wolf, 1989). Similarly, lipopeptides (surfactin, mycosutilin and fengycin) produced by *Bacillus subtilis* have been tested for their ability to induce resistance and protect wheat against powdery mildew (*Blumeria graminis* f.sp. *tritici*) infection (Khong et al., 2012). Under field conditions, surfactin and mycosubtilin provided 41% and 44% protection levels, respectively, when sprayed on wheat leaves at doses corresponding to 4g ha^{-1}. No direct antifungal effect of LPS was observed on *Blumeria graminis* f. sp. *tritici* conidia germination, irrespective of the tested concentrations (Khong et al., 2013). Further, they used RT-qPCR to investigate elicitor and priming effects of surfactin and mycosubtilin on the expression levels of defense-related genes viz., peroxidase encoding genes (*POX2* and *POX381*), lipoxygenase (*LOX*), AOS (allene oxide synthase) and *PR1* in response to pathogen inoculation. Interestingly, only surfactin induced the expression of defense-related genes in response to *Bgt* while mycosubtilin did not exhibit any priming effect (Khong et al., 2013). Thus, depending on the targeted pathogen and on the type of LP, these molecules could exhibit direct antifungal effect and/or defence induction, through either elicitation or potentialisation. In addition to rhizospheric PGPRs, cyclic lipo-polysacharides (CLP) may also be produced from phyllosphere. For example, *Bacillus liquefaciens* isolated from wheat spikes infected with *Fusarium graminearum* also produces CLPs (iturin, plipastatin and surfactin) with strong inhibitory activities against *Fusarium* head blight of wheat (Gong et al., 2015). Further, optical and fluorescence microscopy analyses revealed severe morphological changes in conidial and substantial distortions in *F. graminearum* hyphae treated with iturin A (50 µg/ml) or plipastatin A (100 µg/ml).

Mejri et al. (2017) assessed the biocontrol activity of three cyclic lipopeptides (mycosubtilin, surfactin and fengycin) from *Bacillus subtilis* on wheat against *Zymoseptoria tritici*. Foliar application of these biomolecules at 100 mg L^{-1} concentration on the wheat cultivars Dinosor and Alixan, two days before fungal inoculation, provided significant reduction of disease severity. The best protection (82% disease reduction) was achieved with mycosubtilin+surfactin containing formulations on Dinosor genotype. *In vitro* and *in planta* investigations revealed that mycosubtilin based formulations inhibit fungal growth, with half-maximal inhibitory concentrations of 1.4 mg L^{-1} for both mycosubtilin and mycosubtilin + surfactin and 4.5 mg L^{-1} for mycosubtilin + surfactin + fengycin, thus revealing that the observed efficacy of these products may rely mainly on antagonistic activity. By contrast, surfactin and fengycin had no direct activity on the pathogen, hence suggesting that these lipopeptides act on wheat against *Z. tritici* as resistance inducers rather than as biofungicides. On parallel lines, Mire et al. (2018) demonstrated that surfactin extracted from the *Bacillus amyloliquefaciens* S499 protected wheat against *Zymoseptoria tritici* by triggering both salicylic acid- and jasmonic acid-dependent defense system. Surfactin protected wheat by 70% against *Z. tritici*, similarly to the chemical reference elicitor Bion®50WG. Moreover, *in vitro* biocidal assays revealed no antifungal activities of surfactin towards the pathogen (Mire et al., 2018).

Proteins and Peptides

Cell wall proteins isolated from Pythium oligandrum have been reported to reduced the number of wheat spikelets infected with F. graminearum (Takenaka et al., 2003) and was developed commercially as Bion® by Novartis (now marketed by Syngenta) (Kessmann et al., 1996). Historically, it has been reported most effective against mildew in wheat and could be effective for up to 10 weeks (Görlach et al., 1996). Reignault et al. (2004) tested the effect of Iodus 40®, Milsana®, salicylyl heptanoate, trehalose, and pectic oligosaccharides on wheat to control powdery mildew and showed a reduction in the level of infection with Milsana®, salicylyl heptanoate and trehalose. Iodus 40® is a product of Goëmar (Saint-Malo, France) and is recommended for the control of diseases on wheat.

The syringolin peptide was originally isolated from the bacterium P. syringae pv. syringae (Wäspi et al., 1998). Syringolin A was shown to induce the accumulation of gene transcripts and resistance to powdery mildew infection if applied before inoculation (Wäspi et al., 2001). Further, it eradicated powdery mildew on wheat if applied post-inoculation. This curative effect was accompanied by the induction of cell death and the re-activation of PR-related genes whose transcripts accumulated initially following powdery mildew inoculation but declined as infection progressed. The authors suggested that syringolin A counteracted the suppression of host defense reactions imposed by the pathogen on colonized cells (Wäspi et al., 2001).

Trehalose

Trehalose (TH) is a non-reducing disaccharide commonly found in a wide variety of living organisms, including fungi, and is also associated with the protection of plants against different types of abiotic stresses (Drennan et al., 1993). Trehalose has also been shown to induce resistance. Thus, a reduction in infection intensity of the powdery mildew fungus *Blumeria graminis* f. sp. *tritici* on wheat was observed after treatment with trehalose (Reignault et al., 2001). Wheat plants grown under controlled conditions showed reductions in infection intensity of 50% and 95% respectively, following single spray andthree sprays of a trehalose solution (15 g L^{-1}), respectively, applied 48 hours prior to inoculation with powdery mildew (Reignault et al., 2001). Similarly, Tayeh et al. (2014) investigated the elicitor or priming effect of TH in wheat both before and during a compatible wheat-powdery mildew interaction. In un-inoculated conditions, TH elicited the expression of defense related genes encoding chitinase (*chi*, *chi1* and *chi4* precursor), pathogenesis-related protein 1, as well as oxalate oxidase *(oxo)*. Moreover, lipid metabolism also altered by TR spraying via the upregulation of lipoxygenase (lox) and lipid-transfer protein (ltp)-encoding gene expression (Tayeh et al., 2014).

Silicon

Silicon (Si) alleviate plant disease severity through preventing pathogen penetration via structural reinforcement; by inhibiting pathogen colonization through stimulating systemic acquired resistance; through antimicrobial compound production and enhancing plant resistance by activating multiple signaling pathways and defense-related gene expression (Wang et al., 2017; Remus-Borel et al., 2005). The beneficial effects of Si with regard to plant resistance to disease are attributed to Si accumulation in epidermal tissue, the formation of complexes with organic compounds in cell walls, the induction of phenolic compounds, phytolexin/glucanase/peroxidase production, and regulating pathogenicity or stress-related gene expression to limit pathogen invasion and colonization (Sakr, 2016; Belanger et al., 2003). Bélanger et al. (2003) reported that Si amendments in the form of exogenously supplied nutrient solution or calcium silicate slag protect wheat plants from powdery mildew disease caused by *Blumeria graminis* f.sp. *tritici*. Epidermal cells of plants supplied with Si reacted to fungus attack with papilla formation, production of callose, and release of electron-dense osmiophilic material identified by cytochemical labeling as glycosilated phenolics. Phenolic material not only accumulated along the cell wall, but also was associated with altered integrity of haustorian. Zones of infection by *B. graminis* f.sp. *tritici* in plants supplied with Si were characterized by intense fluorescence and collapse of conidial chains. Further, in wheat plants infected with *Blumeria graminis* f. sp. *tritici*, about 900 genes responding to pathogen infection were altered in control leaves, while few genes were changed by the pathogen in Si-supplied plants, suggesting that Si almost eliminated the stress imposed by the pathogen invasion (Chain et al., 2009). In the pathosystem of *Xanthomonas translucens* pv. *undulosa* and wheat plants, Silva et al. (2010) observed the resistance of plant to bacterial streak and recorded a reduction of 50.2% in the chlorotic leaf area when 0.3 g kg^{-1} of wollastonite (Si source) was added to the soil. Further, they observed increased chitinase activity and tissue lignification, and peroxidase activity with the highest concentration of the total soluble phenolics and lignin-thioglycolic acid derivatives in silicon treated wheat plants. Si application restricted hyphal entry to the first-invaded epidermal cell of wheat leaves infected with *Pyricularia oryzae*, while hyphae successfully invaded several neighboring leaf cells when there was no Si treatment (Sousa et al., 2013). Pagani et al. (2014) conducted field trails to study the effects of furrow application of calcium and magnesium silicate (300 kg ha^{-1}) and foliar application of SiO_2 (30 g l^{-1}) for wheat blast management. Results showed that disease incidence was significantly lower when plants were treated with foliar or furrow silicate. Moreover, Si applications significantly reduced disease and found BRS-264 genotype more responsive to silicon applications compared to BR-18 genotype. Silva et al. (2015) provided evidence at the histochemical level about the involvement of Si in the potentiation of the biosynthetic pathway of flavonoids in the leaves of wheat plants infected with *P. oryzae*. A similar result was found in wheat (*Bipolaris sorokiniana*) pathosystem in which Si supply delayed pathogen ingress into epidermal cells and reduced fungal colonization in foliar tissue (Domiciano et al., 2013).

Phosphites

Phosphites are metal salts of phosphorous acid that have low toxicity and high solubility, show systemic translocation (both upward and downward), and are reported to elicit systemic acquired resistance (Carmona et al. 2018; Deliopoulos et al., 2010; Reuveni, 1997). Potassium phosphite (K_2HPO_3) has been widely used for control of several plant pathogens (Kashyap and Dhiman, 2009; Lovatt and Mikkelsen, 2006; Forster et al. 1998). In wheat, Cruz et al. (2011) has reported that phosphite (1500 mL ha^{-1}), reduced the severity of wheat blast under controlled conditions.

Glycerol

Glycerol is an environmentally friendly, non-toxic, edible and biodegradable sugar alcohol. Li et al. (2016) provided evidence that foliar spray application of glycerol (3%) 1–2 days before powdery mildew infection and subsequent applications once every 4 days was sufficient to stimulate the plant defense responses without causing any significant damage to wheat leaves. Further, exogenous application of glycerol enhanced the Glycerol-3-phosphate (G3P) level, decreased the level of oleic acid (18:1) and led to the induction of expression of *pathogenesis-related* (*PR*) genes (*TaPR-1, TaPR-2, TaPR-3, TaPR-4*, and *TaPR-5*), generation of reactive oxygen species (ROS) before powdery mildew infection, and induced SA accumulation in wheat leaves. Under field conditions, it was noticed that sprayed glycerol significantly reduced the severity of powdery mildew disease and lessened disease-associated kernel weight loss, without any degradation in wheat seed quality. It is also worth mentioning here that glycerol is one of the cosolvent material used in the application of agrochemicals and therefore appears to have significant potential to be applied in wheat field as an environmentally-friendly agricultural chemical to manage crop pathogens.

Biological Inducers

There is a vast literature describing positive effects of microbes on disease supersession by eliciting ISR (Kashyap et al., 2018c; Srivastava et al., 2012). The first evidence for the involvement of IR in wheat disease control by microbe came from work by Kilic-Ekici and Yuen (2003). These researchers experimentally demonstrated the control of leaf spot of wheat caused by *B. sorokiniana* using *Lysobacter enzymogenes* strain C3 on tall fescue and wheat leaves. They observed that application of live and heat killed cells of C3 provided similar protection against *B. sorokiniana* on wheat as on tall fescue. Application of live or heat killed cells of C3 to only the basal portion of a wheat leaf resulted in less germination of *B. sorokiniana* conidia on along the entire length of the treated leaf compared with the distilled water control. In addition, both live and dead C3 cells sprayed onto wheat plants reduced *Bipolaris* leaf spot development compared with distilled water. The induced resistance observed was long lasting and was not pathogen or host specific, with *L. enzymogenes* controlling *B. sorokiniana* on wheat. Moreover, application of C3 either to foliage or roots resulted in significantly higher levels of peroxidase activity in leaves compared with the water treated controls (Kilic-Ekici and Yuen, 2003).

Al-Ani et al. (2011) conducted studies on induced systemic resistance and promotion of wheat and barley plants growth by *Pseudomonas fluorescence*, *Azospirillum irakense*, sea force extract and elsa fungicide (1ml L^{-1}) against barley yellow dwarf virus. The results indicated that *P. fluorescence* was more efficient in reducing disease incidence as compared to *A. irakense*, Sea force extract and Elsa. Moreover, partially resistant lines (cv. IBA 99, Arivate and Karonea) were found more responsive to treatments than the susceptible ones (Hashmia and Kara) (Al-Ani et al., 2011).

Rhizosphere microbe-mediated induction of phenylpropanoid cascade and induced systemic responses against *Bipolaris sorokiniana* in wheat has been demonstrated by Singh et al. (2016). The study revealed that the plants co-inoculated with *B. amyloliquefaciens* B-16 and *T. harzianum* UBSTH-501 up-regulated the phenylpropanoid cascade and manifold increase was recorded in phenylalanine ammonia lyase (PAL), peroxidase (PO), chitinase and other enzymes related to induce systemic resistance.

Singh and Jha (2017) observed that plant growth promoting bacterium *Stenotrophomonas maltophilia* SBP-9 augments resistance against *Fusarium graminearum* in wheat plants by inducing ISR. They noticed that bacterium treatment led to the enhancement in the level of defense enzymes like β-1, 3 glucanase, PAL, PO, and polyphenol oxidase (PPO), which can protect plants from the infection of pathogens.

Samain et al. (2017) demonstrated the potential of *Paenibacillus* sp. strain B2 (PB2) to stimulate plant defense mechanisms via its paenimyxin lipo-polypeptide elicitor. *Paenibacillus* strain B2 conferred more than 59% protection against *Septoria tritici* blotch by inducing systemic resistance which is characterized by the over expression of the pr1, lox, Aos, peroxidase, oxo and gst genes. Paenimyxin conferred 76% local protection characterized by the overexpression of the β-1,3-glucanase (*glu*), lipoxygenase (*lox*), allene oxide synthase (*aos*), phenylalanine ammonia-lyase (*pal*), chalcone synthases (chs), oxalate oxidase (oxo), and glutathione-s-transferase (*gst*) genes, and 82% systemic protection by *chs*. PB2 also stimulates wheat root colonization by *Curtobacterium plantarum* strain EDS, promoting wheat growth.

Bacillus species have been very effective biocontrol agents due to their ability to produce heat and desiccation-resistant spores and to withstand high temperature, unfavorable pH, lack of nutrients or water, and the ease of stable formulation preparation (Solanki et al., 2015; Singh et al. 2014; Solanki et al. 2012). The ability of *Bacillus subtilis* and *Bacillus thuringiensis* to induce systemic resistance in wheat plants against *Septoria nodorum* has been studied by Burkhanova et al. (2017). They showed that strains of *Bacillus* ssp. that possess the capacity for endophytic survival have antagonistic activity against this pathogen *in vitro*. A reduction of *S. nodorum* blotch development on wheat leaves under the influence of *Bacillus* spp. was accompanied by the suppression of catalase activity, an increase in peroxidase activity and H_2O_2 content, and expression of defense related genes such us *PR-1*, *PR-6*, and *PR-9*. Interestingly, they observed that *B. subtilis* 26D induces expression levels of wheat pathogenesis-related (PR) genes which marks a SA-dependent pathway, however, *B. thuringiensis* V-5689 and V-6066 induces a JA/ET-dependent pathway.

The genus *Trichoderma* has gained immense importance in past several decades due to its antagonistic ability against wide range of plant pathogens and growth promotion in crop plants by triggering ISR (Kashyap et al., 2017; Rai et al., 2016a; Rai et al., 2016b; Srivastava et al., 2012; Solanki et al., 2011). Cordo et al. (2007) reported that *Trichoderma* spp. act as an elicitor of wheat plant defense responses against *Septoria tritici*. Coating seeds with *T. harzianum* Th5 was found to be the most efficacious application technique for controlling the leaf blotch of wheat. In addition, the leaf apoplast antifungal proteolytic activity decreased in plants only inoculated with *S. tritici* and increased in those grown from seeds coated with the *T. harzianum*. These results further confirmed that endogenous germin-like protease inhibitor coordinated the proteolytic action and stimulated biochemical systemic induced response against *S. tritici*. On parallel lines, Stocco et al. (2015) reported that native isolates of *T. harzianum* inducting resistance to *Zymoseptoria tritici* on wheat plants. They attributed the reduction of the *Septoria* leaf blotch severity to a systemic induced resistance throughout elicitors with enzymatic function as the serine-protease. This process can be transitory, but strongly potentiates the expression of defense-related proteins when plants are challenged by pathogens at sites distant from the location of the *Trichoderma* strain. Biological control of stem rust of wheat (*Puccinia graminis* f. sp. *tritici*) by using different combinations of *Trichoderma* spp. and arbuscular mycorrhizal (AM) fungi has been reported by El-Sharkawy et al. (2018). Under the natural conditions, application of AM fungi and *Trichoderma* spp. significantly reduced the disease measures, induced the peroxidase and polyphenol oxidase enzymes, increased the total phenol content and improved the wheat growth and yield parameters (El-Sharkawy et al., 2018). Moreover, the synergistic effect of *Trichoderma* spp. and AM fungi in inducing the salicylate, jasmonate and ethylene-regulated defense mechanisms and inhibiting the uredospore germination of *Puccinia graminis* f. sp. *tritici* explains the superiority of the combined treatment (arbuscular mycorrhizal + *T. harzianum* HL1 + *T. viride* HL5) over the other single or combined treatments in regard to the biocontrol efficacy (El-Sharkawy et al., 2018). Overall, these above mentioned results open the way towards the development of biocontrol tools for wheat protection, in order to enhance the sustainability of current agricultural practices.

INDUCED RESISTANCE AS A COMPONENT OF INTEGRATED DISEASE MANAGEMENT (IDM)

Biological control and induced resistance by chemicals and microbes have potential to act as an essential component of an integrated diseases management programme. ASM has been evaluated in combination with fungicides for plant disease control. A tank mixture of ASM and azoxystrobin provided better control of powdery mildew and leaf blotch (*Septoria tritici*) on wheat than that achieved by either component when applied separately (Stadnik and Buchenauer, 1999). Foliar applications of urea with BTH also improved wheat yields without compromising with effective control of BTH, regardless of the increased symptoms of powdery mildew associated with the increase in available nitrogen (Stadnik and Buchenauer, 1999). Similarly, ASM improved efficacy of cypronidil against powdery mildew, but the addition of the plant activator offered no benefit in relation to grain yield compared with fungicide only. In Danish field trials, the use of ASM in an integrated control strategy for cereal diseases enabled a reduction in the number of fungicide applications in three of seven field trials (Jorgensen et al., 1997). In Brazil, Santos et al. (2011) reported that combined application of phosphite with pyraclostrobin + epoxiconazole (400 mL ha^{-1}), pyraclostrobin + epoxiconazole 300 mL ha^{-1}) and pyraclostrobin + epoxiconazole (400 mL ha^{-1}) effectively control leaf rust, powdery mildew, yellow and brown spot in the wheat under glass house conditions.

FACTORS AFFECTING INDUCED RESISTANCE

Wheat genotype is known to affect the expression of induced resistance. For instance, Pagani et al. (2014) examined the influence of host genotype on the expression of induced resistance in wheat against wheat blast pathogen. They observed that wheat cv. BRS 264 is apparently more predisposed to the beneficial effects of Si treatment than BR 18. Similarly, Santos et al. (2011) demonstrated positive effects for the control of *Drechslera tritici-repentis* and *Bipolaris sorokiniana*, but the response to phosphite depends on the crop species and on the pathosystem. As above mentioned findings came from studies employing a single pathogen isolate or a single host genotype which might lead to the wrong conclusions that may affect the effectiveness of elicitors in practice. Newton et al. (2004) demonstrated that bacterial inoculums from previous crop can affect fungal disease development on subsequent non-host crops. This has consequences for agronomy including rotations. While the synergistic enzyme theory may explain the effects on hemi-biotrophic pathogens in their necrotrophic phase, mildew on wheat was also considerably more severe on the heavy bacterial inoculum plots. The mechanism may not be attributable to direct interaction between the bacteria and *Blumeria graminis* inoculum. It could be due to induced susceptibility or even simply a nutritional effect mediated through the plant as increased nitrogen can have very large effects on increasing susceptibility to mildew in cereals. For example, Newton et al. (2000) noted that high nitrogen level increased powdery mildew by 4.58 times compared with a low level under high inoculum pressure conditions. Whatever the explanation, expression of resistance was affected by other organisms in the environment directly or, more likely, indirectly. Both light and humidity have been demonstrated to affect expression of defence-related genes, including those involved in the production of reactive oxygen species (ROS) (Mateo et al., 2004; Zhou et al., 2004). Light will clearly directly affect the chloroplasts and thereby the redox status of cells, and this has been shown to directly affect SAR (Fobert and Després, 2005). In wheat, a yellow rust resistance gene which is expressed better under high light intensity is known (Ash and Rees, 1994). All above mentioned reports demonstrate the fact that the expression of induced resistance can be influenced by host development, varies with plant age or tissue maturity and abiotic stress and highlight the need for such interactions to be taken into account when induced resistance is used in a field situation.

CONSEQUENCES OF RESISTANCE INDUCTION

Induction of resistance is costly in terms of yield penalty and therefore normally only triggered upon actual pathogen recognition (Walters et al., 2013). It has long been recognized that resistance genes and matching virulence may have a cost in terms of fitness to the pathogen, resulting in stabilizing selection (Vanderplank, 1968), and this has been demonstrated experimentally in *A. thaliana* by Tian et al. (2003). Heil et al. (2000) examined the effects of chemically induced resistance on costs in the absence of pathogen pressure. They applied ASM to wheat in the absence of pathogens and found that treated plants had reduced biomass and reduced numbers of ears and grains, with most marked effects under nitrogen-limiting conditions. These results suggesting that use of ASM incurs allocations costs and supports the 'growth-differentiation balance' hypothesis of Herms and Mattson (1992) that assumes a metabolic competition between processes involved in plant growth and those necessary for plant differentiation, such as the synthesis of chemicals for plant defense. Research on ecologically realistic study systems and ecologically intended research in the agricultural systems thus is needed to obtain a reliable risk assessment of ISR in wheat protection

FUTURE CHALLENGES AND DIRECTIONS

Recent research progress in understanding molecular basis for SAR, ISR, priming, and plant immunity indicate a wider role for plant defense activating compounds. Several compounds of biological and non-biological origin are gaining significant momentum in curtailing the use of health hazardous chemicals for wheat disease management. Still, there are several researchable issues that need attention for effective amalgamation of resistance inducers into wheat protection package and for their wider use in farmer's fields, which are enlisted as follows:

- Compounds inducing resistance played important functions in wheat–pathogen interactions. Studies on structures and functions of these compounds in plants, as well as the underlying genetic repertoire of wheat-associated pathogens will help to develop effective disease management tools and to improve disease resistance in wheat plants.
- There is a need for more effective resistance inducers against wheat disease, and the information attained in the last few years on SAR, ISR and RNA silencing mechanisms, and the interactions between them, may guide to develop new molecules with effective action against wheat pathogens. Additionally, deep research efforts is sought to understand the effects of nutrition, doses, timing of application, and genotypic effects post application of resistance inducing compounds.
- An understanding of the molecular changes leading to defense priming and pathogen resistance will provide valuable knowledge for producing disease-resistant crop varieties by exposing parental plants to priming activators, as well as to the development of novel plant protection chemicals that stimulate the plant's inherent disease resistance mechanisms.
- A number of reports emphasized that ISR-inducing PGPR is a useful tool to reduce wheat diseases that are sensitive to JA- and ET-dependent defenses. Thus, integrating ISR-triggering PGPR into disease management programme in conjunction with other strategies will be a worthwhile approach to explore.

REFERENCES

Ahmad, S., Gordon-Weeks, R., Pickett, J. A., & Ton, J. (2010). Natural variation in priming of basal resistance: From evolutionary origin to agricultural exploitation. *Molecular Plant Pathology*, *11*(6), 817–827. doi:10.1111/j.1364-3703.2010.00645.x PMID:21029325

Al-Ani, R. A., Adhab, M. A., El-Muadhidi, M. A., & Al-Fahad, M. A. (2011). Induced systemic resistance and promotion of wheat and barley plants growth by biotic and non-biotic agents against barley yellow dwarf virus. *African Journal of Biotechnology*, *10*(56), 12079–12084. doi:10.5897/AJB11.1921

Al-Maaroof, E., Fayadh, A. H., & Fattah, F. A. (2014). Use of some chemical inducers to improve wheat resistance to *Puccinia striiformis* f. sp. *tritici*. *Ratarstvo i Povrtarstvo*, *51*(2), 83–90. doi:10.5937/ratpov51-5985

Anand, T., Chandrasekaran, A., Kuttalam, S., Raguchander, T., Prakasam, V., & Samiyappan, R. (2007). Association of some plant defense enzyme activities with systemic resistance to early leaf blight and leaf spot induced in tomato plants by azoxystrobin and *Pseudomonas fluorescens*. *Journal of Plant Interactions*, *2*(4), 233–244. doi:10.1080/17429140701708985

Angelova, Z., Georgiev, S., & Roos, W. (2006). Elicitation of Plants. *Biotechnology, Biotechnological Equipment*, *20*(2), 72–83. doi:10.1080/13102818.2006.10817345

Ash, G. J., & Rees, R. G. (1994). Effect of post-inoculation temperature and light-intensity on expression of resistance to stripe rust in some Australian wheat cultivars. *Australian Journal of Agricultural Research*, *45*(7), 1379–1386. doi:10.1071/AR9941379

Belanger, R. R., Benhamou, N., & Menzies, J. G. (2003). Cytological evidence of an active role of silicon in wheat resistance to powdery mildew (Blumeria graminis f. sp tritici). *Phytopathology*, *93*(4), 402–412. doi:10.1094/PHYTO.2003.93.4.402 PMID:18944354

Bhaskara Reddy, M. V., Arul, J., Angers, P., & Couture, L. (1999). Chitosan treatment of wheat seeds induces resistance to Fusarium graminearum and improves seed quality. *Journal of Agricultural and Food Chemistry*, *47*(3), 1208–1216. doi:10.1021/jf981225k PMID:10552439

Bruce, T. J. A. (2010). Tackling the threat to food security caused by crop pests in the new millennium. *Food Security*, *2*(2), 133–141. doi:10.100712571-010-0061-8

Bruce, T. J. A., Smart, L. E., Birch, A. N. E., Blok, V. C., MacKenzie, K., Guerrieri, E., Cascone, P., Luna, E., & Ton, J. (2017). Prospects for plant defence activators and biocontrol in IPM-Concepts and lessons learnt so far. *Crop Protection (Guildford, Surrey)*, *97*, 128–134. doi:10.1016/j.cropro.2016.10.003

Burkhanova, G. F., Veselova, S. V., Sorokan', A. V., Blagova, D. K., Nuzhnaya, T. V., & Maksimov, I. V. (2017). Strains of *Bacillus* ssp. regulate wheat resistance to *Septoria nodorum* Berk. *Applied Biochemistry and Microbiology*, *53*(3), 346–352. doi:10.1134/S0003683817030048

Cai, K., Gao, D., Chen, J., & Shiming, L. (2009). Probing the mechanisms of silicon-mediated pathogen resistance. *Plant Signaling & Behavior*, *4*(1), 1–3. doi:10.4161/psb.4.1.7280 PMID:19568332

Carmona, M. A., Sautua, F. J., Grijalba, P. E., Cassina, M., & Pérez-Hernández, O. (2018). Effect of potassium and manganese phosphites in the control of Pythium damping-off in soybean: A feasible alternative to fungicide seed treatments. *Pest Management Science*, *74*(2), 366–374. doi:10.1002/ps.4714 PMID:28842951

Chain, F., Côté-Beaulieu, C., Belzile, F., Menzies, J., & Bélanger, R. (2009). A comprehensive transcriptomic analysis of the effect of silicon on wheat plants under control and pathogen stress conditions. *Molecular Plant-Microbe Interactions*, *22*(11), 1323–1330. doi:10.1094/MPMI-22-11-1323 PMID:19810802

Chang, S. J., Tzeng, D. D. S., & Li, C. C. (2002). Effect of silicon nutrient on bacterial blight resistance of rice (Oryza sativa L.). *Proceedings of the Second Silicon in Agriculture Conference*, 31–33.

Choudhary, D. K., & Johri, B. N. (2009). Interactions of *Bacillus* spp. and plants – With special reference to induced systemic resistance (ISR). *Microbiological Research*, *164*(5), 493–513. doi:10.1016/j.micres.2008.08.007 PMID:18845426

Conrath, U. (2006). Systemic Acquired Resistance. *Plant Signaling & Behavior*, *1*(4), 179–184. doi:10.4161/psb.1.4.3221 PMID:19521483

Cordo, C. A., Monaco, C. I., Segarra, C. I., Simon, M. R., Mansilla, A. Y., Perelló, A. E., Kripelz, N. I., Bayo, D., & Conde, R. D. (2007). *Trichoderma* spp. as elicitors of wheat plant defense responses against *Septoria tritici*. *Biocontrol Science and Technology*, *17*(7), 687–698. doi:10.1080/09583150701527094

Cortes-Barco, A. M., Goodwin, P. H., & Hsiang, T. (2010). Comparison of induced resistance activated by benzothiadiazole, (2R,3R)-butanediol and an isoparaffin mixture against anthracnose of *Nicotiana benthamiana*. *Plant Pathology*, *59*(4), 643–653. doi:10.1111/j.1365-3059.2010.02283.x

Cruz, M. F. A., Diniz, A. P. C., Rodrigues, F. A., & Barros, E. G. (2011). Foliar application of products on the reduction of wheat blast severity. *Tropical Plant Pathology*, *36*, 424–428. doi:10.1590/S1982-56762011000600014

Deliopoulos, T., Kettlewell, P. S., & Hare, M. C. (2010). Fungal disease suppression by inorganic salts: A review. *Crop Protection (Guildford, Surrey)*, *29*(10), 1059–1075. doi:10.1016/j.cropro.2010.05.011

Domiciano, G. P., Rodrigues, F. A., Guerra, A. M. N., & Vale, F. X. R. (2013). Infection process of *Bipolaris sorokiniana* on wheat leaves is affected by silicon. *Tropical Plant Pathology*, *38*(3), 258–263. doi:10.1590/S1982-56762013005000006

Domiciano, G. P., Rodrigues, F. A., Vale, F. X. R., Filha, M. S. X., Moreira, W. R., Andrade, C. C. L., & Pereira, S. C. (2010). Wheat resistance to spot blotch potentiated by silicon. *Journal of Phytopathology*, *158*(5), 334–343. doi:10.1111/j.1439-0434.2009.01623.x

Drennan, P. M., Smith, M. T., Goldsworthy, D., & van Staden, J. (1993). The occurrence of trehalose in the leaves of the desiccation-tolerant angiosperm Myrothamnus flabellifolius welw. *Journal of Plant Physiology*, *142*(4), 493–496. doi:10.1016/S0176-1617(11)81257-5

Duan, Z., Lv, G., Shen, C., Li, Q., Qin, Z., & Niu, J. (2014). The role of jasmonic acid signalling in wheat (*Triticum aestivum* L.) powdery mildew resistance reaction. *European Journal of Plant Pathology*, *140*(1), 169–183. doi:10.100710658-014-0453-2

Durrant, W. E., & Dong, X. (2004). Systemic acquired resistance. *Annual Review of Phytopathology*, *42*(1), 185–209. doi:10.1146/annurev.phyto.42.040803.140421 PMID:15283665

Dutt, S., Pandey, D., & Kumar, A. (2011). Jasmonate signal induced expression of cystatin genes for providing resistance against Karnal bunt in wheat. *Plant Signaling & Behavior*, *6*(6), 821–830. doi:10.4161/psb.6.6.14743 PMID:21849815

El-Sharkawy, H. H. A., Rashad, Y. M., & Ibrahim, S. A. (2018). Biocontrol of stem rust disease of wheat using arbuscular mycorrhizal fungi and *Trichoderma* spp. *Physiological and Molecular Plant Pathology*, *103*, 84–91. Advance online publication. doi:10.1016/j.pmpp.2018.05.002

Erbs, G., & Newman, M.-A. (2003). The role of lipopolysaccharides in induction of plant defence responses. *Molecular Plant Pathology*, *4*(5), 421–425. doi:10.1046/j.1364-3703.2003.00179.x PMID:20569401

Figueroa, M., Hammond-Kosack, K. E., & Solomon, P. S. (2018). A review of wheat diseases-a field perspective. *Molecular Plant Pathology*, *19*(6), 1523–1536. doi:10.1111/mpp.12618 PMID:29045052

Filha, M. S. X., Rodrigues, F. A., Domiciano, G. P., Oliveira, H. V., Silveira, P. R., & Moreira, W. R. (2011). Wheat resistance to leaf blast mediated by silicon. *Australasian Plant Pathology*, *40*(1), 28–38. doi:10.100713313-010-0010-1

Fobert, P. R., & Després, C. (2005). Redox control of systemic acquired resistance. *Current Opinion in Plant Biology*, *8*(4), 378–382. doi:10.1016/j.pbi.2005.05.003 PMID:15922650

Forster, H., Adaskaveg, J. E., Kim, D. H., & Stanghellini, M. E. (1998). Effect of phosphite on tomato and pepper plants and on susceptibility of pepper to Phytophthora root and crown rot in hydroponic culture. *Plant Disease*, *82*(10), 1165–1170. doi:10.1094/PDIS.1998.82.10.1165 PMID:30856781

Gardner, S. N., Agrawal, A. A., Gressel, J., & Mangel, M. (1999). Strategies to delay the evolution of resistance in pests: Dose rotations and induced plant defenses. *Aspects of Applied Biology*, *53*, 189–196.

Gong, A.-D., Li, H.-P., Yuan, Q.-S., Song, X.-S., Yao, W., He, W.-J., Zhang, J.-B., & Liao, Y.-C. (2015). Antagonistic Mechanism of Iturin A and Plipastatin A from *Bacillus amyloliquefaciens* S76-3 from Wheat Spikes against Fusarium graminearum. *PLoS One*, *10*(2), e0116871. doi:10.1371/journal.pone.0116871 PMID:25689464

Görlach, J., Voirath, S., & Knauf-Beiter, G. (1996). Benzothiadiazole, a novel class of inducers of systemic acquired resistance, activates gene expression and disease resistance in wheat. *The Plant Cell*, *8*(4), 629–643. doi:10.1105/tpc.8.4.629 PMID:8624439

Grant, M., & Lamb, C. (2006). Systemic immunity. *Current Opinion in Plant Biology*, *9*(4), 414–420. doi:10.1016/j.pbi.2006.05.013 PMID:16753329

Griffey, C. A., Das, M. K., & Stromberg, E. L. (1993). Effectiveness of adult plant resistance in reducing grain yield loss to powdery mildew in winter wheat. *Plant Disease*, *77*(6), 618–622. doi:10.1094/PD-77-0618

Guével, M. H., Menzies, J. G., & Bélanger, R. R. (2007). Effect of root and foliar applications of soluble silicon on powdery mildew control and growth of wheat plants. *European Journal of Plant Pathology*, *119*(4), 429–436. doi:10.100710658-007-9181-1

Gutierrez-Barranquero, J. A., Arrebola, E., Bonilla, N., Sarmiento, D., Cazorla, F. M., & de Vicente, A. (2012). Environmentally friendly treatment alternatives to Bordeaux mixture for controlling bacterial apical necrosis (BAN) of mango. *Plant Pathology*, *61*(4), 665–676. doi:10.1111/j.1365-3059.2011.02559.x

Heier, T., Jain, S. K., Kogel, K.-H., & Pons-Kühnemann, J. (2005). Influence of N-fertilization and fungicide strategies on Fusarium head blight severity and mycotoxin content in winter wheat. *Journal of Phytopathology*, *153*(9), 551–557. doi:10.1111/j.1439-0434.2005.01021.x

Heil, M., & Bostock, R. M. (2002). Induced Systemic Resistance (ISR) Against Pathogens in the Context of Induced Plant Defences. *Annals of Botany*, *89*(5), 503–512. doi:10.1093/aob/mcf076 PMID:12099523

Heil, M., Hilpert, A., Kaiser, W., & Linsenmair, K. E. (2000). Reduced growth and seed set following chemical induction of pathogen defence: Does systemic acquired resistance (SAR) incur allocation costs? *Journal of Ecology*, *88*(4), 645–654. doi:10.1046/j.1365-2745.2000.00479.x

Herms, D. A., & Mattson, W. J. (1992). The dilemma of plants to grow or defend. *The Quarterly Review of Biology*, *67*(3), 283–335. doi:10.1086/417659

Hoffland, E., Hakulinem, J., & Van Pelt, J. A. (1996). Comparison of systemic resistance induced by avirulent and nonpathogenic *Pseudomonas* species. *Phytopathology*, *86*(7), 757–762. doi:10.1094/Phyto-86-757

Hofgaard, I. S., Ergon, Å., Henriksen, B., & Tronsmo, A. M. (2010). The effect of potential resistance inducers on development of *Microdochium majus* and *Fusarium culmorum* in winter wheat. *European Journal of Plant Pathology*, *128*(2), 269–281. doi:10.100710658-010-9662-5

Hofgaard, I. S., Ergon, A., Wanner, L. A., & Tronsmo, A. M. (2005). The effect of chitosan and Bion on resistance to pink snow mould in perennial ryegrass and winter wheat. *Journal of Phytopathology*, *153*(2), 108–119. doi:10.1111/j.1439-0434.2005.00937.x

Jasrotia, P., Kashyap, P. L., Bhardwaj, A. K., Kumar, S., & Singh, G. P. (2018). Scope and applications of nanotechnology for wheat production: A review of recent advances. *Wheat and Barley Research*, *10*(1), 137–150. doi:10.25174/2249-4065/2018/76672

Jayaraj, J., Muthukrishnan, S., Liang, G., & Velazhahan, R. (2004). Jasmonic acid and salicylic acid induce accumulation of β-1,3-glucanase and thaumatin-like proteins in wheat and enhance resistance against *Stagonospora nodorum*. *Biologia Plantarum*, *48*(3), 425–430. doi:10.1023/B:BIOP.0000041097.03177.2d

Jørgensen, L. N., Secher, B. J. M., Olesen, J. E., & Mortensen, J. (1997). Need for fungicide treatments when varying agricultural parameters. *Aspects of Applied Biology*, *50*, 285–292.

Joshi, A. K., Kumar, S., Chand, R., & Ortiz-Ferrara, G. (2004). Inheritance of resistance to spot blotch pathogen caused by *Bipolaris sorokiniana* in spring wheat. *Plant Breeding*, *123*(3), 213–219. doi:10.1111/j.1439-0523.2004.00954.x

Joshi, A. K., Ortiz-Ferrara, G., Crossa, J., Singh, G., Alvarado, G., Bhatta, M. R., Duveiller, E., Sharma, R. C., Pandit, D. B., Siddique, A. B., Das, S. Y., Sharma, R. N., & Chand, R. (2007). Associations of environments in South Asia based on spot blotch disease of wheat caused by *Cochliobolus sativus*. *Crop Science*, *47*(3), 1071–1081. doi:10.2135/cropsci2006.07.0477

Karban, R., & Baldwin, I. T. (1997). *Induced responses to herbivory*. Chicago Press. doi:10.7208/chicago/9780226424972.001.0001

Kashyap, P. L., & Dhiman, J. S. (2009). Induction of resistance in cauliflower against *Alternaria* blight using potassium and phosphonic salts. *The Asian and Australasian Journal of Plant Science and Biotechnology*, *3*(1), 66–70.

Kashyap, P. L., Jasrotia, P., Kumar, S., Singh, D. P., & Singh, G. P. (2018a). Identification guide for major diseases and insect-pests of wheat. ICAR-Indian Institute of Wheat and Barley Research, Karnal *Technical Bulletin 18*.

Kashyap, P. L., Kaur, S., & Pannu, P. P. S. (2018b). Induction of systemic tolerance to *Tilletia indica* in wheat by plant defence activators. *Archiv für Phytopathologie und Pflanzenschutz*, *51*(1-2), 1–13. doi:10.1080/03235408.2018.1438778

Kashyap, P. L., Kaur, S., Sanghera, G. S., Kang, S. S., & Pannu, P. P. S. (2011). Novel methods for quarantine detection of Karnal bunt (*Tilletia indica*) of wheat. *Elixir Agriculture, 31*, 1873–1876.

Kashyap, P. L., Kumar, S., Kumar, R. S., Sharma, A., Jasrotia, P., Singh, D. P., & Singh, G. P. (2020). Molecular Diagnostic Assay for Rapid Detection of Flag Smut Fungus (Urocystis agropyri) in Wheat Plants and Field Soil. *Front. Plant Sci., 11*, 1039. doi:10.3389/fpls.2020.01039 PMID:32754183

Kashyap, P. L., Kumar, S., Tripathi, R., Kumar, R. S., Jasrotia, P., Singh, D. P., & Singh, G. P. (2019). Phylogeography and Population Structure Analysis Reveal Diversity by Gene Flow and Mutation in *Ustilago segetum* (Pers.) Roussel *tritici* Causing Loose Smut of Wheat. *Frontiers in Microbiology, 10*, 1072. doi:10.3389/fmicb.2019.01072 PMID:31156587

Kashyap, P. L., Rai, P., Srivastava, A. K., & Kumar, S. (2017). *Trichoderma* for climate resilient agriculture. *World Journal of Microbiology & Biotechnology, 33*(8), 155. doi:10.100711274-017-2319-1 PMID:28695465

Kashyap, P. L., Srivastava, A. K., Tiwari, S. P., & Kumar, S. (2018c). *Microbes for Climate Resilient Agriculture*. John Wiley and Sons. doi:10.1002/9781119276050

Kashyap, P. L., Xiang, X., & Heiden, P. (2015). Chitosan nanoparticle based delivery systems for sustainable agriculture. *International Journal of Biological Macromolecules, 77*, 36–51. doi:10.1016/j.ijbiomac.2015.02.039 PMID:25748851

Kempf, H.-J., & Wolf, G. (1989). *Erwinia herbicola* as a biocontrol agent of *Fusarium culmorum* and *Puccinia recondita* f. sp. *tritici* on wheat. *Phytopathology, 79*(9), 990–994. doi:10.1094/Phyto-79-990

Khong, N. G., Randoux, B., & Deravel, J. (2013). Induction of resistance in wheat by bacterial cyclic lipopeptides. *Communications in Agricultural and Applied Biological Sciences, 78*(3), 479–487. PMID:25151823

Khong, N. G., Randoux, B., & Tayeh, C. (2012). Induction of resistance in wheat against powdery mildew by bacterial cyclic lipopeptides. *Communications in Agricultural and Applied Biological Sciences, 77*(3), 39–51. PMID:23878959

Kilic-Ekici, O., & Yuen, G. Y. (2003). Induced resistance as a mechanism of biological control by Lysobacter enzymogenes strain C3. *Phytopathology, 93*(9), 1103–1110. doi:10.1094/PHYTO.2003.93.9.1103 PMID:18944093

Kim, Y. C., Kim, Y. H., Lee, Y. H., Lee, S. W., Chae, Y.-S., Kang, H.-K., Yun, B.-W., & Hong, J. K. (2013). β-Amino-n-butyric Acid Regulates Seedling Growth and Disease Resistance of Kimchi Cabbage. *The Plant Pathology Journal, 29*(3), 305–316. doi:10.5423/PPJ.OA.12.2012.0191 PMID:25288957

Leijerstam, B. (1972). Studies in powdery mildew on wheat in Sweden. III. Variability of virulence in *Erysiphe graminis* f. sp. *tritici* due to gene recombination and mutation. *Contributions of the National Swedish Institute of Plant Protection, 15*, 231–248.

Li, Y., Song, N., Zhao, C., Li, F., Geng, M., Wang, Y., Liu, W., Xie, C., & Sun, Q. (2016). Application of Glycerol for Induced Powdery Mildew Resistance in Triticum aestivum L. *Frontiers in Physiology, 7*, 413. doi:10.3389/fphys.2016.00413 PMID:27708588

Llorens, E., García-Agustín, P., & Lapeña, L. (2017). Advances in induced resistance by natural compounds: Towards new options for woody crop protection. *Scientia Agrícola*, *74*(1), 90–100. doi:10.1590/1678-992x-2016-0012

Lovatt, C. J., & Mikkelsen, R. L. (2006). Phosphite fertilizers: What are they? Can you use them? What can they do? *Better Crops with Plant Food*, *90*, 11–13.

Lyon, G. (2007). Agents that can elicit induced resistance. In D. Walters, A. Newton, & G. Lyon (Eds.), *Induced Resistance for Plant Defence: A Sustainable Approach to Crop Protection* (pp. 9–29). Blackwell Publishing. doi:10.1002/9780470995983.ch2

Ma, J. F., & Takahashi, E. (2002). *Soil, Fertilizer, and Plant Silicon Research in Japan*. Elsevier Science.

Ma, J. F., & Yamaji, N. (2006). Silicon uptake and accumulation in higher plants. *Trends in Plant Science*, *11*(8), 392–397. doi:10.1016/j.tplants.2006.06.007 PMID:16839801

Mateo, A., Mühlenbock, P., Rustérucci, C., Chang, C. C.-C., Miszalski, Z., Karpinska, B., Parker, J. E., Mullineaux, P. M., & Karpinski, S. (2004). LESION SIMULATING DISEASE 1 is required for acclimation to conditions that promote excess excitation energy. *Plant Physiology*, *136*(1), 2818–2830. doi:10.1104/pp.104.043646 PMID:15347794

Maurhofer, M., Hase, C., Meuwly, P., Metraux, J. P., & Defago, G. (1994). Induction of systemic resistance of tobacco to tobacco necrosis virus by the root-colonizing *Pseudomonas fluorescens* strain CHAO: Influence of the *gacA* gene and of pyoverdine production. *Phytopathology*, *84*(2), 139–146. doi:10.1094/Phyto-84-139

Mejri, S., Siah, A., & Coutte, F. (2017). Biocontrol of the wheat pathogen *Zymoseptoria tritici* using cyclic lipopeptides from *Bacillus subtilis*. *Environmental Science and Pollution Research International*. Advance online publication. doi:10.100711356-017-9241-9 PMID:28634804

Mire, G. L., Siah, A., & Brisset, M.-N. (2018). Surfactin Protects Wheat against *Zymoseptoria tritici* and Activates Both salicylic acid- and jasmonic acid-dependent defense responses. *Agriculture*, *8*(1), 11. doi:10.3390/agriculture8010011

Moldes, C. A., De Lima Filho, O. F., Merini, L. J., Tsai, S. M., & Camiña, J. M. (2016). Occurrence of powdery mildew disease in wheat fertilized with increasing silicon doses: A chemometric analysis of antioxidant response. *Acta Physiologiae Plantarum*, *38*(8), 206. doi:10.100711738-016-2217-4

Molina, A., Görlach, J., Volrath, S., & Ryals, J. (1999). Wheat genes encoding two types of PR-1 proteins are pathogen inducible, but do not respond to activators of systemic acquired resistance. *Molecular Plant-Microbe Interactions*, *12*(1), 53–58. doi:10.1094/MPMI.1999.12.1.53 PMID:9885193

Motallebi, P., Niknam, V., Ebrahimzadeh, H., Enferadi, S. T., & Hashemi, M. (2015). The effect of methyl jasmonate on enzyme activities in wheat genotypes infected by the crown and root rot pathogen *Fusarium culmorum*. *Acta Physiologiae Plantarum*, *37*(11), 237. doi:10.100711738-015-1988-3

Muchembled, J., Lounes-Hadj Sahraoui, A., Grandmoujin-Ferjani, A., & Sancholle, M. (2006). Changes in lipid composition of *Blumeria graminis* f. sp. *tritici* conidia produced on wheat leaves treated with heptanoyl salicylic acid. *Phytochemistry*, *67*(11), 1104–1109. doi:10.1016/j.phytochem.2006.02.025 PMID:16647727

Newton, A., Guy, D., Gaunt, R., & Thomas, W. (2000). The effect of powdery mildew inoculum pressure and fertilizer levels on disease tolerance in spring barley. *Zeitschrift für Pflanzenkrankheiten und Pflanzenschutz*, *107*, 67–73.

Newton, A. C., Toth, I. K., Neave, P., & Hyman, L. J. (2004). Bacterial inoculum from a previous crop affects fungal disease development on subsequent nonhost crops. *The New Phytologist*, *63*(1), 133–138. doi:10.1111/j.1469-8137.2004.01077.x PMID:33873780

Oliveira, M. D. M., Varanda, C. M. R., & Félix, M. R. F. (2016). Induced resistance during the interaction pathogen x plant and the use of resistance inducers. *Phytochemistry Letters*, *15*, 152–158. doi:10.1016/j.phytol.2015.12.011

Ortmann, I., & Moerschbacher, B. M. (2006). Spent growth medium of Pantoea agglomerans primes wheat suspension cells for augmented accumulation of hydrogen peroxide and enhanced peroxidase activity upon elicitation. *Planta*, *224*(4), 963–970. doi:10.100700425-006-0271-7 PMID:16596409

Orzali, L., Forni, C., & Riccioni, L. (2014). Effect of chitosan seed treatment as elicitor of resistance to Fusarium graminearum in wheat. *Seed Science and Technology*, *42*(2), 132–149. doi:10.15258st.2014.42.2.03

Pagani, A. P. S., Dianese, A. C., & Café-Filho, A. C. (2014). Management of wheat blast with synthetic fungicides, partial resistance and silicate and phosphite minerals. *Phytoparasitica*, *42*(5), 609–617. doi:10.100712600-014-0401-x

Park, R. F. (2007). Stem rust of wheat in Australia. *Australian Journal of Agricultural Research*, *58*(6), 558–566. doi:10.1071/AR07117

Pasquer, F., Isidore, E., Zarn, J., & Keller, B. (2005). Specific patterns of changes in wheat gene expression after treatment with three antifungal compounds. *Plant Molecular Biology*, *57*(5), 693–707. doi:10.100711103-005-1728-y PMID:15988564

Pearce, R. B., & Ride, J. P. (1982). Chitin and related compounds as elicitors of the lignification response in wounded wheat leaves. *Physiological Plant Pathology*, *20*(1), 119–123. doi:10.1016/0048-4059(82)90030-3

Pieterse, C. M. J., Zamioudis, C., Berendsen, R. L., Weller, D. M., Van Wees, S. C. M., & Bakker, P. A. H. M. (2014). Induced Systemic Resistance by Beneficial Microbes. *Annual Review of Phytopathology*, *52*(1), 347–375. doi:10.1146/annurev-phyto-082712-102340 PMID:24906124

Rai, S., Kashyap, P. L., Kumar, S., Srivastava, A. K., & Ramteke, P. W. (2016a). Identification, characterization and phylogenetic analysis of antifungal *Trichoderma* from tomato rhizosphere. *SpringerPlus*, *5*(1), 1939. doi:10.118640064-016-3657-4 PMID:27917337

Rai, S., Kashyap, P. L., Kumar, S., Srivastava, A. K., & Ramteke, P. W. (2016b). Comparative analysis of microsatellites in five different antagonistic Trichoderma species for diversity assessment. *World Journal of Microbiology & Biotechnology*, *32*(1), 8. doi:10.100711274-015-1964-5 PMID:26712623

Ranf, S. (2016). Immune Sensing of Lipopolysaccharide in Plants and Animals: Same but Different. *PLoS Pathogens*, *12*(6), e1005596. doi:10.1371/journal.ppat.1005596 PMID:27281177

Reddy, M. V. B., Arul, J., Angers, P., & Couture, L. (1999). Chitosan treatment of wheat seeds induces resistance to *Fusarium graminearum* and improves seed quality. *Journal of Agricultural and Food Chemistry*, *47*(3), 1208–1216. doi:10.1021/jf981225k PMID:10552439

Reignault, P., Cogan, A., Muchembled, J., Lounes-Hadj Sahraoui, A., Durand, R., & Sancholle, M. (2001). Trehalose induces resistance to powdery mildew in wheat. *The New Phytologist*, *149*(3), 519–529. doi:10.1046/j.1469-8137.2001.00035.x PMID:33873340

Reignault, P., Randoux, B., Renard, D., Sanssen, J., & Durand, R. (2004). Defence responses activated by resistance inducers and natural compounds during the wheat/powdery mildew compatible interaction. In *Workshop on Resistance of cereals to biotic stress*. IHAR Radzików.

Remus-Borel, W., Menzies, J. G., & Belanger, R. R. (2005). Silicon induces antifungal compounds in powdery mildew-infected wheat. *Physiological and Molecular Plant Pathology*, *66*(3), 108–115. doi:10.1016/j.pmpp.2005.05.006

Renard-Merlier, D., Randoux, B., Nowak, E., Farcy, F., Durand, R., & Reignault, P. (2007). Iodus 40, salicylic acid, heptanoyl salicylic acid and trehalose exhibit different efficacies and defence targets during a wheat/powdery mildew interaction. *Phytochemistry*, *68*(8), 1156–1164. doi:10.1016/j.phytochem.2007.02.011 PMID:17399750

Reuveni, M. (1997). Post-infection applications of K_3PO_3, phosphorous acid and dimethomorph inhibit development of downy mildew caused by Plasmopara viticola on grapes. *Journal of Small Fruit and Viticulture*, *5*(2), 27–38. doi:10.1300/J065v05n02_03

Sakr, N. (2016). Silicon control of bacterial and viral diseases in plants. *Journal of Plant Protection Research*, *56*(4), 331–336. doi:10.1515/jppr-2016-0052

Samaina, E., Tuinenc, D., Jeandetd, P., Aussenace, T., & Selima, S. (2017). Biological control of *Septoria* leaf blotch and growth promotion in wheat by *Paenibacillus* sp. strain B2 and *Curtobacterium plantarum* strain EDS. *Biological Control*, *114*, 87–96. doi:10.1016/j.biocontrol.2017.07.012

Santos, H. A. A., Pria, M. D., Silva, O. C., & Mio, L. L. M. (2011). Control of wheat diseases using phosphites and acibenzolar-s-methyl alone or associated with piraclostrobina + epoxiconazole. *Semina. Ciências Agrárias*, *32*, 433–442. doi:10.5433/1679-0359.2011v32n2p433

Savadi, S., Prasad, P., Kashyap, P. L., & Bhardwaj, S. C. (2018). Molecular breeding technologies and strategies for rust resistance in wheat (*Triticum aestivum*) for sustained food security. *Plant Pathology*, *67*(4), 771–791. doi:10.1111/ppa.12802

Silva, I. T., Rodrigues, F. A., Oliveira, J. R., Pereira, S. C., Andrade, C. C. L., Silveira, P. R., & Conceição, M. M. (2010). Wheat resistance to bacterial leaf streak mediated by silicon. *Journal of Phytopathology*, *158*(4), 253–262. doi:10.1111/j.1439-0434.2009.01610.x

Silva, W. L. D., Cruz, M. F. A., Fortunato, A. A., & Rodrigues, F. (2015). Histochemical aspects of wheat resistance to leaf blast mediated by silicon. *Scientia Agrícola*, *72*(4), 322–327. doi:10.1590/0103-9016-2014-0221

Singh, D. P. (2017). *Management of wheat and barley diseases*. Apple Academic Press. doi:10.1201/9781315207537

Singh, R. K., Kumar, D. P., Singh, P., Solanki, M. K., Srivastava, S., Kashyap, P. L., Kumar, S., Srivastava, A. K., Singhal, P. K., & Arora, D. K. (2014). Multifarious plant growth promoting characteristics of chickpea rhizosphere associated Bacilli help to suppress soil-borne pathogens. *Plant Growth Regulation*, *73*(1), 91–101. doi:10.100710725-013-9870-z

Singh, R. K., Kumar, D. P., Singh, P., Solanki, M. K., Srivastava, S., Kashyap, P. L., Kumar, S., Srivastava, A. K., Singhal, P. K., & Arora, D. K. (2014). Multifarious plant growth promoting characteristics of chickpea rhizosphere associated Bacilli help to suppress soil-borne pathogens. *Plant Growth Regulation*, *73*(1), 91–101. doi:10.100710725-013-9870-z

Singh, R. P., & Jha, P. N. (2017). The PGPR *Stenotrophomonas maltophilia* SBP-9 augments resistance against biotic and abiotic stress in wheat plants. *Frontiers in Microbiology*, *8*, 1945. doi:10.3389/fmicb.2017.01945 PMID:29062306

Singh, U. B., Malviya, D., Wasiullah, Singh, S., Imran, M., Pathak, N., Alam, M., Rai, J. P., Singh, R. K., Sarma, B. K., Sharma, P. K., & Sharma, A. K. (2016). Compatible salt-tolerant rhizosphere microbe-mediated induction of phenylpropanoid cascade and induced systemic responses against *Bipolaris sorokiniana* (Sacc.) Shoemaker causing spot blotch disease in wheat. (*Triticum aestivum* L.). *Applied Soil Ecology*, *108*, 300–306. doi:10.1016/j.apsoil.2016.09.014

Solanki, M. K., Kumar, S., Pandey, A. K., Srivastava, S., Singh, R. K., Kashyap, P. L., Srivastava, A. K., & Arora, D. K. (2012). Diversity and antagonistic potential of *Bacillus* spp. associated to the rhizosphere of tomato for the management of *Rhizoctonia solani*. *Biocontrol Science and Technology*, *22*(2), 203–217. doi:10.1080/09583157.2011.649713

Solanki, M. K., Singh, N., Singh, R. K., Singh, P., Srivastava, A. K., Kumar, S., Kashyap, P. L., & Arora, D. K. (2011). Plant defense activation and management of tomato root rot by a chitin-fortified Trichoderma/Hypocrea formulation. *Phytoparasitica*, *39*(5), 471–481. doi:10.100712600-011-0188-y

Solanki, M. K., Singh, R. K., Srivastava, S., Kumar, S., Kashyap, P. L., & Srivastava, A. K. (2015). Characterization of antagonistic-potential of two Bacillus strains and their biocontrol activity against *Rhizoctonia solani* in tomato. *Journal of Basic Microbiology*, *55*(1), 82–90. doi:10.1002/jobm.201300528 PMID:24277414

Song, A., Xue, G., Cui, P., Fan, F., Liu, H., Yin, C., Sun, W., & Liang, Y. (2016). The role of silicon in enhancing resistance to bacterial blight of hydroponic- and soil-cultured rice. *Scientific Reports*, *19*(6), 24640. doi:10.1038rep24640 PMID:27091552

Sousa, R. S., Rodrigues, F. A., Schurt, D. A., Souza, N. F. A., & Cruz, M. F. A. (2013). Cytological aspects of the infection process of *Pyricularia oryzae* on leaves of wheat plants supplied with silicon. *Tropical Plant Pathology, 38*(6), 472–477. doi:10.1590/S1982-56762013000600002

Srivastava, A. K., Singh, R. N., Kumar, S., Kashyap, P. L., & Arora, D. K. (2012). Growth promotion and management of *Alternaria* leaf spot in chilli by *Trichoderma harzianum. International Journal of Innovative Horticulture, 1*(2), 158–163.

Stadnik, M. J., & Buchenauer, H. (1999). Control of wheat diseases by a benzothiadiazole-derivative and modern fungicides. *Journal of Plant Diseases and Protection, 106*, 466–475.

Stocco, M. C., Mansilla, A. Y., & Mónaco, C. I. (2015). Native isolates of *Trichoderma harzianum* inducting resistance to *Zymoseptoria tritici* on wheat plants. *Boletín de la Sociedad Argentina de Botánica, 50*(3), 291–301. doi:10.31055/1851.2372.v50.n3.12519

Takenaka, S., Nishio, Z., & Nakamura, Y. (2003). Induction of defense reactions in sugar beet and wheat by treatment with cell wall protein fractions from the mycoparasite *Pythium oligandrum. Phytopathology, 93*(10), 1228–1232. doi:10.1094/PHYTO.2003.93.10.1228 PMID:18944321

Tayeh, C., Randoux, B., Vincent, D., Bourdon, N., & Reignault, P. (2014). Exogenous trehalose induces defenses in wheat before and during a biotic stress caused by powdery mildew. *Phytopathology, 104*(3), 293–305. doi:10.1094/PHYTO-07-13-0191-R PMID:24073639

Thabet, M. (2008). *Induction of resistance in wheat plants against leaf rust caused by* Puccinia triciana [MSc thesis]. Ain Shams University, Faculty of Agriculture.

Thakur, M., & Sohal, B. S. (2013). Role of Elicitors in inducing resistance in plants against pathogen infection: A Review. *ISRN Biochemistry, 762412*, 1–10. Advance online publication. doi:10.1155/2013/762412 PMID:25969762

Tian, D., Traw, M. B., Chen, J. Q., Kreitman, M., & Bergelson, J. (2003). Fitness costs of R-gene-mediated resistance in Arabidopsis thaliana. *Nature, 423*(6935), 74–77. doi:10.1038/nature01588 PMID:12721627

Van der Plank, J. E. (1968). *Disease resistance in plants*. Academic Press.

van Loon, L. C., Bakker, P. A. H. M., & Pieterse, C. M. J. (1998). Systemic resistance induced by rhizosphere bacteria. *Annual Review of Phytopathology, 36*(1), 453–483. doi:10.1146/annurev.phyto.36.1.453 PMID:15012509

van Loon, L. C., Rep, M., & Pieterse, C. M. (2006). Significance of inducible defense-related proteins in infected plants. *Annual Review of Phytopathology, 44*(1), 135–162. doi:10.1146/annurev.phyto.44.070505.143425 PMID:16602946

Včchet, L., & Šerá, B. (2015). Effectiveness of both synthetic compounds and biological extracts against powdery mildew (*Blumeria graminis* f. sp. *tritici*) on winter wheat. *Agrociencia, 49*, 77–85.

Walters, D. R., Newton, A. C., & Lyon, G. D. (2005). Induced resistance: Helping plants to help themselves. *Biologist, 52*, 28–33.

Walters, D. R., Ratsep, J., & Havis, N. D. (2013). Controlling crop diseases using induced resistance: Challenges for the future. *Journal of Experimental Botany*, *64*(5), 1263–1280. doi:10.1093/jxb/ert026 PMID:23386685

Wang, M., Gao, L., Dong, S., Sun, Y., Shen, Q., & Guo, S. (2017). Role of Silicon on Plant–Pathogen Interactions. *Frontiers in Plant Science*, *8*, 701. doi:10.3389/fpls.2017.00701 PMID:28529517

Wäspi, U., Blanc, D., Winkler, T., Ruedi, P., & Dudler, R. (1998). Syringolin, a novel peptide elicitor from Pseudomonas syringae pv. syringae that induces resistance to Pyricularia oryzae in rice. *Molecular Plant-Microbe Interactions*, *11*(8), 727–733. doi:10.1094/MPMI.1998.11.8.727

Wäspi, U., Schweizer, P., & Dudler, R. (2001). Syringolin reprograms wheat to undergo hypersensitive cell death in a compatible interaction with powdery mildew. *The Plant Cell*, *13*(1), 153–161. doi:10.1105/tpc.13.1.153 PMID:11158536

Wiesel, L., Newton, A. C., Elliott, I., Booty, D., Gilroy, E. M., Birch, P. R. J., & Hein, I. (2014). Molecular effects of resistance elicitors from biological origin and their potential for crop protection. *Frontiers in Plant Science*, *5*, 655. doi:10.3389/fpls.2014.00655 PMID:25484886

Yu, G. Y., & Muehlbauer, G. J. (2001). Benzothiadiazole-induced gene expression in wheat spikes does not provide resistance to Fusarium head blight. *Physiological and Molecular Plant Pathology*, *59*(3), 129–136. doi:10.1006/pmpp.2001.0348

Zhang, S., Schisler, D. A., Boehm, M. J., & Slininger, P. J. (2007). Utilization of chemical inducers of resistance and *Cryptococcus flavescens* OH 182.9 to reduce *Fusarium* head blight under greenhouse conditions. *Biological Control*, *42*(3), 308–315. doi:10.1016/j.biocontrol.2007.05.020

Zhou, F., Menke, F. L. H., Yoshioka, K., Moder, W., Shirano, Y., & Klessig, D. F. (2004). High humidity suppresses ssi4-mediated cell death and disease resistance upstream of MAP kinase activation, H_2O_2 production and defense gene expression. *The Plant Journal*, *39*(6), 920–932. doi:10.1111/j.1365-313X.2004.02180.x PMID:15341634

Chapter 26
Cropping Systems:
A Pathway for Sustainable Agriculture

Amit Kumar
Keral Verma Subharti College of Science, Swami Vivekanand Subharti University, Meerut, India

Raj Singh
Maharishi Markandeshwar University (Deemed), India

Permod Kumar
Keral Verma Subharti College of Science, Swami Vivekanand Subharti University, Meerut, India

Rekha Dixit
College of Biotechnology, Sardar Vallabhbhai Patel University of Agriculture and Technology, India

ABSTRACT

The selection of the best crop rotation is very important for the productive property of agriculture. Crop rotation is extremely vital for soil fertility management and weed, insect, and malady management. Legumes are essential in any rotation and will comprise 25% to 45% of the cropland. Mixed cropping is essential for sustainable agriculture. The blessings of sustainable agriculture are ecological balance, low price of cultivation, clean setting, and nutrient food while not chemical. A cropping system will facilitate massive scale crop production from restricted land. Each little and huge farmer will adopt varied inhume cropping systems as a result of it and need low fertiliser and water. The aim of this chapter is to produce essential information of the cropping system for the sustainable agriculture.

INTRODUCTION

Cropping system is a vital part of a farming system. It represent cropping patterns used on a farm and their interaction with farm resources, different farm enterprises and obtainable technology that verify their makeup. Cropping pattern means that the proportion of space underneath varied crops at some extent of your time in a very unit space. It indicates the yearly sequence and spacing of crops and fallow

DOI: 10.4018/978-1-7998-8434-7.ch026

in a region. Crop sequence and crop rotation square measure typically used synonymously. Crop rotation refers to continual succession of crops on an equivalent piece of land either in an every year or over a extended amount of your time (Liebman et al., 1993). Components crops square measure thus chosen in order that soil health isn't impaired.

Modern crop production technology has significantly raised output however has created issues of land degradation, chemical residues in farm turn out, allele erosion, environmentally and pollution. The resource base is degraded and diminished and therefore the quality of the atmosphere sustaining human life is adversely affected. Agricultural production has sustained man and nice civilizations.

The selection of best crop rotation is very important for productive sustainable agriculture (Liebman *et al.,* 1993). Crop rotation is incredibly vital for soil fertility management, weed, and insect and disease management. Legumes are essential in crop rotation and may comprise 35 to 50% of the crop land. A mixed cropping are essential sustainable agriculture (Vilich-Meller, 1992).

Sustainable agriculture substantially differs from modern-day chemical based mostly, agriculture. the most variations between the both are that in fashionable agriculture, artificial chemical containing NPK are used to enrich the soil and chemical pesticides are accustomed management pests whereas in sustainable agriculture farm created yard manure, compost and bio-fertilizers are used (Jackson et al., 2007; Scherr and McNeely,2008). Sustainable agriculture additionally involves agro forestry and multi – level cultivation and integrated agriculture. Crop production and animal husbandry have to be compelled to be used as associate integral system for sustainable agriculture to succeed.

Sustainable agriculture may be a balanced management system of natural resources including soil, wildlife, forests, crops, plant genetic resources and ecosystems while not degradation and to produce food, bread and butter for current and future generation maintaining or improving productivity and scheme services of those resources (Jackson et al., 2007). Sustainable agriculture system must be economically viable each within the short and long run perspectives. property agriculture must stop land degradation and erosion. It has to fill again nutrient and management weeds, pests and diseases through biological and cultural method.

Figure 1. Types of cropping systems

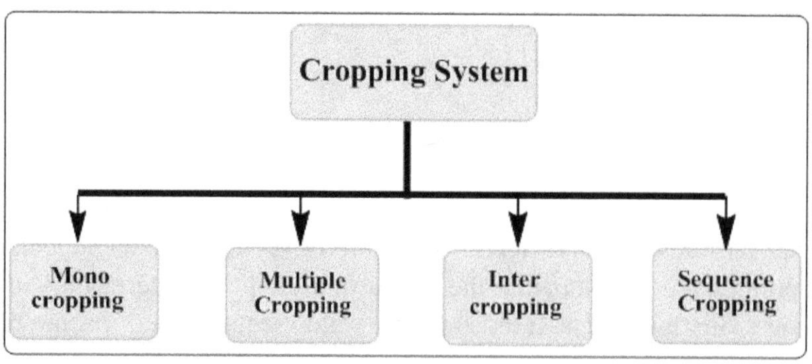

TYPES OF CROPPING SYSTEMS

Based on resources and technology obtainable, four varieties of cropping systems are adopted on farms (Fig 1).

Mono Cropping

Mono cropping or monoculture refers to growing of just one crop on a bit of land year after year. it should ensue to climatologically and socioeconomic conditions or thanks to specialization of a farmer in growing a selected crop. Underneath rainfed conditions, groundnut or cotton are mature year once year due to limitation of rain. In canal irrigated areas, underneath water logged conditions, rice crop is mature because it isn't potential to grow the other crop however major drawbacks of mono-cropping are most water use potency, maintain soil fertility and encourage diseases and peast (Hoshikawa, 1991).

Multiple Cropping

Growing more than two crops on constant piece of land in one year is thought as multiple cropping. it's the intensification of cropping in time and area dimension, i.e., a lot of range of crops at intervals a year and a lot of range of crops on constant piece of land at any given amount. many studies are classified on the idea of crop style, crop management and species composition (Poveda et al, 2008; Maezieux et al 2008 and Lithourgids 2011). It includes intercropping, mixed cropping and sequence cropping.

Intercropping

In intercropping system, two or more crops are growing in the same on constant piece of land with definite row pattern (Gomez and Gomez 1983). As an example, growing fenugreek and wheat in 4:1 quantitative relation i.e., once each four rows of fenugreek, one row of wheat is planted (A. Wasaya et al., 2013). Thus, cropping intensity in area dimension is achieved. Based on the present on plant population used for every crop in intercropping system, it's divided into two sorts i.e. additive series and replacement series. In additive series, that is generally adopted in Asian nation, one crop is planted with 100%of its suggested population in pure stand that is thought because the base crop. Another crop referred to as intercrop is introduced into the base crop by adjusting or dynamical crop . The population of intercrop is a smaller amount than its suggested population in pure stand. In replacement series, each the crops are known as element crops. By scarifying sure proportion of population of single element, another element is introduced. this kind of intercropping is fashionable in western countries. Intercropping is extremely helpful approach for tiny farmers and intercropping of cereals (especially maize) with legumes (especially pigeon pea) are a lot of profitable than the opposite crops (Rusinamhodizi et al., 2012). Intercropping system play an important role in weed suppression, maintain soil fertility, minimum water use potency, high economic profit and atmosphere friendly (fig-2). (Khan et al., 2012) discovered that intercropping system play vital role in weed suppression.

Figure 2. Intercropping system

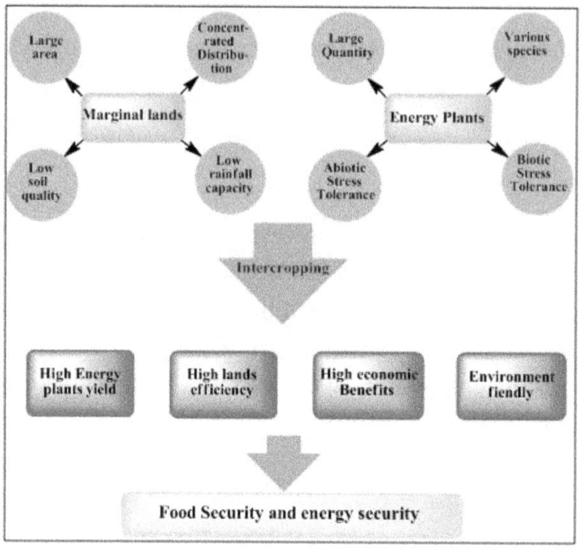

Sequence Cropping

Sequence Cropping is outlined as growing of two or a lot of crops in sequence on constant piece of land during a farming year.depending on the quantity of crops mature during a year, it's known as as double, triple and quadruple cropping involving 2, 3 and 4 crops severally. Sequence cropping give several advantages i.e. nitrogen fixation process, thereby creating extra chemical element for crop production and reducing fertiliser necessities and improved yields and better amino acid levels for the subsequent cereal crop (Cochran et al., 2006). Additionally to those systems, relay cropping, ratoon cropping also are fashionable. Relay cropping refers to planting of the succeeding crop before gather the preceding crop, Ratoon cropping or allocation refers to raising a crop with regrowth starting of roots or stalks once harvest of the crop.

EFFICIENT CROPPING SYSTEMS

Efficient cropping systems for a selected farm rely upon farm resources, farm enterprise and farm technology. The farm resources involve land, labour, water, capital and infrastructure. once land is restricted, intensive cropping is adopted to completely utilize on the available water and labour once comfortable and low-cost labour is accessible, vegetable crops also are enclosed within the cropping system as they need a lot of labour. Capital isn't a constraint. In low rain regions (<750mm/annum) mono-cropping is followed and once rain is quite 750 millimetre intercropping is practiced. With comfortable irrigation water, triple and quadruple cropping is adopted once different climatical factors don't seem to be limiting. Farm enterprises like farming, poultry etc., conjointly influence the sort of copping system. Once the farm enterprise includes farm, the cropping system should contain fodder crops as parts. Amendment in cropping system takes place with the development of technology. The practicability of growing four crop sequence in Gangetic deposit plains gave impetus to multiple cropping.

INTERACTIONS BETWEEN DIFFERENT COMPONENT CROPS

Once crops are mature in association as in entomb cropping, there's interaction between totally different component crops. This interaction is also competitive or non competitive or complementary. once crops are mature in sequence, residual result of the preceding crop influences the succeeding crop. this could be harmful of useful. The venomous chemicals (allelopathic chemical) left within the soil by the roots of helianthus crop inhibit germination of the succeeding crop (Ashrafi et al., 2008). The stubbles of sorghum with high C: N quantitative relation cause immobilization of nitrogen. so inflicting nitrogen element deficiency in early stages of the succeeding crop. The roots of legume crops and their residues add nitrogen element to the soil.

The term interference is employed to explain the result that the presences of one plant has on the atmosphere of another. Interference is divided into 2 removal reactions of one plant on its atmosphere and additive reaction once one thing is additional. once some issue is aloof from the atmosphere, the ensuing response of neighboring species is negative, positive of neutral. Competition among plants is one example for removal interactions. Wherever some issue is additional to the atmosphere, the response of neighboring species is negative, positive or neural. Some such additive interactions are allelopathy and mutualism.

INTERACTIONS IN INTERCROPPING

Plants need growth factors like radiation, water, nutrients and CO2 their growth. In intercropping or mixed cropping, totally different styles of plants vie with each other for the restricted growth factors (fig-3).

Figure 3. Interactions in intercropping

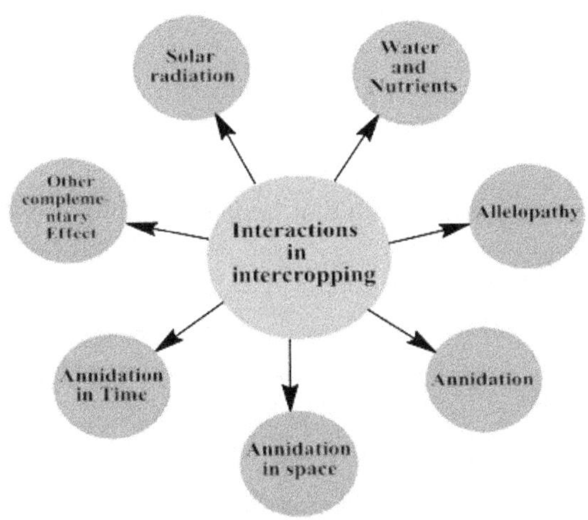

Solar Radiation

The taller crop within the intercropping systems intercepts most of the radiation whereas shorter component suffers. In some intercropping systems, radiation is used with efficiency by each crops. In groundnut + redgram intercropping system, light-weight interception is prolonged as redgram starts growing once the harvest of groundnut (Keshava et al., 2000). If the component crops have totally different growth period, the height demand for light-weight happens at totally different ties. In maize+ greengram intercropping system, greengram flowers in thirty five days once sowing and is harvested sixty fivedays once sowing (mehdi et al., 2013). Peak light-weight demand for maize happens at sixty days after sowing once greengram is prepared for harvest. In such intercrops, there's less competition among component crops and better radiation is intercrops, there's less competition among component crops and better radiation is intercepted in intercropping systems than in pure stands.

Water and Nutrients

Competition for water and nutrients leads to 2 main forms of effects on the less productive or suppressed element (Fusuo et al., 2003). First, the roots of dominated crop could grow less on the perimeters of aggressive element. The suppressed parts adapt to such conditions by magnified capability for uptake. Also, if one a part of the root system is on the depleted aspect, the remaining half shows countervailing activity and vigour. Secondly, plants full of competition for soil factors are possible to own magnified root/shoot quantitative relation.

Allelopathy

Allelopathy is any direct or indirect harmful result that one plant has on another through the discharge of chemical substance or toxins into the root atmosphere (Ashrafi et al., 2008). Some crops is also unsuitable to be mature as intercrops as a result of they'll turn out and eject toxins into the soil that are harmful to different parts.

Annidation

Annidation refers to complementary interaction that happens each in area and time.

Annidation in Space

The canopies of component crops an occupy totally different vertical layers with taller component tolerant to high light and high evaporation demand and shorter component affirmative shade and high ratio. Thus, single component crop helps the another. multistoried cropping in coconut gardens, planting of shade trees in occasional, tea and cocoa plant ations use this principle. Similarly, root systems of the crops exploit nutrients from totally different layer so utilizing the resources with efficiency. Generally, one element with shallow system and another with deep system are elite for intercropping as in monocot genus (shallow) + redgram (deep) intercropping system.

Annidation in Time

Once 2 crops of wide varied period are planted, their peak demands for light-weight and nutrients are possible to occur at totally different periods, so reducing competition. once the first maturing crop is harvested; condition becomes favorable for the late maturing crop. This has been observed to occur in sorghum + redgram, (Egbe et al., 2010) intercropping system.

Other Complementary Effect

In an intercropping system, involving a legume and a non legume, part of the nitrogen fixed in the root nodule of the legume may become available to the non – legume component. The presence of rhizosphere microfflora and mycorrhiza on one species many lead to mobilization and greater availability of nutrient not only to the species concerned, but also to the associated species. Another example is the provision of physical support by one species to the other in intercropping system. Erect crop plants may improve the yield of a climber as in the case of coconut + pepper(Ghosh et al., 2011). The taller component acts as wind barrier protecting the short crop as in maize + groundnut (Hauser et al., 2006).

INTERACTIONS IN SEQUENCE CROPPING

Competition for lightweight, water and nutrients as in mixed crop communities doesn't occur once sole crops are mature in sequence. It happens solely in relay cropping wherever there's a brief span of overlapping between 2 crops during a sequence and also the relay crop experiences the shortage of sunshine. The vital purpose in successive cropping is to extend the use of solar radiation. it's achieved by longer field period and speedy ground coverage. Crops are raised one once another to stay the land occupied by the crop for extended amount. If the crop development is slow a lot of of the radiation reaches the bottom, favouring weed growth and increasing evaporation losses from the soil surface.

In successive cropping, the preceding crop has considerable influence on the succeeding crop principally due to changes in soil conditions, presence of allelopathic chemicals, shift in weeds, and carries over effects of fertilizers, pests and disease (Dore et al., 2011; Ekstrom et al., 2011; Bommarco et al., 2013). Field preparation is tough once rice crop since soil structure is destroyed due to puddling. Crops like sorghum and helianthus leave venomous chemicals within the soil that don't enable germination of succeeding crops (Ashrafi et al., 2008). Phosphorus applied to the previous crop is accessible for the succeeding crop. Weed range and species dissent within the succeeding crop thanks to the result of the previous crop. Wheat crop that follows rice suffers from high density of weed Phalaris minor (Kumar et al., 2013). The pests and diseases in crop stubbles and different residues of the previous crop could infect the next crop.

MANAGEMENT OF INTERCROPPING SYSTEMS

The crops in intercropping system are mature at the same time. Management practices that are followed ought to, therefore, aim to produce favorable atmosphere to all or any the parts, exploit favorable interactions among the component crops and minimise competition among the parts.

Seedbed Preparation

Seedbed preparation for various crops varies depending on the crop. Deep unmoving crops respond to deep plowing whereas for many of the cereals shallow tillage is comfortable. The crops with tiny seed need fine bed, certain crops like cotton and maize are planted on ridges, whereas most of the opposite crops are mature on flat bed. Since quite one crop is concerned in intercropping, the bed preparation is mostly done as per the requirements of base crop. The seedbed for sugarcane, as usual, is created into ridges and furrow. Sugarcane is planted in furrows and intercrops ar planted on ridges. In groundnut + redgram intercropping system, flat seedbed is most well-liked for sowing the crops (Keshava et al., 2000). However, ICRISAT is recommending broadbed and furrows for black soils of semi arid regions for pure crops similarly as intercrops mature underneath rainfed conditions and conjointly in agro – biology, bed preparation is finished severally for component crops. In rice + maize intercropping system, ridges and trenches are shaped (Gounou et al., 2006). Maize is planted on ridges and rice in trenches. In agro forestry, pits are and dug for tree species and a rough bed is ready in inter-space for the introduction of forage crops.

Varieties

The types of element crops elite for intercropping system ought to be less competitive with the base crop and also the peak nutrient demand amount ought to show a discrepancy from the base crop. The difference in period between the parts in intercropping ought to be a minimum duration for thirty days. The short period sorghum hybrids like CSH – half dozen, CSH – nine are appropriate for intercropping with long period redgram varieties like C 11 and LRG 41 due to wider gap between maturity periods (Satheeshkumar et al., 2011). choice of compatible genotypes of component crops increase the complementarily of intercropping system. The varieties elite for intercrop ought to have skinny leaves, tolerant to shading and fewer branching since these crops are typically shaded by the bottom crop. If the base crop is shorter than intercrop, the intercrop ought to be compact with erect branching and its early growth ought to be slow. The characteristics of base crop ought to be as in sole crop.

Sowing

Sowing practices are slightly altered to accommodate intercrop in such some way that it causes less competition to the base crop. Sowing of base crop is finished either as paired row, paired – wider row or skip row plating. The sowing of base crop and intercrop is additionally tired fastened ratios. In paired row planting, two rows of base crop are brought hand yreducing inter-row spacing. The spacing between the 2 pairs of rows are magnified to accommodate the intercrop. As an example, the conventional row spacing of rainfed groundnut is 30 cm (Ramakrishna et al., 2006). The row spacing is reduced to twenty centimeter between the paired row and fifty centimeter spacing is given between 2 pairs of rows. The spacing in combined row designing selected as 25/50cm indicates that the spacing between 2 rows during a pair is 20 cm and among the pairs fifty centimeter. Similarly, cattail millet is planted with row spacing of paired30/60 cm in paired row planting (Rana et al., 2009). These changes in crop pure mathematics don't alter base crop yield, however intercrops are benefited to some extent. The seed drill used for traditional planting contains tines with uniform spacing. The spacing of the tines on the beam of seed drill is to be modified to 20/40 or 30/60 as per demand for sowing in paired rows.

Planting in fastened quantitative relation of intercrops is commonest. The intercropping system of groundnut + redgram is either in 5:1 quantitative relation (Gopal et al., 2010) and sorghum + redgram in 2:1 quantitative relation (Satheeshkumar et al., 2011). In these cases, the conventional three tined or four tined seed drill is used with none modification. the whole pertaining to intercrop row within the hopper is closed with a bit of cloth material. In this row, intercrop is planted with akkadi or Kera. In ancient cropping systems, the component crops are mature with sub optimum population. Yields of intercrops is magnified with higher plant population. For higher yields, base crop population is maintained at its sole crop population and intercrop population is unbroken at eightieth of its sole crop population. Even 100% populations of each the crops are maintained with advantage in some intercropping systems. As an example, sorghum redgram intercropping with p.c population of both crop (180 x 103 and 50 x 103) offers higher yield than lower population of those crops (Satheeshkumar et al., 2011). once the distinction in period of element crops is a smaller amount than thirty days, staggered planting is finished to extend thedistinction in period. The aggressive or dominant crop is planted ten to fifteen days once sowing the dominated crop.

Fertilizer Application

The quantity of nutrients present within the element crops indicate the necessity of fertilizers for the intercropping system. The nutrient uptake is mostly a lot of in intercropping system compared to pure crops. once legumes associated to cereal crop in intercropping system, some of chemical element demand of cereal is supplemented by the legume (Fujita et al., 1992). The quantity is also as tiny as a number of kilograms to 22 kg/ha. Application of upper dose of chemical element of the cereal + legume intercropping system not solely reduces the organic process capability of legume, however conjointly growth of legume is suppressed by aggressive cereals as a result of quick growth of cereals with magnifiedconvenience of nitrogen (Fujita et al., 1992). Cereal + legume intercropping, therefore, is especially advantageous underneath low fertilizer application. Considering all the factors, it's recommended that the the nitrogen does will suggested for base crop as pure crop is comfortable for intercropping system with cereals + legume or legume + legume. With relation to phosphorus and K, one – eighth – one – fourth of the suggested dose of intercrop is additionally additional additionally to suggested dose of base crop to fulfill the additional demand. Basal dose of nitrogen is applied to rows of each parts in cereals + legume intercrop system (Wani et al., 1995).fertilization of nitrogen is finished solely to cereal rows. Phosphorus and K are applied as basal dose to each crop.

Water Requirement

Intercropping systems are typically suggested for rainfed crops to induce stable yields. The entire water employed in intercropping system is nearly constant as for sole crops, however yields are magnified. This water use potency of intercropping is on top of sole crops (walker and Ogindo, 2003). Te element crops dissent on their capability to face up to excess or tough wetness conditions. But the irrigation schedule followed for sole crops is appropriate even for intercropping system.

Weed Management

Weed drawback is a smaller amount in intercropping system compared to their sole crops. The upper plant population used and conjointly complete covering of the soil earlier in intercropping system reduces weed infestation (Matt Liebman and Elizabeth Dyck, 1993). In late maturing crops that are planted on wide rows, presence of early maturing crops helps to hide the vacant inter-row area aded keep weeds underneath check. In sure things, intercrops are used as biological agents to manage weeds. Black gram, greengram, cowpea in sorghum and cowpea in banana cut back weed population and one hand – weeding is avoided by this technique. However, in some intercropping systems like maize + groundnut, weed drawback is analogous to their sole crops (Reddy et al., 2005). the expansion habit of genotypes employed in intercropping contains a nice influence on weed growth (Satheeshkumar et al., 2011). Early crops cover to hide the soil is a lot of vital than speedy increase in plant height.

It's alright illustrious that {different ltotally totally different completely different} species of weeds are related to different crops. This is often due to variations in weed management practices like tillage, herbicides etc., and conjointly due to crop weed interference. For these reasons, weeds present in sole crops are totally different than those in intercropping systems.

Pests and Diseases

Diseases and pests are believed to be less in intercropping system due to crop diversity than in sole crops (Scheer and McNeely,2008). Some plant combination could enhance soil fungi stasis and association through indirect effects on soil organic matter content. The unfold of the diseases is altered by the presence of various crops. Very little leaf of herbaceous plant is a smaller amount once herbaceous plant is protected by maize or sorghum. Because the insect carrying virus initial attack maize or sorghum, infection is a smaller amount on herbaceous plant. Non – host plants in mixture could emit chemicals or odors that have an effect on the pests, thereby protective host plants.

The thought of crop diversification for the management of roundworm population has been applied principally within the variety of decoy and entice crops. Decoy crops are non-host crops that are planted to form nematodes waste their infection potential. This is often full of activating larvae of nematodes within the absence of hosts by the decoy crops (Table-1)

Table 1. Decoy crops for the control of nematodes

S.No.	Crop	Nematode	Decoy crops
1	Tomto	*M.incognita, Pratylenchus alleni*	Castor, *Crysanthemum*, groundnut
2	Soybean	*Pratylenchus sp.*	*Crotalaria spectabilis*
3	Brinjal	*Meloidogyne incognita, M. Javanica*	*Sesamum orientale*

MANAGEMENT OF SEQUENTIAL CROPPING SYSTEMS

Unlike intercropping, crops are grown one after another in sequential cropping and hence, management practices are different.

Seedbed Preparation

The appropriate variety of bed for a selected crop is ready in sequence cropping systems like pudding for rice, ridges and furrows for vegetables, maize and cotton and flat bed for many different crops. However, 2 issues are encountered in seedbed preparation in successive cropping system.

1. The time on the market for bed preparation is a smaller amount in high intensity cropping system. Rains terribly often interfere with land preparation.
2. Due to the result of the previous crop, the sphere might not be in correct condition to hold out field operation.

For example, field preparation after harvesting of rice is difficult. it's principally as a result of soil structure is destroyed throughout pudding (Sharma et al., 1993). The flip – around time, the time between gather of crop to sowing of next crop, is a lot of if rice is that the preceding crop. To avoid this drawback, minimum tillage or zero tillage is adopted. it's a typical observe to sow pulse crop simply before or directly once gather rice crop. In rice – wheat system, the rice stubbles are killed by spraying weed-killer and wheat is planted in plough furrows between stubbles of rice (singh et al., 2000). In indonesia, sorjan system has been developed to beat this drawback. During this system, upland and lowland crops are grown in alternate strips; rice is grown within the flooded strips, whereas maize, beans, mung bean, sorghum, cassava, cowpea or sweet potato are mature within the upland or elevated strips. Throughout the year, cassava or sorghum is mature within the upland areas on residual wetness, whereas a crop of a legume or maize is mature wherever the rice has been harvested (Sattar et al., 2012). A serious constraint within the sorjan system is that the labour needed to make and maintain permanent raised planting areas.

It is insufferable to observe zero or minimum tillage in all sequence cropping systems. If helianthus is that the preceding crop, plowing is crucial to oxidize the allelochemical of helianthus (Narwal et al., 2002). The stubbles of pearl millet and sorghum that contain high C:N raito immobilize nitrogen component. It is, therefore, necessary to get rid of them. They conjointly interfere with running of sowing instrumentation. Typically tillage operations are done at the start of the season (for the primary crop) and later on, zero tillage is practiced for different crops within the system. In rice – rice – greengram system, summer ploughing is finished and later once water is accessible within the rainy season,pudding is finished and initial crop is planted. For the second crop, minimum tillage is finished simply to include crop residues and rice is planted. Greengram is planted as a relay crop within the second rice crop. In cotton – sorghum – finger millet cropping system in garden lands, through field preparation is finished and also the field is laid out into check basins to transplant finger millet. within the next season, cotton is planted among the stubbles for finger millet without land preparation. Weeds are controlled by inter-cultivation in cotton at succeeding stages.

Varieties

Short period types of crops are selected to suit well in multiple cropping systems (Sundara et al., 1990). Image -insensitive varieties are essential for productive sequence cropping system. Most of the high yielding varieties are sensitive and intensive cropping is feasible with the introduction of those varieties.

Sowing

Not like in intercropping system, sowing isn't a retardant, provided there's comfortable time for seedbed preparation. If the right seedbed isn't ready, establishment of crops is tough. Special seeding technique and instrumentation are necessary underneath these conditions. As an example, cotton institution is tough in serious back soils once rice. due to establishment pans within the shallow layer, root penetration is hindered. Comfortable time isn't available for field preparation. If the fields are allowed to dry sufficiently for field preparation, nearly a month time is lost. Thus seedlings are raised either on twisted paddy straw or leaf cups before harvesting of rice crop. After harvesting, a crow bar hole is created up to 25 cm. it's partially stuffed with sand and soil mixture and cotton seedlings are planted.

The establishment of a pulse crop once rice is additionally tough. Broadcasting of seeds in standing rice crop or rice stubbles leads to uneven germination and high seed rate is critical. Crops planted in stubbles are subjected to competition from regenerated stubbles. It is overcome by spraying weedkiller or diquat on stubbles. Delay in sowing or planting is that the commonest drawback in intensive cropping systems. Scenario necessitates transplantation with superannuated seedlings. To reduce yield loss due to transplantation of over-aged seedlings, higher level of N_2 is applied to induce tillering. In rice – wheat system, wheat yields ar reduced significantly once the sowings of wheat are delayed on the far side November (kumar et al., 2013). In such things, transplantation of 35 to 45 days previous seedlings of wheat is finished. farm manure is broadcasted over the sphere to take care of higher soil temperature during December month.

Fertilizer Application

Crucial the fertilizer scheduleis complicated in successive cropping system as many factors have to be considered. The vital factors are; soil supply power, total uptake by crops, residual result of fertilizers, nutrients additional by legume crop, crop residues left on the soil and potency of crops in utilizing the soil and applied nutrient.

Soil Supplying Power

Soil contribution to the crops ought to be illustrious before preferring the quantum of fertilizer application. The results of long term fertility trials discovered that there's no considerable amendment within the soil physical properties and soil deficiency for micronutrients as a consequence of multiple cropping. However, Zn element in light-weight alluvial soils (sandy soils) of Ludhiana and Fe in medium black soils of Coimbatore has been found to be essential. The soil nutrient standing, calculable by soil analysis at the start of the season is altered by growing completely different crops throughout different seasons. The soil supply power increase with legume in rotation,fertiliser application and addition of crop residues. The available N and K in soil once a groundnut crop are higher compared to initial standing of the soil (Suo, 2005). However once pearl millet, only K standing within the soil is improved and there's no amendment in phosphorus standing.

Nutrient Uptake by Crop

The entire quantity of nutrients taken by the crops in one sequence offers a sign of the fertilizer demand of the system (Meng et al., 2005). record approach is followed to illustrious whether or not the quantity of fertiliserapplied is equal, a lot of or less to the entire uptake of nutrients by totally different crops within the system. The balance is obtained by subtracting the fertilizer applied to crops within the system from the nutrients taken by the crop.

Residual Effect of Fertilizers

The extent of residues left over within the soil depends on the type of fertilizer used. due to their quality and solubility, N fertilizers leave no residues after the crop is harvested. N^{15} studies have shown that only one to two percent of N applied to maize was taken the subsequent wheat crop (Russell et al., 2005). However, residues of N occur only previous crop yields are poor. Phosphatic fertilizer and farm manure leave considerable residue within the soil that is beneficial for succeeding crops. Farm manure applied to the previous crops, used only fifty percent of its nutrient and also the rest is accessible for subsequent crops. The residues left by K fertilizers are marginal.

Legume Effect

Legumes add N to the soil within the vary of 20 to 100 kg per hectare. The quantity of N additional depends on the crop and conjointly on the aim that it's mature. Greegram mature for grain contributes fifteen to twenty kilo N per Hectare to the succeeding crop. Cowpea mature for grain and fodder contributes twenty four and thirty kilo N per Hectare severally to the succeeding crop. Inclusion of leguminous green manures within the system adds 40 to 120 kg N per hectare. The availabiltiy of phosphorus is additionally magnified by incorporation of manure crops.

Crop Residues

Crop residues add considerable amount of nutrients to the soil. Cotton planted in finger millet stubbles advantages by 25 to 30 kg N per hectare due to decomposition of finger millet stubbles. Deep rooted crops like cotton and redgram absorb phosphorus and different nutrients from deeper layers. Leaf fall and its succeeding decomposition add phosphorus to the uppermost layers. Crop residues containing high C: N quantitative relation like stubble of sorghum and pearl millet briefly immobilize N. Residues of herbaceous plant crops contains low C: N quantitative relation and that they decompose quickly and release nutrients.

Efficiency of Crops

Crops dissent in their ability to extract and forage nutrients from totally different layers of soils and their capability to utilize them for the assembly of economic product. Jute is a lot of economical crop for utilization of N followed by summer rice, rainy – season rice, maize, potato and groundnut in this order. The order of phosphors economical crops are jute > summer rice > Kharif rice > groundnut > maize. Groundnut is a lot of economical in K utilization and also the different crops so as of potency are maize, jute, summer rice, rainy – season rice, and potato.

WATER MANAGEMENT

Water management for successive cropping is same as for individual crops. There aren't any carry over effects of irrigation as within the case of fertiliser. Depending on the availabiltiy of water, appropriate cropping system should be elected. Rice – rice – rice is economical cropping system for total yield, however it consumes great amount of water particularly throughout summer (Tabbal et al., 2002). The availability of water in summer season is minimum, rather than rice, groundnut is employed within the cropping system. Rice – rice – groundnut cropping system has high water use potency and high returns (yadav et al., 2001). In whet primarily based cropping systems, depending on convenience of water, cropping system is modified. when water is accessible for ten irrigations, the economical cropping system is maize – wheat, for four to six irrigation, blackgram – wheat and for two irrigations fallow – wheat.

With relation to the tactic of irrigation, the layout should to be planned in such some way that's appropriate for many of the crops within the system. In rice – rice – groundnut system, rice is irrigated by flood technique of irrigation, whereas for groundnut border strip technique is appropriate (Huke et al., 1997). In cotton – sorghum – finger millet system, cotton and sorghum is irrigated by furrow technique, whereas for finger millet check – basin technique is adopted. For this technique, the sphere is ordered out into check basins and finger millet is transplanted. after harvested of finger millet, cotton is planted within the check basins in finger millet stubbles. once a amount of thirty days ridges and furrows are created for irrigating cotton. Furrows may well be used for succeeding crop of sorghum.

WEED MANAGEMENT

Weed management practices are followed for the complete system as unit rather than considering weed drawback of individual crops. This is often necessary as a result of weed shift happens due to crops mature within the system and also the carry over effects of weed control ways are felt on succeeding crops (Liebman and janke 1990).

Weeds are dynamic in nature and occasional shifts in weed species occur due to amendment in atmosphere brought either by cultural practices or by weed management techniques. typically broad leafed weeds occur in wheat at later stages and 2, 4-Dichloro acetic acid is applied as post – emergence ferbicide to manage them (Hume, 1982). If weed shift happens, this practice isn't appropriate. As an example, in rice- wheat system, grass (Phalaris minor) may be a menace for wheat crop (Kumar et al., 2013). The weed seeds of many species are rotten and lose viability as soil is submerged for rice however *Phalaris minor* seeds don't lose viability. Wheat planted in rice stubbles is heavily infested with *Phalaris minor*. Weed seeds germinate even before the germination of crop seeds and there's severe competition from these weeds, similarly, in cotton – sorghum – finger millet sequence cropping with zero tillage, weeds are controlled by herbicides in 2 rotations. Weed flora shifts from annual weeds to perennial weeds and *Cynodon dactylon* may be a problematic and dominant weed during this system.

HARVESTING

In sequence cropping, the crop is harvested at physiological maturity rather than at harvest maturity. The fields will them be vacated one week earlier for planting another crop. Due to continuous cropping, the gather tie could coincide with serious rains and special post – harvest operations, like artificial drying, treating the crop with common salt etc. are to be practiced to save lots of the produce.

CONCLUSION

Cropping system is a very important component of a farming system. Farming system represent an applicable combination of farm enterprises i.e. fisheries, forestry and also the mean on the market to the farmer to boost them for increasing profitableness. Cropping pattern means that the proportion of space underneath various crops at some extent of your time during a unit space. In cropping system, crop sequence and crop rotation are typically used. The chapter has introduced the various variety of cropping system for sustainable agriculture. Crop rotation is extremely vital for sustainable agriculture. sustainable agriculture means that a variety of agriculture aimed toward meeting the requirements of this generation while not endangering the resource base of the long run generations. it's nominal dependence of artificial fertilizers, pesticides and antibiotics. it's totally different from trendy agriculture as a result of in trendy agriculture, artificial fertilizers containing NPK are used to enrich the soil and chemical pesticides are used to management pest whereas in property agriculture farm created farm manure, compost, manure and biofertilizers are used.

The vital steps in property agriculture ar water shed management, conservation of genetic resources, integrated nutrient management, integrated weed management and integrated blighter management. It is mainly aimed at obtaining sustainable production with limited or no chemical inputs with preference to farm grown inputs without pollution and minimum damage to natural resources base.

REFERENCES

Anastasios, & Christos, & Christos. (2011). Annual intercrops: An alternative pathway for sustainable agriculture. *Australian Journal of Crop Science*, *5*, 396–410.

Bommarco, R., Kleijn, D., & Potts, S. (2013). Ecological intensification: Harnessing ecosystem services for food security. *Trends in Ecology & Evolution*, *28*(4), 230–238. doi:10.1016/j.tree.2012.10.012 PMID:23153724

Cochran, V., Danielson, J., Kolberg, R., & Miller, P. (2006). Dryland cropping in the Canadian Prairies and the U.S. Northern Great Plains In Dryland Agriculture (2nd ed.). American Society of Agronomy Inc. doi:10.2134/agronmonogr23.2ed.c9

Dore, T., Makowski, D., Malézieux, E., Munier-Jolain, N., Tchamitchian, M., & Tittonell, P. (2011). Facing up to the paradigm of ecological intensification in agronomy: Revisiting methods, concepts and knowledge. *European Journal of Agronomy*, *34*(4), 197–210. doi:10.1016/j.eja.2011.02.006

Egbe, Bar, & Mary. (2010). Pigeonpea/sorghum Intercropping in southern guinea savanna: Effect off Planting Density of Pigeonpea. *Nature and Science*, *8*(11), 156–167.

Ekstrom, G., & Barbara, E. (2011). Pest control in agro-ecosystems: An ecological approach. *Critical Reviews in Plant Sciences*, *30*(1-2), 74–94. doi:10.1080/07352689.2011.554354

Fujita, K., Ofosu-Budu, K. G., & Ogata, S. (1992). Biological nitrogen fixation in mixed legume-cereal cropping systems. *Plant and Soil*, *141*(1-2), 155–175. doi:10.1007/BF00011315

Fusuo, Z., & Long, L. (2003). Using competitive and facilitative interactions in intercropping system enhances crop productivity and nutrient- use efficiency. *Plant and Soil*, *248*(1/2), 305–312. doi:10.1023/A:1022352229863

Ghosh & Bandopadhyay. (2011). Productivity and profitability of coconut based cropping system with fruits and black pepper in West Bengal. *Journal of Crop and Weed*, *7*(2), 134–137.

Gomez, A. A., & Gomez, K. A. (1983). Multiple Cropping in the Humid Tropics of Asia. Academic Press.

Gopal & Jagadeeswar. (2010). Management of Peanut budnecrosis disease in groundnut through intercropping with cereal and pulse crops. Taylor Francis.

Gounou & Schulthess. (2006). Effect of traditional rice/maize intercropping on population densities, crop damage and parasitism of stem-borers in the Ivory Coast. *African Plant Protection*, *12*, 93–102.

Hauser, S., Norgrove, L., & Nkem, J. N. (2006). Groundnut/maize/cassava Intercrop Yield Response to Fallow Age, Cropping Frequency and Crop Plant Desnity on an Ultisol in Southern Cameroon. *Biological Agriculture & Horticulture: An International Journal for Sustainable Production Systems.*, *24*(3), 275–291. doi:10.1080/01448765.2006.9755026

Hoshikawa, K. (1991). Significance of legumes crops in intercropping, the productivity and stability of cropping system. In C. Johanson, K. K. Lee, & K. L. Saharawat (Eds.), *Phosphorus Nutrition of Grain Legume in the Semi Arid Tropics* (pp. 173–176). ICRISAT.

Hume, L. (1982). The long-term effects of fertilizer application and three rotations on weed communities in wheat (after 2 1-22 years at Indian Head. Saskatchewan). *Canadian Journal of Plant Science*, *62*, 741–750. doi:10.4141/cjps82-107

Jackson, L., Pascual, U., & Hodgkin, T. (2007). Utilizing and conserving agrobiodiversity in agricultural landscapes. *Agriculture, Ecosystems & Environment*, *121*(3), 196–210. doi:10.1016/j.agee.2006.12.017

Keshava & Ramachandrappa. (2000). Intercropping of groundnut, cowpea and jowar with redgram. *Current Research*, *29*(4), 49-50.

Khan, Ali, & Hussain. (2012). Impact of maize-legume intercropping on weeds and maize crop. *Pakistan Journal of Weed Science Research*, *18*(1), 127–136.

Kumar, A., & Sengar, S. R. (2013). Effect of delayed sowing on yield and proline content of different wheat cultivars. *Research on Crops*, *14*(2), 409–415.

Kumar, V., Singh, S., Chhokar, R. S., Malik, R. K., Brainard, D. C., & Ladha, J. K. (2013). Weed management strategies to reduce herbicide use in zero – till rice –wheat cropping systems of the Indo Gangetic Plains. *Weed Technology*, *27*(1), 241–254. doi:10.1614/WT-D-12-00069.1

Liebman & Janke. (1990). Sustainable weed management practices. In *Sustainable agriculture in temperate zones*. John Wiley & Sons.

Liebman, M., & Dyck, E. (1993). Crop rotation and intercropping strategies for weed management. *Ecological Applications*, *3*(1), 92–122. doi:10.2307/1941795 PMID:27759234

Mehdi & Khas. (2013). The Influence of Intercropping Maize (Zea mays L.) Green Gram (Vigna Radiata L.) on the Changes of Soil Temperature, Moisture and Nitrogen. *International Journal of Ecosystem*, *3*(2), 13-17.

Meng, Zucong, & Ding. (2005). Carbon contents in soils and crops as affected by long-term fertilization. *Turang Xuebao*, *42*, 769–776.

Narwal, Palaniraj, & Chandra. (2002). Allelopahtic effects of aqueous extracts of Sunflower (Helianthus annus L.) root on some winter oil seed crop. *Geobies*, *29*, 225–228.

Poveda, Gomez, & Martinez. (2008). Diversification practices: Their effect on pest regulation and production. *Revista Colombiana de Entomologia*, *34*, 131–144.

Ramakrishna, A., Tam, H. M., Wani, S. P., & Long, T. D. (2006). Effect of mulch on soil temperature, moisture, weed infestation and yield of groundnut in northern Vietnam. *Field Crops Research*, *95*(2 &3), 115–125. doi:10.1016/j.fcr.2005.01.030

Reddy & Tyagi. (2005). Integrated weed management in maize-groundnut sequential cropping system. *Agricultural Reviews (Karnal)*, *26*(4), 235–248.

Robert & Eleanor. (1997). Area by type of culture: South, Southeast, and East Asia. A revised and updated database. Manila (Philippines). International Rice Research Institute.

Rusinamhodz, L. (2012). Maize–grain legume Intercropping is an attractive option for ecological intensification that reduces climatic risk for smallholder farmers in central Mozambique. *Field Crops Research*, *136*, 12–22. doi:10.1016/j.fcr.2012.07.014

Russell, A., Laird, D. A., Parkin, T. B., & Mallarino, A. P. (2005). Impact of nitrogen fertilization and cropping system on carbon sequestration in Midwestern Mollisols. *Soil Science Society of America Journal*, *69*(2), 413–422. doi:10.2136ssaj2005.0413

Satheesh, Thukkaiyannan, & Ponnuswamy. (2011). Effect of sowing and weed management methods and intercrops on weed control and grain yield of sorghum under intercropping situation. *Crop Res.*, *41*(1-3), 46 – 51.

Sattar & Abedin. (2012). Options for coastal farmers of Bangladesh adapting to Ipacts of climate change. *International conference on Environment, Agriculture and Food Science*, 17 – 21.

Scherr, J. S., & Mcneely, A. J. (2008). Biodiversity conservation and agricultural sustainability: Towards a new paradigm of 'ecoagriculture' landscapes. *Philos Trans Royal Soc B*, *363*(1491), 477–494. doi:10.1098/rstb.2007.2165 PMID:17652072

Sharma & Bhagat. (1993). Puddling and compaction effects on water permeability of texturally different soils. *Journal of the Indian Society of Soil Science*, *41*, 1–6.

Singh & Gangwar. (2000). Alternate tillage and crop residue management strategies in rice–wheat cropping system. Annual Report, Project Directorate for Cropping System Research.

Singh, R. V. (2009). Effect of hybrids, plant density and fertility levels on nutrient concentration, uptake and productivity of pearl millet in semi arid environment. *The South Pacific Journal of Natural Science, 27*(1), 45–48. doi:10.1071/SP09008

Sundara, B., & Subramanian, S. (1990). Changes in soil available NPK in multiple cropping systems based on short duration sugarcane crops relative to a conventional sugarcane cropping system. *Fertilizer Research, 24*(2), 67–76. doi:10.1007/BF01073224

Suo, D. R. (2005). Effects of combined fertilization of chemical and organic fertilizers in a long-term posi- tion experiment. *Ganhan Diqu Nongye Yanjiu, 23*(2), 71–77.

Tabbal, D., Bouman, B. A. M., Bhuiyan, S. I., Sibayan, E. B., & Sattar, M. A. (2002). On farm strategies for reducing water input in irrigated rice: Case studies in the Philippines. *Agricultural Water Management, 56*(2), 93–112. doi:10.1016/S0378-3774(02)00007-0

Vilich-Meller, V. (1992). Mixed cropping of cereals to suppress plant diseases and omit pesticide applications. *Biological Agriculture and Horticulture, 8*(4), 299–308. doi:10.1080/01448765.1992.9754607

Walker, S., & Ogindo, H. O. (2003). The water budget of rainfed maize and bean intercrop. *Physics and Chemistry of the Earth, 28*(20-27), 919–926. doi:10.1016/j.pce.2003.08.018

Wani, P., Chandrapalaiah, S., & Dart, P. J. (1985). Response of Pearl millet Cultivars to Inoculation with Nitrogen Fixing Bacteria. *Experimental Agriculture, 21*(2), 175–182. doi:10.1017/S001447970001245X

Wani, P., Rupela, O. P., & Lee, K. K. (1995). Sustainable agriculture in the semi – arid tropics through biological nitrogen fixation in grain legumes. *Plant and Soil, 174*(1-2), 29–49. doi:10.1007/BF00032240

Wasaya, Ahmad, & Hassan. (2013). Enhancing crop productivity through wheat (*triticum aestivum* l.) - fenugreek intercropping system. *The Journal of Animal & Plant Sciences, 23*(1), 210–215.

Yadav & Abburi. (2001). *Atlas of Cropping Systems in India.* PDCSR Bulletin No. 2001-2, Project Directorate for Cropping Systems Research.

Zoheir, Sadeghi, Mashhadi. (2008). Allelopathic Effects of sunflower (*Helianthus annuus*) on Germination and Growth of Wild Barley (*Hordeum spontaneum*). *Agricultural Technology (Thailand), 4*(1), 219–229.

Chapter 27
Fostering Sustainable Productivity Through Maize Technology Intensification:
Participant Responses

Godfrey Onuwa
Federal College of Forestry, Jos, Nigeria

ABSTRACT

Farmers' efficiency in maize production resulting from adoption of technological innovations has food security implications; low productivity from maize farms has been attributable to non-adoption of improved technologies. Multistage sampling procedure was employed for this study. Primary data collected was analyzed using descriptive statistics. The study revealed that most (78.2%) of the respondents got information on production technologies through their membership of social groups or cooperatives. Production technologies were prevalent among maize farmers in the study area; however, the level of adoption was relatively low. In addition, the constraints identified significantly affected maize farmers' adoption decisions in the study area. Formulation and implementation of policies are strongly recommended to ameliorate adoption constraints.

INTRODUCTION

Maize (Zea mays) is a cereal produced across the world. Maize is one of the most important cereal crop cultivated in the rainforest and derived savannah zones of Nigeria. Maize has now risen to a commercial crop on which many agro-based industries depend on as a raw material (Iken and Amusa, 2004). It is a high yielding crop, easy to process, readily digested with cost advantage when compared to other cereals (IITA, 2001). It plays an important role both in terms of food security and nutritional intake and accounting for about 43% of calorie intake with a daily consumption quantity of 53.20g per capita (Komolafe *et al.*, 2010). It is also used extensively as the main source of calories in animal feeding and

DOI: 10.4018/978-1-7998-8434-7.ch027

feed formulation (Manyong *et al.*, 2010). Apart from feed and food, maize is also useful in medicines and as raw material in agro-industries (Abdulrahaman & Kolawole, 2006). It constitutes a staple food in many regions of the world particularly in developing countries (Abebaw and Abelay, 2001). Maize is a staple food of great economic importance in the sub-Saharan Africa of which Nigeria is inclusive. In Nigeria, it is the third most important cereal crop after sorghum and millet (Ajayi *et al.*, 2008). Maize is used as a staple human food, as feed for livestock and as raw material for many industrial products (Ouma *et al.*, 2002). In most countries, maize is the most important foodstuff and in particular provides the daily bread for indigenous population of rural communities. Frova et al (1999), asserted that maize is one of the world's three most important cereal crop after wheat and rice and it has the widest distribution of any cereal. According to them, the crop is primarily grown for its grain which is consumed as human foods. In some developed countries, maize is also grown for animal feed and as a base for industrial products such as oil, syrup and starch. Maize is the second most important cereal crop in the world in terms of acreage and production. Global production of Maize was about 1040 million MT in the year 2016–2017, where in USA and China contributed about 38 and 23%, respectively. Maize can be cultivated successfully in loamy sand to heavy clay, well aerated, neutral pH soils. As of tropical origin, it is highly sensitive to water stagnation, so avoid the cultivation in low-lying or poor drainage fields. Furthermore, extended low temperature less 5°C severally affects the crop. Optimum range of temperature for better crop growth and yield realization is 25–35°C. Maize crop can be cultivated throughout the year with high yield levels. The average yield of maize in Nigeria is about 2500–3500 kg/ha of threshed grain using recommended fertilizers, varieties, seed-dressing, and improved cultural practices. The total land area planted to maize in Nigeria is above 2.5 million hectares with an estimated yield of about 1.4 metric tons per hectare, though, maize production is still very low especially if considered in relation to the growing food demand of the country. About 20% is processed for secondary uses.

Growth in maize utilization has been driven by the rapidly increasing demand for maize as livestock feed, food and industrial non-food products. The population of Nigeria is expected to grow at a rate of more than 3% per year, while food production is likely to grow at a rate of 2% or less a year. Closing this gap and increasing food production will require intensive agriculture based on use of modern technologies such as the use of improved seed varieties, agrochemicals, management practices, etc. Higher solar radiation received in the northern part of the country relative to the southern part, increases the potential of maize production in the northern zone (NAERLS & FDAE, 2014). Improved agricultural practices are farming practices that have been researched on, tried and found to bring about increased crop yield (Ali-Olubandwa et al. 2010). The practices include use of certified seeds, agronomic practices, fertilizer and agrochemical application among others. Ouma *et al.*, (2002), defines adoption as a decision to make full use of an innovation or new technology as the best opportunity available to the farmers. In view of the uncertainty about the outcome or otherwise profitability of such innovation, a greater effort is required of the farmers so as to decide whether to use the innovation or not. Agricultural productivity in the developing countries like Nigeria continuous to be low and it is generally believed that non-adoption of research results by majority of farmers is the main reason for this situation (Idrisa *et al.*, 2012). Bawa and Ani (2014), found a positive and significant association between ages, farming experience, training received, socio-economic status, cropping intensity, economic motivation, innovativeness, information sources, utilization, adoption, etc. However there has been a fluctuating trend in maize production over the last decades which threatens household. Maize has yielded compelling success stories with the adoption of new technologies that has increased small holder maize production. The diffusion of new technologies in Africa has been more widespread for maize than for other food crops. This implies that technologies adoption can provide a basis for further increase in maize production. Farmers' efficiency in maize production resulting from technological innovations also has some food security implications. Food crises is more serious in sub-Saharan Africa where the attainment of food security is intrinsically linked with poor agricultural techniques which results in decrease in agricultural growth rate. (Ajayi *et al.*, 2008). Despite the economic importance of maize in Nigeria, it has not been produced to meet food and industrial demands. This could be attributed to low productivity from maize farms due to non-adoption of improved technologies for maize production by farmers. Therefore the study intends to evaluate the adoption of production technologies by maize farmers in the study area. To this effect, the following research questions would be asked;

1. What are the sources of information on production technologies among the maize farmers?
2. What were the maize production technologies adopted?
3. What are the constraints of technology adoption among maize farmers in the study area?

METHODOLOGY

Study Area

Toro Local Government Area (LGA) covers a total land area of 6932km^2 and a total population of 350,000 people, projected at 3% growth rate per annum to be 499,586 in 2018(NBS, 2006). The local government shares common boundaries with Plateau in the Southeast, Kaduna in the West and Kano in the North. Toro local government is topographically hilly with altitude of 100m above sea level. It is located on longitude 9°N and 12°E and latitude 8°N and 11°E. It is in the Sudan Savannah zone of Nigeria with an average rainfall ranging between 830mm to 1,100mm annually starting from April to October, with average temperatures of 35°C for lowland and 31°C for highland areas respectively (NBS, 2006). There are three major districts that make up the local government: Toro district with headquarters at Toro, Lame district with headquarters at Gumau and Jama'a district with headquarters at Nabardo. The people of Toro local government is predominantly farmers and the crops grown in the local government include: maize, rice, cowpea, sweet potato, sorghum and soya beans.

Sampling Techniques

A multistage sampling technique was employed to select maize farmers in the study area. At first stage, Toro Local Government Area was selected. The next stage involves a purposive selection of the three district of the Local Government. The three district selected are Toro, Lame and Jama'a. The third stage involves a simple random sampling of two villages from each district. In the last stage, from compiled lists of maize farmers 10% respondents from each village was selected which gave an overall sample size of 101 respondents. Table 1 presents the sample frame distribution.

Method of Data Collection

Data collected for this study was obtained from primary sources. The primary data for this research was collected through the use of well-structured questionnaires.

Table 1. Sample size of the respondents

District	Village	Sample Frame	Sample Size
Toro	Zaranda	189	18
	Takanda	164	16
Lame	Magama gari	150	15
	Babikko	160	16
Jama'a	Zalau	173	17
	Rishi	190	19
TOTAL		1026	101

Source: Field Survey, 2019

Analytical Techniques

Descriptive statistics (such as frequency counts, percentages and means) were used to analyze objectives i, ii and iii.

RESULTS AND DISCUSSION

Source of Information on Maize Production Technology

Table 2. Distribution based on source of information of production technologies

Information source	Frequency	Percentage
Extension contact	23	78.2
Information Communication Technology Social groups or cooperatives	11 67	21.8
Total	101	100.0

Source: Field Survey, 2019

Table 2 revealed that most (78.2%) of the respondents got information on production technologies through their membership of social groups or cooperatives, which suggests that dissemination of agricultural information in the study area was very efficient through farmers membership of social groups or cooperatives; which avails members opportunity to exchange ideas and information that can boost their level of farm productivity.

Maize Production Technology

Table 3. Distribution based on maize production technologies adopted

Technology	Adopted	Not adopted	Ranking
Adjustment in planting date	75 (74.3)	26 (25.7)	1st
Planting method and seed varieties	71 (70.3)	30 (29.7)	2nd
Plant spacing and seed rate	61 (60.4)	40 (39.6)	3rd
Weed management	60 (59.4)	41 (39.6)	4th
Harvesting technology	51 (50.5)	50 (49.5)	5th
Fertilizer application and management	49 (48.5)	52(51.5)	6th
Pest management	43 (42.6)	58(57.4)	7th
Postharvest Processing techniques	41 (40.6)	60(59.4)	8th
Storage methods	35 (34.7)	66 (65.3)	9th
Water management	28 (27.7)	78 (77.3)	10th
Disease control	15 (14.9)	86 (85.1)	11th

Source: Field Survey, 2019; Percentages are in parenthesis

Table 3 revealed the various production technologies adopted by maize farmers in the study area. The significant production technologies adopted among maize farmers include; adjustment in planting date (74.3%); due to occurrence of diverse climatic conditions in the country, planting time varies from place to place. The optimum time to sow the crop depends on availability of irrigation facilities. For example, if irrigation facilities are available, maize crop can be sown during all seasons, planting method and seed varieties (70.3%); Method of planting plays an important role in establishment of crop under given set of conditions. Maize is mainly sown directly through seed by using different methods of tillage & establishment. Recently, resource conservation technologies (RCTs) namely; zero tillage, minimum tillage, ridge/raised bed planting, furrow planting and flat sowing of maize crop have been adopted by farmers in various maize based cropping system and are cost effective and environment friendly. Cultivars or seed varieties can be proffered based on length of growing season, availability of optimum moisture regime etc., plant spacing and seed rate (60.4%); the plant spacing recommended for maize in the Nigerian savannas is 75 × 50 cm, where 2 plants per stand are adopted. Maize may be planted either manually (hand planting) or mechanically. Crop geometry has direct effect on inter and intra-plant competition in field crops. Maize crop can be planted in varied crop geometries depending upon the purpose of cultivation. About 15–20 kg of maize seeds is required to plant a hectare, which is about 2½ acres. Plant 3 seeds per hole and later thin the stand to 2 plants at about 1–2 weeks after germination. When maize is grown as an intercrop, the spacing should be wider than for sole cropped maize. The practice of alternating rows of maize with other crops is highly recommended as it will allow the adoption of recommended packages for each crop. Being a non-tilling crop it cannot compensate for the lost space if proper plant stand is not maintained under field conditions. So maintenance of 60–65,000 plants/ha is pre-requisite for realizing maximum yield. Sowing of the crop should be done 60 × 20–25 cm crop geometry, weed management (59.4%); Good control of weeds in maize is a prerequisite for high yield. Weeds usually compete with the crops for nutrients, sunlight, and water. To permit the maize crop to express its full potential on the field, regular weeding is necessary. Timely weeding is essential, especially at the early stages of growth. Since the crop is shallow-rooted, it is very important to ensure that no mechanical damage is done to the crop roots during this operation. In cases where soil erosion is becoming significant, earthing up or remolding of ridges will be essential to prevent lodging and excessive exposure of the roots to the sun. In many parts of the Nigerian savannas, Striga is a notorious parasitic weed that has continuously posed a great threat to maize production. Under severe infestation, even before the emergence of Striga, the maize plants are usually severely attacked and look yellowish, stunted, and wilting. To effectively control Striga, the various methods need to be implemented in an integrated way and should include rotation with soybean, use of tolerant/resistant varieties, and application of the recommended fertilizer rate, with other cultural practices. Manual weed control weeding may start as much at 2 or 3 weeks after planting, and more than 2 manual weeding may be required. Two manual weeding are recommended to be conducted at 2 weeks and 4–5 weeks after planting. In areas where Imperata cylindrica is a problem. Chemical weed control wherever economically feasible to do so, herbicides can be used to control weeds on a maize farm. Although some have been recommended for sole cropped maize, it is important to note that the effectiveness of the herbicides is enhanced by very good land preparation, whatever implements are used. The increasing problem of lack of appropriate labor has increased the adoption of chemical weed control. Before land preparation, application rate will depend on formulation and label rate. Wait for 10 days after application before land preparation. Maize crop is infested with grassy and broad leaf annual weeds. Among grassy, *Dactyloctenium aegypticum, Eleucine indica, Setaria glauca, Cyanodon dactylon, Cyperus rotundus, Sorghum helepanse, Bracharia rapens* are common. The broad leaf weeds are *Celosia argentia, Commelina bengalensis, Phylanthis niruri, Solanum nigrum, Amaranthus viridis, Trianthema partulacastrum*. Effective weed management strategies have key role in successful maize cultivation. Adoption of weed control practices during the first 6–8 weeks after planting is crucial because maize crop kept weed free for 30–45 days after planting is almost similar in yield as that kept weed free for entire crop period. The annual yield loss in maize because of weed problems is estimated to be approximately 10%. A number of weed management practices can be adopted such as; Non-chemical control (manual weeding, mulching) and chemical control. Non-chemical weed control measures involves physical or cultural methods, such as manual removal of weeds from the maize fields. In cultural method, two hoeing's 15–30 days after sowing is recommended. Mulching is practice of keeping crop residues or plastic sheets on the soil surface within the crop rows to regulate temperature, conserve water as well weed control in field crops. Chemical control; sometimes due to scarce availability of farm labour, the only effective way to control weeds is the use of herbicides. The herbicide should be applied uniformly at recommended rates to minimize residual toxicity to crops sown. Harvesting technology (50.5%); Harvesting when maize is to be consumed fresh, should be conducted when the silk has turned brown (50–70 days after planting). But when grain is needed, harvest as soon as the grain is dry enough (80– 110 days after planting) depending on the variety. During harvest, the cobs can be broken by hand from the plant or the whole plant can be cut with a cutlass. Where the entire plants are harvested, they are often stacked in the field to allow the grain to dry further. In the humid zones, the crop can be dried in a traditional ventilated granary. For use as grain, cobs should be harvested when grains are at about 20% moisture. Whereas to consume as sweet corn, harvesting should be done when tassel starts turning brown and swelling of cob initiates. Other recommended production technologies include; Fertilizer application and management (48.5%); for high yield, adequate and balanced nutrition is important in maize. Ensure proper thinning of the plants to recommended stand density then conduct weeding before fertilizer application. The following fertilizer nutrients are recommended per hectare for maize in the savannas; 100 kg N, 50 kg of single super phosphate (SUPA), and 50 kg of muriate of potash (K20). This recommended fertilizer rate should be applied in two split applications. The first dose (that is 50:50:50) should be applied at planting or within the first week of planting using about 6 bags of NPK 15-15-15. The second dose of N (50 kg, about 2 bags of urea) should be applied between 4 and 5 weeks after planting in the Southern and Northern Guinea savanna zones, but between 3 and 4 weeks after planting for extra-early varieties in the Sudan savanna. All fertilizers should be buried during application. Ensure that the second weeding has been conducted before the second fertilizer dose is applied. Note that delay in the application of the first dose of fertilizer beyond week 1 after planting will drastically reduce the grain yield of early and extra-early varieties. Also, the second dose should not be delayed beyond 4 weeks after planting. Fertilizer should be applied in small quantities of with NPK 1 week after planting. Urea should be applied at 4 weeks after planting. The fertilizer should be poured into a hole made with a stick at a distance of about 5–8 cm away from the maize plant and below the soil surface. Do not leave applied fertilizers exposed on the soil surface. Always cover applied fertilizer with soil. Among the cereal crops, maize in general and specifically hybrids are very responsive to nutrients applied through organic or inorganic means. The rate of application depends on soil nutrient status and cropping system. For realizing required yield, the dose of applied fertilizer should be on par with the soil supplying capacity and crop demand. Also it is reported that the response of maize crop to organic manures is remarkable and a very important option in maize based systems. Integrated fertilizer management may include the following techniques; apply 10–15 t/ha of good quality farmyard manure per hectare to the maize crop; green manure the field, i.e., old green manure crop should be buried and allowed to decompose for about 10 days before sowing of maize; paddy straw compost at 450 kg/ha along with recommended dose of fertilizers can be an alternate to farm yard manure; as a general recommendation, one could apply 120 kg N, 60 kg P_2O_5 and 40 kg K_2O per hectare for hybrids and 80 kg N, 30 kg P_2O_5 and 20 kg K_2O per hectare for composites, pest management (42.6%); Effective pest management includes integrated use of all possible alternates that can be physical, cultural, biological, mechanical or chemical for controlling pests. Cultural control (such as; deep summer plowing helps in destroying resting stage of pests, inter-cropping with legume reduces borer incidence, use of well decomposed farm yard manure and balanced use of fertilizers reduces termite attack. Biological control through use of good quality planting material from reliable source. Mechanical control by cutting and destroying infected plants which ceases further spread. Use of pheromone traps. Set up of light traps. Chemical control: use of synthetic chemicals for the control of insect-pest and diseases, postharvest processing techniques (40.6%); maize shelling can be done by any one of the following methods; Hand shelling: use of internally ribbed tubes, beating cobs with a stick; using a single intake disc (hand-operated); using double intake disc (maybe hand/pedal/engine- operated).The method to use is determined mainly by output, although engine- operated multi-crop threshers used on a contract basis are gradually becoming popular. Grain cleaning. This can be done using the wind for winnowing or with screens of proper size. Maize drying is a vital operation which involves removal of moisture from the cobs/grains. It is carried out because high moisture grain will deteriorate rapidly due to grain respiration and heating, germination of grains, mold (fungal) growth and subsequent incidence of mycotoxins (e.g. aflatoxin) and increase insect multiplication and damage. The optimum moisture content of maize should be 14% or less. Types of drying; Sun drying: it is a popular method of drying grain where spread grain is exposed to direct sunlight until the desired grain moisture content is achieved. It is low energy cost; Smoking: The insect infestation is reduced when hung above the fire as the heat reduces the moisture content and the chemicals in smoke deters insect from laying eggs; Air drying: The maize cobs are hung along the roof of the house to expose it to air and hence the moisture content is minimized. Storage methods (34.7%); the walls and floor of the rumbu or granary must be cleaned thoroughly. Phostoxin tablets or pellets may be used to store more maize in an airtight container or large storage in sacks. Large-scale storage in bags such as pyramids will require a complete cover with tarpaulin after application with 4–6 tablets or pellets per ton, (wrapped in newspaper) in some of the bags. The tablets or pellets could also be enclosed in perforated envelopes instead of newspapers. Such storage should not be disturbed frequently for greater effectiveness. For fumigants to be effective, they must be applied where there is air-tight storage condition in sacks, containers or buildings. For bagged produce, the rate is 1-2 tablets per 100 kg grain sack. The tablet is wrapped in small tissue paper and placed in the middle of the sack. Also, grains must be properly dried; below 12 percent moisture level before storage to avoid mold. Clean granaries and stores before filling with new grains. A farmer can buy these chemicals from reputable agro chemicals dealers or retailers. Whenever phostoxin tablets are used, ensure the store is airtight to avoid the escape of fumes that are poisonous to human beings and animals when inhaled. For non-chemical storage, farmers are advised to store maize in Purdue Improved Crop Storage (PICS) bags. Water management (27.7%); Water requirement of the maize crop varies from 400 to 600 mm. Excess or shortage of moisture can have harmful impact on the crop growth. Proper drainage of standing water and meeting the crop needs at critical stages play a pivotal role in better crop performance. Disease control (14.9%); dressing the seeds prior to planting easily controls most of the diseases (whether soil-or seed-borne). A few pests such as stem borers, and grass hoppers occur occasionally and rarely present any serious challenge. These production technologies adopted significantly affected maize yield and the level of farm productivity. However, the level of adoption of these production technologies among maize farmers was relatively low in the study area.

Constraints of Adoption of Maize Production Technology

Table 4. Distribution based on the constraints of adoption

Constraints	Major (YES)	Minor (NO)	Ranking
High cost of production technology	82 (81.2)	19 (18.8)	1st
Complexity of technology	81 (80.2)	20 (19.8)	2nd
Lack of technology	79 (78.2)	22 (21.8)	3rd
Poor access to technology	73 (72.3)	28 (27.7)	4th
Poor access to agricultural credit	66 (65.3)	35 (34.7)	5th
Inadequate extension contact	65 (64.4)	36 (35.6)	6th
High cost of labour	61 (60.4)	40 (39.6)	7th
Fragmented lands	46 (45.5)	55 (54.5)	8th

Source: Field Survey, 2019; Percentages are in parenthesis

Table 4 revealed that high cost of production technology (81.2%); attributable to non-subsidization of agricultural technology and inputs, poor access to improved maize production technologies (81.2%); attributable to poor extension services, complexity of technology (80.2%); attributable to lack of technical expertise among farmers and stakeholders, lack of technology (78.2%); attributable to inadequate provision and supply of agricultural technology, poor access to technology (72.3%); also attributable to poor access to agricultural input markets, poor access to agricultural credit (65.3%); attributable to farmers poor access to financial institutions, inadequate extension contact (64.4%); attributable to farmers poor access and linkages with extension services, high cost of labour (60.4%); attributable to livelihood diversification activities which result to inadequate farm labour supply, fragmented lands (45.5%); attributable to poor tenure policies. This constraints of adoption of maize production technologies posited by the farmers in the study area were very significant factors that affected their adoption decisions.

CONCLUSION AND RECOMMENDATIONS

This study revealed that membership of social groups or cooperatives was the most significant means of agricultural information dissemination in the study area. Also, several maize production technologies were adopted among by the respondents; however the level of adoption was relatively low. In addition, all the constraints identified significantly affected the adoption of production technologies among maize farmers in the study area. Based on the findings of this study, the following recommendations are made:

1. Formulation and implementation of policies that will subsidize the cost of maize production technologies.
2. Formulation and implementation of policies that will improve farmer's access, supply and technical expertise on maize production technologies.
3. Formulation and implementation of policies that will improve farmer's access to agricultural credit, this will enable them adopt more options of production technologies.
4. Improving farmers access to extension services to improve their access to technology and mitigate any challenges encountered in the adoption of production technologies.
5. Formulation and implementation of policies that will ameliorate labour supply cost
6. Modification of tenure policies to mitigate land fragmentation for maize production.

REFERENCES

Abdulrahaman, A. A., & Kolawole, O. M. (2006). Traditional preparations and uses of maize in Nigeria. Alama J.F (2001). Economic analysis of maize production using two technologies in Kaduna State. *Journal of Arid Agriculture*, *11*, 125–129.

Abebaw, D., & Abelay, K. (2001). Factors Influencing Adoption of High Yielding Maize Varieties in South Eastern Ethiopia. *Quarterly Journal of International Agriculture & Extension Education*, *17*(1), 21–30.

Ali-Olubwanda, A. M., Odero-Wanga, D., Kathuri, N. J., & Shigova, W. A. (2010). Adoption of improved maize production practices among small scale farmers in agricultural reform era: The case of western province of Kenya. *Journal of Integrative Agriculture*, *8*, 160–171.

Bawa, D. B., & Ani, A. O. (2014). Analysis of adoption of improved maize production technology among farmers in Southern Borno, Nigeria. *Research on Humanities and Social Science*, *4*(255), 43–54.

Frova, C., Krajewski, P., Fonzo, N. D., Villa, M., & Sari Gorla, M. (1999). Genetic analysis of drought tolerance maize by molecular markers. *Theoretical and Applied Genetics*, *99*(1-2), 280–288. doi:10.1007001220051233

Idrisa, Y. L., Shehu, H., & Ngandu, M. B. (2012). Effect of adoption of improved maize seed on household food security in Gwoza EGA of Borno State, Nigeria. *Global Journal*, *12*, 20–31.

Iken, J. E., & Amusa, N. A. (2004). Maize research and production in Nigeria. *African Journal of Biotechnology*, *8*, 140–151.

International Institute of Tropical Agriculture. (2001). *Annual Report on Maize*. IITA.

Komolafe, S. E., Adeseji, G. B., & Ajibola, B. O. (2010). Determination of adoption of improved crop practices among women farmers in Ekiti East LGA of Ekiti State, Nigeria. *Journal of Agricultural Research*, *5*(2), 22–33.

Manyong,, V.M., & Makinde,, K.O., & Coulibaly, O. (2005). Economic grains from maize research in West and Central Africa. *Nongxue Xuebao*, *102*(9), 78–89.

National Agricultural Extension Research & Liaison Services (NAERLS) and Federal Department of Agricultural Extension. (2014). *Agricultural Performance Survey 2014 wet season in Nigeria*. Executive summary. Author.

National Bureau of Statistics (NBS). (2006). *Annual abstracts of statistics*. Federal Republic of Nigeria.

Ouma, J. O., Murithi, M. F., Mwangi, W., Macharia, H. G., & Gronte, H. D. (2002). Adoption of maize seed and fertilizer technologies in Embu district, Kenya. *International Journal of Agriculture*, *43*, 98–101.

Chapter 28
Organic Potato Cultivation: A Sustainable Green Technology Opportunity

Satish Kumar Luthra
ICAR-Central Potato Research Institute Regional Station, Modipuram, Meerut, India

Pooja Mankar
ICAR-Central Potato Research Institute Regional Station, Modipuram, Meerut, India

Sanjay Rawal
ICAR-Central Potato Research Institute Regional Station, Modipuram, Meerut, India

Mehi Lal
ICAR-Central Potato Research Institute Regional Station, Modipuram, Meerut, India

ABSTRACT

The extensive use of high-yielding varieties with high demand for fertilizers and pesticides leads to the unjudicious application of agri-inputs that leads to deterioration of precious natural resources. The world over, consumers are becoming serious about food quality, especially chemical residues. Organic farming is rising as a dynamic 'alternate farming system' to produce safe and quality food by environmental equilibrium. The growing demand for organic food is creating new market opportunities for farmers with premium prices. Potato can also play a potential role in improving farm income through organic production. Globally, India is the second-largest potato producer producing around 53 million tonnes from 2.2 million hectares. An organic farming system consisting of an integrated approach is the best technique for achieving higher individual crop and system productivity. The current potato scenario in India suggests its diversified production and utilization in the domestic and export market leading to higher profits by way of organic potato production.

DOI: 10.4018/978-1-7998-8434-7.ch028

INTRODUCTION

Organic agriculture is described as a system of farm design and management to create an ecosystem, which can achieve sustainable productivity without artificial external inputs such as chemicals, fertilizers, pesticides, *etc*. The principal objective of an organic production system is quality food production for humans and animals while maintaining ecological and economic viability. The consumer is becoming more serious about food quality, especially about chemical residues, and this awareness is developing demand for quality food, thereby a new market opportunity for farmers. Farmers can fetch premium prices from such produce and potato is no exception to all these trends and market avenues. Potatoes were introduced as a new crop in India by European settlers and now it is one of the important crops addressing food and nutritional security in India. Potatoes are a significant diet component at the national level as it is consumed fresh and in processed form in numerous ways. The second advance estimate for 2020-21 by the Ministry of Agriculture and Farmers welfare estimates 53.114 million tonnes from an area of 2.247 million hectares (Agricoop, 2021). Approximately 87% contribution comes from the North Indian plains. The current scenario of potatoes in India suggests its diversified production and utilization in the domestic and export market. Therefore, higher profits can be earned in domestic and international markets by way of organic potato production.

This chapter aims to inform the stakeholders, especially the cultivators who are already following organic potato production and those interested in beginning organic farming, about the organic potato production technology. Subjects to be discussed include the national program on Organic farming, conversation requirements, landscapes, diversity in crop production and management, contamination control, along soil and water conservation. We will also describe how selecting the suitable variety, agronomical practices, nutrient and water management, weed control, pest, disease management, and harvest can be adapted to comply with organic production standards while optimizing tuber yield and quality. Recommendations provided in this chapter complement standard crop management practices described in detail throughout the other chapters of this production guide.

The National Program for Organic Production (NPOP)

The National Programme for Organic Production (NPOP), which started in 2001, laid the foundation for the systematic development of the organic agriculture sector in India. This program is managed by the Agriculture and Processed Food Products Export Development Authority (APEDA) under the Ministry of Commerce and Industry. The NPOP provides standards for organic production, systems, criteria, and procedures for accreditation of Certification Bodies, the Logo, and its use regulations. The NPOP proposes to provide an institutional mechanism for implementing the NationalStandards for Organic Production (NSOP). The standards and procedures have been formulated in harmony with other International Standards regulating the import and export of organic products. The Director-General notifies the Standards of Foreign Trade. The NPOP has a system wherein Certification Bodies are accredited with National Accreditation Body (NAB). The NPOP is the mainstay for quality assurance for organic agriculture products in the country. All types of agriculture, horticulture, and non-food crops are grown under the organic certification process. Livestock, aquaculture, animal feed processing and handling, mushroom production, seaweeds, aquatic plants, and greenhouse crop production have also come under the ambit of organic certification.TRACENET is an online application tool managed by APEDA under NPOP to manage the entire organic certification system in the country (APEDA, 2021). The standards and procedures have been formulated in harmony with other International Standards regulating the import and export of organic products (Jaivikbharat, 2021).NPOP has equivalence with organic food regulations of the European Union and Switzerland and a recognition agreement on conformity assessment with USDA. The exports from the country are as per the provisions in NPOP.

ORGANIC FARM CONVERSION REQUIREMENT

A conversion period is an interim period required to establish an organic management system and soil fertility. The start of the conversation period can be counted from the first inspection of the certification agency. The conversion period may or may not be sufficient to develop the organic management system and build-up of soil fertility; however, it is the period during which all necessary actions required to achieve the goal are to be initiated. A clear plan is to be prepared about how to convert a farm into an organic one. The farmer/s should update this to comply with all standards. Requirements mentioned in the standards shall comply in this conversation period. All these requirements shall be applicable from the commencement of the conversion period until its conclusion. At locations where *de facto* requirements prescribed under these standards have been met for several years, a total conversion period is not required, and based on available documentation, the same can be verified. Under such a scenario, inspections shall be carried out at a reasonable time interval before the first harvest.

In potato, an annual crop, the tubers produced can be certified organic when the requirements prescribed under standards have been met during the conversion period of at least two years under organic management before starting the production cycle. The accredited certification bodies shall decide in some instances for extension or reduction of conversion period depending on the past status/use of land and environmental condition. Twelve months reduction in conversion period could be considered for annuals and perennials provided documentary proof has been made available to the certification body that the requirements prescribed under these standards have been met for a period of a minimum of three years or more. This could include the land that has been certified for a minimum of three years and wherein the products approved for use in organic farming as listed in Table 4 have been applied. The certification bodies shall also consider such a reduction in conversion period if it has satisfactory proof to demonstrate that for three years or more, the land has been idle and/or it has been treated with the products approved for use in organic farming as listed in Table 4 and 7. Organic produce of farm shall be sold as "produce of organic agriculture in conversion" or a similar description during the conversion period when the requirements prescribed under these standards have been met for at least twelve months.

Landscape

Organic farming is determined for biodiversity maintenance and natural conservation. The organic farming certification program shall set standards or procedures for a small farm area to facilitate biodiversity and natural conservation. Areas that are managed organically shall facilitate biodiversity. These may be extensive grasslands such as moorlands, reed land or dry land; all areas which are not under rotation and are not heavily manured; extensive pastures, meadows, extensive grassland, extensive orchards, hedges, hedgerows, groups of trees, and/or bushes and forest lines; ecologically rich fallow land or arable land; ecologically diversified extensive field margins; waterways, pools, springs, ditches, wetlands and swamps and other water-rich areas which are not used for intensive agriculture or aqua production.

Diversity In Crop Production And Management Plan

The basis for crop production in organic farming shall consider the structure and fertility of the soil and the surrounding ecosystem to minimize nutrient losses. Where appropriate, organic farms shall be required to maintain sufficient diversity in a manner that considers pressure from insects, weeds, diseases, and other pests while maintaining or increasing soil organic matter, fertility, microbial activity, and general soil health. Soil fertility shall be maintained through the cultivation of legumes or deep-rooted plants and the use of green manures, along with the establishment of a program of crop rotation and nutrient application with organic inputs.

Contamination Control

All relevant measures should be followed to minimize the contamination from outside and within the farm. Buffer zones shall be maintained to prevent contamination from conventional or non-organic farms. The buffer zone should be sufficient in size to prevent the possibility of unintended contact of prohibited substances applied to adjacent conventional land or farms. In case of doubt about contamination, analyze the relevant products and sources like inputs, soil and water etc., under a certification program to measure the contamination level. Polyethene and polypropylene or other polycarbonates coverings such as plastic mulches, fleeces, insect net, and silage wrapping are allowed. These shall be removed from the soil after use and shall not be burnt on the farmland. The use of polychloride-based products is prohibited.

Soil And Water Conservation

Soil and water resources shall be handled sustainably. Relevant measures shall be taken to prevent erosion, salinization of soil, excessive and improper water use, and ground and surface water pollution. Clearing of land by burning organic matter, e.g., slash-and-burn, straw burning, shall be restricted to the minimum. Clearing of primary forest is prohibited. The certification program shall require checking appropriate stocking rates that do not lead to land degradation and ground or surface water pollution.

SUITABLE VARIETIES

ICAR-Central Potato Research Institute has developed 66 varieties suited to different agro-climatic zones of the country. Under organic farming, selection of the suitable variety as per agro-climatic requirements, duration of crop, purpose viz., table purpose or processing, type (red skin, white skin), *etc.,* is crucial. The performance of a crop is affected by its climatic requirements and its purpose of cultivation. Hence the wise selection of variety as per requirement is a must. Usually, potato breeding efforts are focused on conventional production systems with optimum fertilizer and pesticide inputs, but varietal recommendations are rarely specific for organic farming. Genotypes have wider variations in productivity, tolerance to abiotic and biotic stresses, storability and taste under organic farming. In general, foliage senescence is advanced, and tuber number and yield decrease in the organic farming system over inorganic nutrition (Luthra *et al.,* 2017). The potato varieties Kufri Mohan, Kufri Khyati, Kufri Garima, Kufri Sadabahar, Kufri Pushkar and Kufri Anand were found promising for ware production due to better yield potential. For processing purposes, Kufri Himsona and Kufri Chipsona-3 were better varieties for process grade tuber yield. The characteristics and yield performance of potential potato varieties under organic farming are presented in Table 1 and Fig.1.

Table 1. Promising potato varieties under organic farming

Variety	Tuber yield (t/ ha)		Duration (Days)	Resistance	Tuber Characters
	Marketable	Total			
Kufri Khyati	28.4	31.4	70-80	Early and late blight	White cream, ovoid tubers, shallow eyes, cream flesh
Kufri Mohan	28.3	29.9	90-100	Late blight	White-cream ovoid tubers, shallow eyes, white flesh
Kufri Garima	27.0	29.3	90-100	Late blight	Light yellow, ovoid tubers, shallow eyes, light yellow flesh
Kufri Pushkar	26.1	27.9	90-100	Late blight, wart	Light yellow, ovoid tubers, medium-deep eyes, cream flesh
Kufri Sadabahar	26.0	27.0	90-100	Late blight	White cream, ovoid tubers, shallow eyes, white flesh
Kufri Anand	26.0	27.6	90-100	Late blight, wart, tolerant to frost and hopper	White cream, oblong tubers, shallow eyes, white flesh
Kufri Himsona	21.5	26.0	110-120	Late blight and wart	White-cream, round tubers, shallow eyes, cream flesh, suitable for chip making
Kufri Chipsona-3	20.3	23.1	100-110	Late blight	White cream, ovoid tubers, shallow eyes, white flesh, suitable for making chips

Figure 1. Tubers of promising potato varieties under organic farming

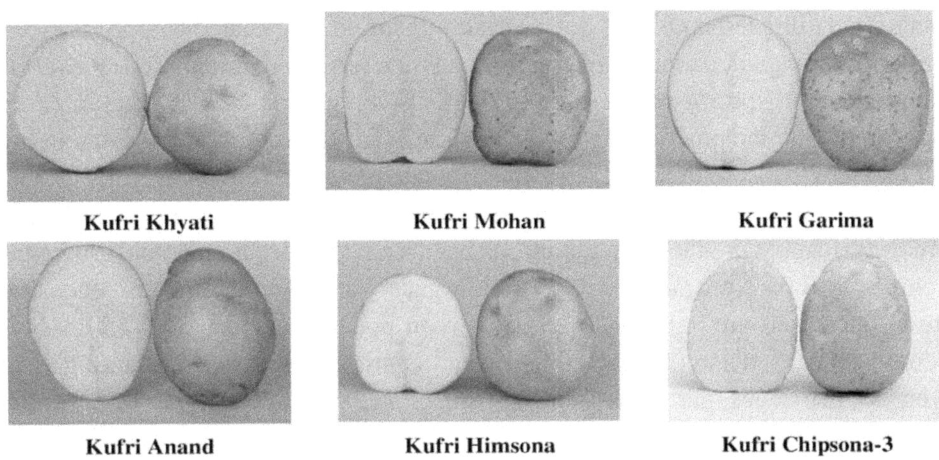

Seed and planting

The NSOP norms state that all seed/ planting material should be certified organic. If such material is not available, then use chemically untreated conventional seed/ planting material. The seed must be procured from a reliable source and replaced within 3-4 years. Use of genetically engineered seed, transgenic plants, or plant material is prohibited in organic farming. Sub-soiling in the initial phase of organic farming is the best method for loosening the soil. It should be performed during May- June for breaking the hardpan just below the plow layer. Later on, organic carbon and fauna activity keep soil sufficiently aerated and loose for better root development. It is recommended to have pre-planting irrigation 7-10 days before field preparation if sufficient soil moisture is not available. Plough the field with a mouldboard plow or disc harrow up to a depth of 20-25 cm, followed by one or two tilling and cross planking. The field should have fine tilth and be leveled for efficient use of irrigation water and proper drainage.

Optimum planting time is when maximum day temperature is below 32 ^{0}C and low night temperatures (< 20 ^{0}C) occur during tuberization, about 25-30 days after planting. Better growth, tuber initiation, and bulking take place during 15-25 ^{0}C. Seed potatoes are withdrawn from the cold store 10-15 days before planting. Seed tubers are kept in a diffused light under shades with proper ventilation for better sprouting. Tubers can be kept in baskets, wooden boxes and plastic trays or spread in a thin layer on the floor. Planting time is adjusted for better tuber initiation and development, which coincides with these temperatures. Row-to-row and plant-to-plant spacing should be kept at 67.5 cm and 20-35 cm. Plant to plant spacing varies with seed size. Medium seed size tubers (35-45 mm) are planted at 20 cm spacing. Planting depth should be kept at 8-10 cm. Bed planting (two/ three-row; base 132 cm) should be followed in micro-irrigation.

Weed Management

Weeds compete for sunlight, space, water and nutrition with the main crop and often serve as alternate hosts for several pests and diseases. Therefore, crop-weed competition is a serious factor affecting yield if timely care is not taken. Early weed competition usually reduces crop yields more than late-season weeds. The most critical period of crop weed competitions is the first 4-6 weeks and 6-8 weeks after planting in plains and hills, respectively. Crop-weed competition is managed by optimum planting date and plant population, companion cropping, crop rotations and fertility manipulation (Rawal et al., 2014). Weeds are suppressed out specifically in initial phases of crop growth by way of crop competition by adopting best crop production practices and making significant crop components grow in favour of the potato crop. It includes vigorous and faster growth of potato plants and having a crop environment such that it always smothers weeds. Weeds emerging out under better crop canopy are generally frail and will not be much harmful to tuber productivity. Potato cultivars having vigorous and rapid growing habits may prove better competitors for weeds as they cover fields quickly and overwhelm these undesirable plants.

Inter-cultural operations are done at 20-25 days when potato plants have 10-15 cm height for reducing soil compaction, better root aeration and weed control. Soil cover at planting is insufficient as irrigation erodes the ridges, particularly in light soils and exposure of stolon to light, making it an aerial stem. Tubers in ridges also become prone to greening, which is not fit for marketing. Therefore, Inter-cultivation is done in the morning; the field is exposed to sunlight for the whole day and weeds are removed manually. Ridging is followed the next day after applying the remaining dose of nitrogen through vermicompost or other organic product. Hand tools like *Khurpa* and spades are used in small fields. Bullocks drawn inter-cultivator and mouldboard plough can also be used in small fields. In mechanical cultivation, tractor-drawn 2-4 row cultivator and ridger are used for this operation. Crop rotation may be done following at least two-year crop rotation in a particular field or having green manure crops like dhaincha, cowpea, etc., for smothering weeds. Two- three-year crop rotation will assist in reducing weed seed bank in the field, while green manuring (45-50 days) shall reduce their intensity as they get buried along with the green manure crops and are decomposed (Rawal et al., 2014). The seedbed should be prepared thoroughly depending upon the soil type of a region. Pre-plant tillage operations for making a proper soil tilth accelerate the faster emergence of the potato plant and destroy weeds and give an edge to the crop. The seedbed should contain sufficient soil moisture and if moisture is low, pre-planting irrigation is advised before cultivation. A competitive edge is given to potato crops by way of optimized planting time. First of all, a variety should be planted at an optimum date and well-sprouted seed tubers should to be used at optimum crop geometry and planting depth. At planting, nutrient inputs should be precisely placed in bands 5-6 cm below seed tubers so that these inputs remain in the crop's root zone for efficient nutrient utilization. Inter-cultivation is done 20- 25 days after planting when plants attain 10-15 cm height. Later on, crop canopy shall cover the field, and weeds do not dominate.

Under organic farming, the aspects of hot weather cultivation, green manuring, mulching, water management, seed and land preparation are essential to keep the weed population under control. Hot weather cultivation can be an economical and environmentally friendly tool for pests, disease, and weed management. The advantage of the hot and dry summer season should be harnessed in a cropping system for desiccating all these menaces. Two-three deep field cultivations in this period are very useful for the control of annuals and also for perennials like *Cynodon dactylon* L. Similarly, soil solarisation may be pretty helpful in specific situations or for premium potato crops like the seed, etc. Soil solarization is done using transparent polyethylene (LLDPE) sheet of 25 to 100 mm thickness for 30 and 40 days during a high-temperature period after giving light irrigation to the field (48 hours before this practice).

Mulching is a very efficient traditional way of controlling weeds, conserving soil moisture, and slow decomposition of organic residues in soil. It is a very proficient way for smothering weed growth during crop season and, more specifically, for annuals. The main objective of mulching is to deprive weeds of solar radiation and thus inhibition of crop growth. Germination of weeds is also hampered, and this practice helps in conserving soil moisture, which facilitates the quick emergence of potato plants. Crop residue, dry straw, dry grasses, pine needles, and other vegetative material may be utilized in this operation. Plastic mulching has also come up as a promising technique for weed control in combination with micro-irrigation

Nutrient Management

As per NSOP norms, only on-farm produced microbial, plant or animal origin biodegradable material should be used as a source of nutrient management. There is a limit for using outside farm-produced microbial, plant, or animal origin biodegradable material. Non-synthetic mineral fertilizers and brought-in biological origin fertilizers should be regarded as supplementary sources and not replacements for nutrient recycling. The objective of nutrient management is to avoid nutrient losses, accumulation of heavy metals and other pollutants. Mineral fertilizers should only be used in a supplementary role to carbon-based materials. Only permitted organic or mineral fertilizers should be used and they should be used on-farm only when other fertility management practices have been optimized. Manures containing human excreta are prohibited to prevent transmission of pests, parasites and infectious agents. Mineral fertilizers shall be applied in their natural composition and shall not be rendered more soluble by chemical treatment. Certification may grant exceptions, but these exceptions shall not include mineral fertilizers containing nitrogen. Restrictions use inputs such as mineral potassium, magnesium fertilizers, trace elements, manures and fertilizers with a relatively high heavy metal content and/or other unwanted substances, e.g., basic slag, rock phosphate and sewage sludge. All synthetic nitrogenous fertilizers are prohibited. Application of well-decomposed biogas slurry, farmyard manure, compost, vermicompost, crop residue recycling and any safe organic source is beneficial for potato crops. Their nutrient composition is provided in Table 4. Examples of nutrient budgeting are mentioned in Tables 2 and 3. The dose of nutrition was optimized by integrating leguminous crop cultivation and the use of cheap available organic farm resources. An organic farming system consisting of an integrated approach (crop residue recycling, microbial treatment of seed compatible with individual crops, farmyard manure and vermicompost) is the best technique for achieving higher individual crop and system productivity. Only a lower rate of nitrogen release in organic farming is the main bottleneck for improving potato yields and this needs to be worked out to bring potato productivity at par with inorganic nutrition.

CROP ROTATION

Preceding and succeeding crops should not be exhaustive like potato crops. Farmers can select potato varieties and adjust crop duration suitable to their overall farming situation. Similar is the case with other crops followed in cropping/ inter-cropping systems. At least one leguminous, vegetables, or green manure crop should be included in potato-based cropping systems. The continuous raising of potato crops should not be followed on the same field to avoid pest build-up, and two to three-year rotation may be adopted. In Institute studies, crop rotation of maize- potato- onion, groundnut- potato- green gram, cowpea (vegetable)- potato- okra and groundnut- potato- maize+ green gram were found promising. Basmati rice- potato- late sown wheat also sustained productivity with an organic source of nutrition and in situ crop residue recycling (Tables 2 and 3).

Green Manuring

Green manures improve potato tuber yield and soil productivity as they supply organic matter and additional nitrogen. Therefore, green manure crops like sun hemp or *dhaincha* should be adjusted in the crop scheme. A leguminous crop with 10-20 tonnes of biomass per hectare adds about 60-100 kg/ha of Nitrogen. Also, it also helps in reducing the incidence of soil-borne diseases like black scurf and common scab etc. Green manure crop should be incorporated in soil 40-50 days after sowing for its proper decomposition.

Composting and Vermicomposting

Composting crop residues and vermicompost preparation is a common and beneficial technique for utilizing farm waste in crop nutrition. This is an invariably indispensable part of organic farming. Several available methods can be used for this activity. The nutrient composition of vermicompost is provided in table 6.

Table 2. Example of doses and method of application for food grain-based organic farming system

Nutrient management	Groundnut	Processing potato	Maize+ green gram
Conventional inorganic system	25 N-50 P_2O_5-50 K_2O kg/ha; Band placement	270 N-80 P_2O_5-150 K_2O kg/ha; Band placement and broadcasting	150 N-80 P_2O_5-80 K_2O kg/ha; Band placement and broadcasting
Integrated organic system	Maize+ *moong* bean stover: Approx. 15-19t/ha; *in situ* incorporation + *Rhizobium, Bacillus subtilis, Trichoderma*; seed treatment & mixed with FYM + FYM (2 t/ha)+vermi-compost (1 t/ha); FYM before sowing, vermi-compost at sowing	Groundnut stover: Approx. 5-10t/ha; *in situ* incorporation + *Azotobacter*, *Bacillus subtilis*, *Trichoderma*; seed treatment & mixed with FYM+ FYM (15 t/ha)+vermi-compost (6 t/ha); FYM before planting, vermi-compost at earthing	Potato haulms: Approx. 4-7t/ha; *in situ* incorporation + *Azotobacter*, *Rhizobium,Bacillus subtilis*, *Trichoderma*; seed treatment & mixed with FYM+ FYM (6 t/ha)+vermi-compost (3 t/ha); FYM before sowing, vermi-compost at 30-35 days
Organic system (Biofertiliser/ microbial formulation): *Rhizobium* (250g/ha seed treatment), *Bacillus subtilis* (250g/ha seed treatment), *Azotobacter* (250g/ha seed treatment), *Trichoderma* (4 kg/ha with 100 kg FYM for mixing in soil)			

Table 3. Example of doses and method of application for vegetable-based organic farming system

Nutrient management	Cowpea	Table purpose potato	Okra
Conventional inorganic system	25 N-50 P_2O_5-25 K_2O kg/ha; Band placement	180 N-80 P_2O_5-100 K_2O kg/ha; Band placement and broadcasting	90 N-50 P_2O_5-75 K_2O kg/ha; Band placement and broadcasting
Integrated organic system	Okra stover: Approx. 7-14 t/ha; *in situ* incorporation + *Rhizobium, Bacillus subtilis, Trichoderma*; seed treatment & mixed with FYM+ FYM (2 t/ha)+vermi-compost (1 t/ha); FYM before sowing, vermi-compost at sowing	Cowpea stover: Approx. 10-16 t/ha; *in situ* incorporation+ *Azotobacter*, *Bacillus subtilis*, *Trichoderma*; seed treatment & mixed with FYM+ FYM (12 t/ha)+vermi-compost (6 t/ha); FYM before planting, vermi-compost at earthing	Potato haulms: Approx. 7-14 t/ha; *in situ* incorporation+ *Azotobacter*, *Bacillus subtilis*, *Trichoderma*; seed treatment & mixed with FYM+ FYM (6 t/ha)+vermi-compost (3 t/ha); FYM & compost before sowing, vermi-compost at first picking
Organic system (Biofertiliser/ microbial formulation): *Rhizobium* (250g/ha seed treatment), *Bacillus subtilis* (250g/ha seed treatment), *Azotobacter* (250g/ha seed treatment), *Trichoderma* (4 kg/ha with 100 kg FYM for mixing in soil)			

Table 4. Products for use in plant nutrition and soil conditioning (Source: NPOP)

Matter Produced on an Organic Farm Unit	
Farmyard & poultry manure, slurry, urine, crop residues and green manure, straw and other mulches	Permitted
Matter Produced Outside the Organic Farm Unit	
Compost from plant residue; Peat without synthetic additives (prohibited for soil conditioning); Sawdust, wood shavings, wood provided it comes permitted from untreated wood	Permitted
Blood, meat, bone and feather meal without preservatives; Compost made from any carbon-based residues (animal excrement including poultry); Farmyard manure, slurry, urine; Fish and fish products without preservatives; Guano; By-products from food and textile industries of biodegradable material of microbial, plant or animal origin without any synthetic additives; Seaweed and seaweed products obtained by physical processes, extraction with water or aqueous acid and/or alkaline solution; Sewage sludge and urban composts from separated sources monitored for contamination; Straw; Vermicasts; Animal charcoal; Compost and spent mushroom and vermiculate substances; Compost from organic household reference; By-products from oil palm, coconut and cocoa (including empty fruit bunch, palm oil mill effluent (pome), coco peat and empty cocoa pods); By-products of industries processing ingredients from organic agriculture	Restricted
Human excrement	Not allowed
Minerals	
Calcified seaweed; Calcium chloride; Calcium carbonate of network origin (chalk, limestone, gypsum and phosphate chalk); Sodium chloride; Magnesium sulfate (Epson salt); Gypsum (calcium sulfate); Clay (bentonite, perlite, zeolite)	Permitted
Basic slag; Calcareous and magnesium rock; Mineral potassium with low chlorine content (e.g., sulfate of potash, kaolinite, sylvinite, patenkali); Natural phosphate (e.g., rock phosphate); Pulverised rock; Trace elements (boron, Fe, Mn, Mo, Zn); Wood ash from untreated wood; Potassium sulfate; Aluminum calcium phosphate; Sulphur; Stone mill	Restricted
Microbial Preparations	
Bacterial preparations (bio-fertilizers), biodynamic preparations, plant preparations, botanical extracts, vermiculite, peat	Permitted

"Restricted: Conditions and procedure for use shall be set by the certification program."

Water Management

Potato is a shallow-rooted crop and sensitive crop to water stress. The stolon formation and tuber initiation stages are the sensitive crop growth stages and water scarcity during this period leads to substantial yield losses. Water management in organic farming is similar to the conventional crop production system. In general, the time interval approach is adopted in the autumn season in the ridge-furrow system where the plots of 8-10 furrows and 6-8 meters in length are prepared, and water is let into them till the water level reaches half of the ridge height. Around 50 mm of water is applied during each irrigation. Irrigation is scheduled at 8-10 days intervals in the initial period when the temperatures are moderate, and the interval is increased to 12-15 days during the winter. Irrigation is stopped 10-12 days prior to haulm cutting for better skin setting.

Precision in irrigation is also an essential component in enhancing crop productivity. It is observed that micro-irrigation has improved overall system productivity. In the case of organic potato cultivation, it is comparable to the conventional system. Water losses are minimum under drip (4-5%) and sprinkler irrigation (15-20%) as compared to flood irrigation (40-50%). In experiments on organic farming, it was observed that more than ninety percent of system productivity could be achieved with a sprinkler or drip irrigation where water was precisely provided in the root zone and nutrient losses were also minimized

(Singh et al., 2018). Under a micro-irrigation system, water applications per cumulative pan evaporation (CPE) are anticipated. Herein, irrigation is started when CPE reaches 15-25 mm, depending upon soil type and variety. However, the recommendations vary from region to region and its best to adopt research recommendations at the regional level. Generally, drip irrigation is done on alternate days at 100-125% CPE for 30-45 minutes and sprinkler irrigation is done twice a week at 125-150 CPE for 90-120 minutes in potato crop (CPRI, 2014).

Table 5. Average nutrient content of bulky organic and green manures

Manure	Percentage content (dry weight)		
	Nitrogen	Phosphoric acid (P_2O_5)	Potash (K_2O)
Bulky organic and green manures			
Farmyard manure	0.95	0.62	2.20
Rural compost	0.75	0.63	1.05
Urban compost	1.35	0.62	1.45
Sesbania	2.01	0.32	2.03
Sunhemp	2.16	0.48	2.11
Cluster bean	1.46	0.25	1.89
Cowpea	2.45	0.56	2.32
Concentrated organic manures			
Castor	4.3	1.8	1.3
Cottonseed cake un-decorticated	3.9	1.8	1.6
Neem cake	5.2	1.0	1.4
Safflower cake un-decorticated	4.9	1.4	1.2
Groundnut cake	7.3	1.5	1.3
Linseed cake	4.9	1.4	1.3
Niger cake	4.7	1.8	1.3
Crop Residues			
Rice straw	0.60	0.24	1.48
Wheat Straw	0.56	0.11	1.52
Maize stalks	0.42	0.45	1.65
Sugarcane trash	0.35	0.10	0.60
Potato haulms	1.90	0.36	2.14
Groundnut straw	2.00	0.38	1.40
Sunflower stalks	1.90	0.61	2.10
Vegetable peas straw	1.40	0.32	1.27
Banana dry	0.61	0.12	1.00
Cotton	0.44	0.10	0.66
Pigeon pea	1.10	0.58	1.28

Table 6. Characteristics of vermicompost prepared from organic wastes

Substrate	C/N ratio	Available Nutrient			Microbial counts
		Nitrogen	P_2O_5	K_2O	
Cow dung					
Original	49.16	0.53	0.003	0.104	-
Vermi composted	16.83	1.20	0.004	0.396	24 x 105
Untreated	24.04	1.00	0.003	0.192	10 x 105
Sugarcane trash					
Original	38.17	0.66	0.080	0.020	-
Vermi composted	16.07	1.40	0.100	0.100	24 x 106
Untreated	30.72	0.80	0.090	0.060	20 x 105

MANAGEMENT OF PESTS AND DISEASES

The potato crop is affected by various diseases caused by fungi, bacteria, viruses and nematodes. The most important diseases are early blight, late blight, leaf spot complex, black scurf, common scab, bacterial wilt, soft rot, charcoal rot, potato virus X, S, M, V, Y, stem necrosis, apical leaf curl virus and root-knot nematode etc. NPOP norms should be considered seriously for the control of pests, diseases and weeds. To ensure minimum losses from these under organic cultivation, emphasis should be on balanced nutrition, suitable variety, fertile soils with high biological activity, crop rotations, green manuring, intercropping, mulching, land preparations etc. Natural enemies of pests and diseases should be protected and encouraged through proper habitat management of hedges, nesting sites etc. Different types of traps, biocontrol agents, chemicals approved under organic production were evaluated for biotic stress management. An ecological equilibrium should be created to bring about a balance in the pest-predator cycle. Products prepared at the farm level from local plants, animals, and microorganisms are permitted for pests, diseases, and weed control. Thermic weed control and physical methods are also permitted. Thermic sterilization of soils to combat pests and diseases is restricted when a proper rotation or renewal of soil cannot occur. Permission from the certification agency is required on case to case basis. All equipment from conventional farming systems should be adequately cleaned and free from residues before being used in the organic field. Use of synthetic herbicides, fungicides, growth regulators, synthetic dyes, insecticides and other pesticides are prohibited. Use of genetically engineered organisms or products is prohibited. Permitted products for weed, pest, and disease control are listed in (Table 7).

Disease Management

Integration of cultural and biological methods is permitted in organic cultivation. As chemical control is prohibited, suitable integrated disease management may be implemented to control one or more potent diseases of the region. Growing resistant cultivars and using disease-free seed is an essential component for the management of all diseases. Green manuring and deep ploughing during summer are also effective in checking such diseases. Tuber damage and injury must be avoided during harvest and post-harvest handling of tubers. Damaged and rotten tubers must be removed from the heap before storage. Late

Table 7. Products for plant pest and disease control (Source: NPOP)

Substances from plant and animal origin	
Plant-based extracts (Neem, garlic, Pongamia etc.); Gelatine; Casein; Extract from mushroom (Shiitake fungus); Extract from *Chlorella*	Permitted
Azadirachta indica (neem oil etc.); Preparation of rotenone from *Derris elliptica, Lonchocarpus, Thephrosia* spp.; Propolis; Preparation of pyrethrins extracted from *Chrysanthemum cinerariaefolium*, containing possibly a synergist, *Pyrethrum cinerafolium*; Preparation from *Quassia amara;* Release of parasite predators of insect pests; Preparation from *Ryania* species; Lecithin; Seaweeds, seaweed meal, seaweed extracts, sea salt and salty water; Fermented product from *Aspergillus*; Natural acids (vinegar)	Restricted
Tobacco/ tea	Not allowed
Minerals	
Clay (bentonite, perlite, vermiculite, zeolite etc.); Sodium bicarbonate	Permitted
Chloride of lime/soda; Copper salts/inorganic salts (Bordeaux mix, copper hydroxide, copper oxychloride) used as a fungicide, maximum 8 kg/ ha/ year depending upon crop and under the supervision of inspection/ certification agency; Diatomaceous earth; Light mineral oils; Permanganate of potash; Lime sulfur (calcium polysulphide); Silicates (sodium silicate, quartz); Sulphur (as a fungicide, acaricide, repellent)	Restricted
Mineral powders (stone meal, silicates)	Not allowed
Microorganisms / Bio-control agents	
Viral preparations (Granulosis viruses, Nuclear polyhydrosis viruses etc.); Fungal preparations (*Trichoderma* species etc.); Bacterial preparations (*Bacillus* species etc.); Parasites, predators and sterilized insects	Permitted
Others	
Softsoap (potassium soap); Homeopathic and ayurvedic preparations; Herbal and biodynamic preparations	Permitted
Carbon dioxide and nitrogen gas	Restricted
Ethyl alcohol	Not allowed
Traps	
Physical methods (chromatic traps, mechanical traps, light traps, sticky traps and pheromones); Mulches, nets, etc.	Permitted

blight is controlled by adjusting planting dates to avoid severe attack and heavy ridges to reduce tuber infection. Field scouting identifies and destroys primary infection patches by removing infected plants after drenching them with suitable restricted or permitted products. Irrigation should be stopped for some time when the weather condition is very congenial.*Trichoderma viride* @ 0.7% in liquid formulation and *Bacillus subtillus* @ 0.25% can be used for its control (Sunania *et al.*2006 & Lal *et al.* 2017). Spraying should be initiated before the appearance of disease and the number of sprays may be more depending upon disease severity. The use of copper oxychloride 50% WP @ 0.2% and copper hydroxide 53.8% DF @ 0.2% have shown lower disease severity, so these can be used for late blight control with approval from the certification body. Some other products, such as *Trichoderma viride* @ 2.0% in dry formulation, *Pseudomonas fluorescens* @ 1.0%, Neem oil @ 6.0%, and fermented neem giri @ 1.0, showed less disease severity of late blight of potato (Lal *et al.*2021). Avoid cultivating solanaceous crops and apply the recommended dose of organic manures/ products for sufficient nitrogen supply for controlling leaf spot complexes. Seed tuber treatment before storage with 3% boric acid for 25-30 minutes followed by drying under the shade, and crop rotation helps avoid several seed/ soil-borne diseases like black scurf, fusarium wilt, dry rot, charcoal rot, bacterial wilt, and common scab. Soil solarization and bio-fumigation with cruciferous plant species (mustard, radish) is also very effective in controlling such diseases. *Trichoderma viride* @ 4.5 g/kg of seed tubers for seed treatment and soil solarization of black scurf infested fields and seed treatment with *T. viride* greatly improved disease control in disease-infested filed (Arora *et al.* 2008). Diseases like charcoal rot and soft rot require potato harvesting before the soil temperatures exceed 28 °C. Trap crops like marigold in alternate rows can reduce root-knot nematodes. Viral diseases are kept in check by field sanitation, regular rouging, and controlling weeds in seed crops.

Pest Management

Aphids, whiteflies, leafhoppers and mites, white grubs, thrips, cutworms, caterpillars, and potato tuber moths are the important potato pests. The integration of eco-friendly methods for reducing vector (aphid) population below critical limits of 20 per 100 compound leaves is helpful in potato crops (Sridhar *et al.*, 2015). Monitoring of aphids is done by employing a yellow water trap (Fig.2). Twelve yellow colored sticky traps per hectare are placed 8-10 days after planting to trap whitefly and aphids. The height of the traps should be adjusted to canopy height. Dried neem leaves are used as mulch in organic plots to reduce the insect population by their decomposition and repellant action. A combination of neem leaves with weekly foliar sprays of horticulture mineral oil @ 0.5% resulted in reducing whiteflies and aphids. Several leaf-cutting caterpillars attack potatoes during December- January. ICAR-VPKAS, Almora, and National Center for Integrated Pest Management (NCIPM), New Delhi, have designed light traps to capture moths at night. Avoid planting the potato crop if temperatures are high to control leafhoppers and mites. Summer deep ploughing and tillage before planting expose larvae of white grubs etc., to predators. Use of liquid culture of EPN (*Heterorhabditis indica*) prepared in water and EPN cultured in *Galleria* cocoons reduce white grub larvae in the soil. Light and frequent irrigation will keep the thrips population in check. Caterpillars can be controlled by using light traps. The potato tuber moth is controlled by heavy ridging, water traps, and potato heaps with dried lantana and eucalyptus leaves

Figure 2. Yellow sticky traps

HARVESTING

Harvest potato tubers at the proper maturity when the skin is appropriately set. Stop irrigation 8-10 days before haulm cutting so that soil has enough moisture at the time of harvest and there is no formation of soil clods. Remove haulms when the crop is fully mature and start digging only after 10-15 days so that tuber skin is fully mature at the time of harvest and heap making. For manual harvesting, use hand tools like spade and *khurpa* to dig out the tubers. The animal-drawn plough is another option, which is faster as compared to the manual method. However, harvesting is delayed, and fields become dry in the case of large farms. So, tractor-operated diggers, multipurpose diggers are fast, economical, cause minor damage to the tubers, and vogue for potato harvestings.

CERTIFICATION OF ORGANIC PRODUCE

The accredited certification bodies in the case of NPOP and Regional Councils in PGS- India is responsible for certifying organic food products. Certification of organic food is essential to ensure that they comply with all the laid out standards. Organic foods should also comply with the requirements for metallic contaminants, aflatoxins, naturally occurring toxic substances (NOTS), and pathogens as specified under the Food Safety and Standards (Contaminants, Toxins and Residues) Regulation, 2011. In the case of residues of insecticides, the residue limit shall be 5% of the maximum limits prescribed under regulations mentioned above or the level of quantification (LoQ), whichever is higher. In addition, organic foods should comply with other standards besides food Safety parameters. Organic foods should comply with the requirements of various parameters covered under the Food Safety and Standards (Food Product Standards and Food Additives) Regulations, 2011.

CONCLUSION

The movement of organic farming is gaining momentum steadily worldwide due to several conclusive ill effects of chemical farming. Unpredictable changes in global climate have also put pressure on the agricultural sector to adopt technologies to mitigate climate change. Therefore, ways and means have to be found out to reduce carbon footprint and this can be made possible by minimum or negligible use of synthetic products and complete recycling of farm resources. Organic farming gives a ray of hope, so all components of the production technology of a system are needed to be developed for any sustainable organic farming system. Development of varieties, rapid availability of nutrients through natural means, organic products for weed control, and crop protection would significantly impact this direction.

Potato crop cannot remain untouched with recent developments of alternative farming as it constitutes the main ingredient of the Indian diet. People prefer fresh potatoes and like its processed products like region-specific dishes, chips, French fries, and flakes, etc. The country is the second-largest producer of potatoes globally, and a significant chunk of this is coming from North Indian plains. Sufficient scope exists in this region of increasing diversified utilization of this crop and organic potatoes can be one of them. States of Punjab, Haryana, Uttar Pradesh, Bihar, West Bengal and Assam have the vast potential of organic potato cultivation. Work on this aspect was going on in the past decade in the Institute, and production technology is now available to have yield levels better than the national average.

REFERENCES

Agricoop. (2021). *Ministry of Agriculture Cooperation and Farmers Welfare*. https://agricoop.nic.in/en/statistics/horticulture-crops-2019-20-second-advance-estimates

Agriculture and Processed Food Products Export Development Authority (APEDA). (n.d.). https://jaivikbharat.fssai.gov.in/standard-npop.php)

Arora, R.K., Sunaina, V., & Somani, A. K. (2008). Biological control of soil and tuber-borne diseases. *Indian Farming*, *58*(9), 44-46.

Lal, M., Chaudhary, S., Rawal, S., Sharma, S., Kumar, M., & Chakrabarti, S. K. (2021). Evaluation of bio-agents and Neem based products against late blight disease (*Phytophthora infestans*) of potato. *Indian Phytopathology*, *74*(1), 181–187. doi:10.100742360-021-00330-6

Lal, M., Yadav, S., Sharma, S., Singh, B. P., & Kaushik, S. K. (2017). Integrated management of late blight of potato. *Journal of Applied and Natural Science*, *9*(3), 1821–1824. doi:10.31018/jans.v9i3.1445

Luthra, S. K., & Sanjay Rawal, V. K. (2017). Gupta. Promising potato genotypes for organic farming systems. *CPRI Newsletter*, *68*(April-June), 2.

Rawal, S., Khan, M. A., Dua, V. K., Singh, S. P., Yadav, S. K., Lal, S. S., & Singh, B. P. (2014). Weed Management in potato. *ICAR-CPRI Technical Bulletin, 91*, 78.

Rawal, S., Luthra, S.K., Malik, K., Lal, M., Mankar, P., Dua, V.K., & Chakrabarti, S.K. (2020). Organic potato Production Technology for North Indian Plains. *ICAR-CPRI Shimla Extension Bulletin, 51*, 28.

Sridhar, J., Venkateswarlu, V., & Nagesh, M. (2015). Aphid. In A manual on diseases and pests of potato-Technical Bulletin No. 101. ICAR-Central Potato Research Institute.

Chapter 29
Scenario of Climate Change on Potato Instects:
Indian Perspective

Anuj Bhatnagar
ICAR-Central Potato Research Institute Regional Station, Modipuram, Meerut, India

V. K. Dua
ICAR-Central Potato Research Institute, Shimla, India

B. P. Singh
ICAR-Central Potato Research Institute, Shimla, India

ABSTRACT

Climate change is not very well understood including how these changes will affect potato crop, insect, disease, natural enemies, and the complex relationship among them. If the climate is warmer, will it increase yield offset losses to pests or will losses to pests outweigh yield advantage from warmer temperatures? It is likely that new insect pests of potato will stabilize in narrower areas and be able attack plants in new regions. It is likely that plants in some regions will be attacked more frequently by certain pests. This chapter explores the effect of climate change on potato insects in India.

INTRODUCTION

There is a growing concern about the global warming caused by emission of gases like carbon dioxide, methane oxide and ozone. Climate change has now become a reality and no more a speculation. Agriculture contributes to climate change both directly through various agricultural practices but also indirectly being a significant user of land and consumer of fossil fuels. Potato is cool climate loving crop. Climatically, there is a short window in which potato has been fitted with appreciable use realized even 80 to 90 days old crop. Moreover a few other crops produce comparatively high yields of nutrients per unit area where land is scare. That's why potatoes highly important for many marginal families of the

DOI: 10.4018/978-1-7998-8434-7.ch029

country. High production levels and highly perishable sale practices enable farmers to be competitive despite the fact of high input cost, labor problem and forced gluts, still potato is an important cash crop and cultivated mainly as an irrigated winter cash crop. Among the various factors related to insect pest and diseases, potato farmers are increasingly forced to face water stress change, changes in rainfall patterns and its intensity, frost and fog etc. These conditions are generally interpreted to climate changes. Potato production is highly dependent on climate and weather factors. All climate change projections are based on simulations and therefore involve uncertainties. However, it is clear that climate change will not only affect the cultivation system but also change the pressure of insect pests of potato. Disease free seed production may also be affected due to changes in the life cycles of aphids. Pests such as mite, potato tuber moth and cutworms may become more serious on potato crop. At present most of the potato seed is grown in Northern plains under no or low aphid period in order to keep it virus free but change in climate, in certain regions, aphids will occurs in greater numbers, resulting threat particularly to virus free seed production. Besides that environmental incidence of extreme weather may reduce the effectiveness of insecticides on targeted insect pests. There is need to scientifically assessed the impact of climate change on the insect pest of potato.

INSECTS AND ENVIRONMENT

Insects are poikilothermic in nature i.e. temperature of insect bodies is almost the same as that of the environment. Therefore, temperature is probably single most important factor which influencing distribution, development, survival, behaviour and its life cycle. Insect life stage, predictions are most often calculated using accumulated degree days from a base temperature and bio fix point. Few researchers believe that effect of temperature on insect largely overwhelms the effects of other environmental factors (Bale *et.al.* 2002). It has been estimated that a 2^0C temperature increaseinsects might experiencing one to five additional life cycles per seasons (Yamamura & Kiritani, 1998). Other researchers have found that moisture and CO_2 effects on insects can be potentially important consideration in a global climate change setting (Hamilton 2005; Coviella and Trumble, 1999; Hunter, 2001). The increase temperature limit their range over wintering, population growth rate, number of generations per annum, length of growing season, crop pest synchroization, intra specific interaction, dispersal, migration and availability of host plants may alter host physiology and resistance against insect pests of potato.

Effect of Rising Temperature on Potato Insects

Elevated temperatures may cause breakdown in temperature sensitive resistance. Increase in food quality due to increase in nitrogen content in plants under stress condition caused by high temperature can results in sudden resurgence of insect pests population. Potato plants become more susceptible to insect pests. Climate change resulting in increasing temperature could impact crop insect pests in several ways. One way temperature effect recent trend to decline insect populations especially in tropical regions, while in temperate regions, warmer temperature will result in more types and higher populations of insects.

Development strategy of insect species can exert different effects (Bale *et.al.,* 2002).Temperature can potentially effect insect survival; population size and life cycle beside that also affect insect physiology and development directly or indirectly through the physiology or existence of hosts.Some soil insects which take several days to complete one life cycle-like white grub, cut worm, potato tuber moth will tend to moderate temperature or variability over the course of their life history. Some potato crop pests like aphids, white fly, leafhoppers, mite and thrips are developed more rapidly during periods of time with suitable temperature. Increased temperatures will accelerate the development of these insects on potato crop possibly resulting in more numbers and generation per year.

Migratory insects such as aphids which are an important insect vector of potato may arrive in Northern plains earlier or the area in which they are able to over winter may be expanded. In *M. persicae,* increase of mean temperatures by 1^{0}C advances the time of progeny migration by two weeks, also well evident in *Acyrothosiphon* spp. Temperature also contributes to morph determination besides differences in responses between species or clones or morphs. *M. persicae*is very sensitive to warmer conditions with a reduced tendency for alate production, lower migratory potential and delay in sexual morph production.*M. persicae*may increase by more number of generations in a year due to lower developmental zero point and low thermal total required for one generation, resulting more infestation. Natural enemy of insects may be responds differently to insect host due to change in temperature. Parasitism could be reduced if host populationsemerge and pass through vulnerable life stages before emergence of parasitoids. Host may pass through vulnerable life stages more quickly at higher temperature, reducing the chance of parasitism. Temperature may change gender ratio of thrips (Lewis, 1997) potentially affecting reproduction rates. Soil inhabiting insects may be more gradually affected by the temperature change than those that are above ground simply because soil provides insulation medium that will tend to buffer temperature changes more than air (Bale *et. al.,* 2002). Lower winter mortality of insects due to warmer winter temperatures could be important in increasing insect population (Harrington *et.al.,* 2001). Insect species tends to decrease with higher latitude (Gaston & Williams, 1996; Andrew & Hughes 2005), meaning that rising temperatures could result in more insect species attacking more hosts in temperate climates (Bale *et.al* 2002). Based on fossil records, it was concluded that the diverging of insect species and their feeding intensity increased historically with increasing temperatures (Bale *et. al.,* 2002).

Findings from Peru suggested that global warming is already affecting the distribution of crop insect pests. Scientist from CIP and the partner institutions are of the opinion that current balance of insect populations is almost certain to be upset. In the opinion of Dr. Aziz Lagnoaui, entomologist from CIP, Lima Peru,and climate change will favor invaders over native species. Some insect pests will become more important while others will decline but the net effect will be to increase the pressure of pests on the crop yields. Experience from CaneteValley in Peru revealed that when white flies are few in number, their feeding on crops pose little damage. But if sprayed, they soon developed resistance and multiply. At high population density, white fly can destroy whole crop stand.

Until recently, only biotype A of *Bemisiatabaci*which causes moderate yield loss was found inCaneteValley. In 2001, however, a more aggressive white fly biotype B was identified. According to farmers, it arrived during EI Nino year when climate in Peru was hotter and drier than usual. But instead of disappearing when weather reverted to normal, the pest apparently adapted to its new environment, where it is now inflicting severe damage to crops. Besides, a new species of whitefly *viz. B. afer*was also detected in the valley. This species is even more aggressive than biotype B of *B. tabaci.* Now it is causing severe crop losses in the valley.

In India,*B. tabaci*was a minor pest till recently. Data on population buildup during the last 20 years revealed that average population of *B. tabaci*was 11 whitefly /100 leaves during 1984 which rose to 24.24/100 leaves in 2004. During this period, average ambient temperature increased by 1.07°C. This indicates that climate warming may lead to whitefly infestation in Indo-Gangetic plains. Increase in *B.tabaci*population led to outbreak of a new viral disease known as apical leaf curl in potato which has since been identified as a Gemini virus. Gemini Viruses are not reported to infect potato crop world over. Therefore, a new dimension has been added to potato seed production in sub-tropics. The infestation of this virus in some of the varieties *viz.* KufriPukhraj, Kufri Anand, Kufri Sutlej reached almost 100% in seed stocks. Consequently, altogether new approach was adopted to manage this virus which included discarding of the old stocks and their replacement with disease-free stocks from tissue culture and integrated management of whitefly which included delayed planting, seed treatment and weekly application of insecticides including summer oil. This disease has since been contained but the danger persists.

Empoascafabae(leaf hopper) is another pest which has assumed significance in early planted crop in sub-tropical plains of India. Its population during 1984 was 16.6 which rose to 23.8 in 2004. The hopper burn damage also increased from 45 to 68% during this intervening period. Sudden warming up of temperature may also lead to flaring up of the pest. This has happened during 2006-07 in Gujarat state. Temperature profile of the region revealed that during crop period right from December to first week of January, ambient temperature was higher and relative humidity was lower than previous years. The increase in temperature was up to 5.8°C during fourth week of December, 2006. This increase in temperature led to faster multiplication of the pest and consequently very heavy (up to 100%) hopper burn damage (Singh *et.al., 2007).*

Similar to hopper damage, mite infestation has also increased in early planted crop. During 1984-85 its damage was 86% which increased to 100% in 2004. Results tend to suggest that in sub-tropical plains of India, *Myzuspersicae*population is on the rise. During 1984-85 mean aphids population/l00 compound leaves were 567 which increased to 653 in 2003-04. On the other hand population of *Aphis gossypii* hasincreased three fold during the last 20 years. Although *A. gossypii*has low vector efficiency but its appearance right from the emergence of the crop and further maintaining its population throughout the crop season may pose serious problems to seed production in sub-tropical plains.

Increased Temperature, Decrease Insect Pest Populations

Temperature increases that causesfarmersnot grow host crop any longer would decrease the population of insect pests specific to potato. The some environmental factors that can impact insect pests and their natural enemies that infect the pests, resulting in increasing attack on insect populations. At higher temperatures, aphids have been shown to be less responsive to the aphid alarm pheromone they release when under attack by natural enemies, resulting in more predation (Awmack *et. al.,*1997).

Effect of Precipitation on Insect Pest Populations

Many insect are sensitive to precipitation and are killed or removed from host plants by heavy rains, theseconsiderations are important when choosing management options for thrips (Reiners and Petzoldt, 2005). Some insects like white grub, termite that overwinter in soil, flooding the soil has been used as a effective measure. It will be predicted more frequent and intense precipitation,forecasted with climate change to negativity impact on these insects. Fungal pathogens of insects are favored by high humidity and their incidence would be increased by climate changes that lengthen periods of high humidity and reduce by those that result in drier conditions.

Effect of CO_2 on Insect Population

In general, CO_2 impact on insects is thought to be indirect impact on insect damage results from changesin the host crop. Some workers have found that rising CO_2 can potentially have important effect on insect pests. Recently free air gas concentration enrichment technology (FACE) was used to create an atmosphere with CO_2 and O_2 concentrations similar to what climate change models predict for the middle of the 21st century. FACE allows for field testing of crop situations with fewer limitations than those conducted in enclosed spaces. In few cases, like Soybeans grown at elevated CO2 atmosphere had 57% more damage from insect then those grown in present atmosphere and required plant protection treatment in order to continue the experiment. It is assumed that increase in the level of simple sugar in the soybeanlevels may have stimulated the additional insects feeding (Hamilton *et.al,* 2005). In another FACE study, it was recorded that insect some time feed more on leaves that have lowered nitrogen content in order to obtain sufficient nitrogen for their metabolism (Coviella and Trumble, 1999; Hunter, 2001). Increased carbon to nitrogen ratio in plant tissue resulting from increased CO_2 levels may slow insect development and increased the length of life stages vulnerable to attack by parasitoids (Coviella and Trumble, 1999). Studies carried out by Bezemer*et. al*., 1998 revealed that aphid abundance was enhanced by the CO_2 and temperature. Parasitism rates remained unchanged in elevated CO_2 but showed an increasing trend in conditions of elevated temperature. Research further revealed that clones of *Myzuspercicae* have emerged that are resistant to three major groups of insecticides. It is feared that global warming could remove the break that winter new places on the number of insecticide-resistant aphids. *Brevicorynebrassicae* reared in *Brassica oleracea*produced significantly less offspring at elevated $CO_{2,}$ whereas, the opposite was found for*M. persicae* on the same host. No response was found for *M. persicae*on*Senecio vulgaris* (Bezemer*et.al*, 1999).

Effect on Farmers

It is likely that farmers will experience extensive impacts on insect management strategies with change in climate.It is assumed that insect will expand their geographic ranges and increase reproduction rates. It is likely that farmers will have more types and higher number of insects to manage. Insecticides and their applications have significant economic costs for growers and environmental cost for society. Additionally, some classes of pesticides like pyrethroids and spinosad have been shown to be less effective in controlling insects at higher temperatures (Musser & Shelton, 2005). It is predicted that additional generations of important insect pests specifically in temperate climate as a result of increased temperatures required more insecticide application to maintain population below economic threshold levels. In order to avoid insecticide resistance is to apply insecticide with a particular mode of action, less frequently with more insecticide applications required the probability of applying a given mode of action insecticide more times in a season will increase, thus increasing the probability of more and more members of insects developing resistance to insecticides specially sucking insects of potato (Shelton *et.al.,* 2001).

A number of cultural practices that can be used by Indian farmers could be affected by change in climate. Although it is not clear whether these practices would be helped, hindered or not affected by the anticipated changes. Using crop rotation as an insect management strategy could be less effective with earlier insect arrival or increased over wintering of insects. However, this can be balanced by change in the earliness of crop planting time, development and harvest. Row covers used for insect exclusion might have to be removed earlier to prevent crop damage by excessive temperatures under the covers would the targeted early insects is also complete their damaging periods earlier or be ready to attack when the row covers were removed.

What Farmers can do to Adopt

Climate change is a slow and gradual process that will give enough opportunity to Indian farmers to adopt them. It is not very well under stood, how these changes will affect potato crop, insect, disease, natural enemies and their complex relationship among them. If climate is warmer, will increases in yield offset losses to pests or losses to pests outweigh yield advantage from warmer temperatures? It is likely that new insect pests of potato will become stabilize in narrower areas and to be able attack on plants in new regions. It is likely that plants in some regions will be attacked more frequently by certain pests. A few pests may be less likely to attack on crop as changes occur. It is likely that we will not know the actual impact of the climate change on pests until they occur. Clearly, it will be important for farmers to be aware of crop pest trends in their regions and flexible in choosing both their management methods in the crop they grow. It is important that the potato farmers closely monitor the occurrence of insect pests in their fields and keep records of their incidence frequency and cost of managing pests over time will be a better position to make decisions about whether it remains economical to continue to grow potato crop with appropriate pest management techniques. It is important that farmers may use integrated pest management practices more precisely such as monitoring of insectvectors, forecasting, keeping records of insect pestincidence and choosing economically and environmentally sound control measures will be most likely to be successful in dealingwith the effects of climate change.

Conditions are more favorable for the proliferation of insect pests in warmer climate. Therefore, warmer winter temperatures may also allow larvae to over winter in areas where they are now limited by cold, thus causing grater infestation during the following crop season. Altered wind patterns may change the spread of air borne insects like thrips and aphids. The possible increases in insect pests infestation may bring about greater use of insecticide to manage them, a situation that will require the further development and application of integrated pest management techniques, some insect pests that are currently present at low levels or that are not considered important one, may be more prevalent.The consequence on population dynamics may lead to substantial increase in the use of agrochemicals. Higher temperature leads to faster completion of insect cycle which indirectly provokes use of insecticides and hence development of resistance. Globally, climate change model predict an increase in probability of intense rainfall, which could result in increase insecticide wash off and reduced insect pest on potato crop. The rates of insecticide application thus have to be modified carefully according to new situations as arise due to change in climatic conditions.

Adaptation Measures for Climate Change

- We have to increased survey and surveillance of insect pests of potato crop and taking up plant protection measures accordingly.
- We have to explore changes in host pest interaction under early, normal and late sowing potato crop in order to recommended optimum sowing dates for reduced pest pressure and increase in yield.
- We have to take-up breeding work of potato cultivars for biotic stresses.

Future Strategies

- Effect of cumulative increase of CO_2 and temperature on biology of insects and mite to be generated on potato crop.
- Efforts on genetic selection and transgenic development for biotic stress management to be intensified on potato crop.
- Forecasting models of potato pest specially aphids to be validated under changing climate scenarios.
- Management of insecticides resistance on sucking pests like aphids, whitefly and hoppers to be increased under changing climate scenarios.
- Studies on change in the pattern of virus –vector relationship on potatounder changing climate scenarios.

CONCLUSION

The precise impact of climate change on insects is somewhat uncertain because some changes may favors insects while other may inhibit the activity of few insects. However, the evidence indicates that there will be an overall increase in the number of out breaks of wider variety of insects. Climate change is likely to alter the balance between insect pests, their natural enemies and their hosts. One of the most important effect of the climate change will be to alter the synchrony between host and insect pests development particularly in spring, but also in autumn the predicted rise in temperature will also generally favors insect development and winter survival, although there will be some exceptions. Few changes have already been observed in the distribution of the insect populations. All these changes will contribute to shifts in geographical distribution of insect pests of potato. Temperate insect species will push further towards the polar regions while tropical and sub-tropical species will encroach on temperate regions. The possible increased use of insecticides resulting from an increase in pest out breaks will likely have negative environment and economic impacts for potato production in the country. The best economic strategy for farmers is to use IPM practices to monitor insect occurrence and their buildup on potato crop.

REFERENCES

Andrew, N. R., & Hughes, L. (2005). Diversity and assemblage structure of phytophagousHemiptera along a latitudinal gradient: Predicting the potential impacts of climate change. *Global Ecology and Biogeography*, *14*(3), 249–262. doi:10.1111/j.1466-822x.2005.00149.x

Awmack, C. S., Woodcock, C. M., & Harrington, R. (1997). Climate change may increase vulnerability of aphids to natural enemies. *Ecological Entomology*, *22*(3), 366–368. doi:10.1046/j.1365-2311.1997.00069.x

Bale, J. S. G. J., Masters, G. J., Hodkinson, I. D., Awmack, C., Bezemer, T. M., Brown, V. K., Butterfield, J., Buse, A., Coulson, J. C., Farrar, J., Good, J. E. G., Harrington, R., Hartley, S., Jones, T. H., Lindroth, R. L., Press, M. C., Symrnioudis, I., Watt, A. D., & Whittaker, J. B. (2002). Herbivory in global climate change research: Direct effects of rising temperatures on insect herbivores. *Global Change Biology*, *8*(1), 1–16. doi:10.1046/j.1365-2486.2002.00451.x

Bezemer, T. M., Jonesh, T. H., & Knight, K. J. (1998). Long term effect of elevated Co_2 and temperature on the population of the Peach Potato aphid *Myzuspercicae* and its parasitoid*Aphidiusmatricariae*. *Oecologia*, *116*(2), 128–135. doi:10.1007004420050571 PMID:28308517

Bezemer, T. M., Knight, K. J., Newington, J. E., & Jones, T. H. (1999). How general are aphid responses to elevated atmospheric CO_2? *Annals of the Entomological Society of America*, *92*(5), 724–730. doi:10.1093/aesa/92.5.724

Coviella, C., & Trumble, J. (1999). Effects of elevated atmospheric carbon dioxide on insect plant interactions. *Conserv. Biol*, *13*(4), 700–712. doi:10.1046/j.1523-1739.1999.98267.x

Gaston, K. J., & Williams, P. H. (1996). Spatial patterns in taxonomic diversity. In *Biodiversity 202-229*. Blackwell Science.

Hamilton, Dermody, Aldea, Zangerl, Rogers, Berenbaum, & Delucia. (2005). Anthropogenic Changes in Tropospheric Composition Increase Susceptibility of Soybean to Insect Herbivory. *Envirn. Entomoi.*, *34*(2), 479-485.

Harrington, R., Fleming, R. A., & Woiwod, I. P. (2001). Climate change impacts on insect management and conservation in temperate regions: Can they be predicted? *Agricultural and Forest Entomology*, *3*(4), 233–240. doi:10.1046/j.1461-9555.2001.00120.x

Hunter, M. D. (2001). Effects on elevated atmospheric carbon dioxide on insect-plant interactions. *Agricultural and Forest Entomology*, *3*(3), 153–159. doi:10.1046/j.1461-9555.2001.00108.x

Lewis, T. (1997). *Thrips as crop pests*. CAB International, Cambridge University Press.

Musser, F. P., & Shelton, A. M. (2005). The influence of post-exposure temperature on the toxicity of insecticides to *Ostrinianubilalis*(Lepidoptera:Crambidae). *Pest Management Science*, *6*(5), 508–510. doi:10.1002/ps.998 PMID:15633207

Reiners, S. C. (Ed.). (2005). Integrated Crop and Pest Management Guidelines for Commercial Vegetable Production. Cornell Cooperative Extension Publication.

Shelton, A. M., Wilsey, W. R., & Soderlund, D. M. (2001). Classification of insecticides and acaricides for resistance management. *NYSAES*.

Singh, Kamlesh, Patel, Patel, Singh, & Singh. (2007). First report of "hopper burn" damage in Gujarat state. *CPRI News Letter*, *37*, 3–4.

Yamamura, K., & Kiritani, K. (1998). A simple method to estimate the potential increase in the number of generations under global warming in temperate zones. *Applied Entomology and Zoology*, *33*(2), 289–298. doi:10.1303/aez.33.289

Chapter 30
Bioinformatics Resources for Sustainable Agriculture

Rekha Dixit
Sardar Vallabhbhai Patel University of Agriculture and Technology, Meerut, India

Neelesh Kapoor
Sardar Vallabhbhai Patel University of Agriculture and Technology, Meerut, India

Naresh Pratap Singh
Sardar Vallabhbhai Patel University of Agriculture and Technology, Meerut, India

Amit Kumar
Swami Vivekanand Subharti University, India

ABSTRACT

Plant breeding and genetic engineering have made significant contributions in improving crops and thus addressing the problem of food security over the last few decades. However, due to continuously growing population, increasing urbanization, environmental pollution, land degradation, and global climate change, food security is still a major concern worldwide. More attention needs to be paid to sustainable crop production for ensuring enough food for current and future generations. Emergences of genomic technologies holds promise for improving crop productivity and provide solutions to the food shortage. Recent advances in high throughput technologies have led to generation of a tremendous amount of genomics data from plants that can be put to a variety of uses. Computational tools are essential to analyze this complex data generated from sequencing of genomes, determine complete set of genes, analyze transcripts and proteins encoded by an organism, and understand their interactions that bring out changes in phenotypes.

DOI: 10.4018/978-1-7998-8434-7.ch030

INTRODUCTION

Bioinformatics is an interdisciplinary field involving Microarray, Genomics, Proteomics, Protein modeling Bio Grid computing, Molecular biology, system biology, protein ligand docking and structure prediction of macromolecule (Fig-1). Bioinformatics overlaps other areas of research that are designated informatics and computational biology. Informatics has traditionally been a discipline in which mathematicians, computer scientists, statisticians, and engineers develop technologies for supporting information management in fields like health care. Bioinformatics is now involved in these activities by organizing biological data related to genomes with a view to applying this information in agriculture, pharmacology, and other commercial applications.

Figure 1. Role of bioinformatics in various field

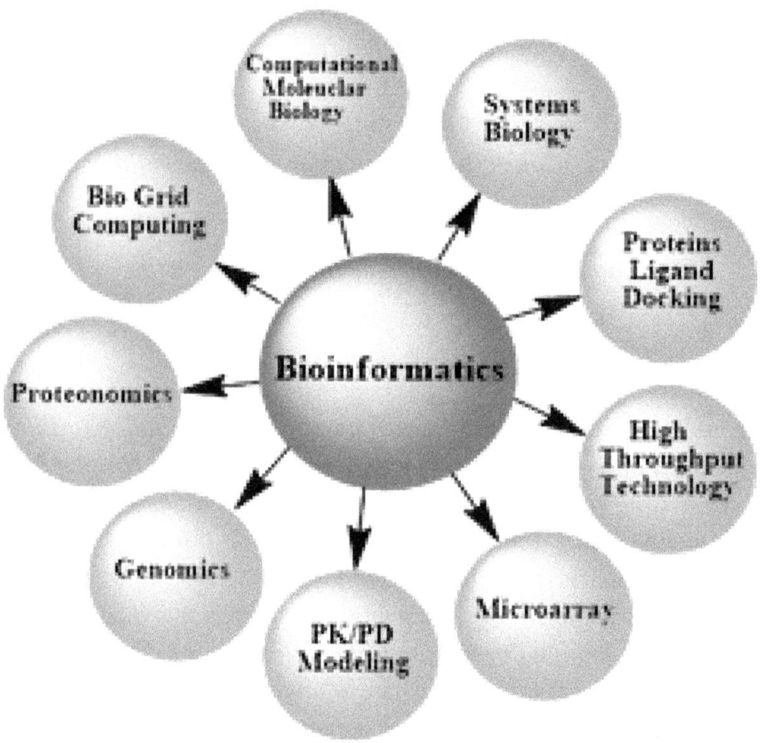

In recent years sustainable agriculture has assumed a great significance in face of ever-increasing human population, global climate change and environmental problems (Brown and Funk 2008, Turner *et al.* 2009), newly emerging insect pests, abiotic stress resulting in poor yields etc. It aims at meeting the needs of present generation without eroding the natural resource base of the future generations. Modern agriculture is based on use of synthetic fertilizer and chemical pesticides and has caused enormous damage to land and water bodies. In addition effluents of industries constantly degrade arable land and pollute water bodies. There is loss of biological diversity and erosion of germplasm resources due to monoculture. Genetically modified crops too promote large scale monoculture at expense of diversity. The conventional plant breeding practices may not able to ensure the sustainable production of crop plants under such circumstances. To resolve these problems there is urgent need to breed plants on the basis of better molecular understanding of gene function and regulation of gene expression responsible for increase in crop yield and stress tolerance. Advent of genomic era in plant biotechnology have lead to identification of the genes related to crop yields, crop nutritional quality, and tolerance to biotic and abiotic stresses. Recent advances made in the area of crop genomics and bioinformatics offer substantial opportunities for enhancing the effectiveness of classical plant breeding programs. Bioinformatic information and web sites have become important for crop scientists in gene data mining, and linking this knowledge to its biological significance (Mochida and Shinozaki 2010).

Genomic revolution is providing insights into genes and regulatory networks of complex agronomic traits. Genomics refers to mapping, sequencing and analysis of genome and function of the genome. Structural genomics represent the initial phase of genome analysis and leads to the construction of high resolution genetic, physical and transcript maps of an organism. The ultimate physical map of an organism is its complete sequence. Structural genomics includes the different techniques and analytical tools such as Linkage analysis, Molecular cytogenetics, physical mapping, EST sequencing etc.

Functional genomics refers to the development and application of genome system wide experimental approaches to assess gene function by making use of the information and reagents provided by structural genomics. It is characterized by high throughput experimental methodologies combined with statistical and computational analysis of the results.

Proteomics is the study of proteomes. Proteome refers to all proteins produced by a species and it varies with time. Proteomics is a rapidly evolving field and there has been a very fast development in proteomics technologies like mass spectrometry for rapid and quantitative measurements of proteins in a complex mixture. Proteomics has led to the development of protein sequence databases and has immense application in healthcare, drug reaseach and diagnostics.

Microarrays are the transcriptional profiles of most genes within a genome can be analyzed using microarrays and quantitative analysis of microarray data. Transcriptional profile can help to generate gene expression data that can be used to define a cell type or condition. There is a huge amount of data produced in any microarray experiment and hence computational methods are essential to conduct this analysis.

Microarrays utilize the preferential binding of complementary single – stranded nucleic acid sequences. Microarray is built using a glass slide, on to which cDNA molecules are attached at fixed locations. There are a very large number of spots on an array, each containing a huge number of identical DNA molecules, of lengths from twenty to hundreds of nucleotides. Gene expression monitoring and SNP detection are the two most important application of microarray technology.

NEW GENERATION SEQUENCING (NGS) FOR CROP ENVIRONMENT

Arrival of NGS technologies has revolutionized the field of genomics and have redefined discovery in biological sciences. These new technologies are capable of performing millions of sequencing reactions in parallel and have potential to detect rare mutations, transcripts and biomarkers. Among the "second generation" technologies, the 454(http://454.com/products-solutions/systemfeatures.asp), Illumina (Illumina, http://www.illumina.com) platforms are already being deployed for sequencing crop genomes.

Others, like Solid (Applied Biosystems, http://www.appliedbiosystems.com/technologies), have been less exploited in plants. Roche/454's GS FLX Titanim, can generate 0.45 Gb raw data per run with average read length of 330 bp. Illumina Solexa's GA instrument is capable of generating 18 Gb data with average read length of 75 bp in 18 days. Applied Biosystems SOLid system generates about 10 Gb data per run with average read length of 25 bp. Moreover, new, "third generation" platforms are being developed and incorporated to sequencing projects, such as PacBio RS (Pacific Biosciences, http://www.pacificbiosciences.com), Helicos (Helicos, http://www.helicosbio.com), or Ion Torrent (Life Technologies, http://www.iontorrent.com). Helicos platform is a single molecule sequencing platform with read length of about 32 bp only but the data generated in one run is maximum up to 37 Gb. *Single Molecule Real time (SMRT) sequencing* is released by Pacific Biosciences. Its read length is 1kb which can go up to 10kb. These Platforms are generating large amount of sequencing data. Next generation sequencing strategies are generating information equivalent to billions of bases in a day at reduced cost and high speed. Computational tools are required for downstream applications as storing, organizing, visualizing and analyzing the massive data. Management and analysis of this data has been a challenge (Batley and Edwards, 2009). Consequent to technological advancement in genome sequencing, novel Bioinformatic capabilities for analysis and management of huge data are being developing (Lai *et al* 2012, Mochida, K. and Shinozaki, K. (2010). A wide array of state of art bioinformatic tools for alignment of sequence reads to a reference, base-calling and/or polymorphism detection, *denovo* assembly from paired or unpaired reads, structural variant detection and genome browsing are available. The first step for analyzing the data is to align the NGS reads to a reference genome. Traditional tools for aligning capillary reads are not efficient for this huge amount of data. For alignment of short sequence length of Solexa, SOLiD and Helicos reads, novel fast short read aligner are required .Bioinformatic software tools for short read alignment as Bowtie, Maq, SSAHA2, SOAP2, BWA. offer advantages of speed and accuracy. If reference genome sequence does not exist, *de novo* assembly of short reads is required. The short reads present many challenges for *De novo* assembly (Jonathan Butler *et al* 2008). In the last two years, many algorithms have been proposed for the *de novo* assembly, especially for bacterial genomes, including AbySS (Simpson *et al.*, 2009), ALLPATHS (Butler *et al.*, 2008), Velvet (Zerbino & Birney, 2008) and SOAP denovo (http://soap.genomics.org.cn). These are capable of assembling short reads reliably by removing low quality bases to improve assembly accuracy. For crop improvement, knowledge of molecular mechanism underlying host pathogen interaction could contribute towards development of biotic stress tolerant plant. A number of plant pathogenic bacteria and fungi have either been sequenced in by sequencing platform including *Agrobacterium tumefaciens*, *Magnaporthe grisea*, *Pseudomonas syringae* pv. tomato *Xanthomonas campestris* etc. Assembly of bacterial genomes from ion PGM system is established using Floton assembler. Tools that allow for the visualization of the alignment or assembly of short read data include EagleView (Huang W, Marth G, 2008), MapView (Bao H *et al* 2009), the Text Alignment Viewer of SAMtools, MaqView, Tablet (Milne I,2010) and IGV. A variety of software tools are available for gene prediction, protein structure prediction and the detection of structural variants. Analysis of data generated by these high-throughput sequencing technologies has greatly impacted agricultural biotechnology research. These have contributed to our understanding of crop genomes identification of genetic diversity, mining candidate genes and molecular markers for beneficial traits. These resources provide better understanding of basic processes relevant to crop production, response to environment which will go a long way in achieving targeted objectives in breeding programs to improve the performance and productivity of crops. Thus sustainable agriculture is possible owing to vast array of information available or crop plants.

GENOME SEQUENCING PROJECTS

Recent technological advances in sequencing technologies are making large crop genome projects successful and providing resources for increased understanding of plant biological processes, crop improvement through genomics assisted plant breeding and development of novel sequence based approaches for crop improvement (Edwards and Batley, 2010). Published Crop genomes include *Oryza sativa, Populus trichocarpa (*Tuskan *et al* 2006), *Vitis vinifera, Carica papaya, Sorghum bicolor, Cucumis sativus, Zea mayes,* (Schnable P *et al* 2009) *Glycine max, Ricinus communis, Malus x domestica, Jatropha curcas,Theobroma cacao, Fragaria vesca, Phoenix dactylifera, Solanum tuberosum, Brassica rapa, Cannabis sativa, Cajanus cajan, Solanum lycopersicum, Cucumis melo, Linum usitatissimum, Musa acuminate, Gossypium raimondii, Azadirachta indica, Hordeum vulgare, Triticum aestivum, Pyrus bretschneideri, Citrullus lanatus, Prunus mume, Cicer arietinum, Hevea brasiliensis, Prunus persica and Picea abies.* The Norway Spruce (Picea abies), the first gymnosperm to be sequenced completely is also the largest being at 19,600 Mb (Nystedt *et al.* 2013). 454 were used first time for sequencing of grape genome in 2007. Sequencing of papaya genome gave insight into evolution of angiospermic genomes (Ming R *et al,* 2008). Large size of sorghum genome as compared ton rice was found to be due to long terminal repeat (LTR) retrotransposons (Paterson *et al* 2009). Advances in sequencing technologies and computational biology helped researchers to sequence this giant genome. Illumina platform has been used extensively for completion of genome plant sequence projects. However, SOLiD platform is being used for two crops only tomato and strawberry.

International Rice Genome Sequencing Project

Rice Genome was the first crop genome to be sequenced entirely (Goff S.A. *et al.* 2002, International Rice Genome sequencing project, 2005). Publication of high quality rice genome sequence annotated by searching the non-redundant protein database using BLASTX software (Altschul *et al.*, 1997), searching the rice EST database using BLASTN software (Altschul *et al.*, 1997), and scanning the sequence with GenScan (Burge C, Karlin S, 1997) (trained for maize) to predict open reading frames and with Splice Predictor (Kleffe J *et al,* 1996) to project exon/intron splice sites.

International Wheat Genome Sequencing Consortium (IWGSC)

Large 17 giga basepairs (Gbp) genome of the bread wheat genome was sequenced using whole-genome shotgun sequencing and was published in Nature journal in 2012 (Rachel Brenchley, *et al.*2012. Highest quality genome sequence *ie* of 21 wheat chromosomes of the bread wheat variety *Chinese Spring* will enable breeders to develop better wheat varieties.

Wheat genome is the most widely cultivated crop species. The genome is complex being hexaploid as wheat originated from hybridization between cultivated tetraploid emmer wheat (AABB, *Triticum dicoccoides*) and diploid goat grass (DD, *Aegilops tauschii*).

. Based on the sequence analysis 107,891 genes and more than 4 million molecular markers in wheat have been identified (IWGSC (2018). Several classes of genes relevant to crop productivity, stress tolerance, photosynthesis and metabolism have been identified.

The assembled genome sequences provide foundation for breeding of crop species by focusing on the challenges of ecological and economic importance as adaptation to changing climate, insect resistance, abiotic stress tolerance. The genome sequences are becoming available at the amazing pace due to Next generation sequencing technologies and increasingly becoming valuable tool to understand important crop traits such as fruit ripening, grain traits and flowering time adaptation.

Due to the immense contribution agriculture makes to ensure food security, global economy it is essential to grow new, suitably adapted crop plants for both now and for the coming generations. The genome projects opens up the avenues for genome-assisted breeding, developing genome –wide markers for stress tolerance, desirable growth, yield, quality traits, and biodiversity and conservation studies and thus are of paramount importance for sustainable agriculture.

WEB-BASED SITES AND GENOME DATABASES

Emergence of inexpensive next generation sequencing platforms is generating vast quantity of sequence data. This data is an essential resource for accelerating comprehension of molecular mechanism underlying biological processes in crop plants and thus providing leads for designing novel strategies for crop improvement. The information is made available to crop science research community by systemically organizing the information in biological databases. These hold huge records that are periodically updated by new data generated in particular field. Genbank of National Center for Biotechnology Information (NCBI), is one of the earliest primary biological data repository (Benson *et al.,* 1997). In addition European Molecular Biology Laboratory (EMBL) and the DNA Databank of Japan (DDBJ), are biological databases for submission of raw DNA and protein sequences.

Plant Genome databases are dedicated to genomic data of specific crops. These databases also incorporate many levels and types of information as physical maps, genomic maps, QTL, molecular markers, mutants, germplasm accessions, expression data etc.For example genome database, Integrated Rice Genome Explorer (INE), has been developed to integrate map and sequence information into a basic database for rice genomics (Sakata K *et al* 2000). Graingenes (http://wheat.pw.usda.gov/) for *triticea*, oats, and sugarcane; MaizeGDB (https://www.maizegdb.org/) for maize; Oryzabase (http://www.shigen.nig.ac.jp/rice/oryzabase/) for rice; BRAD (https://brassicadb.org/brad/) for *Brassica* crops; Legume information system (http://www.comparative-legumes.org/) for legumes; SOL Genomics Network (SGN) (https://solgenomics.net/) for *Solanaceae* crop species, Castor Bean Genome Database, CGPDB, SoyBase for soybean(http://www.soybase.org), are the important databases for genome information mining.

Some databases strive to integrate genome data for multiple plant species.

Gramene

Curated resource for genetic, genomic and comparative genomics data for the major crop species, including rice, maize, wheat and many other plant (mainly grass) species (Ware *et al,* 2002). Holds genomic information from all grass species including genome sequence information. This integrated data resource facilitates comparative and functional genomics in crops and model plant species. It hosts a genome browser, BLAST tool, Plant reactome for browsing metabolic pathways and structural analysis of proteins

Phytozome

Phytozome created by Department of Energy's Joint Genome Institute, provides access to fifty-eight sequenced and annotated green plant genomes, fifty-two of which have been clustered into gene families at 15 evolutionarily significant nodes. Each gene has been annotated with PFAM, KOG, KEGG, PANTHER and GO assignments. This is data resource for comparative genomic studies amongst other green plants.

Plant Markers

A genetic marker web database that contains molecular markers such as SNP, SSR and conserved ortholog set cosmid (COS) markers and primers from various plant resources (Heesacker *et al.* 2008).

PRGDB (Plant Resistance Genetic Data Base)

It is the plant resistance gene database a valuable resource of genes involved in plant defense mechanisms.

NCBI—BioProject

The NCBI site provides genome sequences and information for many plant species (Viridiplantae) designed to facilitate comparative genomic studies.

NCBI—Entrez

Entrez facility provides a gateway to DNA and protein sequences, genome mapping data, 3 – D macromolecular structures, the Pub Med bibliographic database and the like. A search for a particular gene in either database will allow smooth transition to the genome it comes from, the protein sequence it encodes, its structure, bibliographic reference and equivalent entries for all related genes. Tracks over 800 whole genome projects from biological organisms, and the 115 species of Viridiplantae; including agronomically important crops for food and fruit, medicine and a number of green algae.

Plant GDB

It is the primary resource for comparative plant genomics. The goal of the PlantGDB web site is to establish the basis for identifying sets of genes common to all plants or specific to particular species by integrating a number of bioinformatics tools that facilitate gene prediction and cross- species comparisons (Dong *et al,* 2005).

Barley Base

It is the Expression profiling database for analysis and visualization of microarray data.

Meta Crop

This Resource is used for information about diverse metabolic pathways in crop plants.

Genevestigator

It is Most useful database that contains microarray data from different species.

Green PhylDB

GreenPhylDB (http://greenphyl.cirad.fr) is a comprehensive platform designed to facilitate comparative functional genomics in Oryza sativa and Arabidopsis thaliana genomes (Conte *et al,*2008).

UNCOVERING THE COMPLEXITY OF TRANSCRIPTOMES WITH RNA-SEQ

Microarrays offer hybridization based platform for expression profiling (Aharoni and Vorst, 2002). They are capable of measuring tens of thousands of different transcripts in the same reaction.Several web resources and software packages are available for microarrays analysis as the Bioconductor (http://www.bioconductor.org/help/workflows/oligo-arrays/) or MeV (ttp://www.tm4.org/mev/). Expression profiling data is made accessible to research community through databases as GEO (https://www.ncbi.nlm.nih.gov/geo/), ArrayExpress (https://www.ebi.ac.uk/arrayexpress/)

RNA-seq (massively parallel sequencing of cDNA) has revolutionized transcript profiling and has rapidly emerged as alternative powerful techniques to microarray. In addition to analysis of differential gene expression it offers advantage of detection of sequence polymorphism. In RNA-seq cDNA that has been derived from an RNA sample is sequenced using next-generation sequencing methods. NGS produces millions of short reads. High quality sequence reads are retained for further analysis. These reads are then assembled *denovo* or mapped with reference genome using algorithms. The most common use of transcriptome profiling is in the search of differentially expressed genes (DEG). Differentially expressed genes are subjected to Ontology analysis to identify the Biological Process, Cellular Component, and Molecular Function in differentially expressed genes.RNA-seq offer number of advantages over the microarray for differential expression analysis. Bioinformatics software, such as Cufflinks/Cuffdiff, facilitate transcriptome analysis by systematic analysis from the aligned reads to the differential expression results. After identifying differentially expressed genes, Biological pathway analysis can be performed in up and down regulated genes using ExPATH tool. Biological pathway analysis reveals the involvement of differentially expressed genes in various Biological pathways important for plants.

CONCLUSION

Global warming and growing population both are creating pressure on food security. Conventional breeding technique is not enough to produce sufficient food for growing population in all over the world. But now a day's bioinformatics play essential role for crop improvement. Bioinformatics is interdisciplinary science of interpreting data using information technology and computer science. Bioinformatics is used to organize biological data to help the researcher's access information, add new information arising out of experiments and modify existing information. Bioinformatics play an important role in the field of agriculture science where it can be used for enhancing the nutritional value, productivity of agriculture product and development of diseases diagnostic tool. Bioinformatics tools are required for assembling of biological information into databases, data mining, analysis and annotation of this information. Furthermore, Computational tools are essential to analyze this complex data generated from sequencing of genomes, determination of complete set of genes, analysis of transcripts and proteins encoded by an organism, understanding of their interactions that bring out changes in phenotype.

REFERENCES

Aharoni, A., & Vorst, O. (2002). DNA microarrays for functional plant genomics. *Plant Molecular Biology*, *48*(1/2), 99–118. doi:10.1023/A:1013734019946 PMID:11860216

Altschul, S. F., Madden, T. L., Schäffer, A. A., Zhang, J., Zhang, Z., Miller, W., & Lipman, D. J. (1997). Gapped BLAST and PSI-BLAST: A new generation of protein database search programs. *Nucleic Acids Research*, *25*(17), 3389–3402. doi:10.1093/nar/25.17.3389 PMID:9254694

Appels, R., Eversole, K., Stein, N., Feuillet, C., Keller, B., Rogers, J., Pozniak, C. J., Choulet, F., Distelfeld, A., Poland, J., Ronen, G., Sharpe, A. G., Barad, O., Baruch, K., Keeble-Gagnère, G., Mascher, M., Ben-Zvi, G., Josselin, A.-A., Himmelbach, A., ... Wang, L.IWGSC. (2018). Shifting the limits in wheat research and breeding using a fully annotated reference genome. *Science*, *361*(6403), eaar7191. Advance online publication. doi:10.1126cience.aar7191 PMID:30115783

Bao, H., Guo, H., Wang, J., Zhou, R., Lu, X., & Shi, S. (2009). MapView: Visualization of short reads alignment on a desktop computer. *Bioinformatics (Oxford, England)*, *25*(12), 1554–1555. doi:10.1093/bioinformatics/btp255 PMID:19369497

Batley, J., & Edwards, D. (2009). Genome sequence data: Management, storage, and visualization. *BioTechniques*, *46*(5), 333–334, 336. doi:10.2144/000113134 PMID:19480628

Benson, D. A., Boguski, M. S., Lipman, D. J., & Ostell, J. (1997). GenBank. *Nucleic Acids Research*, *25*(1), 1–6. doi:10.1093/nar/25.1.1 PMID:9016491

Brenchley, R., Spannagl, M., Pfeifer, M., Barker, G. L. A., D'Amore, R., Allen, A. M., McKenzie, N., Kramer, M., Kerhornou, A., Bolser, D., Kay, S., Waite, D., Trick, M., Bancroft, I., Gu, Y., Huo, N., Luo, M.-C., Sehgal, S., Gill, B., ... Hall, N. (2012). Analysis of the bread wheat genome using whole-genome shotgun sequencing. *Nature*, *491*(7426), 705–710. doi:10.1038/nature11650 PMID:23192148

Brown, M. E., & Funk, C. C. (2008). Climate. Food security under climate change. *Science*, *319*(5863), 580–581. doi:10.1126cience.1154102 PMID:18239116

Burge, C., & Karlin, S. (1997). Prediction of complete gene structures in human genomic DNA. *Journal of Molecular Biology*, *268*(1), 78–94. doi:10.1006/jmbi.1997.0951 PMID:9149143

Butler, J., MacCallum, I., Kleber, M., Shlyakhter, I. A., Belmonte, M. K., Lander, E. S., Nusbaum, C., & Jaffe, D. B. (2008). ALLPATHS: De novo assembly of whole-genome shotgun microreads. *Genome Research*, *18*(5), 810–820. doi:10.1101/gr.7337908 PMID:18340039

Conte, M. G., Gaillard, S., Lanau, N., Rouard, M., & Perin, C. (2008). Green PhylDB: A database for plant comparative genomics. *Nucleic Acids Research*, *36*(Database), 991–998. doi:10.1093/nar/gkm934 PMID:17986457

Dong, Q., Lawrence, C. J., Schlueter, S. D., Wilkerson, M. D., Kurtz, S., Lushbough, C., & Brendel, V. (2005). Comparative plant genomics resources at Plant GDB. *Plant Physiology*, *139*(2), 610–618. doi:10.1104/pp.104.059212 PMID:16219921

Edwards, D., & Batley, J. (2010). Plant genome sequencing: Applications for crop improvement. *Plant Biotechnology Journal*, *8*(1), 2–9. doi:10.1111/j.1467-7652.2009.00459.x PMID:19906089

Goff, S. A., Ricke, D., Lan, T.-H., Presting, G., Wang, R., Dunn, M., Glazebrook, J., Sessions, A., Oeller, P., Varma, H., Hadley, D., Hutchison, D., Martin, C., Katagiri, F., Lange, B. M., Moughamer, T., Xia, Y., Budworth, P., Zhong, J., ... Briggs, S. (2002). A draft sequence of the rice genome (*Oryza sativa* L. ssp. *japonica*). *Science*, *296*(5565), 92–100. doi:10.1126cience.1068275 PMID:11935018

Heesacker, A., Kishore, V. K., Gao, W., Tang, S., Kolkman, J. M., Gingle, A., Matvienko, M., Kozik, A., Michelmore, R. M., Lai, Z., Rieseberg, L. H., & Knapp, S. J. (2008). SSRs and INDELs mined from the sunflower EST database: Abundance, polymorphisms, and cross-taxa utility. *Theoretical and Applied Genetics*, *117*(7), 1021–1029. doi:10.100700122-008-0841-0 PMID:18633591

Huang, W., & Marth, G. (2008). EagleView: A genome assembly viewer for next-generation sequencing technologies. *Genome Research*, *18*(9), 1538–1543. doi:10.1101/gr.076067.108 PMID:18550804

Kleffe, J., Hermann, K., Vahrson, W., Wittig, B., & Brendel, V. (1996). Logitlinear models for the prediction of splice sites in plant pre-mRNA sequences. *Nucleic Acids Research*, *24*(23), 4709–4718. doi:10.1093/nar/24.23.4709 PMID:8972857

Lai, W. K., Bard, J. E., & Buck, M. J. (2012). ArchTEx: Accurate extraction and visualization of next-generation sequence data. *Bioinformatics (Oxford, England)*, *28*(7), 1021–1023. doi:10.1093/bioinformatics/bts063 PMID:22302569

Milne, I., Bayer, M., Cardle, L., Shaw, P., Stephen, G., Wright, F., & Marshall, D. (2010). Tablet—Next generation sequence assembly visualization. *Bioinformatics (Oxford, England)*, *26*(3), 401–402. doi:10.1093/bioinformatics/btp666 PMID:19965881

Ming, R., Hou, S., Feng, Y., Yu, Q., Dionne-Laporte, A., Saw, J. H., Senin, P., Wang, W., Ly, B. V., Lewis, K. L., Salzberg, S. L., Feng, L., Jones, M. R., Skelton, R. L., Murray, J. E., Chen, C., Qian, W., Shen, J., Du, P., ... Alam, M. (2008). The draft genome of the transgenic tropical fruit tree papaya (Carica papaya linnaeus). *Nature*, *452*(7190), 991–996. doi:10.1038/nature06856 PMID:18432245

Mochida, K., & Shinozaki, K. (2010). Genomics and bioinformatics resources for crop improvement. *Plant & Cell Physiology*, *51*(4), 497–523. doi:10.1093/pcp/pcq027 PMID:20208064

Nystedt, B., Street, N. R., Wetterbom, A., Zuccolo, A., Lin, Y. C., Scofield, D. G., Vezzi, F., Delhomme, N., Giacomello, S., Alexeyenko, A., Vicedomini, R., Sahlin, K., Sherwood, E., Elfstrand, M., Gramzow, L., Holmberg, K., Hällman, J., Keech, O., Klasson, L., ... Jansson, S. (2013). The Norway spruce genome sequence and conifer genome evolution. *Nature*, *497*(7451), 579–584. doi:10.1038/nature12211 PMID:23698360

Paterson, A. H., Bowers, J. E., Bruggmann, R., Dubchak, I., Grimwood, J., Gundlach, H., Haberer, G., Hellsten, U., Mitros, T., Poliakov, A., Schmutz, J., Spannagl, M., Tang, H., Wang, X., Wicker, T., Bharti, A. K., Chapman, J., Feltus, F. A., Gowik, U., ... Rokhsar, D. S. (2009). The Sorghum bicolor genome and the diversification of grasses. *Nature*, *457*(7229), 551–556. doi:10.1038/nature07723 PMID:19189423

Sakata, K., Antonio, B. A., Mukai, Y., Nagasaki, H., Sakai, Y., Makino, K., & Sasaki, T. (2000). INE: A rice genome database with an integrated map view. *Nucleic Acids Research*, *28*(1), 97–101. doi:10.1093/nar/28.1.97 PMID:10592192

Sasaki, T.International Rice Genome Sequencing Project. (2005). The map-based sequence of the rice genome. *Nature*, *436*(7052), 793–800. doi:10.1038/nature03895 PMID:16100779

Schmutz, J., Cannon, S. B., Schlueter, J., Ma, J., Mitros, T., Nelson, W., Hyten, D. L., Song, Q., Thelen, J. J., Cheng, J., Xu, D., Hellsten, U., May, G. D., Yu, Y., Sakurai, T., Umezawa, T., Bhattacharyya, M. K., Sandhu, D., Valliyodan, B., ... Jackson, S. A. (2010). Genome sequence of the palaeopolyploid soybean. *Nature*, *463*(7278), 178–183. doi:10.1038/nature08670 PMID:20075913

Schnable, P., Ware, D., Fulton, R., Stein, J. C., Wei, F., Pasternak, S., Liang, C., Zhang, J., Fulton, L., Graves, T. A., Minx, P., Reily, A. D., Courtney, L., Kruchowski, S. S., Tomlinson, C., Strong, C., Delehaunty, K., Fronick, C., Courtney, B., ... Wilson, R. K. (2009). The B73 maize genome: Complexity, diversity, and dynamics. *Science*, *326*(5956), 1112–1115. doi:10.1126cience.1178534 PMID:19965430

Simpson, J. T., Wong, K., Jackman, S. D., Schein, J. E., Jones, S. J., & Birol, I. (2009). ABySS: A parallel assembler for short read sequence data. *Genome Research*, *19*(6), 1117–1123. doi:10.1101/gr.089532.108 PMID:19251739

Turner, W. R., Oppenheimer, M., & Wilcove, D. S. (2009). A force to fight global warming. *Nature*, *462*(7271), 278–279. doi:10.1038/462278a PMID:19924191

Tuskan, G. A., Difazio, S., Jansson, S., Bohlmann, J., Grigoriev, I., Hellsten, U., Putnam, N., Ralph, S., Rombauts, S., Salamov, A., Schein, J., Sterck, L., Aerts, A., Bhalerao, R. R., Bhalerao, R. P., Blaudez, D., Boerjan, W., Brun, A., Brunner, A., ... Rokhsar, D. (2006). The genome of black cottonwood, Populus trichocarpa (Torr. & Gray). *Science*, *313*(5793), 1596–1604. doi:10.1126cience.1128691 PMID:16973872

Ware, D. H., Jaiswal, P., Ni, J., Yap, I. V., Pan, X., Clark, K. Y., Teytelman, L., Schmidt, S. C., Zhao, W., Chang, K., Cartinhour, S., Stein, L. D., & McCouch, S. R. (2002). Gramene, a tool for grass genomics. *Plant Physiology*, *130*(4), 1606–1613. doi:10.1104/pp.015248 PMID:12481044

Zerbino, D. R., & Birney, E. (2008). Velvet: Algorithms for de novo short read assembly using de Bruijn graphs. *Genome Research*, *18*(5), 821–829. doi:10.1101/gr.074492.107 PMID:18349386

Chapter 31
A Perspective on RNAi-Based Technology for Crop Productivity and Protection

Pradeep Kumar Shukla
Sam Higginbottom University of Agriculture, Technology, and Sciences, India

Suman Sanju
Sam Higginbottom University of Agriculture, Technology, and Sciences, India

Pragati Misra
Sam Higginbottom University of Agriculture, Technology, and Sciences, India

Pramod W. Ramteke
Sam Higginbottom University of Agriculture, Technology, and Sciences, India

Bir P. Singh
ICAR-Central Potato Research Institute, India

Aditi Thakur
College of Horticulture and Forestry, India

Ashu Singh
Sardar Vallabhbhai Patel University of Agriculture and Technology, India

ABSTRACT

The RNAi-mediated technology has emerged as a powerful platform for developing eco-friendly and stable molecular tools for crop improvement. Both siRNA and miRNA are found in all the kingdoms of eukaryotic life and have been employed for crop improvement by enhancing plant nutritional value, development of seedless fruits, and enhancement of shelf life in various plants. It has also been exploited for elimination of anti-nutritional factors and allergens in the crop plants and in floricultural plants to improve and modify the flower color. RNAi-based host plant-mediated gene silencing technique can be used to combat plant pathogens such as viruses, bacteria, nematodes, fungal, and oomycetes plant pathogens of various agronomically important crop plants. RNA silencing technologies have been used in vegetable plants to enhance pest and disease resistance, alter architectural changes in plant and flowering time, improve commercial traits of fruits and flowers, improve nutritional content, eliminate toxic compounds and allergens, and promote a greater industrial product.

DOI: 10.4018/978-1-7998-8434-7.ch031

INTRODUCTION

To achieve higher yields and sustainable cultivation for global food security, present crops require significant upgrades. Conventional breeding has been used in various crop plants to enhance the productivity. Moreover, due to linkage drag crossing obstacles, uncertainty associated with traditional breeding is tedious and long. (Tammu *et al.* 2021, Rodewald and Trognitz 2013). To get a desired target of crop improvement, there is a thorough need to align the conventional crop breeding with new modern-day technologies. Another challenge revolving successful crop improvement is minimizing crop damage from diseases and pests, which is generally accomplished by the use of pesticides and chemicals. But these hazardous chemicals come with their share of destruction to the environment and it is also a recent environmental responsibility to reduce their usage as a part of global environmental agenda. As evident from recent progress in transgenic engineering, which itself provides a substantial prospect for modifying food crop to meet the current world demands but this technology is not fully accepted by the consumers and it's an ecological ethical concern. We have also exhausted our gene pools due to concurrent cross breeding as a natural consequence of domestication and breeding of crop plants. additionally, due to specialization of the gene activity most of the genes no matter having beneficial effect in one plant might be not so useful in other or may be disruptive in other plant tissues. Breeding has its own share of glitches where gene transfer is easily likeably lead to transfer of the gene to non-target organisms. RNAi-mediated resistance is used to enhance the potential of gene exchange between organisms. The RNAi construct is used to silence the gene of interest in RNAi-mediated gene silencing. RNAi has proved to be very potent and promising strategy for crop improvement.

RNAi-mediated gene silencing has proven to be a useful platform for producing environmentally friendly and stable molecular tools for crop development. This genetic revolution can be used to decrease the transcription of certain genes that are involved in biotic and abiotic stress. The RNAi approach will be used to increase unique features in plants, such as resistance to disease. The host organism's gene of interest can be repressed throughout the organism or in specific tissues (Zhang et al. 2015). RNAi technique can be described as a green chemistry, biosafe, and renewable technology since it eliminates some of the risks associated with the creation of transgenic plants containing first generation designs (binary vector and sense and antisense genes).

History of Discovery: RNA Interference

The main function of RNA interference (RNAi) is its transcriptional regulation mechanism that suppresses gene expression by using sequence-specific RNA to reduce gene expression at the transcriptional or post-transcriptional levels. The RNA interference mechanism most likely developed to rid cells of undesirable foreign genes. It's a common feature of gene control systems in all eukaryotes, and it involves the use of short RNAs to suppress gene expression (sRNA). TGS (transcriptional gene silencing) is the repression of transcription caused by promoter deactivation, whereas PTS is the repression of transcription caused by transcript degradation (PTGS). And due to this RNA synthesis is reduced in transcriptional gene silencing (TGS) as a result of of promoter methylation, (Pélissier et al. 1999, Zhong et al., 2013) while in post-transcriptional gene silencing (PTGS) sequence-specific RNA degradation leads to silencing of target gene (Vanyushin 2006). RNA interference is a favoured method in organisms because it is the most precise activity at the genomic level for controlling gene expression. The antisense RNA approach (asRNA) was utilized for gene silencing before the introduction of RNAi in the development of plant biotechnology for the inhibition of gene expression. (Romay 2017). RNA silencing or gene silencing was first discovered accidently in petunia plants while attempting the over expression of *chalcone synthase (ChsA)* gene to produce purple-colored petals. The introduction of the ChsA transgene, however, resulted in the suppression of endogenous gene expression, resulting in white petunia petals (van der Krol *et al.* 1990). This was termed as co–suppression. Further studies suggested that this phenomenon involved degradation of the transcript but not reduced transcription. This précised gene silencing was triggered by partial duplex mRNA resulting in degradation of transcripts. Since this mechanism was involving post-transcriptional mRNA degradation, it was termed post-transcriptional gene silencing in plants. The post-transcriptional silencing has attracted much attention because of the discovery of RNA interference (Fire *et al.* 1998).

PTGS or RNAi had been discovered in plants, but it soon became apparent that RNAi was present in practically all eukaryotic species. Co-suppression or PTGS in plants, quelling in fungi, RNAi in animals, and transposable element silencing have all been described as phenotypically different but mechanistically comparable forms of RNAi. (Wagner and Sun 1998, Chandler and vaucheret et al. 2001, Cogoni and Macino 1999). Fire et al. (1998) showed that double-stranded RNA (dsRNA) was the silencing trigger in gene silencing while working with the nematode *Caenorhabditis elegans.* Andrew Fire and Craig Mello discovered that the dsRNA injected into *C. elegans* triggered the silencing of genes due to sequence similarity to that of dsRNA, six years after the discovery of co-suppression and quelling. They coined the term RNA interference (RNAi) to describe the mechanism behind previously known endogenous gene silencing. In 2006, Fire and Mello were awarded the Nobel Prize in Physiology or Medicine for this discovery.

Mode of Action

RNA silencing, also known as post-transcriptional gene silencing, is initiated by long double-stranded RNA (dsRNA) or hairpin-shaped RNAs. The precursor dsRNA is processed or cleaved into short 21-25 nucleotide small interfering RNA duplexes by the cleavage activity of the enzyme Dicer, a ribonuclease III type enzyme (siRNAs) (Strauss 1999, Preall et al. 2006). The RNA-induced suppression complex is loaded with the resultant dsRNAs/siRNAs (RISC). In an ATP-dependent process, the active RISC unwinds the siRNA duplexes to produce the 'antisense or guide strand.' AGO and P-element induced wimpy testis (PIWI) proteins, as well as siRNA, are found in the RISC complex (Hunter et al. 2013, Vaucheret 2008). The AGO/PIWI proteins are the key locations for RNAi, and these proteins are found in the cytoplasm's P-body regions (Pauley et al. 2006, Macfarlane 2010). In the dsRNAs/siRNAs duplex, the "guide strand" directs silence, whilst the "passenger strand" is destroyed by an RNA helicase activity (Kusaba 2004). The dsRNAs/siRNAs are separated into single stranded RNAs in association with RISC complex. The antisense/guide strand of dsRNA/siRNA molecule is incorporated into a nuclease containing RISC complex (Kusaba 2004). The guide strand uses base-pairing interaction to target the homologous transcript, resulting in either transcript destruction or translational suppression (Shen et al.2021). RNAi's primary biological purpose is to regulate cell gene activity at the PTGS level. This can be accomplished by inhibiting mRNA translation or by degrading mRNA directly (Hannon 2002). Furthermore, the components of the RNAi pathway (i.e. Dicer, siRNA, and RISC) may be responsible for genome order and structure through RNA-induced histone modification, which may quiet gene activity at the pre-transcriptional stage (Schmid and Jensen 2008).

The Start-up Phase

The type III endonuclease enzyme Dicer is responsible for the creation of siRNAs from hundreds or thousands of base pair long dsRNA in the initiation stage. In plants, a protein called DCL1 (DICER-LIKE 1) performs both the initiation and effector phases in the nucleus (Fukudome and Fukuhara 2017). The Dicer enzyme recognises double-stranded RNA (dsRNA) and breaks it down into short interfering RNA fragments of 21-23 nucleotides (siRNAs). Dicer's Piwi-Argonaute-Zwille (PAZ) domain identifies the substrate and determines the size of the siRNA product. The Dicer enzyme is an ATP-dependent enzyme with a preference for the stem-loop form of RNA (precursor of miRNA). The Dicer enzyme, which forms a heterodimeric complex with Dicer, also requires a double-stranded RNA binding protein partner (R2D2). R2D2 plays a vital role in development and is responsible for the Dicer enzyme's stability. These findings suggest that Dicer and R2D2's dsRNA binding domains are required for the heterodimeric Dcr-2/R2D2 complex to bind and load siRNA into the RISC complex. The RNA helicase domain, a DEAD box, DUF283, Piwi-Argonaute-Zwille (PAZ) domain at the N-terminus, RNase III, and dsRBD motif at the C-terminus are among the domains found in Dicer family members (Sontheimer 2005)

The Effector Phase

The effector phase of the RNAi mechanism is where the silencing complex is developed. The siRNA duplex is unwound by the RISC. The RNA-induced silencing complex include the guide strands, which are the single strands of siRNA and miRNA duplexes (RISC). The respective thermodynamic stabilities of the two duplex ends determine the strand selection, according to many in vitro and in vivo experiments. The guide strand will be favoured over the passenger strand, which has a less stable base-paired 5' terminal end (Tomari and Zamore 2005). Despite the fact that the RNAse-III domain of Dicer and the Piwi domain of Ago proteins bind by direct contact, dsRNA processing and target-mRNA degradation are biochemically separate activities (Carbonell and Daros 2017, Hammond et al. 2001). RISC looks for complementary mRNA in the cell and inactivates it using its RNAse activity, either by cleavage or by translation repression via homology-dependent degradation. RISC is a ribonucleoprotein complex with two major components: single-stranded siRNAs and ribonucleases from the argonaute family, which constitute the RISC's active components. They snip the target mRNA in the guide strand's direction. As a result, argonautes boost RISC's slicer activity. The target mRNA is cleaved at a single spot by the antisense/guide strand's 5' end (dubbed the'seed sequence') attaching to the target mRNA sequence. The RISC hydrolyses the target mRNA 10 to 11 nucleotides proximal of the guide strand and mRNA nucleotide pairing (Elbashir et al. 2001). This task necessitates a high to perfect level of complementarity between the target mRNA and the 5' end of the guide strand (Long et al., 2008, Din and Barozai 2014). This process does not require ATP. Mg2+ ions are required for the hydrolysis of target mRNA. The cleavage reaction is catalysed by a subclass of argonaute proteins known as the PIWI domain. The PIWI domain is a Mg2+-dependent endoribonuclease with structural similarity to RNAse H. It hydrolyses the RNA strand in RNA-DNA hybrids. A straightforward representation of siRNA-mediated gene silencing pathway is portrayed in Fig.1.

Because an A-form helix must be created with the target RNA at the centre of a siRNA guide strand for cleavage to occur, complementarities between the small RNA and its target mRNA define the small RNA-directed target cleavage or inhibition of mRNA translation. The siRNA/miRNA remains intact after mRNA cleavage and can facilitate identification.

Figure 1. Current model of post-transcriptional gene silencing

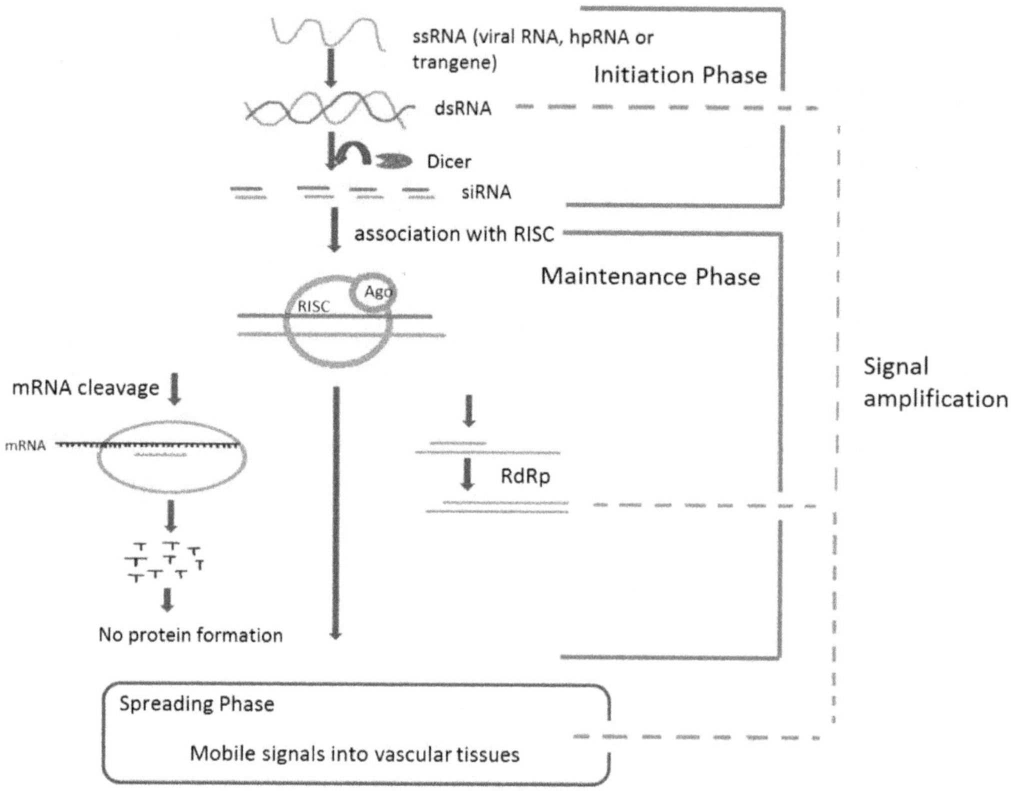

The initiation phase can be triggered by a viral RNA-directed RNA polymerase (RdRP) that makes a complementary RNA strand, using the single-stranded as a template, by a transgenic antisense RNA (asRNA) or by a hairpin RNA (hpRNA). Double stranded RNA (dsRNA) is recognized by the Dicer enzyme and is chopped into pieces of 21-23 nucleotides, called small interfering RNAs (siRNAs). In the maintenance phase, RNA-induced silencing complex (RISC) unwinds the siRNAs, triggers the surveillance mechanism to find siRNA complementary RNA in the cell, and inactivates them with its RNase activity. Alternatively, siRNAs can serve as primers for an RdRP to make double-stranded RNA during the silencing signal amplification, in a feedback fashion. Moreover, siRNAs are believed to constitute (at least partially) mobile signals that spread the gene silencing mechanism to other parts of the plant, through vascular tissues. ssRNA = single-stranded RNA. (Modified from Benedito et al. 2004)

The RISC core is composed of PAZ PIWI Domain proteins, the highly conserved super family of proteins. Upon activation RISC complex move on to target more mRNA transcript copies and the gene silencing get amplified. It has been observed that the "primary siRNAs" diced from dsRNAs are used as primers by the RNA-dependent RNA polymerase enzyme (RdRp) to amplify more dsRNA and abundant "secondary siRNAs" are created (Sijen et al. 2001). Moreover, the gene silencing mechanism spread from one to other parts of the plant through the siRNAs that are believed to constitute (at least partially) the mobile signals (Benedito et al. 2004). This has been observed that locally initiated silencing can be inherited and spread systemically in worms and plants to distant parts of the organism (Boerjan et al.

1994). Systemic spread of siRNA signals in the plants has been demonstrated through various studies. Increased resistance against Colorado potato beetle actin gene has been observed in potato through foliar application of dsRNA (San Miguel and Scott 2016). Similar results were obtained in tobacco plants upon foliar application of dsRNA against Tobacco mosaic virus (TMV) p126 (silencing suppressor) and coat protein genes (Konakalla et al. 2016). Therefore, the uptake and systemic spread of silencing signals in the form of dsRNA or siRNAs in plants result in the increased resistance for diseases as demonstrated above and through several other studies also. However, RNAi efficiency is clearly dependent on dsRNA dose and on insect species that possess different abilities of dsRNA degradation (Shou Yan et al., 2020).

Small RNAs

Small RNAs are the non-coding RNAs (ncRNA), the RNA molecules that do not code for any protein but play regulatory function in the cell. These include both large and small RNAs, rRNAs, tRNAs and also sRNA classes. Small RNAs have different characteristics and sizes ranging from 16-40 nucleotides. Until now, many classes of sRNAs have been described based on their biogenesis, interacting protein complexes, functions and targets. However, the silencing pathways of different sRNAs are inter-connected with each other for substrate or accessory proteins. Different classes of small RNAs play an active role in RNAi. Despite the fact that a number of new sRNA classes have been discovered, the focus of RNA biology research remains on the three main classes of sRNA: siRNA, miRNA and piRNA. Both siRNA and miRNA are found in all the kingdoms of eukaryotic life and is the product of cleavage of dsRNA. Both siRNA and miRNA are important gene expression regulators but they differ in their precursor structures, biogenesis pathways and mode of action (Axtell 2013). The siRNA originate from exogenous and endogenous sources, whereas miRNA originate from endogenous transcripts that form stem-loop structures. Until now, Piwi-interacting RNAs (piRNAs) are found in animals only. They are derived from single-stranded precursors and regulate transposable elements in the germline (Malone and Hannon 2009).

siRNA

siRNAs are 21 nucleotides small RNAs derived from long, linear, double-stranded RNAs that may be introduced dsRNA, transgenes or viruses (Table1). siRNAs identify their targets with perfect complementarity after being processed by RNAase III-like activity into duplexes with 2-3 nucleotides overhangs. Primarily siRNAs were discovered in virus-induced gene silencing experiments in plants (Carthew and Sontheimer 2009). Based upon their source of origin (dsRNA), siRNAs are categorized as exo-siRNAs and endo-siRNAs.

The exogenous siRNAs:

- Source of Origin: Foreign nucleic acids such as mRNAs, transposons, viruses or heterochromatic DNA having higher nucleotide divergence in related species.
- In plants, besides above mentioned precursors, exogenous siRNA may also originate from single-stranded sense transcripts of transgenes. These transgenes are converted to dsRNA by RDR6, belonging to the RNA-dependent RNA polymerase family. They silence genes through Ago-mediated cleavage, nearly identical to those from which they originated (Bartel and Bartel 2003, Pyott and Molnar 2015).

- Source of Origin: In animals, exogenous dsRNA originate from the transgene or viral RNAs (Qu et al. 2005).

The endogenous siRNAs:

- Endogenous siRNAs are produced by the action of enzyme RNA-dependent RNA polymerase (RdRp) from the endogenous genome loci. The endogenous siRNAs originate from convergent mRNA transcripts, naturally occurring sense-antisense pairs, pseudogene derived antisense transcript duplexes, and sense mRNA from their cognate genes and also hairpin RNAs.
- However in organisms lacking RdRp, such as *Drosophila* and mouse, the precursors for endo-siRNAs are long hairpin structures generated by the transcription of palindromic sequences, mobile genetic elements, overlapping RNAs formed by convergent transcription and cognate pseudogenes (Czech et al. 2008, Nilsen 2008).
- The endo siRNAs silence transposons by inhibiting the transposable elements (TEs) from causing insertional mutagenesis (Malone and Hannon 2009). The siRNAs act as defenders of genome integrity and respond to foreign or invasive nucleic acids such as transgenes, viruses and transposons.

There are a number of subclasses of siRNA that are derived from different pathways, are of different sizes, and rely on distinct proteins to exert their function. The siRNAs are the first line of defense against invading viruses and transposons and aberrant transcription products.

To create the RNAi silencing construct for stable gene silencing, the partial target gene sequence of 200-300bp is oriented in sense and anti-sense direction interspersed by an intron which act as a spacer. This is also called as hairpin RNAi construct the whole cassette is controlled by a suitable promoter. As a result, the dsRNA is expressed in transgenic plants in the form of siRNAs by the activity of enzyme DICER for transient gene silencing (Yogindran and Rajam, 2015, Rajam, 2020).

The siRNAs-guided cleavage of target RNA by AGO proteins may be recognized by RNA-dependent RNA polymerase (RdRp). RdRp amplifies the dsRNA and Suppressor of Gene silencing (SGS) stabilizes the dsRNA substrate for enzyme Dicer (DCLs) and hence the secondary siRNAs are produced and the RNA silencing process is reinforced (Peragine et al. 2004). The siRNAs guide the post transcriptional gene silencing (PTGS) and transcriptional gene silencing (TGS) using the components described above. In PTGS, the mRNA is degraded and transcription is suppressed while in TGS, the homologous DNA is targeted causing DNA methylation and hence transcription is suppressed.

miRNA

microRNAs are endogenous 21-24 nucleotides regulatory small non-coding RNAs that are expressed in all eukaryotes (Table1). The miRNAs are transcribed by their own encoding genes in the host genome. The precursors for miRNAs are pri-miRNA which is the capped and polyadenylated transcripts that contain a local hairpin structure (Jones-Rhoades and Bartel 2004). This pri-miRNA eventually gets processed into pre-miRNA and later into miRNA: miRNA* duplexes. From these stem-loop precursors, an active and single-stranded stable miRNA is liberated. The precursor hairpin structures are processed into 21-24 nucleotide sRNAs *i.e.* miRNAs by RNase III enzymes which are Dicer in plants and Dicer and Drosha family proteins in animals. miRNA are found in plants and animals (Ambros et al. 2003,

Bartel 2004) and were first identified in *Caenorhabditis elegans* (Wightman et al. 1993, Lee et al. 1993). Later on, miRNAs were found in several eukaryotes (Bartel 2004). They target mRNA transcripts by blocking mRNA translation at PTGS level. In addition, they regulate gene expression through epigenetic changes in the target RNA by DNA and histone methylation. The miRNAs regulate most biological process in both animals and plants (Ghildiyal and Zamore 2009) that includes gene expression, growth, development, general metabolic pathways, stress responses, transcription factors, transport pathways and host-microbe interactions (Bartel 2004, Stefani and Slack 2008, Katiyar-Agarwal and Jin 2010). The miRNAs regulate the developmental timings, therefore also called as small temporal RNAs (Chen et al. 2004). Within related organisms, miRNAs show high nucleotide identity and they are used by the cell to regulate various genes. A single target gene can be controlled by different miRNAs and vice versa, a single miRNA can control many targets RNAs (Ray et al. 2005, Pillai 2005).

Piwi-Interacting (pi) RNAs

Piwi-interacting RNAs (piRNAs) are single-stranded 25–31 nt small RNAs generated from Dicer-dependent pathways (Table 1). These piRNAs have recently been detected in rat, mouse and human testes. They have been shown to form a Piwi-interacting RNA complex (piRC) with the association of Piwi protein (a subclass of Argonaute proteins) and the human RecQ1 protein. These Piwi interacting RNA complexes are thought to regulate the genome inside developing sperm cells (Carthew 2006).

Table 1. The three major types of small RNAs

	siRNA	miRNA	Piwi-RNA
Size	20-25 nucleotides	21-24 nucleotides	21-35 nucleotides
Origin	dsRNA from Foreign invaders such as transposons, viruses	Endogenous DNA	Long single stranded RNA transcripts of active transposable elements, piRNA clusters
Functional unit	siRNA	miRNA	PiRNA–mediated Pi-RNAs
Roles	Defense against viruses and transposable elements	Regulation of gene expression	Suppression of germ line transposon expression, fight viral infections, regulate gene expression

Applications of RNAi in Crop Improvement

RNAi approach has nowadays emerged as a potential and novel technique to revolutionize the agriculture biotechnology. RNAi-based host plant-mediated gene silencing technique can be used to combat biotic stresses of various agronomically important crop plants, abiotic stresses, nutritional enhancement through bio-fortification as well as bio-elimination, elimination or reduction of food allergen contents, elimination of toxic compounds in foods, altered morphology of plants, production of seedless varieties, increased synthesis of secondary metabolites etc . The technique followed in this method is the down-regulation of the desired gene(s) by the uptake of dsRNA/siRNA produced by transgenic plants which contain the hairpin RNA specific to target gene sequence. The technique takes advantage of the heritable and stable RNAi phenotypes in plants.

Quality Improvement and Nutrient Enhancement

RNAi technique has been harnessed for improvement of quality and nutritional value in various plants. It has also been exploited for elimination of anti-nutritional factors and allergens in the crop plants.

Toxins and Allergens Elimination

Certain allergens are present in the foods cause allergic responses when consumed like apple mango and peanuts etc. So, these allergens need to be eliminated from the foods to make them fit for consumption:

- The expression of major apple allergen Mald1 has been reduced significantly through RNAi (Gilisen et al. 2005). Mald1 belongs to the group of pathogenesis-related protein PR10.
- Similarly, the expression of tomato allergen Lyce3 has been reduced using RNAi (Le et al. 2006).
- The content of allergen Ara h 2 which is the most allergic protein in peanut has been reduced up to 25% through RNAi technology (Dodo et al. 2008).
- In cassava plants, the content of a cyanogenic substance- linamarin (Siritungam and Sayre 2003). Through RNAi the production of enzyme cytochrome P450 has been suppressed that in turn inhibit the production of linamrin and lotaustralin and as a result up to 92% of cyanogenic glucosides are reduced from tubers of cassava (Jorgensen et al., 2005).
- Tearless onion have been produced by suppression of lfs (lacrymatory factor synthase) gene by hpRNAi method through 35S CaMV promoter (Eady et al., 2008). Lfs enzyme convert 1-propenyl sulfenic acid to propanthial S-oxide (tear inducing lachrymatory factorLF).
- In cotton, the compound Gossypol from the seeds has been reduced through RNAi technology. The cotton seeds are rich source of proteins and hence elimination of Gossypol from cotton seeds can make them fit for human consumption (Risco and Chase, 1997).

Development of Seedless Fruits

RNAi technique also enables us to develop seedless fruits and enhancement of shelf life and thus in reducing the inedible waste due to spoilage of fruits and vegetables. The rate of ethylene production was significantly inhibited in the ripened fruits of transgenic plants leading to prolonged shelf life through gene silencing. The fruit quality of apple has been improved by utilizing RNAi approach by enhancing shelf life (Dandekar et al. 2004). The delayed ripening of tomato has been achieved through RNAi (Gupta et al. 2013). By down-regulating a *chalcone synthase* gene in tomato, a gene involved in first step of flavonoid biosynthesis, seedless fruits have been developed (Schijlen et al. 2007). Parthenocarpy is a highly favorable character and it enhances the market value of the fruits.

Improvement of Nutritional Content

Over two-third population across the globe are deprived of nutrition rich diet as they lack one or more nutritional contents in their food (White and Broadley, 2009). RNAi technology can also be harnessed in enriching the dietary elements with required nutrients. In maize, the expression of 22kDa maize *zein* storage protein, the protein with poor lysine content has been knocked out through RNAi generating a dominant high lysine maize variant (Segal et al. 2003). RNAi also generates quality and normal maize seeds with the high level of lysine-rich proteins (Tang et al. 2004). The potato starch content has been altered using RNAi methodology. The amylase content of potato can be decreased or increased either by suppressing the *Granule Bound Starch Synthase* (GBSS) gene or by simultaneously suppressing both starch branching enzymes A and B (Schwall et al. 2000).

The glutenin content of a rice variety called LGC-1 has been produced using RNAi technology by Kusaba et al. 2003 which is beneficial for the patients unable to digest glutenin. They reduced the level of glutenin through GluB hairpin RNA method.

Transgenic soybean has been produced by down regulation of omega -3-fatty acid desaturase using hairpin RNA using glycine promoter for seed specific silencing that in turn reduced the production of alpha–linolenic acid. This improved the flavor and stability of soybean oil with no need of hydrogenation (Flores et al. 2008).

RNAi technology has also been employed to generate tomato with improved carotenoid and flavonoid levels (Davuluei et al. 2005) and in the production of coffee varieties with very low caffeine content (Van Uyen 2006). RNAi has been utilized to increase the carotenoid content of rapeseed (*Brassica napus*) by down- regulating the expression of *lycopene epsilon cyclase* (ε-*CYC*) gene (Yu et al. 2007). Various fruits have been improved in their quality with the help of RNAi technique. *e.g.* transgenic hypoallergic apples have been produced through RNAi (Gillsen et al. 2005). In citrus, the down-regulation of *pectin methylesterase* genes is projected as the possible solution for undesirable separation of juice into clear serum and particular phase (Guo et al. 2005).

Bio-pharming

The use of genetic engineering and transient expression systems such as virus infection, agro-infiltration and magnification for production of potential pharmaceutical compounds is called as bio-pharming or molecular pharming (Obembe et al., 2011). The applications of bio-pharming are in biosynthesis of secondary metabolites and proteins which are costly in the market but are necessary for human consumption.

Altered Architecture

The plant architecture can be manipulated as per requirement through RNAi technology. The concerned genes can be manipulated to enhance the crop yield, plant height, inflorescence, branching etc. The high yielding rice plants have been generated by manipulating the plant architecture through RNAi. RNAi mediated suppression of GA20-oxidase (*OsGA20ox2*) gene resulted in semi-dwarf plants from a taller rice variety QX1 and exhibited higher panicle length, number of seeds per panicle and higher test (1000 grain) weight (Qiao et al. 2007).

Transgenic poplar with significantly enhanced axillary meristem outgrowth, reduced stem lignin content (upto 30%) as compared to wild type and with shorter internode length has been created using miRNA technique under the control of cauliflower mosaic virus 35S promoter (Rubinelli et al., 2013). Similarly, in jute plants, the quantity of lignin has been regulated through the down regulation of key enzyme *4- coumarate:-CoA ligase* (4-cl) in the early stages of lignin biosynthesis. Thus, jute varieties for low lignin content have been developed allowing easier, environmentally friendly and cost-effective processing of fiber (Williams et al. 2004).

RNAi-mediated gene silencing has also been used in floricultural plants to improve and modify the flower colour, *e.g.* Genetin flowers (Nishihara et al. 2006), blue carnation flowers (Tanaka et al. 2005), petunia, torenia and tobacco plants (Tsuda et al. 2004, Nishihara et al. 2005, Nakamura et al. 2006, Nakatsuka et al. 2007a, Nakatsuka et al. 2007b, Nakatsuka et al. 2008). RNAi mediated silencing of tobacco gene Chalcone isomerase (CHI) reduced the pigmentation of flower petals as well as change of flavonoid components (Nishihara et al., 2005)

Recently the new addition in the applications of RNAi include the non – browning Arctic apples and non - browning Innate potatoes which are developed by two private firms harnessing the RNAi methodology through suppression of certain genes. The enzyme polyphenoloxidases (PPOs) is the key enzyme responsible for browning of apples which has been silenced through RNAi to obtain non-browning Arctic apples by a Canadian firm Okanagan Specialty Fruits Inc. (OSF) developed in Australia. Similarly, the non – browning Innate potatoes are developed by J.R. Sipmplot company by silencing four different proteins through RNAi. Both, Arctic apples and Innate potatoes, have been granted approval for commercial sale in the United States (Kumar et al. 2017).

Role of RNAi in Biotic and Abiotic Stress

RNA interference technique has been exploited to develop resistance to various plant pathogens such as viruses, fungi, bacteria, nematodes etc. by improving their defense mechanisms. To attain the pathogen derived resistance in plants, the pathogen derived genes or gene sequences are transformed into the plants targeting blockage of specific life or infection cycle stages of the pathogen. Some of the examples are cited below for various pathogens.

Virus

Viruses are the most important pathogens in agriculture as they are source of various crop diseases and are difficult to control. However, RNAi has enabled to combat successfully virus induced crop diseases in the past few years. Virus Induced Gene Silencing has been extensively used RNA silencing methods to regulate the plant viruses (Senthilkumar et al. 2011). Targeting of viral coat protein for generating resistant plants against viruses using RNAi is found to be very effective. This method has been used for many virus resistant plants such as potato virus Y(PVY)-resistant potato (Missiou et al., 2004), Plum-pox virus (PPV)-resistant *Nicotiana benthamiana* and *Prunus domestica* (Hily, et al., 2007), cucumber green mottle mosaic virus (CGMMV)-resistant *N. benthamiana* (Kamachi et al., 2007) etc. Banana bract mosaic virus resistant varieties of banana have been successfully generated through RNAi (Rodoni et al. 1999). The utilization of RNAi technology has also been utilized for inducing immunity reaction against several other viruses in different plants for examples Bean golden mosaic virus (Bernstein et al. 2001), Rice dwarf virus (Takumi et al. 2008), Tobacco rattle virus (Andre et al. 2009), Tomato leaf curl virus (Shelly et al. 2010) and Citrus Tristeza Virus (Nuria et al. 2011). Using RNAi technology, barley variety resistant to barley yellow dwarf virus has been developed in barley (Wang et al. 2000).

Fungi

RNAi technique has also been used for the production of resistant crop plants against the fungal pathogens. Kadotani et al. (2003), first of all carried out the gene silencing in fungus *Magnaporthe grisea* the rice blast fungus by using the gene green florescent protein. RNAi has been used in ascomycetes *Fusarium graminearum* (*Gibberella zeae*) and *Aspergillus flavus* (McDonald et al. 2005). RNAi is also effective against obligate parasites *Blumeria* (Nowara et al., 2010) and *Puccinia* (Yin et al., 2011), respectively. The RNAi gene silencing technique has also been successfully used against *Puccinia triticina* in wheat (Panwar et al. 2013), *Fusarium* wilt in banana (Ghag et al. 2014), downy mildew of lettuce (Govindarajulu et al. 2015), *Fusarium culmorum* in wheat (Chen et al. 2016). Dual silencing of *Bc-Dcl1* and *Bc-Dcl2* genes in wide variety of plants such as tomato, grapes, onion, rose, lettuce and strawberry was apttempted by Wang et al. (2016). Exogenous application of synthetic hpRNA or the use of transgenic Arabidopsis and tomato plants overexpressing hpRNAs resulted in significant reduction of fungal pathogenecity and growth.

Bacteria

Bacterial diseases are difficult to control as they spread very fast and hence, prevention is the easiest alternative in bacterial diseases. One of the remarkable examples of bacterial disease management was production of transgenic *Arabidopsis thaliana* and *Lycopersicon esculentum* transformed with RNAi constructs targeting the bacterial oncogenes *iaaM* and *ipt* targeting the process of tumourogensis (gall formation) conferring resistance to crown gall disease (Escobar et al. 2001). Expression of these oncogenes is a prerequisite for wild type tumor formation. This was the first report to manage a major bacteria disease through RNAi. The effective management of the pathogens that multiply rapidly causing tumors such as *Albugo candida*, *Synchytrium endobioticum*, and *Erwinia amylovora* among others can be attempted through RNAi.

Nematodes and Insect Resistance

Plant nematodes are the most important plant pathogens causing huge damage to crop plants. Numerous genes have been characterized in the nematode *Caenorhabditis elegans* using the technique RNAi since the discovery of RNA interference by Fire et al. (1998) using siRNA duplexes in the nematode (Kamath et al. 2003). Plant nematologists greatly harnessed RNAi technique for nematode management in agriculture. Steves et al. (2006) reported reduction in the developing SCN females using RNAi in soybean plants. In nematodes, the down regulation of target genes is attempted through *in vitro* dsRNA ingestion and *in planta* delivery of dsRNA to the feeding nematodes (Sindhu et al. 2009). It has been reported that the genes from the intestine, both subventral and dorsal oesophageal glands, female reproductive system and sperms can be silenced through ingested dsRNA in the nematodes (Urwin et al. 2002, Lilley et al. 2005, Steeves et al. 2006, Huang et al. 2006)

Similarly, this technique has been applied to produce other disease free crop plants. Cotton plants through RNAi have been developed which were resistant to insects and pathogen to that of wild type cotton and also produced seeds with higher nutritional value (Tang et al. 2007). RNAi has also been used in pest control through sequence specificity. RNAi does not have any detrimental effect on the non target insects and nematodes, birds, reptiles, fish or mammals, as in the case of pesticides. All transgenic plant-mediated insect control technology is based on the plant producing protein derived from a specific type of bacteria, *Bacillus thuringiensis* or Bt. RNAi and Bt technology can be combined to boost insect control and enhance product performance. Through this combination, resistance development against Bt proteins can also be checked (Goldstein et al. 2009). A recent technique of utilizing RNAi involves a spray of nano-sized degradable clay that protects plants from specific disease-causing pathogens through sustained release of dsRNA and silencing of homologous RNA on topical application (Mitter et al. 2017).

Abiotic Stress

Abiotic stress adversely affects the growth and yield of plants and is a serious hazard to the well-being of plants. In the past few years RNAi has been used in several crops in abiotic stress stimulation due to its specificity and sequence centered gene silencing. The miRNAs are regulatory in function and vital for development and stress response in plants by deleteriously affecting post-transcriptional gene expression. Functional characterization of miRNAs is performed through high-throughput sequencing at the global genome-level in plants (Wang et al. 2011b). Through several studies in different crops it has been found that in *Arabidopsis* miR159 is drought responsive (Achard et al. 2004), in rice 16 miRNAs involved in down-regulation and 14miRNAs in up-regulation, identified based on microarray profiling (Liu et al. 2008) and in maize miR474 has been shown to up-regulated in drought stress (Wei et al. 2009). Transgenic rice plants tolerant to drought stress were developed by silencing of receptor for activated C-kinase1 (*RACK1*; Li et al. 2009). In legume plant *Medicago trunculata*, up-regulation of 22 members of 4 miRNA families and down-regulation of 10 members of 6 miRNA families was observed by Wang et al. (2011b) in response to drought stress. These drought responsive miRNAs play important role in developmental and regulatory processes in plants such as cross adaptation transcription, protein degradation etc.

Likewise, numerous regulatory miRNAs have been reported in salt stressed plants through different studies. The impending production of crop plants is reduced significantly due to low availability of irrigation water quality that leads to salt salinity which is dangerous in plants. Under salt stress, the results of microarray profiling indicated the members of miRNA groups' miR396, miR156, miR167 and miR164 were up-regulated while miR474, miR162, miR395 and miR168 groups were up-regulated in maize (Ding et al. 2009). The miR398 was down-regulated and the miRNAs such as miR397, miR156, miR394, miR158, miR393, miR159, miR319, miR165, miR171, miR167, miR169, miR168 and miR398 were up-regulated in response to salinity stress in *Arabidopsis* (Liu et al. 2008). The miRNA393 was strongly up-regulated in response to various abiotic stresses *i.e.* dehydration, high salinity, cold and abscisic acid (ABA) exposures. In addition, miR319c, miR389a, miR397b and miR402 were also found to be regulated in *Arabidopsis* upon exposure of varying degrees of abiotic stresses Sunkar and Zhu (2004).

The cold stress has been studied through miRNA expression profiling in the plants *Arabidopsis*, *Brachypodium*, and *Populus* (Liu et al. 2008, Arenas-Huertero et al. 2009, Zhang et al. 2009). In all the three species studied for cold stress, up-regulation of miR169 and miR397 was observed and miR172 was also up-regulated in all except *Populus*.

The effect of heat stress responses has been studied in wheat in which 9 miRNAs were identified as heat responsive out of 32 families of miRNA illustrated. Under heat stress, the miRNAs *i.e.* miR827, miR156, miR169, miR159, miR168, miR160, miR166 and miR393 were up-regulated whereas miR172 was down-regulated (Xin et al. 2010). Several other successful attempts have been made on the crop plants to modify them for higher yield and productivity as well as improved nutritional quality.

The legume plants have the ability to fix atmospheric nitrogen in symbiotic association with bacteria. *Asnodf32*, encodes a nodule-specific cysteine proteinase in *Astragalus sinicus* and is silenced through *Agrobacterium rhizogenes*-mediated RNAi. The silencing of *Asnodf32* by Li et al. (2008b) delayed root nodule and bacteroid senescence in legume plants and developed enlarged nodules with prolonged period for active nitrogen fixation. Hence, a functional symbiotic interaction may be regulated through secreted RNA-binding peptides.

The role of miRNAs in Al^{3+} toxicity and tolerance has been studied by Chen et al. (2012) and they concluded that after 4 h of Al^{3+} treatment 18 miRNAs were responsive, 4 miRNAs of 4 families were responsive at both 4 h and 24 h of Al^{3+} treatment.

CURRENT CHALLENGES AND FUTURE PROSPECTIVE

To fulfill the demand of nutritious diet for healthy world, the bio-fortified crops with nutritionally essential minerals, vitamins, antioxidants, amino acids and fatty acids needs to be developed. The transgenic crops are the promising solution to the growing challenge of food shortage but there are certain shortcomings that needed to be eliminated. The high cost of production of transgenic crops is the biggest hindrance in the field application. The low production cost with high precision is prerequisite for the field application of transgenic crops. The resistance developed by the insect pests against the transgenic crops either through mutations of target gene or through the genes of RNAi machinery or lower dsRNA uptake is another challenge in commercialization of transgenic crops (Zhu and Palli, 2020). Another major challenge in RNAi research is the selection of the target sequences for a particular disease and effective *in vivo* siRNA delivery to the particular cell types which needed to be addressed through utmost caution and precision. In addition, the multiple target approach can be utilized to increase the effectiveness of RNAi methodology. The non specific binding of siRNA in nontarget organisms other than pests may also lead to some lethal effects. However, with the development of some computational design tools the non-target and off-target effects can be completely evaluated with precision but they should be used with appropriate bioassays to guard the exposed nontarget organisms.

Hence, it is concluded that RNAi is a very potential and promising method to combat the emerging phytopathogenic fungi, nematodes, insects and other pests for durable and long lasting resistance in crop plants, to increase yield of cereals and grains, plant architecture alterations, nutritional improvement, alteration of metabolic pathways for prolonged shelf life, removal of allergens and toxic substances, development of seedless fruits and male sterile lines and flower colour modulation. It seems to be specific and effective tool for reverse genetics. RNAi technology has emerged to combat plant pathogens and holds great potential in future for crop improvement by enhancing productivity and nutritional value. The potential of RNAi-based technology has been utilized in various plants till date and can be utilized for many other purposes in future.

REFERENCES

Achard, P., Herr, A., Baulcombe, D. C., & Harberd, N. P. (2004). Modulation of floral development by gibberellin-regulated microRNA. *Development*, *131*(14), 3357–3365. doi:10.1242/dev.01206 PMID:15226253

Ambros, V., & Lee, R. (2003). RC, Lavanway A, Williams PT, Jewell D. "Micro RNAs and other tiny endogenous RNAs in *C. elegans*. *Current Biology*, *13*, 807–818. doi:10.1016/S0960-9822(03)00287-2 PMID:12747828

Andre, C. V., Suma, C., & Gregory, B. M. (2009). Virus-induced gene silencing (VIGS) in *Nicotiana benthamiana* and tomato. *Journal of Visualized Experiments*, *28*, 1292–1301. PMID:19516240

Arenas-Huertero, C., Pérez, B., Rabanal, F., Blanco-Melo, D., De la Rosa, C., Estrada-Navarrete, G., Sanchez, F., Covarrubias, A. A., & Reyes, J. L. (2009). Conserved and novel miRNAs in the legume *Phaseolus vulgaris* in response to stress. *Plant Molecular Biology*, *70*(4), 385–401. doi:10.100711103-009-9480-3 PMID:19353277

Axtell, M., Westholm, J., & Lai, E. (2011). Vive la difference: Biogenesis and evolution of microRNAs in plants and animals. *Genome Biology*, *12*(4), 221. doi:10.1186/gb-2011-12-4-221 PMID:21554756

Bartel, B., & Bartel, D. P. (2003). MicroRNAs: At the root of plant development? *Plant Physiology*, *132*(2), 709–717. doi:10.1104/pp.103.023630 PMID:12805599

Bartel, D. P. (2004). MicroRNAs: Genomics, biogenesis, mechanism, and function. *Cell*, *116*(2), 281–297. doi:10.1016/S0092-8674(04)00045-5 PMID:14744438

Benedito, V. A., Visser, P. B., Angenent, G. C., & Kerns, F. A. (2004). The potential of virus-induced gene silencing for speeding up functional characterization of plant genes. *Genetics and Molecular Research*, *3*, 323–341. PMID:15614725

Boerjan, W., Bauw, G., Van Montagu, M., & Inzé, D. (1994). Distinct phenotypes generated by overexpression and suppression of *S*-adenosyl-l-methionine reveal developmental patterns of gene silencing in tobacco. *The Plant Cell*, *6*, 1401–1414. PMID:7994174

Carbonell, A., & Daros, J. A. (2017). Artificial microRNAs and synthetic transacting small interfering RNAs interfere with viroid infection. *Molecular Plant Pathology*, *18*(5), 746–753. doi:10.1111/mpp.12529 PMID:28026103

Carthew, R. W. (2006). Gene regulation by microRNAs. *Current Opinion in Genetics & Development*, *16*(2), 203–208. doi:10.1016/j.gde.2006.02.012 PMID:16503132

Carthew, R. W., & Sontheimer, E. J. (2009). Origins and mechanisms of miRNAs and siRNAs. *Cell*, *136*(4), 642–655. doi:10.1016/j.cell.2009.01.035 PMID:19239886

Chandler, V. L., & Vaucheret, H. (2001). Gene activation and gene silencing. *Plant Physiology*, *125*(1), 145–148. doi:10.1104/pp.125.1.145 PMID:11154319

Chen, C. Z., Li, L., Lodish, H. F., & Bartel, D. P. (2004). MicroRNAs modulate hematopoietic lineage differentiation. *Science*, *303*(5654), 83–86. doi:10.1126cience.1091903 PMID:14657504

Chen, L., Wang, T., Zhao, M., Tian, Q., & Zhang, W. H. (2012). Identification of aluminum-responsive microRNAs in *Medicago truncatula* by genome-wide high-throughput sequencing. *Planta*, *235*(2), 375–386. doi:10.100700425-011-1514-9 PMID:21909758

Chen, W., Kastner, C., & Nowara, D. (2016). Host-induced silencing of *Fusarium culmorum* genes protects wheat from infection. *Journal of Experimental Botany*. Advance online publication. doi:10.1093/jxb/erw263(2016) PMID:27540093

Cogoni, C., & Macino, G. (1999). Post transcriptional gene silencing in *Neurospora* by a RecQ DNA helicase. *Science*, *286*(5448), 2342–2344. doi:10.1126cience.286.5448.2342 PMID:10600745

Czech, B., Malone, C. D., Zhou, R., Stark, A., Schlingeheyde, C., Dus, M., Perrimon, N., Kellis, M., Wohlschlegel, J. A., Sachidanandam, R., Hannon, G. J., & Brennecke, J. (2008). An endogenous small interfering RNA pathway in *Drosophila. Nature, 453*(7196), 798–802. doi:10.1038/nature07007 PMID:18463631

Dandekar, A. M., Teo, G., Defilippi, B. G., Uratsu, S. L., Passey, A. J., Kader, A. A., Stow, J. R., Colgan, R. J., & James, D. J. (2004). Effect of down-regulation of ethylene biosynthesis on fruit flavor complex in apple fruit. *Transgenic Research, 13*(4), 373–384. doi:10.1023/B:TRAG.0000040037.90435.45 PMID:15517996

Davuluri, G. R., Tuinen, A., & Fraser, P. D. (2005). Fruit-specific RNAi-mediated suppression of DET1 enhances carotenoid and flavonoid content in tomatoes. *Nature Biotechnology, 23*(7), 890–895. doi:10.1038/nbt1108 PMID:15951803

Din, M., & Barozai, M.Y. (2014). Profiling microRNAs and their targets in an important fleshy fruit: tomato (Solanum lycopersicum). *Gene, 10*(535), 198-203.

Ding, D., Zhang, L., Wang, H., Liu, Z., Zhang, Z., & Zheng, Y. (2009). Differential expression of miRNAs in response to salt stress in maize roots. *Annals of Botany, 103*(1), 29–38. doi:10.1093/aob/mcn205 PMID:18952624

Dodo, H. W., Konan, K. N., Chen, F. C., Egnin, M., & Viquez, O. M. (2008). Alle viating peanut allergy using genetic engineering: The silencing of the immunodominant allergen Arah2 leads to its significant reduction and a decrease in peanut allergenicity. *Plant Biotechnology Journal, 6*(2), 135–145. doi:10.1111/j.1467-7652.2007.00292.x PMID:17784907

Eady, C. C., Kamoi, T., Kato, M., Porter, N. G., Davis, S., Shaw, M., Kamoi, A., & Imai, S. (2008). Silencing onion lachrymatory factor synthase causes a significant change in the sulfur secondary metabolite profile. *Plant Physiology, 147*(4), 2096–2106. doi:10.1104/pp.108.123273 PMID:18583530

Elbashir, S. M., Lendeckel, W., & Tuschl, T. (2001). RNA interference is mediated by 21-and22-nucleotide RNAs. *Genes & Development, 15*(2), 188–200. doi:10.1101/gad.862301 PMID:11157775

Escobar, M. A., Civerolo, E. L., Summerfelt, K. R., & Dandekar, A. M. (2001). RNAi-mediated oncogene silencing confers resistance to Crown Gall Tumorigenesis. *Proceedings of the National Academy of Sciences of the United States of America, 98*(23), 13437–13442. doi:10.1073/pnas.241276898 PMID:11687652

Flores, T., Karpova, O., Su, X., Zeng, P., Bilyeu, K., Sleper, D. A., Nguyen, H. T., & Zhang, Z. J. (2008). Silencing of GmFAD3 gene by siRNA leads to low alpha-linolenic acids (18:3) of fad3-mutant phenotype in soybean [*Glycine max* (Merr.)]. *Transgenic Research, 17*(5), 839–850. doi:10.100711248-008-9167-6 PMID:18256901

Fukudome, A., & Fukuhara, T. (2017). Plant dicer-like proteins: Double-stranded RNA-cleaving enzymes for small RNA biogenesis. *Journal of Plant Research, 130*(1), 33–44. doi:10.100710265-016-0877-1 PMID:27885504

Ghag, S. B., Shekhawat, U. K. S., & Ganapathi, T. R. (2014). Host-induced post-transcriptional hairpin RNA-mediated gene silencing of vital fungal genes confers efficient resistance against *Fusarium* wilt in banana. *Plant Biotechnology Journal, 12*(5), 541–553. doi:10.1111/pbi.12158 PMID:24476152

Ghildiyal, M., & Zamore, P. D. (2009). Small silencing RNAs: An expanding universe. *Nature Reviews. Genetics*, *10*(2), 94–108. doi:10.1038/nrg2504 PMID:19148191

Gillsen, L. J., Bolhaar, S. T., & Matos, C. I. (2005). Silencing of major apple allergen Mald1 by using the RNA interference approach. *The Journal of Allergy and Clinical Immunology*, *115*(2), 364–369. doi:10.1016/j.jaci.2004.10.014 PMID:15696096

Goldstein, E., Paur, K., Fraser, C., Kenah, E., Wallinga, J., & Lipsitch, M. (2009). Reproductive numbers, epidemic spread and control in a community of households. *Mathematical Biosciences*, *221*(1), 11–25. doi:10.1016/j.mbs.2009.06.002 PMID:19559715

Govindarajulu, M., Epstein, L., Wroblewski, T., & Michelmore, R. W. (2015). Host-induced gene silencing inhibits the biotrophic pathogen causing downy mildew of lettuce. *Plant Biotechnology Journal*, *13*(7), 875–883. doi:10.1111/pbi.12307 PMID:25487781

Guo, W., Duan, Y., Olivares-Fuster, O., Wu, Z., Arias, C. R., Burns, J. K., & Grosser, J. W. (2005). Protoplast transformation and regeneration of transgenic Valencia sweet orange plants containing a juice quality-related pectin methylesterase. *Plant Cell Reports*, *24*(8), 482–486. doi:10.100700299-005-0952-x PMID:15875190

Gupta, A., Pal, R. K., & Rajama, M. V. (2013). Delayed ripening and improved fruit processing quality in tomato by RNAi-mediated silencing of three homologs of 1-aminopropane-1-carboxylate synthase gene. *Journal of Plant Physiology*, *170*(11), 987–995. doi:10.1016/j.jplph.2013.02.003 PMID:23507024

Hammond, S. M., Candy, A. A., & Hannon, G. J. (2001). Post transcriptional gene silencing by double-stranded RNA. *Nature Reviews. Genetics*, *2*(2), 110–119. doi:10.1038/35052556 PMID:11253050

Hannon, G. J. (2002). RNA interference. *Nature*, *418*(6894), 244–251. doi:10.1038/418244a PMID:12110901

Hily, J. M., Ravelonandro, M., Damsteegt, V., Basset, C., Petri, C., Liu, Z., & Scorza, R. (2007). Plum pox virus coat protein gene intron-hairpin-RNA (ihpRNA) constructs provide resistance to plum pox virus in *Nicotiana bethamiana* and *Prunus domestica. Journal of the American Society for Horticultural Science*, *132*(6), 850–858. doi:10.21273/JASHS.132.6.850

Holmquis, G., & Ashley, T. (2006). Chromosome organization and chromatin modification: Influence on genome function and evolution. *Cytogenetic and Genome Research*, *114*(2), 96–125. doi:10.1159/000093326 PMID:16825762

Huang, G., Allen, R., Davis, E. L., Baum, T. J., & Hussey, R. S. (2006). Engineering broad root-knot resistance in transgenic plants by RNAi silencing of a conserved and essential root-knot nematode parasitism gene. *Proceedings of the National Academy of Sciences of the United States of America*, *103*(39), 4302–14306. doi:10.1073/pnas.0604698103 PMID:16985000

(2013). Hunter, Lydia J R; Westwood, Jack H; Heath, Geraldine; Macaulay, Keith; Smith, Alison G. "Regulation of RNA-Dependent RNA Polymerase 1 and Isochorismate Synthase Gene Expression in Arabidopsis. *PLoS One*, *8*, 6.

Jones-Rhoades, M. W., & Bartel, D. P. (2004). Computational identification of plant microRNAs and their targets, including a stress-induced miRNA. *Molecular Cell*, *14*(6), 787–799. doi:10.1016/j.molcel.2004.05.027 PMID:15200956

Jørgensen, K., Bak, S., Busk, P. K., Sørensen, C., Olsen, C. E., Puonti-Kaerlas, J., & Moller, B. L. (2005). Cassava plants with a depleted cyanogenic glucoside content in leaves and tubers. *Plant Physiology*, *139*(1), 363–374. doi:10.1104/pp.105.065904 PMID:16126856

Kadotani, N., Nakayashiki, H., Tosa, Y., & Mayama, S. (2003). RNA silencing in the pathogenic fungus *Magnaporthe oryzae*. *Molecular Plant-Microbe Interactions*, *16*(9), 769–776. doi:10.1094/MPMI.2003.16.9.769 PMID:12971600

Kamachi, S., Mochizuki, A., Nishiguchi, M., & Tabei, Y. (2007). Transgenic *Nicotiana benthamiana* plants resistant to cucumber green mottle mosaic virus based on RNA silencing. *Plant Cell Reports*, *26*(8), 1283–1288. doi:10.100700299-007-0358-z PMID:17453214

Katiyar-Agarwal, S., & Jin, H. (2010). Role of small RNAs in host-microbe interactions. *Annual Review of Phytopathology*, *48*(1), 225–246. doi:10.1146/annurev-phyto-073009-114457 PMID:20687832

Konakalla, N. C., Kaldis, A., Berbati, M., Masarapu, H., & Voloudakis, A. E. (2016). Exogenous application of double-stranded RNA molecules from TMV p126 and CP genes confers resistance against TMV in tobacco. *Planta*, *244*(4), 961–969. doi:10.100700425-016-2567-6 PMID:27456838

Kumar, P., Mundiyara, R., & Jitarwal, R. C. (2017). RNA Interference: New Approach of Gene Silencing in Plants. *International Journal of Pure and Applied Biosciences*, *5*(3), 421–425. doi:10.18782/2320-7051.5066

Kusaba, M. (2004). RNA interference in plants. *Current Opinion in Biotechnology*, *15*(2), 139–143. doi:10.1016/j.copbio.2004.02.004 PMID:15081052

Kusaba, M., Miyahara, K., Iida, S., Fukuoka, H., Takano, T., Sassa, H., Nishimura, M., & Nishio, T. (2003). Low glutelin content 1: A dominant mutation that suppresses the glutelin multigene family via RNA silencing in rice. *The Plant Cell*, *15*(6), 1455–1467. doi:10.1105/tpc.011452 PMID:12782736

Le, L., Lorenz, Y., Scheurer, S., Fotisch, K., Enrique, E., Bartra, J., Biemelt, S., Vieths, S., & Sonnewald, U. (2006). Design of tomato fruits with reduced allergenicity by dsRNAi-mediated inhibition of ns-LTP (Lyce3) expression. *Plant Biotechnology Journal*, *4*(2), 231–242. doi:10.1111/j.1467-7652.2005.00175.x PMID:17177799

Lee, R. C., Feinbaum, R. L., & Ambros, V. (1993). The *C. elegans* heterochronic gene *lin-4* encodes small RNAs with antisense complementarity to *lin-14*. *Cell*, *75*(5), 843–854. doi:10.1016/0092-8674(93)90529-Y PMID:8252621

Li, D. H., Liu, H., Yang, Y., Zhen, P., & Liang, J. (2009). Down-regulated expression of RACK1 gene by RNA interference enhances drought tolerance in rice. *Rice Science*, *16*(1), 14–20. doi:10.1016/S1672-6308(08)60051-7

Li, Y., Zhou, L., Li, Y., Chen, D., Tan, X., Lei, L., & Zhou, J. A. (2008b). nodule-specific plant cysteine proteinase, AsNODF32, is involved in nodule senescence and nitrogen fixation activity of the green manure legume *Astragalus sinicus. The New Phytologist*, *180*(1), 185–192. doi:10.1111/j.1469-8137.2008.02562.x PMID:18643938

Lilley, C. J., Goodchild, S. A., Atkinson, H. J., & Urwin, P. E. (2005). Cloning and characterization of a *Heterodera glycines* minopeptidase cDNA. *International Journal for Parasitology*, *35*(14), 1577–1585. doi:10.1016/j.ijpara.2005.07.017 PMID:16216247

Liu, H. H., Tian, X., Li, Y. J., Wu, C.-A., & Zheng, C.-C. (2008). Microarray-based analysis of stress-regulated microRNAs in *Arabidopsis thaliana. RNA (New York, N.Y.)*, *14*(5), 836–843. doi:10.1261/rna.895308 PMID:18356539

Liu, Z., Bos, J. I., Armstrong, M., Whisson, S. C., da Cunha, L., Torto-Alalibo, T., Win, J., Avrova, A. O., Wright, F., Birch, P. R. J., & Kamoun, S. (2005). Patterns of diversifying selection in the phytotoxin-like *scr74* gene family of *Phytophthora infestans. Molecular Biology and Evolution*, *22*(3), 659–672. doi:10.1093/molbev/msi049 PMID:15548752

Long, D., Chan, C. Y., & Ding, Y. (2008). Analysis of microRNA-target interactions by a target structure based hybridization model. *Pacific Symposium on Biocomputing*, 64–74. PMID:18232104

Macfarlane, L.-A., & Paul, R. (2010). MicroRNA: Biogenesis, Function and Role in Cancer. *Current Genomics*, *11*(7), 537–561. doi:10.2174/138920210793175895 PMID:21532838

Malone, C., & Hannon, G. J. (2009). Small RNAs as guardians of the genome. *Cell*, *136*(4), 656–668. doi:10.1016/j.cell.2009.01.045 PMID:19239887

Margis, R., Fusaro, A. F., Smith, N. A., Curtin, S. J., Watson, J. M., Finnegan, E. J., & Waterhouse, P. M. (2006). The evolution and diversification of Dicers in plants. *FEBS Letters*, *580*(10), 2442–2450. doi:10.1016/j.febslet.2006.03.072 PMID:16638569

McDonald, T., Brown, D., Keller, N. P., & Hammond, T. M. (2005). RNA silencing of mycotoxin production in *Aspergillus* and *Fusarium* species. *Molecular Plant-Microbe Interactions*, *18*(6), 539–545. doi:10.1094/MPMI-18-0539 PMID:15986923

Missiou, A., Kalantidis, K., Boutla, A., Tzortzakaki, S., Tabler, M., & Tsagris, M. (2004). Generation of transgenic potato plants highly resistant to potato virus Y (PVY) through RNA silencing. *Molecular Breeding*, *14*(2), 185–197. doi:10.1023/B:MOLB.0000038006.32812.52

Mitter, N., Worrall, E. A., Robinson, K. E., Li, P., Jain, R. G., Taochy, C., Fletcher, S. J., Carroll, B. J., Lu, G. Q., & Xu, Z. P. (2017). Clay nanosheets for topical delivery of RNAi for sustained protection against plant viruses. *Nature Plants*, *3*(2), 16207. doi:10.1038/nplants.2016.207 PMID:28067898

Nakamura, N., Fukuchi-Mizutani, M., Miyazaki, K., Suzuki, K., & Tanaka, Y. (2006). RNAi suppression of the anthocyanidin synthase gene in Torenia hybrida yields white flowers with higher frequency and better stability than antisense and sense suppression. *Plant Biotechnology (Sheffield, England)*, *23*(1), 13–18. doi:10.5511/plantbiotechnology.23.13

Nakatsuka, A., Mizuta, D., Kii, Y., Miyajima, I., & Kobayashi, N. (2008). Isolation and expression analysis of flavonoid biosynthesis genes in evergreen azalea. *Scientia Horticulturae*, *118*(4), 314–320. doi:10.1016/j.scienta.2008.06.016

Nakatsuka, T., Abe, Y., Kakizaki, Y., Yamamura, S., & Nishihara, M. (2007a). Production of red-flowered plants by genetic engineering of multiple flavonoid biosynthetic genes. *Plant Cell Reports*, *26*(11), 1951–1959. doi:10.100700299-007-0401-0 PMID:17639403

Nakatsuka, T., Chetsadaporn, P., Yamamura, S., & Nishihara, M. (2007b). Induction of differential flower pigmentation patterns by RNAi using promoters with distinct tissue-specific activity. *Plant Biotechnology Reports*, *1*(4), 251–257. doi:10.100711816-007-0039-y

Nielsen, E., Cheung, A. Y., & Ueda, T. (2008). The regulatory RAB and ARF GTPases for vesicular trafficking. *Plant Physiology*, *147*(4), 1516–1526. doi:10.1104/pp.108.121798 PMID:18678743

Nishihara, M., Nakatsuka, T., Hosokawa, K., Yokoi, T., Abe, Y., Mishiba, K., & Yamamura, S. (2006). Dominant inheritance of white-flowered and herbicide-resistant traits in trangenic gentian plants. *Plant Biotechnology (Sheffield, England)*, *23*(1), 25–31. doi:10.5511/plantbiotechnology.23.25

Nishihara, M., Nakatsuka, T., & Yamamura, S. (2005). Flavonoid components and flower color change in transgenic tobacco plants by suppression of *chalcone isomerase* gene. *FEBS Letters*, *579*(27), 6074–6078. doi:10.1016/j.febslet.2005.09.073 PMID:16226261

Nowara, D., Gay, A., Lacomme, C., Shaw, J., Ridout, C., Douchkov, D., Hensel, G., Kumlehn, J., & Schweizer, P. (2010). HIGS: Host-induced gene silencing in the obligate biotrophic fungal pathogen *Blumeria graminis*. *The Plant Cell*, *22*(9), 3130–3141. doi:10.1105/tpc.110.077040 PMID:20884801

Nuria, S., Carmen, F., & Sinda, C. (2011). RNAi mediated protection against citrus *Tristeza* virus in Transgenic citrus plants. *Non Coding RNAs in Plants, 8*, 447–460.

Obembe, O. O., Popoola, J. O., Leelavathi, S., & Reddy, S. V. (2011). Advances in plant molecular farming. *Biotechnology Advances*, *29*(2), 210–222. doi:10.1016/j.biotechadv.2010.11.004 PMID:21115109

Panwar, V., McCallum, B., & Bakkeren, G. (2013). Endogenous silencing of *Puccinia triticina* pathogenicity genes through *in planta*-expressed sequences leads to the suppression of rust diseases on wheat. *The Plant Journal*, *73*(3), 521–532. doi:10.1111/tpj.12047 PMID:23110316

Pauley, K. M., Eystathioy, T., Jakymiw, A., Hamel, J. C., Fritzler, M. J., & Chan, E. K. (2006). Formation of GW bodies is a consequence of microRNA genesis. *EMBO Reports*, *7*(9), 904–910. doi:10.1038j.embor.7400783 PMID:16906129

Pélissier, T., Thalmeir, S., Kempe, D., Sänger, H.L., & Wassenegger, M. (1999). Heavy de novo methylation at symmetrical and non-symmetrical sites is a hallmark of RNA-directed DNA methylation. *Nucleic Acids Res., 27*(7), 1625-34. doi:10.1093/nar/27.7.1625

Peragine, A., Yoshikawa, M., Wu, G., Albrecht, H. L., & Poethig, R. S. (2004). *SGS3* and *SGS2/SDE1/RDR6* are required for juvenile development and the production of *trans*-acting siRNAs in *Arabidopsis*. *Genes & Development*, *18*(19), 2368–2379. doi:10.1101/gad.1231804 PMID:15466488

Pillai, R. S. (2005). MicroRNA function: Multiple mechanisms for a tiny RNA? *RNA*, *11*(12), 1753–1761. doi:10.1261/rna.2248605 PMID:16314451

Preall, J. B., He, Z., Gorra, J. M., & Sontheimer, E. J. (2006). Short interfering RNA strand selection is independent of dsRNA processing polarity during RNAi in Drosophila. *Current Biology*, *16*(5), 530–535. doi:10.1016/j.cub.2006.01.061 PMID:16527750

Pyott, D. E., & Molnar, A. (2015). Going mobile: Non-cell-autonomous small RNAs shape the genetic landscape of plants. *Plant Biotechnology Journal*, *13*(3), 306–318. doi:10.1111/pbi.12353 PMID:25756494

Qiao, F., Yang, Q., Wang, C. L., Fan, Y.-L., Wu, X.-F., & Zhao, K.-J. (2007). Modification of plant height via RNAi suppression of OsGA20 ox2 gene in rice. *Euphytica*, *158*(1-2), 35–45. doi:10.100710681-007-9422-6

Qu, F., Ye, X., Hou, G., Sato, S., Clemente, T. E., & Morris, T. J. (2005). RDR6 has a broad-spectrum but temperature-dependent antiviral defense role in *Nicotiana benthamiana. Journal of Virology*, *79*(24), 15209–15217. doi:10.1128/JVI.79.24.15209-15217.2005 PMID:16306592

Ray, B. D., Jarori, G. K., Raghunathan, V., Yan, H., & Nageswara Rao, B. D. (2005). Conformations of nucleotides bound to wild type and Y78F mutant yeast guanylate kinase: Proton two-dimensional transferred NOESY measurements. *Biochemistry*, *44*(42), 13762–13770. doi:10.1021/bi0509088 PMID:16229466

Risco, C. A., & Chase, C. C. Jr. (1997). Handbook of plant and fungal toxicants. *CRC Press*.

Rodewald, J., & Trognitz, B. (2013). *Solanum* resistance genes against *Phytophthora infestans* and their corresponding avirulence genes. *Molecular Plant Pathology*, *14*(7), 740–757. doi:10.1111/mpp.12036 PMID:23710878

Rodoni, B. C., Dale, J. L., & Harding, R. M. (1999). Characterization and expression of the coat protein-coding region of the banana bract mosaic potyvirus, development of diagnostic assays and detection of the virus in banana plants from five countries in Southeast Asia. *Archives of Virology*, *144*(9), 1725–1737. doi:10.1007007050050700 PMID:10542022

Romay, G., & Bragard, C. (2017). Antiviral Defenses in Plants through Genome Editing. *Frontiers in Microbiology*, *8*, 47. doi:10.3389/fmicb.2017.00047 PMID:28167937

San Miguel, K., & Scott, J. G. (2016). The next generation of insecticides: dsRNA is stable as a foliar-applied insecticide. *Pest Management Science*, *72*(4), 801–809. doi:10.1002/ps.4056 PMID:26097110

Schijlen, E. G. W. M., deVos, R. C. H., Martens, S., Jonker, H. H., Rosin, F. M., Molthoff, J. W., Tikunov, Y. M., Angenent, G. C., van Tunen, A. J., & Bovy, A. G. (2007). RNA Interference silencing of *chalcone synthase*, the first step in the flavonoid biosynthesis pathway, leads to parthenocarpic tomato fruits. *Plant Physiology*, *144*(3), 1520–1530. doi:10.1104/pp.107.100305 PMID:17478633

Schmid, M., & Jensen, T. H. (2008). The exosome: A multipurpose RNA-decay machine. *Trends in Biochemical Sciences*, *33*(10), 501–510. doi:10.1016/j.tibs.2008.07.003 PMID:18786828

Schwall, G. P., Safford, R., Westcott, R. J., Jeffcoat, R., Tayal, A., Shi, Y.-C., Gidley, M. J., & Jobling, S. A. (2000). Production of very high-amylose potato starch by inhibition of SBE A and B. *Nature Biotechnology*, *18*(5), 551–554. doi:10.1038/75427 PMID:10802625

Segal, G., Song, R., & Messing, J. (2003). A new opaque variant of maize by a single dominant RNA-interference-inducing transgene. *Genetics*, *165*(1), 387–397. doi:10.1093/genetics/165.1.387 PMID:14504244

Sen, G., & Blau, H. M. (2005). Argonaute 2/RISC resides in sites of mammalian mRNA decay known as cytoplasmic bodies. *Nature Cell Biology*, *7*(6), 633–636. doi:10.1038/ncb1265 PMID:15908945

Senthilkumar, M., & Mysore, K. S. (2011). Virus-induced gene silencing can persist for more than two years and also be transmitted to progeny seedlings in *Nicotiana benthamiana* and tomato. *Plant Biotechnology Journal*, *9*(7), 797–806. doi:10.1111/j.1467-7652.2011.00589.x PMID:21265998

Shelly Praveen, S., Ramesh, V., & Anil, K. M. (2010). Silencing potential of viral derived RNAi constructs in tomato leaf curl virus-AC4 gene suppression in tomato. *Transgenic Research*, *19*(1), 45–55. doi:10.100711248-009-9291-y PMID:19548101

Shen, Y., Xie, Y., Yu, X., Zhang, S., Wen, Q., Ye, G., & Guo, J. (2021).Clinical diagnostic values of transfer RNA-derived fragment tRF-19-3L7L73JD and its effects on the growth of gastric cancer cells. *J Cancer, 12*(11), 3230-3238.

Sijen, T., Fleenor, J., Simmer, F., Thijssen, K. L., Parrish, S., Timmons, L., Plasterk, R. H. A., & Fire, A. (2001). On the role of RNA amplification in dsRNA-triggered gene silencing. *Cell*, *107*(4), 465–476. doi:10.1016/S0092-8674(01)00576-1 PMID:11719187

Sindhu, S., Maier, T., Mitchum, M. G., Hussey, R. S., Davis, E. L., & Baum, T. J. (2009). Effective and specific in Plant RNAi in cyst nematodes: Expression interference of four parasitism genes reduces parasitic success. *Journal of Experimental Botany*, *60*(1), 315–324. doi:10.1093/jxb/ern289 PMID:19015219

Siritungam, D., & Sayre, R. (2003). Generation of cyanogen-free transgenic cassava. *Planta*, *217*(3), 367–373. doi:10.100700425-003-1005-8 PMID:14520563

Sontheimer, E. J. (2005). Assembly and function of RNA silencing complexes. *Nature Reviews. Molecular Cell Biology*, *6*(2), 127–138. doi:10.1038/nrm1568 PMID:15654322

Stefani, G., & Slack, F. J. (2008). Small non-coding RNAs in animal development. *Nature Reviews. Molecular Cell Biology*, *9*(3), 219–230. doi:10.1038/nrm2347 PMID:18270516

Steves, R. M., Todd, T. C., Essig, J. S., & Trick, H. N. (2006). Transgenic soybeans expressing siRNAs specific to a major sperm protein gene suppress *Heterodera glycines* reproduction. *Functional Plant Biology*, *33*(11), 991–999. doi:10.1071/FP06130 PMID:32689310

Strauss E. (1999). Molecular biology. Candidate 'gene silencers' found. *Science, 286*(5441), 886.

Sunkar, R., & Zhu, J. K. (2004). Novel and stress-regulated micro RNAs and other small RNAs from Arabidopsis. *The Plant Cell*, *16*(8), 2001–2019. doi:10.1105/tpc.104.022830 PMID:15258262

Takumi, S., Motoyasu, Y., & Taiyun, W. (2008). Silencing by RNAi of the gene for Pns12, a viroplasm matrix protein of *Rice dwarf viruso*, results in strong resistance of transgenic rice plants to the *virus. Plant Biotechnology Journal*, *7*, 24–32.

Tammu, R. N., Nuringtyas, T. S., & Daryono, B. S. (2021). Daryono.B.S. "Colchicine effects on the ploidy level and morphological characters of Katokkon pepper (*Capsicum annuum* L.) from North Toraja, Indonesia. *Journal of Genetic Engineering and Biotechnology*, *19*(1), 1. doi:10.118643141-021-00131-4 PMID:33580394

Tanaka, K. N., Mukae, H., Dewar, M., van Breugel, M., James, E. K., Prescott, A. R., Antony, C., & Tanaka, T. U. (2005). Molecular mechanisms of kinetochore capture by spindle microtubules. *Nature*, *434*(7036), 987–994. doi:10.1038/nature03483 PMID:15846338

Tang, L., Kwon, S. Y., & Kwak, S. S. (2004). Selection of transgenic potato plants expressing both CuZnSOD and APX in chloroplasts with enhanced tolerance to oxidative stress. *Korean Journal of Plant Biotechnology*, *31*(2), 109–113. doi:10.5010/JPB.2004.31.2.109

Tang, Y.-Y., Ma, Y., Wang, J., Fan, Y., Feng, S., Lu, Q., Yu, Q., Sui, D., Rothbart, M. K., Fan, M., & Posner, M. I. (2007). Short-term meditation training improves attention and self-regulation. *Proceedings of the National Academy of Sciences of the United States of America*, *104*(43), 17152–17156. doi:10.1073/pnas.0707678104 PMID:17940025

Tomari, Y., & Zamore, P. (2005). Perspective: Machines for RNAi. *Genes & Development*, *19*(5), 517–529. doi:10.1101/gad.1284105 PMID:15741316

Tsuda, S., Fukui, Y., Nakamura, N., Katsumoto, Y., Yonekura-Sakakibara, K., Fukuchi-Mizutani, M., Ohira, K., Ueyama, Y., Ohkawa, H., Holton, T. A., Kusumi, T., & Tanaka, Y. (2004). Flower color modification of *Petunia 83ybrid* commercial varieties by metabolic engineering. *Plant Biotechnology (Sheffield, England)*, *21*(5), 377–386. doi:10.5511/plantbiotechnology.21.377

Urwin, P. E., Lilley, C. J., & Atkinson, H. J. (2002). Ingestion of double-stranded RNA by pre-parasitic juvenile cyst nematodes leads to RNA interference. *Molecular Plant-Microbe Interactions*, *15*(8), 747–752. doi:10.1094/MPMI.2002.15.8.747 PMID:12182331

van der Krol, A. R., & Mur, L. A. (1990). M Beld, J N Mol, A R Stuitje. "Flavonoid genes in petunia: Addition of a limited number of gene copies may lead to a suppression of gene expression. *The Plant Cell*, *4*(2), 291–299. PMID:2152117

Van Uyen, N. (2006). Novel approaches in plants breeding RNAi technology. *Proceedings of International Workshop on Biotechnology in Agriculture*, 12-16.

Vanyushin, B. F. (2006). DNA methylation in plants. *Current Topics in Microbiology and Immunology*, *301*, 67–122. doi:10.1007/3-540-31390-7_4 PMID:16570846

Vaucheret, H. (2008). Plant ARGONAUTES. *Trends in Plant Science*, *13*(7), 350–358. doi:10.1016/j.tplants.2008.04.007 PMID:18508405

Wagner, R.W., & Sun, L. (1998). Double-stranded RNA poses puzzle. *Nature, 391*(6669), 744-5.

Wang, M., Abbott, D. C., & Waterhouse, P. M. (2000). A single copy of a virus- derived transgene encoding hairpin RNA gives immunity to barley yellow dwarf virus. *Molecular Plant Pathology*, *1*(6), 347–356. doi:10.1046/j.1364-3703.2000.00038.x PMID:20572982

Wang, M., Weiberg, A., Lin, F.-M., Thomma, B. P. H. J., Huang, H.-D., & Jin, H. (2016). Bidirectional cross-kingdom RNAi and fungal uptake of external RNAs confer plant protection. *Nature Plants*, *2*(10), 16151. doi:10.1038/nplants.2016.151 PMID:27643635

Wang, T., Chen, L., Zhao, M., Tian, Q., & Zhang, W. H. (2011b). Identification of drought-responsive microRNAs in *Medicago truncatula* by genome-wide high-throughput sequencing. *BMC Genomics*, *12*(1), 367. doi:10.1186/1471-2164-12-367 PMID:21762498

Wei, L., Zhang, D., Xiang, F., & Zhang, Z. (2009). Differentially expressed miRNAs potentially involved in the regulation of defense mechanism to drought stress in maize seedlings. *International Journal of Plant Sciences*, *170*(8), 979–989. doi:10.1086/605122

White, P. J., & Broadley, M. R. (2009). Biofortification of crops with seven mineral elements often lacking in human diets—Iron, zinc, copper, calcium, magnesium, selenium and iodine. *The New Phytologist*, *182*(1), 49–84. doi:10.1111/j.1469-8137.2008.02738.x PMID:19192191

Wightman, B., Ha, I., & Ruvkun, G. (1993). Post transcriptional regulation of the heterochronic gene lin-4 mediates temporal formation in *C. elegans*. *Cell*, *75*(5), 855–862. doi:10.1016/0092-8674(93)90530-4 PMID:8252622

Williams, M. M. II, Walsh, D. B., & Boydston, R. A. (2004). Integrating arthropod herbivory and reduced herbicide use for weed management. *Weed Science*, *52*(6), 1018–1025. doi:10.1614/WS-04-034R

Xiang, S., Fruehauf, J., & Li, C. J. (2006). Short hairpin RNA-expressing bacteria elicit RNA interference in mammals. *Nature Biotechnology*, *24*(6), 697–702. doi:10.1038/nbt1211 PMID:16699500

Xin, M., Wang, Y., Yao, Y., Xie, C., Peng, H., Ni, Z., & Sun, Q. (2010). Diverse set of microRNAs are responsive to powdery mildew infection and heat stress in wheat (*Triticum aestivum* L.). *BMC Plant Biology*, *10*(1), 123. doi:10.1186/1471-2229-10-123 PMID:20573268

Yan, S., Ren, B., Zeng, B., & Shen, J. (2020). Improving RNAi efficiency for pest control in crop species. *BioTechniques*, *68*(5), 283–290. doi:10.2144/btn-2019-0171 PMID:32202134

Yin, C., Jurgenson, J. E., & Hulbert, S. H. (2011). Development of a host-induced RNAi system in the wheat stripe rust fungus *Puccinia striiformis* f. sp. *tritici*. *Molecular Plant-Microbe Interactions*, *24*(5), 554–561. doi:10.1094/MPMI-10-10-0229 PMID:21190437

Yu, B., Lydiate, D. J., Young, L. W., Schäfer, U. A., & Hannoufa, A. (2007). Enhancing the carotenoid content of *Brassica napus* seeds by downregulating lycopene epsilon cyclase. *Transgenic Research*, *17*(4), 573–585. doi:10.100711248-007-9131-x PMID:17851775

Zhang, J., Xu, Y., Huan, Q., & Chong, K. (2009). Deep sequencing of *Brachypodium* small RNAs at the global genome level identifies microRNAs involved in cold stress response. *BMC Genomics*, *10*(1), 449. doi:10.1186/1471-2164-10-449 PMID:19772667

Zhang, T., Jin, Y., Zhao, J.-H., Gao, F., Zhou, B.-J., Fang, Y.-Y., & Guo, H.-S. (2016). Host-induced gene silencing of the target gene in fungal cells confers effective resistance to the cotton wilt disease pathogen *Verticillium dahliae*. *Molecular Plant*, *9*(6), 939–942. doi:10.1016/j.molp.2016.02.008 PMID:26925819

Zhong, S. H., Liu, J. Z., Jin, H., Lin, L., Li, Q., Chen, Y., Yuan, Y.-X., Wang, Z.-Y., Huang, H., Qi, Y.-J., Chen, X.-Y., Vaucheret, H., Chory, J., Li, J., & He, Z.-H. (2013). Warm temperatures induce transgenerational epigenetic release of RNA silencing by inhibiting siRNA biogenesis in *Arabidopsis. Proceedings of the National Academy of Sciences of the United States of America, 110*(22), 9171–9176. doi:10.1073/pnas.1219655110 PMID:23686579

Zhu, K.Y., & Palli, S.R. (2020). Mechanisms, applications, and challenges of insect RNA interference. *Annu. Rev. Entomol., 65*, 14.1–14.19.

Chapter 32
Organic Techniques in Vegetable Crop Production for Sustainable Agriculture

Shahnaz Mufti
Sher-e-Kashmir University of Agricultural Sciences and Technology, India

Sumati Narayan
Sher-e-Kashmir University of Agricultural Sciences and Technology, India

Aijaz Ahmad Malik
Sher-e-Kashmir University of Agricultural Sciences and Technology, India

ABSTRACT

Sustainable crop production is a way of growing or raising food in an ecologically and ethically responsible manner. Sustainable crop production is in contrast to industrial crop production, which generally relies upon monocropping (growing only one crop in a large area of land), intensive application of commercial fertilizers, heavy use of pesticides, and other inputs that are damaging to the environment, to communities, and to farm workers. Organic farming is one of the several approaches found to meet the objectives of sustainable agriculture. Organic vegetable cultivation offers one of the most sustainable farming systems with the recurring benefits to long term soil health and provide a lasting stability in production by importing better resistance against various biotic and abiotic stresses. Most of organic products are growing at a faster rate (20%) as compared to conventional ones (5%). This growth rate is highest in Japan, USA, Australia, and EU.

INTRODUCTION

Organic farming methods combine scientific knowledge of ecology and modern technology with traditional farming practices based on naturally occurring biological processes. The major farming techniques of organic farming include:

DOI: 10.4018/978-1-7998-8434-7.ch032

1. Crop Rotation
2. Green Manuring
3. Composting
4. Vermicomposting
5. Mulching
6. Plant Protection Practices
    a. Cultural practices
    b. Inter cropping and trap crops
    c. Mixed Cropping
    d. Growing disease resistant varieties

Other natural practices include

- Flooding
- Exclusion of pathogen
- Application of organic amendments
- Physical methods
- Application of biocontrol agents
- Using Biodyanamic liquid pesticides
- Manipulation in Sowing Dates
- Elimination of Host Plants Summer Ploughing

CROP ROTATION

Crop rotation is the systematic planting of different crops rather than the same vegetable or members of the same family of vegetables in a particular order over several years in the same growing space. Crop rotation will benefit the vegetable crops in two ways: it will prevent the build –up of soil – borne pests and diseases; second it will allow for the replenishment and efficient use of soil nutrients. The development and implementation of well-designed crop rotations is central to the success of organic production systems (Stockdale 2000). To minimize pest and disease problems and to help renew soil nutrients, members of the same plant family should not be planted in the same part of the field more than once every three or four years. Vegetable insect pests tend to feed on similar plants and members of the same plant family e.g., an insect pest that attacks and eats cabbage lay its eggs before it dies. If cabbage or a member of the cabbage family is planted in the same spot the next year, the eggs of the insect will hatch and the pests will find exactly the food they need to continue the pest life cycle. Soil borne diseases, fungi, bacteria and viruses can also be hosted by specific plants as well. Removing host plants or alternating unrelated plants in the field can break the cycle of pests and diseases. Crop rotation also helps to prevent soil nutrients from being depleted. Vegetables draw upon a wide range of soil nutrients for growth: nitrogen, phosphorus and potassium are the key or major soil nutrients. Members of the same vegetable family usually draw the same nutrients from the soil. Crop rotation will prevent the soil from wearing out. Heavy N,P,K feeding crops such as tomatoes are rotated with soil building crops such as beans which add nitrogen to the soil and then with the light feeding crops such as onion.

Major Plant Families used in Crop Rotation:

Onion Family

(Alliaceae, Amaryllidaceae): Garlic, onions, leek, shallots. These are light feeders. Plant these after heavy feeders. Follow these crops with legumes.

Cabbage Family

(Brassicaceae): Broccoli, Brussels sprouts, cabbage, cauliflower, kale, kohlrabi, radish, turnip. These are heavy feeders. These crops should follow legumes. After these crops allow the garden to go fallow for season or plant a cover crop or add plenty of compost and organic matter to the garden.

Carrot Family

(Umbelliferae): Carrot, celery, coriander, parsley. These are light to moderate feeders. These crops can follow any other group.

Squash Family

(Cucurbitaceae): Cucumbers, melons, summer and winter squash, pumpkins, watermelon. These crops are heavy feeders. Follow these crops with legumes.

Bean Family

(Leguminosae): Beans, peas. These crops enrich the soil. Plant these crops before or after any other crop family.

Tomato Family

(Nightshade Family): tomatoes, potatoes, Eggplant. These crops are heavy feeders. Follow these crops with legumes.

GREEN MANURING

The positive role of green manuring has been known since ancient times. Importance of this soil ameliorating practice is increasing in recent years because of high cost of chemical fertilizers, increased risk of environmental pollution and need of sustainable cropping systems. Green manuring crops include:

1. Legumes (clover family) - which have the ability to fix nitrogen.
2. Non-legumes- they don't fix nitrogen but provides suitable amount of organic matter and retain nutrients that might otherwise be leached.

Benefits of Green manuring include:

1. **Improve soil physical, chemical and biological properties and consequently crop yields:** Once incorporated, the green manure provide a pool of fresh organic matter and there are numerous examples where using green manures increases soil organic matter in comparison to treatments where only inorganic fertilizers are applied (Shepherd et al., 2002). This organic matter provides food to soil microorganisms, encouraging an increase in their number and activity (Campbell et al., 1991; N' Dayegamuye & Tran, 2001). The more succulent a green manure is the more rapidly it will decay once incorporated and the less effect it will have on long term soil organic matter. Older woody green manures with a higher C: N ratios break down more slowly and breakdown is reported to decrease to a very slow rate when C: N ratio is above 16:1 (Enwezor, 1996). Legumes which have a low C:N ratio, break down rapidly so have little effect on long term soil organic matter but give a larger short term boost to soil structure as they have a large short term effect on soil microorganisms. Green manures can also play a part in reducing soil erosion both by wind and rain (Cransberg & McFarlane, 1994).
2. **Minimize nitrate leaching risk**: Green manures vary in their ability to reduce leaching. Grazing rye is particularly effective in leaching by an average of 95%.
3. **Provision of Nitrogen for succeeding crops**: Of all the nutrient elements, nitrogen is the most labile in soil and the one most likely to be affected by green manures. This is because it exists in so many different forms that are inter-converted by a range of biological processes – some of these forms are prone to losses (by leaching or gaseous emissions of ammonia, nitrogen or nitrous oxides). For some crops (eg cauliflowers) it is particularly important that sufficient nitrogen is available at certain growth stages to ensure that the plants produce yields of a marketable quality and correct management of green manures can be used to manipulate its availability. Large amounts of nitrogen are added to the soil by a successful green manure (e.g., an overwintered crop of vetch may accumulate up to 200kg N/ha by early May). Non legumes (e.g., grazing rye or mustard) may add as much nitrogen to the soil as legumes but less of this is made available in the short term. This is because they have a lower nitrogen content (1-2%) and so more carbon is added as well (i.e., these plants have a higher C:N ratio). This means that nitrogen is released more slowly as they are decomposed by microorganisms. Furthermore, the nitrogen concentration in these crops usually decreases as the plants age and prepare to set seed so the date of incorporation is crucial e.g., rye incorporated at the end of April will release nitrogen (although not as much as vetch) but if left until May the decomposing organisms will actually have to compete with plants for soil nitrate in order to tackle the residues. This is sometimes called 'nitrogen robbery'. In the longer term, the nitrogen conserved by cover crops, especially the slower mineralizing non legumes, is prone to leaching in the first winter after incorporation if the land remains uncropped; rotation design should take account of this. Green manures have their biggest effect on soil mineral nitrogen within the first year after incorporation and so this is the crucial period when the cropping regimes need to be planned to make effective use of it.
4. **Effects on the availability of other nutrients**: Green manures are effective in preventing nitrogen losses from organic rotations. However inputs of other nutrients, such as P and K, are often limited in organic systems, so it is important that these nutrients are made available to crops and their losses minimized which can be achieved by incorporating green manures into the soil.

5. **Pest, disease and weed control effects** Weed suppression can be one of the key benefits of green manures, but again there can be a negative effect if weed control in the green manure has been poor, or the green manure itself becomes a weed in subsequent crops.
6. **Pollen and nectar source for predators**: Flowering green manures can also act as pollen and nectar sources for predatory insects such as ladybirds, hoverflies, lacewings and parasitic wasps. Commonly used green manures such as crimson clover (Tillman *et al.*, 1999), phacelia (Hickman & Wratten, 1996; Denys & Tscharntke, 2002) and buckwheat (Bowie et al., 1995) have all been reported to act as attractants to predatory insects when grown in conjunction with other crops. A number of disadvantages can also be identified in green manuring. Some of which are mentioned below:
    - Direct costs of seed and extra cultivations
    - Lost opportunities for cash cropping.
    - Extra work at busy times of the year.
    - Exacerbated pest and disease problems (due to the 'green bridge' effect).
    - Potential for the green manures to become weeds.

COMPOSTING

One of the easiest and most widely used practice in organic farming which promotes greater soil and plant health. Composting is a form of waste disposal where organic wastes decompose naturally under oxygen rich conditions. Any type of plant material can be used for composting. These materials can include leaves, lawn, straw or kitchen scraps. Crushed egg shells and coffee grinds can also be used for composting. Adding egg shells to the compost will add calcium to it which is an important nutrient that helps to build the plant cell wall. In case, of some vegetables like tomatoes and squash fruits which are susceptible to blossom end rot can be prevented by adding the egg shells to the compost Composts perform many functions when added to soil such as it can increase soil organic matter and influence wide range of functional soil properties and have direct influence on crop performance .The use of good quality compost can reduce regular fertilizer use thereby saving money for commercial growers.

VERMICULTURE TECHNOLOGY

Vermiculture appears to be a novel sustainable technology for waste treatment which has a great future in the field of organic farming. In this technology nontoxic organic residues of plant and animal wastes are used as substrates to produce vermicastings. Organic wastes are converted into usable form of nutrients by bio-oxidation and stabilization with the synergism of earthworms and microorganisms. Vermicompost has a great potential as plant growth media and also served as soil conditioner. It is highly rich in nutrients and beneficial microbes. In addition, during the vermicomposting process, a honey brown colored leachate is produced by the action of earthworms called vermiwash (Aalok *et al.* 2008, Gopal *et al.* 2010). It also contains macro and micronutrients along with microbes which have important plant growth promoting attributes. Thus, the organic waste can be excellently recycled and converted into useful vermiproducts that have been proved to be essential component in plants growth and productivity and also maintaining the soil health. It possesses a great potential to promote growth of various crops and directly or indirectly inhibit the growth and multiplication of plant pathogens (Kale 1998, Lalitha *et al.*, 2000, Zularisam *et al.*, 2010, Kaur *et al.*, 2015). It also contains auxin, gibberllic acid and cytokinins like substances that enhance its application to plants for their growth. Application of vermiproducts increases the soil organic matter (SOM) to above 3-5% which acts as a 'glue' to bind 'soil particles' into aggregates and improve the soil structure, infiltration, air porosity, nutrient and water holding capacity. Organic carbon in soil plays an important role in improving soil health. Application of vermiproducts compensates the loss of soil carbon thus maximize production of crop over time. The SOM (soil organic matter) decomposes over time results in the accumulation of more stable carbon compound called humus which improves mineral breakdown and provide continues supply of nutrient to plants. Stable and highly mature composts contain longlasting form of carbon called as 'humates' or 'humic and fulvic acids' which are very important for maintaining soil health and fertility (Compost Australia 2011). The technology of vermiculture can effectively manage the waste. This process allows composting the degradable materials and at the same time utilizing their products to enhance crop production and thus eliminates the use of chemical fertilizers. The emphasis on chemical fertilizers has led to the long term imbalance in soil pH and its fertility, which has caused irreparable damage to the ecosystem. To cope with these incisive problems, the vermiculture technology has become main stream all over the world because it works in harmony with nature rather than against it. It has come up with a new hope to sustain agriculture in an ecofriendly manner which provides us good quality of food without any harmful chemical constituents in it.

MULCHING

The practice of applying mulches for the production of vegetables can be dated back to thousands of years (Lightfoot, 1994). Typically mulching involves placing a layer of material on the soil around the crop of interest to modify the growing environment to improve crop productivity. The primary purpose for using mulches is for weed suppression in the crop to be grown. Mulches typically function by blocking light or creating environmental conditions which can prevent germination or suppress weed growth shortly after germination. However, numerous other benefits are often obtained including: increased earliness, moisture conservation, temperature regulation of the root zone and above-ground growing environment, reduced nutrient leaching, altered insect and disease pressures, and, in some instances, reduced soil compaction or improved soil organic matter (Lamont, 2005; Ngouajio and McGiffen, 2004). The use of mulches typically results in higher yields and quality in vegetable crops enhancing profitability for the grower.

Types of Mulches

Organic-based Mulches

Such as plant waste, straw, sawdust, and manure have been used to a great extent for vegetable production. Traditionally, organic mulches have consisted of materials which are locally plentiful. When applied as a thick layer, organic-based mulches can effectively suppress weeds and increase soil moisture levels (Diaz-Perez et al., 2004). However, research dated back to the late 19th century has shown the variable results of organic mulches on yield. In areas with warm temperatures and limitations on water for plant growth, straw mulches have been found to positively affect growth and yields of several cucurbits

(Emerson, 1993). However, when used in cool climates, the addition of straw mulch, while beneficial for controlling weeds, has been shown to retard the growth of warm season vegetables and decrease yields (Rowe-Dutton, 1997). Nonetheless, organic mulches remain popular due to their low cost and ready availability.

Paper-based Mulches

Utilized through much of the early 20th century with positive results. Thompson and Platenius (2001) reported positive results with paper mulches in several vegetable crops including pepper (*Capsicum* spp.), tomato and muskmelon (*Cucumis melo* L.). Paper mulches controlled weeds, increased soil temperatures and moisture levels, resulting in greater yields (Thompson and Platenius, 1931). By altering the growing environment, black-paper mulches affected root distribution of several vegetable crops when compared to bare-ground production (Knavel and Mohr, 1967). The use of paper mulches, particularly those impregnated with asphalt or tar, suppressed weeds, conserved moisture, and warmed the soil, increasing yields in most warm-season crops. However, issues with cost and durability led to the development of alternative mulches.

Degradable Mulch Films

The degradable mulch must be flexible, lightweight, prevent light transmittance, and degrade in a timely manner after harvest. Exposure to light, temperature, and moisture can influence degradation (Kyrikou and Briassoulis, 2007). Normalizing degradation rates between growing regions with vastly different climates is a challenge as well. Some crops will quickly form a canopy shading mulches and thus delaying degradation; while others do not. Developing mulch that will degrade on-demand at a competitive cost is a challenge. Degradable plastic mulches are often labeled as biodegradable. Degradable mulches should break down into small brittle pieces and would not interfere with subsequent planting. Halley et al. (2001) developed a mulch film using modified-starch polymers that performed as well as conventional polyethylene mulch for pepper production. This mulch withstood 14 weeks of water exposure and remained largely stable during crop production. However, just two weeks after the mulch was plowed into the soil, it was visually undetectable, with composting trials indicating that the mulch completely degraded to carbon dioxide and water after 45 days (Halley *et al.,* 2001). Paper is a renewable resource that readily biodegrades. Newspaper-based mulches represent an available and cost effective resource and have been frequently trialed; though they often deteriorate rapidly under field conditions, reducing effectiveness (Shogren, 2001). Shredded newspapers have been successfully used as a weed suppressing mulch in organic high tunnel cucumber production (Sanchez et al., 2008). A high-tunnel environment (no wind or rain) is conducive for using newspaper mulches.

Synthetic Mulches

Includes polyethylene transparent and multicolor (natural, white, red, blue, yellow, silver, black, opaque, grey, smoke green), translucent polyethylene, perforated polyethylene, spun bonded polypropylene and photo selective coloured films.

Inorganic Mulches

Reflective aluminium foils, asphalt sprays, wax, chalk powder, oil emulsion, kaolin, lime, bitumen, riverstone, metal strips, lime sprays and anti – fogging covers.

PLANT PROTECTION PRACTICES

Plant protection practices in Organic farming include:

Cultural Practices

Include crop rotation, intercropping, trap crops, mixed cropping, use of resistant varieties.

Intercropping and Trap Crops

Intercropping is a practice to grow suitable crops along with main crop. This helps to break the pattern of growing the same crop year after year. This helps to break the pattern of growing same crop year after year and limits insect infestation. Natural enemies of insect pests tend to be more abundant in intercrops than in monocrops as they find better distribution of nectar and pollen sources. Diverse planting not only control adult insects from laying eggs but also release volatile allele chemicals from particular crops. Cabbage along with carrot or tomato is an important intercrop combination to manage diamond back moth. Similarly raising trap crops as inter and border crops is an important cropping system approach. The trap crop is attractive to insects compared to the main crop. It provides protection by preventing the insects from attacking the main crop. Mustard along with cabbage is a trap crop for the management of diamond back moth, aphids and leaf webber. Marigold is a good trap crop for the boll worm besides it also attracts the adults of leaf minor which lay eggs on its leaves. Maize can also be used a trap crop to attract fruit fly adults in vegetable cultivation where the incidence of fruit fly is a major problem.

Mixed Cropping

It is defined as growing of two or more crops in the same field during the same period. This system of cropping intensifies the farm system and ensures increased productivity and biodiversity of field. This system of cropping is often considered as a promising technique for sustainable crop production as it has multifunctional roles and higher productivity. It is an important tool for insects and pest management in organic farming. It reduces phytophagous pests by encouraging the buildup of natural enemies due to:

1. Increased ground cover, particularly important for diurnal insect pests.
2. Greater temporal and spatial distribution of nectar and pollen
3. Pest's ability to find host plants by conferring associational resistance by the non – host plant masking the odors of the host plant.

Growing Disease Resistant Varieties

For low external input organic farming, resistant crops represent an important alternative to pesticides. Exploiting the diversity and variability in the host genetic constitution for resistance against a pathogen in a crop is the best strategy for disease management without application of hazardous pesticides. It can individually restrict the incidence of a particular disease in a crop. Successful disease establishment depends on the compatible gene for gene interaction between a host and a pathogen. Resistant varieties tend to remain disease free for a long period of time owing to morphological manifestation of their genetic constitution in form of leaf and stem toughness, time of maturity, nutrient content, plant architecture, growth habit which can deter growth of pathogen, their reproduction and host preference. Care need to be taken to include more than on resistant variety in a region to dissipate selection pressure on the pathogen.

Other organic techniques for plant protection in organic farming include:

1. **Flooding**: It can be used to manage soil borne insects. If the field is infested with termites flooding will help to manage them. Similarly, many insects like cut worms, potato tuber moth and root grubs can't withstand flooding. It also prevents many insects from laying eggs because of the destruction of their habitats.
2. **Exclusion of pathogen**: Preventing the potent and viable disease propagules to interact with the host results in reduction in disease incidence. Use of disease-free seeds and planting material would prevent seed borne disease, management of vectors, and in situ destruction of soil borne pathogens through soil solarisation or Anaerobic soil disinfestations (ASD) involves the incorporation of fresh organic material in moist soil under airtight plastic for 3–6 weeks, depending on the outside temperature (van Bruggen et al., 2016).
3. **Application of organic amendments**: Soils with low microbial diversity promote establishment of plant pathogenic organisms. Healthy soil is the mainstay of organic agriculture. Improved soil biological activity is known to play a key role in suppressing weeds, pests and diseases (IFOAM, 1998). Improving soil health through use of cover crops, green manures, animal manures to fertilize the soil not only helps in restricting soil borne pathogens but also maximizes biological activity and maintains long-term soil health. Application of composts and organic amendments tends to increase quantity and diversity of soil microbial diversity and consecutive disease suppressiveness. Organic amendments are biodegradable and are generally available on the farmer's fields. Neem cake used for soil amendment @ 0.25 to 0.5 t/ha contributes significantly in control of nematodes and soil borne pathogens. Soils rich in organic matter are high on soil biodiversity with abundance of beneficial soil microorganisms.
4. **Physical methods**: Soil solarisation of nursery beds reduces soil borne inoculum. Hot water/steam treatment of seeds/planting material has been successful in many crops (Cohen et al., 2005).
5. **Application of bio control agents**: Microbial bio-control agents isolated from native environments are relatively safe, host specific and do not disturb other biotic systems (Srinivas and Ramakrishna, 2005). They are ideal for both short and long term pest suppression and are also compatible with most other control methods. Their mechanisms of action include competition, antagonism, antibiosis, enhanced nutrient uptake, induction of host resistance (Kloepper *et al.*, 1997) etc. Unlike chemical pesticides, they are harmless to humans and other non-target organisms, they do not leave chemical residues on crops, are easy and safe to dispose of and do not contaminate water systems. Commercial bio-fungicides are containing beneficial living organisms are can be locally

produced and used for pest management in organic farming. These are available as powders for seed treatments, as granules for soil application, and as suspensions for root drenches and foliar sprays. Biological control agents like *Trichoderma* spp, *Pseudomonas* spp and *Bacillus* spp. have proven their worth in managing a range of plant diseases.

6. **Using Biodyanamic liquid pesticides:** Are prepared by using different materials like cow dung, urine, leaves of plants having pesticidal properties (neem, pongamia, Datura, stinging needle, Calatropis, caster etc). Besides pesticidal properties, it contains macro and micronutrients which promotes the plant health.
7. **Manipulation in Sowing Dates:** It helps in avoidance of egg laying period of certain insects and establishment of plants before the insect incidence. It also facilitates the early maturation of the crop before the insect incidence.
8. **Elimination of Host Plants:** The host plants are very attractive to many insects and serve as focal point for future infestation. Unless they are destroyed, the perpetuation of pest problem will continue as these plants supply food to support life cycle of pests of proceeding crops. ix. **Summer ploughing:** Deep ploughing during summer destroys the weeds, eggs and insects which are in their initial stages.

REFERENCES

Aalok, A., Tripath, A. K., & Soni, P. (2008). Vermicomposting: A better option for organic solid waste management. *Journal of Human Ecology (Delhi, India)*, *24*(1), 59–64. doi:10.1080/09709274.2008.11906100

Bowie, M., Wratten, S. D., & White, A. J. (1995). Agronomy and phenology of "companion plants" of potential for enhancement of insect biological control. *New Zealand Journal of Crop and Horticultural Science*, *23*(4), 423–427. doi:10.1080/01140671.1995.9513919

Campbell, C. A., Biederbeck, V. O., Zentner, R. P., & Lafond, G. P. (1991). Effect of crop rotations and cultural practices on soil organic matter, microbial biomass and respiration in a thin Black Chernozem. *Canadian Journal of Soil Science*, *71*(3), 363–376. doi:10.4141/cjss91-035

Cohen, M. R., Yamasaki, H., & Mazzola, M. (2005). *Brassica napus* seed meal soil amendment modifies microbial community structure, nitric oxide production and incidence of Rhizoctonia root rot. *Soil Biology & Biochemistry*, *37*(7), 1215–1227. doi:10.1016/j.soilbio.2004.11.027

Compost Australia. (2011). *Compost for soils*. Compost Australia. http://www.compostforsoils.com.au

Cransberg, L., & McFarlane, D. J. (1994). Can perennial pastures provide the basis for a sustainable farming system in southern Australia. *New Zealand Journal of Agricultural Research*, *37*(3), 287–294. doi:10.1080/00288233.1994.9513067

Denys, C., & Tscharntke, T. (2002). Plant-insect communities and predator-prey ratios in field margin strips, adjacent crop fields, and fallows. *Oecologia*, *130*(2), 315–324. doi:10.1007004420100796 PMID:28547156

Diaz-Perez, J. C., Randle, W. M., Boyhan, G., Walcott, R. W., Giddings, D., Bertrand, D., Sanders, H. F., & Gitaitis, R. D. (2004). Effects of mulch and irrigation system on sweet onion: 1. Bolting, plant growth, and bulb yield and quality. *Journal of the American Society for Horticultural Science, 129*(2), 218–224. doi:10.21273/JASHS.129.2.0218

Emerson, R. A. (1993). *Experiments in mulching garden vegetables.* University of Nebraska, Agricultural Experiment Station of Nebraska.

Enwezor, W. O. (1996). The mineralization of nitrogen and phosphorus in organic materials of varying C:N and C:P ratios. *Plant and Soil, 44*(1), 237–240. doi:10.1007/BF00016972

Gopal, M., Gupta, A., Palaniswami, C., Dhanapal, R., & Thomas, G. V. (2010). Coconut leaf vermiwash: A bio-liquid from coconut leaf vermicompost for improving the crop production capacities of soil. *Current Science, 98*, 1202–1210.

Halley, P., Rutgers, R., Coombs, S., Kettels, J., Gralton, J., Christie, G., Jenkins, M., Beh, H., Griffin, K., Jayasekara, R., & Lonergan, G. (2001). Developing biodegradable mulch films from starch based polymers. *Starch, 53*(8), 362–367. doi:10.1002/1521-379X(200108)53:8<362::AID-STAR362>3.0.CO;2-J

Hickman, J. M., & Wratten, S. D. (1996). Use of *Phacelia tanacetifolia* strips to enhance biological control of aphids by hoverfly larvae in cereal fields. *Journal of Economic Entomology, 89*(4), 832–840. doi:10.1093/jee/89.4.832

IFOAM. (1998). *IFOAM Basic Standards for Organic Production and Processing*. IFOAM Publications.

Kale, R. D. (1998). *Earthworm cinderella of organic farming*. Prism Book Pvt. Ltd.

Kaur, P., Bhardwaj, M., & Babbar, I. (2015). Effect of Vermicompost and vermiwash on growth of vegetables. *Research Journal of Animal. Veterinary and Fishery Sciences, 3*, 9–12.

Kloepper, J. W., Tuzun, S., Zehnder, G. W., & Wei, G. (1997). Multiple disease protection by rhizobacteria that induce systemic resistance – historic precedence. *Phytopathology, 87*(2), 136–137. doi:10.1094/PHYTO.1997.87.2.136 PMID:18945131

Lalitha, R., Fathima, K., & Ismail, S. A. (2000). Impact of biopesticides and microbial fertilizers on productivity and growth of Abelmoschus esculentus. *Vasundhara The Earth, 1-2*, 4-9.

Lamont, W. J. (2005). Plastics: Modifying the microclimate for the production of vegetable crops. *HortTechnology, 15*(3), 477–481. doi:10.21273/HORTTECH.15.3.0477

Lightfoot, D. R. (1994). Morphology and ecology of lithic-mulch agriculture. *Geographical Review, 84*(2), 172–185. doi:10.2307/215329

N'Dayegamiye, A., & Tran, T. S. (2001). Effects of green manures on soil organic matter and wheat yields and nutrition. *Canadian Journal of Soil Science, 81*(4), 371–382. doi:10.4141/S00-034

Ngouajio, M., & McGiffen, M. E. (2004). Sustainable Vegetable production: Effects of cropping systems on weed and insect population dynamics. *Acta Horticulturae*, (638), 77–83. doi:10.17660/ActaHortic.2004.638.8

Rowe-Dutton, P. (1997). *The mulching of vegetables*. Commonwealth Bureau of Horticulture and Plantation Crops.

Sanchez, E., Lamont, W. J. Jr, & Orzolele, M. D. (2008). Newspaper mulches for suppressing weeds for organic high-tunnel cucumber production. *HortTechnology*, *18*(1), 154–157. doi:10.21273/HORTTECH.18.1.154

Shepherd, M. A., Harrison, R., & Webb, J. (2002). Managing soil organic matter-implications for soil structure on organic farms. *Soil Use and Management*, *18*(3), 284–292. doi:10.1079/SUM2002134

Shogren, R. L. (2001). Biodegradable mulches from renewable resources. *Journal of Sustainable Agriculture*, *16*(4), 33–47. doi:10.1300/J064v16n04_05

Srinivas, P., & Ramakrishna, G. (2005). Biological management of rice seed borne pathogens by native biocontrol agents. *Annals of Plant Protection Sciences*, *13*(2), 422–426.

Thompson, H., & Platenius, H. (1931). Results of paper mulch experiments with vegetable crops. *Proceedings of the American Society for Horticultural Science*, *28*, 305–309.

Tillman, P. G., Schomberg, H. H., Phatak, S., Mullinix, B., Timper, P., Lachnicht, S. L., & Olson, D. M. (2004). Influence of cover crops on insect pests and predators in conservation-tillage cotton. *Journal of Economic Entomology*, *97*(4), 1217–1232. doi:10.1093/jee/97.4.1217 PMID:15384330

Van Bruggen, A. H. C., Gamliel, A., & Finckh, M. R. (2016). Plant disease management in organic farming systems. *Pest Management Science*, *72*(1), 30–44. doi:10.1002/ps.4145 PMID:26331771

Zularisam, A. W., Zahirah, Z. S., Zakaria, I., Syukiri, M. M., Anwar, A., & Sakinah, M. (2010). Production of biofertilizer from vermicomposting process of municipal sewage sludge. *Journal of Applied Sciences*, *10*(7), 580–584. doi:10.3923/jas.2010.580.584

Chapter 33
Nutrigenomics and Green Technologies:
The Flavor of Future

Shalani Gupta
Sardar Vallabhbhai Patel University of Agriculture and Technology, Meerut, India

Varsha Rani
Sardar Vallabhbhai Patel University of Agriculture and Technology, Meerut, India

ABSTRACT

Green technologies can operate in diversified areas, and at the same time, these technologies contribute to produce more healthy and nutritious food. Biotechnology, the greenest tool of green technology, has the potential to increase the agricultural production, as well as to genetically modify food or its nutritional components to support health and to reduce the risk of diet-related diseases. Recent advances in molecular biology coupled with the wealth of information from 'omics technologies' have fostered the emergence of a multidisciplinary field of science 'Nuturigenomics' that tends to unfold the role of nutrition on gene expression. Amalgamation of both green technologies and nutrigenomics will help to solve problems arising due to negative effects of food, nutrition, and environment. The chapter provides an overview of green technology and its impact on nutrition and health through the study of nutrigenomics.

INTRODUCTION

Green Revolution and other technological innovation in the field of agriculture has been the major driving force to increase agricultural productivity to more than three folds between 1960 and 2015. A remarkable change in the process of industrialization and globalization of food and agriculture has been noticed during the same period (FAO, 2017). In the earlier days, the primary aims of agricultural technologies were focused on increasing production and crop productivity. Today the challenges before agriculture are comparatively higher and much more complex. Rapid population growth and increased human activities have resulted in the overexploitation of the environment, and has threatened the ability

DOI: 10.4018/978-1-7998-8434-7.ch033

of the agriculture sector to provide food as well as nutrition for the people (Welch et al, 1999). There are increasing concerns that the agricultural production system must be supported by modern technologies that are ecologically sound, eco-friendly and economically viable.

Satisfying increased demands on agriculture with existing farming practices is likely to pose more competition for natural resources and increased use of agrochemicals. This resource intensive farming system have resulted in massive deforestation, soil degradation, nutrient depletion, high levels of greenhouse gas emissions, loss of biodiversity, water scarcity, carbon foot-print, natural resource depletion etc which are considered as major challenges to the sustainability of the world's agriculture. For sustainability there is a need to increase food production while at the same time it is required to decrease the negative impacts of agriculture on land, water, and climate (Foley et al, 2011).

There is requirement of sustainable as well as innovative agriculture systems that protect and enhance the natural resource base, and at the same time increasing productivity (Murgai et al., 2001). A sustainable agricultural system is that where high quality of food is produced without degrading the quality of environment. Indiscriminate use of various agrochemicals such as insceticides, pesticides, weedicides, chemical fertilizers etc. may cause undesirable effects not only to the agricultural ecosystem but also to human health due to their persistent nature. Global warming and rising temperature is also a major detrimental factor for crop yield as well as the nutritional quality of the crop (Leeuwis and Hall, 2010). At higher temperatures plants develop more quickly, leaving less time to accumulate human nutrients such as sugars, fat, and protein (Fedoroff et al, 2010). Similarly, rising CO_2 levels are also predicted to lower the nutritional quality of certain crops, lowering mineral and protein content (Myers et al, 2014). Interestingly research have shown that future increases to atmospheric CO_2 will also impact the nutritional protein content of crops (Medek et al, 2017) as well as minerals such as zinc and iron (Medek et al, 2017; Myers et al, 2014; Myers et al, 2017) Hence, to maintain or improve an individual's nutritional profile, it will remain important to make available additional varieties of nutrient-rich foods.Thus, eco-friendly as well as economically viable technological improvements are needed which apart from curtailing our dependency on chemical based agro products and agricultural fossil fuel, would help address undesirable effects of climate change and the intensification of natural hazards (CEO Council, 2008). In this regard the use of green technology plays a vital role as its having the potential to meet the requirements of sustainable agricultural (Ghadiyali and Manish, 2012. Dwivedi et al 2017).

Green technology:Global collaborative encyclopedia, Wikipedia, defined Green technology as ''the application of one or more of environmental science, green chemistry, environmental monitoring and electronic devices to monitor, model and conserve the natural environment and resources, and to curb the negative impacts of human involvement'' (http://en.wikipedia.org). The Organization for EconomicCo-operation and Development (OECD) has defined the term ''green growth''as ''the pursuitof economic growth and development, while preventingenvironmental degradation, biodiversity loss and unsustainable natural resource use'' (Diretto, 2007).

Green Technology (GT) is eco-friendly technology that reduces environmental damage (OECD, 2010). Various environmental technologies, used for monitoring and assessing environmental pollution, their prevention, control and remediation, are taken in under green technologies to reduce the human impact on the environment and create ways of sustainable development (Acemoglu et al, 2012; Soni. 2015). Green technologies are sustainable technologies which will not create footprint when used for various processes/applications. Green inventions often involve - renewable resources, energy efficiency, recycling, safety and health concerns etc. Green technologies support the use of natural organic resources and avoid production of green gasses and aim to reduce waste, pollution, and use of fossil fuel in agriculture. They also consume less resource and do not support any kind of environmental degradation in agricultural processes. They support automation of every process, hence improves the lifestyle of the people and contribute for human comfortability.

Different Approaches for Green Technology

Green technology is a combination of various technologies that can operate in diversified areas such as genetic engineering, IPM, organic farming, biofuel, ecoforestry, renewable energy, and solid waste management etc. All these ground- breaking technologies are having a great potential to drastically modify modern agriculture by keeping pace with the sustainability. More over these technologies can contribute towards the production of high-quality, organic and environmentally safe food. Small-scale farmers would also be get benefited with the use of these advanced technologies due to increase in their income by producing a high-quality, marketable farm produce. However, it is neither viable nor required to adopt all the available technologies at one time without considering their specific strengths and weaknesses.Organization for Economic Cooperation and Development (OECD, 2011) grouped these wide range of approaches for green technology applicable in the field of agriculture into four major groups -

1. **New Science and Generic Technologies with Green Potential:** Major technologies included in this group are mainly biotechnology or Genetically modified (GM) or Genetically Engineered (GE) crops (Conway, 1999; Arundel and Sawaya, 2009)., Communication Technology and its applications/ Precision Farming techniques/ Precision Agriculture Information and Techniques for sustainable Bio Production/ Biofortification (Bouis and Saltzman, 2017) of major crop plants. Use of these advanced technological innovation can improve the environmental performance of farming systems Technologies, included in this group can reduce and protect ground or surface waters, reduce the use of known toxins in agricultural production, conserve natural habitats, reduce nutrient loads in soils, lower loss of gaseous nitrogen and reduce the amount of non-renewable energy used in the agricultural cropping cycle.
2. **Green technologies for Farming System Innovations:** All those technologies that can increase agricultural productivity and profitability, having capabilities of reducing environmental degradation as well as conserving natural resources and simultaneously can reduce adverse environmental impacts are included in this group. Green technologies in this group includes Integrated Pest Management (IPM) (Singh and Sharma, 2004), Systems of Rice Intensification (SRI) (Prasad, 2006)., Organic Production Systems (Scialabba and Müller-Lindenlauf, 2010;IFOAM, 2010), Conservation Agriculture/zero tillage/Crop Rotation (Trigo et al, 2009), Water Management Systems (OECD, 2010b), Natural Resource Management (management of natural resources such as land, water, soil, plants and animals) (De Tourdonnet et al, 2010), Urban/Peri-Urban Agriculture in the form of Micro-Gardens, Roof Gardens/ food mills etc (DFID, 2006).
3. **Green technologies for Integrated National Green Regimes:** The technologies those are considered as part of national/regional green agenda like Biofuels, agrotourism, use of renewable energies, Integrated Food Energy Systems (IFES) etc. These technologies mainly focusing to save energy and its renewal uses and simultaneously addressing the production of environmental safe food and energy (Chel and Kaushik, 2011; Ministry of New and Renewable Energy, Annual Report, 2015-16).

4. **Cross-cutting techniques toward greenness:** The fourth group is a cross-cutting mode examines that under what circumstances market or policy-driven mechanisms are most suited to drive innovation in pursuit of a green agenda. These includes Policy-Driven Mechanisms, Voluntary Sustainability Initiatives (includes market-based approaches having ability to generate new markets and investment for sustainable practice) (Vojtech, 2010; Hall et al, 2010).

Choice for a Suitable Green Technology for Sustainable Agriculture

Since a lot of different advanced, technologies are available identifying appropriate technology suitable for sustainable agriculture is a specific challenge in their specific use. To remain sustainable, the green technology should be efficient, practical, cost effective and free from pollution. For adopting a suitable green technology for sustainable agriculture integrates three major goals- environmental health, economic profitability, and social and economic equity. Thus, the sustainability factor of these different groups of approaches should be looked at their ability to maintain acceptable levels of production over a long period of time, without degrading the environment. In this context, agricultural sustainability should seek to maximize food production within constraints of profitability. A paradigm shift is occurring from the current production/productivity goals to developing nutritionally enhanced and resource use-efficient crops. Biotechnological tools are considered as to be one of the 'greenest' tools available to help farmers better provide the food, fuel and fiber to serve a growing population.

It can be argued that the major twin problems of food insecurity and dire poverty in Africa will not go away until water security is achieved, to allow rural populations to have adequate year round water supplies for safe drinking and irrigated agriculture. As priorities, improving crop and animal genetics to cope with environmental and biotic stresses, such as water deficits, heat and animal diseases, should be part of an integrated holistic approach for enhancing agricultural sustainability and poverty eradication. Application of nutritional genomics for biofortification of food staples has great potential to reduce hunger and malnutrition on the continent. To attain these benefits, increased educational and research efforts are needed by government agencies, charities and foundations for development, enhancement and/or adoption of genomics, computational biology and bioinformatics tools in agriculture, food and healthcare delivery systems. Furthermore, partnerships need to be forged between African research institutions and international, public and private sector genomics and breeding efforts. The New Partnership for Africa's Development (NEPAD) can play a key role in providing stewardship of this effort, by facilitating the generation and use of cutting-edge science and technology by th

Role of Biotechnology as 'Greenest' Tools of GT

Biotechnology is a field of applied biology that has a high potential for green agriculture but its use has proved highly controversial.Biotechnology provides tools and technologies that provide solutions to many of today's global ecological problems/ environmental challenges and could result in an emerging bioeconomy(Arundel and Sawaya, 2009; OECD, 2009c). The potential environmental benefits offered by biotechnology throughout the world are also very significant.Agricultural biotechnology provides solutions tovarious environmental challenges -

1. Biotechnology is now globally used to grow herbicide and insecticide/pesticides resistant crop plants, allowing the farmers to almost completely eliminate plowing on their fields (no till farming),

in turn enhancing soil health, improved moisture content retention, reducing soil erosion, limiting CO2 emissions, reducing the amount of insecticide/pesticides that need to be used and decreased herbicide runoff. This can help reduce carbon emissions and soil erosion, thereby reducing the negative impact on the environment. Based on the amount of farm land that has been converted from full-till to reduced and zero-till systems, biotechnology has resulted in an 8,053 million kilogram reduction in CO_2 emissions from agriculture, which is equivalent to 3,579,298 average-sized family cars (Brookes and Barfoot, 2008). Global pesticide applications decreased six percent in the 10 years after biotechnology derived crops were first introduced, eliminating 379 million pounds of pesticide applications (OECD, 2009c).

2. Biotechnology derived crops are improving water quality both through less herbicide and pesticide in runoff from fields, and in the future also through reducing phosphorus excretion in livestock by using biotechologicallyproduced feed that contains reduced levels of phytate.
3. Increasing production yields, thereby reducing pressures to force more land, often marginal and highly erodible land, into production.

IMPACT OF BIOTECHNOLOGY ON HUMAN HEALTH

Food security and adequate nutrition are two crucial aspects of sustainability. In addition to addressing hunger through increasing production and availability of food, lately, attention has focused on developing nutrient-rich crops in response to consumer demand for high nutritional quality foods. Substantial improvement of the nutritional content of food will likely require new germplasm generatd using advanced biotechnological approaches. To increase agricultural productivity to a large extent, there is need to develop new crop germplasm wit henhanced nutritional content. Biotechnology may help address critical nutritional deficiencies by developing a second generation of GE crops which are also called as nutritionally enhanced GE crops. These nutritionally enhanced GE crops can address nutritional requirements directly by contributing to multipoint intervention strategies (Yuan et al, 2011). Plant components of dietary interest can be broadly divided into four main categories- i) macronutrients (proteins, carbohydrates, lipids [oils], fiber); ii) micronutrients (vitamins, minerals, functional metabolites); iii) antinutrients (substances such as phytate that limit the bioavailability of nutrients); and iv) allergens (intolerances and toxins). From a health perspective, the first two dietary components need to be enhanced and the latter two to be limited or removed.

Diet- related factors, which predominantly reflect interactions among micronutrients and other organic compounds, have a profound effect on the efficacy of a nutritionally enhanced crop (Graham et al, 2001). Such factors include the chemical form of the nutrient (Combs, 2001; Monsen and Balintfy, 1982) leading to improved nutritional and/or health characteristics. Genetically modified food and feed, obtained either through extensive genetic modifications/ incorporation of a complete metabolic pathways (such as rice containing beta-carotene, soybeans with enhanced oleic acid content, or tomato with increased concentration of flavonoids) or through incorporation of a new gene producing a novel protein into a crop, there may be chances of having an allergic reaction to that protein(s) in human beings (Byrne et al, 2004).

Additionally, these genetically or metabolically engineered plants, with desired quality traits, in general could potentially lead to unexpected or unintended changes in concentration of toxic substances or nutrient composition of plants. Although progress in dissecting metabolic pathways and our ability to manipulate gene expression in genetically modified (GM) plants has progressed at a greater pace, attempts to use these tools for engineering plant metabolism have not quite kept pace. The success of this approach depends on the ability to account for the changes formed by these modified dietary metabolite on host metabolism. Thus, the ultimate success of any genetic modification in plants can only be judged by its impact on human health and nutrition after deployment (Konig et al, 2004) leading to the application of these nutritionally GM products highly controversial, as many consumers remain skeptical because of socio-psychological factors associated with potential food risks.

Nutrition, Health, and Global Food Demand

The paradigm of agricultural development based on maximizing grain yield in major cereal crops (Stevenson et al., 2013) led to increased production of high-yielding bread wheat, rice, and maize and replacement of other more nutrient-rich cereals (DeFries, et al. 2015). Between 1961 and 2013, the land area planted with wheat, risce, and maize increased from 66% to 79% of all cereals (FAO, 2015), while the land area planted with other cereals, having higher nutrient content, such as small millets, barley, oats, rye, sorghum etc. declined from 33% to 19%. As a result, between 1961 and 2011, the energy density of the cereal supply remained constant but the protein, iron, and zinc contents in the global cereal supply declined by 4%, 19%, and 5% respectively, with an overall decline of the nutrient-to-calories ratio (DeFries, et al. 2015).Nutritional quality has been compromised by the emphasis on edible yield and through the loss of biodiversity (Charrondiére et al.,, 2013) caused by the introduction of high-yielding uniform cultivars and breeds: today 95% of the world's calories come from 30 species, but almost half of the global calorie demand is supplied, as noted above, by three crops, namely maize, rice, and wheat. Of 30 000 edible species, only about 150 are grown. This loss of diversity alone has had significant negative health consequences (WHO, 2015). Therefore, to meet the growing demand for safe and nutritious food, it is urgent to promote agro-biodiversity on farms or through local production. We can now grow enough food for today's needs and even those for the world's population in 2050. Persistent and widespread malnutrition and hunger remain a huge challenge in many parts of the world. The current world population of 7.6 billion is expected to reach 8.6 billion in 2030 and 9.8 billion in 2050, with roughly 83 million people being added to the world's population every year (UN, DESA https://www.un.org, 2017). Estimates indicate agricultural production must as much as double to meet projected demands for food (Foley et al, 2011). More food is needed and food quality must be improved, particularly regarding nutrient content (Tester et al, 2010). It is time when the focus is on 'healthy foods from healthy soils'. The current rate of progress will not be enough to eradicate hunger by 2030, and not even by 2050. However, while the importance of the linkages between agriculture and nutrition are increasingly recognized, food and nutrition security cannot be achieved through increased agricultural production alone (Fan and Pandya-Lorch, 2012). An important step in this direction is a new path for agricultural development, where agricultural growth is used not only to increase food production but also to enhance nutrition and health (Shenggen Fan,2015). It is necessary to produce sufficient, safe, and nutritionally enhanced food as now the challenge is one of health (FAO, 2018).

The notion that quantity and quality cannot progress together is not true in several crops for micro-nutrients such as iron, zinc, and vitamin A.Although subject to debate (Newell-McGloughlin, M. 2008), a number of nutrition studies have linked diet with certain human health maladies–cancer, osteoporosis, diabetes, hypertension, cardiovascular health, and age/lifestyle-related diseases (Block et al, 1992; Potter and Steinmetz, 1996; Botella-Pavía and Rodriguez-Conceptíon, 2006; Mattoo et al, 2010). Diet fortification with vitamins and antioxidants is a solution to enhanc human nutrition, but the potential of a nutrient is realized through its bioavailability to human beings (Rao and Agarwal, 1995; Milner, 2008). Nutrition needs to be personalized for it to be effective long-term. Diagnostic labs that assess genetic information, as well as functional biomarkers, can be utilized for this purpose.

Nutrients and Gene Interactions

There has been an increasing focus on using diet to maintain health. Fundamental to better health is understanding nutrient- gene interactions in growth and development and in disease prevention.The dietary constituents can affect the structure and expression of gene(s) and have impact on transcription as well as on DNA structure. The genetic makeup can be a major factor contributing to the degree of a diet's effects on health and disease. Although genes are critical for determining expression and function, nutrition modifies the extent to which different genes are expressed and thereby modulates whether individual will be able to attain the potential established by their genotype/genetic background or not. The nutrients- gene interactions can be grouped in three categories-

1. **Direct interactions:** Here nutrients act as a ligand for transcription factor receptors. and have direct relationship with DNA to modify genetic expression.
2. **Epigenetic interactions:** In epigenetic interactions, the nutrients can alter the structure of DNA (DNA methylation and chromatin remodeling), and thus affecting gene expression
3. **Genetic variation:** In the genetic variations the single nucleotide polymorphism (SNPs), within humans that relate to the variations between humans in their response to diet, can alter the expression or functionality of genes.

The diverse tissue and organ-specific effects of bioactive dietary components can be assessed with help of different tools, including transcriptome (gene-expression patterns); proteome (protein-expression patterns, including posttranslational modifications); metabolome (metabolite profiles); as well as epigenome (organization and modification of the chromatin). All thesetools to studythe impact of nutrient –gene interaction were not available at earlier time and are only now becomingavailable after discovery of "omics technology" enabling the development of Nutrigenomics. The knowledge in this field not only offer new insight into how genetics shape our nutritionalneeds but also significantly improve understanding of the molecular mechanism of thefood components.

NUTRIGENOMICS

Recent advances in molecular biology combined with the wealth of information generated by the Human Genome Project have fostered the emergence of 'Nutrigenomics', which is one of the most exciting new areas in the food and wellness.This new wave of molecular nutrition, which unfolds the mystery behind nutrient-gene interactions, may provide the strategies for developing safe and effective nutritherapy for individuals (Afman and Muller, 2006; Ferguson, 2006).

Nutrigenomics is the study of how DNA and our genetic code affect ourneed for certain nutrients and help maintain optimal heath throughout our life. The concept relies on the fact that foods we eat provide a feedback stimulus to our DNA and trigger our genes to respond accordingly. Since it studies the interactions between different types of food and the genome, it can aid in selecting foodstuffs that are vital to health (Ouhtit, 2014). Knowing the right amount and the type of specific nutrients that our body requires would positively impact the management of our health (German, 2005). Nutrigenomics is addressing why some people can control disease with diet, whereas others require drugs. Therefore, nutrigenomics is expected to significantly contribute to personalised medicine (Miggiano and De Sanctis, 2006).

Nutrigenomics has the potential to impact various parts of the food components and food chain including the genetic modification of the crop plants and animal feeds.The Nutrigenomics study represents a novel and powerful approach in nutritional sciences which not only elucidates the underlying molecular mechanisms but also being used to determine the extent and effectiveness of the genetic modification introduced(Davidovich-Rikanati et al, 2007; Direttoet al, 2007). Simultaneously, it evaluates the health and safety concerns related to the consumption of genetically modified dietary nutrients (Catchpoleet al, 2005; Dixon et al, 2006; Kutchan, 2005).

Thus, nutrigenomics mainly focuses on the effect of nutrients on the genome, proteome, and metabolome and has the potential in disease prevention, disease mitigation and ways in treating chronic disease (Davis. 2012).

Definition of Nutrigenomics

Initially nutrigenomics referred to the study of the effects of nutrients on the expression of an individual's genetic make up. The term 'Nutrigenomics' was used by Della Penna (1999), who defined Nutrigenomics as the general approach to the gene discovery that is most applicable to the compounds of nutritional importance that are synthesized or accumulated by plants or other organisms. The term "nutrigenomics" was first described by Pelegrin (2001) and then, by Van Ommen and Stierum (2002). Later, Nutrigenomics was defined as the study of how genes and gene products interact with dietary chemicals to alter phenotype and, conversely, how genes and their products metabolize nutrients (Muller and Kersten, 2003; Kaput et al, 2005a). An international group of biologists, ethicists and sociologists (Kaput et al, 2005b) defined Nutrigenomics as 'a multidisciplinary approach for the comprehensive investigation of the influence of diet and individual genetic variation as risk factors for chronic diseases'. In 2006, Mariman (Mariman, 2006) defined Nutrigenomics as the study of the effect of nutrition or dietary components on the transcriptome of cells and tissues. Sales et al. (2014) defined Nutrigenomics as an emerging science which investigates a certain area of nutrition that uses molecular tools to search, access and understand the several responses obtained through a certain diet applied between individual and population groups. Recently this definition has been broadened to encompass nutritional factors that protect the genome from damage (Wellen and Hotamisligil, 2005).

Thus, the early definition of Nutrigenomics referred only to studies on the effects of nutrients/bioactive food on gene expression of an individual. Nowadays, this definition expanded and recently, Nutrigenomics seeks to understand the influence of dietary components on the Genome, Transcriptome, Proteome, and Metabolome (Ronteltap et al, 2009; Cozzolino, 2012; Cozzolino and Cominetti, 2013).

Different Phases of Development of Nutrigenomics

Role of nutrition on human health was noticed way back to 400 BC. It was well understood by the quote of Hippocrates "to let food be thy medicine and medicine be thy food." Development on research on nutrigenomics acan be grouped into following five distinct developmental phases/periods-

1. Period of Analytical Chemistry: This was the first phase in nutritional research that began around 1700 AD (Cruz et al, 2003). How food was metabolized by the body to generate water, carbon dioxide, and energy, was first discovered during this period known as period of analytical chemistry (Vasconcelos, 2010), when the role of various macromolecules (such as carbohydrates, proteins, lipids, etc) was identified in releasing the heat (Cruz et al, 2003).
2. Chemical and Analytical period of Nutrition: The period between the 18th and 20th centuries was called as Chemical and Analytical period of Nutrition. During this period of nutritional research, important discoveries were made on food metabolism and their relation with energy production, including its relevance on breathing and oxidation (Fujii et al, 2010).
3. Biological period: The 19th century was known as the "Biological period." During this period emphasis was led down on the role of nutrients in the development and prevention of chronic diseases, such as cancer, cardiovascular, neurodegenerative, and bone metabolism disorders (Vasconcelos, 2010). During the 20th century, Nutritional Science focused on finding role of vitamins (Ghoshal et al, 2003) and minerals (Menon et al, 2010), defining their use and preventing the diseases (Gopalan, 1992; Godbole et al, 2009; Rao, 2001; Naushad et al, 2010) that they caused due to their deficiency. Later, nutrition related health problems of the developed world shifted to over nutrition, obesity (Raj et al, 2007) and type-2 diabetes (Mohan et al, 2007a; Mohan et al, 2007b). To prevent the development of these types of disease, nutrition research is investigating on how nutrition can optimize and maintain cellular, tissue, organ and whole body homeostasis. This requires understanding the mode of action of these nutrients at the molecular level which in turn involves a massive amount of nutrient-related interactions at gene, protein and metabolic levels.
4. Genomic Period: This period was started soon after the completion of the first draft of the human genome in 2001. With the advances of genome sequencing and the decoding of the human genome (Lander et al, 2001), nutrition research shifted to Molecular Biology and Genetics and as new area of research in nutrition science was emerged which is known as 'Nutrigenomics' (Peregrin, 2001). By this time, the researches in nutritional research were focused on identifying and understanding molecular-level interactions between ingested bioactive nutrients, and other dietary compounds. Thus, Nutrigenomics studies would provide a basis for the development of safe and effective diet therapies to rationally and specifically improve health in individuals or subgroups of the population (Peregrin, 2001). The advent of 'omics' technologies have provided the tools for the understanding of diet-gene interaction at the molecular level. This lead the Nutrigenomics towards the study of effect of nutrients on genome, transcriptomic, proteome as well as on metabolome. Thus, Nutrigenomics involves in characterization of gene products, their physiological function and their interactions i.e. how chemical molecules in food can affect gene expression in an individual and that this influence may vary between individuals with resulting health consequences.
5. Post-Genomic period: The present phase of nutritional research is called post- genomic period which is characterized by the integration of three fields: biological, social, and environmental, where scientific discoveries on nutritional pathophysiology and metabolism are included (Fialho et al, 2008; Norheim et al, 2012).

Scientific Disciplines Contributing to Nutrigenomics

Nutrigenomics is an integrative, multidisciplinary science that seeks to provide a genetic and molecular understanding for how common dietary chemicals affect the balance between health and disease by changing the structure and/or expression of an individual's genotype. This is the result of three factors that have led to a growing realization that the effects of nutrition on health and disease cannot be understood without a profound understanding of how nutrients act at the molecular level. By understanding the interaction between nutrients and the genome, better dietary regimens may be designed to address the treatment of important diet related diseases . Nutrigenomics uses many types of tools to understand effect of nutrients on health as well as to identify disease risk and progression.Such tools include all those methods thatmeasure the whole genome, modifications to the genome, transcriptome,proteome, metabolome, genetic variation at single nucleotide level and all the data pertaining to the methodsthat improve diet and health. These tools help identifying genes, proteins, or metabolites that change in the state of prediseases, as well as assisting on recognizing and characterizing the pathways regulated by nutrients or bioactive compoundsin foods (Fialho et al, 2008; Lorenzo, 2008; Liu and Qian, 2011; Dalmiel, et al., 2012).

- **Transcriptomics:** In nutrition research, transcriptomics is a very valuable way of beginning to understand how nutritional exposure influences gene expression on a genomic scale. To identify differentially expressed genes under different conditions techniques of transcriptomics are being utilized. This leads to the understanding of the genes or pathways associated with a particular nutrient or diet. Transcriptom analysis is an essential component in understanding normalcy and disease at many levels. Different modern techniques like Expressed sequence Tag (EST), Serial Analysis of Gene Expression (SAGE), DNA microarray, Exon array, Tiling arrays etc. have efficiently been used to measure the transcriptome analysis.
- **Proteomics:** investigate differential protein expression under different conditions. Apart from analyzing gene expression in tissues/ organs, proteomics can also be used for the identification of presence or absence of certain key proteins. This can give information about the early stages of disease by identifying new protein biomarkers. Proteomics uses a set of technologies designed to study the expression of proteins. However, the tools that measure the proteome are comparatively less developed, these utilizes methods based on gel electrophoresis, chromatography and mass spectrometry and protein profiling, protein chip, reverse phased protein microarry etc. (Cozzolino, 2012; Costa and Rosa, 2011; Dalmiel, et al., 2012; Mahan and Scott Stump, 2005).
- **Metabolomics:** examines global patterns of metabolites present in the cell or in body fluids in response to specific dietary environmental and genetic exposures (Liu and Qian, 2011; Norheim et al, 2012). The tools that measure the metabolome are also less developed and research advances in this area may facilitate the understanding of how the genotype is related to the phenotype of an individual. The tools that measure the metabolome include methods based on Nuclear magnetic resonance imaging as well as gas and liquid chromatography. Once the metabolome allows about theknowledge of the arrangements and metabolic disorders caused due to a person's diet it will be applied to cure those changes/disease (Rocha, et al, 2006; Ong and Rogero, 2009; Ronteltap et al, 2009; Cozzolino, 2012;, Dalmiel, et al., 2012).
- **Epigenetics:** which means "above the genetics", is the study of modifications to the genome which are copied from one cell generation to the next but which do not involve changes to the primary DNA sequence.DNA methylation and changes to chromatin packaging of DNA by post-translational modifications of histone proteins(Klose and Bird, 2006; Talbert and Henikoff, 2006), regulation by non-coding RNAs, such as microRNAs (Jirtle and Skinner, 2007), are the most

extensively investigated mechanisms. The expression profiles of epigenetically controlled genes reveal early biomarkers for disease.

- **Bioinformatics:** Recent advances in "omics" research have resulted in the creation of large datasets that were generated by consortiums and centers, small datasets that were generated by individual investigators, and bioinformatics tools for mining these datasets. It is important for nutrigenomics studies to take full advantage of the analysis tools to interrogate datasets for information relevant to genomics, epigenomics, transcriptomics, proteomics, and metabolomics

Aims of Nutrigenomics

Nutrigenomics examines relationships among genes, diet, and health. Specifically, nutrigenomic research is "the study of how foods affect the expression of genetic information in an individual and how an individual's genetic makeup affects the metabolism and response to nutrients and other bioactive components in food."(Kaput and Dawson, 2007). Since, different diets bring forth a variety of gene and protein expression and metabolite production, nutrigenomics described such diet-induced patterns and their effects as "signature dietary patterns". The main aim of nutrigenomics is to understand the mechanism of this signature dietary patterns by amalgamating all the available technologies for the study. Subsequently, this can be used to determine naturally occurring chemical agents in food that could prevent the onset of diseases such as obesity, type-2 diabetes and cancer (Ordovas and Mooser .2004; Muller and Kersten, 2003; Kauwell, 2005; Stover, 2006). . To achieve this long term goal:

1. The genes and gene variants that are affected by varying amounts of nutrients are first identified, and their role in regulation is then studied.
2. Since every individual has their own unique genetic makeup (genotype) and the effect of nutrient variation is distinct for everyone, genotypes associated with diet-related disease are then identified.
3. Increasing the understanding about the influence of nutrients/bioactive food components on health and overall metabolism and maintenance of a constant internal environment of the body(homeostasis).
4. demonstrating the positive effects of functional foods on health, modify and improve dietary pattern for the treatment or prevention of disease at group and individual levels and further use of this understanding to prevent the onset of chronic conditions, especially cancer, obesity and type II diabetes; for educating people and promoting their use, particularly among those with inadequate nutrition.
5. Eventually, nutrigenomics will lead to evidence-based dietary intervention strategies for restoring health and fitness and for preventing diet-related disease.
6. As nutrigenomics research reveals more details about the roles of various nutrients, considering these findings into account for future research and development the agricultural biotechnology industry will be able to modify nutrition related traits in nutritionally enhanced GM crops.

Challenges Faced by Nutrigenomics

Applying the concepts of nutrigenomics to individuals is challenging because of the diversity of genetically individual responses to the complexity of foods, and the study design of health or disease processes. While some DNA is common to all people, there are variables within genes known as polymorphisms. When they occur they may be responsible for differences in the protein product of that gene, proteins that play key roles in important body processes. Current research has determined how genetic variation affects our body's ability to utilize nutrients and remove toxins. One's health is a result of how these processes interacts with environmental and lifestyle factors such as diet, exercise, stress, smoking andalcohol. By learning about the specific nature of genes, we will find out how their variations (polymorphisms) affect the way in which our body manages its nutrition and detoxification processes. By testing for these variations one can learn if these body processes are below their optimum level. Every individual has a predetermined susceptibility to disease based on his/her genetic profile As with our understanding of the very nature of DNA, nutrigenomics will find its way into our life by evolution not by revolution.

Although there is growing expectation that nutrigenomic research will improve individual and group health through personalized nutrition **(Lampe, 2006; Kaput, 2007)** the field faces several methodological challenges, including: defining and measuring dietary intake; shifting outcomes of interest from actual diseased condition to biomarkers identification indicating early stages of disease incidences; analyzing phenotypic, genetic, molecular, dietary, and clinical data simultaneously to describe for the full effects of food. These challenges contribute to incoherent findings across genetic association studies and complicate the development of nutrigenomic interventions to improve individual health.Further, the completion of several large genome projects has markedly altered the research agenda by drawing attention to the importance of genes in human nutrition, and has provided a wealth of new genetic information to be explored.

The Expected Benefits from Nutrigenomics

The expected benefits from nutrigenomics are tremendous and cover:

- a better understanding of the toxicity and safety profile of macro- and micronutrients/ bioactive component of the food
- useful in the prevention of certain diet-associated diseases
- the enjoyment of otherwise less healthy food by individuals whose health is not likely to be affected
- the avoidance of unhealty dietary supplements that are routinely used by certain people
- the prevention of diseases and an increase in life expectancy.
- Identifying rate-limiting steps in synthesis could provide targets for genetically engineering biochemical pathways to produce improved amounts of compounds and new compounds.
- Targeted expression will be used to channel metabolic flow into new pathways, while gene-silencing tools can reduce or eliminate undesirable compounds or traits or switch off genes to increase the amount of desirable products (Liu et al., 2002; Herman et al., 2003; Davies, 2007)

CONCLUSION

In the past, food production has been increased almost to double through substantial use of synthetic fertilizer, pesticides, and irrigation, all at significant environmental cost. Future production of nutrition-dense food will require next-generation crop production systems with decreased reliance on synthetic fertilizer and pesticide using techniques based on gre en technology. Morenutritiousfoodsarealsoneed of the time.A substantial improvement of the nutritional content of food has been claimed to be achieved using advanced biotechnological approaches. Thus, it would be appropriate to mention thattoday we live in a nutritional environment that is quite different from the environments to which we had genetically adapted. Therefore, to improve our knowledge on the agriculture-nutrition-health nexus, we need to learn more about the health and nutritional impacts of different technological benefits of agricultural sub sector development. Differences in genetic makeup (genotype), as well as nutrient deficiencies, are factors in various diseases, including cancer, cardiovascular, digestive and inflammatory diseases. Nutrigenomics is addressing why some people can control disease with diet, whereas others require drugs. Therefore, nutrigenomics is expected to significantly contribute to personalised medicine. Important steps include the development and dissemination of more nutritious, biofortified food, availability of advance techniques to study the impact of different components of food on human health and their future applications. Furthermore, knowing the right amount and the type of specific nutrients that our body requires would positively impact the management of our health. Although studies conducted in this field have been costly and difficult, the results are deemed to be helpful for the advancement of medical technologies pertaining to health care. Once the required knowledge is attained this can also be utilizedto design agricultural strategies to minimize risks and maximize the benefits to nutrition and health across the entire value chain, from production to consumption.

REFERENCES

A Smarter Shade of Green: How Innovative Technologies Are Saving Energy, Time and Money. (2008). Technology CEO Council.

Acemoglu, D., Aghion, P., Bursztyn, L., & Hemous, D. (2012). The environment and directed technical change. *The American Economic Review*, 131–166.

Afman, L., & Müller, M. (2006). Nutrigenomics: From Molecular Nutrition to Prevention of Disease. *Journal of the Academy of Nutrition and Dietetics*, *106*(4), 569–576.

Arundel, A., & Sawaya, D. (2009). *Biotechnologies in Agriculture and Related Natural Resources to 2015. OECD Journal.*

Avensblog. (2016). *Nutrigenomics – Eating for better genes*. Avensblog.

Block, G., Patterson, B., & Subar, A. (1992). Fruit, vegetable and cancer prevention: A review of the epidemiological evidence. *Nutrition and Cancer*, 1–29.

Botella-Pavía, P., & Rodriguez-Conceptíon, M. (2006). Carotenoid biotechnology in plants for nutritionally improved foods. *Plant Physiology*, *126*, 369–381.

Bouis, H. E., & Saltzman, A. (2017). Improving nutrition through biofortification: A review of evidence from HarvestPlus, 2003 through 2016. *Global Food Security*, *12*, 49–58.

Brookes & Barfoot. (2008). *GM crops: Global socio-economic and environmental impacts 1996-2006*. PG Economics Ltd.

Byrne, P., Sarah, W., Harrington, J., & Fuller, L. (2004). *Transgenic Crops: An Introduction and Resource Guide*. Department of Soil and Crop Sciences, Colorado State University.

Catchpole, G. S., Beckman, M., & Enot, D. P. (2005). Hierarchical metabolomics demonstrates substantial compositional similarity between genetically modified and conventional potato crops. *Proceedings of the National Academy of Sciences of the United States of America*, *102*(40), 14458–14462.

Charrondiere, & Stadylmayr, & Rittenschober. (2013). FAO/INFOODS food consumption database for biodiversity. *Food Chemistry*, *140*(3), 408–412.

Chel, A., & Kaushik, G. (2011). Renewable energy for sustainable agriculture. *Agronomy for Sustainable Development*, *31*(1), 91–118. doi:10.1051/agro/2010029

Combs, G. F. (2001). Selenium in global food systems. *British Journal of Nutrition*, *85*(5), 517–547.

Conway. (1999). *GM Foods can Benefit the Developing Countries*. Academic Press.

Costa, N. M. B., & Rosa, C. O. B. (2011). *Functional Foods: Bioactive Components and Physiological Effects. 1 Reprint*. Rúbio.

Cozzolino, S. M. F. (2012). *The Bioavailability of Nutrients* (4th ed.). Monole.

Cozzolino, S. M. F., & Cominetti, C. (2013). *Biochemical and Physiological Bases of Nutrition in Different Stages of Life in Health and Disease* (1st ed.). Monole.

Cruz, I. B. M., Taufer, M., & Schwanke, C. H. A. (2003). Genomics in the era of aging and its potential application in gerontology and geriatrics. In A. C. A. Souza (Ed.), Institute of Geriatrics and Gerontology PUCRS: The Cradle of Academic Geriatrics In Brazil (pp. 83–84). Academic Press.

Dalmiel, L., Vargas, T., & Molina, A. R. (2012). Nutritional genomics for the characterization of the effect of bioactive molecules in lipid metabolism and related pathways. *Electrophoresis Journal*, *33*(15), 2266–2289.

Davidovich-Rikanati, R., Sitrit, Y., & Tadmor, Y. (2007). Enrichment of tomato flavor by diversion of the early plastidial terpenoid pathway. *Nature Biotechnology*, *25*, 899–901.

Davis, U. C. (2012). *Nutrigenomics is a multidisciplinary science*. The NCMHD Center of Excellence for Nutritional Genomics.

De Lorenzo, D. (2008). Nutrigenomics: For a personalized nutrition. Journal Digitalis, 22, 1–3.

De Tourdonnet, S., Triomphe, B., & Scopel, E. (2010). *Ecological, Technical and Social Innovation Processes in Conservation Agriculture*. Paper presented at Innovation and Sustainable Development in Agriculture and Food (ISDA), Montpellier, France.

DeFries, R., Franzo, J., Remans, R., Palm, C., Wood, S., & Anderman, T. L. (2015). Metrics for land-scarce agriculture. *Science*, *349*(6245), 238–240.

DFID. (2006). *A Synthesis of Urban and Peri-Urban Agricultural Research Commissioned by the RNRRS Programme*. Department for International Development.

Diretto, G., Al-Babili, S., Tavazza, R., Papacchioli, V., & Giuliano, G. (2007). Metabolic engineering of potato carotenoid content through tuber-specific overexpression of a bacterial mini-pathway. *PLoS One*, *2*, e350.

Dixon, R. A., Gang, D. R., & Charlton, A. J. (2006). Applications of metabolomics in agriculture. *Journal of Agricultural and Food Chemistry*, *54*(24), 8984–8994.

Dwivedi, N., Kumar, P., Dwivedi, S. K., Kumar, P., Sonowal, K., & Singh, J. (2017). Green technology: An eco-friendly approach towards sustainable agriculture. (G-. *Journal of Environmental Science and Technology*, *4*(4).

Fan, S. (2015). *How agriculture can improve health and nutrition*. World Economic Forum Report.

Fan, S., & Pandya-Lorch, R. (2012). *Reshaping Agriculture for Nutrition and Health*. International Food Policy Research Institute.

FAO, IFAD, UNICEF, WFP, & WHO. (2018). The State of Food Security and Nutrition in the World 2017. FAO.

Ferguson, L. R. (2006). Nutrigenomics: Integrating genomic approaches into nutrition research. *Molecular Diagnosis & Therapy*, *10*(2), 101–108.

Fialho, E., Moreno, F. S., & Ong, T. P. P. (2008). Nutrition in the post-genomics: Fundamentals and applications of omics tools. *The Journal of Nutrition*, *21*(6), 757–766.

Fogg-Johnson, N., & Kaput, J. (2003). Nutrigenomics: An Emerging Scientific Discipline. *Food Technology*, *57*(4), 60–67.

Foley, J. A., Ramankutty, N., & Brauman, K. A. (2011). Solutions for a cultivated planet. *Nature*, *478*, 337–342.

Fujii, T. M. M., Medeiros, R., & Yamada, R. (2010). Nutrigenomics and nutrigenetics: Important concepts for the science of nutrition. *Journal of the Brazilian Society of Food and Nutrition*, *35*(1), 149166.

German, J. B. (2005). Genetic dietetics: Nutrigenomics and the future of dietetics practice. *Journal of the American Dietetic Association*, *2005*, 530–531.

Ghadiyali, T. R., & Manish, M. (2012). Contribution of green technology in Sustainable development of Agriculture sector. *Journal of Environmental Research and Development*, *7*(1A), 590–596.

Ghoshal, S., Pasham, S., Odom, D. B., Furr, H. C., & McGrane, M. M. (2003). Vitamin A depletion is associated with low phosphoenolpyruvate carboxykinase mRNA levels during late fetal development and at birth in mice. *The Journal of Nutrition*, *133*, 2131–2136.

Godbole, K., Deshmukh, U., & Yajnik, C. (2009). Nutrigenetic determinants of neural tube defects in India. *Indian Pediatrics*, *46*(6), 467–475.

Godfray, H. C. J., Beddington, J. R., & Crute, I. R. (2010). Food security: The challenge of feeding 9 billion people. *Science*, *237*, 812–818.

Gopalan, C. (1992). The contribution of nutrition research to the control of under nutrition: The Indian experience. *Annual Review of Nutrition*, *12*, 1–17.

Graham, R. D., Welch, R. M., & Bouis, H. F. (2001). Addressing micronutrient malnutrition through enhancing the nutritional quality of staple foods: Principles, perspectives and knowledge gaps. *Advances in Agronomy*, *70*, 77142.

Gupta, S. (2018). Nutrigenomics: The Science to Unfold the Impact of GM Food on Health. *Food Sci Nutr Technol*, *3*(3), 000150.

Hall, A., Clark, N., & Frost, A. (2010). *Bottom-up, Bottom-line: Development-Relevant Enterprises in East Africa and their Significance for Agricultural Innovation*. RIU Discussion Paper Series. Research Into Use.

Hall, A., & Dorai, K. (2011). *The greening of agriculture: agricultural innovation and sustainable growth*. Report of on OECD Workshop on Agriculture and Green Growth.

IFOAM. (2010). *Organic Agriculture Worldwide: The Main Results of the FiBL-IFOAM Survey 2010*. Presentation at the BioFach Congress, International Federation of Organic Agriculture Movements.

Jaenisch, R., & Bird, A. (2003). Epigenetic regulation of gene expression: How the genome integrates intrinsic and environmental signals. *Nature Genetics Suppl.*, *33*, 245–254.

Jirtle, R. L., & Skinner, M. K. (2007). Environmental epigenomics and disease susceptibility. *Nature Reviews. Genetics*, *8*(4), 253–262.

Julie Casper, C. (2018). *Epigenetics, Nutrigenomics, and Genetic Testing*. nutritionalbalancing.org

Kaput, J. (2007). Developing the promise of nutrigenomics through complete science and international collaborations. *Forum of Nutrition*, *60*, 209–223.

Kaput, J., & Dawson, K. (2007). Complexity of type 2 diabetes mellitus data sets emerging from nutrigenomic research: A case for dimensionality reduction. *Mutation Research*, *622*(1–2), 19–32.

Kaput, J., Ordovas, J. M., & Ferguson, L. (2005a). Nutritional genomics: The next frontier in the post-genomic era. *Physiological Genomics*, *16*(2), 166–177.

Kaput, J., Ordovas, J. M., & Ferguson, L. (2005b). The case for strategic international alliances to harness nutritional genomics for public and personal health. *British Journal of Nutrition*, *94*(5), 623–632.

Kauwell, G. P. A. (2005). Emerging concepts in nutrigenomics: A preview of what is to come. *Nutrition in Clinical Practice*, *20*, 75–87.

Klose, R. J., & Bird, A. P. (2006). Genomic DNA methylation: The mark and its mediators. *Trends in Biochemical Sciences*, *31*(2), 89–97.

Konig, A., Cockburn, A., Crevel, R. W. R., Debruyne, E., Grafstroem, R., Hammerling, U., Kimber, I., Knudsen, I., Kuiper, H. A., Peijnenburg, A. A. C. M., Penninks, A. H., Poulsen, M., Schauzu, M., & Wal, J. M. (2004). Assessment of the safety of food derived from genetically modified GM crops. *Food and Chemical Toxicology*, *42*(7), 1047–1088. doi:10.1016/j.fct.2004.02.019 PMID:15123382

Kutchan, T. M. (2005). Predictive metabolic engineering in plants: Still full of surprises. *Trends in Biotechnology*, *23*(8), 381–383.

Lamp, J. W. (2006). For debate: Investment in nutrigenomics will advance the role of nutrition in public health. *Cancer Epidemiology, Biomarkers & Prevention, 15*, 2329–2330.

Lander, E. S., Linton, L. M., & Birren, B. (2001). Initial sequencing and analysis of the human genome. *Nature, 409*, 860–921.

Leeuwis, C., & Hall, A. (2010). Facing the Challenges of Climate Change and Food Security: The Role of Research, Extension and Communication Institutions. Report Commissioned by the Research and Extension Branch of the United Nations Food and Agriculture Organization.

Liu, B., & Qian, S. B. (2011). Translational regulation in nutrigenomics. *American Society for Nutrition, 2*, 511–519.

Mahan, L. K., & Scott Stump, S. (2005). *Food* (6th ed.). Nutrition & Diet Therapy.

Malkaram, S. A., Hassan, Y. I., & Zempleni, J. (2012). Bioinformatics tools in nutrition research. (American Society for Nutrition. *Advances in Nutrition, 3*, 654–665.

Mariman, E. C. (2006). Nutrigenomics and nutrigenetics: The 'omics' revolution in nutritional science. *Biotechnology and Applied Biochemistry, 44*(3), 119–128.

Mathers, J. (2005). The Science Of Nutrigenomics. In NUTRIGENOMICS - Report of a workshop hosted by The Nuffield Trust and organized by the Public Health Genetics Unit on 5 February 2004. The Nuffield Trust.

Mattoo, A. K., Shukla, V., Fatima, T., Handa, A. K., & Yachha, S. K. (2010). Genetic engineering to enhance crop-based phytonutrients (nutraceuticals) to alleviate diet-related diseases. In M. T. Giardi, G. Rea, & B. Berra (Eds.), *Bio-Farms for Nutraceuticals: Functional Food and Safety Control by Biosensors* (pp. 122–143). Springer. doi:10.1007/978-1-4419-7347-4_10

Medek, D. E., Schwartz, J., & Myers, S. S. (2017). Estimated effects of future atmospheric CO2 concentrations on protein intake and the risk of protein deficiency by country and region. *Environmental Health Perspectives, 128*, 8.

Menon, B., Harinarayan, C. V., & Raj, M. N. (2010). Prevalence of low dietary calcium intake in patients with epilepsy: A study from South India. *Neurology India, 58*, 209–212.

Miggiano, G. A., & De Sanctis, R. (2006). Nutritional genomics: Toward a personalized diet US National Library of Medicine, National institutes of health. *La Clinica Terapeutica, 157*, 355–361.

Milner, J. A. (2008). Nutrition and cancer: Essential elements for a roadmap. *Cancer Letters, 269*, 189–198.

Ministry of New and Renewable Energy. (n.d.). *Annual Report, 2015-16*. Author.

Mohan, V., Sandeep, S., Deepa, R., Shah, B., & Varghese, C. (2007a). Epidemiology of type 2 diabetes: Indian scenario. *The Indian Journal of Medical Research, 125*(3), 217–230.

Mohan, V., Sudha, V., & Radhika, G. (2007b). Gene-environment interactions and the diabetes epidemic in India. *Forum of Nutrition, 60*, 118–126.

Monsen, E. R., & Balintfy, J. (1982). Calculating dietary iron bioavailability: Refinement and computerization. *Journal of the American Dietetic Association*, *80*(4), 307–311.

Muller, M., & Kersten, S. (2003). Nutrigenomics: Goals and strategies. *Nature Reviews. Genetics*, *4*(4), 315–322.

Murgai, R., Ali, M., & Byerlee, D. (2001). Productivity Growth and Sustainability in Post-Green Revolution Agriculture: The Case of the Indian and Pakistan Punjabs. *The World Bank Research Observer*, *16*(2).

Myers, S. S., Smith, M. R., Guth, S., Golden, C. D., Vaitla, B., & Mueller, N. D. (2017). Climate change and global food systems: Potential impacts on food security and undernutrition. *Annual Review of Public Health*, *38*, 259–277.

Myers, S. S., Zanobetti, A., & Kloog, I. (2014). Increasing CO 2 threatens human nutrition. *Nature*, *510*, 139–142.

Naushad, S. M., Radha, A., & Devi, R. (2010). Role of parental folate pathway single nucleotide polymorphisms in altering the susceptibility to neural tube defects in South India. *Journal of Perinatal Medicine*, *38*(1), 63–69.

Newell-McGloughlin, M. (2008). Nutritionally improved agricultural crops. *Plant Physiology*, 939–953.

Norheim, F., Gjelstad, I. M. F., & Hjorth, M. (2012). Molecular nutrition research-the modern way of performing nutritional science. *Journal Nutrients*, *4*(12), 1898–1944.

OECD. (2010). *Interim Report on the Green Growth Strategy: Implementing our Commitment for a Sustainable Future*. OECD.

OECD. (2010b). *Sustainable Management of Water Resources in Agriculture*. OECD.

Ong & Rogero. (2009). Nutrigenomics: Importance of nutrient-gene interaction for health promotion. *Journal of the ABESO, 40*.

Ordovas, J. M., & Mooser, V. (2004). Nutrigenomics and nutrigenetics. *Current Opinion in Lipidology*, *15*, 101–108.

Ouhtit, A. (2014). Nutrigenomics From promise to practice. *Sultan Qaboos University Medical Journal*, *14*(1), e1–e3. doi:10.12816/0003328 PMID:24516738

Peregrin, T. (2001). The new frontier of nutrition science: Nutrigenomics. *Journal of the American Dietetic Association*, *101*(11), 1306.

Potter, J. D., & Steinmetz, K. (1996). Vegetables, fruit and phytoestrogens as preventive agents. *IARC Scientific Publications*, 61–90.

Raj, M., Sundaram, K. R., & Murli Paul, A. S. (2007). Deepa and R. K. Kumar. Obesity in Indian children: Time trends and relationship with hypertension. *The National Medical Journal of India*, *20*(6), 288–293.

Rao, A. V., & Agarwal, S. (1995). Bioavailability and in vivo antioxidant properties of lycopene from tomato products and their possible role in the prevention of cancer. *Nutrition and Cancer*, *31*(3), 199–203. doi:10.1080/01635589809514703 PMID:9795972

Rao, S. (2001). Nutritional status of Indian population. *Journal of Biosciences*, *26*(4), 481–489.

Rocha, T. L. R., Evaristo, R. G. S., & Silva, L. P. (2006). *Metabolomics: Applications and Perspectives* (1st ed.). EMBRAPA.

Ronteltap, A., Trijp, H. V., Berezowska, A., & Goossens, J. (2013). Nutrigenomics-based personalised nutritional advice: In search of a business model. *Genes & Nutrition*, *8*(2), 153–163. doi:10.100712263-012-0308-4 PMID:22903899

Ronteltap, A., van Trijp, J. C. M., & Renes, R. J. (2009). Consumer acceptance of Nutrigenomics-based personalised nutrition. *British Journal of Nutrition*, *101*(1), 132–144. doi:10.1017/S0007114508992552 PMID:18479587

Ross, M. (1999). Welch and, Robin D. Graham. A New Paradigm for World Agriculture: Meeting Human Needs. Productive, Sustainable, Nutritious. *Field Crops Research*, *60*, 1–10.

Sales, N. M. R., Pelegrini, P. B., & Goersch, M. C. (2014). Nutrigenomics: Definitions and Advances of This New Science. *Journal of Nutrition and Metabolism.*

Scialabba & Müller-Lindenlauf. (2010). Organic agriculture and climate change. *Renewable Agriculture and Food Systems*, *25*(2), 158–169.

Shambu Prasad, C. (2006). *System of Rice Intensification in India: Innovation History and Institutional Challenges.* WWF-International-ICRISAT Dialogue Project.

Singh, A., & Sharma, O. P. (2004). Integrated Pest Management for Sustainable Agriculture. In Integrated Pest Management in Indian Agriculture. National Centre for Agricultural Economics and Policy Research (NCAP).

Soni, G. D. (2015). Advantages of green technology. *International Journal of Research – Granthaalayah*, *3*, 2–5.

Stevenson, J. R., Villoria, N., Byerlee, D., Kelley, T., & Maredia, M. (2013). Green revolution research saved an estimated 18 to 27 million hectares from being brought into agricultural production. *Proceedings of the National Academy of Sciences of the United States of America*, *110*(21), 8363–8368.

Stover, P. J. (2006). Influence of human genetic variation on nutritional requirements. *The American Journal of Clinical Nutrition*, *83*, 436S–442S.

Talbert, P. B., & Henikoff, S. (2006). Spreading of silent chromatin: Inaction at a distance. *Nature Reviews. Genetics*, *7*(10), 793–803.

Tester, M., & Langridge, P. (2010). Breeding technologies to increase crop production in a changing world. *Science*, *327*, 818–822.

Trigo, E., Cap, E., Malach, V., & Villarreal, F. (2009). *The Case of Zero-Tillage Technology in Argentina.* IFPRI Discussion Paper 00915. International Food Policy Research Institute.

Van Ommen, B., & Stierum, R. (2002). Nutrigenomics: Exploiting systems biology in the nutrition and health arena. *Current Opinion in Biotechnology*, *13*(5), 517–521.

Vasconcelos, F. A. G. (2010). The science of nutrition in transit: From nutrition and dietetics to nutrigenomics. *Revista de Nutrição*, *23*(6), 935–945.

Vojtech, V. (2010). *Policy Measures Addressing Agrienvironmental Issues*. OECD Food, Agriculture and Fisheries Working Papers, No. 24. OECD Publishing.

Wellen, K. E., & Hotamisligil, G. S. (2005). Inflammation, stress, and diabetes. *The Journal of Clinical Investigation*, *115*(5), 1111–1119.

WHO. (2015). *World Health Organization and Secretariat of the Convention of Biological Diversity Connecting Global Priorities: Biodiversity and Human Health: A Status of Knowledge Review*. WHO.

Wiwanitkit, V. (2012). Database and tools for nutrigenomics: A brief summary. *J Med Nutr Nutraceut.*, *1*, 87–91.

Yuan, D., Bassie, L., & Sabalza, M. (2011). Potential impact of plant biotechnology on the millennium development goals. *Plant Cell Reports*, *30*(3), 249–265.

Compilation of References

A decade of EU-funded GMO research 2001-2010. (2010). European Commission Directorate-General for Research and Innovation Biotechnologies, Agriculture, Food. Publications Office of the European Union.

A Smarter Shade of Green: How Innovative Technologies Are Saving Energy, Time and Money. (2008). Technology CEO Council.

Aalok, A., Tripath, A. K., & Soni, P. (2008). Vermicomposting: A better option for organic solid waste management. *Journal of Human Ecology (Delhi, India)*, *24*(1), 59–64. doi:10.1080/09709274.2008.11906100

Abbas, S. S., Haneef, M., & Lohani, M. (2016). Nanomaterials used as a plants growth enhancer: An update. *International Journal of Pharmaceutical Sciences Review and Research*, *5*, 17–23.

Abbo, A. S. H., Idris, M. O., & ElBalla, M. M. A. (2009). The response of tea tree oil as a biofungicide against early blight disease in tomato crop (Solanum lycopersicum) in Sudan. In *Conference on International Research on Food Security, Natural Resource Management and Rural Development* (pp. 1-9). Academic Press.

Abdelrahman, H., Cocozza, C., Olk, D. C., Ventrella, D., Montemurro, F., & Miano, T. (2020). Changes in labile fractions of soil organic matter during the conversion to organic farming. *Journal of Soil Science and Plant Nutrition*, 1–10.

Abdulrahaman, A. A., & Kolawole, O. M. (2006). Traditional preparations and uses of maize in Nigeria. Alama J.F (2001). Economic analysis of maize production using two technologies in Kaduna State. *Journal of Arid Agriculture*, *11*, 125–129.

Abebaw, D., & Abelay, K. (2001). Factors Influencing Adoption of High Yielding Maize Varieties in South Eastern Ethiopia. *Quarterly Journal of International Agriculture & Extension Education*, *17*(1), 21–30.

Abou-Shanab, R. A., Angle, J. S., Delorme, T. A., Chaney, R. L., Berkum, P. V., Moawad, H., Ghanem, K., & Ghozlan, H. A. (2003). Rhizobacterial effects on nickel extraction from soil and uptake by *Alyssum murale*. *The New Phytologist*, *158*(1), 219–224. doi:10.1046/j.1469-8137.2003.00721.x

Abou-Shanab, R., Ghanem, N., Ghanem, K., & Al-Kolaibe, A. (2007). Phytoremediation potential of crop and wild plants for multimetal contaminated soils. *Research Journal of Agriculture and Biological Sciences*, *3*, 370–376.

Abraham, E., Deepa, B., Pothan, L. A., Jacob, M., Thomas, S., & Anandjiwala, R. (2020). Extraction of nanocellulose fibrils from lignocellulosic filaments: A novel methodology. *Starch Polymers*, *86*, 1468–1475.

Acemoglu, D., Aghion, P., Bursztyn, L., & Hemous, D. (2012). The environment and directed technical change. *The American Economic Review*, 131–166.

Achard, P., Herr, A., Baulcombe, D. C., & Harberd, N. P. (2004). Modulation of floral development by gibberellin-regulated microRNA. *Development*, *131*(14), 3357–3365. doi:10.1242/dev.01206 PMID:15226253

Adekalu, K. O., Olorunfeli, I. A., & Osunbitan, J. A. (2007). Grass mulching effect on infiltration, surface runoff and soil loss of three agricultural soils in Nigeria. *Bioresource Technology*, *98*, 912–917. PMID:16678407

Adisarwanto, T., Utomo, W. H., Kirchhof, G., & So, H. B. (1996). Composition of food legume crops to different soil management practices after rainfed lowland rice East Jawa. *Proc. Management of Clay Soils for Rainfed Lowland Rice-Based Cropping Systems*, 142–147.

Adriano, S. K., & Shah, B. A. (2020). Danger of arsenic pollution in groundwater influencing the Ganga alluvial plain. *Ecological Health Perspectives*, *112*, 19–20.

Afman, L., & Müller, M. (2006). Nutrigenomics: From Molecular Nutrition to Prevention of Disease. *Journal of the Academy of Nutrition and Dietetics*, *106*(4), 569–576.

AFNOR - Association Française de Normalisation. (1996). XP T 90-337: Détermination de l'inhibition de la croissance de Lemna minor. Author.

Afsharian, S., & Shiea, B. A. M. (2020). A Review of Tidal and Wave Energy in Southern Waters of Iran. *Journal of Marine Science*, *2*(3). doi:10.30564/jms.v2i3.1761

Afzal, A., & Bano, A. (2008). *Rhizobium* and phosphate solubilizing bacteria improve the yield and phosphorus uptake in wheat (*Triticum aestivum* L.). *International Journal of Agriculture and Biology*, *10*, 85–88.

Agarwal, S., Satyavada, A., Patra, P., & Kumar, R. (2008). Strengthening functional community–provider linkages: Lessons from the Indore urban health programme. *Global Public Health: An International Journal for Research, Policy and Practice*, *3*(3), 308–325. doi:10.1080/17441690701592957

Aggarwal, A., Kadian, N., Tanwar, A., Yadav, A., & Gupta, K. K. (2011). Role of arbuscular mycorrhizal fungi (AMF) in global sustainable development. *Journal of Applied and Natural Science*, *3*(2), 340–351. doi:10.31018/jans.v3i2.211

Agricoop. (2021). *Ministry of Agriculture Cooperation and Farmers Welfare.* https://agricoop.nic.in/en/statistics/horticulture-crops-2019-20-second-advance-estimates

Agricultural Biotechnology (A Lot More than Just GM Crops). (2014). International Service for the Acquisition of Agri-Biotech Applications (ISAAA) Resources Publications.

Agriculture and Processed Food Products Export Development Authority (APEDA). (n.d.). https://jaivikbharat.fssai.gov.in/standard-npop.php)

Aharoni, A., & Vorst, O. (2002). DNA microarrays for functional plant genomics. *Plant Molecular Biology*, *48*(1/2), 99–118. doi:10.1023/A:1013734019946 PMID:11860216

Ahemad, M., & Kibret, M. (2014). Mechanisms and applications of plant growth promoting rhizobacteria: Current perspective. *Journal of King Saud University. Science*, *26*(1), 1–20. doi:10.1016/j.jksus.2013.05.001

Ahirwar, N. K., Singh, R., Chaurasia, S., Chandra, R., & Ramana, S. (2020). Effective role of beneficial microbes in achieving the sustainable agriculture and eco-friendly environment development goals: A review. *Frontiers in Microbiology*, *5*, 111–123.

Ahmad, F., Ahmad, I., & Khan, M. S. (2008). Screening of free-living rhizospheric bacteria for their multiple plant growth promoting activities. *Microbiological Research*, *163*(2), 173–181. doi:10.1016/j.micres.2006.04.001 PMID:16735107

Ahmad, S., Gordon-Weeks, R., Pickett, J. A., & Ton, J. (2010). Natural variation in priming of basal resistance: From evolutionary origin to agricultural exploitation. *Molecular Plant Pathology*, *11*(6), 817–827. doi:10.1111/j.1364-3703.2010.00645.x PMID:21029325

Ahn, Y., & Logan, B. E. (2012). A multi-electrode continuous flow microbial fuel cell with separator electrode assembly design. *Applied Microbiology and Biotechnology*, *93*(5), 2241–2248. doi:10.100700253-012-3916-4 PMID:22314518

Ajayan, P. M., Schadler, L. S., & Braun, P. V. (2003). Nanocomposite Science and Technology. Wiley-VCH VerlagGmb-Hand Co. doi:10.1002/3527602127

Akshay, S. M., Kushwaha, H. L., Kumar, A., Bhowmik, A., & Singh, A. K. (2021). Pulsating air pollinator for green-house cultivation. *JSIR*, (80), 477–485.

Al-Ani, R. A., Adhab, M. A., El-Muadhidi, M. A., & Al-Fahad, M. A. (2011). Induced systemic resistance and promotion of wheat and barley plants growth by biotic and non-biotic agents against barley yellow dwarf virus. *African Journal of Biotechnology*, *10*(56), 12079–12084. doi:10.5897/AJB11.1921

Alfadul, S. M., Altahir, O. S., & Khan, M. A. (2017). Application of nanotechnology in the field of food production. *Academia Journal of Scientific Research*, *5*, 143–154.

Alfaverh, F., Denai, M., & Sun, Y. (2020). Demand response strategy based on reinforcement learning and fuzzy reasoning for home energy management. *IEEE Access: Practical Innovations, Open Solutions*, *8*, 39310–39321. doi:10.1109/ACCESS.2020.2974286

Ali, A., Medhi, D. N., Dekamedhi, B., & Baroova, R. (1995). Effect of rice straw in combination with diVerent levels of nitrogen, phosphorus and potash on transplanted rice. *Journal of Agricultural Soil Science Society of North East India*, *8*, 248–250.

Aliero, M. S., Qureshi, K. N., Pasha, M. F., & Jeon, G. (2021). Smart Home Energy Management Systems in Internet of Things networks for green cities demands and services. *Environmental Technology & Innovation*, 101443.

Ali, H., Khan, E., & Sajad, M. A. (2013). Phytoremediation of heavy metals-concepts and applications. *Chemosphere*, *91*(7), 869–881. doi:10.1016/j.chemosphere.2013.01.075 PMID:23466085

Ali-Olubwanda, A. M., Odero-Wanga, D., Kathuri, N. J., & Shigova, W. A. (2010). Adoption of improved maize production practices among small scale farmers in agricultural reform era: The case of western province of Kenya. *Journal of Integrative Agriculture*, *8*, 160–171.

Allen, L. H. Jr, Boote, K. J., Jones, J. W., Jones, P. H., Valle, R. R., Acock, B., Rogers, H. H., & Dahlman, R. C. (1987). Response of vegetation to rising carbon dioxide: Photosynthesis, biomass, and seed yield of soybean. *Global Biogeochemical Cycles*, *1*(1), 1–14. doi:10.1029/GB001i001p00001

Allito, B. B., Nana, E. M., & Alemneh, A. A. (2015). *Rhizobia* strain and host-legume interaction effects on nitrogen fixation and yield of grain legume: A review. *Molecular Soil Biology*, *6*, 1–6.

Alloway, B. J. (2008). Zinc in soils and plant nutrition. Academic Press.

Al-Maaroof, E., Fayadh, A. H., & Fattah, F. A. (2014). Use of some chemical inducers to improve wheat resistance to *Puccinia striiformis* f. sp. *tritici*. *Ratarstvo i Povrtarstvo*, *51*(2), 83–90. doi:10.5937/ratpov51-5985

Almas, A., Singh, B. R., & Salbu, B. (1999). Mobility of Cadmium-109 and Zinc-65 in soil influenced by equilibration time, temperature, and organic matter. *Journal of Environmental Quality*, *28*(6), 1742–1750. doi:10.2134/jeq1999.00472425002800060008x

Almsatar, T. (2020). Environmental issues of biomass-burning in sub-Saharan African countries. In *Biomass Burning in Sub-Saharan Africa* (pp. 1–14). Springer. doi:10.1007/978-94-007-0808-2_1

Aloo, B. N., Makumba, B. A., & Mbega, E. R. (2019). The potential of Bacilli rhizobacteria for sustainable crop production and environmental sustainability. *Microbiological Research, 219*, 26–39. doi:10.1016/j.micres.2018.10.011 PMID:30642464

Alroe, H. F., Byrne, J., & Glover, L. (2006). *Organic agriculture and ecological justice: Ethics and practice. In Global Development of Organic Agriculture. Challenges and Prospects*. CABI Publishing.

Altieri, M. A., & Nicholls, C. I. (2020). Agroecology and the reconstruction of a post-COVID-19 agriculture. *The Journal of Peasant Studies*, *47*(5), 881–898.

Altman, A. (1999). Plant biotechnology in the 21st century: the challenges ahead: Review, EJB. *Electronic Journal of Biotechnology*, *2*(2), 51–55. doi:10.2225/vol2-issue2-fulltext-1

Altschul, S. F., Madden, T. L., Schäffer, A. A., Zhang, J., Zhang, Z., Miller, W., & Lipman, D. J. (1997). Gapped BLAST and PSI-BLAST: A new generation of protein database search programs. *Nucleic Acids Research*, *25*(17), 3389–3402. doi:10.1093/nar/25.17.3389 PMID:9254694

Ambros, V., & Lee, R. (2003). RC, Lavanway A, Williams PT, Jewell D. "Micro RNAs and other tiny endogenous RNAs in *C. elegans. Current Biology*, *13*, 807–818. doi:10.1016/S0960-9822(03)00287-2 PMID:12747828

Amico, E. D., Cavalca, L., & Andreoni, V. (2008). Improvement of *Brassica napus* growth under cadmium stress by cadmium-resistant rhizobacteria. *Soil Biology & Biochemistry*, *40*(1), 74–84. doi:10.1016/j.soilbio.2007.06.024

Anandraj, M., Venugopal, M.N., Veena, S.S., Kumar, A., & Sarma, Y.R. (2001). Ecofriendly management of disease of species. *Indian Species*, *38*, 28-31.

Anand, T., Chandrasekaran, A., Kuttalam, S., Raguchander, T., Prakasam, V., & Samiyappan, R. (2007). Association of some plant defense enzyme activities with systemic resistance to early leaf blight and leaf spot induced in tomato plants by azoxystrobin and *Pseudomonas fluorescens. Journal of Plant Interactions*, *2*(4), 233–244. doi:10.1080/17429140701708985

Anastasios, & Christos, & Christos. (2011). Annual intercrops: An alternative pathway for sustainable agriculture. *Australian Journal of Crop Science*, *5*, 396–410.

Anastas, P. T., & Warner, J. C. (1998). *Green Chemistry: Theory and Practice*. Oxford University Press.

Anderson, A. J., & Dawes, E. A. (1990). Occurrence, metabolism, metabolic role, and industrial uses of bacterial polyhyroxyalkanoates. *Microbiological Reviews*, *54*(4), 450–472. doi:10.1128/mr.54.4.450-472.1990 PMID:2087222

Anderson, D., Gnanasambandam, A., Mills, E., O'Shea, M., Nielsen, L., & Brumbley, S. (2011). Synthesis of short-chain length/medium chain length polyhydroxyalkanoate (PHA) copolymers in peroxisomes of transgenic sugarcane plants. *Tropical Plant Biology*, *4*(3), 170–184. doi:10.100712042-011-9080-7

Andre, C. V., Suma, C., & Gregory, B. M. (2009). Virus-induced gene silencing (VIGS) in *Nicotiana benthamiana* and tomato. *Journal of Visualized Experiments*, *28*, 1292–1301. PMID:19516240

Andrew, N. R., & Hughes, L. (2005). Diversity and assemblage structure of phytophagousHemiptera along a latitudinal gradient: Predicting the potential impacts of climate change. *Global Ecology and Biogeography*, *14*(3), 249–262. doi:10.1111/j.1466-822x.2005.00149.x

Andrieu, N., Howland, F., Acosta-Alba, I., Le Coq, J. F., Osorio-Garcia, A. M., Martinez-Baron, D., ... Chia, E. (2019). Co-designing climate-smart farming systems with local stakeholders: A methodological framework for achieving large-scale change. *Frontiers in Sustainable Food Systems*, *3*, 37.

Angelova, Z., Georgiev, S., & Roos, W. (2006). Elicitation of Plants. *Biotechnology, Biotechnological Equipment*, *20*(2), 72–83. doi:10.1080/13102818.2006.10817345

Anjali, C. H., Sharma, Y., Mukherjee, A., & Chandrasekaran, N. (2012). Neem oil (*Azadirachtaindica*) nanoemulsion a potent larvicidal agent against *Culexquinquefasciatus*. *Pest Management Science*, *68*(2), 158–163. doi:10.1002/ps.2233 PMID:21726037

Anjum, A., Zuber, M., Zia, K. M., Noreen, A., Anjum, M. N., & Tabasum, S. (2016). Microbial production of polyhydroxyalkanoates (PHAs) and its copolymers: A review of recent advancements. *International Journal of Biological Macromolecules*, *89*, 161–174. doi:10.1016/j.ijbiomac.2016.04.069 PMID:27126172

Anser, M. K., Yousaf, Z., Zaman, K., Nassani, A. A., Alotaibi, S. M., Jambari, H., Khan, A., & Kabbani, A. (2020). Determination of resource curse hypothesis in mediation of financial development and clean energy sources: Go-for-green resource policies. *Resources Policy*, *66*, 101640. doi:10.1016/j.resourpol.2020.101640

Appels, R., Eversole, K., Stein, N., Feuillet, C., Keller, B., Rogers, J., Pozniak, C. J., Choulet, F., Distelfeld, A., Poland, J., Ronen, G., Sharpe, A. G., Barad, O., Baruch, K., Keeble-Gagnère, G., Mascher, M., Ben-Zvi, G., Josselin, A.-A., Himmelbach, A., ... Wang, L.IWGSC. (2018). Shifting the limits in wheat research and breeding using a fully annotated reference genome. *Science*, *361*(6403), eaar7191. Advance online publication. doi:10.1126cience.aar7191 PMID:30115783

Arai, Y., Nakashita, H., Doi, Y., & Yamaguchi, I. (2001). Plastid targeting of polyhydroxybutyrate biosynthetic pathway in tobacco. *Plant Biotechnology (Sheffield, England)*, *18*(4), 289–293. doi:10.5511/plantbiotechnology.18.289

Arai, Y., Nakashita, H., Suzuki, Y., Kobayashi, Y., Shimizu, T., Yasuda, M., Doi, Y., & Yamaguchi, I. (2002). Synthesis of a novel class of polyhydroxyalkanoates in *Arabidopsis* peroxisomes and their use in monitoring short-chain-length intermediates of β-oxidation. *Plant & Cell Physiology*, *43*(5), 555–562. doi:10.1093/pcp/pcf068 PMID:12040103

Arai, Y., Shikanai, T., Doi, Y., Yoshida, S., Yamaguchi, I., & Nakashita, H. (2004). Production of polyhydroxybutyrate by polycistronic expression of bacterial genes in tobacco plastid. *Plant & Cell Physiology*, *45*(9), 1176–1184. doi:10.1093/pcp/pch139 PMID:15509840

Are, A. K., Anuradha, K., Ramaiah, B., & Grando, S. (2015). Recent advances in Sorghum biofortification research. *In book. Plant Breeding Reviews*, *39*, 89–124.

Arenas-Huertero, C., Pérez, B., Rabanal, F., Blanco-Melo, D., De la Rosa, C., Estrada-Navarrete, G., Sanchez, F., Covarrubias, A. A., & Reyes, J. L. (2009). Conserved and novel miRNAs in the legume *Phaseolus vulgaris* in response to stress. *Plant Molecular Biology*, *70*(4), 385–401. doi:10.100711103-009-9480-3 PMID:19353277

Arends, J., Speeckaert, J., Blondeel, E., De Vrieze, J., Boeckx, P., Verstraete, W., Rabaey, K., & Boon, N. (2014). Greenhouse gas emissions from rice microcosms amended with a plant microbial fuel cell. *Applied Microbiology and Biotechnology*, *98*(7), 3205–3217. doi:10.100700253-013-5328-5 PMID:24201892

Aroonsrimorakot, S., Laiphrakpam, M., & Paisantanakij, W. (2020). Solar panel energy technology for sustainable agriculture farming: A review. *Agricultural Technology (Thailand)*, *16*(3), 553–562.

Arora, R.K., Sunaina, V., & Somani, A. K. (2008). Biological control of soil and tuber-borne diseases. *Indian Farming*, *58*(9), 44-46.

Arsenio, S. K., & Arsenio, V.M.L. (2010). *Ecological Crisis Our Responsibility let us amend our lives*. C & E Publishing Inc.

Arundel, A., & Sawaya, D. (2009). *Biotechnologies in Agriculture and Related Natural Resources to 2015. OECD Journal.*

Arun, J., Gopinath, K. P., Sivaramakrishnan, R., SundarRajan, P., Malolan, R., & Pugazhendhi, A. (2020). Technical insights into the production of green fuel from CO2 sequestered algal biomass: A conceptual review on green energy. *The Science of the Total Environment*, 142636. PMID:33065504

Asada, H., & Matsumoto, J. (2009). Effects of rainfallvariation on rice production in the Ganges-Brahmaputra Basin. *Climate Research*, *38*, 249–260. doi:10.3354/cr00785

Asgher, M., Bashir, F., & Hafiz, M. N. I. (2014). A comprehensive ligninolytic pre-treatment approach from lignocellulose green biotechnology to produce bio-ethanol. *Chemical Engineering Research & Design*, *92*(8), 1571–1578. doi:10.1016/j.cherd.2013.09.003

Ash, G. J., & Rees, R. G. (1994). Effect of post-inoculation temperature and light-intensity on expression of resistance to stripe rust in some Australian wheat cultivars. *Australian Journal of Agricultural Research*, *45*(7), 1379–1386. doi:10.1071/AR9941379

Ashraf, S., Ali, Q., Zahir, Z. A., Ashraf, S., & Asghar, H. N. (2019). Phytoremediation: Environmentally sustainable way for reclamation of heavy metal polluted soils. *Ecotoxicology and Environmental Safety*, *174*, 714–727. doi:10.1016/j.ecoenv.2019.02.068 PMID:30878808

Aspelin, L., & Grube, A. H. (1999). Pesticides Industry Sales and Usage, 1996 and 1997, U.S. *Environmental Protection Agency.*, *733*, 99–001.

ASTM - American Society for Testing and Materials. (2000). Standard Guide for Conducting Static Toxicity Test With Lemna gibba G3. E 1415-91 (Reapproved 1998). In Annual Book of ASTM Standards, Vol. 11.05 Biological Effects and Environmental Fate; Biotechnology; Pesticides. ASTM.

Atkinson, D., & McKinlay, R. G. (1997). Crop protection and its integration within sustainable farming systems. *Agriculture, Ecosystems & Environment*, *64*(2), 87–93. doi:10.1016/S0167-8809(97)00026-1

Auffan, M., Rose, J., Bottero, J. Y., Lowry, G. V., Jolivet, J.-P., & Wiesner, M. R. (2009). Towards a definition of inorganic nanoparticles from an environmental, health and safety perspective. *Nature Nanotechnology*, *4*(10), 634–664. doi:10.1038/nnano.2009.242 PMID:19809453

Auskalniene, O., Psibisauskiene, G., Auskalnis, A., & Kadzys, A. (2010). Cultivar and plant density influence on weediness in spring barley crops. *Zemdirb Agric*, *97*, 53–60.

Avensblog. (2016). *Nutrigenomics – Eating for better genes*. Avensblog.

Awmack, C. S., Woodcock, C. M., & Harrington, R. (1997). Climate change may increase vulnerability of aphids to natural enemies. *Ecological Entomology*, *22*(3), 366–368. doi:10.1046/j.1365-2311.1997.00069.x

Axtell, M., Westholm, J., & Lai, E. (2011). Vive la difference: Biogenesis and evolution of microRNAs in plants and animals. *Genome Biology*, *12*(4), 221. doi:10.1186/gb-2011-12-4-221 PMID:21554756

Ayana, B., Suin, S., & Khatua, B. B. (2014). Highly exfoliated eco-friendly thermoplastic starch (TPS)/Poly(lactic acid) (PLA)/clay nanocomposites using unmodified nanoclay. *Carbohydrate Polymers*, *110*, 430–439. doi:10.1016/j.carbpol.2014.04.024 PMID:24906776

Azad, B. S., Singh, H., & Dubey, A. N. (1988). Control of Phalaris minor in wheat by cultural and chemical methods. *Pesticides*, *22*, 41–42.

Azeez, G. (2000). *The biodiversity benefits from organic faming*. Soil Association/WWFUK.

Aziz, G., Bajsa, N., & Haghjou, T. (2012). Abundance, diversity and prospecting of culturable phosphate solubilizing bacteria on soils under crop-pasture rotations in a no-tillage regime in Uruguay. *Applied Soil Ecology*, *61*, 320–326. doi:10.1016/j.apsoil.2011.10.004

Aziz, N., Faraz, M., Pandey, R., Shakir, M., Fatma, T., Varma, A., Barman, I., & Prasad, R. (2015). Facile algae-derived route to biogenic silver nanoparticles: Synthesis, antibacterial and photocatalytic properties. *Langmuir*, *31*(42), 11605–11612. doi:10.1021/acs.langmuir.5b03081 PMID:26447769

Aziz, N., Pandey, R., Barman, I., & Prasad, R. (2016). Leveraging the attributes of *Mucorhiemalis*-derived silver nanoparticles for a synergistic broad-spectrum antimicrobial platform. *Frontiers in Microbiology*, *7*, 1984. doi:10.3389/fmicb.2016.01984 PMID:28018316

Babu, R., Rojas, N. P., & Prasanna, B. M. (2015). Biofortified Maize - A Genetic Avenue for Nutritional Security. *In book. Translational Genomics for Crop Breeding*, *2*, 161–176.

Baharlouei, K., Pazira, E., & Solhi, M. (2011). Evaluation of inoculation of plant growth-promoting rhizobacteria on cadmium. In *International Conference on Environmental Science & Technology*. IACSIT Press.

Baier, A. H. (2012). Organic certification of farms and Business producing Agricultural Products. *USDA Organic*, 1-8. www.attra.ncat.org

Baker, A. J. M., McGrath, S. P., Reeves, R. D., & Smith, J. A. C. 2000. Metal hyperaccumulator plants: A review of the ecology and physiology of a biological resource for phytoremediation of metal-polluted soils. In Phytoremediation of Contaminated Soil and Water. Lewis Publishers.

Bakhtiari, M., Moaveni, P., & Sani, B. (2015). The effect of iron nanoparticles spraying time and concentration on wheat. *Biological Forum: An International Journal*, *7*, 679–683.

Bale, J. S. G. J., Masters, G. J., Hodkinson, I. D., Awmack, C., Bezemer, T. M., Brown, V. K., Butterfield, J., Buse, A., Coulson, J. C., Farrar, J., Good, J. E. G., Harrington, R., Hartley, S., Jones, T. H., Lindroth, R. L., Press, M. C., Symrnioudis, I., Watt, A. D., & Whittaker, J. B. (2002). Herbivory in global climate change research: Direct effects of rising temperatures on insect herbivores. *Global Change Biology*, *8*(1), 1–16. doi:10.1046/j.1365-2486.2002.00451.x

Bandel, V. A., Dziernia, G., Stanford, G., & Legg, J. O. (1975). N behavious under no-tillage and conventional corn culture. First year results using unlabeled N fertilizer. *Agronomy Journal*, *67*, 782–786.

Bani, A., Echevarria, G., Sulçe, S., Morel, J. L., & Mullai, A. (2007). In-situ phytoextraction of Ni by a native population of *Alyssum murale* on a ultramafic site (Albania). *Plant and Soil*, *293*(1-2), 79–89. doi:10.100711104-007-9245-1

Banting, D., Doshi, H., Missios, A., Currie, A., & Verrati, M. (2005). Report on the Environmental Benefits and Costs of Green Roof Technology for the City of Toronto. Prepared for City of Toronto and Ontario centres of Excellence – Earth and Environmental Technologies (OCE-Etech).

Banuelos, G. S. (2000). Factors influencing field phytoremediation of selenium-laden soils. In N. Terry & G. Banuelos (Eds.), *Lewis Publishers* (pp. 41–60). CRC.

Bao, H., Guo, H., Wang, J., Zhou, R., Lu, X., & Shi, S. (2009). MapView: Visualization of short reads alignment on a desktop computer. *Bioinformatics (Oxford, England)*, *25*(12), 1554–1555. doi:10.1093/bioinformatics/btp255 PMID:19369497

Baraibar, B., Westerman, P. R., Carrion, E., & Recasens, J. (2009). Effects of tillage and irrigation in cereal fields on weed seed removal by seed predators. *Journal of Applied Ecology*, *46*(2), 380–387. doi:10.1111/j.1365-2664.2009.01614.x

Barberi, P., & Lo Cascio, B. (2001). Long-term tillage and crop rotation effects on weed seed bank size and composition. *Weed Research*, *41*(4), 325–340. doi:10.1046/j.1365-3180.2001.00241.x

Bardin, S. D., Huang, H. C., Pinto, J., Amundsen, E. J., & Erickson, R. S. (2004). Biological control of *Pythium* damping-off of pea and sugar beet by *Rhizobium leguminosarum* bv. *viceae*. *Canadian Journal of Botany*, *82*(3), 291–296. doi:10.1139/b04-003

Barker, A. V., & Pilbeam, D. J. (2006). *Handbook of plant nutrition*. CRC Press.

Barnes, M., Correll, R., & Stevens, D. P. 2003. A Simple spreadsheet estimating low-effect concentrations and associated logistic dose response curves. *Abstracts SETAC Asia/pacific Australasian Society for Ecotoxicology Meeting: Solutions to Pollution, Christchurch, New Zealand, September 28-October 1*, 156.

Bartel, B., & Bartel, D. P. (2003). MicroRNAs: At the root of plant development? *Plant Physiology*, *132*(2), 709–717. doi:10.1104/pp.103.023630 PMID:12805599

Bartel, D. P. (2004). MicroRNAs: Genomics, biogenesis, mechanism, and function. *Cell*, *116*(2), 281–297. doi:10.1016/S0092-8674(04)00045-5 PMID:14744438

Bashan, Y., de Bashan, L. E., Prabhu, S. R., & Hernandez, J. P. (2014). Advances in plant growth-promoting bacterial inoculant technology: Formulations and practical perspectives. *Plant and Soil*, *378*(1-2), 1–33. doi:10.100711104-013-1956-x

Basker, S., & Narmatha Bai, V. (2006). Micropropagation of Coelogyne stricta (D.Don) Schltr. via Pseudobulb Segment Cultures. *Tropical and Subtropical Agroecosystems*, *6*, 31–35.

Batish, D. R., Singh, H. P., Setia, N., Kohli, R. K., Kaur, S., & Yadav, S. S. (2007). Alternative control of littleseed canary grass using eucalypt oil. *Agronomy for Sustainable Development*, *27*(3), 171–177. doi:10.1051/agro:2007008

Batley, J., & Edwards, D. (2009). Genome sequence data: Management, storage, and visualization. *BioTechniques*, *46*(5), 333–334, 336. doi:10.2144/000113134 PMID:19480628

Baudo, M. M., Lyons, R., Powers, S., Pastori, G. M., Edwards, K. J., Floldsworth, M. J., & Shewry, P. R. (2006). Transgenesis has lessimpact on the transcriptome of wheat grain than conventional breeding. *Plant Biotechnology Journal*, *4*(4), 369–380. doi:10.1111/j.1467-7652.2006.00193.x PMID:17177803

Baum, C., Hrynkiewicz, K., Leinweber, P., & Meissner, R. (2006). Heavy metal mobilization and uptake by mycorrhizal and nonmycorrhizal willows (*Salix x dasyclados*). *Journal of Plant Nutrition and Soil Science*, *169*(4), 516–522. doi:10.1002/jpln.200521925

Bawa, D. B., & Ani, A. O. (2014). Analysis of adoption of improved maize production technology among farmers in Southern Borno, Nigeria. *Research on Humanities and Social Science*, *4*(255), 43–54.

Beaton, J. D., Hasegawa, M., Jiung-Chang, X., Keng, J. C. W., & Halstead, E. H. (1992, September). Influence of intensive long-term fertilization on properties of paddy soils and sustainable yields. In *Proceedings of International Symposium on Paddy Soils* (pp. 246-251). Academic Press.

Bekchanov, M., & Mirzabaev, A. (2018). Circular economy of composting in Sri Lanka: Opportunities and challenges for reducing waste related pollution and improving soil health. *Journal of Cleaner Production*, *202*, 1107–1119.

Belanger, R. R., Benhamou, N., & Menzies, J. G. (2003). Cytological evidence of an active role of silicon in wheat resistance to powdery mildew (Blumeria graminis f. sp tritici). *Phytopathology*, *93*(4), 402–412. doi:10.1094/PHYTO.2003.93.4.402 PMID:18944354

Benedito, V. A., Visser, P. B., Angenent, G. C., & Kerns, F. A. (2004). The potential of virus-induced gene silencing for speeding up functional characterization of plant genes. *Genetics and Molecular Research*, *3*, 323–341. PMID:15614725

Bengtsson, J., Ahnstrom, J., & Weibull, A.-C. (2005). The effects of organic agriculture on biodiversity and abundance: A meta-analysis. *Journal of Applied Ecology*, *42*(2), 261–269. doi:10.1111/j.1365-2664.2005.01005.x

Benson, D. A., Boguski, M. S., Lipman, D. J., & Ostell, J. (1997). GenBank. *Nucleic Acids Research*, *25*(1), 1–6. doi:10.1093/nar/25.1.1 PMID:9016491

Berlioz, S., Molina-Boisseau, S., Nishiyama, Y., & Heux, L. (2018). Gas-Phase Surface Esterification of Cellulose Microfibrils and Whiskers. *Biomacromolecules*, *10*(8), 2144–2151. doi:10.1021/bm900319k PMID:19572699

Bessette, D. L., & Mills, S. B. (2021). Farmers vs. lakers: Agriculture, amenity, and community in predicting opposition to United States wind energy development. *Energy Research & Social Science*, *72*, 101873. doi:10.1016/j.erss.2020.101873

Bezemer, T. M., Jonesh, T. H., & Knight, K. J. (1998). Long term effect of elevated Co_2 and temperature on the population of the Peach Potato aphid *Myzuspercicae* and its parasitoid*Aphidiusmatricariae. Oecologia*, *116*(2), 128–135. doi:10.1007004420050571 PMID:28308517

Bezemer, T. M., Knight, K. J., Newington, J. E., & Jones, T. H. (1999). How general are aphid responses to elevated atmospheric CO_2? *Annals of the Entomological Society of America*, *92*(5), 724–730. doi:10.1093/aesa/92.5.724

BGS (British Geological Survey)/DPHE. (2018). *Arsenic pollution of groundwater in Bangladesh 2: Final report*. BGS Technical Report WC/00/19.

Bhan, V. M., & Kumar, S. (1997). Integrated management of Phalaris minor in rice–wheat ecosystems in India. *Proceedings of international conference on ecological agriculture: towards sustainable development,* 400-415.

Bhandari, A. L., Sood, A., Sharma, K. N., & Rana, D. S. (1992). Integrated nutrient management in a rice-wheat system. *Journal of the Indian Society of Soil Science*, *40*(4), 742–747.

Bhardwaj, D., Ansari, M. W., Sahoo, R. K., & Tuteja, N. (2014). Biofertilizers function as key player in sustainable agriculture by improving soil fertility, plant tolerance and crop productivity. *Microbial Cell Factories*, *13*(1), 66. doi:10.1186/1475-2859-13-66 PMID:24885352

Bhargava, Dilta, Gupta, Dhiman, & Modgil. (2013). Studies on micropropagation of gerbera (Gerbera jamesonii Bolus). *Indian Journal of Applied Research*, *3*, 8–11. doi:10.15373/2249555X/NOV2013/4

Bhaskara Reddy, M. V., Arul, J., Angers, P., & Couture, L. (1999). Chitosan treatment of wheat seeds induces resistance to Fusarium graminearum and improves seed quality. *Journal of Agricultural and Food Chemistry*, *47*(3), 1208–1216. doi:10.1021/jf981225k PMID:10552439

Bhattacharya, P., Ahmad, K. M., Hasan, M. A., Broms, S., Fogelstrom, J., Jacks, G., Sracek, O., & Routh, J. (2018). Versatility of arsenic in groundwater in a piece of Brahmanbaria area, NE Bangladesh. *Acta Hydrochimica et Hydrobiologica*, *31*, 97–107.

Bhattacharyva, P., & Chakraborty, G. (2005). Current Status of Organic Farming in India and other Countries. *Indian Journal of Fertilisers*, *1*(9), 111-123.

Bhattacharyya, A., Duraisamy, P., & Govindarajan, M. (2016). Nano-biofungicides: emerging trend in insect pest control. In R. Prasad (Ed.), *Advances & Applications through Fungal Nanobiotechnology* (pp. 307–319). Springer. doi:10.1007/978-3-319-42990-8_15

Biotechnology and Management of Bioresources Division (BMBD) Brochure. (n.d.). The Energy and Resources Institute (TERI).

Blake, D. M., Maness, P.-C., Huang, Z., Wolfrum, E. J., Jacoby, W. A., & Huang, J. (1999). Application of the photocatalytic chemistry of titanium dioxide to disinfection and the killing of cancer cells. *Separation and Purification Methods*, *28*(1), 1–50. doi:10.1080/03602549909351643

Block, G., Patterson, B., & Subar, A. (1992). Fruit, vegetable and cancer prevention: A review of the epidemiological evidence. *Nutrition and Cancer*, 1–29.

BMLF. (1995). *Organic Farming in Austria*. Wien: BMLF

Bockus, W. W., Webster, R. K., Wick, C. M., & Jackson, L. F. (1979). Rice residue disposal influences overwintering inoculum level of *Sclerotiumoryzae* and stem rot severity. *Phytopathology*, *69*(8), 862–865. doi:10.1094/Phyto-69-862

Bodek, I., Lyman, W. J., Reehl, W. F., & Rosenblatt, D. H. (2018). *Natural Inorganic Chemistry: Properties, Processes and Estimation Methods*. Pergamon Press.

Boerjan, W., Bauw, G., Van Montagu, M., & Inzé, D. (1994). Distinct phenotypes generated by overexpression and suppression of *S*-adenosyl-l-methionine reveal developmental patterns of gene silencing in tobacco. *The Plant Cell*, *6*, 1401–1414. PMID:7994174

Bohmert, K., Balbo, I., Kopka, J., Mittendorf, V., Nawrath, C., Poirier, Y., Tischendorf, G., Trethewey, R. N., & Willmitzer, L. (2000). Transgenic *Arabidopsis* plants can accumulate polyhydroxybutyrate to up to 4% of their fresh weight. *Planta*, *211*(6), 841–845. doi:10.1007004250000350 PMID:11144269

Bohmert, K., Balbo, I., Steinbuchel, A., Tischendorf, G., & Willmitzer, L. (2002). Constitutive expression of the beta-ketothiolase gene in transgenic plants: A major obstacle for obtaining polyhydroxybutyrate-producing plants. *Plant Physiology*, *128*(4), 1282–1290. doi:10.1104/pp.010615 PMID:11950977

Bohmert-Tatarev, K., McAvoy, S., Daughtry, S., Peoples, O. P., & Snell, K. D. (2011). High levels of bioplastic are produced in fertile transplastomic tobacco plants engineered with a synthetic operon for the production of polyhydroxybutyrate. *Plant Physiology*, *155*(4), 1690–1708. doi:10.1104/pp.110.169581 PMID:21325565

Bommarco, R., Kleijn, D., & Potts, S. G. (2013). Ecological intensification: Harnessing ecosystem services for food security. *Trends in Ecology & Evolution*, *28*(4), 230–238.

Bondeson, D. (2018). Mathew An and Oksman K. Advancement of the Isolation of Nanocrystals From Microcrystalline Cellulose by Acid Hydrolysis. *Cellulose (London, England)*, *13*, 171–180. doi:10.100710570-006-9061-4

Boschetti, L., Roy, D., Barbosa, P., Boca, R., & Justice, C. O. (2007). A MODIS assessment of the summer 2007 extent burned in Greece. *International Journal of Remote Sensing*, *29*(8), 2433–2436. doi:10.1080/01431160701874561

Botella-Pavía, P., & Rodriguez-Conceptíon, M. (2006). Carotenoid biotechnology in plants for nutritionally improved foods. *Plant Physiology*, *126*, 369–381.

Bouis, H. E., & Saltzman, A. (2017). Improving nutrition through biofortification: A review of evidence from HarvestPlus, 2003 through 2016. *Global Food Security*, *12*, 49–58. doi:10.1016/j.gfs.2017.01.009 PMID:28580239

Bowie, M., Wratten, S. D., & White, A. J. (1995). Agronomy and phenology of "companion plants" of potential for enhancement of insect biological control. *New Zealand Journal of Crop and Horticultural Science*, *23*(4), 423–427. doi:10.1080/01140671.1995.9513919

Brainard, D. C., Bellinde, R. R., Hahn, R. R., & Shah, D. A. (2008). Crop rotation, cover crop, and weed management effects on weed seed banks and yields in snap bean, sweet corn, and cabbage. *Weed Science*, *56*(3), 434–441. doi:10.1614/WS-07-107.1

Brain, P., & Cousens, R. (1989). An equation to describe dose-responses where there is stimulation of growth at low doses. *Weed Research*, *29*, 93–96.

Brandt, K., & Mølgaard, J. P. (2001). Organic agriculture: Does it enhance or reduce the nutritional value of plant foods. *Journal of the Science of Food and Agriculture*, *81*(9), 924–931. doi:10.1002/jsfa.903

Branton, D., Deamer, D. W., Marziali, A., Bayley, H., Benner, S. A., Butler, T., Di Ventra, M., Garaj, S., Hibbs, A., Huang, X., Jovanovich, S. B., Krstic, P. S., Lindsay, S., Ling, X. S., Mastrangelo, C. H., Meller, A., Oliver, J. S., Pershin, Y. V., Ramsey, J. M., ... Schloss, J. A. (2008). The potential and challenges of nanopore sequencing. *Nature Biotechnology*, *10*(10), 1146–1153. doi:10.1038/nbt.1495 PMID:18846088

Brenchley, R., Spannagl, M., Pfeifer, M., Barker, G. L. A., D'Amore, R., Allen, A. M., McKenzie, N., Kramer, M., Kerhornou, A., Bolser, D., Kay, S., Waite, D., Trick, M., Bancroft, I., Gu, Y., Huo, N., Luo, M.-C., Sehgal, S., Gill, B., ... Hall, N. (2012). Analysis of the bread wheat genome using whole-genome shotgun sequencing. *Nature*, *491*(7426), 705–710. doi:10.1038/nature11650 PMID:23192148

Briones, M. J. I., & Schmidt, O. (2017). Conventional tillage decreases the abundance and biomass of earthworms and alters their community structure in a global meta-analysis. *Global Change Biology*, *23*(10), 4396–4419.

Brookes & Barfoot. (2008). *GM crops: Global socio-economic and environmental impacts 1996-2006*. PG Economics Ltd.

Brownlie, S. (1990). Roof Gardens A Review. Peterborough Nature Conservancy Council.

Brown, M. E., & Funk, C. C. (2008). Climate. Food security under climate change. *Science*, *319*(5863), 580–581. doi:10.1126cience.1154102 PMID:18239116

Bruce, R. D., & Versteeg, D. J. (1992). A statistical procedure for modelling continuous toxicity data. *Environmental Toxicology and Chemistry*, *11*, 1485–1494.

Bruce, T. J. A. (2010). Tackling the threat to food security caused by crop pests in the new millennium. *Food Security*, *2*(2), 133–141. doi:10.100712571-010-0061-8

Bruce, T. J. A., Smart, L. E., Birch, A. N. E., Blok, V. C., MacKenzie, K., Guerrieri, E., Cascone, P., Luna, E., & Ton, J. (2017). Prospects for plant defence activators and biocontrol in IPM-Concepts and lessons learnt so far. *Crop Protection (Guildford, Surrey)*, *97*, 128–134. doi:10.1016/j.cropro.2016.10.003

Brumbley, S. M., Petrasovits, L. A., Bonaventura, P. A., O'Shea, M. J., Purnell, M. P., & Nielsen, L. K. (2003). Production of polyhydroxyalkanoates in sugarcane. *Proc. of Int. Soc. of Sugarcane Tech., Mol. Biol. Workshop*, *4*, 31.

Budhar, M. N., & Palaniappan, S. (1994). Effect of incorporation of preceding cotton (Gossypium species) on succeeding rice (Oryza sativa). *Indian Journal of Agronomy*, *39*(3), 461–462.

Bulc, T. G. (2006). Long term performance of a constructed wetland for landfill leachate treatment. *Ecological Engineering*, *26*(4), 365–374. doi:10.1016/j.ecoleng.2006.01.003

Burchi, F., Fanzo, J., & Frison, E. (2011). The role of food and nutrition system approaches in tackling Hidden Hunger. *International Journal of Environmental Research and Public Health*, *8*(2), 358–373. doi:10.3390/ijerph8020358 PMID:21556191

Buresh, R. J., & De Datta, S. K. (1991). Nitrogen dynamics and management in rice-legume cropping systems. *Advances in Agronomy*, *45*, 1–59.

Buresh, R. J., Pampolino, M. F., & Witt, C. (2010). Field-specific potassium and phosphorus balances and fertilizer requirements for irrigated rice-based cropping systems. *Plant and Soil*, *335*(1), 35–64.

Burge, C., & Karlin, S. (1997). Prediction of complete gene structures in human genomic DNA. *Journal of Molecular Biology*, *268*(1), 78–94. doi:10.1006/jmbi.1997.0951 PMID:9149143

Burkhanova, G. F., Veselova, S. V., Sorokan', A. V., Blagova, D. K., Nuzhnaya, T. V., & Maksimov, I. V. (2017). Strains of *Bacillus* ssp. regulate wheat resistance to *Septoria nodorum* Berk. *Applied Biochemistry and Microbiology*, *53*(3), 346–352. doi:10.1134/S0003683817030048

Burns, R. C., & Hardy, R. W. (2012). Nitrogen fixation in bacteria and higher plants. Springer Science & Business Media, 21, 192.

Butler, G., Nielsen, J. H., Slots, T., Seal, C., Eyre, M. D., Sanderson, R., & Leifert, C. (2008). Fatty acid and fat-soluble antioxidant concentrations in milk from high- and low-input conventional and organic systems: Seasonal variation. *Journal of the Science of Food and Agriculture*, *88*(8), 1431–1441. doi:10.1002/jsfa.3235

Butler, J., MacCallum, I., Kleber, M., Shlyakhter, I. A., Belmonte, M. K., Lander, E. S., Nusbaum, C., & Jaffe, D. B. (2008). ALLPATHS: De novo assembly of whole-genome shotgun microreads. *Genome Research*, *18*(5), 810–820. doi:10.1101/gr.7337908 PMID:18340039

Byrne, P., Sarah, W., Harrington, J., & Fuller, L. (2004). *Transgenic Crops: An Introduction and Resource Guide*. Department of Soil and Crop Sciences, Colorado State University.

Bystrzejewska-Piotrowska, G., Asztemborska, M., & Steborowski, R. (2012). Application of neutron activation for investigation of Fe_3O_4 nanoparticles accumulation by plants. *Nukleonika*, *57*, 427–430.

Cadisch, G., & Giller, K. E. (1997). *Driven by nature: plant residue quality and decomposition*. CAB International.

Cai, K., Gao, D., Chen, J., & Shiming, L. (2009). Probing the mechanisms of silicon-mediated pathogen resistance. *Plant Signaling & Behavior*, *4*(1), 1–3. doi:10.4161/psb.4.1.7280 PMID:19568332

Cai, R., Van, G. M., Aw, P. K., & Itoh, K. (2006). Itoh, K. Solar-driven self-cleaning coating for a painted surface. *Comptes Rendus. Chimie*, *9*(5-6), 829–835. doi:10.1016/j.crci.2005.04.007

Campbell, B. M., Thornton, P., Zougmoré, R., van Asten, P., & Lipper, L. (2014). Sustainable intensification: What is its role in climate smart agriculture? *Current Opinion in Environmental Sustainability*, *8*, 39–43. doi:10.1016/j.cosust.2014.07.002

Campbell, C. A., Biederbeck, V. O., Zentner, R. P., & Lafond, G. P. (1991). Effect of crop rotations and cultural practices on soil organic matter, microbial biomass and respiration in a thin Black Chernozem. *Canadian Journal of Soil Science*, *71*(3), 363–376. doi:10.4141/cjss91-035

Campiglia, E., Radicetti, E., & Mancinelli, R. (2018). Floristic composition and species diversity of weed community after 10 years of different cropping systems and soil tillage in a Mediterranean environment. *Weed Research*, *58*(4), 273–283.

Camprubí, A., Estaún, V., El Bakali, M. A., Garcia-Figueres, F., & Calvet, C. (2007). Alternative strawberry production using solarization, metham sodium and beneficial soil microbes as plant protection methods. *Agronomy for Sustainable Development*, *27*(3), 179–184. doi:10.1051/agro:2007007

Cao, X., Wang, H., Li, X. Q., Fang, Z., & Li, X. N. (2017). Enhanced degradation of azo dye by a stacked microbial fuel cell-biofilm electrode reactor coupled system. *Bioresource Technology*, *227*, 273–278. doi:10.1016/j.biortech.2016.12.043 PMID:28040648

Cappuyns, V., Alian, V., Vassilieva, E., & Swennen, R. (2014). pH dependent leaching behaviour of Zn, Cd, Pb, Cu and As from mining wastes and slags: Kinetics and mineralogical control. *Waste and Biomass Valorization*, *5*(3), 355–368. doi:10.100712649-013-9274-3

Carbonell, A., & Daros, J. A. (2017). Artificial microRNAs and synthetic transacting small interfering RNAs interfere with viroid infection. *Molecular Plant Pathology*, *18*(5), 746–753. doi:10.1111/mpp.12529 PMID:28026103

Carmona, M. A., Sautua, F. J., Grijalba, P. E., Cassina, M., & Pérez-Hernández, O. (2018). Effect of potassium and manganese phosphites in the control of Pythium damping-off in soybean: A feasible alternative to fungicide seed treatments. *Pest Management Science*, *74*(2), 366–374. doi:10.1002/ps.4714 PMID:28842951

Carthew, R. W. (2006). Gene regulation by microRNAs. *Current Opinion in Genetics & Development*, *16*(2), 203–208. doi:10.1016/j.gde.2006.02.012 PMID:16503132

Carthew, R. W., & Sontheimer, E. J. (2009). Origins and mechanisms of miRNAs and siRNAs. *Cell*, *136*(4), 642–655. doi:10.1016/j.cell.2009.01.035 PMID:19239886

Carvajal-Muñoz, J. S., & Carmona-Garcia, C. E. (2012). Benefits and limitations of biofertilization in agricultural practices. *Livestock Research for Rural Development*, *24*, 43.

Carvalho, J. A., Costa, F. S., Veras, C. A. G., Sandberg, D. V., Alvarado, E. C., Gielow, R., Serra, A. M., & Santos, J. C. (2004). Biomass fire consumption and carbon release rates of rainforest-clearing experiments conducted in northern Mato Grosso, Brazil. *Journal of Geophysical Research*, *106*(D16), 17877–17887. doi:10.1029/2000JD900791

Cassman, K. G., & Grassini, P. (2020). A global perspective on sustainable intensification research. *Nature Sustainability*, *3*(4), 262–268.

Catchpole, G. S., Beckman, M., & Enot, D. P. (2005). Hierarchical metabolomics demonstrates substantial compositional similarity between genetically modified and conventional potato crops. *Proceedings of the National Academy of Sciences of the United States of America*, *102*(40), 14458–14462.

Catroux, G., Hartmann, A., & Revellin, C. (2001). Trends in rhizobial inoculant production and use. *Plant and Soil*, *230*(1), 21–30. doi:10.1023/A:1004777115628

Celestino, M. D. R., Font, R., Rojas, R. M., & Bailon, A. D. H. (2006). Uptake of Lead and Zinc by wild plants growing on contaminated soils. *Industrial Crops and Products*, *24*(3), 230–237. doi:10.1016/j.indcrop.2006.06.013

Chain, F., Côté-Beaulieu, C., Belzile, F., Menzies, J., & Bélanger, R. (2009). A comprehensive transcriptomic analysis of the effect of silicon on wheat plants under control and pathogen stress conditions. *Molecular Plant-Microbe Interactions*, *22*(11), 1323–1330. doi:10.1094/MPMI-22-11-1323 PMID:19810802

Chakraborti, D., Rahman, M. M., Paul, K., Chowdhury, U. K., Sengupta, M. K., Lodh, D., Chanda, C. R., Saha, K. C., & Mukherjee, S. C. (2017). Arsenic catastrophe in the Indian subcontinent: What exercises have been realized? *Talanta*, *58*, 3–22. doi:10.1016/S0039-9140(02)00270-9 PMID:18968730

Chakraborty, S. N., Chakraborty, L., Agrawal, S., & Narula, G. K. (2010). Next generation protein rich potato by expressing a seed protein gene AmA1 as a result of proteome rebalancing in transgenic tuber. *Proceedings of the National Academy of Sciences of the United States of America*, *107*(41), 17533–17538. doi:10.1073/pnas.1006265107 PMID:20855595

Chamberlain, D. E., Wilson, J. D., & Fuller, R. J. (1999). A comparison of bird populations on organic and conventional farm systems in southern Britain. *Biological Conservation*, *88*(3), 307–320.

Chandler, V. L., & Vaucheret, H. (2001). Gene activation and gene silencing. *Plant Physiology*, *125*(1), 145–148. doi:10.1104/pp.125.1.145 PMID:11154319

Chandrasekaran, Raju, Varutharaju, Asiarn, & Shajahan. (2015). Improved Agrobacterium rhizogenes-mediated hairy root culture system of Withania somnifera (L.) Dunal using sonication and heat treatment. *Biotech, 3*, 297.

Chang, H., Jwo, C. S., Lo, C. H., Tsung, T. T., Kao, M. J., & Lin, H. M. (2005). Rheology of CuO nanoparticle suspension prepared by ASNSS. *Reviews on Advanced Materials Science*, *10*, 128–132.

Chang, S. J., Tzeng, D. D. S., & Li, C. C. (2002). Effect of silicon nutrient on bacterial blight resistance of rice (Oryza sativa L.). *Proceedings of the Second Silicon in Agriculture Conference*, 31–33.

Charrondiere, & Stadylmayr, & Rittenschober. (2013). FAO/INFOODS food consumption database for biodiversity. *Food Chemistry*, *140*(3), 408–412.

Chatterjee, B. N., & Mondal, S. S. (1996). Potassium nutrition under intensive cropping. *Journal of Potassium Research*, *12*(4), 358–364.

Chauhan, B. S., & Johnson, D. E. (2009). Influence of tillage systems on weed seedling emergence pattern in rainfed rice. *Soil & Tillage Research*, *106*(1), 15–21. doi:10.1016/j.still.2009.10.004

Chauhan, B. S., & Johnson, D. E. (2010). Implications of narrow crop row spacing and delayed Echinochloa colona and Echinochloa crusgalli emergence for weed growth and crop yield loss in aerobic rice. *Field Crops Research*, *117*(2-3), 177–182. doi:10.1016/j.fcr.2010.02.014

Chauhan, B. S., Singh, R. G., & Mahajan, G. (2012). Ecology and management of weeds under conservation agriculture: A review. *Crop Protection (Guildford, Surrey)*, *38*, 57–65.

Chauhan, B. S., Yadav, A., & Malik, R. K. (2001). Competitive wheat genotypes under zero tillage-an important tool to manage resistant *Phalaris minor. Indian Journal of Weed Science*, *33*, 75–76.

Cheke, R. A., & Tratalos, J. A. (2007). Migration, patchiness and population processes illustrated by two migrant pests. *Bioscience*, *57*(2), 145–154. doi:10.1641/B570209

Chel, A., & Kaushik, G. (2011). Renewable energy for sustainable agriculture. *Agronomy for Sustainable Development*, *31*(1), 91–118. doi:10.1051/agro/2010029

Chen, J.-H. (2006). The combined use of chemical and organic fertilizers and/or biofertilizer for crop growth and soil fertility. In *Proceeding of International Workshop on Sustained Management of the Soil-Rhizosphere System for Efficient Crop Production and Fertilizer Use*. Land Development Department.

Chen, C. Z., Li, L., Lodish, H. F., & Bartel, D. P. (2004). MicroRNAs modulate hematopoietic lineage differentiation. *Science*, *303*(5654), 83–86. doi:10.1126cience.1091903 PMID:14657504

Cheney, C. (2004). A busy spring for rooftops in New York City. *Journal of Green Roof Infrastructure Monitor*, *6*(1), 16–20.

Cheng, G., Varanasi, P., Li, C., Liu, H., Melnichenko, Y.B., Simmons, B.A., Kent, M.S., & Singh, S. (2017). *Progress of Cellulose Crystalline Structure and Surface Morphology of Biomass as a Function of Ionic Liquid Pretreatment and Its Relation to Enzymatic Hydrolysis*. doi:10.1021/bm101240z

Cheng, X., Chen, G., & Rodrigue, W. R. (2009). Micro- and nanotechnology for viral detection. *Analytical and Bioanalytical Chemistry*, *393*(2), 487–501. doi:10.100700216-008-2514-x PMID:19052733

Chen, H. D., & Yada, R. (2011). Nanotechnologies in agriculture: New tools for sustainable development. *Trends in Food Science & Technology*, *22*(11), 585–594. doi:10.1016/j.tifs.2011.09.004

Chen, J., Lin, N., Huang, J., & Dufresne, A. (2017). Profoundly alkynyl-functionalization of cellulose nanocrystals and progressed nanocomposites there of by means of snap science. *Polymer Chemistry*, *6*, 4385–4395. doi:10.1039/C5PY00367A

Chen, L., Wang, T., Zhao, M., Tian, Q., & Zhang, W. H. (2012). Identification of aluminum-responsive microRNAs in *Medicago truncatula* by genome-wide high-throughput sequencing. *Planta*, *235*(2), 375–386. doi:10.100700425-011-1514-9 PMID:21909758

Chen, W., Kastner, C., & Nowara, D. (2016). Host-induced silencing of *Fusarium culmorum* genes protects wheat from infection. *Journal of Experimental Botany*. Advance online publication. doi:10.1093/jxb/erw263(2016) PMID:27540093

Chen, Y. N., Chai, L. Y., & Shu, Y. D. (2017). Investigation of arsenic (V) adsorption on bone roast from watery arrangement. *Diary of Hazardous Materials*, *160*, 168–172. doi:10.1016/j.jhazmat.2008.02.120

Chen, Y., Wang, Y., Wu, W., Lin, Q., & Xue, S. (2006). Impacts of chelate-assisted phytoremediation on microbial community composition in the rhizosphere of a copper accumulator and non-accumulator. *The Science of the Total Environment*, *356*(1-3), 247–255. doi:10.1016/j.scitotenv.2005.04.028 PMID:15935447

Chen, Z., Huang, Y. C., Liang, J. H., Zhao, F., & Zhu, Y. G. (2012). A novel sediment microbial fuel cell with a biocathode in the rice rhizosphere. *Bioresource Technology*, *108*, 55–59. doi:10.1016/j.biortech.2011.10.040 PMID:22265978

Chhimwal, M., Narayan, M., & Srivastava, R. K. (2017). Electricity generation from microbial fuel cell by using different bio-wastes as substrate. *Environment and Ecology*, *35*(4A), 2958–2964.

Chibuike, G. U., & Obiora, S. C. (2014). Heavy metal polluted soils: Effect on plants and bioremediation methods. *Applied and Environmental Soil Science*, *2014*, 752708. doi:10.1155/2014/752708

Chigbo, C., & Batty, L. (2015). Chelate-assisted phytoremediation of cu-pyrene-contaminated soil using Z. *mays*. *Water, Air, and Soil Pollution*, *226*(3), 1–10. doi:10.100711270-014-2277-2

Chilton, M., Chyatte, M., & Breaux, J. (2007). The negative effects of poverty & food insecurity on child development. *The Indian Journal of Medical Research*, *126*, 262–272. PMID:18032801

Chinnamuthu, C., & Boopathi, P. M. (2009). Nanotechnology and agroecosystem. *Madras Agricultural Journal*, *96*, 17–31.

Chirinos, J., Leal, A., & Montilla, J. (2006). Uso de insumos biológicos como alternativa para la agricultura sostenible en la zona sur del Estado Anzoátegui. *Revista Digital Ceniap Hoy*, *11*, 1–7.

Choudhary, B., Gheysen, G., Buysse, J., van der Meer, P., & Burssens, S. (2014). Regulatory options for genetically modified crops in India [Review]. *Plant Biotechnology Journal*, *12*(2), 135–146. doi:10.1111/pbi.12155 PMID:24460889

Choudhary, D. K., & Johri, B. N. (2009). Interactions of *Bacillus* spp. and plants – With special reference to induced systemic resistance (ISR). *Microbiological Research*, *164*(5), 493–513. doi:10.1016/j.micres.2008.08.007 PMID:18845426

Choudhury, M. A., & Kennedy, I. R. (2004). Prospects and potentials for system of biological nitrogen fixation in sustainable rice production. *Biology and Fertility of Soils*, *39*(4), 219–227. doi:10.100700374-003-0706-2

Chouhan, R. S., Vinayaka, A. C., & Thakur, M. S. (2010). Thiol-stabilized luminescent CdTe quantum dot as biological fluorescent probe for sensitive detection of methyl parathion by a fluoro-immune chromatographic technique. *Analytical and Bioanalytical Chemistry*, *397*(4), 1467–1475. doi:10.100700216-009-3433-1 PMID:20101498

Chowdhury, M. S., Rahman, K. S., Selvanathan, V., Nuthammachot, N., Suklueng, M., Mostafaeipour, A., ... Techato, K. (2020). Current trends and prospects of tidal energy technology. *Environment, Development and Sustainability*, 1–16. PMID:33041645

Chugh, G., Siddique, K. H. M., & Solaiman, Z. M. (2021). Nanobiotechnology for agriculture: Smart technology for combating nutrient deficiencies with nanotoxicity challenges. *Sustainability*, *13*(4), 1781. doi:10.3390u13041781

Clark, M. S., Horwath, W. R., Shennan, C., Scow, K. M., Lantini, W. T., & Ferris, H. (1999). Nitrogen, weeds and water as yield-limiting factors in conventional, low-input, and organic tomato systems. *Agriculture, Ecosystems & Environment*, *73*(3), 257–270. doi:10.1016/S0167-8809(99)00057-2

Claus, D., Dietze, H., Gerth, A., Grosser, W., & Hebner, A. (2007). Application of agronomic practice improves phytoextraction on a multipolluted site. *Journal of Environmental Engineering and Landscape Management*, *15*(4), 208–212. doi:10.3846/16486897.2007.9636932

Clemente, R., Walker, D. J., & Bernal, M. P. (2005). Uptake of heavy metals and As by *Brassica juncea* grown in a contaminated soil in Aznalcollar (Spain): The effect of soil amendments. *Environmental Pollution*, *138*(1), 46–58. doi:10.1016/j.envpol.2005.02.019 PMID:15894412

Clemente, Z., Grillo, R., Jonsson, M., Santos, N. Z. P., Feitosa, L. O., Lima, R., & Fraceto, L. F. (2014). Ecotoxicological evaluation of poly(ε-caprolactone) nanocapsules containing triazine herbicides. *Journal of Nanoscience and Nanotechnology*, *14*(7), 4911–4917. doi:10.1166/jnn.2014.8681 PMID:24757962

Cochran, V., Danielson, J., Kolberg, R., & Miller, P. (2006). Dryland cropping in the Canadian Prairies and the U.S. Northern Great Plains In Dryland Agriculture (2nd ed.). American Society of Agronomy Inc. doi:10.2134/agronmonogr23.2ed.c9

Cogoni, C., & Macino, G. (1999). Post transcriptional gene silencing in *Neurospora* by a RecQ DNA helicase. *Science*, *286*(5448), 2342–2344. doi:10.1126cience.286.5448.2342 PMID:10600745

Cohen, M. R., Yamasaki, H., & Mazzola, M. (2005). *Brassica napus* seed meal soil amendment modifies microbial community structure, nitric oxide production and incidence of Rhizoctonia root rot. *Soil Biology & Biochemistry*, *37*(7), 1215–1227. doi:10.1016/j.soilbio.2004.11.027

Combs, G. F. (2001). Selenium in global food systems. *British Journal of Nutrition*, *85*(5), 517–547.

Compost Australia. (2011). *Compost for soils*. Compost Australia. http://www.compostforsoils.com.au

Conrath, U. (2006). Systemic Acquired Resistance. *Plant Signaling & Behavior*, *1*(4), 179–184. doi:10.4161/psb.1.4.3221 PMID:19521483

Conte, M. G., Gaillard, S., Lanau, N., Rouard, M., & Perin, C. (2008). Green PhylDB: A database for plant comparative genomics. *Nucleic Acids Research*, *36*(Database), 991–998. doi:10.1093/nar/gkm934 PMID:17986457

Conway. (1999). *GM Foods can Benefit the Developing Countries*. Academic Press.

Cook, H. F., Valdes, G. S., & Lee, H. C. (2006). Mulch effects on rainfall interception, soil physical characteristics and temperature under Zea mays L. *Soil & Tillage Research*, *91*(1-2), 227–235.

Corbella, C., Garfí, M., & Puigagut, J. (2014). Vertical redox profiles in treatment wetlands as function of hydraulic regime and macrophytes presence: Surveying the optimal scenario for microbial fuel cell implementation. *The Science of the Total Environment*, *470*, 754–758. doi:10.1016/j.scitotenv.2013.09.068 PMID:24184552

Corbella, C., Guivernau, M., Viñas, M., & Puigagut, J. (2015). Operational, design and microbial aspects related to power production with microbial fuel cells implemented in constructed wetlands. *Water Research*, *84*, 232–242. doi:10.1016/j.watres.2015.06.005 PMID:26253894

Cordo, C. A., Monaco, C. I., Segarra, C. I., Simon, M. R., Mansilla, A. Y., Perelló, A. E., Kripelz, N. I., Bayo, D., & Conde, R. D. (2007). *Trichoderma* spp. as elicitors of wheat plant defense responses against *Septoria tritici*. *Biocontrol Science and Technology*, *17*(7), 687–698. doi:10.1080/09583150701527094

Corpoica, Gobernación de Antioquia, & FAO. (2007). *Manual de buenas prácticas agrícolas en la producción de fríjol voluble*. Medellín: FAO.

Corradini, E., Moura, M. R., & Mattoso, L. H. C. (2010). A preliminary study of the incorporation of NPK fertilizer into chitosan nanoparticles. *Express Polymer Letters*, *4*(8), 509–515. doi:10.3144/expresspolymlett.2010.64

Cortes-Barco, A. M., Goodwin, P. H., & Hsiang, T. (2010). Comparison of induced resistance activated by benzothiadiazole, (2R,3R)-butanediol and an isoparaffin mixture against anthracnose of *Nicotiana benthamiana*. *Plant Pathology*, *59*(4), 643–653. doi:10.1111/j.1365-3059.2010.02283.x

Cortes-Lobos, R. (2013). *Nanotechnology research in the US agri-food sectoral system of innovation: towards sustainable development* [Dissertation]. Georgia Institute of Technology.

Costa, N. M. B., & Rosa, C. O. B. (2011). *Functional Foods: Bioactive Components and Physiological Effects. 1 Reprint*. Rúbio.

Coviella, C., & Trumble, J. (1999). Effects of elevated atmospheric carbon dioxide on insect plant interactions. *Conserv. Biol*, *13*(4), 700–712. doi:10.1046/j.1523-1739.1999.98267.x

Cox, A., Venkatachalam, P., Sahi, S., & Sharma, N. (2017). Reprint of: silver and titanium dioxide nanoparticle toxicity in plants: a review of current research. *Plant Physiology and Biochemistry*, *110*, 33–49. doi:10.1016/j.plaphy.2016.08.007 PMID:27569179

Cozzolino, S. M. F. (2012). *The Bioavailability of Nutrients* (4th ed.). Monole.

Cozzolino, S. M. F., & Cominetti, C. (2013). *Biochemical and Physiological Bases of Nutrition in Different Stages of Life in Health and Disease* (1st ed.). Monole.

Cransberg, L., & McFarlane, D. J. (1994). Can perennial pastures provide the basis for a sustainable farming system in southern Australia. *New Zealand Journal of Agricultural Research*, *37*(3), 287–294. doi:10.1080/00288233.1994.9513067

Cromar, H. E., Murphy, S. D., & Swanton, C. J. (1999). Influence of tillage and crop residue on post dispersal predation of weed seeds. *Weed Science*, *47*(2), 184–194. doi:10.1017/S0043174500091608

Crop Protection Manager. (1997). Insect Management- Insecticides Do Have a Role in IPM. Author.

Cruz, I. B. M., Taufer, M., & Schwanke, C. H. A. (2003). Genomics in the era of aging and its potential application in gerontology and geriatrics. In A. C. A. Souza (Ed.), Institute of Geriatrics and Gerontology PUCRS: The Cradle of Academic Geriatrics In Brazil (pp. 83–84). Academic Press.

Cruz, M. F. A., Diniz, A. P. C., Rodrigues, F. A., & Barros, E. G. (2011). Foliar application of products on the reduction of wheat blast severity. *Tropical Plant Pathology*, *36*, 424–428. doi:10.1590/S1982-56762011000600014

Cruz, R., & Husain, T. (2008). Effect of vesicular arbuscular mycorrhiza (VAM) fungi inoculation on coppicing ability and drought resistance of Senna spectabilis. *Pakistan Journal of Botany*, *40*(5), 2217–2224.

Cui, H., Qi, L., Gao, S., & Shang, J. (2017). Strong adsorption of arsenic species by indistinct zirconium oxide nanoparticles. *Journal of Industrial and Engineering Chemistry*, *18*(4), 1418–1427. doi:10.1016/j.jiec.2012.01.045

Cullen, W. R., & Reimer, K. J. (2017). *Arsenic speciation in the climate. Substance Reviews, 89, 713-764.*

Cutlip, J. (2006). Green roofs a sustainable technology. *Docstoc Report*, 23.

Czech, B., Malone, C. D., Zhou, R., Stark, A., Schlingeheyde, C., Dus, M., Perrimon, N., Kellis, M., Wohlschlegel, J. A., Sachidanandam, R., Hannon, G. J., & Brennecke, J. (2008). An endogenous small interfering RNA pathway in *Drosophila. Nature*, *453*(7196), 798–802. doi:10.1038/nature07007 PMID:18463631

Dagoumas, A. S., Polemis, M. L., & Soursou, S. E. (2020). Revisiting the impact of energy prices on economic growth: Lessons learned from the European Union. *Economic Analysis and Policy*, *66*, 85–95. doi:10.1016/j.eap.2020.02.013

DalCorso, G., Fasani, E., Manara, A., Visioli, G., & Furini, A. (2019). Heavy metal pollutions: State of the art and innovation in phytoremediation. *International Journal of Molecular Sciences*, *20*(14), 3412. doi:10.3390/ijms20143412 PMID:31336773

Dalmiel, L., Vargas, T., & Molina, A. R. (2012). Nutritional genomics for the characterization of the effect of bioactive molecules in lipid metabolism and related pathways. *Electrophoresis Journal*, *33*(15), 2266–2289.

Dalton, D. A., Ma, C., Shrestha, S., Kitin, P., & Strauss, S. H. (2011). Trade-offs between biomass growth and inducible biosynthesis of polyhydroxybutyrate in transgenic poplar. *Plant Biotechnology Journal*, *9*(7), 759–767. doi:10.1111/j.1467-7652.2010.00585.x PMID:21265995

Dandekar, A. M., Teo, G., Defilippi, B. G., Uratsu, S. L., Passey, A. J., Kader, A. A., Stow, J. R., Colgan, R. J., & James, D. J. (2004). Effect of down-regulation of ethylene biosynthesis on fruit flavor complex in apple fruit. *Transgenic Research*, *13*(4), 373–384. doi:10.1023/B:TRAG.0000040037.90435.45 PMID:15517996

Dang, J. C., Kumaria, S., Kumar, S., & Tandon, P. (2011). Micropropagation of Ilex- khasiana, a critically endangered and endemic holly of Northeast India. *AoB Plants*, *2011*, 1–7. doi:10.1093/aobpla/plr012 PMID:22476482

Darn, V. B. H., Doan, H. D., Dang-Vu, T., & Lohi, A. (2017). Harmony and energy of biosorption of cadmium(II) and copper(II) particles by wheat straw. *Bioresource Technology*, *100*, 211–219.

Darwin, R., Tsigas, M., Lewandrowski, J., & Raneses, A. (1995). World agriculture and climate change: economic adaptations. Agricultural Economic Report 703, US Department of Agriculture.

Darwin, R. F. (1999). A Farmer's View of the Ricardian Approach to Measuring Agricultural Effects of Climatic Change. *Climatic Change*, *41*(3-4).

Das, S., Bhattacharya, A., Debnath, N., Datta, A., & Goswami, A. (2013). Nanoparticle-induced morphological transition of *Bombyxmori* nucleopolyhedrovirus: A novel method to treat silkworm grasserie disease. *Applied Microbiology and Biotechnology*, *97*(13), 6019–6030. doi:10.100700253-013-4868-z PMID:23588933

Dastager, S. G., Deepa, C. K., & Pandey, A. (2010). Isolation and characterization of novel plant growth promoting *Micrococcus* sp NII-0909 and its interaction with cowpea. *Plant Physiology and Biochemistry*, *48*(12), 987–992. doi:10.1016/j.plaphy.2010.09.006 PMID:20951599

Davídková, V. (2020, December). Geothermal Potential of Mining Water in the Bílina Mines. *IOP Conference Series. Earth and Environmental Science*, *609*(1), 012063. doi:10.1088/1755-1315/609/1/012063

Davidovich-Rikanati, R., Sitrit, Y., & Tadmor, Y. (2007). Enrichment of tomato flavor by diversion of the early plastidial terpenoid pathway. *Nature Biotechnology*, *25*, 899–901.

Davis, U. C. (2012). *Nutrigenomics is a multidisciplinary science*. The NCMHD Center of Excellence for Nutritional Genomics.

Davuluri, G. R., Tuinen, A., & Fraser, P. D. (2005). Fruit-specific RNAi-mediated suppression of DET1 enhances carotenoid and flavonoid content in tomatoes. *Nature Biotechnology*, *23*(7), 890–895. doi:10.1038/nbt1108 PMID:15951803

De Koning, G. (1995). Physical properties of bacterial poly (R)-3-hydroxyalkanoates). *Canadian Journal of Microbiology, 41*(13), 303–309. doi:10.1139/m95-201

De Lorenzo, D. (2008). Nutrigenomics: For a personalized nutrition. Journal Digitalis, 22, 1–3.

De Pinto, A., Cenacchi, N., Kwon, H. Y., Koo, J., & Dunston, S. (2020). Climate smart agriculture and global food-crop production. *PLoS One, 15*(4), e0231764.

De Tourdonnet, S., Triomphe, B., & Scopel, E. (2010). *Ecological, Technical and Social Innovation Processes in Conservation Agriculture.* Paper presented at Innovation and Sustainable Development in Agriculture and Food (ISDA), Montpellier, France.

De, A., Bose, R., Kumar, A., & Mozumdar, S. (2014). Worldwide pesticide use. In *Targeted delivery of pesticides using biodegradable polymeric nanoparticles* (pp. 5–6). Springer. doi:10.1007/978-81-322-1689-6_2

Debnath, N., Das, S., Seth, D., Chandra, R., Bhattacharya, S. C., & Goswami, A. (2011). Entomotoxic effect of silica nanoparticles against *Sitophilus oryzae* (L.). *Journal of Pest Science, 84*(1), 99–105. doi:10.100710340-010-0332-3

DeFries, R., Franzo, J., Remans, R., Palm, C., Wood, S., & Anderman, T. L. (2015). Metrics for land-scarce agriculture. *Science, 349*(6245), 238–240.

Deka, B., & Arjuna, M. (2014). Effect of Plant Growth Regulators on in vitro propagation of Tagetes erecta Linn [Review]. *Indian Journal of Basic and Applied Medical Research, 3*, 15–23.

Delbaere, B., Mikos, V., & Pulleman, M. (2014). European Policy Review: Functional agrobiodiversity supporting sustainable agriculture. *Journal for Nature Conservation, 22*(3), 193–194.

Deliopoulos, T., Kettlewell, P. S., & Hare, M. C. (2010). Fungal disease suppression by inorganic salts: A review. *Crop Protection (Guildford, Surrey), 29*(10), 1059–1075. doi:10.1016/j.cropro.2010.05.011

Dell'Anna, F. (2021). Green jobs and energy efficiency as strategies for economic growth and the reduction of environmental impacts. *Energy Policy, 149*, 112031. doi:10.1016/j.enpol.2020.112031

Delong, E. F. (2009). The microbial ocean from genomes to biomes. *Nature, 459*(7244), 200–206. doi:10.1038/nature08059 PMID:19444206

Denardo, J. C., Jarret, A. R., Manbeck, H. B., Beattie, D. J., & Berghage, R. D. (2003). Stormwater detention and retention abilities of green roofs. In *World water and environmental resources congress and related symposia.* American Society of Civil Engineers. 10.1061/40685(2003)310

Denman, K. L., Astuk, S. N., & Zimmerman, A. S. (2007). Couplings between Changes in the Climate System and Biogeochemistry. In *Climate Change 2007: The Physical Science Basis, Contribution of Working Group I to the Fourth Assessment Report of the Intergovernmental Panel on Climate Change.* Cambridge University Press.

Denys, C., & Tscharntke, T. (2002). Plant-insect communities and predator-prey ratios in field margin strips, adjacent crop fields, and fallows. *Oecologia, 130*(2), 315–324. doi:10.1007004420100796 PMID:28547156

Departemen, A. R. I. Syaamil Al-Qur'an. (2004). *Environmental Benefits of Organic Farming.* www.organicabout.com

De, R., Rao, Y. Y., & Ali, W. (1983). Grain and fodder legumes as preceding crops affecting the yield and N economy of rice. *The Journal of Agricultural Science, 101*(2), 463–466.

DeRosa, M. C., Monreal, C., Schnitzer, M., Walsh, R., & Sultan, Y. (2010). Nanotechnology in fertilizers. *Nature Nanotechnology, 5*(2), 91. doi:10.1038/nnano.2010.2 PMID:20130583

Devi, S. I., Somkuwar, B., Potshangbam, M., & Talukdar, N. C. (2012). Genetic characterization of Burkholderia cepacia strain from Northeast India: A potential bio-control agent. *Advances in Bioscience and Biotechnology*, *3*(8), 1179–1188. doi:10.4236/abb.2012.38144

Dexter, A. R. (1997). Physical properties of tilled soils. *Soil & Tillage Research*, *43*(1-2), 41–63.

DFID. (2006). *A Synthesis of Urban and Peri-Urban Agricultural Research Commissioned by the RNRRS Programme*. Department for International Development.

Dhankher, O. P., Shasti, N. A., Rosen, B. P., Fuhrmann, M., & Meagher, R. B. (2003). Increased cadmium tolerance and accumulation by plants expressing bacterial arsenate reductase. *The New Phytologist*, *159*, 431–441.

Di Simine, C. D., Sayed, J. A., & Gadd, G. M. (1998). Solubilization of zinc phosphate by a strain of *Pseudomonas fluorescens* isolated from a forest soil. *Biology and Fertility of Soils*, *28*(1), 87–94. doi:10.1007003740050467

Diaz-Perez, J. C., Randle, W. M., Boyhan, G., Walcott, R. W., Giddings, D., Bertrand, D., Sanders, H. F., & Gitaitis, R. D. (2004). Effects of mulch and irrigation system on sweet onion: 1. Bolting, plant growth, and bulb yield and quality. *Journal of the American Society for Horticultural Science*, *129*(2), 218–224. doi:10.21273/JASHS.129.2.0218

Dikshit, A., Shukla, S. K., & Mishra, R. K. (2013). *Exploring nanomaterials with PGPR in current agricultural scenario*. Lap Lambert Academic Publishing.

Dimkpa, C. O. (2014). Can nanotechnology deliver the promised benefits without negatively impacting soil microbial life? *Journal of Basic Microbiology*, *54*(9), 889–904. doi:10.1002/jobm.201400298 PMID:24913194

Dimkpa, C. O., & Bindraban, P. S. (2017). Nanofertilizers: New products for the industry? *Journal of Agricultural and Food Chemistry*. Advance online publication. doi:10.1021/acs.jafc.7b02150 PMID:28535672

Din, M., & Barozai, M.Y. (2014). Profiling microRNAs and their targets in an important fleshy fruit: tomato (Solanum lycopersicum). *Gene, 10*(535), 198-203.

Ding, D., Zhang, L., Wang, H., Liu, Z., Zhang, Z., & Zheng, Y. (2009). Differential expression of miRNAs in response to salt stress in maize roots. *Annals of Botany*, *103*(1), 29–38. doi:10.1093/aob/mcn205 PMID:18952624

Diretto, G., Al-Babili, S., Tavazza, R., Papacchioli, V., & Giuliano, G. (2007). Metabolic engineering of potato carotenoid content through tuber-specific overexpression of a bacterial mini-pathway. *PLoS One*, *2*, e350.

Dixit, R., Wasiullah, Malaviya, D., Pandiyan, K., Singh, U., Sahu, A., Shukla, R., Singh, B., Rai, J., Sharma, P., Lade, H., & Paul, D. (2015). Bioremediation of heavy metals from soil and aquatic environment: An overview of principles and criteria of fundamental processes. *Sustainability*, *7*(2), 2189–2212. doi:10.3390u7022189

Dixit, S. (2020). Solar technologies and their implementations: A review. *Materials Today: Proceedings*, *28*, 2137–2148. doi:10.1016/j.matpr.2020.04.134

Dixon, R. A., Gang, D. R., & Charlton, A. J. (2006). Applications of metabolomics in agriculture. *Journal of Agricultural and Food Chemistry*, *54*(24), 8984–8994.

Dobbs, T. L., & Smolik, J. D. (1996). Productivity and profitability of conventional and alternative farming systems: A long-term on-farm paired comparison. *Journal of Sustainable Agriculture*, *91*(1), 63–79. doi:10.1300/J064v09n01_06

Dobermann, A., & Witt, C. (2000). The potential impact of crop intensification on carbon and nitrogen cycling in intensive rice systems. *Carbon and nitrogen dynamics in flooded soils*, 1-25.

Dobrogojski, J., Spychalski, M., Lucinski, R., & Borek, S. (2018). Transgenic plants as a source of polyhydroxyalkanoates. *Acta Physiologiae Plantarum*, *40*(9), 162–178. doi:10.100711738-018-2742-4

Dodo, H. W., Konan, K. N., Chen, F. C., Egnin, M., & Viquez, O. M. (2008). Alle viating peanut allergy using genetic engineering: The silencing of the immunodominant allergen Arah2 leads to its significant reduction and a decrease in peanut allergenicity. *Plant Biotechnology Journal*, *6*(2), 135–145. doi:10.1111/j.1467-7652.2007.00292.x PMID:17784907

Dogan, K., Kamail, C. I., Mustafa, G. M., & Ali, C. (2011). Effect of different soil tillage methods on rhizobial nodulation, biomass and nitrogen content of second crop soybean. *African Journal of Microbiological Research*, *5*(20), 3186–3194. doi:10.5897/AJMR11.165

Doherty, L., Zhao, Y., Zhao, X., Hu, Y., Hao, X., Xu, L., & Liu, R. (2015). A review of a recently emerged technology: Constructed wetland–microbial fuel cells. *Water Research*, *85*, 38–45. doi:10.1016/j.watres.2015.08.016 PMID:26295937

Domiciano, G. P., Rodrigues, F. A., Guerra, A. M. N., & Vale, F. X. R. (2013). Infection process of *Bipolaris sorokiniana* on wheat leaves is affected by silicon. *Tropical Plant Pathology*, *38*(3), 258–263. doi:10.1590/S1982-56762013005000006

Domiciano, G. P., Rodrigues, F. A., Vale, F. X. R., Filha, M. S. X., Moreira, W. R., Andrade, C. C. L., & Pereira, S. C. (2010). Wheat resistance to spot blotch potentiated by silicon. *Journal of Phytopathology*, *158*(5), 334–343. doi:10.1111/j.1439-0434.2009.01623.x

Dong, Q., Lawrence, C. J., Schlueter, S. D., Wilkerson, M. D., Kurtz, S., Lushbough, C., & Brendel, V. (2005). Comparative plant genomics resources at Plant GDB. *Plant Physiology*, *139*(2), 610–618. doi:10.1104/pp.104.059212 PMID:16219921

Dore, T., Makowski, D., Malézieux, E., Munier-Jolain, N., Tchamitchian, M., & Tittonell, P. (2011). Facing up to the paradigm of ecological intensification in agronomy: Revisiting methods, concepts and knowledge. *European Journal of Agronomy*, *34*(4), 197–210. doi:10.1016/j.eja.2011.02.006

dos Santos, R.M., Diaz, P.A.E, Lobo, L.L.B., & Rigobelo, E.C. (2020). Use of plant growth-promoting rhizobacteria in maize and sugarcane: characteristics and applications. *Frontiers in Sustainable Food Systems*, *4*(136), 1-15.

Dreistadt, S. H. (2004). *Pests of Landscape Trees and Shrubs: An Integrated Pest Management Guide, Publication 3359*. University of California Division of Agriculture and Natural Resources.

Drennan, P. M., Smith, M. T., Goldsworthy, D., & van Staden, J. (1993). The occurrence of trehalose in the leaves of the desiccation-tolerant angiosperm Myrothamnus flabellifolius welw. *Journal of Plant Physiology*, *142*(4), 493–496. doi:10.1016/S0176-1617(11)81257-5

Drinkwater, L. E., Wagoner, P., & Sarrantonio, M. (1998). Legume-based cropping systems have reduced carbon and nitrogen losses. *Nature*, *396*(6708), 262–265. doi:10.1038/24376

Duan, Z., Lv, G., Shen, C., Li, Q., Qin, Z., & Niu, J. (2014). The role of jasmonic acid signalling in wheat (*Triticum aestivum* L.) powdery mildew resistance reaction. *European Journal of Plant Pathology*, *140*(1), 169–183. doi:10.100710658-014-0453-2

Dubey, R. K., & Nidhi, S. (2014). Organic Farming: An Eco-Friendly Technology and Its Importance And Opportunities in the Sustainable Development. *International Journal of Innovative Research in Science, Engineering and Technology*, *3*, 3.

Dubock, A. (2014). The politics of Golden Rice. *GM Crops and Food: Biotechnology in Agriculture and the Food Chain*, *5*(3), 210–222. doi:10.4161/21645698.2014.967570 PMID:25437240

Dubock, A. (2014). The present status of Golden Rice. *Huazhong Nongye Daxue Xuebao*, *33*(6).

Duhan, J. S., Kumar, R., Kumar, N., Kaur, P., Nehra, K., & Duhan, S. (2017). Nanotechnology: The new perspective in precision agriculture. *Biotechnology Reports (Amsterdam, Netherlands)*, *15*, 11–23. doi:10.1016/j.btre.2017.03.002 PMID:28603692

Dunnett, C. W. (1955). A multiple comparisons procedure for comparing several treatments with a control. *Journal of the American Statistical Association*, *50*(272), 1096–1121. doi:10.1080/01621459.1955.10501294

Dunnett, C. W. (1964). New tables for multiple comparisons with a control. *Biometrics*, *20*(3), 482–491. doi:10.2307/2528490

Dunnett, N., & Kingsbury, N. (2004). *Planting green roofs and living walls. Academic Press.*

Durrant, W. E., & Dong, X. (2004). Systemic acquired resistance. *Annual Review of Phytopathology*, *42*(1), 185–209. doi:10.1146/annurev.phyto.42.040803.140421 PMID:15283665

Dutt, S., Pandey, D., & Kumar, A. (2011). Jasmonate signal induced expression of cystatin genes for providing resistance against Karnal bunt in wheat. *Plant Signaling & Behavior*, *6*(6), 821–830. doi:10.4161/psb.6.6.14743 PMID:21849815

Du, W., Tan, W., Peralta-Videa, J. R., Gardea-Torresdey, J. L., Ji, R., Yin, Y., & Guo, H. (2017). Interaction of metal oxide nanoparticles with higher terrestrial plants: Physiological and biochemical aspects. *Plant Physiology and Biochemistry*, *110*, 210–225. doi:10.1016/j.plaphy.2016.04.024 PMID:27137632

Dwivedi, N., Kumar, P., Dwivedi, S. K., Kumar, P., Sonowal, K., & Singh, J. (2017). Green technology: An eco-friendly approach towards sustainable agriculture. (G-. *Journal of Environmental Science and Technology*, *4*(4).

Dwyer, J. T., Wiemer, K. L., Dary, O., Keen, C. L., King, J. C., Miller, K. B., Philbert, M. A., Tarasuk, V., Taylor, C. L., Gaine, P. C., Jarvis, A. B., & Bailey, R. L. (2015). Fortification and Health: Challenges and Opportunities. *Advances in Nutrition*, *6*(1), 124–131. doi:10.3945/an.114.007443 PMID:25593151

Eady, C. C., Kamoi, T., Kato, M., Porter, N. G., Davis, S., Shaw, M., Kamoi, A., & Imai, S. (2008). Silencing onion lachrymatory factor synthase causes a significant change in the sulfur secondary metabolite profile. *Plant Physiology*, *147*(4), 2096–2106. doi:10.1104/pp.108.123273 PMID:18583530

Easterling, W. E., Aggarwal, P. K., Batima, P., Brander, K. M., Erda, L., Howden, S. M., Kirilenko, A., Morton, J., Soussana, J. F., Schmidhuber, J., & Tubiello, F. N. (2007). Food, fibre and forest products. Climate Change 2007: Impacts, adaptation and vulnerability. Contribution of working group II to the fourth assessment report of Intergovernmental Panel on Climate Change, 273-313.

Ebbs, S. D., Lasat, M., Brady, D. J. M., Cornish, J., Gordon, R., & Kochian, V. (1997). Phytoextraction of cadmium and zinc from a contaminated soil. *Journal of Environmental Quality*, *26*, 1424–1430.

EC. (2012). *Communication from the Commission to the European Parliament, the Council, the European Economic and Social Committee and the Committee of the Regions. 'A European strategy for key enabling technologies-A bridge to growth and jobs'*. https://eur-lex.europa.eu/LexUriServ/ LexUriServ.do?uri=COM: 2012:0341:FIN:EN:PDF

Edwards, S. (2007). The impact of compost use on crop yields in Tigray, Ethiopia. Institute for Sustainable Development (ISD). In *Proceedings of the International Conference: Organic Agriculture and Food Security*. FAO.

Edwards, D., & Batley, J. (2010). Plant genome sequencing: Applications for crop improvement. *Plant Biotechnology Journal*, *8*(1), 2–9. doi:10.1111/j.1467-7652.2009.00459.x PMID:19906089

Egamberdieva, D., & Lugtenberg, B. (2014). Use of Plant Growth-Promoting Rhizobacteria to Alleviate Salinity Stress in Plants. In *Use of Microbes for the Alleviation of Soil Stresses* (Vol. 1, pp. 73–96). Springer. doi:10.1007/978-1-4614-9466-9_4

Egamberdiyeva, D. (2007). The effect of plant growth promoting bacteria on growth and nutrient uptake of maize in two different soils. *Applied Soil Ecology*, *36*(2-3), 184–189. doi:10.1016/j.apsoil.2007.02.005

Egbe, Bar, & Mary. (2010). Pigeonpea/sorghum Intercropping in southern guinea savanna: Effect off Planting Density of Pigeonpea. *Nature and Science*, *8*(11), 156–167.

Ekstrom, G., & Barbara, E. (2011). Pest control in agro-ecosystems: An ecological approach. *Critical Reviews in Plant Sciences*, *30*(1-2), 74–94. doi:10.1080/07352689.2011.554354

El Janati, M., Akkal-Corfini, N., Bouaziz, A., Oukarroum, A., Robin, P., Sabri, A., ... Thomas, Z. (2021). Benefits of Circular Agriculture for Cropping Systems and Soil Fertility in Oases. *Sustainability*, *13*(9), 4713.

Elayaraja, M., & Vijai, C. (2020). Organic farming in India: Benefits and Challenges. *Eur. J. Mol. Clin. Med.*, *7*(11), 3021–3029.

Elbashir, S. M., Lendeckel, W., & Tuschl, T. (2001). RNA interference is mediated by 21-and22-nucleotide RNAs. *Genes & Development*, *15*(2), 188–200. doi:10.1101/gad.862301 PMID:11157775

Elevitch, C. R., Mazaroli, D. N., & Ragone, D. (2018). Agroforestry standards for regenerative agriculture. *Sustainability*, *10*(9), 3337.

El-Ramady, H. R. (2014). Integrated nutrient management and postharvest of crops. *Sustainable Agriculture Research*, *13*, 163–274.

El-Sharkawy, H. H. A., Rashad, Y. M., & Ibrahim, S. A. (2018). Biocontrol of stem rust disease of wheat using arbuscular mycorrhizal fungi and *Trichoderma* spp. *Physiological and Molecular Plant Pathology*, *103*, 84–91. Advance online publication. doi:10.1016/j.pmpp.2018.05.002

Emadian, S. E. (2017). *Physiological responses of loblolly pine (Finustaeda L.) to silicon and water stress* [PhD thesis]. Texas A & M Univ, College Station, TX, USA.

Emerson, R. A. (1993). *Experiments in mulching garden vegetables.* University of Nebraska, Agricultural Experiment Station of Nebraska.

Endo, N., Yoshida, K., Akiyoshi, M., & Manji, S. (2006). Hybrid fiber production: A wood and plastic combination in transgenic rice and *Tamarix* made by accumulating poly-3-hydroxybutyrate. *Plant Biotechnology (Sheffield, England)*, *23*(1), 99–109. doi:10.5511/plantbiotechnology.23.99

Engler, N., & Krarti, M. (2021). Review of energy efficiency in controlled environment agriculture. *Renewable & Sustainable Energy Reviews*, *141*, 110786. doi:10.1016/j.rser.2021.110786

Environment Canada. (1993). *Proposed Guidelines for Registration of Chemical Pesticides: NonTarget Plant Testing and Evaluation.* Canadian Wildlife Service, Technical Report Series No. 145.

Environment Canada. (1999). Biological Test Method: Test for Measuring the Inhibition of Growth Using the Freshwater Macrophyte, Lemna minor. EPS 1/RM/37. Author.

Enwezor, W. O. (1996). The mineralization of nitrogen and phosphorus in organic materials of varying C:N and C:P ratios. *Plant and Soil*, *44*(1), 237–240. doi:10.1007/BF00016972

Erbs, G., & Newman, M.-A. (2003). The role of lipopolysaccharides in induction of plant defence responses. *Molecular Plant Pathology*, *4*(5), 421–425. doi:10.1046/j.1364-3703.2003.00179.x PMID:20569401

Eriksson, J. E. (1989). The influence of pH, soil type and time on adsorption and uptake by plants of Cd added to the soil. *Water, Air, and Soil Pollution*, *48*(3-4), 317–335. doi:10.1007/BF00283334

Erisman, J. W., Eekeren, N. V., Doorn, A. V., Geertsema, W., & Polman, N. (2017). Measures for nature-inclusive agriculture. *Alterra-rapport-Wageningen University and Research Centre*, (2821).

Erisman, J. W., van Eekeren, N., de Wit, J., Koopmans, C., Cuijpers, W., Oerlemans, N., & Koks, B. J. (2016). Agriculture and biodiversity: A better balance benefits both. *AIMS Agriculture and Food*, *1*(2), 157–174.

Escobar, M. A., Civerolo, E. L., Summerfelt, K. R., & Dandekar, A. M. (2001). RNAi-mediated oncogene silencing confers resistance to Crown Gall Tumorigenesis. *Proceedings of the National Academy of Sciences of the United States of America*, *98*(23), 13437–13442. doi:10.1073/pnas.241276898 PMID:11687652

Essefi, E. (2020). On the Geochemistry and Mineralogy of the Anthropocene. *Int J Water Wastewater Treat*, *6*(2).

Essefi, E. (2021). Record of the Anthropocene-Great Acceleration along a core from the coast of Sfax, southeastern Tunisia. Turkish Journal of Earth Sciences Fthenakis, V. M., Kim, H. C., & Alsema, E. (2008). Emissions from photovoltaic life cycles. *Environmental Science & Technology*, *42*(6), 2168–2174.

Fabian, Y., Sandau, N., Bruggisser, O. T., Aebi, A., Kehrli, P., Rohr, R. P., ... Bersier, L. F. (2013). The importance of landscape and spatial structure for hymenopteran-based food webs in an agro-ecosystem. *Journal of Animal Ecology*, *82*(6), 1203–1214.

Facknath, S., & Lalljee, B. (2000). *Allelopathic strategies for eco-friendly crop protection*. Springer. doi:10.1007/978-94-011-4173-4_3

Fajer, E. D., & Bazzaz, F. A. (1992). Is carbon dioxide a 'good' greenhouse gas? Effects of increasing carbon dioxide on ecological systems. *Global Environmental Change*, *2*(4), 301–310. doi:10.1016/0959-3780(92)90047-B

Fan, S. (2015). *How agriculture can improve health and nutrition*. World Economic Forum Report.

Fang, Z., Cao, X., Li, X., Wang, H., & Li, X. (2017). Electrode and azo dye decolorization performance in microbial-fuel-cell-coupled constructed wetlands with different electrode size during long-term wastewater treatment. *Bioresource Technology*, *238*, 450–460. doi:10.1016/j.biortech.2017.04.075 PMID:28463809

Fang, Z., Song, H. L., Cang, N., & Li, X. N. (2013). Performance of microbial fuel cell coupled constructed wetland system for decolorization of azo dye and bioelectricity generation. *Bioresource Technology*, *144*, 165–171. doi:10.1016/j.biortech.2013.06.073 PMID:23867535

Fang, Z., Song, H. L., Cang, N., & Li, X. N. (2015). Electricity production from Azo dye wastewater using a microbial fuel cell coupled constructed wetland operating under different operating conditions. *Biosensors & Bioelectronics*, *68*, 135–141. doi:10.1016/j.bios.2014.12.047 PMID:25562740

Fan, S., & Pandya-Lorch, R. (2012). *Reshaping Agriculture for Nutrition and Health*. International Food Policy Research Institute.

Fan, Y., Han, S. K., & Liu, H. (2012). Improved performance of CEA microbial fuel cells with increased reactor size. *Energy & Environmental Science*, *5*(8), 8273–8280. doi:10.1039/c2ee21964f

FAO Study-World Agriculture. (2015). *Towards 2015/2030 Summary Report*. www.fao.org

FAO, IFAD, UNICEF, WFP, & WHO. (2018). The State of Food Security and Nutrition in the World 2017. FAO.

FAO. (2020). *Responding to the impact of the COVID-19 outbreak on food value chains through efficient logistics*. Report of activity No.166 from 04.06.2020 to 06.07.2020. https://www.fao.org/fsnforum/activities/discussions/covid19_logistics

FAO/LEG. (1996). Regional compendium on aquaculture and inland fisheries legislation (Asian Region). FAO Development Law Service. FAO.

Fasim, F., Ahmed, N., Parsons, R., & Gadd, G. M. (2002). Solubilization of zinc salts by a bacterium isolated from the air environment of a tannery. *FEMS Microbiology Letters*, *213*(1), 1–6. doi:10.1111/j.1574-6968.2002.tb11277.x PMID:12127480

Fasusi, O.A., Cruz, C., & Babalola, O.O. (2021). Agricultural sustainability: microbial biofertilizers in rhizosphere management. *Agriculture, 11*(163), 1-19.

Fay, P. (1983). *The blue-greens*. Edward Arnold.

Ferguson, L. R. (2006). Nutrigenomics: Integrating genomic approaches into nutrition research. *Molecular Diagnosis & Therapy*, *10*(2), 101–108.

Fialho, E., Moreno, F. S., & Ong, T. P. P. (2008). Nutrition in the post-genomics: Fundamentals and applications of omics tools. *The Journal of Nutrition*, *21*(6), 757–766.

Figueroa, M., Hammond-Kosack, K. E., & Solomon, P. S. (2018). A review of wheat diseases-a field perspective. *Molecular Plant Pathology*, *19*(6), 1523–1536. doi:10.1111/mpp.12618 PMID:29045052

Filha, M. S. X., Rodrigues, F. A., Domiciano, G. P., Oliveira, H. V., Silveira, P. R., & Moreira, W. R. (2011). Wheat resistance to leaf blast mediated by silicon. *Australasian Plant Pathology*, *40*(1), 28–38. doi:10.100713313-010-0010-1

Finamore, A., Britti, M. S., Roselli, M., Bellovino, D., Gaetani, S., & Mengheri, E. (2004). Novel Approach for food safety evaluation. *Journal of Agricultural and Food Chemistry*, *52*(24), 7425–7431. doi:10.1021/jf049097p PMID:15563230

Firth, A. G., Baker, B. H., Brooks, J. P., Smith, R., Iglay, R. B., & Davis, J. B. (2020). Low external input sustainable agriculture: Winter flooding in rice fields increases bird use, fecal matter and soil health, reducing fertilizer requirements. *Agriculture, Ecosystems & Environment*, *300*, 106962.

Fischer, R. A., & Connor, D. J. (2018). Issues for cropping and agricultural science in the next 20 years. *Field Crops Research*, *222*, 121–142.

Flores, T., Karpova, O., Su, X., Zeng, P., Bilyeu, K., Sleper, D. A., Nguyen, H. T., & Zhang, Z. J. (2008). Silencing of GmFAD3 gene by siRNA leads to low alpha-linolenic acids (18:3) of fad3-mutant phenotype in soybean [*Glycine max* (Merr.)]. *Transgenic Research*, *17*(5), 839–850. doi:10.100711248-008-9167-6 PMID:18256901

Fobert, P. R., & Després, C. (2005). Redox control of systemic acquired resistance. *Current Opinion in Plant Biology*, *8*(4), 378–382. doi:10.1016/j.pbi.2005.05.003 PMID:15922650

Fogg-Johnson, N., & Kaput, J. (2003). Nutrigenomics: An Emerging Scientific Discipline. *Food Technology*, *57*(4), 60–67.

Foley, J. A., Ramankutty, N., & Brauman, K. A. (2011). Solutions for a cultivated planet. *Nature*, *478*, 337–342.

Food and Agriculture Organization of the United Nations (FAO). (1988). *Report of the FAO Council, 94th Session*. FAO.

Forster, H., Adaskaveg, J. E., Kim, D. H., & Stanghellini, M. E. (1998). Effect of phosphite on tomato and pepper plants and on susceptibility of pepper to Phytophthora root and crown rot in hydroponic culture. *Plant Disease*, *82*(10), 1165–1170. doi:10.1094/PDIS.1998.82.10.1165 PMID:30856781

Frey-Klett, P., Burlinson, P., Deveau, A., Barret, M., Tarkka, M., & Sarniguet, A. (2011). Bacterial-fungal interactions: Hyphens between agricultural, clinical, environmental, and food microbiologists. *Microbiology and Molecular Biology Reviews*, *75*(4), 583–609. doi:10.1128/MMBR.00020-11 PMID:22126995

Frick & Bonn. (2016). *The world of organic agriculture.* FiBL & IFOAM.

Frova, C., Krajewski, P., Fonzo, N. D., Villa, M., & Sari Gorla, M. (1999). Genetic analysis of drought tolerance maize by molecular markers. *Theoretical and Applied Genetics*, *99*(1-2), 280–288. doi:10.1007001220051233

Fujii, T. M. M., Medeiros, R., & Yamada, R. (2010). Nutrigenomics and nutrigenetics: Important concepts for the science of nutrition. *Journal of the Brazilian Society of Food and Nutrition*, *35*(1), 149166.

Fujita, K., Ofosu-Budu, K. G., & Ogata, S. (1992). Biological nitrogen fixation in mixed legume-cereal cropping systems. *Plant and Soil*, *141*(1-2), 155–175. doi:10.1007/BF00011315

Fukudome, A., & Fukuhara, T. (2017). Plant dicer-like proteins: Double-stranded RNA-cleaving enzymes for small RNA biogenesis. *Journal of Plant Research*, *130*(1), 33–44. doi:10.100710265-016-0877-1 PMID:27885504

Funk, C. C., & Brown, M. E. (2009). Declining per capita agricultural production and warming oceans threaten food security? *Food Security*, *1*(3), 271–289. doi:10.100712571-009-0026-y

Fusuo, Z., & Long, L. (2003). Using competitive and facilitative interactions in intercropping system enhances crop productivity and nutrient- use efficiency. *Plant and Soil*, *248*(1/2), 305–312. doi:10.1023/A:1022352229863

Gabriela, F., Casanovas, E. M., & Quillehauquy, V. (2015). *Azospirillum* inoculation effects on growth, product quality and storage life of lettuce plants grown under salt stress. *Scientia Horticulturae*, *195*, 154–162. doi:10.1016/j.scienta.2015.09.015

Gangwar, K. S., Singh, K. K., Sharma, S. K., & Tomar, O. K. (2006). Alternative tillage and crop residue management in wheat after rice in sandy loam soils of Indo-Gangetic Plains. *Soil & Tillage Research*, *88*(1-2), 242–252. doi:10.1016/j.still.2005.06.015

Gao, J. (2020). Green Energy Strategic Management for Service of Quality Composition in the Internet of Things Environment. *Complexity*, *2020*, 2020. doi:10.1155/2020/6678612

Garcia-Casal, M. N., Pena-Rosas, J. P., & Giyose, B. (2016). Staple crops biofortified with increased vitamins and minerals: Considerations for a public health strategy. *Annals of the New York Academy of Sciences*, *1390*(1), 3–13. doi:10.1111/nyas.13293 PMID:27936288

Gardner, S. N., Agrawal, A. A., Gressel, J., & Mangel, M. (1999). Strategies to delay the evolution of resistance in pests: Dose rotations and induced plant defenses. *Aspects of Applied Biology*, *53*, 189–196.

Garg, M., Sharma, N., Sharma, S., Kapoor, P., Kumar, A., Chunduri, V., & Arora, P. (2018). Biofortified crops generated by Breeding, Agronomy, and Transgenic approaches are improving lives of millions of people around the world. *Frontiers in Nutrition*, *5*, 12. doi:10.3389/fnut.2018.00012 PMID:29492405

Garibaldi, L. A., Pérez-Méndez, N., Garratt, M. P., Gemmill-Herren, B., Miguez, F. E., & Dicks, L. V. (2019). Policies for ecological intensification of crop production. *Trends in Ecology & Evolution*, *34*(4), 282–286.

Gaston, K. J., & Williams, P. H. (1996). Spatial patterns in taxonomic diversity. In *Biodiversity 202-229*. Blackwell Science.

Gaur, A. C. (2001). Organic manure: A basic input in organic farming. *Indian Farming*, *51*(3), 3–11.

Gaur, A. C. (2006). *Handbook of organic Farming and Biofertilizers*. Ambica Book Publication.

Gealy, D. R., Anders, M., Watkins, B., & Duke, S. (2014). Crop performance and weed suppression by weed-suppressive rice cultivars in furrow- and flood-irrigated systems under reduced herbicide inputs. *Weed Science*, *62*(2), 303–320. doi:10.1614/WS-D-13-00104.1

Gedge, D., & Frith, M. (2004). Green roofs benefits and cost implications. Birmingham City Council.

Gerhardt, K. E., Gerwing, P. D., & Greenberg, B. M. (2017). Opinion: Taking phytoremediation from proven technology to accepted practice. *Plant Science*, *256*, 170–185. doi:10.1016/j.plantsci.2016.11.016 PMID:28167031

German, J. B. (2005). Genetic dietetics: Nutrigenomics and the future of dietetics practice. *Journal of the American Dietetic Association*, *2005*, 530–531.

Ghadiyali, T. R., & Manish, M. (2012). Contribution of green technology in Sustainable development of Agriculture sector. *Journal of Environmental Research and Development*, *7*(1A), 590–596.

Ghag, S. B., Shekhawat, U. K. S., & Ganapathi, T. R. (2014). Host-induced post-transcriptional hairpin RNA-mediated gene silencing of vital fungal genes confers efficient resistance against *Fusarium* wilt in banana. *Plant Biotechnology Journal*, *12*(5), 541–553. doi:10.1111/pbi.12158 PMID:24476152

Ghany, T. M. A., Alawlaqi, M. M., & Al Abboud, M. A. (2013). Role of biofertilizers in agriculture: A brief review. *Mycopathologia*, *11*, 95–101.

Gharsalli, N., Essefi, E., Baydoun, R., & Yaich, C. (2020). The Anthropocene and great acceleration as controversial epoch of human-induced activities: case study of the Halk El Menjel Wetland, Eastern Tunisia. *Applied Ecology and Environmental Research*, *18*(3), 4137–4166. doi:10.15666/aeer/1803_41374166

Ghildiyal, M., & Zamore, P. D. (2009). Small silencing RNAs: An expanding universe. *Nature Reviews. Genetics*, *10*(2), 94–108. doi:10.1038/nrg2504 PMID:19148191

Gholami, L., Karimi, N., & Kavian, A. (2017). Soil bioengineering Methods used in water management and stabilization slopes steep. *Ecohydrology*, *4*(1), 149–162.

Ghormade, V., Deshpande, M. V., & Paknikar, K. M. (2011). Perspectives for nano-biotechnology enabled protection and nutrition of plants. *Biotechnology Advances*, *29*(6), 792–803. doi:10.1016/j.biotechadv.2011.06.007 PMID:21729746

Ghosh & Bandopadhyay. (2011). Productivity and profitability of coconut based cropping system with fruits and black pepper in West Bengal. *Journal of Crop and Weed*, *7*(2), 134–137.

Ghoshal, S., Pasham, S., Odom, D. B., Furr, H. C., & McGrane, M. M. (2003). Vitamin A depletion is associated with low phosphoenolpyruvate carboxykinase mRNA levels during late fetal development and at birth in mice. *The Journal of Nutrition*, *133*, 2131–2136.

Ghuman, R. S., Brar, L. S., & Walia, U. S. (2008). Role of variety and plant geometry on weed management in transplanted rice (Oryza sativa L.). *Indian Journal of Weed Science*, *40*, 137–141.

Giller, K. E., Cadisch, G., Ehaliotis, C., Adams, E., Sakala, W. D., & Mafongoya, P. L. (1997). Building soil nitrogen capital in Africa. *Replenishing Soil Fertility in Africa, 51*, 151-192.

Giller, K. E., Hijbeek, R., Andersson, J. A., & Sumberg, J. (2021). Regenerative Agriculture: An agronomic perspective. *Outlook on Agriculture*, *50*(1), 13–25.

Gill, K. S., Arshad, M. A., Chivundu, B. K., Phiri, B., & Gumbo, M. (1992). Influence of residue mulch, tillage and cultural practices on weed mass and corn yield from three field experiments. *Soil & Tillage Research*, *24*(3), 211–223. doi:10.1016/0167-1987(92)90088-S

Gillsen, L. J., Bolhaar, S. T., & Matos, C. I. (2005). Silencing of major apple allergen Mald1 by using the RNA interference approach. *The Journal of Allergy and Clinical Immunology*, *115*(2), 364–369. doi:10.1016/j.jaci.2004.10.014 PMID:15696096

Ginneken, L. V., Meers, E., Guisson, R., Ruttens, A., Elst, K., Tack, F. M. G., Vangronsveld, J., Diels, L., & Dejonghe, W. (2007). Phytoremediation for heavy metal-contaminated soils combined with bioenergy production. *Journal of Environmental Engineering and Landscape Management*, *15*(4), 227–236. doi:10.3846/16486897.2007.9636935

Glick, B.R. (2012). Plant growth-promoting bacteria: mechanisms and applications. *Scientifica*, 15.

Gliessman, S. (2013). Agroecology: Growing the roots of resistance. *Agroecology and Sustainable Food Systems*, *37*(1), 19–31.

Gliessman, S. R. (2018). Defining agroecology. *Agroecology and Sustainable Food Systems*, *42*, 599–600.

Godbole, K., Deshmukh, U., & Yajnik, C. (2009). Nutrigenetic determinants of neural tube defects in India. *Indian Pediatrics*, *46*(6), 467–475.

Godfray, H. C. J., Beddington, J. R., & Crute, I. R. (2010). Food security: The challenge of feeding 9 billion people. *Science*, *237*, 812–818.

Goff, S. A., Ricke, D., Lan, T.-H., Presting, G., Wang, R., Dunn, M., Glazebrook, J., Sessions, A., Oeller, P., Varma, H., Hadley, D., Hutchison, D., Martin, C., Katagiri, F., Lange, B. M., Moughamer, T., Xia, Y., Budworth, P., Zhong, J., ... Briggs, S. (2002). A draft sequence of the rice genome (*Oryza sativa* L. ssp. *japonica*). *Science*, *296*(5565), 92–100. doi:10.1126cience.1068275 PMID:11935018

Goldstein, E., Paur, K., Fraser, C., Kenah, E., Wallinga, J., & Lipsitch, M. (2009). Reproductive numbers, epidemic spread and control in a community of households. *Mathematical Biosciences*, *221*(1), 11–25. doi:10.1016/j.mbs.2009.06.002 PMID:19559715

Gomez, A. A., & Gomez, K. A. (1983). Multiple Cropping in the Humid Tropics of Asia. Academic Press.

Gomiero, T., Pimentel, D., & Paoletti, M. G. (2011). Environmental impact of different agricultural management practices: Conventional vs. organic agriculture. *Critical Reviews in Plant Sciences*, *30*(1-2), 95–124. doi:10.1080/07352689.2011.554355

Gong, A.-D., Li, H.-P., Yuan, Q.-S., Song, X.-S., Yao, W., He, W.-J., Zhang, J.-B., & Liao, Y.-C. (2015). Antagonistic Mechanism of Iturin A and Plipastatin A from *Bacillus amyloliquefaciens* S76-3 from Wheat Spikes against Fusarium graminearum. *PLoS One*, *10*(2), e0116871. doi:10.1371/journal.pone.0116871 PMID:25689464

Gopal & Jagadeeswar. (2010). Management of Peanut budnecrosis disease in groundnut through intercropping with cereal and pulse crops. Taylor Francis.

Gopalan, C. (1992). The contribution of nutrition research to the control of under nutrition: The Indian experience. *Annual Review of Nutrition*, *12*, 1–17.

Gopal, M., Gupta, A., Palaniswami, C., Dhanapal, R., & Thomas, G. V. (2010). Coconut leaf vermiwash: A bio-liquid from coconut leaf vermicompost for improving the crop production capacities of soil. *Current Science*, *98*, 1202–1210.

Görlach, J., Voirath, S., & Knauf-Beiter, G. (1996). Benzothiadiazole, a novel class of inducers of systemic acquired resistance, activates gene expression and disease resistance in wheat. *The Plant Cell*, *8*(4), 629–643. doi:10.1105/tpc.8.4.629 PMID:8624439

Goswami, A., Roy, I., Sengupta, S., & Debnath, N. (2010). Novel applications of solid and liquid formulations of nanoparticles against insect pests and pathogens. *Thin Solid Films*, *519*(3), 1252–1257. doi:10.1016/j.tsf.2010.08.079

Goswami, B. K., & Mittal, A. (2004). Management of root-knot nematode infecting tomato by *Trichodermaviride* and *Paecilomyceslilacinus*. *Indian Phytopathology*, *57*, 235–236.

Gottwald, F., & Stein-Bachinger, K. (2016). *Landwirtschaft für Artenvielfalt-Ein Naturschutzstandard fürkologisch bewirtschaftete Betriebe. 2*. WWF-Deutschland. www. landwirtschaft-artenvielfalt. de

Gounou & Schulthess. (2006). Effect of traditional rice/maize intercropping on population densities, crop damage and parasitism of stem-borers in the Ivory Coast. *African Plant Protection*, *12*, 93–102.

Govaerts, B., Verhulst, N., Castellanos-Navarrete, A., Sayre, K. D., Dixon, J., & Dendooven, L. (2009). Conservation agriculture and soil carbon sequestration: Between myth and farmer reality. *Critical Reviews in Plant Sciences*, *28*(3), 97–122.

Govindarajulu, M., Epstein, L., Wroblewski, T., & Michelmore, R. W. (2015). Host-induced gene silencing inhibits the biotrophic pathogen causing downy mildew of lettuce. *Plant Biotechnology Journal*, *13*(7), 875–883. doi:10.1111/pbi.12307 PMID:25487781

Graham, R. D., Welch, R. M., & Bouis, H. E. (2001). Addressing micronutrient malnutrition through enhancing the nutritional quality of staple foods: Principles, perspectives and knowledge gaps. *Advances in Agronomy*, *70*, 77–142. doi:10.1016/S0065-2113(01)70004-1

Grant, M., & Lamb, C. (2006). Systemic immunity. *Current Opinion in Plant Biology*, *9*(4), 414–420. doi:10.1016/j.pbi.2006.05.013 PMID:16753329

Gray, E. J., & Smith, D. L. (2005). Intracellular and extracellular PGPR: Commonalities and distinctions in the plant-bacterium signaling processes. *Soil Biology & Biochemistry*, *37*(3), 395–412. doi:10.1016/j.soilbio.2004.08.030

Green Biotechnology and Climate Change. (2009). Factshect produced by EuropaBio. The European Association for Bioindustries.

Green, J. M., & Beestman, G. B. (2007). Recently patented and commercialized formulation and adjuvant technology. *Crop Protection (Guildford, Surrey)*, *26*(3), 320–327. doi:10.1016/j.cropro.2005.04.018

Gregory, P. J., Johnson, S. N., Newton, A. C., & Ingram, J. S. I. (2009). Integrating pests and pathogens into the climate change/food security debate. *Journal of Experimental Botany*, *60*, 2827–2838.

Griffey, C. A., Das, M. K., & Stromberg, E. L. (1993). Effectiveness of adult plant resistance in reducing grain yield loss to powdery mildew in winter wheat. *Plant Disease*, *77*(6), 618–622. doi:10.1094/PD-77-0618

Grillo, R., Mattos, B. D., & Antunes, D. R. (2021). Foliage adhesion and interactions with particulate delivery systems for plant nanobionics and intelligent agriculture. Vol. 37, Nano Today. *Elsevier B.V.*, *2021*, 101078.

Gruère, G. P. (2012). Implications of nanotechnology growth in food and agriculture in OECD countries. *Food Policy*, *37*(2), 191–198. doi:10.1016/j.foodpol.2012.01.001

Guével, M. H., Menzies, J. G., & Bélanger, R. R. (2007). Effect of root and foliar applications of soluble silicon on powdery mildew control and growth of wheat plants. *European Journal of Plant Pathology*, *119*(4), 429–436. doi:10.100710658-007-9181-1

Guidelines for Hariyali. (2003). https://megsoil.gov.in/docs/HARAYALI_ Guidelines.pdf

Gumel, A. M., Annuar, M. S. M., & Chisti, Y. (2013). Recent advances in the production, recovery and applications of polyhydroxyalkanoates. *Journal of Polymers and the Environment*, *21*(2), 580–605. doi:10.100710924-012-0527-1

Guo, W., Duan, Y., Olivares-Fuster, O., Wu, Z., Arias, C. R., Burns, J. K., & Grosser, J. W. (2005). Protoplast transformation and regeneration of transgenic Valencia sweet orange plants containing a juice quality-related pectin methylesterase. *Plant Cell Reports*, *24*(8), 482–486. doi:10.100700299-005-0952-x PMID:15875190

Gupta Raj, K., & Seth, A. (2007). A review of resource conservation technologies for sustainable management of the rice–wheat cropping systems of the Indo-Gangetic Plains. *Crop Protection (Guildford, Surrey)*, *26*, 436–447.

Gupta, A., Pal, R. K., & Rajama, M. V. (2013). Delayed ripening and improved fruit processing quality in tomato by RNAi-mediated silencing of three homologs of 1-aminopropane-1-carboxylate synthase gene. *Journal of Plant Physiology*, *170*(11), 987–995. doi:10.1016/j.jplph.2013.02.003 PMID:23507024

Gupta, N., Jain, S. K., Tomar, B. S., Anand, A., Singh, J., & Singh, A. K. (2021). Influence of fruit load per vine on seed quality in cucumber (Cucumis sativus L.) grown under open field and protected environments. *Indian Journal of Horticulture*, *78*(1), 93-100.

Gupta, S. (2010). Biopesticides: An eco-friendly approach for pest control. *Journal of Biopesticides*, *3*(1), 186–188.

Gupta, S. (2018). Nutrigenomics: The Science to Unfold the Impact of GM Food on Health. *Food Sci Nutr Technol*, *3*(3), 000150.

Gutierrez-Barranquero, J. A., Arrebola, E., Bonilla, N., Sarmiento, D., Cazorla, F. M., & de Vicente, A. (2012). Environmentally friendly treatment alternatives to Bordeaux mixture for controlling bacterial apical necrosis (BAN) of mango. *Plant Pathology*, *61*(4), 665–676. doi:10.1111/j.1365-3059.2011.02559.x

Güven, A., & Aydemir, A. (2020). Dams. In *Risk Assessment of Dams* (pp. 1–14). Springer. doi:10.1007/978-3-030-47139-2_1

Habel, J. C., Ulrich, W., Biburger, N., Seibold, S., & Schmitt, T. (2019). Agricultural intensification drives butterfly decline. *Insect Conservation and Diversity*, *12*(4), 289–295.

Hahn, J. J., Eschenlauer, A. C., Sleytr, U. B., Somers, D. A., & Srienc, F. (1999). Peroxisomes as sites for synthesis of polyhydroxyalkanoates in transgenic plants. *Biotechnology Progress*, *15*(6), 1053–1057. doi:10.1021/bp990118n PMID:10585189

Halberg, N. (2012). Assessment of the environmental sustainability of organic farming: Definitions, indicators and the major challenges. *Canadian Journal of Plant Science*, *92*(6), 981–996.

Halberg, N., Alrøe, H. F., & Kristensen, E. S. (2005). Synthesis: Perspectives for organic agriculture in a global context. *Global Development of Organic Agriculture: Challenges and Prospects,CAB Int.*, *12*, 343–367.

Hall, A., & Dorai, K. (2011). *The greening of agriculture: agricultural innovation and sustainable growth*. Report of on OECD Workshop on Agriculture and Green Growth.

Hall, A., Clark, N., & Frost, A. (2010). *Bottom-up, Bottom-line: Development-Relevant Enterprises in East Africa and their Significance for Agricultural Innovation*. RIU Discussion Paper Series. Research Into Use.

Halley, P., Rutgers, R., Coombs, S., Kettels, J., Gralton, J., Christie, G., Jenkins, M., Beh, H., Griffin, K., Jayasekara, R., & Lonergan, G. (2001). Developing biodegradable mulch films from starch based polymers. *Starch*, *53*(8), 362–367. doi:10.1002/1521-379X(200108)53:8<362::AID-STAR362>3.0.CO;2-J

Hamilton, Dermody, Aldea, Zangerl, Rogers, Berenbaum, & Delucia. (2005). Anthropogenic Changes in Tropospheric Composition Increase Susceptibility of Soybean to Insect Herbivory. *Envirn. Entomoi.*, *34*(2), 479-485.

Hammer, D. A. (Ed.). (2020). *Constructed wetlands for wastewater treatment: municipal, industrial and agricultural*. CRC Press. doi:10.1201/9781003069850

Hammond, S. M., Candy, A. A., & Hannon, G. J. (2001). Post transcriptional gene silencing by double-stranded RNA. *Nature Reviews. Genetics*, *2*(2), 110–119. doi:10.1038/35052556 PMID:11253050

Handayanto, E., Giller, K. E., & Cadisch, G. (1997). Regulating N release from legume tree prunings by mixing residues of different quality. *Soil Biology & Biochemistry*, *29*(9-10), 1417–1426.

Han, H. S., Supanjani, E., & Lee, K. D. (2006). Effect of co-inoculation with phosphate and potassium solubilizing bacteria on mineral uptake and growth of pepper and cucumber. *Plant, Soil and Environment*, *52*(3), 130–136. doi:10.17221/3356-PSE

Hannon, G. J. (2002). RNA interference. *Nature*, *418*(6894), 244–251. doi:10.1038/418244a PMID:12110901

Han, Y., Zhang, L., Gu, J., Zhao, J., & Fu, J. (2018). Citric acid and EDTA on the growth, photosynthetic properties and heavy metal accumulation of *Iris halophila Pall.* cultivated in Pb mine tailings. *International Biodeterioration & Biodegradation*, *128*, 15–21. doi:10.1016/j.ibiod.2016.05.011

Haoliang, L., Chongling, Y., & Jingchun, L. (2007). Low-molecular-weight organic acids exuded by Mangrove (*Kandelia candel* (L.) Druce) roots and their effect on cadmium species change in the rhizosphere. *Environmental and Experimental Botany*, *61*(2), 159–549. doi:10.1016/j.envexpbot.2007.05.007

Harding, K. L., Aguayo, V. M., Namirembe, G., & Webb, P. (2017). Determinants of anemia among women and children in Nepal and Pakistan: An analysis of recent national survey data. *Maternal and Child Nutrition*, *4*(1), e12478. PMID:28857410

Hardman, R. (2006). A toxicologic review of quantum dots: Toxicity depends on physicochemical and environmental factors. *Environmental Health Perspectives*, *114*(2), 165–172. doi:10.1289/ehp.8284 PMID:16451849

Harrington, R., Fleming, R. A., & Woiwod, I. P. (2001). Climate change impacts on insect management and conservation in temperate regions: Can they be predicted? *Agricultural and Forest Entomology*, *3*(4), 233–240. doi:10.1046/j.1461-9555.2001.00120.x

Harrison, H. W. (2000). *Roofs and Roofing*. Building Research Establishment Report 16.

Hasaneen, M. N. A., Abdel-Aziz, H. M. M., El-Bialy, D. M. A., & Omer, A. M. (2014). Preparation of chitosan nanoparticles for loading with NPK fertilizer. *African Journal of Biotechnology*, *13*(31), 3158–3164. doi:10.5897/AJB2014.13699

Hasan, M. M., Uddin, M. N., Ara-Sharmeen, F. I., Alharby, H., Alzahrani, Y., & Hakeem, K. R. (2019). Assisting phytoremediation of heavy metals using chemical amendments. *Plants*, *8*, 295. doi:10.3390/plants8090295 PMID:31438507

Hasan, M., Saver, N., & Singh, A. K. (2018). *Hydroponics Technology for Horticultural Crops. TB-ICN No.188/2018. IARI.*

Hatfield, J. L., & Prueger, J. H. (1996). Microclimate effects of crop residues on biological processes. *Theoretical and Applied Climatology*, *54*(1), 47–59.

Hattori, H., Kuniyasu, K., Chiba, K., & Chino, M. (2006). Effect of chloride application and low soil pH on cadmium uptake from soil by plants. *Soil Science and Plant Nutrition*, *52*(1), 89–94. doi:10.1111/j.1747-0765.2006.00007.x

Hauser, S., Norgrove, L., & Nkem, J. N. (2006). Groundnut/maize/cassava Intercrop Yield Response to Fallow Age, Cropping Frequency and Crop Plant Desnity on an Ultisol in Southern Cameroon. *Biological Agriculture & Horticulture: An International Journal for Sustainable Production Systems.*, *24*(3), 275–291. doi:10.1080/01448765.2006.9755026

Heaton, S. (2002). Assessing organic food quality: Is it better for you? *Proceedings of the UK Organic Research Conference*, 55-60.

Heesacker, A., Kishore, V. K., Gao, W., Tang, S., Kolkman, J. M., Gingle, A., Matvienko, M., Kozik, A., Michelmore, R. M., Lai, Z., Rieseberg, L. H., & Knapp, S. J. (2008). SSRs and INDELs mined from the sunflower EST database: Abundance, polymorphisms, and cross-taxa utility. *Theoretical and Applied Genetics, 117*(7), 1021–1029. doi:10.100700122-008-0841-0 PMID:18633591

Heier, T., Jain, S. K., Kogel, K.-H., & Pons-Kühnemann, J. (2005). Influence of N-fertilization and fungicide strategies on Fusarium head blight severity and mycotoxin content in winter wheat. *Journal of Phytopathology, 153*(9), 551–557. doi:10.1111/j.1439-0434.2005.01021.x

Heil, M., & Bostock, R. M. (2002). Induced Systemic Resistance (ISR) Against Pathogens in the Context of Induced Plant Defences. *Annals of Botany, 89*(5), 503–512. doi:10.1093/aob/mcf076 PMID:12099523

Heil, M., Hilpert, A., Kaiser, W., & Linsenmair, K. E. (2000). Reduced growth and seed set following chemical induction of pathogen defence: Does systemic acquired resistance (SAR) incur allocation costs? *Journal of Ecology, 88*(4), 645–654. doi:10.1046/j.1365-2745.2000.00479.x

Heldbjerg, H., Fox, A. D., Levin, G., & Nyegaard, T. (2016). The decline of the Starling Sturnus vulgaris in Denmark is related to changes in grassland extent and intensity of cattle grazing. *Agriculture, Ecosystems & Environment, 230*, 24–31.

Helder, M., Strik, D. P., Hamelers, H. V., & Buisman, C. J. (2012). The flat-plate plant-microbial fuel cell: The effect of a new design on internal resistances. *Biotechnology for Biofuels, 5*(1), 1–11. doi:10.1186/1754-6834-5-70 PMID:22998846

Helga, W., & Julia, L. (2017). BIOFACH, "Organic agriculture worldwide." In *Current Statistics*. FiBL.

Hellin, J., & Fisher, E. (2019). Climate-smart agriculture and non-agricultural livelihood transformation. *Climate (Basel), 7*(4), 48.

Herms, D. A., & Mattson, W. J. (1992). The dilemma of plants to grow or defend. *The Quarterly Review of Biology, 67*(3), 283–335. doi:10.1086/417659

Herrmann, L., & Lesueur, D. (2013). Challenges of formulation and quality of biofertilizers for successful inoculation. *Applied Microbiology and Biotechnology, 97*(20), 8859–8873. doi:10.100700253-013-5228-8 PMID:24037408

Herzig, R., Nehnevajova, E., Pfistner, C., Schwitzguebel, J. P., Ricci, A., & Keller, C. (2014). Feasibility of labile Zn phytoextraction using enhanced tobacco and sunflower: Results of five- and one-year field scale experiments in Switzerland. *International Journal of Phytoremediation, 16*(7-8), 735–754. doi:10.1080/15226514.2013.856846 PMID:24933882

Hickman, J. M., & Wratten, S. D. (1996). Use of *Phacelia tanacetifolia* strips to enhance biological control of aphids by hoverfly larvae in cereal fields. *Journal of Economic Entomology, 89*(4), 832–840. doi:10.1093/jee/89.4.832

Hiloidhari, M., Das, D., & Baruah, D. C. (2014). Bioenergy potential from crop residue biomass in India. *Renewable & Sustainable Energy Reviews, 32*, 504–512.

Hily, J. M., Ravelonandro, M., Damsteegt, V., Basset, C., Petri, C., Liu, Z., & Scorza, R. (2007). Plum pox virus coat protein gene intron-hairpin-RNA (ihpRNA) constructs provide resistance to plum pox virus in *Nicotiana bethamiana* and *Prunus domestica. Journal of the American Society for Horticultural Science, 132*(6), 850–858. doi:10.21273/JASHS.132.6.850

Hinsinger, P., Plassard, C., & Jaillard, B. (2006). Rhizosphere: A new frontier for soil biogeochemistry. *Journal of Geochemical Exploration, 88*(1-3), 210–213. doi:10.1016/j.gexplo.2005.08.041

Hirschfeld, S., & Van Acker, R. (2021). ecosystem services in permaculture systems. *Agroecology and Sustainable Food Systems, 45*(6), 794–816.

Hoarcă, I. C., Enescu, F. M., & Bizon, N. (2020). Comparative study regarding the integration of photovoltaic sources in agriculture. In *2020 12th International Conference on Electronics, Computers and Artificial Intelligence (ECAI)* (pp. 1-6). IEEE. 10.1109/ECAI50035.2020.9223176

Hoffland, E., Hakulinem, J., & Van Pelt, J. A. (1996). Comparison of systemic resistance induced by avirulent and nonpathogenic *Pseudomonas* species. *Phytopathology*, *86*(7), 757–762. doi:10.1094/Phyto-86-757

Hoffmann, M., Holtze, E. M., & Wiesner, M. R. (2007). Reactive oxygen species generation on nanoparticulate material. In M. R. Wiesner & J. Y. Bottero (Eds.), *Environmental Nanotechnology. Applications & Impacts of Nanomaterials* (pp. 155–203). McGraw Hill.

Hofgaard, I. S., Ergon, Å., Henriksen, B., & Tronsmo, A. M. (2010). The effect of potential resistance inducers on development of *Microdochium majus* and *Fusarium culmorum* in winter wheat. *European Journal of Plant Pathology*, *128*(2), 269–281. doi:10.100710658-010-9662-5

Hofgaard, I. S., Ergon, A., Wanner, L. A., & Tronsmo, A. M. (2005). The effect of chitosan and Bion on resistance to pink snow mould in perennial ryegrass and winter wheat. *Journal of Phytopathology*, *153*(2), 108–119. doi:10.1111/j.1439-0434.2005.00937.x

Hole, D. G., Perkins, A. J., Wilson, J. D., Alexander, I. H., Grice, P. V., & Evans, A. D. (2005). Does organic farming benefit biodiversity? *Biological Conservation*, *122*(1), 113–130.

Holmgren, D. (2020). *Essence of permaculture*. Melliodora Publishing.

Holmquis, G., & Ashley, T. (2006). Chromosome organization and chromatin modification: Influence on genome function and evolution. *Cytogenetic and Genome Research*, *114*(2), 96–125. doi:10.1159/000093326 PMID:16825762

Honorato-Salazar, J. A., & Sadhukhan, J. (2020). Annual biomass variation of agriculture crops and forestry residues, and seasonality of crop residues for energy production in Mexico. *Food and Bioproducts Processing*, *119*, 1–19. doi:10.1016/j.fbp.2019.10.005

Hoshikawa, K. (1991). Significance of legumes crops in intercropping, the productivity and stability of cropping system. In C. Johanson, K. K. Lee, & K. L. Saharawat (Eds.), *Phosphorus Nutrition of Grain Legume in the Semi Arid Tropics* (pp. 173–176). ICRISAT.

Hossain, M. D. (2009) Nutrient and Residue Management for Improving Productivity and N Use Efficiency of Rice-Wheat-Mungbean Systems in Bangladesh. *Proceedings of the International Plant Nutrition Colloquium XVI*. https://escholarship.org/uc/item/2r29j9t0

Houghton, J. T., Jenkins, G. J., & Ephramus, J. J. (Eds.). (1990). *Climate Change: The IPCC Scientific Assessment*. Cambridge University Press.

Houmiel, K.L., Slater, S., Broyles, D. P. (1999). (β-hydroxybutyrate) production in oilseed leukoplasts of *Brassica napus*. *Planta, 209*, 547-550.

Houng, K. H., & Hwa, C. (1975). The effect of crop resdiues on the growth of following crops II. EVect of decomposing rice straw and bagasse on the nutrient uptake of rice seedlings. *Journal of Chinese Agricultural Chemistry Society*, *13*, 412–415.

Huang, G., Allen, R., Davis, E. L., Baum, T. J., & Hussey, R. S. (2006). Engineering broad root-knot resistance in transgenic plants by RNAi silencing of a conserved and essential root-knot nematode parasitism gene. *Proceedings of the National Academy of Sciences of the United States of America*, *103*(39), 4302–14306. doi:10.1073/pnas.0604698103 PMID:16985000

Huang, W., & Marth, G. (2008). EagleView: A genome assembly viewer for next-generation sequencing technologies. *Genome Research*, *18*(9), 1538–1543. doi:10.1101/gr.076067.108 PMID:18550804

Hubenova, Y., & Mitov, M. (2012). Conversion of solar energy into electricity by using duckweed in direct photosynthetic plant fuel cell. *Bioelectrochemistry (Amsterdam, Netherlands)*, *87*, 185–191. doi:10.1016/j.bioelechem.2012.02.008 PMID:22445768

Huber, B. (2007). Regulation and Organic Certification. *Disampaikan dalam workshop Biofach.*

Huber, M., & Van, de V.L. (2009). *Overview of research linking organic production methods and health effects in the lab, in animals and in humans*. Department of Health Care and Nutrition, Louis Bolk Institute.

Huber, D. M., & Graham, R. D. (1999). The role of nutrition in crop resistance and tolerance to disease. In Z. Rengel (Ed.), *Mineral Nutrition of Crops Fundamental Mechanisms and Implications* (pp. 205–226). Food Product Press.

Huber, D. M., & McCay-Buis, T. S. (1993). A multiple component analysis of the take-all disease of cereals. *Plant Disease*, *77*(5), 437–447. doi:10.1094/PD-77-0437

Hübler, C., & Rolfes, R. (2021). Analysis of the influence of climate change on the fatigue lifetime of offshore wind turbines using imprecise probabilities. *Wind Energy (Chichester, England)*, *24*(3), 275–289. doi:10.1002/we.2572

Huebert, D. B., & Shay, J. M. (1993). Considerations in the assessment of toxicity using duckweeds. *Environmental Toxicology and Chemistry*, *12*(3), 481–483. doi:10.1002/etc.5620120309

Hu, J. L., Chan, P. T., Wu, F. Y., Wu, S. C., Zhang, J. H., Lin, X. G., & Wong, M. H. (2013). Arbuscular mycorrhizal fungi induce differential Cd and P acquisition by Alfred stonecrop (*Sedum alfredii* Hance) and upland kangkong (*Ipomoea aquatica* Forsk.) in an intercropping system. *Applied Soil Ecology*, *63*, 29–35. doi:10.1016/j.apsoil.2012.09.002

Hume, L. (1982). The long-term effects of fertilizer application and three rotations on weed communities in wheat (after 2 1-22 years at Indian Head. Saskatchewan). *Canadian Journal of Plant Science*, *62*, 741–750. doi:10.4141/cjps82-107

Hunter, M. D. (2001). Effects on elevated atmospheric carbon dioxide on insect-plant interactions. *Agricultural and Forest Entomology*, *3*(3), 153–159. doi:10.1046/j.1461-9555.2001.00108.x

Hunt, J. R., Celestina, C., & Kirkegaard, J. A. (2020). The realities of climate change, conservation agriculture and soil carbon sequestration. *Global Change Biology*, *26*(6), 3188–3189.

Ibeawuchi, I. I., Obiefuna, J. C., & Iwuanyanwu, U. P. (2015). Low External Input Agricultural Farming System for the Increase in Productivity of Resource Poor Farmers. *Soil and Water, 5*(2).

ICAR. (2006). Handbook of Agriculture. Indian Council of Agricultural Research.

ICAR. (2012) Degraded and Wastelands of India-Status and Special Distribution. Indian Council of Agricultural Research & National Academy of Agricultural Sciences.

Idrisa, Y. L., Shehu, H., & Ngandu, M. B. (2012). Effect of adoption of improved maize seed on household food security in Gwoza EGA of Borno State, Nigeria. *Global Journal*, *12*, 20–31.

IFCAM. (2009). *The World of Organic Agriculture – Statistics & Emerging Trends 2008*. https://www.soel.de/fachtheraaii

IFOAM Norms, . (2002). *IFOAM Basic Standards - IFOAM Accreditation Criteria*. IFOAM.

IFOAM. (1998). *IFOAM Basic Standards for Organic Production and Processing*. IFOAM Publications.

IFOAM. (2000). *Basic Standards for Organic Production and Processing*. IFOAM General Assembly. http://www.ifoam.org/

IFOAM. (2003a). Pest and disease management in organic management. A Latin American perspective. IFOAM.

IFOAM. (2003b). *Training manual for organic agriculture in the tropics*. IFOAM.

IFOAM. (2005). *The IFOAM Norms for organic production and processing including IFOAM Basic Standards and IFOAM Accreditation Criteria*. IFOAM.

IFOAM. (2010). *Organic Agriculture Worldwide: The Main Results of the FiBL-IFOAM Survey 2010.* Presentation at the BioFach Congress, International Federation of Organic Agriculture Movements.

IIRR & ACT. (2005). *Conservation agriculture: A manual for farmers and extension workers in Africa*. International Institute of Rural Reconstruction and African Conservation Tillage Network.

Iken, J. E., & Amusa, N. A. (2004). Maize research and production in Nigeria. *African Journal of Biotechnology*, *8*, 140–151.

Ilyas, N., Bano, A., Iqbal, S., & Raja, N. I. (2012). Physiological, biochemical and molecular characterization of *Azospirillum* sp. isolated from maize under water stress. *Pakistan Journal of Botany*, *44*, 71–80.

International Institute of Tropical Agriculture. (2001). *Annual Report on Maize*. IITA.

Ion, A. C., Ion, I., & Culetu, A. (2010). Carbon-based nanomaterials: Environmental applications. *University Politehnica of Bucharest*, *38*, 129–132.

IPCC. (2007). *Climate Change 2007: The Physical Science Basis. Contribution of Working Group I to the Fourth Assessment Report of the Intergovernmental Panel on Climate Change* (S. Solomon, D. Qin, M. Manning, Z. Chen, M. Marquis, K. B. Averyt, M. Tignor, & H. L. Miller, Eds.). Cambridge University Press.

Iqbal, J., Cheema, Z. A., & An, M. (2007). Intercropping of field crops in cotton for the management of purple nutsedge (*Cyperus rotundus* L.). *Plant and Soil*, *300*(1-2), 163–171. doi:10.100711104-007-9400-8

Iqbal, M., Puschenreiter, M., Oburger, E., Santner, J., & Wenzel, W. W. (2012). Sulfur-aided phytoextraction of Cd and Zn by *Salix smithiana* combined with *in situ* metal immobilization by gravel sludge and red mud. *Environmental Pollution*, *170*, 222–231. doi:10.1016/j.envpol.2012.07.008 PMID:22842051

Isbell, F., Reich, P. B., Tilman, D., Hobbie, S. E., Polasky, S., & Binder, S. (2013). Nutrient enrichment, biodiversity loss, and consequent declines in ecosystem productivity. *Proceedings of the National Academy of Sciences of the United States of America*, *110*(29), 11911–11916.

Ismunadji, M. (1978). Utilization of cereal crop residues and its agricultural significance in Indonesia. Central Research Institute for Agriculture.

ISO CD 2009. Water Quality - Determination of the Toxic Effect of Water Constituents and Waste Water to Duckweed (Lemna minor) - Duckweed Growth Inhibition Test.

ITRC (Interstate Technology & Regulatory Council). (2009). *Phytotechnology Technical and Regulatory Guidance and Decision Trees*. Author.

Jackson, L., Pascual, U., & Hodgkin, T. (2007). Utilizing and conserving agrobiodiversity in agricultural landscapes. *Agriculture, Ecosystems & Environment*, *121*(3), 196–210. doi:10.1016/j.agee.2006.12.017

Jacob, J. M., Karthik, C., Saratale, R. G., Kumar, S. S., Prabakar, D., Kadirvelu, K., & Pugazhendhi, A. (2018). Biological approaches to tackle heavy metal pollution: A survey of literature. *Journal of Environmental Management*, *217*, 56–70. doi:10.1016/j.jenvman.2018.03.077 PMID:29597108

Jaenisch, R., & Bird, A. (2003). Epigenetic regulation of gene expression: How the genome integrates intrinsic and environmental signals. *Nature Genetics Suppl.*, *33*, 245–254.

Jain, M., Solomon, D., Capnerhurst, H., Arnold, A., Elliott, A., Kinzer, A. T., ... Weinstein, C. (2020). How much can sustainable intensification increase yields across South Asia? A systematic review of the evidence. *Environmental Research Letters*, *15*(8), 083004.

Janssen, J., Weyens, N., Croes, S., Beckers, B., Meiresonne, L., Peteghem, P. V., Carleer, R., & Vangronsveld, J. (2015). Phytoremediation of metal contaminated soil using Willow: Exploiting plant-associated bacteria to improve biomass production and metal uptake. *International Journal of Phytoremediation*, *17*(11), 1123–1136. doi:10.1080/15226514.2015.1045129 PMID:25942689

Jansson, C., Faiola, C., Wingler, A., Zhu, X. G., Kravchenko, A., de Graaff, M. A., ... Beckles, D. M. (2021). Crops for Carbon Farming. *Frontiers in Plant Science*, *12*, 938.

Jasrotia, P., Kashyap, P. L., Bhardwaj, A. K., Kumar, S., & Singh, G. P. (2018). Scope and applications of nanotechnology for wheat production: A review of recent advances. *Wheat and Barley Research*, *10*(1), 137–150. doi:10.25174/2249-4065/2018/76672

Jat, M. L., Singh, R. G., & Saharawat, Y. S. (2009). Innovations through conservation agriculture: progress and prospects of participatory approach in the Indo-Gangetic plains. *4th World Congress on Conservation Agriculture*, 60-64.

Jat, M. L., Singh, S., Rai, H. K., Chhokar, R. S., Sharma, S. K., & Gupta, R. K. (2005). Furrow irrigated raised bed planting technique for diversification of rice–wheat system of Indo-Gangetic plains. *J Jpn Assoc Int Coop Agric For*, *28*, 25–42.

Javed, M. S., Ma, T., Jurasz, J., & Amin, M. Y. (2020). Solar and wind power generation systems with pumped hydro storage: Review and future perspectives. *Renewable Energy*, *148*, 176–192. doi:10.1016/j.renene.2019.11.157

Jayaraj, J., Muthukrishnan, S., Liang, G., & Velazhahan, R. (2004). Jasmonic acid and salicylic acid induce accumulation of β-1,3-glucanase and thaumatin-like proteins in wheat and enhance resistance against *Stagonospora nodorum. Biologia Plantarum*, *48*(3), 425–430. doi:10.1023/B:BIOP.0000041097.03177.2d

Jayaram, D., Chatterjee, B. N., & Mondal, S. S. (1990). Effect of FYM, crop residues and fertilizers management in sustaining productivity under intensive cropping. *Journal of Potash Research*, *6*, 172–179.

Jayceta Ghosh, R. D. (2014). An Assessment of Genetic Relatedness between Soybean [Glycine max (L.) Merrill] Cultivars Using SSR Markers. *American Journal of Plant Sciences*, *5*(20), 3089–3096. doi:10.4236/ajps.2014.520325

Jelin, J., Dhanrajan, M. S., & Mariappan, V. (2013). Assessment of compost as a bio-fertilizer for the growth of paddy. *Journal of Environmental Biology*, *34*, 975–979. PMID:24555324

Jenrick, K. (2005). *Green roofs: a horticultural perspective*. Ritak Vitabuc Gardens.

Jensen & Weiser. (1971). *Crop Rotations for Profit*. Extension Bulletin, No. 14. North Dakota State University.

Jha, M. N., & Prasad, A. N. (2006). Efficacy of new inexpensive cyanobacterial biofertilizer including its shelf-life. *World Journal of Microbiology & Biotechnology*, *22*(1), 73–79. doi:10.100711274-005-7024-9

Jimenez, J. (2010). *Social policy and Social change*. Sage.

Jirtle, R. L., & Skinner, M. K. (2007). Environmental epigenomics and disease susceptibility. *Nature Reviews. Genetics*, *8*(4), 253–262.

Jitendra, K., Patel, N., Singh, R., Sahoo, P. K., Susama, S., Sehgal, V. K., Marwaha, S., & Singh, A. K. (2021). Development and evaluation of automation system for irrigation scheduling in Broccoli crop. *Indian Journal of Agricultural Sciences*, *91*(5), 221–225.

John, M. E., & Keller, G. (1996). Metabolic pathway engineering in cotton: Biosynthesis of polyhydroxybutyrate in fiber cells. *Proceedings of the National Academy of Sciences of the United States of America*, *93*(23), 12768–12773. doi:10.1073/pnas.93.23.12768 PMID:11038522

John, P. S., Prasad, R., Pandey, R. K., Buresh, R. J., & Prasad, R. (1989). Lowland rice response to urea following three cowpea cropping systems. *Agronomy Journal*, *81*, 853–857.

Johnston, J. (1995). Rooftop Wildlife. *Enact*, 3.

Johnston, J., & Newton, J. (1993). *Building green- A guide to using plants on roofs, walls and pavements*. London Ecology Unit.

Jones-Rhoades, M. W., & Bartel, D. P. (2004). Computational identification of plant microRNAs and their targets, including a stress-induced miRNA. *Molecular Cell*, *14*(6), 787–799. doi:10.1016/j.molcel.2004.05.027 PMID:15200956

Jørgensen, K., Bak, S., Busk, P. K., Sørensen, C., Olsen, C. E., Puonti-Kaerlas, J., & Moller, B. L. (2005). Cassava plants with a depleted cyanogenic glucoside content in leaves and tubers. *Plant Physiology*, *139*(1), 363–374. doi:10.1104/pp.105.065904 PMID:16126856

Jørgensen, L. N., Secher, B. J. M., Olesen, J. E., & Mortensen, J. (1997). Need for fungicide treatments when varying agricultural parameters. *Aspects of Applied Biology*, *50*, 285–292.

Joshi, A. K., Kumar, S., Chand, R., & Ortiz-Ferrara, G. (2004). Inheritance of resistance to spot blotch pathogen caused by *Bipolaris sorokiniana* in spring wheat. *Plant Breeding*, *123*(3), 213–219. doi:10.1111/j.1439-0523.2004.00954.x

Joshi, A. K., Ortiz-Ferrara, G., Crossa, J., Singh, G., Alvarado, G., Bhatta, M. R., Duveiller, E., Sharma, R. C., Pandit, D. B., Siddique, A. B., Das, S. Y., Sharma, R. N., & Chand, R. (2007). Associations of environments in South Asia based on spot blotch disease of wheat caused by *Cochliobolus sativus*. *Crop Science*, *47*(3), 1071–1081. doi:10.2135/cropsci2006.07.0477

Joshi, P. K., Jha, A. K., Wani, S. P., Sreedevi, T. K., & Shaheen, F. A. (2008). *Impact of Watershed Program and Conditions for Success: A Meta-Analysis Approach*. Global Theme on Agroecosystems, International Crops Research Institute for the Semi-Arid Tropics and National Centre for Agricultural Economics and Policy Research.

Julie Casper, C. (2018). *Epigenetics, Nutrigenomics, and Genetic Testing*. nutritionalbalancing.org

Kadotani, N., Nakayashiki, H., Tosa, Y., & Mayama, S. (2003). RNA silencing in the pathogenic fungus *Magnaporthe oryzae*. *Molecular Plant-Microbe Interactions*, *16*(9), 769–776. doi:10.1094/MPMI.2003.16.9.769 PMID:12971600

Kah, M. (2015). Nanopesticides and nanofertilizers: Emerging contaminants or opportunities for risk mitigation? *Frontiers in Chemistry*, *3*, 64. doi:10.3389/fchem.2015.00064 PMID:26636068

Kaiser, M., Bernauer, M., Sunstein, C. R., & Reisch, L. A. (2020). The power of green defaults: The impact of regional variation of opt-out tariffs on green energy demand in Germany. *Ecological Economics*, *174*, 106685. doi:10.1016/j.ecolecon.2020.106685

Kale, R. D. (1998). *Earthworm cinderella of organic farming*. Prism Book Pvt. Ltd.

Kalyan, S. (2005). *Development of sustainable farming system model for the irrigated agro-ecosystem of eastern UP, ICAR, adhoc project. Final annual report, department of agronomy, Institute of agriculture science*. BHU.

Kamachi, S., Mochizuki, A., Nishiguchi, M., & Tabei, Y. (2007). Transgenic *Nicotiana benthamiana* plants resistant to cucumber green mottle mosaic virus based on RNA silencing. *Plant Cell Reports*, *26*(8), 1283–1288. doi:10.100700299-007-0358-z PMID:17453214

Kandasamy, S., & Prema, R. S. (2015). Methods of synthesis of nanoparticles and its applications. *Journal of Chemical and Pharmaceutical Research*, *7*, 278–285.

Kannaujia, P. K., Asrey, R., Singh, A. K., & Varghese, E. (2019). Postharvest treatments to reduce chilling injury in summer squash (Cucurbita Pepo) fruits during storage. *International Journal of Agricultural Sustainability*, *89*(10), 1633–1637.

Kapruwan, S., Bakshi, M., & Kaur, M. (2014). Effect of growth regulators on the in vitro multiplication of Dendrocalamus Hamiltonii. *International Journal of Engineering Research and Applications*, *4*, 83–86.

Kapur, M. L., & Kanwar, R. S. (1994). Sugarbeet top as a source of N for maize and rice grown in rotation with sugarbeet in subtropical north-west India. *Tropical Agriculturist*, *71*, 12–16.

Kaput, J. (2007). Developing the promise of nutrigenomics through complete science and international collaborations. *Forum of Nutrition*, *60*, 209–223.

Kaput, J., & Dawson, K. (2007). Complexity of type 2 diabetes mellitus data sets emerging from nutrigenomic research: A case for dimensionality reduction. *Mutation Research*, *622*(1–2), 19–32.

Kaput, J., Ordovas, J. M., & Ferguson, L. (2005a). Nutritional genomics: The next frontier in the postgenomic era. *Physiological Genomics*, *16*(2), 166–177.

Kaput, J., Ordovas, J. M., & Ferguson, L. (2005b). The case for strategic international alliances to harness nutritional genomics for public and personal health. *British Journal of Nutrition*, *94*(5), 623–632.

Karadağ, H. (2018). Biodynamic Agriculture in Turkey; Istafil Biodynamic Organic Farmland Sample. In *4th International Conference on Engineering and Natural Science. Book of Abstracts* (*Vol. 116*). Academic Press.

Karasawa, T., Takebe, M., Sato, F., Komada, M., Nagaoka, K., Takenaka, M., ... Kato, N. (2015). Trends of lettuce and carrot yields and soil enzyme activities during transition from conventional to organic farming in an Andosol. *Soil Science and Plant Nutrition*, *61*(2), 295–311.

Karban, R., & Baldwin, I. T. (1997). *Induced responses to herbivory*. Chicago Press. doi:10.7208/chicago/9780226424972.001.0001

Karthikeyan, N., Prasanna, R., Lata, N., & Kaushik, B. D. (2007). Evaluating the potential of plant growth promoting cyanobacteria as inoculants for wheat. *European Journal of Soil Biology*, *43*(1), 23–30. doi:10.1016/j.ejsobi.2006.11.001

Kashem, M. A., & Singh, B. R. (2002). The effect of fertilizer additions on the solubility and plant-availability of Cd, Ni and Zn in soil. *Nutrient Cycling in Agroecosystems*, *62*(3), 287–296. doi:10.1023/A:1021226201136

Kashyap, P. L., & Dhiman, J. S. (2009). Induction of resistance in cauliflower against *Alternaria* blight using potassium and phosphonic salts. *The Asian and Australasian Journal of Plant Science and Biotechnology*, *3*(1), 66–70.

Kashyap, P. L., Jasrotia, P., Kumar, S., Singh, D. P., & Singh, G. P. (2018a). Identification guide for major diseases and insect-pests of wheat. ICAR-Indian Institute of Wheat and Barley Research, Karnal *Technical Bulletin 18.*

Kashyap, P. L., Kaur, S., & Pannu, P. P. S. (2018b). Induction of systemic tolerance to *Tilletia indica* in wheat by plant defence activators. *Archiv für Phytopathologie und Pflanzenschutz*, *51*(1-2), 1–13. doi:10.1080/03235408.2018.1438778

Kashyap, P. L., Kaur, S., Sanghera, G. S., Kang, S. S., & Pannu, P. P. S. (2011). Novel methods for quarantine detection of Karnal bunt (*Tilletia indica*) of wheat. *Elixir Agriculture*, *31*, 1873–1876.

Kashyap, P. L., Kumar, S., Kumar, R. S., Sharma, A., Jasrotia, P., Singh, D. P., & Singh, G. P. (2020). Molecular Diagnostic Assay for Rapid Detection of Flag Smut Fungus (Urocystis agropyri) in Wheat Plants and Field Soil. *Front. Plant Sci.*, *11*, 1039. doi:10.3389/fpls.2020.01039 PMID:32754183

Kashyap, P. L., Kumar, S., Tripathi, R., Kumar, R. S., Jasrotia, P., Singh, D. P., & Singh, G. P. (2019). Phylogeography and Population Structure Analysis Reveal Diversity by Gene Flow and Mutation in *Ustilago segetum* (Pers.) Roussel *tritici* Causing Loose Smut of Wheat. *Frontiers in Microbiology*, *10*, 1072. doi:10.3389/fmicb.2019.01072 PMID:31156587

Kashyap, P. L., Rai, P., Srivastava, A. K., & Kumar, S. (2017). *Trichoderma* for climate resilient agriculture. *World Journal of Microbiology & Biotechnology*, *33*(8), 155. doi:10.100711274-017-2319-1 PMID:28695465

Kashyap, P. L., Srivastava, A. K., Tiwari, S. P., & Kumar, S. (2018c). *Microbes for Climate Resilient Agriculture*. John Wiley and Sons. doi:10.1002/9781119276050

Kashyap, P. L., Xiang, X., & Heiden, P. (2015). Chitosan nanoparticle based delivery systems for sustainable agriculture. *International Journal of Biological Macromolecules*, *77*, 36–51. doi:10.1016/j.ijbiomac.2015.02.039 PMID:25748851

Kassam, A., Friedrich, T., & Derpsch, R. (2019). Global spread of conservation agriculture. *The International Journal of Environmental Studies*, *76*(1), 29–51.

Katiyar-Agarwal, S., & Jin, H. (2010). Role of small RNAs in host-microbe interactions. *Annual Review of Phytopathology*, *48*(1), 225–246. doi:10.1146/annurev-phyto-073009-114457 PMID:20687832

Kaur, H., Anand, M., & Goyal, D. (2011). Establishment of an efficient protocol for micropropagation of stem explants of Tylophora indica, an important medicinal plant. *African Journal of Biotechnology*, *10*, 6928–6932.

Kaur, P., Bhardwaj, M., & Babbar, I. (2015). Effect of Vermicompost and vermiwash on growth of vegetables. *Research Journal of Animal. Veterinary and Fishery Sciences*, *3*, 9–12.

Kaushik, B. D. (2013). Developments in cyanobacterial biofertilizer. *Proceeding of Indian Natural Science Academy*, *80*(2), 379–388. doi:10.16943/ptinsa/2014/v80i2/55115

Kauwell, G. P. A. (2005). Emerging concepts in nutrigenomics: A preview of what is to come. *Nutrition in Clinical Practice*, *20*, 75–87.

Kavinandan, S. K., Gupta, J. P., & Mahapatra, I. C. (1987). Studies on residue management and biofertilizers in rice-wheat sequence. *Indian Journal of Agronomy*, *32*, 278–279.

Kayser, A., Wenger, K., Keller, A., Attinger, W., Felix, H. R., Gupta, S. K., & Schulin, R. (2003). Enhancement of phytoextraction of Zn, Cd and Cu from calcareous soil: The use of NTA and sulfur amendments. *Environmental Science & Technology*, *34*(9), 1778–1783. doi:10.1021/es990697s

Kempf, H.-J., & Wolf, G. (1989). *Erwinia herbicola* as a biocontrol agent of *Fusarium culmorum* and *Puccinia recondita* f. sp. *tritici* on wheat. *Phytopathology*, *79*(9), 990–994. doi:10.1094/Phyto-79-990

Kennedy, I. R., Choudhury, A. T. M. A., & Kecskés, M. L. (2004). Non-symbiotic bacterial diazotrophs in crop-farming systems: Can their potential for plant growth promotion be better exploited? *Soil Biology & Biochemistry*, *36*(8), 1229–1244. doi:10.1016/j.soilbio.2004.04.006

Keshava & Ramachandrappa. (2000). Intercropping of groundnut, cowpea and jowar with redgram. *Current Research*, *29*(4), 49-50.

Ke, Y. C., & Stroeve, P. (2005). *Polymer-layered silicate and silica nanocomposites* (1st ed.). Elsevier. doi:10.1016/B978-044451570-4/50004-5

Khadse, V. A., Ghatol, P. U., Thakare, P. D., & Bhale, V. M. (2021). *Adoption of biodynamic organic agriculture practices on farmer's field of Vidarbha region*. Academic Press.

Khaliq, A., Matloob, A., Irshad, M. S., Tanveer, A., & Zamir, M. S. I. (2010). Organic weed management in maize through integration of allelopathic crop residues. *Pakistan Journal of Weed Science Research*, *16*, 409–420.

Khan, Ali, & Hussain. (2012). Impact of maize-legume intercropping on weeds and maize crop. *Pakistan Journal of Weed Science Research*, *18*(1), 127–136.

Khande, R., Sushil, K. S., Ramesh, A., & Mahaveer, P. S. (2017). Zinc solubilizing *Bacillus* strains that modulate growth, yield and zinc biofortification of soybean and wheat. *Rhizosphere*, *4*, 126–138. doi:10.1016/j.rhisph.2017.09.002

Khan, M. S., & Zaidi, A. (2007). Synergistic effects of the inoculation with plant growth promoting rhizobacteria and arbuscular mycorrhizal fungus on the performance of wheat. *Turkish Journal of Agriculture and Forestry*, *31*, 355–362.

Khan, M. S., Zaidi, A., Ahemad, M., Oves, M., & Wani, P. A. (2010). Plant growth promotion by phosphate solubilizing fungi- current perspective. *Archives of Agronomy and Soil Science*, *56*(1), 73–98. doi:10.1080/03650340902806469

Khan, S. A. R., Yu, Z., Belhadi, A., & Mardani, A. (2020). Investigating the effects of renewable energy on international trade and environmental quality. *Journal of Environmental Management*, *272*, 111089. doi:10.1016/j.jenvman.2020.111089 PMID:32854892

Khodakovskaya, M., Dervishi, E., Mahmood, M., Xu, Y., Li, Z., Watanabe, F., & Biris, A. S. (2009). Carbon nanotubes are able to penetrate plant seed coat and dramatically affect seed germination and plant growth. *ACS Nano*, *3*(10), 3221–3227. doi:10.1021/nn900887m PMID:19772305

Khoie, R., Bose, A., & Saltsman, J. (2020). The carbon emissions of wind power; a study of emissions of windmill in the panhandle of Texas. In *Conference proceedings of SOLAR* (pp. 24-25). 10.18086olar.2020.01.14

Khong, N. G., Randoux, B., & Deravel, J. (2013). Induction of resistance in wheat by bacterial cyclic lipopeptides. *Communications in Agricultural and Applied Biological Sciences*, *78*(3), 479–487. PMID:25151823

Khong, N. G., Randoux, B., & Tayeh, C. (2012). Induction of resistance in wheat against powdery mildew by bacterial cyclic lipopeptides. *Communications in Agricultural and Applied Biological Sciences*, *77*(3), 39–51. PMID:23878959

Khot, L. R., Sankaran, S., Maja, J. M., Ehsani, R., & Schuster, E. W. (2012). Applications of nanomaterials in agricultural production and crop protection: A review. *Crop Protection (Guildford, Surrey)*, *35*, 64–70. doi:10.1016/j.cropro.2012.01.007

Khush, G. S. (2008). Biofortification of Crops for Reducing Malnutrition. *Proc. Indian Natn. Sci. Acad.*, *74*, 21–25.

Kidd, P., Mench, M., Alvarez-Lopez, V., Bert, V., Dimitriou, I., Friesl-Hanl, W., Herzig, R., Olga Janssen, J., Kolbas, A., Müller, I., Neu, S., Renella, G., Ruttens, A., Vangronsveld, J., & Puschenreiter, M. (2015). Agronomic practices for improving gentle remediation of trace element-contaminated soils. *International Journal of Phytoremediation*, *17*(11), 1005–1037. doi:10.1080/15226514.2014.1003788 PMID:25581041

Kilic-Ekici, O., & Yuen, G. Y. (2003). Induced resistance as a mechanism of biological control by Lysobacter enzymogenes strain C3. *Phytopathology*, *93*(9), 1103–1110. doi:10.1094/PHYTO.2003.93.9.1103 PMID:18944093

Kim, Y. B., & Lenz, R. W. (2001). Polysters from microorganisms. *Advances in Biochemical Engineering/Biotechnology*, *71*, 51–79. doi:10.1007/3-540-40021-4_2 PMID:11217417

Kim, Y. C., Kim, Y. H., Lee, Y. H., Lee, S. W., Chae, Y.-S., Kang, H.-K., Yun, B.-W., & Hong, J. K. (2013). β-Amino-n-butyric Acid Regulates Seedling Growth and Disease Resistance of Kimchi Cabbage. *The Plant Pathology Journal*, *29*(3), 305–316. doi:10.5423/PPJ.OA.12.2012.0191 PMID:25288957

Kiraly, Z. (1976). Plant disease resistance as influenced by biochemical effects of nutrients and fertilizers. *Fertilizer Use and Plant Health, Proceedings of Colloquium 12.*

Kleffe, J., Hermann, K., Vahrson, W., Wittig, B., & Brendel, V. (1996). Logitlinear models for the prediction of splice sites in plant pre-mRNA sequences. *Nucleic Acids Research, 24*(23), 4709–4718. doi:10.1093/nar/24.23.4709 PMID:8972857

Kleijn, D., Bommarco, R., Fijen, T. P., Garibaldi, L. A., Potts, S. G., & van der Putten, W. H. (2019). Ecological intensification: Bridging the gap between science and practice. *Trends in Ecology & Evolution, 34*(2), 154–166.

Kloepper, J. W., Tuzun, S., Zehnder, G. W., & Wei, G. (1997). Multiple disease protection by rhizobacteria that induce systemic resistance – historic precedence. *Phytopathology, 87*(2), 136–137. doi:10.1094/PHYTO.1997.87.2.136 PMID:18945131

Klose, R. J., & Bird, A. P. (2006). Genomic DNA methylation: The mark and its mediators. *Trends in Biochemical Sciences, 31*(2), 89–97.

Kogan, M. (1998). Integrated Pest Management: Historical Perspectives and Contemporary Developments. *Annual Review of Entomology, 43*(1), 243–270. doi:10.1146/annurev.ento.43.1.243 PMID:9444752

Kogel, K. H., Franken, P., & Huckelhoven, R. (2006). Endophyte or parasite- what decides? *Current Opinion in Plant Biology, 9*(4), 358–363. doi:10.1016/j.pbi.2006.05.001 PMID:16713330

Kohler, M., Schmidt, G. F. W., Wilhelm Grimme, F., Laar, M., Lúcia de Assunção Paiva, V., & Tavares, S. (2002). Green roofs in temperate climates and in the hot-humid tropics-far beyond the aesthetics. *Environmental Management and Health, 13*(4), 382–391. doi:10.1108/09566160210439297

Kolbas, A. (2012). *Phenotypic traits and development of plants exposed to trace elements; use for phytoremediation and biomonitoring* (PhD thesis). University of Bordeaux.

Komolafe, S. E., Adeseji, G. B., & Ajibola, B. O. (2010). Determination of adoption of improved crop practices among women farmers in Ekiti East LGA of Ekiti State, Nigeria. *Journal of Agricultural Research, 5*(2), 22–33.

Konakalla, N. C., Kaldis, A., Berbati, M., Masarapu, H., & Voloudakis, A. E. (2016). Exogenous application of double-stranded RNA molecules from TMV p126 and CP genes confers resistance against TMV in tobacco. *Planta, 244*(4), 961–969. doi:10.100700425-016-2567-6 PMID:27456838

Konig, A., Cockburn, A., Crevel, R. W. R., Debruyne, E., Grafstroem, R., Hammerling, U., Kimber, I., Knudsen, I., Kuiper, H. A., Peijnenburg, A. A. C. M., Penninks, A. H., Poulsen, M., Schauzu, M., & Wal, J. M. (2004). Assessment of the safety of food derived from genetically modified GM crops. *Food and Chemical Toxicology, 42*(7), 1047–1088. doi:10.1016/j.fct.2004.02.019 PMID:15123382

Kosseva, M. R., & Rusbandi, R. (2018). Trends in the biomanufacture of polyhydroxyalkanoates with focus on downstream processing. *International Journal of Biological Macromolecules, 107*, 762–778. doi:10.1016/j.ijbiomac.2017.09.054 PMID:28928063

Kour, D., Rana, K.L., & Yadav, A.N. (2020). Microbial biofertilizers: bioresources and eco-friendly technologies for agricultural and environmental sustainability. *Biocatalysis and Agricultural Biotechnology, 23*, 1-11.

Kourtz, L., Dillon, K., Daughtry, S., Madison, L. L., Peoples, O., & Snell, K. D. (2005). A novel thiolase-reductase gene fusion promotes the production of polyhydroxybutyrate in *Arabidopsis. Plant Biotechnology Journal, 3*(4), 435–447. doi:10.1111/j.1467-7652.2005.00136.x PMID:17173631

Kourtz, L., Dillon, K., Daughtry, S., Madison, L. L., Peoples, O., & Snell, K. D. (2007). Chemically inducible expression of the PHB biosynthetic pathway in *Arabidopsis. Transgenic Research, 16*(6), 759–769. doi:10.100711248-007-9067-1 PMID:17279436

Kovács-Hostyánszki, A., Espíndola, A., Vanbergen, A. J., Settele, J., Kremen, C., & Dicks, L. V. (2017). Ecological intensification to mitigate impacts of conventional intensive land use on pollinators and pollination. *Ecology Letters, 20*(5), 673–689.

Kovochich, M., Xia, T., & Xu, J. (2005). Principles and procedures to assess nanoparticles. *Environmental Science & Technology, 39*, 1250–1256.

Kragt, M. E., Blackmore, L., Capon, T., Robinson, C. J., Torabi, N., & Wilson, K. A. (2014). *What are the barriers to adopting carbon farming practices?* (No. 1784-2016-141870).

Kratochvil, R. (2005). Organic Farming: An approach to make agriculture more sustainable. *Jahrbuch der Österreichischen Gesellschaft für Agrarökonomie, 12*, 245–260.

Krebs, J., & Bach, S. (2018). Permaculture—Scientific evidence of principles for the agroecological design of farming systems. *Sustainability, 10*(9), 3218.

Kristiansen, P. (2006). Organic agriculture: opportunities and challenges. *Organic Agriculture: A Global Perspective, 18*, 421-441.

Kucey, R. M. N. (1988). Effect of *Penicillium billai* on the solubility and uptake of P and micronutrients from soil by wheat. *Canadian Journal of Soil Science, 68*, 261–270. doi:10.4141/cjss88-026

Kuepper, G. (2010). *A brief overview of the history and philosophy of organic agriculture*. Kerr Center of Sustainable Agriculture. www.kerrcenter.com

Kulkarni, D. B., Sakhale, B. K., & Giri, N. A. (2018). A potential review on millet grain processing. *International Journal of Nursing Sciences, 3*(1), 1018.

Kulkarni, K. R., & Pandey, R. K. (1988). Annual legumes for food and as green manure in a rice-based cropping system. In *Sustainable Agriculture-Green Manure in Rice Farming* (pp. 280–289). Int. Rice Res. Inst.

Kulma, A., Skorkowska-Telichowska, K., Kostyna, K., Szatkowskia, M., Skała, J., Drulis-Kawa, Z., Preisnera, M., Zuk, M., Szperlika, J., Wang, Y. F., & Szopaa, J. (2015). New flax producing bioplastic fibers for medical purposes. *Industrial Crops and Products, 68*, 80–90. doi:10.1016/j.indcrop.2014.09.013

Kumar, A., Choudhary, A. K., Pooniya, V., Suri, V. K., & Singh, U. (2016). Soil Factors Associated with Micronutrient Acquisition in Crops- Biofortification Perspective. *Biofortification of Food Crops*, 158-176.

Kumar, J., Shukla, S. M., Bhat, V., & Gupta, S. (2005). In Vitro Plant Regeneration and Genetic Transformation of Dichanthium annulatum. DNA and Cell Biology, 24, 670-679.

Kumar, T. S. (n.d.). Review on biofertilizers. *IJCRCPS, 1*, 17-23.

Kumar, A., & Sengar, S. R. (2013). Effect of delayed sowing on yield and proline content of different wheat cultivars. *Research on Crops, 14*(2), 409–415.

Kumar, D., Sengar, R. S., Malik, N., Yadav, M. K., Gupta, S., Chand, P., Singh, G., & Kumar, P. (2020). In vitro evaluation to intensify the differential morphogenetic response through plant growth regulators and antibiotic supplementation in sugarcane. *Plant Physiol. Rep., 25*(2), 335–346. doi:10.100740502-020-00516-6

Kumar, D., Sengar, R. S., Yadav, M. K., anchand, P., Singh, G., & Gupta, S. (2019). Evaluation of Sterilant Effect on In-vitro Culture Establishment in Sugarcane Variety Co 0118. *International Journal of Current Microbiology and Applied Sciences*, *8*(07), 1226–1233. doi:10.20546/ijcmas.2019.807.146

Kumar, J., Singh, D., Ghosh, P., & Kumar, A. (2017). Endophytic and epiphytic modes of microbial interactions and benefits. In *Plant-Microbe Interactions in Agro-ecological Perspectives* (vol. 2, pp. 227–253). doi:10.1007/978-981-10-5813-4_12

Kumar, K. A., Saket, R. K., Rajendran, R., & Eslamian, S. (2020). Design and development of green energy conversion system using waste water and organic kitchen wastes. *International Journal of Hydrology Science and Technology*, *10*(4), 315–333. doi:10.1504/IJHST.2020.108292

Kumar, K. K., Kumar, K. R., Ashrit, R. G., Deshpande, N. R., & Hansen, J. W. (2004). Climate impacts on Indian agriculture. *International Journal of Climatology*, *24*, 1375–1393.

Kumar, K., & Goh, K. M. (2000). Crop residues and management practices: Effects on soil quality, soil nitrogen dynamics, crop yields, and nitrogen recovery. *Advances in Agronomy*, *68*, 197–319.

Kumar, M., Shirohi, H. S., Singh, B., Tomer, B. S., & Singh, A. K. (2020). Effect of insect-proof net house on flowering and seeds quality of cucumber (Cucumis sativus L.). *Indian Journal of Agricultural Sciences*, *90*(6), 1202–1204.

Kumar, N. (Ed.). (2020). *Biotechnology for Biofuels: A Sustainable Green Energy Solution*. Springer Nature. doi:10.1007/978-981-15-3761-5

Kumar, P., Mundiyara, R., & Jitarwal, R. C. (2017). RNA Interference: New Approach of Gene Silencing in Plants. *International Journal of Pure and Applied Biosciences*, *5*(3), 421–425. doi:10.18782/2320-7051.5066

Kumar, S. P. (1989). Use of *Azolla* in Asian Agriculture. *Applied Agricultural Research*, *4*, 149–161.

Kumar, V. (1994). Nitrogen economy in Indian mustard through use of *Azotobacterchroococcum*. *Crop Research*, *8*, 449–452.

Kumar, V., Singh, S., Chhokar, R. S., Malik, R. K., Brainard, D. C., & Ladha, J. K. (2013). Weed management strategies to reduce herbicide use in zero – till rice –wheat cropping systems of the Indo Gangetic Plains. *Weed Technology*, *27*(1), 241–254. doi:10.1614/WT-D-12-00069.1

Kurgat, B. K., Lamanna, C., Kimaro, A., Namoi, N., Manda, L., & Rosenstock, T. S. (2020). Adoption of climate-smart agriculture technologies in Tanzania. *Front. Sustain. Food Syst*, *4*(3), 55.

Kuriqi, A., Pinheiro, A. N., Sordo-Ward, A., Bejarano, M. D., & Garrote, L. (2021). Ecological impacts of run-of-river hydropower plants—Current status and future prospects on the brink of energy transition. *Renewable & Sustainable Energy Reviews*, *142*, 110833. doi:10.1016/j.rser.2021.110833

Kusaba, M. (2004). RNA interference in plants. *Current Opinion in Biotechnology*, *15*(2), 139–143. doi:10.1016/j.copbio.2004.02.004 PMID:15081052

Kusaba, M., Miyahara, K., Iida, S., Fukuoka, H., Takano, T., Sassa, H., Nishimura, M., & Nishio, T. (2003). Low glutelin content 1: A dominant mutation that suppresses the glutelin multigene family via RNA silencing in rice. *The Plant Cell*, *15*(6), 1455–1467. doi:10.1105/tpc.011452 PMID:12782736

Küstermann, B., Kainz, M., & Hülsbergen, K. J. (2008). Modeling carbon cycles and estimation of greenhouse gas emissions from organic and conventional farming systems. *Renewable Agriculture and Food Systems*, *23*(1), 38–52. doi:10.1017/S1742170507002062

Kutchan, T. M. (2005). Predictive metabolic engineering in plants: Still full of surprises. *Trends in Biotechnology*, *23*(8), 381–383.

Kuzemko, C., Bradshaw, M., Bridge, G., Goldthau, A., Jewell, J., Overland, I., & Westphal, K. (2020). Covid-19 and the politics of sustainable energy transitions. *Energy Research & Social Science*, *68*, 101685. doi:10.1016/j.erss.2020.101685 PMID:32839704

Laborie, M. P. G. (2009). Bacterial cellulose and its polymeric nanocomposites. In L. A. Lucia & O. J. Rojas (Eds.), *The Nanoscience & Technology ofRenewable Biomaterials*. Wiley. doi:10.1002/9781444307474.ch9

Lai, W. K., Bard, J. E., & Buck, M. J. (2012). ArchTEx: Accurate extraction and visualization of next-generation sequence data. *Bioinformatics (Oxford, England)*, *28*(7), 1021–1023. doi:10.1093/bioinformatics/bts063 PMID:22302569

Lakshmanan, R., Prasad, R., & Jain, M. C. (2005). Yield and uptake of micronutrients by rice as influenced by variety and nitrogen fertilization. *Archives of Agronomy and Soil Science*, *51*, 1–14.

Lalitha, R., Fathima, K., & Ismail, S. A. (2000). Impact of biopesticides and microbial fertilizers on productivity and growth of Abelmoschus esculentus. *Vasundhara The Earth*, *1-2*, 4-9.

Lal, M., Chaudhary, S., Rawal, S., Sharma, S., Kumar, M., & Chakrabarti, S. K. (2021). Evaluation of bio-agents and Neem based products against late blight disease (*Phytophthora infestans*) of potato. *Indian Phytopathology*, *74*(1), 181–187. doi:10.100742360-021-00330-6

Lal, M., Yadav, S., Sharma, S., Singh, B. P., & Kaushik, S. K. (2017). Integrated management of late blight of potato. *Journal of Applied and Natural Science*, *9*(3), 1821–1824. doi:10.31018/jans.v9i3.1445

Lal, R. (2005). World crop residues production and implications of its use as a biofuel. *Environment International*, *31*, 575–584. PMID:15788197

Lal, R. (2006). Enhancing crop yields in the developing countries through restoration of the soil organic carbon pool in agricultural lands. *Land Degradation & Development*, *17*(2), 197–209. doi:10.1002/ldr.696

Lal, R. (2020). Regenerative agriculture for food and climate. *Journal of Soil and Water Conservation*, *75*(5), 123A–124A.

Lamichhane, J. R., & Reay-Jones, F. P. F. (2021). Editorial: Impacts of COVID-19 on global plant health and crop protection and the resulting effect on global food security and safety. *Crop Protection (Guildford, Surrey)*, *139*, 105383. doi:10.1016/j.cropro.2020.105383 PMID:32921863

Lamont, W. J. (2005). Plastics: Modifying the microclimate for the production of vegetable crops. *HortTechnology*, *15*(3), 477–481. doi:10.21273/HORTTECH.15.3.0477

Lamp, J. W. (2006). For debate: Investment in nutrigenomics will advance the role of nutrition in public health. *Cancer Epidemiology, Biomarkers & Prevention*, *15*, 2329–2330.

Lampkin, N. H. (1994). *Organic Farming: Sustainable Agriculture in Practice. An International Perspective*. Cab International.

Lander, E. S., Linton, L. M., & Birren, B. (2001). Initial sequencing and analysis of the human genome. *Nature*, *409*, 860–921.

Landis, D. A. (2017). Designing agricultural landscapes for biodiversity-based ecosystem services. *Basic and Applied Ecology*, *18*, 1–12.

Langer, I., Santner, J., Krpata, D., Fitz, W. J., Wenzel, W. W., & Schweiger, P. F. (2012). Ectomycorrhizal impact on Zn accumulation of *Populus tremula* L. grown in metalliferous soil with increasing levels of Zn concentration. *Plant and Soil*, *355*(1-2), 283–297. doi:10.100711104-011-1098-y

Lanning, F. C., Ponnaiya, B. W. X., & Crumpton, C. F. (1958). The chemical nature of silica in plants. *Plant Physiology*, *33*(5), 339–343. doi:10.1104/pp.33.5.339 PMID:16655143

Lasat, M. M. (2000). Phytoextraction of metals from contaminated soil: A review of plant/soil/metal interaction and assessment of pertinent agronomic issues. *Journal of Hazardous Substance Research*, *2*, 1–25.

Lasat, M. M. (2002). Phytoremediation of toxic metals: A review of biological mechanisms. *Journal of Environmental Quality*, *31*, 109–120. PMID:11837415

Lauwers, A. M., & Heinen, W. (1974). Biodegradation and utilization of silica and quartz. *Archives of Microbiology*, *95*(1), 67–78. doi:10.1007/BF02451749 PMID:4365644

Lead, J. R., & Wilkinson, K. J. (2006). Natural aquatic colloids: Current knowledge and future trends. *Environmental Chemistry*, *3*, 159–171. doi:10.1071/EN06025

Lebeau, T., Braud, A., & Jezequel, K. (2008). Performance of bioaugmentation-assisted phytoextraction applied to metal contaminated soils: A review. *Environmental Pollution*, *153*(3), 497–522. doi:10.1016/j.envpol.2007.09.015 PMID:17981382

Lebedeva, E., Nazarenko, A. V., & Vedenyapina, E. G. (1999). Soil micromycetes in the area of "Severstal" metallurgical works (Cherepovets, Russia). *Botanica Lithuanica*, *3*, 107–110.

Lee, G. E., & Lee, J. H. (2021). Spatial correlation analysis using the indicators of the anthropocene focusing on atmospheric pollution: A case study of Seoul. *Ecological Indicators*, *125*, 107535. doi:10.1016/j.ecolind.2021.107535

Lee, H. (2017). Transgenic Pro-Vitamin A Biofortified Crops for Improving Vitamin A Deficiency and Their Challenges. *The Open Agriculture Journal*, *11*(1), 11–23. doi:10.2174/1874331501711010011

Lee, R. C., Feinbaum, R. L., & Ambros, V. (1993). The *C. elegans* heterochronic gene *lin-4* encodes small RNAs with antisense complementarity to *lin-14*. *Cell*, *75*(5), 843–854. doi:10.1016/0092-8674(93)90529-Y PMID:8252621

Lee, S., Moon, J. S., Ko, T. S., Petros, D., Goldsbrough, P. B., & Korban, S. S. (2003). Overexpression of *Arabidopsis* phytochelatin synthase paradoxically leads to hypersensitivity to cadmium stress. *Plant Physiology*, *131*(2), 656–663. doi:10.1104/pp.014118 PMID:12586889

Leeuwis, C., & Hall, A. (2010). Facing the Challenges of Climate Change and Food Security: The Role of Research, Extension and Communication Institutions. Report Commissioned by the Research and Extension Branch of the United Nations Food and Agriculture Organization.

Lehmann, A., Veresoglou, S. D., Leifheit, E. F., & Rillig, M. C. (2014). Arbuscular mycorrhizal influence on zinc nutrition in crop plants- a meta-analysis. *Soil Biology & Biochemistry*, *69*, 123–131. doi:10.1016/j.soilbio.2013.11.001

Lehmann, J., Gaunt, J., & Rondo, M. (2006). Bio-char sequestration in terrestrial ecosystems—A review. *Mitigation and Adaptation Strategies for Global Change*, *11*, 403–427.

Leijerstam, B. (1972). Studies in powdery mildew on wheat in Sweden. III. Variability of virulence in *Erysiphe graminis* f. sp. *tritici* due to gene recombination and mutation. *Contributions of the National Swedish Institute of Plant Protection*, *15*, 231–248.

Le, L., Lorenz, Y., Scheurer, S., Fotisch, K., Enrique, E., Bartra, J., Biemelt, S., Vieths, S., & Sonnewald, U. (2006). Design of tomato fruits with reduced allergenicity by dsRNAi-mediated inhibition of ns-LTP (Lyce3) expression. *Plant Biotechnology Journal*, *4*(2), 231–242. doi:10.1111/j.1467-7652.2005.00175.x PMID:17177799

Lemaux, P. G. (2008). Genetically engineered plants and foods: A scientist's analysis of the issues. *Annual Review of Plant Biology*, *59*(1), 771–812. doi:10.1146/annurev.arplant.58.032806.103840 PMID:18284373

Lenssen, A. W. (2008). Planting date and preplant weed management influence yield, water use, and weed seed production in herbicide-free forage barley. *Weed Technology*, *22*(3), 486–492. doi:10.1614/WT-08-009.1

Lenz, R. W., & Marchessault, R. H. (2005). Bacterial Polyesters: Biosynthesis, Biodegradable Plastics and Biotechnology. *Biomacromol*, *6*(1), 1–8. doi:10.1021/bm049700c

Lewis, T. (1997). *Thrips as crop pests.* CAB International, Cambridge University Press.

Li, B., Feng, Y., Xia, X., & Feng, M. (2021). Evaluation of China's Circular Agriculture Performance and Analysis of the Driving Factors. *Sustainability*, *13*(4), 1643.

Lichtenberg, E. M., Kennedy, C. M., Kremen, C., Batary, P., Berendse, F., Bommarco, R., ... Crowder, D. W. (2017). A global synthesis of the effects of diversified farming systems on arthropod diversity within fields and across agricultural landscapes. *Global Change Biology*, *23*(11), 4946–4957.

Lichter, K., Govaerts, B., & Six, J. (2008). Aggregation and C and N contents of soil organic matter fractions in the permanent raised-bed planting system in the Highlands of Central Mexico. *Plant and Soil*, *305*, 237–252.

Li, D. H., Liu, H., Yang, Y., Zhen, P., & Liang, J. (2009). Down-regulated expression of RACK1 gene by RNA interference enhances drought tolerance in rice. *Rice Science*, *16*(1), 14–20. doi:10.1016/S1672-6308(08)60051-7

Liebman & Janke. (1990). Sustainable weed management practices. In *Sustainable agriculture in temperate zones*. John Wiley & Sons.

Liebman, M., & Dyck, E. (1993). Crop rotation and intercropping strategies for weed management. *Ecological Applications*, *3*(1), 92–122. doi:10.2307/1941795 PMID:27759234

Lightfoot, D. R. (1994). Morphology and ecology of lithic-mulch agriculture. *Geographical Review*, *84*(2), 172–185. doi:10.2307/215329

Li, J. T., Liao, B., Dai, Z. Y., Zhu, R., & Shu, W. S. (2009). Phytoextraction of Cd-contaminated soil by carambola (*Averrhoa carambola*) in field trials. *Chemosphere*, *76*(9), 1233–1239. doi:10.1016/j.chemosphere.2009.05.042 PMID:19541343

Lilley, C. J., Goodchild, S. A., Atkinson, H. J., & Urwin, P. E. (2005). Cloning and characterization of a *Heterodera glycines* minopeptidase cDNA. *International Journal for Parasitology*, *35*(14), 1577–1585. doi:10.1016/j.ijpara.2005.07.017 PMID:16216247

Lindenthal, T., Vogl, C. und Heb, J. (1996). Integrale Schwerpun ktthemen und Methodikkriterien der Forschung im okologischen Landbau [Integral focus topics and methodological criteria of research in oological agriculture]. In *Erstellung eines Strategiepapiers furs die Forschungsforderung. Forderungsdienst.* Wien: BMWFK und BMLF.

Liu, B., & Qian, S. B. (2011). Translational regulation in nutrigenomics. *American Society for Nutrition*, *2*, 511–519.

Liu, H. H., Tian, X., Li, Y. J., Wu, C.-A., & Zheng, C.-C. (2008). Microarray-based analysis of stress-regulated microRNAs in *Arabidopsis thaliana. RNA (New York, N.Y.)*, *14*(5), 836–843. doi:10.1261/rna.895308 PMID:18356539

Liu, H., Cheng, S., Huang, L., & Logan, B. E. (2008). Scale-up of membrane-free single-chamber microbial fuel cells. *Journal of Power Sources*, *179*(1), 274–279. doi:10.1016/j.jpowsour.2007.12.120

Liu, R., & Lal, R. (2015). Potentials of engineered nanoparticles as fertilizers for increasing agronomic productions. *The Science of the Total Environment*, *514*, 131–139. doi:10.1016/j.scitotenv.2015.01.104 PMID:25659311

Liu, S., Song, H., Li, X., & Yang, F. (2013). Power generation enhancement by utilizing plant photosynthate in microbial fuel cell coupled constructed wetland system. *International Journal of Photoenergy*, *2013*, 2013. doi:10.1155/2013/172010

Liu, S., Song, H., Wei, S., Yang, F., & Li, X. (2014). Bio-cathode materials evaluation and configuration optimization for power output of vertical subsurface flow constructed wetland—Microbial fuel cell systems. *Bioresource Technology*, *166*, 575–583. doi:10.1016/j.biortech.2014.05.104 PMID:24956029

Liu, Z., Bos, J. I., Armstrong, M., Whisson, S. C., da Cunha, L., Torto-Alalibo, T., Win, J., Avrova, A. O., Wright, F., Birch, P. R. J., & Kamoun, S. (2005). Patterns of diversifying selection in the phytotoxin-like *scr74* gene family of *Phytophthora infestans*. *Molecular Biology and Evolution*, *22*(3), 659–672. doi:10.1093/molbev/msi049 PMID:15548752

Li, X., Zhao, L., Yu, J., Liu, X., Zhang, X., Liu, H., & Zhou, W. (2020). Water splitting: From electrode to green energy system. *Nano-Micro Letters*, *12*(1), 1–29. doi:10.100740820-020-00469-3 PMID:34138146

Li, Y., Song, N., Zhao, C., Li, F., Geng, M., Wang, Y., Liu, W., Xie, C., & Sun, Q. (2016). Application of Glycerol for Induced Powdery Mildew Resistance in Triticum aestivum L. *Frontiers in Physiology*, *7*, 413. doi:10.3389/fphys.2016.00413 PMID:27708588

Li, Y., Zhou, L., Li, Y., Chen, D., Tan, X., Lei, L., & Zhou, J. A. (2008b). nodule-specific plant cysteine proteinase, AsNODF32, is involved in nodule senescence and nitrogen fixation activity of the green manure legume *Astragalus sinicus*. *The New Phytologist*, *180*(1), 185–192. doi:10.1111/j.1469-8137.2008.02562.x PMID:18643938

Llorens, E., García-Agustín, P., & Lapeña, L. (2017). Advances in induced resistance by natural compounds: Towards new options for woody crop protection. *Scientia Agrícola*, *74*(1), 90–100. doi:10.1590/1678-992x-2016-0012

Lockhart, W. L., Billeck, B. N., & Baron, C. L. (1989). Bioassays with a floating plant (Lemna minor) for effects of sprayed and dissolved glyphosate. *Hydrobiologia*, *118/119*(1), 353–359. doi:10.1007/BF00027800

Lockman, P., Oyewumi, M., Koziara, J., Roder, K. E., Mumper, R. J., & Allen, D. D. (2003). Brain uptake of thiamine-coated nanoparticles. *Journal of Controlled Release*, *93*(3), 271–282. doi:10.1016/j.jconrel.2003.08.006 PMID:14644577

Lodriche, S. S., Soltani, S., & Mirzazadeh, R. (2013). *Silicon nanocarrier for delivery of drug, pesticides and herbicides, and for waste water treatment*. United States Patent US 20130225412 A1.

Logan, B. E. (2008). *Microbial Fuel Cell*. John Wiley & Sons.

Loli, O. O., & Chuguizuta, F. (1993). Effect of crop rotations and incorporation of residues on yield of dry season rice (*Oryza sativa* L.). *Proceedings of the 11th Latin-American Congress and the 2nd Cuban Congress of Soil Science*, 1418–1424.

Long, D., Chan, C. Y., & Ding, Y. (2008). Analysis of microRNA-target interactions by a target structure based hybridization model. *Pacific Symposium on Biocomputing*, 64–74. PMID:18232104

Lori, M., Symnaczik, S., Mäder, P., De Deyn, G., & Gattinger, A. (2017). Organic farming enhances soil microbial abundance and activity—A meta-analysis and meta-regression. *PLoS One*, *12*(7), e0180442.

Lossl, A., Eibl, C., Harloff, H. J., Jung, C., & Koop, H. U. (2003). Polyester synthesis in transplastomic tobacco (*Nicotiana tabacum* L.): Significant contents of polyhydroxybutyrate are associated with growth reduction. *Plant Cell Reports*, *21*(9), 891–899. doi:10.100700299-003-0610-0 PMID:12789507

Lovatt, C. J., & Mikkelsen, R. L. (2006). Phosphite fertilizers: What are they? Can you use them? What can they do? *Better Crops with Plant Food*, *90*, 11–13.

Lu, L., Xing, D., & Ren, Z. J. (2015). Microbial community structure accompanied with electricity production in a constructed wetland plant microbial fuel cell. *Bioresource Technology*, *195*, 115–121. doi:10.1016/j.biortech.2015.05.098 PMID:26066972

Luthra, S. K., & Sanjay Rawal, V. K. (2017). Gupta. Promising potato genotypes for organic farming systems. *CPRI Newsletter*, *68*(April-June), 2.

Lyon, G. (2007). Agents that can elicit induced resistance. In D. Walters, A. Newton, & G. Lyon (Eds.), *Induced Resistance for Plant Defence: A Sustainable Approach to Crop Protection* (pp. 9–29). Blackwell Publishing. doi:10.1002/9780470995983.ch2

Macfarlane, L.-A., & Paul, R. (2010). MicroRNA: Biogenesis, Function and Role in Cancer. *Current Genomics*, *11*(7), 537–561. doi:10.2174/138920210793175895 PMID:21532838

Macivor, J. S., Ranalli, M. A., & Lundholm, J. T. (2011). Performance of dryland and wetland plant species on extensive green roofs. *Annals of Botany*, *107*(4), 671–679. doi:10.1093/aob/mcr007 PMID:21292676

Mahajan, G., Brar, L. S., & Walia, U. S. (2002). Minor response in wheat in relation to planting dates, tillage and herbicides. *Indian Journal of Weed Science*, *34*, 213–215.

Mahan, L. K., & Scott Stump, S. (2005). *Food* (6th ed.). Nutrition & Diet Therapy.

Mahanty, T., Bhattacharjee, S., Goswami, M., Bhattacharyya, P., Das, B., Ghosh, A., & Tribedi, P. (2017). Biofertilizers : A potential approach for sustainable agriculture development. *Environmental Science and Pollution Research International*, *24*(4), 3315–3335. doi:10.100711356-016-8104-0 PMID:27888482

Mahapatra & Richard. (2002). Jhabua: A Green Miracle. In Making Water Everybody's Business - Practice and Policy of Water Harvesting Centre for Science and Environment. Academic Press.

Mahapatra, B. S., Mishra, A., & Kumar, A. (2004). *Green manuring: a basic concept and its role in weed control. Training manual: advances in weed management*. G.B. Pant University of Agriculture and Technology.

Mahawer, H., Presanna, R., Gogoi, R., & Singh, A. K. (2020). *Differential modes of disease suppression elicited by silver nanoparticles alone and augmented with calothrix elenkinii against leaf blight in tomato*. European Jr. of Plant Pathology., doi:10.100710658-02021-w

Mahdi, S. S., Talat, M. A., Hussain, D. M., Hamid, A., & Ahmad, L. (2012). Soil phosphorus fixation chemistry and role of phosphate solubilizing bacteria in enhancing its efficiency for sustainable cropping- a review. *Journal of Pure & Applied Microbiology*, *6*, 1–7.

Maillet, F., Poinsot, V., André, O., Puech-Pagès, V., Haouy, A., Gueunier, M., Cromer, L., Giraudet, D., Formey, D., Niebel, A., Martinez, E. A., Driguez, H., Bécard, G., & Dénarié, J. (2011). Fungal lipochitooligosaccharide symbiotic signals in arbuscular mycorrhiza. *Nature*, *469*(7328), 58–63. doi:10.1038/nature09622 PMID:21209659

Ma, J. F., & Takahashi, E. (2002). *Soil, Fertilizer, and Plant Silicon Research in Japan*. Elsevier Science.

Ma, J. F., & Yamaji, N. (2006). Silicon uptake and accumulation in higher plants. *Trends in Plant Science*, *11*(8), 392–397. doi:10.1016/j.tplants.2006.06.007 PMID:16839801

Maleva, M., Borisova, G., Koshcheeva, O., & Sinenko, O. (2017). Biofertilizer based on silicate solubilizing bacteria improves photosynthetic function of *Brassica juncea*. *AGROFOR International Journal*, *2*, 13–19.

Malhi, S. S., & Lemke, R. (2007). Tillage, crop residue and N fertilizer effects on crop yield, nutrient uptake, soil quality and nitrous oxide gas emissions in a second 4-year rotation cycle. *Soil & Tillage Research*, *96*(1-2), 269–283. doi:10.1016/j.still.2007.06.011

Malik, M. R., Yang, W., Patterson, N., Tang, J., Wellinghoff, R. L., Preuss, M. L., Burkitt, C., Sharma, N., Ji, Y., Jez, J. M., Peoples, O. P., Jaworski, J. G., Cahoon, E. B., & Snell, K. D. (2015). Production of high levels of poly-3-hydroxybutyrate in plastids of *Camelina sativa* seeds. *Plant Biotechnology Journal*, *13*(5), 675–688. doi:10.1111/pbi.12290 PMID:25418911

Malik, N., Kumar, D., & Babu, B. K. (2020). Analysis of genetic divergence and population structure through microsatellite markers in normal and quality protein maize genotypes from NW Himalayan region of India. *Vegetos*, *33*(1), 194–202. doi:10.100742535-020-00100-1

Malik, R. K., Balyan, R. S., & Bhan, V. M. (1988). Effect of herbicides and planting dates of weeds in chickpea. *Indian Journal of Weed Science*, *20*, 75–81.

Malkaram, S. A., Hassan, Y. I., & Zempleni, J. (2012). Bioinformatics tools in nutrition research. (American Society for Nutrition. *Advances in Nutrition*, *3*, 654–665.

Mallikarjuna, M. G., & Nepolian, T. (2014). Genetic variability and correlation of kernel micronutrients among exotic quality protein maize inbreads and their utility in breeding programme. *Indian Journal of Genetics and Plant Breeding*, *74*(2), 166–173. doi:10.5958/0975-6906.2014.00152.7

Malone, C., & Hannon, G. J. (2009). Small RNAs as guardians of the genome. *Cell*, *136*(4), 656–668. doi:10.1016/j.cell.2009.01.045 PMID:19239887

Malusa, E., Sas-Paszt, L., & Ciesielska, J. (2012). Technologies for beneficial microorganisms inocula used as biofertilizers. *TheScientificWorldJournal*, *2012*, 491206. doi:10.1100/2012/491206 PMID:22547984

Manyong,, V.M., & Makinde,, K.O., & Coulibaly, O. (2005). Economic grains from maize research in West and Central Africa. *Nongxue Xuebao*, *102*(9), 78–89.

Margis, R., Fusaro, A. F., Smith, N. A., Curtin, S. J., Watson, J. M., Finnegan, E. J., & Waterhouse, P. M. (2006). The evolution and diversification of Dicers in plants. *FEBS Letters*, *580*(10), 2442–2450. doi:10.1016/j.febslet.2006.03.072 PMID:16638569

Mariman, E. C. (2006). Nutrigenomics and nutrigenetics: The 'omics' revolution in nutritional science. *Biotechnology and Applied Biochemistry*, *44*(3), 119–128.

Marmiroli, M., Visioli, G., Maestri, E., & Marmiroli, N. (2011). Correlating SNP genotype with the phenotypic response to exposure to cadmium in populus spp. *Environmental Science & Technology*, *45*(10), 4497–4505. doi:10.1021/es103708k PMID:21491876

Martens, D. C., & Westermann, D. T. (1991). Fertilizer applications for correcting micronutrient deficiencies. In B. J. Alloway (Ed.), *Micronutrient Deficiencies in Global Crop Production* (pp. 549–553). Springer.

Martin, X. M., Sumathi, C. S., & Kannan, V. R. (2011). Influence of agrochemicals and *Azotobacter* sp. application on soil fertility in relation to maize growth under nursery conditions. *Eurasian Journal of Biosciences*, *5*, 19–28. doi:10.5053/ejobios.2011.5.0.3

Marzbani, P., Afrouzi, Y. M., & Omidvar, A. (2015). The effect of nano-zinc oxide on particleboard decay resistance. *Maderas. Ciencia y Tecnología*, *17*(ahead), 63–68. doi:10.4067/S0718-221X2015005000007

Masood, F., Yasin, T., & Hameed, A. (2014). Polyhydroxyalkanoates—What are the uses? Current challenges and perspectives. *Critical Reviews in Biotechnology*, *35*(4), 514–521. doi:10.3109/07388551.2014.913548 PMID:24963700

Mateo, A., Mühlenbock, P., Rustérucci, C., Chang, C. C.-C., Miszalski, Z., Karpinska, B., Parker, J. E., Mullineaux, P. M., & Karpinski, S. (2004). LESION SIMULATING DISEASE 1 is required for acclimation to conditions that promote excess excitation energy. *Plant Physiology*, *136*(1), 2818–2830. doi:10.1104/pp.104.043646 PMID:15347794

Mathers, J. (2005). The Science Of Nutrigenomics. In NUTRIGENOMICS - Report of a workshop hosted by The Nuffield Trust and organized by the Public Health Genetics Unit on 5 February 2004. The Nuffield Trust.

Mathew, A. P., Laborie, M. P., & Oksman, K. (2009). Cross-linked chitosan/chitin crystal nanocomposites with improved permeation selectivity and pH stability. *Biomacromolecules*, *10*(6), 1627–1632. doi:10.1021/bm9002199 PMID:19388678

Mathivanan, R., Umavathi, S., Ramasamy, P. K., & Thangam, Y. (2015). Influence of vermicompost on the activity of the plant growth regulators in the leaves of the Indian butter bean plant, *Dolichos lab lab* L. *International Journal of Advanced Research in Biological Sciences*, *2*, 84–89.

Matsumoto, K., Aoki, E., Takase, K., Doi, Y., & Taguchi, S. (2006a). *In vivo* and *in vitro* characterization of Ser477X mutations in polyhydroxyalkanoate (PHA) synthase 1 from *Pseudomonas* sp 61-3: Effects of beneficial mutations on enzymatic activity, substrate specificity, and molecular weight of PHA. *Biomacromolec*, *7*(8), 2436–2442. doi:10.1021/bm0602029 PMID:16903693

Matsumoto, K., Arai, Y., Nagao, R., Murata, T., Takase, K., Nakashita, H., Taguchi, S., Shimada, H., & Doi, Y. (2006b). Synthesis of short-chain-length/medium-chain-length polyhydroxyalkanoate (PHA) copolymers in peroxisome of the transgenic *Arabidopsis thaliana* harboring the PHA synthase gene from *Pseudomonas* sp. 61-3. *Journal of Polymers and the Environment*, *14*(4), 369–374. doi:10.100710924-006-0035-2

Matsumoto, K., Morimoto, K., Gohda, A., Shimada, H., & Taguchi, S. (2011). Improved polyhydroxybutyrate (PHB) production in transgenic tobacco by enhancing translation efficiency of bacterial PHB biosynthetic genes. *Journal of Bioscience and Bioengineering*, *111*(4), 485–488. doi:10.1016/j.jbiosc.2010.11.020 PMID:21185778

Matsumoto, K., Murata, T., Nagao, R., Nomura, C. T., Arai, S., Arai, Y., Takase, K., Nakashita, H., Taguchi, S., & Shimada, H. (2009). Production of short-chain-length/medium-chain-length polyhydroxyalkanoate (PHA) copolymer in the plastid of *Arabidopsis thaliana* using an engineered 3-ketoacyl-acyl carrier protein synthase III. *Biomacromolec*, *10*(4), 686–690. doi:10.1021/bm8013878 PMID:19265441

Matsumoto, K., Nagao, R., Murata, T., Arai, Y., Kichise, T., Nakashita, H., Taguchi, S., Shimada, H., & Doi, Y. (2005). Enhancement of poly(3-hydroxybutyrate-co-3-hydroxyvalerate) production in the transgenic *Arabidopsis thaliana* by the *in vitro* evolved highly active mutants of polyhydroxyalkanoate (PHA) synthase from *Aeromonas caviae. Biomacromolecules*, *69*(4), 2126–2130. doi:10.1021/bm050113g PMID:16004454

Mattoo, A. K., Shukla, V., Fatima, T., Handa, A. K., & Yachha, S. K. (2010). Genetic engineering to enhance crop-based phytonutrients (nutraceuticals) to alleviate diet-related diseases. In M. T. Giardi, G. Rea, & B. Berra (Eds.), *Bio-Farms for Nutraceuticals: Functional Food and Safety Control by Biosensors* (pp. 122–143). Springer. doi:10.1007/978-1-4419-7347-4_10

Maurhofer, M., Hase, C., Meuwly, P., Metraux, J. P., & Defago, G. (1994). Induction of systemic resistance of tobacco to tobacco necrosis virus by the root-colonizing *Pseudomonas fluorescens* strain CHAO: Influence of the *gacA* gene and of pyoverdine production. *Phytopathology*, *84*(2), 139–146. doi:10.1094/Phyto-84-139

Ma, Y., Rajkumar, M., & Freitas, H. (2009). Inoculation of plant growth promoting bacterium *Achromobacter xylosoxidans* strain Ax10 for the improvement of copper phytoextraction by *Brassica juncea*. *Journal of Environmental Management*, *90*(2), 831–837. doi:10.1016/j.jenvman.2008.01.014 PMID:18329785

Mayer, J. E., & Potrykus, I. (2011). 'Golden Rice' and Biofortification - Their Potential to Save Lives Is Being Hampered by Overzealous Regulation. *Acta Horticulturae*, (941), 21–34. doi:10.17660/ActaHortic.2012.941.1

Mazid, M., & Khan, T. A. (2015). Future of bio-fertilizers in Indian agriculture: An overview. *Irish Journal of Agricultural and Food Research*, *3*, 10–23.

Mburu, S. M., Wakiru, J. M., Muchiri, P. N., & Pintelon, L. (2021). Maintenance Impact on Geothermal Drilling Operations. A Case Study Approach. *American Journal of Industrial and Business Management*, *11*(2), 183–201. doi:10.4236/ajibm.2021.112012

McComb, R. B., Bowers, G. N. Jr, & Posen, S. (2013). *Alkaline phosphatase*. Springer Science and Business Media.

McDonald, T., Brown, D., Keller, N. P., & Hammond, T. M. (2005). RNA silencing of mycotoxin production in *Aspergillus* and *Fusarium* species. *Molecular Plant-Microbe Interactions*, *18*(6), 539–545. doi:10.1094/MPMI-18-0539 PMID:15986923

McIntyre, D., Herren, H. R., Wakhungu, J., & Watson, R. T. (2009). *International Assessment of Agricultural Knowledge, Science and Technology for Development (IAASTD): Global Report*. Island Press.

McMurray, T. A., Dunlop, P. S. M., & Byrne, J. A. (2006). The photocatalytic degradation of atrazine on nanoparticulate TiO2 films. *Journal of Photochemistry and Photobiology A Chemistry*, *182*(1), 43–51. doi:10.1016/j.jphotochem.2006.01.010

McQualter, R. B., Fong Chong, B., O'Shea, M., Meyer, K., van Dyk, D. E., Viitanen, P. V., & Brumbley, S. M. (2005). Initial evaluation of sugarcane as a production platform for a *p*-hydroxybenzoic acid. *Plant Biotechnology Journal*, *3*(1), 29–41. doi:10.1111/j.1467-7652.2004.00095.x PMID:17168897

McQualter, R. B., Petrasovits, L. A., Gebbie, L. K., Schweitzer, D., Blackman, D. M., Chrysanthopoulos, P., Hodson, M. P., Plan, M. R., Riches, J. D., Snell, K. D., Brumbley, S. M., & Nielsen, L. K. (2015). The use of an acetoacetyl-CoA synthase in place of a β-ketothiolase enhances poly-3-hydroxybutyrate production in sugarcane mesophyll cells. *Plant Biotechnology Journal*, *13*(5), 700–707. doi:10.1111/pbi.12298 PMID:25532451

Medek, D. E., Schwartz, J., & Myers, S. S. (2017). Estimated effects of future atmospheric CO2 concentrations on protein intake and the risk of protein deficiency by country and region. *Environmental Health Perspectives*, *128*, 8.

Meenakshi, J. V., Johnson, N. L., Manyong, V. M., Groote, H. D., Javelosa, J., & Yanggen, D. R. (2010). How cost-effective is biofortification in combating micronutrient malnutrition? An ex ante assessment. *World Development*, *38*(1), 64–75. doi:10.1016/j.worlddev.2009.03.014

Meena, R. P., Meena, H. P., & Meena, R. S. (2013). Organic farming: Concept and components. *Popular Kheti*, *1*(4), 5–14.

Mehdi & Khas. (2013). The Influence of Intercropping Maize (Zea mays L.) Green Gram (Vigna Radiata L.) on the Changes of Soil Temperature, Moisture and Nitrogen. *International Journal of Ecosystem*, *3*(2), 13-17.

Mehdipour-Moghaddam, M. J., Emtiazi, G., & Salehi, Z. (2012). Enhanced auxin production by *Azospirillum* pure cultures from plant root exudates. *Journal of Agricultural Science and Technology*, *14*, 985–994.

Mehnaz, S. (2015). *Azospirillum*: a biofertilizer for every crop. In *Plant microbes symbiosis: applied facets* (pp. 297–314). Springer.

Mejri, S., Siah, A., & Coutte, F. (2017). Biocontrol of the wheat pathogen *Zymoseptoria tritici* using cyclic lipopeptides from *Bacillus subtilis*. *Environmental Science and Pollution Research International*. Advance online publication. doi:10.100711356-017-9241-9 PMID:28634804

Memon, J. R., Memon, S. Q., Bhangera, M. I., Memon, G. Z., Turki, A. E., & Allen, G. C. (2008). Characterization of banana peel by scanning electron microscopy and FT-IR spectroscopy and its use for cadmium removal. *Colloids and Surfaces. B, Biointerfaces*, *66*(2), 260–265. doi:10.1016/j.colsurfb.2008.07.001 PMID:18760572

Mench, M., Lepp, N., Bert, V., Schwitzguébel, J.P., Gawronski, S., Schröder, P., & Vangronsveld, J. (2010). Successes and limitations of phytotechnologies at field scale: outcomes, assessment and outlook from COST Action 859. *Journal of Soil and Sediment Contamination, 10*, 1039-1070.

Mench, M. J. (1998). Cadmium availability to plants in relation to major long-term changes in agronomy systems. *Agriculture, Ecosystems & Environment*, *67*(2-3), 175–187. doi:10.1016/S0167-8809(97)00117-5

Mendoza, T. C. (2002). *An Evaluation study of plan international sustainable agriculture program in Calapan*. Oriental Mindoro.

Meng, Zucong, & Ding. (2005). Carbon contents in soils and crops as affected by long-term fertilization. *Turang Xuebao*, *42*, 769–776.

Menon, B., Harinarayan, C. V., & Raj, M. N. (2010). Prevalence of low dietary calcium intake in patients with epilepsy: A study from South India. *Neurology India*, *58*, 209–212.

Menzel, G., Harloff, H. J., & Jung, C. (2003). Expression of bacterial poly (3-hydroxybutyrate) synthesis genes in hairy roots of sugar beet (*Beta vulgaris* L.). *Applied Microbiology and Biotechnology*, *60*(5), 571–576. doi:10.100700253-002-1152-z PMID:12536258

Mier y Terán Giménez Cacho, M., Giraldo, O. F., Aldasoro, M., Morales, H., Ferguson, B. G., Rosset, P., ... Campos, C. (2018). Bringing agroecology to scale: Key drivers and emblematic cases. *Agroecology and Sustainable Food Systems*, *42*(6), 637–665.

Miggiano, G. A., & De Sanctis, R. (2006). Nutritional genomics: Toward a personalized diet US National Library of Medicine, National institutes of health. *La Clinica Terapeutica*, *157*, 355–361.

Milani, N., McLaughlin, M. J., Stacey, S. P., Kirby, J. K., Hettiarachchi, G. M., Beak, D. G., & Cornelis, G. (2012). Dissolution kinetics of macronutrient fertilizers coated with manufactured zinc oxide nanoparticles. *Journal of Agricultural and Food Chemistry*, *60*(16), 3991–3998. doi:10.1021/jf205191y PMID:22480134

Miller, H. (2017). How College Students Are Being Misled About 'Sustainable' Agriculture. *National Review*.

Miller, R. G. Jr. (1986). *Beyond ANOVA, basics of applied statistics*. John Wiley & Sons.

Mills, T. (2006). Composting cafeteria residuals with earthworm. *BioCycle*, *47*, 54–55.

Milne, I., Bayer, M., Cardle, L., Shaw, P., Stephen, G., Wright, F., & Marshall, D. (2010). Tablet—Next generation sequence assembly visualization. *Bioinformatics (Oxford, England)*, *26*(3), 401–402. doi:10.1093/bioinformatics/btp666 PMID:19965881

Milner, J. A. (2008). Nutrition and cancer: Essential elements for a roadmap. *Cancer Letters*, *269*, 189–198.

Minghui, L., Wen, Z., Yu, X. I. A., & Yongsheng, G. (2011). Study on removal efficiencies of pollutant from constructed wetland in aquiculture waste water around Poyang Lake. *Procedia Environmental Sciences*, *10*, 2444–2448. doi:10.1016/j.proenv.2011.09.380

Ming, R., Hou, S., Feng, Y., Yu, Q., Dionne-Laporte, A., Saw, J. H., Senin, P., Wang, W., Ly, B. V., Lewis, K. L., Salzberg, S. L., Feng, L., Jones, M. R., Skelton, R. L., Murray, J. E., Chen, C., Qian, W., Shen, J., Du, P., ... Alam, M. (2008). The draft genome of the transgenic tropical fruit tree papaya (Carica papaya linnaeus). *Nature*, *452*(7190), 991–996. doi:10.1038/nature06856 PMID:18432245

Ministry of New and Renewable Energy. (n.d.). *Annual Report, 2015-16.* Author.

Mire, G. L., Siah, A., & Brisset, M.-N. (2018). Surfactin Protects Wheat against *Zymoseptoria tritici* and Activates Both salicylic acid- and jasmonic acid-dependent defense responses. *Agriculture*, *8*(1), 11. doi:10.3390/agriculture8010011

Mishra, D. J., Singh, R., Mishra, U. K., & Kumar, S. S. (2013). Role of bio-fertilizer in organic agriculture: A review. *Research Journal of Recent Sciences*, *2*, 39–41.

Mishra, D., Rajvir, S., Mishra, U., & Kumar, S. S. (2013). Role of bio-fertilizer in organic agriculture: A review. *Research Journal of Recent Sciences ISSN*, *2277*, 2502.

Mishra, J. S., & Singh, V. P. (2008). Competitive ability of wheat (Triticum aestivum) cultivars against wild oat (Avena ludoviciana) infestation under zero tillage. *Indian Journal of Agricultural Sciences*, *78*, 838–841.

Mishra, P., & Dash, D. (2014). Rejuvenation of biofertilizer for sustainable agriculture and economic development. *Consilience: The Journal of Sustainable Development*, *11*, 41–61.

Miśkiewicz, R. (2021). The Impact of Innovation and Information Technology on Greenhouse Gas Emissions: A Case of the Visegrád Countries. *Journal of Risk and Financial Management*, *14*(2), 59. doi:10.3390/jrfm14020059

Missiou, A., Kalantidis, K., Boutla, A., Tzortzakaki, S., Tabler, M., & Tsagris, M. (2004). Generation of transgenic potato plants highly resistant to potato virus Y (PVY) through RNA silencing. *Molecular Breeding*, *14*(2), 185–197. doi:10.1023/B:MOLB.0000038006.32812.52

Mitsky, T. A., Slater, S., Reiser, S. E., Hao, M., & Houmiel, K. L. (2003). *Multigene expression vectors for the biosynthesis of products via multienzyme biological pathways.* US Patent Application 2003/0182678.

Mittal & Decker. (2013). Application of biotechnology for biofuels: Transforming biomass to biofuels. *Biotech, 3*, 341-343.

Mittal, D., Kaur, G., Singh, P., Yadav, K., & Ali, S. A. (2020). Nanoparticle-based sustainable agriculture and food science: Recent advances and future outlook. *Front. Nanotechnol.*, *2*, 579954. doi:10.3389/fnano.2020.579954

Mittendorf, V., Robertson, E. J., Leech, R. M., Kruger, N., Steinbuchel, A., & Poirier, Y. (1998). Synthesis of medium-chain-length polyhydroxyalkanoates in *Arabidopsis thaliana* using intermediates of peroxisomal fatty acid β-oxidation. *Proceedings of the National Academy of Sciences of the United States of America*, *95*(23), 13397–13402. doi:10.1073/pnas.95.23.13397 PMID:9811811

Mitter, N., Worrall, E. A., Robinson, K. E., Li, P., Jain, R. G., Taochy, C., Fletcher, S. J., Carroll, B. J., Lu, G. Q., & Xu, Z. P. (2017). Clay nanosheets for topical delivery of RNAi for sustained protection against plant viruses. *Nature Plants*, *3*(2), 16207. doi:10.1038/nplants.2016.207 PMID:28067898

MNRE (Ministry of New and Renewable Energy Resources). (2009). *Govt. of India*. www.mnre.gov.in/biomassrsources

Moaveni, P., & Kheiri, T. (2011). TiO2nano particles affected on maize (*Zea mays* L). In 2nd *International Conference on Agricultural & Animal Science*. IACSIT Press.

Mochida, K., & Shinozaki, K. (2010). Genomics and bioinformatics resources for crop improvement. *Plant & Cell Physiology*, *51*(4), 497–523. doi:10.1093/pcp/pcq027 PMID:20208064

Mohammadi, K., & Sohrabi, Y. (2012). Bacterial biofertilizers for sustainable crop production: A review. *Journal of Agricultural and Biological Science*, *7*, 307–316.

Mohan, V., Sandeep, S., Deepa, R., Shah, B., & Varghese, C. (2007a). Epidemiology of type 2 diabetes: Indian scenario. *The Indian Journal of Medical Research*, *125*(3), 217–230.

Mohan, V., Sudha, V., & Radhika, G. (2007b). Gene-environment interactions and the diabetes epidemic in India. *Forum of Nutrition*, *60*, 118–126.

Mohapatra, B., Verma, D. K., Sen, A., Panda, B. B., & Asthir, B. (2013). Bio-fertilizers- a gateway to sustainable agriculture. *Popular Kheti*, *1*, 97–106.

Mohler, C. L., & Calloway, M. B. (1992). Effects of tillage and mulch on the emergence and survival of weeds in sweet corn. *Journal of Applied Ecology*, *29*(1), 21–34. doi:10.2307/2404343

Mohsin, M., Kuittinen, S., Salam, M. M. A., Peräniemi, S., Laine, S., Pulkkinen, P., Kaipiainen, E., Vepsäläinen, J., & Pappinen, A. (2019). Chelate-assisted phytoextraction: Growth and ecophysiological responses by *Salix schwerinii ELWolf* grown in artificially polluted soils. *Journal of Geochemical Exploration*, *205*, 205. doi:10.1016/j.gexplo.2019.106335

Mokrani, S., & El-hafid, N. (2021) Microbes Associated with Crops: Functional attributes for crop productivity. Sustainable Development and Biodiversity, 27. doi:10.1007/978-3-030-73507-4_2

Moldes, C. A., De Lima Filho, O. F., Merini, L. J., Tsai, S. M., & Camiña, J. M. (2016). Occurrence of powdery mildew disease in wheat fertilized with increasing silicon doses: A chemometric analysis of antioxidant response. *Acta Physiologiae Plantarum*, *38*(8), 206. doi:10.100711738-016-2217-4

Molina, A., Görlach, J., Volrath, S., & Ryals, J. (1999). Wheat genes encoding two types of PR-1 proteins are pathogen inducible, but do not respond to activators of systemic acquired resistance. *Molecular Plant-Microbe Interactions*, *12*(1), 53–58. doi:10.1094/MPMI.1999.12.1.53 PMID:9885193

Mondal, S. S., & Banerjee, W. (2020). Design of Green Energy Based Charge Controller for Domestic Uses. In *Energy Systems, Drives and Automations* (pp. 173–185). Springer. doi:10.1007/978-981-15-5089-8_16

Mondelaers, K., Aertsens, J., & Van Huylenbroeck, G. (2009). A meta-analysis of the differences in environmental impacts between organic and conventional farming. *British Food Journal*.

Monsen, E. R., & Balintfy, J. (1982). Calculating dietary iron bioavailability: Refinement and computerization. *Journal of the American Dietetic Association*, *80*(4), 307–311.

Montiel-Rozas, M. M., Madejon, E., & Madejon, P. (2016). Effect of heavy metals and organic matter on root exudates (low molecular weight organic acids) of herbaceous species: An assessment in sand and soil conditions under different levels of contamination. *Environmental Pollution*, *216*, 273–281. doi:10.1016/j.envpol.2016.05.080 PMID:27267743

Moore, M. N. (2006). Do nanoparticles present ecotoxicological risks for the health of the aquatic environment? *Environment International*, *32*, 967–976.

Moraditochaee, M., Azarpour, E., & Bozorgi, H. R. (2014). Study effects of bio- fertilizers, nitrogen fertilizer and farmyard manure on yield and physiochemical properties of soil in lentil farming. *International Journal of Biosciences*, *4*, 41–48.

Motallebi, P., Niknam, V., Ebrahimzadeh, H., Enferadi, S. T., & Hashemi, M. (2015). The effect of methyl jasmonate on enzyme activities in wheat genotypes infected by the crown and root rot pathogen *Fusarium culmorum*. *Acta Physiologiae Plantarum*, *37*(11), 237. doi:10.100711738-015-1988-3

Motsara & Bisoyi. (2001). Crop demonstration on biofertilizer. National Biofertilizer Development Centre.

Mozejko-Ciesielska, J., & Kiewisz, R. (2016). Bacterial polyhydroxyalkanoates: Still fabulous? *Microbiological Research*, *192*, 271–282. doi:10.1016/j.micres.2016.07.010 PMID:27664746

Muchembled, J., Lounes-Hadj Sahraoui, A., Grandmoujin-Ferjani, A., & Sancholle, M. (2006). Changes in lipid composition of *Blumeria graminis* f. sp. *tritici* conidia produced on wheat leaves treated with heptanoyl salicylic acid. *Phytochemistry*, *67*(11), 1104–1109. doi:10.1016/j.phytochem.2006.02.025 PMID:16647727

Mukhopadhyay, S. S. (2014). Nanotechnology in agriculture: Prospects and constraints. *Nanotechnology, Science and Applications*, *7*, 63–71. doi:10.2147/NSA.S39409 PMID:25187699

Mullen, Orden, & Gulati. (2005). *Agricultural policies in India*. International Food Policy Research Institute (IFPRI).

Muller, M., & Kersten, S. (2003). Nutrigenomics: Goals and strategies. *Nature Reviews. Genetics*, *4*(4), 315–322.

Murgai, R., Ali, M., & Byerlee, D. (2001). Productivity Growth and Sustainability in Post-Green Revolution Agriculture: The Case of the Indian and Pakistan Punjabs. *The World Bank Research Observer*, *16*(2).

Musibau, H. O., Nepal, R., Vespignani, J., & Yanotti, M. (2020). The moderating role of green energy and energy-innovation in environmental Kuznets: Insights from quantile-quantile analysis. *Globalization and Monetary Policy Institute Working Paper*, (385).

Musser, F. P., & Shelton, A. M. (2005). The influence of post-exposure temperature on the toxicity of insecticides to *Ost rinianubilalis*(Lepidoptera:Crambidae). *Pest Management Science*, *6*(5), 508–510. doi:10.1002/ps.998 PMID:15633207

Muthusamy, V. (2015). Biofortification of maize: An Indian perspective. *Indian J. Genet.*, *75*(1), 1–22. doi:10.5958/0975-6906.2015.00001.2

Muthusamy, V., Hossain, F., Thirunavukkarasu, N., Choudhary, M., Saha, S., Bhat, J. S., Prasanna, B. M., & Gupta, H. S. (2014). Development of β-Carotene rich Maize hybrids through Marker-Assisted Introgression of *β-carotene hydroxylase* Allele. *PLoS One*, *9*(12), e113583. doi:10.1371/journal.pone.0113583 PMID:25486271

Myers, S. S., Smith, M. R., Guth, S., Golden, C. D., Vaitla, B., & Mueller, N. D. (2017). Climate change and global food systems: Potential impacts on food security and undernutrition. *Annual Review of Public Health*, *38*, 259–277.

Myers, S. S., Zanobetti, A., & Kloog, I. (2014). Increasing CO 2 threatens human nutrition. *Nature*, *510*, 139–142.

N'Dayegamiye, A., & Tran, T. S. (2001). Effects of green manures on soil organic matter and wheat yields and nutrition. *Canadian Journal of Soil Science*, *81*(4), 371–382. doi:10.4141/S00-034

Nabi, A., Narayan, S., Afroza, B., Mushtaq, F., Mufti, S., Ummyiah, H. M., & Magray, M. M. (2017). Biodynamic farming in vegetables. *Journal of Pharmacognosy and Phytochemistry*, *6*(6), 212–219.

Naderi, M. R., & Abedi, A. (2012). Application of nanotechnology in agriculture and refinement of environmental pollutants. *Journal of Nanotechnology*, *11*(1), 18–26.

Naik, K., Mishra, S., Srichandan, H., Singh, P.K., & Sarangi, P.K. (2019). Plant growth promoting microbes: Potential link to sustainable agriculture and environment. *Biocatalysis and Agricultural Biotechnology, 21*, 1-12.

Naiman, R (1994). Watershed Management: Balancing Sustainability and Environmental Change. In *Technology & Engineering*. Springer.

Naiman, A. D., Latrónico, A., & de Salamone, I. E. (2009). Inoculation of wheat with *Azospirillum brasilense* and *Pseudomonas fluorescens*: Impact on the production and culturable rhizosphere microflora. *European Journal of Soil Biology*, *45*(1), 44–51. doi:10.1016/j.ejsobi.2008.11.001

Nakamura, N., Fukuchi-Mizutani, M., Miyazaki, K., Suzuki, K., & Tanaka, Y. (2006). RNAi suppression of the anthocyanidin synthase gene in Torenia hybrida yields white flowers with higher frequency and better stability than antisense and sense suppression. *Plant Biotechnology (Sheffield, England)*, *23*(1), 13–18. doi:10.5511/plantbiotechnology.23.13

Nakashita, H., Arai, Y., Shikanai, T., Doi, Y., & Yamaguchi, I. (2001a). Introduction of bacterial metabolism into higher plants by polycistronic transgene expression. *Bioscience, Biotechnology, and Biochemistry*, *65*(7), 1688–1691. doi:10.1271/bbb.65.1688 PMID:11515561

Nakashita, H., Arai, Y., Suzuki, Y., & Yamaguchi, I. (2001b). Molecular breeding of transgenic tobacco plants which accumulate polyhydroxyalkanoates. *RIKEN Rev*, *42*, 67–70.

Nakashita, H., Arai, Y., Yoshioka, K., Fukui, T., Doi, Y., Usami, R., Horikoshi, K., & Yamaguchi, I. (1999). Production of biodegradable polyester by a transgenic tobacco. *Bioscience, Biotechnology, and Biochemistry*, *63*(5), 870–874. doi:10.1271/bbb.63.870 PMID:27385570

Nakatsuka, A., Mizuta, D., Kii, Y., Miyajima, I., & Kobayashi, N. (2008). Isolation and expression analysis of flavonoid biosynthesis genes in evergreen azalea. *Scientia Horticulturae*, *118*(4), 314–320. doi:10.1016/j.scienta.2008.06.016

Nakatsuka, T., Abe, Y., Kakizaki, Y., Yamamura, S., & Nishihara, M. (2007a). Production of red-flowered plants by genetic engineering of multiple flavonoid biosynthetic genes. *Plant Cell Reports*, *26*(11), 1951–1959. doi:10.100700299-007-0401-0 PMID:17639403

Nakatsuka, T., Chetsadaporn, P., Yamamura, S., & Nishihara, M. (2007b). Induction of differential flower pigmentation patterns by RNAi using promoters with distinct tissue-specific activity. *Plant Biotechnology Reports*, *1*(4), 251–257. doi:10.100711816-007-0039-y

Namaya, P. (2011). *Towards the biofortification of banana fruit for enhanced micronutrient content* (PhD thesis). Queensland University of Technology.

Nambiar, K. K. M., Soni, P. N., Vats, M. R., Sehgal, D. K., & Mehta, D. K. (1992). *AICRP on long term fertilizer experiments, Annual Reports 1987-88 and 1988-89*. IARI.

Nandwani, D., & Nwosisi, S. (2016). Global trends in organic agriculture. In *Organic farming for sustainable agriculture* (pp. 1–35). Springer.

Nannoni, F., Rossi, S., & Protano, G. (2014). Soil properties and metal accumulation by earthworms in the Siena urban area. *Applied Soil Ecology*, *77*, 9–17. doi:10.1016/j.apsoil.2014.01.004

Narayanan, S. (2005). Organic Farming. In *India: Relevance, Problems and Constraints, Occasional, Paper No. 38*. Department of Economic Analysis and Research National Bank for Agriculture and Rural Development.

Narayan, M., Solanki, P., & Srivastava, R. K. (2018). Treatment of sewage (domestic wastewater or municipal wastewater) and electricity production by integrating constructed wetland with microbial fuel cell. In *Sewage*. IntechOpen. doi:10.5772/intechopen.75658

Narimani, H., Rahimi, M. M., Ahmadikhan, A., & Vaezi, B. (2010). Study on the effects of foliar spray of micronutrient on yield and yield components of durum wheat. *Applied Scientific Research*, *2*, 168–176.

Narwal, Palaniraj, & Chandra. (2002). Allelopahtic effects of aqueous extracts of Sunflower (Helianthus annus L.) root on some winter oil seed crop. *Geobies*, *29*, 225–228.

Natarajan, N., & Sivasubramaniam, K. (2007). Nanotechnology application in seed management. In *Application of Nanotechnology in Agriculture*. Tamil Nadu Agricultural University.

National Agricultural Extension Research & Liaison Services (NAERLS) and Federal Department of Agricultural Extension. (2014). *Agricultural Performance Survey 2014 wet season in Nigeria*. Executive summary. Author.

National Bureau of Statistics (NBS). (2006). *Annual abstracts of statistics*. Federal Republic of Nigeria.

Naureen, Z., Aqeel, M., & Hassan, M. N. (2015). Isolation and screening of silicate bacteria from various habitats for biological control of phytopathogenic fungi. *American Journal of Plant Sciences*, *6*, 2850–2859. doi:10.4236/ajps.2015.618282

Naushad, S. M., Radha, A., & Devi, R. (2010). Role of parental folate pathway single nucleotide polymorphisms in altering the susceptibility to neural tube defects in South India. *Journal of Perinatal Medicine*, *38*(1), 63–69.

Nawrath, C., Poirier, Y., & Somerville, C. (1994). Targeting of the polyhydroxybutyrate biosynthetic-pathway to the plastids of *Arabidopsis thaliana* results in high-levels of polymer accumulation. *Proceedings of the National Academy of Sciences of the United States of America*, *91*(26), 12760–12764. doi:10.1073/pnas.91.26.12760 PMID:7809117

Nayak, A., Panda, S., Basu, A., & Dhal, N. (2018). Enhancement of toxic Cr (VI), Fe, and other heavy metals phytoremediation by the synergistic combination of native *Bacillus cereus* strain and *Vetiveria zizanioides* L. *International Journal of Phytoremediation*, *20*(7), 682–691. doi:10.1080/15226514.2017.1413332 PMID:29723050

Nehnevajova, E., Herzig, R., Bourigault, C., Bangerter, S., & Schwitzguebel, J. P. (2009). Stability of enhanced yield and metal uptake by sunflowers mutants for improved phytoremediation. *International Journal of Phytoremediation*, *11*(4), 329–346. doi:10.1080/15226510802565394

Nel, A., Xia, T., Mädler, L., & Li, N. (2006). Toxic potential of materials at nanolevel. *Science*, *311*(5761), 622–627. doi:10.1126cience.1114397 PMID:16456071

Nelson, G. C., Rosegrant, M. W., Koo, J., Robertson, R., Sulser, T., Zhu, T., Ringler, C., & … . (2009). *Climate change impact on agriculture and costs of adaptation*. International Food Policy Research Institute.

Nelson, M. L. (2004). Plant growth promoting rhizobacteria (PGPR): Prospects for new inoculants. *Crop Management*, *2*(1), 301–305. doi:10.1094/CM-2004-0301-05-RV

Nesci, A., Bluma, R., & Etcheverry, M. (2005). *In vitro* selection of maize rhizobacteria to study potential biological control of *Aspergillus*section *Flavi*. *European Journal of Plant Pathology*, *113*, 1–13. doi:10.100710658-005-5548-3

Newell-McGloughlin, M. (2008). Nutritionally improved agricultural crops. *Plant Physiology*, 939–953.

Newman, J. A. (2004). Climate change and cereal aphids: The relative effects of increasing CO2 and temperature on aphid population dynamics. *Global Change Biology*, *10*, 5–15.

Newton, J. (2007). *Building greener: guidance on the use of green roofs, green walls and complementary features on buildings*. CIRIA Report 16.

Newton, A. C., Toth, I. K., Neave, P., & Hyman, L. J. (2004). Bacterial inoculum from a previous crop affects fungal disease development on subsequent nonhost crops. *The New Phytologist*, *63*(1), 133–138. doi:10.1111/j.1469-8137.2004.01077.x PMID:33873780

Newton, A., Guy, D., Gaunt, R., & Thomas, W. (2000). The effect of powdery mildew inoculum pressure and fertilizer levels on disease tolerance in spring barley. *Zeitschrift für Pflanzenkrankheiten und Pflanzenschutz*, *107*, 67–73.

Ngouajio, M., & McGiffen, M. E. (2004). Sustainable Vegetable production: Effects of cropping systems on weed and insect population dynamics. *Acta Horticulturae*, (638), 77–83. doi:10.17660/ActaHortic.2004.638.8

Nguyen, T. A. H., Ngo, H. H., Guo, W. S., Zhang, J., Liang, S., Yue, Q. Y., Li, Q., & Nguyen, T. V. (2013). Applicability of agricultural waste and byproducts for adsorptive removal of heavy metals from wastewater. *Bioresource Technology, 148*, 574–585. doi:10.1016/j.biortech.2013.08.124

Nicewarner-Pena, S. R., Freeman, R. G., Reiss, B. D. H., He, L., Peña, D. J., Walton, I. D., Cromer, R., Keating, C. D., & Natan, M. J. (2001). Submicrometer metallic barcodes. *Science, 294*(5540), 137–141. doi:10.1126cience.294.5540.137 PMID:11588257

Nicholls, C. I., & Altieri, M. A. (2018). Pathways for the amplification of agroecology. *Agroecology and Sustainable Food Systems, 42*(10), 1170–1193.

Nielsen, E., Cheung, A. Y., & Ueda, T. (2008). The regulatory RAB and ARF GTPases for vesicular trafficking. *Plant Physiology, 147*(4), 1516–1526. doi:10.1104/pp.108.121798 PMID:18678743

Niggli, U., Slabe, A., Schmid, O., Halberg, N., & Schlueter, M. (2008). Vision for an organic food and farming research agenda to 2025. *Organic Knowledge for the Future.*

Niggli, U. (2014). *How will organic farming look like in 20 years and to which challenges will it have to respond to? In TIPI Workshop.* BIOFACH.

Niggli, U., Fließbach, A., Hepperly, P., & Scialabba, N. (2009). *Low greenhouse gas agriculture: Mitigation and adaptation potential of sustainable farming systems.* FAO.

Nishihara, M., Nakatsuka, T., Hosokawa, K., Yokoi, T., Abe, Y., Mishiba, K., & Yamamura, S. (2006). Dominant inheritance of white-flowered and herbicide-resistant traits in trangenic gentian plants. *Plant Biotechnology (Sheffield, England), 23*(1), 25–31. doi:10.5511/plantbiotechnology.23.25

Nishihara, M., Nakatsuka, T., & Yamamura, S. (2005). Flavonoid components and flower color change in transgenic tobacco plants by suppression of *chalcone isomerase* gene. *FEBS Letters, 579*(27), 6074–6078. doi:10.1016/j.febslet.2005.09.073 PMID:16226261

Njoroge, J. W. (1994). *Field notes on organic farming.* Kenya Institute of Organic Farming.

Norberg-King, T. J. (1988). *An interpolation estimate for chronic toxicity: The ICp approach.* National Effluent Toxicity Assessment Center Technical Report 05-88.

Norheim, F., Gjelstad, I. M. F., & Hjorth, M. (2012). Molecular nutrition research-the modern way of performing nutritional science. *Journal Nutrients, 4*(12), 1898–1944.

Nosheen, S., Ajmal, I., & Song, Y. (2021). Microbes as biofertilizers, a potential approach for sustainable crop production. *Sustainability, 13*, 1-20.

Nowak, D. J., & Crane, D. E. (1998). The urban forest effects (UFORE) model Quantifying Urban Forest Structure and Functions. *Integrated Tools Proceedings*, 1-7. https://nrs.fs.fed.us/pubs/gtr/other/gtr_nc212/gtr_nc212co.pdf

Nowara, D., Gay, A., Lacomme, C., Shaw, J., Ridout, C., Douchkov, D., Hensel, G., Kumlehn, J., & Schweizer, P. (2010). HIGS: Host-induced gene silencing in the obligate biotrophic fungal pathogen *Blumeria graminis. The Plant Cell, 22*(9), 3130–3141. doi:10.1105/tpc.110.077040 PMID:20884801

Nuria, S., Carmen, F., & Sinda, C. (2011). RNAi mediated protection against citrus *Tristeza* virus in Transgenic citrus plants. *Non Coding RNAs in Plants, 8*, 447–460.

Nyquist, J., & Greger, M. (2009). A field study of constructed wetlands for preventing and treating acid mine drainage. *Ecological Engineering, 35*(5), 630–642. doi:10.1016/j.ecoleng.2008.10.018

Nystedt, B., Street, N. R., Wetterbom, A., Zuccolo, A., Lin, Y. C., Scofield, D. G., Vezzi, F., Delhomme, N., Giacomello, S., Alexeyenko, A., Vicedomini, R., Sahlin, K., Sherwood, E., Elfstrand, M., Gramzow, L., Holmberg, K., Hällman, J., Keech, O., Klasson, L., ... Jansson, S. (2013). The Norway spruce genome sequence and conifer genome evolution. *Nature*, *497*(7451), 579–584. doi:10.1038/nature12211 PMID:23698360

Obembe, O. O., Popoola, J. O., Leelavathi, S., & Reddy, S. V. (2011). Advances in plant molecular farming. *Biotechnology Advances*, *29*(2), 210–222. doi:10.1016/j.biotechadv.2010.11.004 PMID:21115109

Oberdörster, E. (2004). Manufactured nanomaterials (fullerens, C60) induce oxidative stress in the brain of juvenile largemouth bass. *Environmental Health Perspectives*, *112*(10), 1058–1062. doi:10.1289/ehp.7021 PMID:15238277

Oberson, A., Fardeau, J. C., Besson, J. M., & Sticher, H. (1993). Soil phosphorus dynamics in cropping systems managed according to conventional and biological agricultural methods. *Biology and Fertility of Soils*, *16*(2), 111–117. doi:10.1007/BF00369411

OECD. (2000). *Guidance Document on Aquatic Toxicity Testing of Difficult Substances and Mixtures.* OECD Environmental Health and Safety Publications, Series on Testing and Assessment No.23.

OECD. (2010). *Interim Report on the Green Growth Strategy: Implementing our Commitment for a Sustainable Future*. OECD.

OECD. (2010b). *Sustainable Management of Water Resources in Agriculture*. OECD.

Oechel, W. C., Cowles, S., Gurlke, N., Hastings, S. J., Lawrence, B., Prudhomme, T., Reichers, G., Strain, B., Tissue, D., & Vourlitis, G. L. (1994). Transient nature of CO_2 fertilization in Arctic tundra. *Nature*, *371*, 500–503.

Ogbo, F. C. (2010). Conversion of cassava wastes for biofertilizer production using phosphate solubilizing fungi. *Bioresource Technology*, *101*(11), 4120–4124. doi:10.1016/j.biortech.2009.12.057 PMID:20138509

Ogero, Mburugu, Mwangi, Ngugi, & Ombori. (2012). Low Cost Tissue Culture Technology in the Regeneration of Sweet Potato (Ipomoea batatas (L) Lam). *Research Journal of Biology*, *2*, 51-58.

Oke, T. R. (1987). *Boundary Layer Climates* (2nd ed.). Methuen.

Oliveira, M. D. M., Varanda, C. M. R., & Félix, M. R. F. (2016). Induced resistance during the interaction pathogen x plant and the use of resistance inducers. *Phytochemistry Letters*, *15*, 152–158. doi:10.1016/j.phytol.2015.12.011

Olson, D. G., McBride, J. E., Shaw, J. A., & Lynd, L. R. (2012). Recent progress in consolidated bioprocessing. *Current Opinion in Biotechnology*, *23*(3), 396–405. doi:10.1016/j.copbio.2011.11.026 PMID:22176748

Ombodi, A., & Saigusa, M. (2000). Broadcast application versus band application of polyolefin coated fertilizer on green peppers grown on andisol. *Journal of Plant Nutrition*, *23*(10), 1485–1493. doi:10.1080/01904160009382116

Onder, D., Onder, S., Daghan, S., & Uygu, V. (2016). Influence of different irrigation level and different nickel (ni) doses on phytoremediative capacity of *Tagetes erecta* l. *Fresenius Environmental Bulletin*, *25*, 191–199.

Ong & Rogero. (2009). Nutrigenomics: Importance of nutrient-gene interaction for health promotion. *Journal of the ABESO, 40.*

Onmura, S., Matsumoto, M., & Hokoi, S. (2001). Study on the evaporative cooling effect of roof lawn gardens. *Energy and Building*, *33*(7), 653–666. doi:10.1016/S0378-7788(00)00134-1

Oon, Y. L., Ong, S. A., Ho, L. N., Wong, Y. S., Dahalan, F. A., Oon, Y. S., Lehl, H. K., & Thung, W. E. (2016). Synergistic effect of up-flow constructed wetland and microbial fuel cell for simultaneous wastewater treatment and energy recovery. *Bioresource Technology*, *203*, 190–197. doi:10.1016/j.biortech.2015.12.011 PMID:26724550

Oon, Y. L., Ong, S. A., Ho, L. N., Wong, Y. S., Oon, Y. S., Lehl, H. K., & Thung, W. E. (2015). Hybrid system up-flow constructed wetland integrated with microbial fuel cell for simultaneous wastewater treatment and electricity generation. *Bioresource Technology*, *186*, 270–275. doi:10.1016/j.biortech.2015.03.014 PMID:25836035

Ordovas, J. M., & Mooser, V. (2004). Nutrigenomics and nutrigenetics. *Current Opinion in Lipidology*, *15*, 101–108.

Ortmann, I., & Moerschbacher, B. M. (2006). Spent growth medium of Pantoea agglomerans primes wheat suspension cells for augmented accumulation of hydrogen peroxide and enhanced peroxidase activity upon elicitation. *Planta*, *224*(4), 963–970. doi:10.100700425-006-0271-7 PMID:16596409

Orzali, L., Forni, C., & Riccioni, L. (2014). Effect of chitosan seed treatment as elicitor of resistance to Fusarium graminearum in wheat. *Seed Science and Technology*, *42*(2), 132–149. doi:10.15258st.2014.42.2.03

Osmundson, T. (1999). *Roof gardens: History, design and construction*. W.W. Norton & Company.

Ouhtit, A. (2014). Nutrigenomics From promise to practice. *Sultan Qaboos University Medical Journal*, *14*(1), e1–e3. doi:10.12816/0003328 PMID:24516738

Ouma, J. O., Murithi, M. F., Mwangi, W., Macharia, H. G., & Gronte, H. D. (2002). Adoption of maize seed and fertilizer technologies in Embu district, Kenya. *International Journal of Agriculture*, *43*, 98–101.

Owens, G., Bhattacharya, P., & Nadebaum, P. (Eds.). (2018). *Managing arsenic in the climate: From soil to human wellbeing*. CSIRO Publishing.

Pagani, A. P. S., Dianese, A. C., & Café-Filho, A. C. (2014). Management of wheat blast with synthetic fungicides, partial resistance and silicate and phosphite minerals. *Phytoparasitica*, *42*(5), 609–617. doi:10.100712600-014-0401-x

Palmer, W. E., Bromley, P. T., & Brandenburg, R. L. (2007). *Wildlife & pesticides - Peanuts*. North Carolina Cooperative Extension Service.

Palys, M. J., Wang, H., Zhang, Q., & Daoutidis, P. (2021). Renewable ammonia for sustainable energy and agriculture: Vision and systems engineering opportunities. *Current Opinion in Chemical Engineering*, *31*, 100667. doi:10.1016/j.coche.2020.100667

Pamungkas, A., & Jumnahdi, M. (2020, November). The Study of Tidal Currents as Renewable Energy in Kelabat Bay, Bangka Belitung. *IOP Conference Series. Earth and Environmental Science*, *599*(1), 012001. doi:10.1088/1755-1315/599/1/012001

Pandey, S., Shrivastava, A. K., Rai, R., & Rai, L. C. (2013). Molecular characterization of Alr1105 a novel arsenate reductase of the diazotrophic cyanobacterium *Anabaena* sp. PCC7120 and decoding its role in abiotic stress management in *Escherichia coli. Plant Molecular Biology*, *83*(4-5), 417–432. doi:10.100711103-013-0100-x PMID:23836391

Panwar, V., McCallum, B., & Bakkeren, G. (2013). Endogenous silencing of *Puccinia triticina* pathogenicity genes through *in planta*-expressed sequences leads to the suppression of rust diseases on wheat. *The Plant Journal*, *73*(3), 521–532. doi:10.1111/tpj.12047 PMID:23110316

Paramo, L. A., Feregrino-Pérez, A. A., Guevara, R., Mendoza, S., & Esquivel, K. (2020). Nanoparticles in agroindustry: Applications, toxicity, challenges, and trends. *Nanomaterials (Basel, Switzerland)*, *10*(9), 1654. doi:10.3390/nano10091654 PMID:32842495

Parker, D. T. (1962). Decomposition in the field of buried and surface-applied cornstalk residue. *Soil Science Society of America Journal*, *26*, 559–562.

Park, R. F. (2007). Stem rust of wheat in Australia. *Australian Journal of Agricultural Research*, *58*(6), 558–566. doi:10.1071/AR07117

Parry, M. L., Rosenzweig, C., Iglesias, A., Livermore, M., & Fischer, G. (2004). Effects of climate change on global food production under SRES emissions and socio-economic scenarios. *Global Environmental Change*, *14*, 53–67.

Parthasarathy, R. P., Birthal, P. S., Reddy, B. V. S., Rai, K. N., & Ramesh, S. (2006). Diagnostics of Sorghum and Pearl Millet Grains-based Nutrition in India. *International Sorghum and Millets Newsletter*, *47*, 93–96.

Parveez, G. K. A., Bahariah, B., Ayub, N. H., Masani, M. Y. A., Rasid, O. A., Tarmizi, A. H., & Ishak, Z. (2015). Production of polyhydroxybutyrate in oil palm (*Elaeis guineensis* Jacq.) mediated by microprojectile bombardment of PHB biosynthesis genes into embryogenic calli. *Front Plant Sci*, *6*, 598. doi:10.3389/fpls.2015.00598 PMID:26322053

Pasquer, F., Isidore, E., Zarn, J., & Keller, B. (2005). Specific patterns of changes in wheat gene expression after treatment with three antifungal compounds. *Plant Molecular Biology*, *57*(5), 693–707. doi:10.100711103-005-1728-y PMID:15988564

Paterson, A. H., Bowers, J. E., Bruggmann, R., Dubchak, I., Grimwood, J., Gundlach, H., Haberer, G., Hellsten, U., Mitros, T., Poliakov, A., Schmutz, J., Spannagl, M., Tang, H., Wang, X., Wicker, T., Bharti, A. K., Chapman, J., Feltus, F. A., Gowik, U., ... Rokhsar, D. S. (2009). The Sorghum bicolor genome and the diversification of grasses. *Nature*, *457*(7229), 551–556. doi:10.1038/nature07723 PMID:19189423

Pathak, H., Sankhyan, S., Dubey, D. S., Bhatia, A., & Jain, N. (2013). Dry direct-seeding of rice for mitigating greenhouse gas emission: Field experimentation and simulation. *Paddy and Water Environment*, *11*(1-4), 593–601. doi:10.100710333-012-0352-0

Pathak, H., & Wassmann, R. (2007). Introducing greenhouse gas mitigation as a development objective in rice-based agriculture: I Generation of technical coefficients. *Agricultural Systems*, *94*, 807–825.

Patra, J. K., & Baek, K. H. (2017). Antibacterial activity and synergistic antibacterial potential of biosynthesized silver nanoparticles against foodborne pathogenic bacteria along with its anticandidal and antioxidant effects. *Frontiers in Microbiology*, *8*, 167. doi:10.3389/fmicb.2017.00167 PMID:28261161

Patterson, N., Tang, J., Cahoon, E. B., Jaworski, J. G., Wang, W., Peoples, O. P., & Snell, K. D. (2011). *Generation of high polyhydroxybutryate producing oilseeds.* International Patent Application WO/2011/034945.

Paul, K., Acram, T., & John, R. (2006). *Organic agriculture a global perspective.* Csiro Publishing.

Paul, D., & Nair, S. (2008). Stress adaptations in a plant growth promoting Rhizobacterium (PGPR) with increasing salinity in the coastal agricultural soils. *Journal of Basic Microbiology*, *48*(5), 1–7. doi:10.1002/jobm.200700365 PMID:18702073

Pauley, K. M., Eystathioy, T., Jakymiw, A., Hamel, J. C., Fritzler, M. J., & Chan, E. K. (2006). Formation of GW bodies is a consequence of microRNA genesis. *EMBO Reports*, *7*(9), 904–910. doi:10.1038j.embor.7400783 PMID:16906129

Pearce, R. B., & Ride, J. P. (1982). Chitin and related compounds as elicitors of the lignification response in wounded wheat leaves. *Physiological Plant Pathology*, *20*(1), 119–123. doi:10.1016/0048-4059(82)90030-3

Peck, S., & Khun, M. (2001). Design the guidelines of green roof system. Canada Mortgage and Housing Corporation. http://www.cmhc-sghl.gc.ca/en/imquaf/himu/himu_002.cfm

Peck, S., Callaghan, C., Bass, B., & Kuhn, M. (1999). *Greenbacks from green roofs: forging a new industry in Canada.* Status report on benefits, barriers and opportunities for green roof and vertical garden technology diffusion.

Peck, S., & Kuhn, M. (2001). *Design guidelines for green roofs*. Canada Mortgage and Housing Corporation. https://www.cmhc-schl.gc.ca/

Peera, S. K. P. G., Balasubramaniam, P., & Mahendran, P. P. (2016). Effect of silicate solubilizing bacteria and fly ash on silicon uptake and yield of rice under lowland ecosystem. *Journal of Applied and Natural Science*, *8*(1), 55–59. doi:10.31018/jans.v8i1.746

Pélissier, T., Thalmeir, S., Kempe, D., Sänger, H.L., & Wassenegger, M. (1999). Heavy de novo methylation at symmetrical and non-symmetrical sites is a hallmark of RNA-directed DNA methylation. *Nucleic Acids Res.*, *27*(7), 1625-34. doi:10.1093/nar/27.7.1625

Pellegrino, E., & Bedini, S. (2014). Enhancing ecosystem services in sustainable agriculture: Biofertilization and biofortification of chickpea (*Cicer arietinum* L.) by arbuscular mycorrhizal fungi. *Soil Biology & Biochemistry*, *68*, 429–439. doi:10.1016/j.soilbio.2013.09.030

Peragine, A., Yoshikawa, M., Wu, G., Albrecht, H. L., & Poethig, R. S. (2004). *SGS3* and *SGS2/SDE1/RDR6* are required for juvenile development and the production of *trans*-acting siRNAs in *Arabidopsis*. *Genes & Development*, *18*(19), 2368–2379. doi:10.1101/gad.1231804 PMID:15466488

Peregrin, T. (2001). The new frontier of nutrition science: Nutrigenomics. *Journal of the American Dietetic Association*, *101*(11), 1306.

Pérez-de-Luque, A., & Rubiales, D. (2009). Nanotechnology for parasitic plant control. *Pest Management Science*, *65*(5), 540–545. doi:10.1002/ps.1732 PMID:19255973

Perlatti, B., Bergo, P. L. S., & Silva, M. F. G. (2013). Polymeric nanoparticle-based insecticides: a controlled release purpose for agrochemicals. In S. Trdan (Ed.), *Insecticides-Development of Safer & More Effective Technologies* (pp. 523–550). InTech Open. doi:10.5772/53355

Perveen, S., Khan, M. S., & Zaidi, A. (2002). Effect of rhizospheric microorganisms on growth and yield of greengram (*Phaseolus radiatus*). *Indian Journal of Agricultural Sciences*, *72*, 421–423.

Petrasovits, L. A., Purnell, M. P., Nielsen, L. K., & Brumbley, S. M. (2007). Production of polyhydroxybutyrate in sugarcane. *Plant Biotechnology Journal*, *5*(1), 162–172. doi:10.1111/j.1467-7652.2006.00229.x PMID:17207265

Petrasovits, L. A., Zhao, L., McQualter, R. B., Snell, K. D., Somleva, M. N., Patterson, N. A., Nielsen, L. K., & Brumbley, S. M. (2012). Enhanced polyhydroxybutyrate production in transgenic sugarcane. *Plant Biotechnology Journal*, *10*(5), 569–578. doi:10.1111/j.1467-7652.2012.00686.x PMID:22369516

Pfiffner, L., & Mäder, P. (1997). Effects of biodynamic, organic and conventional production systems on earthworm populations. Entomological Research in Organic Agriculture, 15, 3–10.

Philip, D. (2011). *Mangiferaindica* leaf-assisted biosynthesis of well dispersed silver nanoparticles. *SpectrochimicaActa Part A*, *78*(1), 327–331. doi:10.1016/j.saa.2010.10.015 PMID:21030295

Pickering, B., Grams, C. M., & Pfenninger, S. (2020). Sub-national variability of wind power generation in complex terrain and its correlation with large-scale meteorology. *Environmental Research Letters*, *15*(4), 044025. doi:10.1088/1748-9326/ab70bd

Pieterse, C. M. J., Zamioudis, C., Berendsen, R. L., Weller, D. M., Van Wees, S. C. M., & Bakker, P. A. H. M. (2014). Induced Systemic Resistance by Beneficial Microbes. *Annual Review of Phytopathology*, *52*(1), 347–375. doi:10.1146/annurev-phyto-082712-102340 PMID:24906124

Pillai, R. S. (2005). MicroRNA function: Multiple mechanisms for a tiny RNA? *RNA*, *11*(12), 1753–1761. doi:10.1261/rna.2248605 PMID:16314451

Plaster, E. (2000). *La ciencia del suelo y su manejo*. Thompson Editores.

Plieninger, T., Torralba, M., Hartel, T., & Fagerholm, N. (2019). Perceived ecosystem services synergies, trade-offs, and bundles in European high nature value farming landscapes. *Landscape Ecology*, *34*(7), 1565–1581.

Poirier, Y., & Gruys, K.J. (2001). Production of PHAs in transgenic plants. *Biopolyesters,* 401-435.

Poirier, Y., Dennis, D. E., Klomparens, K., & Somerville, C. (1992). Polyhydroxybutyrate a biodegradable thermoplastic, produced in transgenic plants. *Science*, *256*(5056), 520–523. doi:10.1126cience.256.5056.520 PMID:17787950

Poirier, Y., Somerville, C., Schechtman, L. A., Satkowski, M. M., & Noda, I. (1995). Synthesis of high-molecular-weight poly ([R]-(-)-3-hydroxybutyrate) in transgenic *Arabidopsis thaliana* plant cells. *International Journal of Biological Macromolecules*, *17*(1), 7–12. doi:10.1016/0141-8130(95)93511-U PMID:7772565

Ponnamperuma, F. N. (1984). Straw as a Source of Nutrients for Wet-Land Rice. In Organic Matter and Rice. IRRI.

Ponzio, C., Gangatharan, R., & Neri, D. (2013). Organic and biodynamic agriculture: A review in relation to sustainability. *International Journal of Plant and Soil Science*, 95–110.

Popkin, R. (1995). Good news for waste watchers: Recycling, composting show results for the future. *Environmental Protection Agency Journal*, *21*, 188–190.

Potrykus, I. (2017). The GMO-crop potential for more, and more nutritious food is blocked by unjustified regulation. *Journal of Innovation & Knowledge*, *2*(2), 90–96. doi:10.1016/j.jik.2017.03.003

Potter, J. D., & Steinmetz, K. (1996). Vegetables, fruit and phytoestrogens as preventive agents. *IARC Scientific Publications*, 61–90.

Poveda, Gomez, & Martinez. (2008). Diversification practices: Their effect on pest regulation and production. *Revista Colombiana de Entomologia*, *34*, 131–144.

Pradhan, S., Patra, P., Das, S., Chandra, S., Mitra, S., Dey, K. K., Akbar, S., Palit, P., & Goswami, A. (2013). Photochemical modulation of biosafe manganese nanoparticles on *Vignaradiata*: A detailed molecular, biochemical, and biophysical study. *Environmental Science & Technology*, *47*(22), 13122–13131. doi:10.1021/es402659t PMID:24144189

Prajapati, U., Asrey, R., Varghese, E., Singh, A. K., & Singh, M. P. (2021). Effects of Post harvest ultraviolet-C treatment on shelf-life and quality of bitter gourd fruit during storage. *Elsevier Journal on Food Packaging and Self Life*, *28*(100665), 1–8.

Prakash, P., Kumar, P., Kar, A., Singh, A.K., & Anbukani, P. (2019). Progress and performance of protected cultivation in Maharashtra. *Indian Journal of Economics and Development, 15*(4), 555-563.

Prakash, P., Kumar, P., Kar, A., & Singh, A. K. (2020). Status and impact of protected, cultivation of horticultural crops in, Maharashtra. *Indian Journal of Horticulture*, *77*(3), 518–526.

Prasad, A. S. (2013). Discovery of Human Zinc Deficiency: Its Impact on Human Health and Disease. *Advances in Nutrition*, *4*(2), 176–190. doi:10.3945/an.112.003210 PMID:23493534

Prasad, P. V. V., Boote, K. J., Allen, L. H., & Thomas, J. M. G. (2003). Super-optimal temperatures are detrimental to peanut (*Arachis hypogea* L.) reproductive processes and yield at both ambient and elevated carbon dioxide. *Global Change Biology*, *9*, 1775–1787.

Prasad, R. (2014). Synthesis of silver nanoparticles in photosynthetic plants. *Journal of Nanoparticles*, *2014*, 963961. doi:10.1155/2014/963961

Prasad, R. C., & Prasad, B. N. (2001). Cyanobacteria as a source biofertilizer for sustainable agriculture in Nepal. *Botanica Orientalis. Journal of Plant Sciences*, 127–133.

Prasad, R., Bhattacharyya, A., & Nguyen, Q. D. (2017). Nanotechnology in sustainable agriculture: Recent developments, challenges, and perspectives. *Frontiers in Microbiology*, *8*, 1014. doi:10.3389/fmicb.2017.01014 PMID:28676790

Prasad, R., John, P. S., George, M., Singh, S., & Sharma, S. N. (1990). Effect of lentil residue management on the productivity and NPK removal by lentil-rice double cropping. *Lens Newsletter*, *17*(1), 5–9.

Prasad, R., Kumar, V., & Prasad, K. S. (2014). Nanotechnology in sustainable agriculture: Present concerns and future aspects. *African Journal of Biotechnology*, *13*(6), 705–713. doi:10.5897/AJBX2013.13554

Prasad, R., & Palaniappan, S. P. (1987). Pulse crop residue as N source in rice-based cropping system. *International Rice Research Newsletter*, *12*(1), 31.

Prasad, S., Singh, A., Korres, N. E., Rathore, D., Sevda, S., & Pant, D. (2020). Sustainable utilization of crop residues for energy generation: A life cycle assessment (LCA) perspective. *Bioresource Technology*, *303*, 122964. doi:10.1016/j.biortech.2020.122964 PMID:32061494

Praveen, S., Narayan, M., Rabha, A. K., & Srivastava, R. K. (2018). Assessment of Cadmium Scavenging Potential of Canna indica L. *Bulletin of Environmental Contamination and Toxicology*, *101*(4), 446–450. doi:10.100700128-018-2416-3 PMID:30116850

Praveen, S., Rabha, A. K., Maitreyie, N., & Srivastava, R. K. (2018). Relative comparison for phytoremediation potential of Canna and Pistia for wastewater recycling. *Environment and Ecology*, *36*(1A), 316–320.

Preall, J. B., He, Z., Gorra, J. M., & Sontheimer, E. J. (2006). Short interfering RNA strand selection is independent of dsRNA processing polarity during RNAi in Drosophila. *Current Biology*, *16*(5), 530–535. doi:10.1016/j.cub.2006.01.061 PMID:16527750

Premi, O. P., Kandpal, B. K., Rathore, S. S., Shekhawat, K., & Chauhan, J. S. (2013). Green manuring, mustard residues recycling and fertilizer application affects productivity and sustainability of Indian mustard (*Brassica juncea* L.) in Indian semi arid tropics. *Industrial Crops and Products*, *41*, 423–429. doi:10.1016/j.indcrop.2012.04.030

Premi, O. P., Rathore, S. S., Shekhawat, K., Kandpal, B. K., & Chauhan, J. S. (2012). Sustainability of fallow – Indian mustard system as influenced by green manure, mustard-straw recycling and fertilizer application. *Indian Journal of Agronomy*, *57*(3), 229–234.

Prestele, R., Hirsch, A. L., Davin, E. L., Seneviratne, S. I., & Verburg, P. H. (2018). A spatially explicit representation of conservation agriculture for application in global change studies. *Global Change Biology*, *24*(9), 4038–4053.

Pretty, J., & Bharucha, Z. P. (2014). Sustainable intensification in agricultural systems. *Annals of Botany*, *114*(8), 1571–1596.

Price, A. J., Stoll, M. E., Bergtold, J. S., Arriaga, F. J., Balkcom, K. S., Kornecki, T. S., & Raper, R. L. (2008). Effect of cover crop extracts on cotton and radish radicle elongation. *Communications in Biometry and Crop Science*, *3*, 60–66.

Puig, S., Coma, M., Desloover, J., Boon, N., Colprim, J., & Balaguer, M. D. (2012). Autotrophic denitrification in microbial fuel cells treating low ionic strength waters. *Environmental Science & Technology*, *46*(4), 2309–2315. doi:10.1021/es2030609 PMID:22257136

Pulleman, M., Jongmans, A., Marinissen, J., & Bouma, J. (2003). Effects of organic versus conventional arable farming on soil structure and organic matter dynamics in a marine loam in the Netherlands. *Soil Use and Management*, *19*(2), 157–165. doi:10.1079/SUM2003186

Puoci, F., Lemma, F., & Spizzirri, U. G. (2008). Polymer in agriculture: A review. *American Journal of Agricultural and Biological Sciences*, *3*(1), 299–314. doi:10.3844/ajabssp.2008.299.314

Purnell, M. P., Petrasovits, L. A., Nielsen, L. K., & Brumbley, D. W. (2007). Spatiotemporal characterisation of polyhydroxybutyrate accumulation in sugarcane. *Plant Biotechnology Journal*, *5*(1), 173–184. doi:10.1111/j.1467-7652.2006.00230.x PMID:17207266

Pursuit and Promotion of Science: The Indian Experience. (2001). Indian National Science Academy (INSA).

Puschenreiter, M., Wittstock, F., Friesl-Hanl, W., & Wenzel, W. W. (2013). Predictability of the Zn and Cd phytoextraction efficiency of a *Salix smithiana* clone by DGT and conventional bioavailability assays. *Plant and Soil*, *369*(1-2), 531–541. doi:10.100711104-013-1597-0

Pyott, D. E., & Molnar, A. (2015). Going mobile: Non-cell-autonomous small RNAs shape the genetic landscape of plants. *Plant Biotechnology Journal*, *13*(3), 306–318. doi:10.1111/pbi.12353 PMID:25756494

Qiao, F., Yang, Q., Wang, C. L., Fan, Y.-L., Wu, X.-F., & Zhao, K.-J. (2007). Modification of plant height via RNAi suppression of OsGA20 ox2 gene in rice. *Euphytica*, *158*(1-2), 35–45. doi:10.100710681-007-9422-6

Qu, F., Ye, X., Hou, G., Sato, S., Clemente, T. E., & Morris, T. J. (2005). RDR6 has a broad-spectrum but temperature-dependent antiviral defense role in *Nicotiana benthamiana*. *Journal of Virology*, *79*(24), 15209–15217. doi:10.1128/JVI.79.24.15209-15217.2005 PMID:16306592

Quinn, J. E., Awada, T., Trindade, F., Fulginiti, L., & Perrin, R. (2017). Combining habitat loss and agricultural intensification improves our understanding of drivers of change in avian abundance in a North American cropland anthrome. *Ecology and Evolution*, *7*(3), 803–814.

Rabb, R.L., & Guthrie, F.E. (1970). *Concepts of pest management*. Academic Press.

Ragaei, M., & Sabry, A. H. (2014). Nanotechnology for insect pest control. *International Journal of Science, Environment and Technology*, *3*, 528–545.

Rahmann, G. (2011). Biodiversity and organic farming: what do we know? *vTI Agriculture and Forestry Research, 3*, 189-208.

Raiesh, Sabana, Rachana, Rahman, Jerard, & Karun. (n.d.). Genetic relationship and diversity among coconut (*Cocos nucifera* L.) accessions revealed through SCoT analysis. *Biotech, 3*, 304.

Rai, M., & Ingle, A. (2012). Role of nanotechnology in agriculture with special reference to management of insect pests. *Applied Microbiology and Biotechnology*, *94*(2), 287–293. doi:10.100700253-012-3969-4 PMID:22388570

Rai, S., Kashyap, P. L., Kumar, S., Srivastava, A. K., & Ramteke, P. W. (2016a). Identification, characterization and phylogenetic analysis of antifungal *Trichoderma* from tomato rhizosphere. *SpringerPlus*, *5*(1), 1939. doi:10.118640064-016-3657-4 PMID:27917337

Rai, S., Kashyap, P. L., Kumar, S., Srivastava, A. K., & Ramteke, P. W. (2016b). Comparative analysis of microsatellites in five different antagonistic Trichoderma species for diversity assessment. *World Journal of Microbiology & Biotechnology*, *32*(1), 8. doi:10.100711274-015-1964-5 PMID:26712623

Rai, V., Acharya, S., & Dey, N. (2012). Implications of nanobiosensors in agriculture. *Journal of Biomaterials and Nanobiotechnology*, *3*(02), 315–324. doi:10.4236/jbnb.2012.322039

Raja, N. (2013). Biopesticides and biofertilizers : Ecofriendly sources for sustainable agriculture. *Journal of Biofertilizers & Biopesticides*, *4*(1), 1–2. doi:10.4172/2155-6202.1000e112

Rajasekaran, S., Ganesh Shankar, K., Jayakumar, K., Rajesh, M., Bhaaskaran, C., & Sundaramoorthy, P. (2012). Biofertilizers current status of Indian agriculture. *J. Environ. Bioenergy*, *4*(3), 176.

Rajkumar, M., & Freitas, H. (2008). Influence of metal resistant-plant growth-promoting bacteria on the growth of *Ricinus communis* in soil contaminated with heavy metals. *Chemosphere*, *71*(5), 834–842. doi:10.1016/j.chemosphere.2007.11.038 PMID:18164365

Raj, M., Sundaram, K. R., & Murli Paul, A. S. (2007). Deepa and R. K. Kumar. Obesity in Indian children: Time trends and relationship with hypertension. *The National Medical Journal of India*, *20*(6), 288–293.

Raju, K. V., Aziz, A., Sundaram, M. S. S., Sekher, M., Wani, S. P., & Sreedevi, T. K. (2008). *Guidelines for Planning and Implementation of Watershed Development Program in India: A Review*. Global Theme on Agroecosystems Report no. 48. Monograph. International Crops Research Institute for the Semi-Arid Tropics.

Raliya, R., Saharan, V., Dimkpa, C., & Biswas, P. (2017). Nanofertilizer for precision and sustainable agriculture: Current state and future perspectives. *Journal of Agricultural and Food Chemistry*. Advance online publication. doi:10.1021/acs.jafc.7b02178 PMID:28835103

Rama Subbu, R., Chandraprabha, A., & Sevugaperumal, R. (2008). Invitro Clonal propagation of vulnerable medicinal plant, Saraca asoca (Roxb.) De Wilde. *Natural Product Radiance*, *7*, 338–341.

Ramachandran, V. K., East, A. K., Karunakaran, R., Downie, J. A., & Poole, S. P. (2011). Adaptation of *Rhizobium leguminosarum* to pea, alfalfa and sugar beet rhizosphere investigated by comparative transcriptomics. *Genome Biology*, *12*(10), 106–109. doi:10.1186/gb-2011-12-10-r106 PMID:22018401

Ramakrishna, A., Tam, H. M., Wani, S. P., & Long, T. D. (2006). Effect of mulch on soil temperature, moisture, weed infestation and yield of groundnut in northern Vietnam. *Field Crops Research*, *95*(2 &3), 115–125. doi:10.1016/j.fcr.2005.01.030

Ramesh, S. A., Choimes, S., & Schachtman, D. P. (2004). Over-Expression of an Arabidopsis Zinc Transporter in Hordeum Vulgare Increases Short-Term Zinc Uptake after Zinc Deprivation and Seed Zinc Content. *Plant Molecular Biology*, *54*(3), 373–385. doi:10.1023/B:PLAN.0000036370.70912.34 PMID:15284493

Ramez, N. (2013). *Greener than green: Biotech and the future of agriculture (Part I)*. http://www.geneticliteracyprojectorg/2013/04/221greener-than-green-biotech-andthefuture-of-agriculture/

Ramirez, V. J., & Munnecke, D. E. (1988). Effect of solar heating and soil amendments of cruciferous residues on *Fusarium oxysporum* sp. conglutinans and other organisms. *Phytopathology*, *78*(3), 289–295. doi:10.1094/Phyto-78-289

Rana, S., & Kalaichelvan, P. T. (2011). Antibacterial effects of metal nanoparticles. *Advanced Biotech*, *2*, 21–23.

Rana, S., & Kalaichelvan, P. T. (2013). Ecotoxicity of nanoparticles. *ISRN Toxicology*, 574648. PMID:23724300

Ranf, S. (2016). Immune Sensing of Lipopolysaccharide in Plants and Animals: Same but Different. *PLoS Pathogens*, *12*(6), e1005596. doi:10.1371/journal.ppat.1005596 PMID:27281177

Ranyaphia, R. A., Maoa, A. A., & Borthakurb, S. K. (2011). Direct organogencsis from leaf and internode explants of in vitro raised wintergreen plant (Gaultheria fragrantissima). *Science Asia*, *37*(3), 186–194. doi:10.2306cienceas ia1513-1874.2011.37.186

Rao, A. V., & Agarwal, S. (1995). Bioavailability and in vivo antioxidant properties of lycopene from tomato products and their possible role in the prevention of cancer. *Nutrition and Cancer*, *31*(3), 199–203. doi:10.1080/01635589809514703 PMID:9795972

Rao, E. P., & Puttana, K. (2001). Nitrate, agriculture and environment. *Current Science*, (9), 1163–1168.

Rao, P. P., Birthal, P. S., Reddy, B. V., Rai, K. N., & Ramesh, S. (2006). Diagnostics of sorghum and pearl millet grains-based nutrition in India. *Int. Sorghum Millets News*, *47*, 93–96.

Rao, S. (1982). Biofertilizer. In N. S. Subba Rao (Ed.), *Advances in Agricultural Microbiology* (pp. 219–242). Oxford & IBH Pub. Co.

Rao, S. (1993). *Biofertilizers in Agriculture*. Oxford and IBH Publishing Co.

Rao, S. (2001). Nutritional status of Indian population. *Journal of Biosciences*, *26*(4), 481–489.

Rao, V. S. (2000). *Principles of weed science* (2nd ed.). Science Publishers Inc Enfield. doi:10.1201/9781482279603

Rasmussen, P. E., & Collins, H. P. (1991). Long-term impacts of tillage, fertilizer and crop residue on soil organic matter in temperate semi arid regions. *Advances in Agronomy*, *45*, 93–134. doi:10.1016/S0065-2113(08)60039-5

Rathore, S. S., Shekhawat, K., Dass, A., Kandpal, B. K., & Singh, V. K. (2017). Phytoremediation mechanism in Indian mustard (*Brassica juncea*) and its enhancement through agronomic interventions. *Proceedings of the National Academy of Sciences, 2017. Section B: Biological Sciences.*

Rathore, A. L., Pal, A. R., & Sahu, K. K. (1998). Tillage and mulching eVects on water use, root growth and yield of rainfed mustard and chickpea grown after lowland rice. *Journal of the Science of Food and Agriculture*, *78*, 149–161.

Rawal, S., Khan, M. A., Dua, V. K., Singh, S. P., Yadav, S. K., Lal, S. S., & Singh, B. P. (2014). Weed Management in potato. *ICAR-CPRI Technical Bulletin, 91*, 78.

Rawal, S., Luthra, S.K., Malik, K., Lal, M., Mankar, P., Dua, V.K., & Chakrabarti, S.K. (2020). Organic potato Production Technology for North Indian Plains. *ICAR-CPRI Shimla Extension Bulletin, 51*, 28.

Ray, B. D., Jarori, G. K., Raghunathan, V., Yan, H., & Nageswara Rao, B. D. (2005). Conformations of nucleotides bound to wild type and Y78F mutant yeast guanylate kinase: Proton two-dimensional transferred NOESY measurements. *Biochemistry*, *44*(42), 13762–13770. doi:10.1021/bi0509088 PMID:16229466

Reddy & Tyagi. (2005). Integrated weed management in maize-groundnut sequential cropping system. *Agricultural Reviews (Karnal)*, *26*(4), 235–248.

Reddy, M. K., Srivastava, A., Kumar, S., Kumar, R., Chawda, N., Ebert, A. W., & Vishwakanna, M. (2014). Chilli (Capsicum Annum L.) Breeding In India: An Overview. *SABRAO Journal of Breeding and Genetics*, *46*, 160–173.

Regenerative Agricultural Technologies. (1990 *Trainors Kit*. International Institute of Rural Reconstruction.

Reicosky, D. C., & Wilts, A. R. (2005). Crop-residue management. Encyclopedia of Soils in the Environment, 334–338. doi:10.1016/b0-12-348530-4/00254-x

Reignault, P., Randoux, B., Renard, D., Sanssen, J., & Durand, R. (2004). Defence responses activated by resistance inducers and natural compounds during the wheat/powdery mildew compatible interaction. In *Workshop on Resistance of cereals to biotic stress*. IHAR Radzików.

Reignault, P., Cogan, A., Muchembled, J., Lounes-Hadj Sahraoui, A., Durand, R., & Sancholle, M. (2001). Trehalose induces resistance to powdery mildew in wheat. *The New Phytologist*, *149*(3), 519–529. doi:10.1046/j.1469-8137.2001.00035.x PMID:33873340

Reijnders, L. (2006). Cleaner nanotechnology and hazard reduction of manufactured nanoparticles. *Journal of Cleaner Production*, *14*, 124–133.

Reiners, S. C. (Ed.). (2005). Integrated Crop and Pest Management Guidelines for Commercial Vegetable Production. Cornell Cooperative Extension Publication.

Rekhi, R. S., & Meelu, O. P. (1983). Effect of complementary use of moong straw and inorganic fertilizer N on the nitrogen availability and yield of rice. *Oryza*, *20*, 125–129.

Remus-Borel, W., Menzies, J. G., & Belanger, R. R. (2005). Silicon induces antifungal compounds in powdery mildew-infected wheat. *Physiological and Molecular Plant Pathology*, *66*(3), 108–115. doi:10.1016/j.pmpp.2005.05.006

Renard-Merlier, D., Randoux, B., Nowak, E., Farcy, F., Durand, R., & Reignault, P. (2007). Iodus 40, salicylic acid, heptanoyl salicylic acid and trehalose exhibit different efficacies and defence targets during a wheat/powdery mildew interaction. *Phytochemistry*, *68*(8), 1156–1164. doi:10.1016/j.phytochem.2007.02.011 PMID:17399750

Reuveni, M. (1997). Post-infection applications of K_3PO_3, phosphorous acid and dimethomorph inhibit development of downy mildew caused by Plasmopara viticola on grapes. *Journal of Small Fruit and Viticulture*, *5*(2), 27–38. doi:10.1300/J065v05n02_03

Riddle, W. C., Gillespie, T. J., & Swanton, C. J. (1996). Rye mulch characterization for the purpose of microclimatic modelling. *Agricultural Meteorology*, *78*, 67–81.

Risco, C. A., & Chase, C. C. Jr. (1997). Handbook of plant and fungal toxicants. *CRC Press*.

Rivera-Cruz, M., Trujillo, A., Córdova, G., Kohler, J., Caravaca, F., & Roldán, A. (2008). Poultry manure and banana waste are effective bio-fertilizer carriers for promoting plant growth and soil sustainability in banana crops. *Soil Biology & Biochemistry*, *40*(12), 3092–3095. doi:10.1016/j.soilbio.2008.09.003

Robert & Eleanor. (1997). Area by type of culture: South, Southeast, and East Asia. A revised and updated database. Manila (Philippines). International Rice Research Institute.

Rocha, T. L. R., Evaristo, R. G. S., & Silva, L. P. (2006). *Metabolomics: Applications and Perspectives* (1st ed.). EMBRAPA.

Rockström, J., & Sukhdev, P. (2016). *How food connects all the SDGs. Keynote Speech: Prof. Johan Rockström & CEO Pavan Sukhdev*. Stockholm Resilience Centre, Stockholm University.

Rodewald, J., & Trognitz, B. (2013). *Solanum* resistance genes against *Phytophthora infestans* and their corresponding avirulence genes. *Molecular Plant Pathology*, *14*(7), 740–757. doi:10.1111/mpp.12036 PMID:23710878

Rodoni, B. C., Dale, J. L., & Harding, R. M. (1999). Characterization and expression of the coat protein-coding region of the banana bract mosaic potyvirus, development of diagnostic assays and detection of the virus in banana plants from five countries in Southeast Asia. *Archives of Virology*, *144*(9), 1725–1737. doi:10.1007007050050700 PMID:10542022

Rodr'ıguez, H., & Fraga, R. (1999). Phosphate solubilizing bacteria and their role in plant growth promotion. *Biotechnology Advances*, *17*(4-5), 319–339. doi:10.1016/S0734-9750(99)00014-2 PMID:14538133

Rodrignes, F. A., & Datnoff, L. E. (2005). Silicon and rice disease management. *Fitopatologia Brasileira*, *30*(5), 457–469. doi:10.1590/S0100-41582005000500001

Romano, A., Raemakers, K., Bernardi, J., Visser, R., & Mooibroek, H. (2003). Transgene organization in potato after particle bombardment-mediated co-transformation using plasmids and gene cassettes. *Transgenic Research*, *12*(4), 461–473. doi:10.1023/A:1024267906219 PMID:12885167

Romano, A., Van der Plas, L. H. W., Witholt, B., Eggink, G., & Mooibroek, H. (2005). Expression of poly-3-(R)-hydroxyalkanoate (PHA) polymerase and acyl-CoA-transacylase in plastids of transgenic potato leads to the synthesis of a hydrophobic polymer, presumably medium-chain-length PHAs. *Planta*, *3*(220), 455–464. doi:10.100700425-004-1349-8 PMID:15351883

Romanyà, J., & Casals, P. (2019). Biological nitrogen fixation response to soil fertility is species-dependent in annual legumes. *Journal of Soil Science and Plant Nutrition*, 1–11.

Romay, G., & Bragard, C. (2017). Antiviral Defenses in Plants through Genome Editing. *Frontiers in Microbiology*, *8*, 47. doi:10.3389/fmicb.2017.00047 PMID:28167937

Ronteltap, A., Trijp, H. V., Berezowska, A., & Goossens, J. (2013). Nutrigenomics-based personalised nutritional advice: In search of a business model. *Genes & Nutrition*, *8*(2), 153–163. doi:10.100712263-012-0308-4 PMID:22903899

Ronteltap, A., van Trijp, J. C. M., & Renes, R. J. (2009). Consumer acceptance of Nutrigenomics-based personalised nutrition. *British Journal of Nutrition*, *101*(1), 132–144. doi:10.1017/S0007114508992552 PMID:18479587

Ross, M. (1999). Welch and, Robin D. Graham. A New Paradigm for World Agriculture: Meeting Human Needs. Productive, Sustainable, Nutritious. *Field Crops Research*, *60*, 1–10.

Rowe-Dutton, P. (1997). *The mulching of vegetables*. Commonwealth Bureau of Horticulture and Plantation Crops.

Roy & Mandal. (2005). Anther culture response in indica rice and variations in major agronomic characters among the androclones of a scented cultivar, Karnal local. *African Journal of Biotechnology*, *4*, 235–240.

Roychowdhury, D. E., Paul, M. A., & Banerjee, S. K. (2014). A review on the effects of biofertilizers and biopesticides on rice and tea cultivation and productivity. *International Journal of Science. Engineering & Technology*, *2*, 96–106.

Roychowdhury, D., Paul, M., & Banerjee, S. K. (2014). A review on the effects of biofertilizers and biopesticides on rice and tea cultivation and productivity. *International Journal of Science, Engineering and Technology*, *2*(8), 96–105.

Roy, S., Khasa, D. P., & Greer, C. W. (2007). Combining alders, frankiae, and mycorrhizae for the revegetation and remediation of contaminated ecosystems. *Canadian Journal of Botany*, *85*(3), 237–251. doi:10.1139/B07-017

Ruiz-Sanchez, M., Aroca, R., Munoz, Y., Polon, R., & Ruiz-Lozano, J. M. (2010). The arbuscular mycorrhizal symbiosis enhances the photosynthetic efficiency and the antioxidative response of rice plants subjected to drought stress. *Journal of Plant Physiology*, *167*(11), 862–869. doi:10.1016/j.jplph.2010.01.018 PMID:20227134

Runhaar, H. (2017). Governing the transformation towards 'nature-inclusive' agriculture: Insights from the Netherlands. *International Journal of Agricultural Sustainability*, *15*(4), 340–349.

Rusinamhodz, L. (2012). Maize–grain legume Intercropping is an attractive option for ecological intensification that reduces climatic risk for smallholder farmers in central Mozambique. *Field Crops Research*, *136*, 12–22. doi:10.1016/j.fcr.2012.07.014

Russell, A., Laird, D. A., Parkin, T. B., & Mallarino, A. P. (2005). Impact of nitrogen fertilization and cropping system on carbon sequestration in Midwestern Mollisols. *Soil Science Society of America Journal*, *69*(2), 413–422. doi:10.2136ssaj2005.0413

Ryan, J. N., & Elimelech, M. (1996). Colloid mobilization and transport in groundwater. *Colloids and Surfaces. A, Physicochemical and Engineering Aspects*, *107*, 1–56. doi:10.1016/0927-7757(95)03384-X

Sabbagh, F., & Muhamad, I. (2017). Production of polyhydroxyalkanoate as secondary metabolite with main focus on sustainable energy. *Renewable & Sustainable Energy Reviews*, *72*, 95–104. doi:10.1016/j.rser.2016.11.012

Sabir, N., & Singh, A.K. (2020). Home and Smart Grading for Health and Environmental Improvement. In *Gardening for Health and Environment*. India Kitchens Garden Association.

Sabir, S., Arshad, M., & Chaudhari, S. K. (2014). Zinc oxide nanoparticles for revolutionizing agriculture: Synthesis and applications. *TheScientificWorldJournal*, *2014*, 8. doi:10.1155/2014/925494 PMID:25436235

Sadeghzadeh, B. (2013). A review of zinc nutrition and plant breeding. *Journal of Soil Science and Plant Nutrition*, *13*(ahead), 905–927. doi:10.4067/S0718-95162013005000072

Safarpour, H., Safarnejad, M. R., Tabatabaei, M., Mohsenifar, A., Rad, F., Basirat, M., Shahryari, F., & Hasanzadeh, F. (2012). Development of a quantum dots FRET-based biosensor for efficient detection of *Polymyxabetae*. *Canadian Journal of Plant Pathology*, *34*(4), 507–515. doi:10.1080/07060661.2012.709885

Sahu, D., Priyadarshani, I., & Rath, B. (2012). Cyanobacteria- as potential biofertilizer. *CIBTech Journal of Microbiology*, *1*, 20–26.

Saikia, D. N., Sarma, J., Dutta, P. K., & Baruah, D. K. (2011). Sustainable productivity of tea through conservation of bio-mass, addition of bio-fertilizers and reduction of inorganic fertilizer. *Two and a Bud, 58*, 109-117.

Sailor, D. J. (2008). *A green roof model for building energy simulation programs*. Portland State University. doi:10.1016/j.enbuild.2008.02.001

Sailor, D. J., Elley, T. B., & Gibson, M. (2011). Exploring the building energy impacts of green roof design decisions – a modeling study of buildings in four distinct climates. *Journal of Building Physics*, *35*(4), 372–391. doi:10.1177/1744259111420076

Sakata, K., Antonio, B. A., Mukai, Y., Nagasaki, H., Sakai, Y., Makino, K., & Sasaki, T. (2000). INE: A rice genome database with an integrated map view. *Nucleic Acids Research*, *28*(1), 97–101. doi:10.1093/nar/28.1.97 PMID:10592192

Sakhuja, P. K., & Jain, R. K. (2001). Nematode diseases of vegetable crops and their management. In T. S. Thind (Ed.), *Diseases of Fruits and Vegetables and their Management* (pp. 439–459). Kalayani Pub.

Sakr, N. (2016). Silicon control of bacterial and viral diseases in plants. *Journal of Plant Protection Research*, *56*(4), 331–336. doi:10.1515/jppr-2016-0052

Sales, N. M. R., Pelegrini, P. B., & Goersch, M. C. (2014). Nutrigenomics: Definitions and Advances of This New Science. *Journal of Nutrition and Metabolism*.

Salim, M. (1995). Rice crop residue use for crop production. In Organic Recycling in Asia and the Pacific. United Nations.

Salt, D. E., Smith, R. D., & Raskin, I. (1998). Phytoremediation. *Annual Review of Plant Physiology and Plant Molecular Biology*, *49*(1), 643–668. doi:10.1146/annurev.arplant.49.1.643 PMID:15012249

Saltzman, A., Birol, E., Bouis, H. E., Boy, E., De Moura, F. F., Islam, Y., & Pfeiffer, W. (2013). Biofortification: Progress toward a more nourishing future. *Global Food Security*, *2*(1), 9–17. doi:10.1016/j.gfs.2012.12.003

Samaina, E., Tuinenc, D., Jeandetd, P., Aussenace, T., & Selima, S. (2017). Biological control of *Septoria* leaf blotch and growth promotion in wheat by *Paenibacillus* sp. strain B2 and *Curtobacterium plantarum* strain EDS. *Biological Control, 114*, 87–96. doi:10.1016/j.biocontrol.2017.07.012

San Miguel, K., & Scott, J. G. (2016). The next generation of insecticides: dsRNA is stable as a foliar-applied insecticide. *Pest Management Science, 72*(4), 801–809. doi:10.1002/ps.4056 PMID:26097110

Sanchez, E., Lamont, W. J. Jr, & Orzolele, M. D. (2008). Newspaper mulches for suppressing weeds for organic high-tunnel cucumber production. *HortTechnology, 18*(1), 154–157. doi:10.21273/HORTTECH.18.1.154

Sangakkara, P., Bandaranayake, U., & Gyjanayake, J. N. (2008). Organic matter addition in organic farming over dry season. IFOAM and the Consorzio Modena Bio.

Sangakkau, U. K. (1987). Legume for intensification of rice-based farming systems of mid country Sri Lanka. In *Food legumes and Improvement for Asian Farming System pp* (pp. 222–223). Int. Rice Res. Inst.

Santi, C., Bogusz, D., & Franche, C. (2013). Biological nitrogen fixation in non-legume plants. *Annals of Botany, 111*(5), 743–767. doi:10.1093/aob/mct048 PMID:23478942

Santos, H. A. A., Pria, M. D., Silva, O. C., & Mio, L. L. M. (2011). Control of wheat diseases using phosphites and acibenzolar-s-methyl alone or associated with piraclostrobina + epoxiconazole. *Semina. Ciências Agrárias, 32*, 433–442. doi:10.5433/1679-0359.2011v32n2p433

Santos, R. E., Pinto-Coelho, R. M., Drumond, M. A., Fonseca, R., & Zanchi, F. B. (2020). Damming Amazon Rivers: Environmental impacts of hydroelectric dams on Brazil's Madeira River according to local fishers' perception. *Ambio, 49*(10), 1612–1628. doi:10.100713280-020-01316-w PMID:31994028

Sanyal, B., Mondal, S. S., & Chatterjee, B. N. (1993). Fertilizer management with bulky organic manure in rice-potato-groundnut sequence for sustaining productivity. *Journal of Potash Research, 9*, 218–227.

Saraswat, S., & Rai, J. P. N. (2011a). Prospective application of *Leucaena leucocephala* for phytoextraction of Cd and Zn and nitrogen fixation in metal polluted soils. *International Journal of Phytoremediation, 13*(3), 1–18. doi:10.1080/15226514.2010.483259 PMID:21598792

Saraswat, S., & Rai, J. P. N. (2011b). Complexation and detoxification of Zn and Cd in metal accumulating plants. *Reviews in Environmental Science and Biotechnology, 10*(4), 327–339. doi:10.100711157-011-9250-y

Saravanan, V. S., Kumar, M. R., & Sa, T. M. (2011). Microbial zinc solubilization and their role on plants. In D. K. Maheshwari (Ed.), *Bacteria in Agrobiology, Plant Nutrient Management* (pp. 47–63). Springer. doi:10.1007/978-3-642-21061-7_3

Saravanan, V. S., Madhaiyan, M., & Thangaraju, M. (2007). Solubilization of zinc compounds by the diazotrophic, plant growth promoting bacterium *Gluconacetobacter diazotrophicus. Chemosphere, 66*(9), 1794–1798. doi:10.1016/j.chemosphere.2006.07.067 PMID:16956644

Saruul, P., Srienc, F., Somers, D. A., & Samac, D. A. (2002). Production of a biodegradable plastic polymer, poly-beta-hydroxybutyrate in transgenic alfalfa. *Crop Science, 42*, 919–927.

Sasaki, T.International Rice Genome Sequencing Project. (2005). The map-based sequence of the rice genome. *Nature, 436*(7052), 793–800. doi:10.1038/nature03895 PMID:16100779

Satapanajaru, T., Anurakpongsatorn, P., Pengthamkeerati, P., & Boparai, H. (2008). Remediation of atrazine-contaminated soil and water by nanozerovalent iron. *Water, Air, and Soil Pollution, 192*(1-4), 349–359. doi:10.100711270-008-9661-8

Satheesh, Thukkaiyannan, & Ponnuswamy. (2011). Effect of sowing and weed management methods and intercrops on weed control and grain yield of sorghum under intercropping situation. *Crop Res., 41*(1-3), 46 – 51.

Sattar & Abedin. (2012). Options for coastal farmers of Bangladesh adapting to Ipacts of climate change. *International conference on Environment, Agriculture and Food Science*, 17 – 21.

Savadi, S., Prasad, P., Kashyap, P. L., & Bhardwaj, S. C. (2018). Molecular breeding technologies and strategies for rust resistance in wheat (*Triticum aestivum*) for sustained food security. *Plant Pathology, 67*(4), 771–791. doi:10.1111/ppa.12802

Savage, S. (2015). *The lower productivity of organic farming: A new analysis and its big implications*. Academic Press.

Savci, S. (2012). An agricultural pollutant: Chemical fertilizer. *International Journal of Environmental Sciences and Development, 3*, 73–80. doi:10.7763/IJESD.2012.V3.191

Saver, N., Singh, A. K., & Hasan, M. (2021). Greenhouse Agriculture: Production and Protections. Pentimer Publications.

Scherr, J. S., & Mcneely, A. J. (2008). Biodiversity conservation and agricultural sustainability: Towards a new paradigm of 'ecoagriculture' landscapes. *Philos Trans Royal Soc B, 363*(1491), 477–494. doi:10.1098/rstb.2007.2165 PMID:17652072

Schijlen, E. G. W. M., deVos, R. C. H., Martens, S., Jonker, H. H., Rosin, F. M., Molthoff, J. W., Tikunov, Y. M., Angenent, G. C., van Tunen, A. J., & Bovy, A. G. (2007). RNA Interference silencing of *chalcone synthase*, the first step in the flavonoid biosynthesis pathway, leads to parthenocarpic tomato fruits. *Plant Physiology, 144*(3), 1520–1530. doi:10.1104/pp.107.100305 PMID:17478633

Schmid, M., & Jensen, T. H. (2008). The exosome: A multipurpose RNA-decay machine. *Trends in Biochemical Sciences, 33*(10), 501–510. doi:10.1016/j.tibs.2008.07.003 PMID:18786828

Schmutz, J., Cannon, S. B., Schlueter, J., Ma, J., Mitros, T., Nelson, W., Hyten, D. L., Song, Q., Thelen, J. J., Cheng, J., Xu, D., Hellsten, U., May, G. D., Yu, Y., Sakurai, T., Umezawa, T., Bhattacharyya, M. K., Sandhu, D., Valliyodan, B., ... Jackson, S. A. (2010). Genome sequence of the palaeopolyploid soybean. *Nature, 463*(7278), 178–183. doi:10.1038/nature08670 PMID:20075913

Schnable, P., Ware, D., Fulton, R., Stein, J. C., Wei, F., Pasternak, S., Liang, C., Zhang, J., Fulton, L., Graves, T. A., Minx, P., Reily, A. D., Courtney, L., Kruchowski, S. S., Tomlinson, C., Strong, C., Delehaunty, K., Fronick, C., Courtney, B., ... Wilson, R. K. (2009). The B73 maize genome: Complexity, diversity, and dynamics. *Science, 326*(5956), 1112–1115. doi:10.1126cience.1178534 PMID:19965430

Schnell, J. A., Treyvaud-Amiguet, V., Arnason, J. T., & Johnson, D. A. (2012). Expression of polyhydroxybutyric acid as a model for metabolic engineering of soybean seed coats. *Transgenic Research, 4*(21), 895–899. doi:10.100711248-011-9575-x PMID:22183687

Schwall, G. P., Safford, R., Westcott, R. J., Jeffcoat, R., Tayal, A., Shi, Y.-C., Gidley, M. J., & Jobling, S. A. (2000). Production of very high-amylose potato starch by inhibition of SBE A and B. *Nature Biotechnology, 18*(5), 551–554. doi:10.1038/75427 PMID:10802625

Scialabba & Müller-Lindenlauf. (2010). Organic agriculture and climate change. *Renewable Agriculture and Food Systems, 25*(2), 158–169.

Scott, N., & Chen, H. (2013). Nanoscale science and engineering for agriculture and food systems. *Industrial Biotechnology (New Rochelle, N.Y.), 9*(1), 17–18. doi:10.1089/ind.2013.1555

Scrivens, S. (1980a). Roof gardens: Construction. *Architects Journal*, 445-449.

Seabridge, K. (2015). *Organic farming statistics 2014*. Department of Environment, Food and Rular Affairs.

Segal, G., Song, R., & Messing, J. (2003). A new opaque variant of maize by a single dominant RNA-interference-inducing transgene. *Genetics*, *165*(1), 387–397. doi:10.1093/genetics/165.1.387 PMID:14504244

Sekhon, B. S. (2014). Nanotechnology in agri-food production: An overview. *Nanotechnology, Science and Applications*, *7*, 31–53. doi:10.2147/NSA.S39406 PMID:24966671

Sell, J., Kayser, A., Schulin, R., & Brunner, I. (2005). Contribution of ectomycorrhizal fungi to cadmium uptake of poplars and willows from a heavily polluted soil. *Plant and Soil*, *277*(1-2), 245–253. doi:10.100711104-005-7084-5

Seneviratne, S. I., Phipps, S. J., Pitman, A. J., Hirsch, A. L., Davin, E. L., Donat, M. G., ... Kravitz, B. (2018). Land radiative management as contributor to regional-scale climate adaptation and mitigation. *Nature Geoscience*, *11*(2), 88–96.

Sen, G., & Blau, H. M. (2005). Argonaute 2/RISC resides in sites of mammalian mRNA decay known as cytoplasmic bodies. *Nature Cell Biology*, *7*(6), 633–636. doi:10.1038/ncb1265 PMID:15908945

Senior, P. J., & Dawes, E. A. (1973). The regulation of poly ß-hydroxybutyrate metabolism in *Azotobacter beijerinckii. The Biochemical Journal*, *134*(1), 225–238. doi:10.1042/bj1340225 PMID:4723225

Senthilkumar, M., & Mysore, K. S. (2011). Virus-induced gene silencing can persist for more than two years and also be transmitted to progeny seedlings in *Nicotiana benthamiana* and tomato. *Plant Biotechnology Journal*, *9*(7), 797–806. doi:10.1111/j.1467-7652.2011.00589.x PMID:21265998

Sessitsch, A., Kuffner, M., Kidd, P., Vangronsveld, J., Wenzel, W., Fallmann, K., & Puschenreiter, M. (2013). The role of plant-associated bacteria in the mobilization and phytoextraction of trace elements in contaminated soils. *Soil Biology & Biochemistry*, *60*, 182–194. doi:10.1016/j.soilbio.2013.01.012 PMID:23645938

Shambu Prasad, C. (2006). *System of Rice Intensification in India: Innovation History and Institutional Challenges.* WWF-International-ICRISAT Dialogue Project.

Shamseldin, A. (2013). The role of different genes involved in symbiotic nitrogen fixation-review. *Global Journal of Biotechnology & Biochemistry*, *8*, 84–94.

Sharma & Bhagat. (1993). Puddling and compaction effects on water permeability of texturally different soils. *Journal of the Indian Society of Soil Science*, *41*, 1–6.

Sharma, A. R., & Mitra, B. N. (1992). Integrated nitrogen management in rice (*Oryza sativa*)- wheat (*Triticum aestivum*) cropping system. *Indian Journal of Agricultural Sciences*, *62*, 70–72.

Sharma, A. R., Singh, R., Dhyani, S. K., & Dube, R. K. (2010). Effect of live mulching with annual legumes on performance of maize (Zea mays) and residual effect on following wheat (Triticum aestivum). *Indian Journal of Agronomy*, *55*, 177–184.

Sharma, H.K., Singh, A.K., & Lal, J. (2014). Effect of Methom Sodium on Root-knot nematode Infestation in tomato under protected Cultivation. *I.J. of Nimatology*, *43*(2), 236–237.

Sharma, L., Srivastava, J. K., & Singh, A. K. (2016). Biodegradable polyhydroxyalkanoate thermoplastics substituting xenobiotic plastics: a way forward for sustainable environment. In A. Singh, S. Prasad, & R. Singh (Eds.), *Plant responses to xenobiotics* (pp. 317–346). Springer. doi:10.1007/978-981-10-2860-1_14

Sharma, P. K., De Datta, S. K., & Redulla, C. A. (1989). Effect of percolation rate on nutrient kinetics and rice yield in tropical rice soils. II. Role of straw amendments. *Plant and Soil*, *119*, 121–126.

Sharma, P., Sardana, V., & Kandola, S. S. (2011). Response of groundnut (*Arachis hypogaea* L.) to *Rhizobium* inoculation. *Libyan Agriculture Research Centre Journal Interaction*, *2*, 101–104.

Sharma, R. (2002). *Watershed Development Adaptation Strategy for Climate Change. Paper presented in South Asia expert workshop on Adaptation to Climate Change for Agricultural Productivity, organized by the Government of India,* UNEP and CGIAR.

Sharma, S., Duveiller, E., Basnet, R., Karki, C. B., & Sharma, R. C. (2005). Effect of potash fertilization on helminthosporium leaf blight severity in wheat, and associated increases in grain yield and kernel weight. *Field Crops Research, 93*(2-3), 142–150. doi:10.1016/j.fcr.2004.09.016

Sharma, S., & Shahzad, A. (2014). An overview on Decalepsis: A Genus of Woody Medicinal Climbers. *Journal of Plant Science Research, 1*, 104.

Sharmila, R. C. (2010). *Nutrient release pattern of nano-fertilizer formulations* (PhD thesis). Tamil Nadu Agricultural University, Coimbatore, Tamil Nadu, India.

Shaterabadi, M., Jirdehi, M. A., Amiri, N., & Omidi, S. (2020). Enhancement the economical and environmental aspects of plus-zero energy buildings integrated with INVELOX turbines. *Renewable Energy, 153*, 1355–1367. doi:10.1016/j.renene.2020.02.089

Shekhawat, K., Rathore, S. S., & Dass, A. (2017). Weed menace and management strategies for enhancing oilseed brassicas production in the Indian sub-continent: A review. *Crop Protection (Guildford, Surrey)*. Advance online publication. doi:10.1016/j.cropro.2017.02.017

Shekhawat, K., Rathore, S. S., & Kandpal, B. K. (2016). Crop establishment techniques affect productivity, sustainability, and soil health under mustard-based cropping systems of Indian semi-arid regions. *Soil & Tillage Research, 158*, 137–146.

Shelly Praveen, S., Ramesh, V., & Anil, K. M. (2010). Silencing potential of viral derived RNAi constructs in tomato leaf curl virus-AC4 gene suppression in tomato. *Transgenic Research, 19*(1), 45–55. doi:10.100711248-009-9291-y PMID:19548101

Shelton, A. M., Wilsey, W. R., & Soderlund, D. M. (2001). Classification of insecticides and acaricides for resistance management. *NYSAES*.

Shen, Y., Xie, Y., Yu, X., Zhang, S., Wen, Q., Ye, G., & Guo, J. (2021).Clinical diagnostic values of transfer RNA-derived fragment tRF-19-3L7L73JD and its effects on the growth of gastric cancer cells. *J Cancer, 12*(11), 3230-3238.

Sheng, X. F., & Xia, J. J. (2006). Improvement of rape (*Brassica napus*) plant growth and cadmium uptake by cadmium-resistant bacteria. *Chemosphere, 64*(6), 1036–1042. doi:10.1016/j.chemosphere.2006.01.051 PMID:16516946

Sheng, X. F., Xia, J. J., Jiang, C. Y., He, L. Y., & Qian, M. (2008). Characterization of heavy metal-resistant endophytic bacteria from rape (*Brassica napus*) roots and their potential in promoting the growth and lead accumulation of rape. *Environmental Pollution, 156*(3), 1164–1170. doi:10.1016/j.envpol.2008.04.007 PMID:18490091

Shen, J., Li, C., Mi, G., Li, L., Yuan, L., Jiang, R., & Zhang, F. (2013). Maximizing root/rhizosphere efficiency to improve crop productivity and nutrient use efficiency in intensive agriculture of China. *Journal of Experimental Botany, 64*(5), 1181–1192. doi:10.1093/jxb/ers342 PMID:23255279

Shepherd, M. A., Harrison, R., & Webb, J. (2002). Managing soil organic matter-implications for soil structure on organic farms. *Soil Use and Management, 18*(3), 284–292. doi:10.1079/SUM2002134

Shogren, R. L. (2001). Biodegradable mulches from renewable resources. *Journal of Sustainable Agriculture, 16*(4), 33–47. doi:10.1300/J064v16n04_05

Shukla, H. (2015). Watershed Management: Its Role In Environmental Planning And Management. *IOSR Journal of Environmental Science, Toxicology, and Food Technology, 1*(5), 8-11.

Shukla, S. K., Kumar, R., & Mishra, R. K. (2015). Prediction and validation of gold nanoparticles (GNPs) on plant growth promoting rhizobacteria (PGPR): A step toward development of nano-biofertilizers. *Nano Reviews*, *4*, 439–448.

Shyla, K. K., Natarajan, N., & Nakkeeran, S. (2014). Antifungal activity of zinc oxide, silver and titanium dioxide nanoparticles against *Macrophominaphaseolina. Journal of Mycology and Plant Pathology*, *44*, 268–273.

Sidhu, H.S., Singh, M., & Humphreys, E. (2007). The Happy Seeder enables direct drilling of wheat into rice stubble. *Australian Journal of Experimental Agriculture*, *47*, 844–854.

Sidhu, H.S., & Singh, M., Singh, Y., & Blackwell, J. (2015). Development and evaluation of the Turbo Happy Seeder for sowing wheat into heavy rice residues in NW India. *Field Crops Research*, *184*, 201–212.

Sijen, T., Fleenor, J., Simmer, F., Thijssen, K. L., Parrish, S., Timmons, L., Plasterk, R. H. A., & Fire, A. (2001). On the role of RNA amplification in dsRNA-triggered gene silencing. *Cell*, *107*(4), 465–476. doi:10.1016/S0092-8674(01)00576-1 PMID:11719187

Sikka, D. R., & Gadgil, S. (1980). On the Maximum Cloud Zone and the ITCZ over Indian, Longitudes during the Southwest Monsoon. *Monthly Weather Review*, *108*(11), 1840–1853. doi:10.1175/1520-0493(1980)108<1840:OTMCZA>2.0.CO;2

Silva, I. T., Rodrigues, F. A., Oliveira, J. R., Pereira, S. C., Andrade, C. C. L., Silveira, P. R., & Conceição, M. M. (2010). Wheat resistance to bacterial leaf streak mediated by silicon. *Journal of Phytopathology*, *158*(4), 253–262. doi:10.1111/j.1439-0434.2009.01610.x

Silva, W. L. D., Cruz, M. F. A., Fortunato, A. A., & Rodrigues, F. (2015). Histochemical aspects of wheat resistance to leaf blast mediated by silicon. *Scientia Agrícola*, *72*(4), 322–327. doi:10.1590/0103-9016-2014-0221

Silwana, T. T., Lucas, E. O., & Olaniyan, A. B. (2007). The effect of inorganic and organic fertilizers on the growth and development of component crop in maize/ bean intercrop in eastern cape of south Africa. *Journal of Food Agriculture and Environment*, *5*(1), 262–272.

Simpson, J. T., Wong, K., Jackman, S. D., Schein, J. E., Jones, S. J., & Birol, I. (2009). ABySS: A parallel assembler for short read sequence data. *Genome Research*, *19*(6), 1117–1123. doi:10.1101/gr.089532.108 PMID:19251739

Sims, Whitehouse, & Lacey. (1999). *The OECD Lemna Growth Inhibition Test. Development and Ring-testing of draft OECD Test Guideline*. R&D Technical Report EMA 003. WRc plc - Environment Agency.

Sindhu, S., Maier, T., Mitchum, M. G., Hussey, R. S., Davis, E. L., & Baum, T. J. (2009). Effective and specific in Plant RNAi in cyst nematodes: Expression interference of four parasitism genes reduces parasitic success. *Journal of Experimental Botany*, *60*(1), 315–324. doi:10.1093/jxb/ern289 PMID:19015219

Sing, A. K., Shri Dhar Saber, N., Patel, N., Singh, V., Singh, A., & Kumar, A. (2021). Response of Long melon on spacing and pruning methods under insect proof net house during off-season in the plain reason of India. *Indian Journal of Agriculture Sciences, 91*(5), 226-30.

Singh & Gangwar. (2000). Alternate tillage and crop residue management strategies in rice–wheat cropping system. Annual Report, Project Directorate for Cropping System Research.

Singh & Mungali. (2011). *Water Pollution and Management*. Academic Press.

Singh, A. K., & Sabir, N. (2019). Design and size of Greenhouse Structure in India: Crops and Regional Suitability. Sping the Future of Horticulture, 371-392.

Singh, A. K., Shrivastva, R., & Mathala, M. J. (2004). Use of plastic in agriculture. A Hindi-Bulitin-IARI, TB-ICN:H-6/2004.

Singh, A. K., Singh, V. K., Mehta, K. S., Singh, B., Sinthu, S. S., Kumar, R., Kumar, M., & Singh, V. (2012). Socio-Economic Empowerment of Hill Women through Protected Cultivation Technology. *Global conference on women in Agriculture*, 21-22.

Singh, A., & Sharma, O. P. (2004). Integrated Pest Management for Sustainable Agriculture. In Integrated Pest Management in Indian Agriculture. National Centre for Agricultural Economics and Policy Research (NCAP).

Singh, A.K. (2019). Producing Long Melon in Off-Season. *Indian Horticulture*, 11-13.

Singh, A.K. (2021). *Protected Cultivation of Vegetables under Polyhouse structure*. UDWDP.

Singh, G. (2021). Lockdown cash crunch effects organic farming. *VillageSquare*. https://www.villagesquare.in

Singh, Jaiswal, & Jaiswal. (2014). Thidiazuron-Induced Anatomical Changes and Direct Shoot Morphogenesis In Dendrocalainus strictus nees. *Canadian Journal of Pure and Applied Sciences*, *8*, 2901–2904.

Singh, K. N, Sharma, I. P., & Srivastava, V. C., (2001). Effect on FYM fertilizer and plant density on productivity of rice-what sequence. *Journal of Research, Birsa Agriculture University*, *13*(2), 159-162.

Singh, Kamlesh, Patel, Patel, Singh, & Singh. (2007). First report of "hopper burn" damage in Gujarat state. *CPRI News Letter*, *37*, 3–4.

Singh, R. K., Singh, A. K., & Gautam, U. S. (2021). Protected cultivation for an increase in farmer's income in Bundelkhand region of UP. New Age Protected Cultivation Magazine, 7(2), 44-48.

Singh, R. V. (Ed.). (2000). Watershed planning and management. Yash Publishing House.

Singh, Y., Singh, D., & Tripathi, R. P. (1996). Crop residue management in rice-wheat cropping system. In *Abstracts of Poster Sessions, 2nd International Crop Science Congress*. National Academy of Agricultural Sciences.

Singh. (2004). Long-term effects of organic inputs on yield and soil fertility in rice-wheat rotation. *Soil Society of American Journal.*, *68*(3), 845–853.

Singh. (2005). Crop residue management for nutrient cycling and improving soil productivity in rice-based cropping systems in the tropics. *Advances in Agronomy*, *85*, 269–407.

Singh. (2010). *Options for Effective Utilization of Crop Residues. Research Bulletin 3/2010*. Punjab Agricultural University.

Singh, A. K. (2007a). *Drip-irrigation system for hilly areas. KVK.*

Singh, A. K. (2007b). *Polyhouse Cultivation in hilly areas. KVK.*

Singh, A. K. (2013). Traditional knowledge based adaptations of bio-mulching for ginger crop management: Empirical learning with resource-poor farmers under rain-fed agro-eco-system of western Himalaya. *Indian Journal of Traditional Knowledge*.

Singh, A. K. (2017). Horticulture based Plasticulture cum covered cultivation for livelihood security. *Indian Horticulture*, *62*(5), 68–74.

Singh, A. K. (2017). Jal-Sanchyan and Sichai Pravendhan me Sanrakkchhit Kheti ka Yogdan. *Kuroochhetra*, *64*(1), 61–63.

Singh, A. K. (2018). *Mulching Technology for Vegetables production in Hilly Area. UDWDP.*

Singh, A. K. (2018). Protected Cultivation Technologies: A Sustainable Solution. *ISRAEL: An Indian Prospective in Agriculture Word*, *4*(5), 60–68.

Singh, A. K., Rawat, S. S., Singh, V. K., Mehta, K. S., Singh, B., Sindhu, S. S., Hasan, M., Rakesh, K., & Mukul, K. (2012). Rain Water Harvesting: A Boon for Protected Horticulture in the Peri -Urban Area of Hills. *1st Biennial International Congress on Urban Green Spaces*, 151-152.

Singh, A. K., Saver, N., Jat, G. S., Singh, J., Singh, V., Singh, A., Kumar, A., & Kumar, J. (2021). Effect of different spacing and pruning methods on growth, yield and economics of Long melon during off-season under naturally ventilated polyhouse. *Indian Journal of Agriculture Sciences, 91*(6), 421-24.

Singh, B., & Rengel, Z. (2007). The role of crop residues in improving soil fertility. In *Nutrient cycling in terrestrial ecosystems* (pp. 183–214). Springer.

Singh, D. P. (2017). *Management of wheat and barley diseases*. Apple Academic Press. doi:10.1201/9781315207537

Singh, D., Ghosh, P., Kumar, J., & Kumar, A. (2019). Plant Growth-Promoting Rhizobacteria (PGPRs): functions and benefits. In *Microbial Interventions in Agriculture and Environment* (pp. 205–227). doi:10.1007/978-981-13-8383-0_7

Singh, G., Brar, M. S., & Malhi, S. S. (2007). Decontamination of chromium by farm yard manure application in spinach grown in two texturally different Cr-contaminated soils. *Journal of Plant Nutrition*, *30*(2), 289–308. doi:10.1080/01904160601118125

Singh, J. S., Kumar, A., Rai, A. N., & Singh, D. P. (2016). Cyanobacteria: A precious bio-resource in agriculture, ecosystem, and environmental sustainability. *Frontiers in Microbiology*, *7*, 1–19. doi:10.3389/fmicb.2016.00529 PMID:27148218

Singh, M. (2021). Organic farming for sustainable agriculture. *Indian Journal of Organic Farming*, *1*(1).

Singh, O. K. (2021). Development of a solar cooking system suitable for indoor cooking and its exergy and enviroeconomic analyses. *Solar Energy*, *217*, 223–234. doi:10.1016/j.solener.2021.02.007

Singh, R. K., Kumar, D. P., Singh, P., Solanki, M. K., Srivastava, S., Kashyap, P. L., Kumar, S., Srivastava, A. K., Singhal, P. K., & Arora, D. K. (2014). Multifarious plant growth promoting characteristics of chickpea rhizosphere associated Bacilli help to suppress soil-borne pathogens. *Plant Growth Regulation*, *73*(1), 91–101. doi:10.100710725-013-9870-z

Singh, R. P., Handa, R., & Manchanda, G. (2021). Nanoparticles in sustainable agriculture: An emerging opportunity. *Journal of Controlled Release*, *329*, 1234–1248. doi:10.1016/j.jconrel.2020.10.051 PMID:33122001

Singh, R. P., & Jha, P. N. (2017). The PGPR *Stenotrophomonas maltophilia* SBP-9 augments resistance against biotic and abiotic stress in wheat plants. *Frontiers in Microbiology*, *8*, 1945. doi:10.3389/fmicb.2017.01945 PMID:29062306

Singh, R. V. (2009). Effect of hybrids, plant density and fertility levels on nutrient concentration, uptake and productivity of pearl millet in semi arid environment. *The South Pacific Journal of Natural Science*, *27*(1), 45–48. doi:10.1071/SP09008

Singh, S., Gautam, A., Sharma, A., & Batra, A. (2010). *Centella asiatica* (L.): A Plant with Immense Medicinal Potential but Threatened. *International Journal of Pharmaceutical Sciences Review and Research*, *4*, 9–17.

Singh, S., Ladha, J. K., Gupta, R. K., Bhushan, L., Rao, A. N., Sivaprasad, B., & Singh, P. P. (2007). Evaluation of mulching, intercropping with Sesbania and herbicide use for weed management in dry-seeded rice (Oryza sativa L.). *Crop Protection (Guildford, Surrey)*, *26*(4), 518–524. doi:10.1016/j.cropro.2006.04.024

Singh, S., Vishwakarma, K., Singh, S., Sharma, S., Dubey, N. K., Singh, V. K., Liu, S., Tripathi, D. K., & Chauhan, D. K. (2017). Understanding the plant and nanoparticle interface at transcriptomic and proteomic level: A concentric overview. *Plant Gene*, *11*, 265–272. Advance online publication. doi:10.1016/j.plgene.2017.03.006

Singh, U. B., Malviya, D., Wasiullah, Singh, S., Imran, M., Pathak, N., Alam, M., Rai, J. P., Singh, R. K., Sarma, B. K., Sharma, P. K., & Sharma, A. K. (2016). Compatible salt-tolerant rhizosphere microbe-mediated induction of phenylpropanoid cascade and induced systemic responses against *Bipolaris sorokiniana* (Sacc.) Shoemaker causing spot blotch disease in wheat. (*Triticum aestivum* L.). *Applied Soil Ecology*, *108*, 300–306. doi:10.1016/j.apsoil.2016.09.014

Singh, Y. (2003). *Crop residue management in rice-wheat system. Addressing resource conservation issues in rice-wheat Consortium for the Indo-Gangetic Plains*. CIMMYT.

Sinha, S., & Mukherjee, S. K. (2008). Cadmium-induced siderophore production by a high Cd-resistant bacterial strain relieved Cd toxicity in plants through root colonization. *Current Microbiology*, *56*(1), 55–60. doi:10.100700284-007-9038-z PMID:17899260

Siritungam, D., & Sayre, R. (2003). Generation of cyanogen-free transgenic cassava. *Planta*, *217*(3), 367–373. doi:10.100700425-003-1005-8 PMID:14520563

Sisworo, W. H., Mitrosuhardjo, M. M., & Rasjid, H. (1990). The relative roles of N fixation, crop residues and soil in supplying N in multiple cropping systems in a humid, tropical upland cropping system. *Plant and Soil*, *121*, 73–82.

Skordoulis, M., Ntanos, S., & Arabatzis, G. (2020). Socioeconomic evaluation of green energy investments. *International Journal of Energy Sector Management*.

Slater, S., Mitsky, T. A., Houmiel, K. L., Hao, M., Reiser, S. E., Taylor, N. B., Tran, M., Valentin, H. E., Rodriguez, D. J., Stone, D. A., Padgette, S. R., Kishore, G., & Gruys, K. J. (1999). Metabolic engineering of *Arabidopsis* and *Brassica* for poly (3-hydroxybutyrateco-3-hydroxyvalerate) copolymer production. *Nature Biotechnology*, *17*(10), 1011–1016. doi:10.1038/13711 PMID:10504704

Smical, A. I., Hotea, V., Oros, V., Juhasz, J., & Pop, E. (2008). Studies on transfer and bioaccumulation of heavy metals from soil into lettuce. *Environmental Engineering and Management Journal*, *7*(5), 609–615. doi:10.30638/eemj.2008.085

Smith, B. E., Richards, R. L., & Newton, W. E. (2013). Catalysts for nitrogen fixation: nitrogenases, relevant chemical models and commercial processes. Springer Science & Business Media.

Smith, S. M. (1996). Biological control with Trichogramma: Advances, successes, and potential of their use. Annual Review of Entomology, 375–406. doi:10.1146/annurev.en.41.010196.002111

Sokal & Rohlf. (1981). *Biometry. The principles and practice of statistics in biological research* (2nd ed.). W.H. Freeman and Company.

Solanki, M. K., Kumar, S., Pandey, A. K., Srivastava, S., Singh, R. K., Kashyap, P. L., Srivastava, A. K., & Arora, D. K. (2012). Diversity and antagonistic potential of *Bacillus* spp. associated to the rhizosphere of tomato for the management of *Rhizoctonia solani*. *Biocontrol Science and Technology*, *22*(2), 203–217. doi:10.1080/09583157.2011.649713

Solanki, M. K., Singh, N., Singh, R. K., Singh, P., Srivastava, A. K., Kumar, S., Kashyap, P. L., & Arora, D. K. (2011). Plant defense activation and management of tomato root rot by a chitin-fortified Trichoderma/Hypocrea formulation. *Phytoparasitica*, *39*(5), 471–481. doi:10.100712600-011-0188-y

Solanki, M. K., Singh, R. K., Srivastava, S., Kumar, S., Kashyap, P. L., & Srivastava, A. K. (2015). Characterization of antagonistic-potential of two Bacillus strains and their biocontrol activity against *Rhizoctonia solani* in tomato. *Journal of Basic Microbiology*, *55*(1), 82–90. doi:10.1002/jobm.201300528 PMID:24277414

Someswar & Barman. (2014). In Vitro Micropropagation of Dendrobium chtysanthum Wall. ex Lindl. -A Threatened Orchid. Scholars Academic Journal of BIosciences, 2, 39-42.

Somleva, M., & Ali, A. (2010). *Propagation of transgenic plants*. International Patent Application WO/2010/102220.

Somleva, M., Chinnapen, H., Ali, A., Snell, K. D., Peoples, O. P., Patterson, N., Tang, J., & Bohmert-Tatarev, K. (2012). *Increasing carbon flow for polyhydroxybutyrate production in biomass crops.* US Patent Application 2012/0060413.

Somleva, M. N., Snell, K. D., Beaulieu, J. J., Peoples, O. P., Garrison, B. R., & Patterson Nii, A. (2008). Production of polyhydroxybutyrate in switchgrass, a value-added co-product in an important lignocellulosic biomass crop. *Plant Biotechnology Journal*, *6*(7), 663–678. doi:10.1111/j.1467-7652.2008.00350.x PMID:18498309

Sommer, A. (2008). Vitamin A deficiency and clinical disease: An historical overview. *Clinical Nutrition (Edinburgh, Lothian)*, *138*, 1835–1839. PMID:18806089

Sommer, S. G., Petersen, S. O., & Møller, H. B. (2004). Algorithms for calculating methane and nitrous oxide emission from manure management. *Nutrient Cycling in Agroecosystems*, *69*(2), 143–154. doi:10.1023/B:FRES.0000029678.25083.fa

Song, A., Xue, G., Cui, P., Fan, F., Liu, H., Yin, C., Sun, W., & Liang, Y. (2016). The role of silicon in enhancing resistance to bacterial blight of hydroponic- and soil-cultured rice. *Scientific Reports*, *19*(6), 24640. doi:10.1038rep24640 PMID:27091552

Song, H., Zhang, S., Long, X., Yang, X., Li, H., & Xiang, W. (2017). Optimization of bioelectricity generation in constructed wetland-coupled microbial fuel cell systems. *Water (Basel)*, *9*(3), 185. doi:10.3390/w9030185

Soni, G. D. (2015). Advantages of green technology. *International Journal of Research – Granthaalayah*, *3*, 2–5.

Sontheimer, E. J. (2005). Assembly and function of RNA silencing complexes. *Nature Reviews. Molecular Cell Biology*, *6*(2), 127–138. doi:10.1038/nrm1568 PMID:15654322

Souchie, E. L., Saggin-Ju'nior, O. J., Silva, E. M. R., Campello, E. F. C., Azco'n, R., & Barea, J. M. (2006). Communities of P-solubilizing bacteria, fungi and arbuscular mycorrhizal fungi in grass pasture and secondary forest of Paraty, RJ- Brazil. *Anais da Academia Brasileira de Ciências*, *78*(1), 183–193. doi:10.1590/S0001-37652006000100016 PMID:16532216

Soumya, K. M. (2015). Organic Farming: An effective way to promote sustainable Agriculture Development in India. *Journal of the Humanities and Social Sciences*, *20*(6), 31–36.

Souri, Z., & Karimi, N. (2017). Enhanced Phytoextraction by As Hyperaccumulator *Isatis cappadocica* Spiked with Sodium Nitroprusside. *Soil & Sediment Contamination*, *26*(4), 457–468. doi:10.1080/15320383.2017.1326457

Sousa, R. S., Rodrigues, F. A., Schurt, D. A., Souza, N. F. A., & Cruz, M. F. A. (2013). Cytological aspects of the infection process of *Pyricularia oryzae* on leaves of wheat plants supplied with silicon. *Tropical Plant Pathology*, *38*(6), 472–477. doi:10.1590/S1982-56762013000600002

Sridhar, J., Venkateswarlu, V., & Nagesh, M. (2015). Aphid. In A manual on diseases and pests of potato-Technical Bulletin No. 101. ICAR-Central Potato Research Institute.

Srikanth, S., & Mohan, S. V. (2012). Influence of terminal electron acceptor availability to the anodic oxidation on the electrogenic activity of microbial fuel cell (MFC). *Bioresource Technology*, *123*, 480–487. doi:10.1016/j.biortech.2012.07.049 PMID:22940358

Srinivas, P., & Ramakrishna, G. (2005). Biological management of rice seed borne pathogens by native biocontrol agents. *Annals of Plant Protection Sciences*, *13*(2), 422–426.

Srivastava, A. K., Singh, R. N., Kumar, S., Kashyap, P. L., & Arora, D. K. (2012). Growth promotion and management of *Alternaria* leaf spot in chilli by *Trichoderma harzianum. International Journal of Innovative Horticulture*, *1*(2), 158–163.

Srivastava, P., Yadav, A. K., & Mishra, B. K. (2015). The effects of microbial fuel cell integration into constructed wetland on the performance of constructed wetland. *Bioresource Technology*, *195*, 223–230. doi:10.1016/j.biortech.2015.05.072 PMID:26144020

Sriyanaa, I. (2020). Watershed management index based on the village watershed model (VWM) approach towards sustainability. *International Soil and Water Conservation Research*, *8*(1), 35–46. doi:10.1016/j.iswcr.2020.01.003

SSI - Swedish Standards Institute. (1995). Water quality - Determination of growth inhibition (7- d) Lemna minor, duckweed. SS 02 82 13. Author.

Stadnik, M. J., & Buchenauer, H. (1999). Control of wheat diseases by a benzothiadiazole-derivative and modern fungicides. *Journal of Plant Diseases and Protection*, *106*, 466–475.

Stanley, J. T., & Johnson, S. N. (2008). Climate change impacts on root herbivores. In S. N. Johnson & P. J. Murray (Eds.), *Root Feeders: an ecosystem perspective*. CABI.

Stapleton, J. J., & Banuelos, G. S. (2009). Biomass crops can be used for biological disinfestation and remediation of soils and water. *California Agriculture*, *63*(1), 41–46. doi:10.3733/ca.v063n01p41

Steenhoudt, O., & Vanderleyden, J. (2000). *Azospirillum*, a free living nitrogen-fixing bacterium closely associated with grasses: Genetic, biochemical and ecological aspects. *FEMS Microbiology Reviews*, *24*(4), 487–506. doi:10.1111/j.1574-6976.2000.tb00552.x PMID:10978548

Stefani, G., & Slack, F. J. (2008). Small non-coding RNAs in animal development. *Nature Reviews. Molecular Cell Biology*, *9*(3), 219–230. doi:10.1038/nrm2347 PMID:18270516

Steinbuchel, A., & Valentin, H.E. (1995). Diversity of bacterial polyhydroxyalkanoic acids. *FEMS Microbiol Lett, 128*, 219-228.

Stern, V. M., Smith, R. F., van den Bosch, R., & Hagen, K. S. (1959). The integrated control concept. *Hilgardia*, *29*(2), 81–101. doi:10.3733/hilg.v29n02p081

Stevenson, J. R., Villoria, N., Byerlee, D., Kelley, T., & Maredia, M. (2013). Green revolution research saved an estimated 18 to 27 million hectares from being brought into agricultural production. *Proceedings of the National Academy of Sciences of the United States of America*, *110*(21), 8363–8368.

Steves, R. M., Todd, T. C., Essig, J. S., & Trick, H. N. (2006). Transgenic soybeans expressing siRNAs specific to a major sperm protein gene suppress *Heterodera glycines* reproduction. *Functional Plant Biology*, *33*(11), 991–999. doi:10.1071/FP06130 PMID:32689310

Stewart, C. P., Dewey, K. G., & Ashoran, P. (2010). The undernutrition epidemic: An urgent health priority. *Lancet*, *375*(9711), 282. doi:10.1016/S0140-6736(10)60132-8 PMID:20109955

Stigter, K., Ramesh, K., & Upadhyay, P. K. (2018). Mulching for microclimate modifications in farming–an overview. *Indian Journal of Agronomy*, *63*(3), 89–97.

Stocco, M. C., Mansilla, A. Y., & Mónaco, C. I. (2015). Native isolates of *Trichoderma harzianum* inducting resistance to *Zymoseptoria tritici* on wheat plants. *Boletín de la Sociedad Argentina de Botánica*, *50*(3), 291–301. doi:10.31055/1851.2372.v50.n3.12519

Stoeckli, S., Birrer, S., Zellweger-Fischer, J., Balmer, O., Jenny, M., & Pfiffner, L. (2017). Quantifying the extent to which farmers can influence biodiversity on their farms. *Agriculture, Ecosystems & Environment*, *237*, 224–233.

Stoll, G. (1986). Natural crop protection based on local farm resources in the tropics and subtropics. Intermediate Technology Publications.

Stover, P. J. (2006). Influence of human genetic variation on nutritional requirements. *The American Journal of Clinical Nutrition*, *83*, 436S–442S.

Strategies for Agricultural Research in the North-East. (2001). National Academy of Agricultural Sciences.

Strauss E. (1999). Molecular biology. Candidate 'gene silencers' found. *Science, 286*(5441), 886.

Strik, D. P., Timmers, R. A., Helder, M., Steinbusch, K. J., Hamelers, H. V., & Buisman, C. J. (2011). Microbial solar cells: Applying photosynthetic and electrochemically active organisms. *Trends in Biotechnology*, *29*(1), 41–49. doi:10.1016/j.tibtech.2010.10.001 PMID:21067833

Stroo, H. F., & Jencks, E. M. (1982). Enzyme activity and respiration in minesoils. *Soil Science Society of America Journal*, *46*(3), 548–553. doi:10.2136ssaj1982.03615995004600030021x

Study on The Indian Green Biotech Market and EU Collaboration Opportunities. (2012). *European Business and Technology Centre (EBTC)/Euro India Research Centre*.

Subramanian, A. (2017). *Transforming Indian Agriculture: By Loving Some Agriculture Less and the Rest More*. Lecture at National Academy of Agricultural Sciences.

Subramanian, K. S., & Tarafdar, J. C. (2011). Prospects of nanotechnology in Indian farming. *Indian Journal of Agricultural Sciences*, *81*, 887–893.

Sudesh, K., Abe, H., & Doi, Y. (2000). Synthesis, structure and properties of polyhydroxyalkanoates: Biological polyesters. *Progress in Polymer Science*, *25*(10), 1503–1555. doi:10.1016/S0079-6700(00)00035-6

Sujanya, S., & Chandra, S. (2011). Effect of part replacement of chemical fertilizers with organic and bio-organic agents in ground nut, *Arachis hypogea. Journal of Algal Biomass Utilization*, *2*, 38–41.

Suma, Keshavachandran, & Nybe. (2008). Agrobacterium tumefaciens mediated transformation and regeneration of ginger (*Zingiber officinale Rosc.*). *Journal of Tropical Agriculture*, *46*, 38–44.

Suman, R., Javaid, M., Nandan, D., Bahl, S., & Haleem, A. (2020). Electricity Generation Through Water Supply Pipes in High Rise Buildings. *Journal of Industrial Integration and Management*, 1-20.

Suman, G. (2010). Biopesticides: An eco-friendly approach for pest control. *Journal of Biopesticides*, *3*(1), 186–188.

Suman, J., Uhlik, O., Viktorova, J., & Macek, T. (2018). Phytoextraction of heavy metals: A promising tool for clean-up of polluted environment? *Frontiers in Plant Science*, *9*, 1476. doi:10.3389/fpls.2018.01476 PMID:30459775

Sundara, B., & Subramanian, S. (1990). Changes in soil available NPK in multiple cropping systems based on short duration sugarcane crops relative to a conventional sugarcane cropping system. *Fertilizer Research*, *24*(2), 67–76. doi:10.1007/BF01073224

Sunkar, R., & Zhu, J. K. (2004). Novel and stress-regulated micro RNAs and other small RNAs from Arabidopsis. *The Plant Cell*, *16*(8), 2001–2019. doi:10.1105/tpc.104.022830 PMID:15258262

Suo, D. R. (2005). Effects of combined fertilization of chemical and organic fertilizers in a long-term posi- tion experiment. *Ganhan Diqu Nongye Yanjiu*, *23*(2), 71–77.

Sureshkumar, S., Tamilkumar, P., Senthil, N., Nagarajan, P., Thangavelu, A. U., Raveendran, M., Vellaikumar, S., Ganesan, K. N., Balagopal, R., Vijayalakshmi, G., & Shobana, V. (2014). Marker assisted selection of low phytic acid trait in maize (*Zea mays* L.). *Hereditas*, *151*(1), 20–27. doi:10.1111/j.1601-5223.2013.00030.x PMID:24627970

Suriyamongkol, P., Weselake, R., Narine, S., Moloney, M., & Shah, S. (2007). Biotechnological approaches for the production of polyhydroxyalkanoates in microorganisms and plants—A review. *Biotechnology Advances*, *25*(2), 148–175. doi:10.1016/j.biotechadv.2006.11.007 PMID:17222526

Suryawanshi, S. P., & Kamble, A. (2012) Watershed management in India. *Proceeding of International Conference SWRDM-2012.*

Susanto. (2015). Sustainable organic farming for Environmental Health: A social development model. *International Journal of Scientific & Technology Research, 4*(5), 196-211.

Susarla, S., Medina, V. F., & Mc Cutcheon, S. C. (2002). Phytoremediation: An ecological solution to organic chemical contamination. *Ecological Engineering*, *18*(5), 647–658. doi:10.1016/S0925-8574(02)00026-5

Sustainable agriculture practices and technologies: Guidelines for farmers Africa 2000 network. (1997). UNDP.

Su, X. L., & Li, Y. (2004). Quantum dot biolabelling coupled with immunomagnetic separation for detection of *Escherichia coli* 0157:H7.2004. *Analytical Chemistry*, *76*(16), 4806–4810. doi:10.1021/ac049442+ PMID:15307792

Suyamto, S. (1993). Direct and residual eVect of potassium fertilizer on rice-maize cropping rotation on Vertisols. *Indonesian Journal of Crop Science*, *8*(2), 29–38.

Suzaki, T., Yoro, E., & Kawaguchi, M. (2015). Leguminous plants: Inventors of root nodules to accommodate symbiotic bacteria. *International Review of Cell and Molecular Biology*, *316*, 111–158. doi:10.1016/bs.ircmb.2015.01.004 PMID:25805123

Suzuki, Y., Kurano, M., Arai, Y., Nakashita, H., Doi, Y., Usami, R., Horikoshi, K., & Yamaguchi, I. (2002). Enzyme inhibitors to increase poly-3-hydroxybutyrate production by transgenic tobacco. *Bioscience, Biotechnology, and Biochemistry*, *66*(12), 2537–2542. doi:10.1271/bbb.66.2537 PMID:12596845

Swapana, L. A. (2013). Development of biofertilizers and its Future Perspective. *Journal of Pharmaceutics*, (4), 327–332.

Tabbal, D., Bouman, B. A. M., Bhuiyan, S. I., Sibayan, E. B., & Sattar, M. A. (2002). On farm strategies for reducing water input in irrigated rice: Case studies in the Philippines. *Agricultural Water Management*, *56*(2), 93–112. doi:10.1016/S0378-3774(02)00007-0

Takenaka, S., Nishio, Z., & Nakamura, Y. (2003). Induction of defense reactions in sugar beet and wheat by treatment with cell wall protein fractions from the mycoparasite *Pythium oligandrum*. *Phytopathology*, *93*(10), 1228–1232. doi:10.1094/PHYTO.2003.93.10.1228 PMID:18944321

Takumi, S., Motoyasu, Y., & Taiyun, W. (2008). Silencing by RNAi of the gene for Pns12, a viroplasm matrix protein of *Rice dwarf viruso*, results in strong resistance of transgenic rice plants to the *virus*. *Plant Biotechnology Journal*, *7*, 24–32.

Talbert, P. B., & Henikoff, S. (2006). Spreading of silent chromatin: Inaction at a distance. *Nature Reviews. Genetics*, *7*(10), 793–803.

Taleshi, S., Dhumal, K., Alipour, A., Espahbodi, K., & Ghasemi, O. (2009). Impact of alder (*Alnus subcordata*) in fertility of forest soil. *Research Journal of Environmental Sciences*, *3*(6), 640–644. doi:10.3923/rjes.2009.640.644

Tamak, J. C., Narwal, S. S., & Singh, S. (1993). Effect of rice residues incorporated in soil on seedling emergence, growth and yield of lentil (*Lens esculanta*). *Legume Research*, *16*, 77–78.

Tammu, R. N., Nuringtyas, T. S., & Daryono, B. S. (2021). Daryono.B.S. "Colchicine effects on the ploidy level and morphological characters of Katokkon pepper (*Capsicum annuum* L.) from North Toraja, Indonesia. *Journal of Genetic Engineering and Biotechnology*, *19*(1), 1. doi:10.118643141-021-00131-4 PMID:33580394

Tanaka, K. N., Mukae, H., Dewar, M., van Breugel, M., James, E. K., Prescott, A. R., Antony, C., & Tanaka, T. U. (2005). Molecular mechanisms of kinetochore capture by spindle microtubules. *Nature*, *434*(7036), 987–994. doi:10.1038/nature03483 PMID:15846338

Tandy, S., Healey, J. R., Nason, M. A., Williamson, J. C., & Jones, D. L. (2009). Remediation of metal polluted mine soil with compost: Co-composting versus incorporation. *Environmental Pollution*, *157*(2), 690–697. doi:10.1016/j.envpol.2008.08.006 PMID:18819736

Tang, L., Kwon, S. Y., & Kwak, S. S. (2004). Selection of transgenic potato plants expressing both CuZnSOD and APX in chloroplasts with enhanced tolerance to oxidative stress. *Korean Journal of Plant Biotechnology*, *31*(2), 109–113. doi:10.5010/JPB.2004.31.2.109

Tang, Y.-Y., Ma, Y., Wang, J., Fan, Y., Feng, S., Lu, Q., Yu, Q., Sui, D., Rothbart, M. K., Fan, M., & Posner, M. I. (2007). Short-term meditation training improves attention and self-regulation. *Proceedings of the National Academy of Sciences of the United States of America*, *104*(43), 17152–17156. doi:10.1073/pnas.0707678104 PMID:17940025

Tanu, P. A., & Adholeya, A. (2004). Effect of different organic manures/composts on the herbage and essential oil yield of *Cymbopogon winterianus* and their influence on the native AM population in a marginal alfisol. *Bioresource Technology*, *92*, 311-319.

Tao, S., Pang, R., Chen, C., Ren, X., & Hu, S. (2012). Synthesis, characterization and slow release properties of O-naphthylacetyl chitosan. *Carbohydrate Polymers*, *88*(4), 1189–1194. doi:10.1016/j.carbpol.2012.01.076

Taskin, H., Baktemur, G., Kurul, M., & Buyiikalaca, S. (2013). Use of Tissue Culture Techniques for Producing Virus-Free Plant in Garlic and Their Identification through Real-Time PCR. *TheScientificWorldJournal*, *2013*, 1–5. doi:10.1155/2013/781282 PMID:23935432

Tayeh, C., Randoux, B., Vincent, D., Bourdon, N., & Reignault, P. (2014). Exogenous trehalose induces defenses in wheat before and during a biotic stress caused by powdery mildew. *Phytopathology*, *104*(3), 293–305. doi:10.1094/PHYTO-07-13-0191-R PMID:24073639

Tejavathi, D. H., & Indira, M. N. (2013). Regeneration of Shoots from Leaf Callus Cultures of Dryntaria cordata (1) willd ex Roem and Schult. *Indian Journal of Fundamental and Applied Life Sciences*, *3*, 111–115.

Teodor, A. H., & Bruce, B. D. (2020). Putting photosystem I to work: Truly green energy. *Trends in Biotechnology*, *38*(12), 1329–1342. doi:10.1016/j.tibtech.2020.04.004 PMID:32448469

Teodoro, S., Micaela, B., & David, K. W. (2010). Novel use of nano-structured alumina as an insecticide. *Pest Management Science*, *66*(6), 577–579. doi:10.1002/ps.1915 PMID:20127753

Tester, M., & Langridge, P. (2010). Breeding technologies to increase crop production in a changing world. *Science*, *327*, 818–822.

Thabet, M. (2008). *Induction of resistance in wheat plants against leaf rust caused by* Puccinia triciana [MSc thesis]. Ain Shams University, Faculty of Agriculture.

Thakur, M., & Sohal, B. S. (2013). Role of Elicitors in inducing resistance in plants against pathogen infection: A Review. *ISRN Biochemistry*, *762412*, 1–10. Advance online publication. doi:10.1155/2013/762412 PMID:25969762

The Children's Health Act of 2000, PL 106-310, Title V, Section 511, enacted October 17, 2000. Codified in: Title 42, U.S.Code, Section 300w- 3(a)(1)(E).

Thomas, J., & Kintrup, J. (1991). Pyrethroids as household insecticides: Analysis, indoor exposure and persistence. *Fresenius' Journal of Analytical Chemistry*, *340*(7), 446–453. doi:10.1007/BF00322420

Thomas, O. E. (2021). Mathematical Modelling and Simulation of Tidal Energy. *International Journal of Engineering Science and Application*, *5*(1), 38–46.

Thompson, H., & Platenius, H. (1931). Results of paper mulch experiments with vegetable crops. *Proceedings of the American Society for Horticultural Science*, *28*, 305–309.

Thontadarya, T. S., & Jairao, K. (1980). Bhendi as trap crop in the integrated control of cotton bollworms. *Andhra Agriculture Journal*, *27*, 100–101.

Tian, D., Traw, M. B., Chen, J. Q., Kreitman, M., & Bergelson, J. (2003). Fitness costs of R-gene-mediated resistance in Arabidopsis thaliana. *Nature*, *423*(6935), 74–77. doi:10.1038/nature01588 PMID:12721627

Tien, T., Gaskins, M., & Hubbel, D. (1979). Plant growth substances produced by *Azospirillum brasilese* and their effect on growth of pearl millet (*Pennisetum americanum* L.). *Applied and Environmental Microbiology*, *37*(5), 1016–1024. doi:10.1128/aem.37.5.1016-1024.1979 PMID:16345372

Tilak, K. V. B. R., & Manoharachary, C. (2016). Eco-friendly Plant Growth Promoting Rhizobacteria for Crop Improvement. *Microbial Inoculants in Sustainable Agricultural Productivity*, 297-309.

Tilbrook, K., Gebbie, L., Schenk, P. M., Poirier, Y., & Brumbley, S. M. (2011). Peroxisomal polyhydroxyalkanoate biosynthesis is a promising strategy for bioplastic production in high biomass crops. *Plant Biotechnology Journal*, *9*(9), 958–969. doi:10.1111/j.1467-7652.2011.00600.x PMID:21447054

Tillman, P. G., Schomberg, H. H., Phatak, S., Mullinix, B., Timper, P., Lachnicht, S. L., & Olson, D. M. (2004). Influence of cover crops on insect pests and predators in conservation-tillage cotton. *Journal of Economic Entomology*, *97*(4), 1217–1232. doi:10.1093/jee/97.4.1217 PMID:15384330

Tilman, D., Cassman, K. G., Matson, P. A., Naylor, R., & Polasky, S. (2002). Agricultural sustainability and intensive production practices. *Nature*, *418*(6898), 671–677. doi:10.1038/nature01014 PMID:12167873

Timmers, R. A., Rothballer, M., Strik, D. P., Engel, M., Schulz, S., Schloter, M., Hartmann, A., Hamelers, B., & Buisman, C. (2012). Microbial community structure elucidates performance of Glyceria maxima plant microbial fuel cell. *Applied Microbiology and Biotechnology*, *94*(2), 537–548. doi:10.100700253-012-3894-6 PMID:22361855

Tisserant, E., Malbreil, M., Kuo, A., Kohler, A., Symeonidi, A., Balestrini, R., Charron, P., Duensing, N., Frei dit Frey, N., Gianinazzi-Pearson, V., Gilbert, L. B., Handa, Y., Herr, J. R., Hijri, M., Koul, R., Kawaguchi, M., Krajinski, F., Lammers, P. J., Masclaux, F. G., ... Martin, F. (2013). Genome of an arbuscular mycorrhizal fungus provides insight into the oldest plant symbiosis. *Proceedings of the National Academy of Sciences of the United States of America*, *110*(50), 20117–20122. doi:10.1073/pnas.1313452110 PMID:24277808

Tittonell, P. (2014). Ecological intensification of agriculture—Sustainable by nature. *Current Opinion in Environmental Sustainability*, *8*, 53–61.

Tiwari, K. N. (1995). Urvarak aur Khad. Indian Council of Agricultural Research.

Tomari, Y., & Zamore, P. (2005). Perspective: Machines for RNAi. *Genes & Development*, *19*(5), 517–529. doi:10.1101/gad.1284105 PMID:15741316

Torney, F., Trewyn, B. G., Lin, V. S. Y., & Wang, K. (2007). Mesoporous Silica Nanoparticles Deliver DNA and Chemicals into Plants. *Nature Nanotechnology*, *2*(5), 295–300. doi:10.1038/nnano.2007.108 PMID:18654287

Trenbath, B. R. (1993). Intercropping for the management of pests and diseases. *Field Crops Research*, *34*, 381–405.

Trenkel, M. E. (1997). *Controlled-release and stabilized fertilizers in agriculture*. International Fertilizer Industry Association.

Trigo, E., Cap, E., Malach, V., & Villarreal, F. (2009). *The Case of Zero-Tillage Technology in Argentina*. IFPRI Discussion Paper 00915. International Food Policy Research Institute.

Tripathi, D.K., Tripathi, A., & Shweta, S.S. (2017c). Uptake, accumulation and toxicity of silver nanoparticle in autotrophic plants, and heterotrophic microbes: a concentric review. *Frontiers in Microbiology, 8*, 7.

Tripathi, D. K., Mishra, R. K., Singh, S., Singh, S., Vishwakarma, K., Sharma, S., Singh, V. P., Singh, P. K., Prasad, S. M., Dubey, N. K., Pandey, A. C., Sahi, S., & Chauhan, D. K. (2017a). Nitric oxide ameliorates zinc oxide nanoparticles phytotoxicity in wheat seedlings: Implication of the ascorbate-glutathione cycle. *Frontiers in Plant Science*, *8*, 1. doi:10.3389/fpls.2017.00001 PMID:28220127

Tripathi, D. K., Singh, S., Singh, S., Srivastava, P. K., Singh, V. P., Singh, S., Prasad, S. M., Singh, P. K., Dubey, N. K., Pandey, A. C., & Chauhan, D. K. (2017b). Nitric oxide alleviates silver nanoparticles (AgNps)-induced phytotoxicity in *Pisumsativum* seedlings. *Plant Physiology and Biochemistry*, *110*, 167–177. doi:10.1016/j.plaphy.2016.06.015 PMID:27449300

Tripathi, V. (2016). Organic farming: A new avenue for rural entrepreneurs for sustainable business. *Int J Educ Management Stud.*, *6*(1), 8–10.

Tripathy, S. K., & Maharana, M. (2017). Quality protein maize (QPM): Genetic basis and breeding perspective. *Tropical Plant Research*, *4*(1), 145–152. doi:10.22271/tpr.2017.v4.i1.021

Tsang, C. H. A., Huang, H., Xuan, J., Wang, H., & Leung, D. Y. C. (2020). Graphene materials in green energy applications: Recent development and future perspective. *Renewable & Sustainable Energy Reviews*, *120*, 109656. doi:10.1016/j.rser.2019.109656

Tseng, M. C., Roel, A., Deambrosi, E., Terra, J. A., Zorrilla, G., Riccetto, S., & Pittelkow, C. M. (2020). Towards actionable research frameworks for sustainable intensification in high-yielding rice systems. *Scientific Reports*, *10*(1), 1–13.

Tsuda, S., Fukui, Y., Nakamura, N., Katsumoto, Y., Yonekura-Sakakibara, K., Fukuchi-Mizutani, M., Ohira, K., Ueyama, Y., Ohkawa, H., Holton, T. A., Kusumi, T., & Tanaka, Y. (2004). Flower color modification of *Petunia 83ybrid* commercial varieties by metabolic engineering. *Plant Biotechnology (Sheffield, England)*, *21*(5), 377–386. doi:10.5511/plantbiotechnology.21.377

Tuan, T., & Thuiet, R. (1979). *Use of Azolla in rice production in Vietnam. Nitrogen and Rice*. IRRI.

Tuck, S. L., Winqvist, C., Mota, F., Ahnström, J., Turnbull, L. A., & Bengtsson, J. (2014). Land-use intensity and the effects of organic farming on biodiversity: A hierarchical meta-analysis. *Journal of Applied Ecology*, *51*(3), 746–755.

Türker, O. C., & Yakar, A. (2017). A hybrid constructed wetland combined with microbial fuel cell for boron (B) removal and bioelectric production. *Ecological Engineering*, *102*, 411–421. doi:10.1016/j.ecoleng.2017.02.034

Turner, W. R., Oppenheimer, M., & Wilcove, D. S. (2009). A force to fight global warming. *Nature*, *462*(7271), 278–279. doi:10.1038/462278a PMID:19924191

Tuskan, G. A., Difazio, S., Jansson, S., Bohlmann, J., Grigoriev, I., Hellsten, U., Putnam, N., Ralph, S., Rombauts, S., Salamov, A., Schein, J., Sterck, L., Aerts, A., Bhalerao, R. R., Bhalerao, R. P., Blaudez, D., Boerjan, W., Brun, A., Brunner, A., ... Rokhsar, D. (2006). The genome of black cottonwood, Populus trichocarpa (Torr. & Gray). *Science*, *313*(5793), 1596–1604. doi:10.1126cience.1128691 PMID:16973872

U.S. EPA. (2000). Constructed Wetlands Treatment of Municipal Wastewaters. United States Environmental Protection Agency.

Ullah, A., Heng, S., Munis, M. F. H., Fahad, S., & Yang, X. (2015). Phytoremediation of heavy metals assisted by plant growth promoting (PGP) bacteria: A review. *Environmental and Experimental Botany*, *117*, 28–40. doi:10.1016/j.envexpbot.2015.05.001

Ummyiah, H. M., Jabeen, N., Afroza, B., Mushtaq, F., Mufti, S., & Nabi, A. (2017). Certification of organic products. *International Journal of Current Microbiology and Applied Sciences*, *6*(11), 2300–2308. doi:10.20546/ijcmas.2017.611.273

Urwin, P. E., Lilley, C. J., & Atkinson, H. J. (2002). Ingestion of double-stranded RNA by pre-parasitic juvenile cyst nematodes leads to RNA interference. *Molecular Plant-Microbe Interactions*, *15*(8), 747–752. doi:10.1094/MPMI.2002.15.8.747 PMID:12182331

USEPA - United States Environmental Protection Agency. (1996). OPPTS 850.4400 Aquatic Plant Toxicity Test Using Lemna spp., "Public draft". EPA 712-C-96-156.

Vaarst, M. (2010). Organic Farming as a development strategy: Who are interested and who are not. *Journal of Sustainable Development*, *3*(1), 38–50. doi:10.5539/jsd.v3n1p38

Vaghasiya & Shiroya. (n.d.). *A Review Green biotechnology - a help to the environment.* http://www pharmatutor.org/articles/review-green-biotechnology-helpenvironment

Valančius, K., & Mikučionienė, R. (2020). Solar energy as a tool of renovating soviet-type multi apartment buildings. *Solar Energy*, *198*, 93–100. doi:10.1016/j.solener.2020.01.046

Valentin, H. E., Broyles, D. L., Casagrande, L. A., Colburn, S. M., Creely, W. L., DeLaquil, P. A., Felton, H. M., Gonzalez, K. A., Houmiel, K. L., Lutke, K., Mahadeo, D. A., Mitsky, T. A., Padgette, S. R., Reiser, S. E., Slater, S., Stark, D. M., Stock, R. T., Stone, D. A., Taylor, N. B., ... Gruys, K. J. (1999). PHA production, from bacteria to plants. *International Journal of Biological Macromolecules*, *25*(1-3), 303–306. doi:10.1016/S0141-8130(99)00045-8 PMID:10416678

Valkama, E., Kunypiyaeva, G., Zhapayev, R., Karabayev, M., Zhusupbekov, E., Perego, A., ... Acutis, M. (2020). Can conservation agriculture increase soil carbon sequestration? A modelling approach. *Geoderma*, *369*, 114298.

Vamerali, T., Marchiol, L., Bandiera, M., Fellet, G., Dickinson, N. M., Lucchini, P., Mosca, G., & Zerbi, G. (2012). Advances in agronomic management of phytoremediation: Methods and results from a 10-year study of metal-polluted soils. *Italian Journal of Agronomy*, *7*(4), 323–330. doi:10.4081/ija.2012.e42

Vamvakaki, V., & Chaniotakis, N. A. (2007). Pesticide detection with a liposome-based nanobiosensor. *Biosensors & Bioelectronics*, *22*(12), 2848–2853. doi:10.1016/j.bios.2006.11.024 PMID:17223333

Van Bruggen, A. H. C., Gamliel, A., & Finckh, M. R. (2016). Plant disease management in organic farming systems. *Pest Management Science*, *72*(1), 30–44. doi:10.1002/ps.4145 PMID:26331771

Van Damme, V., Hoedekie, A., & Viaene, N. (2005). Long-term efficacy of *Pochoniachlamydosporia* for management of *Meloidogynejavanica* in glasshouse crops. *Nematology*, *7*, 741–745.

van der Krol, A. R., & Mur, L. A. (1990). M Beld, J N Mol, A R Stuitje. "Flavonoid genes in petunia: Addition of a limited number of gene copies may lead to a suppression of gene expression. *The Plant Cell*, *4*(2), 291–299. PMID:2152117

Van der Plank, J. E. (1968). *Disease resistance in plants*. Academic Press.

van der Sloot, H. A., van Zomeren, A., Meeussen, J. C. L., Seignette, P., & Bleijerveld, R. (2007). Test method selection, validation against field data, and predictive modelling for impact evaluation of stabilised waste disposal. *Journal of Hazardous Materials*, *141*(2), 354–369. doi:10.1016/j.jhazmat.2006.05.106 PMID:16889893

Van der walle, G.A.M., De Koning, G.J.M., Weusthuis, R.A., & Eggink, G. (2001). Properties, modifications and applications of biopolyesters. *Adv Bioch Eng/Biotech, 71*, 263-291.

Van Duivenbooden, N. (1992). *Sustainability in terms of nutrient elements with special reference to West Africa. Agrobiological Research Centre*. CABODLO.

Van Ewijk & Hoekstra. (1993). Alternatives for the no-observed-effect-level. *Environmental Toxicology and Chemistry*, *12*, 187–194.

van Loon, L. C., Bakker, P. A. H. M., & Pieterse, C. M. J. (1998). Systemic resistance induced by rhizosphere bacteria. *Annual Review of Phytopathology*, *36*(1), 453–483. doi:10.1146/annurev.phyto.36.1.453 PMID:15012509

van Loon, L. C., Rep, M., & Pieterse, C. M. (2006). Significance of inducible defense-related proteins in infected plants. *Annual Review of Phytopathology*, *44*(1), 135–162. doi:10.1146/annurev.phyto.44.070505.143425 PMID:16602946

Van Ommen, B., & Stierum, R. (2002). Nutrigenomics: Exploiting systems biology in the nutrition and health arena. *Current Opinion in Biotechnology*, *13*(5), 517–521.

Van Uyen, N. (2006). Novel approaches in plants breeding RNAi technology. *Proceedings of International Workshop on Biotechnology in Agriculture*, 12-16.

Vandergheynst, J. S., Scher, H. B., Gou, H. Y., & Schultz, D. L. (2007). Water in oil emulsions that improve the storage and delivery of the biolarvacide*Lagenidiumgiganteum*. *Biological Control*, *52*, 207–229.

Vandergheynst, J. S., Scher, H., & Hong-Yun, G. (2006). Design of formulations for improved biological control agent viability and sequestration during storage. *Industrial Biotechnology (New Rochelle, N.Y.)*, *2*(3), 213–219. doi:10.1089/ind.2006.2.213

Vanegas, R. (2003). *La transición: la búsqueda del cambio hacia sistemas sustentables de producción agropecuaria. Foro Agrario «Sistemas Agrícolas Sustentables*. Quito Universidad Central del Ecuador.

Vangronsveld, J., Herzig, R., Weyens, N., Boulet, J., Adriaensen, K., Ruttens, A., Thewys, T., Vassilev, A., Meers, E., Nehnevajova, E., van der Lelie, D., & Mench, M. (2009). Phytoremediation of contaminated soils and groundwater: Lessons from the field. *Environmental Science and Pollution Research International*, *16*(7), 765–794. doi:10.100711356-009-0213-6 PMID:19557448

Vannini, C., Domingo, G., Onelli, E., De Mattia, F., Bruni, I., Marsoni, M., & Bracale, M. (2014). Phytotoxic and genotoxic effects of silver nanoparticles exposure on germinating wheat seedlings. *Journal of Plant Physiology*, *171*(13), 1142–1148. doi:10.1016/j.jplph.2014.05.002 PMID:24973586

Vanyushin, B. F. (2006). DNA methylation in plants. *Current Topics in Microbiology and Immunology*, *301*, 67–122. doi:10.1007/3-540-31390-7_4 PMID:16570846

Varade, Wankhade, Yadav, & Khandare. (2011). Watershed management as a tool for changing the Kaleidoscope of central India: A case study from Jhabua district of Madhya Pradesh, India. *Nature Environment and Pollution Technology*, *10*(5), 589–594.

Vasanthi, N., Saleena, L. M., & Raj, S. A. (2012). Silicon in day today life. *WorldAppliedSciencesJournal*, *17*, 1425–1440.

Vasanthi, N., Saleena, L. M., & Raj, S. A. (2016). Silica solubilization potential of certain bacterial species in the presence of different silicate minerals. *Silicon, 10*(2), 267–275. doi:10.100712633-016-9438-4

Vasconcelos, F. A. G. (2010). The science of nutrition in transit: From nutrition and dietetics to nutrigenomics. *Revista de Nutrição, 23*(6), 935–945.

Vats, S., & Sharma, V. (2011). *Organic farming history and techniques*. eVaidya Jiwellness Pvt. Ltd.

Vaucheret, H. (2008). Plant ARGONAUTES. *Trends in Plant Science, 13*(7), 350–358. doi:10.1016/j.tplants.2008.04.007 PMID:18508405

Věchet, L., & Šerá, B. (2015). Effectiveness of both synthetic compounds and biological extracts against powdery mildew (*Blumeria graminis* f. sp. *tritici*) on winter wheat. *Agrociencia, 49*, 77–85.

Velmurugan, N., Gnana Kumar, G., & Sub Han, S. (2009). Synthesis and characterization of potential fungicidal silver nano-sized particles and chitosan membrane containing silver particles. *Iranian Polymer Journal, 18*, 383–392.

Venkataraman, G. S., & Neelakantan, S. (1967). Effect of cellular constituents of the nitrogen fixing blue-green alga *Cylindrospermum musciola* on the root growth of rice seedlings. *The Journal of General and Applied Microbiology, 13*(1), 53–61. doi:10.2323/jgam.13.53

Venkateswarlu, B., Rao, A. V., Raina, P., & Ahmad, N. (1984). Evaluation of phosphorus solubilization by microorganisms isolated from arid soil. *Journal of the Indian Society of Soil Science, 32*, 273–277.

Verdejo-Lucas, S., Sorribas, F. J., Ornat, C., & Galeano, M. (2003). Evaluating *Pochoniachlamydosporia* in a double cropping system of lettuce and tomato in plastic houses infested with *Meloidogynejavanica. Plant Pathology, 52*(4), 521–528. doi:10.1046/j.1365-3059.2003.00873.x

Verhagen, A., Dietz, A. J., Ruben, R., Van Dijk, H., De Jong, A., Zaal, F., De Bruin, M., & Van Keulen, H. (2004). Climate change and food security in the drylands of West Africa. In A. J. Dietz, R. Ruben, & A. Verhagen (Eds.), *The Impact of Climate Change on Drylands with a focus on West Africa* (pp. 365–384). Kluwer Academic Publishers.

Verma, J. P., Yadav, J., Tiwari, K. N., Lavakush, S. V., & Singh, V. (2010). Impact of plant growth promoting rhizobacteria on crop production. *International Journal of Agricultural Research, 5*(11), 954–983. doi:10.3923/ijar.2010.954.983

Vessey, J. K. (2003). Plant growth promoting rhizobacteria as biofertilizers. *Plant and Soil, 255*(2), 571–586. doi:10.1023/A:1026037216893

Vidotti, M., Carvalhal, R. F., Mendes, R. K., Ferreira, D. C. M., & Kubota, L. T. (2011). Biosensors based on gold nanostructures. *Journal of the Brazilian Chemical Society, 22*(1), 3–20. doi:10.1590/S0103-50532011000100002

Vijayakumar, P. S., Abhilash, O. U., Khan, B. M., & Prasad, B. L. V. (2010). Nanogold-loaded sharp-edged carbon bullets as plant gene carriers. *Advanced Functional Materials, 20*(15), 2416–2423. doi:10.1002/adfm.200901883

Vilich-Meller, V. (1992). Mixed cropping of cereals to suppress plant diseases and omit pesticide applications. *Biological Agriculture and Horticulture, 8*(4), 299–308. doi:10.1080/01448765.1992.9754607

Villasenor, J., Capilla, P., Rodrigo, M. A., Canizares, P., & Fernandez, F. J. (2013). Operation of a horizontal subsurface flow constructed wetland–microbial fuel cell treating wastewater under different organic loading rates. *Water Research, 47*(17), 6731–6738. doi:10.1016/j.watres.2013.09.005 PMID:24074815

Voigt, C. C. (2021). Insect fatalities at wind turbines as biodiversity sinks. *Conservation Science and Practice, 3*(5), 366. doi:10.1111/csp2.366

Vojtech, V. (2010). *Policy Measures Addressing Agrienvironmental Issues*. OECD Food, Agriculture and Fisheries Working Papers, No. 24. OECD Publishing.

Vrhovšek, D., Kukanja, V., & Bulc, T. (1996). Constructed wetland (CW) for industrial waste water treatment. *Water Research*, *30*(10), 2287–2292. doi:10.1016/0043-1354(96)00114-5

Vuichard, P., Stauch, A., & Wüstenhagen, R. (2020). Keep it local and low-key: Social acceptance of alpine solar power projects. *Renewable & Sustainable Energy Reviews*, 110516.

Vyslouzilova, M., Tlustos, P., & Szakova, J. (2003). Cadmium and zinc phytoextraction potential of seven clones of *Salix* spp. planted on heavy metal contaminated soils. *Plant, Soil and Environment*, *49*(12), 542–547. doi:10.17221/4191-PSE

Wagner, R.W., & Sun, L. (1998). Double-stranded RNA poses puzzle. *Nature, 391*(6669), 744-5.

Wakatsuki, T. (1995). Metal oxidoreduction by microbial cells. *Journal of Industrial Microbiology*, *14*(2), 169–177. doi:10.1007/BF01569900 PMID:7766210

Wakelin, S. A., Warren, R. A., Harvey, P. R., & Ryder, M. H. (2004). Phosphate solubilization by *Penicillium* sp. closely associated with wheat roots. *Biology and Fertility of Soils*, *40*(1), 36–43. doi:10.100700374-004-0750-6

Walbridge, C. T. (1977). *A flow-through testing procedure with duckweed (Lemna minor L.)*. US EPA Report No. EPA600/3-77 108.

Walker, S., & Ogindo, H. O. (2003). The water budget of rainfed maize and bean intercrop. *Physics and Chemistry of the Earth*, *28*(20-27), 919–926. doi:10.1016/j.pce.2003.08.018

Walters, D. R., Newton, A. C., & Lyon, G. D. (2005). Induced resistance: Helping plants to help themselves. *Biologist*, *52*, 28–33.

Walters, D. R., Ratsep, J., & Havis, N. D. (2013). Controlling crop diseases using induced resistance: Challenges for the future. *Journal of Experimental Botany*, *64*(5), 1263–1280. doi:10.1093/jxb/ert026 PMID:23386685

Walters, S. A., & Young, B. G. (2008). Utility of winter rye living mulch for weed management in zucchini squash production. *Weed Technology*, *22*(4), 724–728. doi:10.1614/WT-08-035.1

Wang, A. S., Angle, J. S., Chaney, R. L., Delorme, T. A., & Reeves, R. D. (2006). Soil pH effects on uptake of Cd and Zn by *Thlaspi caerulescens*. *Plant and Soil*, *281*(1-2), 325–337. doi:10.100711104-005-4642-9

Wang, J., Song, X., Wang, Y., Bai, J., Li, M., Dong, G., Lin, F., Lv, Y., & Yan, D. (2017). Bioenergy generation and rhizodegradation as affected by microbial community distribution in a coupled constructed wetland-microbial fuel cell system associated with three macrophytes. *The Science of the Total Environment*, *607*, 53–62. doi:10.1016/j.scitotenv.2017.06.243 PMID:28686895

Wang, J., Song, X., Wang, Y., Zhao, Z., Wang, B., & Yan, D. (2017). Effects of electrode material and substrate concentration on the bioenergy output and wastewater treatment in air-cathode microbial fuel cell integrating with constructed wetland. *Ecological Engineering*, *99*, 191–198. doi:10.1016/j.ecoleng.2016.11.015

Wang, M., Abbott, D. C., & Waterhouse, P. M. (2000). A single copy of a virus- derived transgene encoding hairpin RNA gives immunity to barley yellow dwarf virus. *Molecular Plant Pathology*, *1*(6), 347–356. doi:10.1046/j.1364-3703.2000.00038.x PMID:20572982

Wang, M., Gao, L., Dong, S., Sun, Y., Shen, Q., & Guo, S. (2017). Role of Silicon on Plant–Pathogen Interactions. *Frontiers in Plant Science*, *8*, 701. doi:10.3389/fpls.2017.00701 PMID:28529517

Wang, M., Weiberg, A., Lin, F.-M., Thomma, B. P. H. J., Huang, H.-D., & Jin, H. (2016). Bidirectional cross-kingdom RNAi and fungal uptake of external RNAs confer plant protection. *Nature Plants*, *2*(10), 16151. doi:10.1038/nplants.2016.151 PMID:27643635

Wang, R., van Dorp, L., Shaw, L. P., Bradley, P., Wang, Q., Wang, X., ... Balloux, F. (2018). The global distribution and spread of the mobilized colistin resistance gene mcr-1. *Nature Communications*, *9*(1), 1–9.

Wang, T., Chen, L., Zhao, M., Tian, Q., & Zhang, W. H. (2011b). Identification of drought-responsive microRNAs in *Medicago truncatula* by genome-wide high-throughput sequencing. *BMC Genomics*, *12*(1), 367. doi:10.1186/1471-2164-12-367 PMID:21762498

Wang, Y. H., Wu, Z. Y., Zhang, X. H., Chen, G. Q., Wu, Q., Huang, C. L., & Yang, Q. (2005). Synthesis of medium-chain-length-polyhydroxyalkanoates in tobacco via chloroplast genetic engineering. *Chinese Science Bulletin*, *50*(11), 1113–1120. doi:10.1360/982005-130

Wang, Y. P., Li, Q. B., Hui, W., Shi, J. Y., Lin, Q., Chen, X. C., & Chen, Y. X. (2008). Effect of sulphur on soil Cu/Zn availability and microbial community composition. *Journal of Hazardous Materials*, *159*(2-3), 385–389. doi:10.1016/j.jhazmat.2008.02.029 PMID:18394797

Wani, S. P., & Garg, K. K. (2009). Watershed Management Concept and Principles. In *Best-bet Options for Integrated Watershed Management Proceedings of the Comprehensive Assessment of Watershed Programs in India*. ICRISAT.

Wani, S. P., & Ramakrishna, Y. S. (2005). Sustainable management of rainwater through integrated watershed approach for improved rural livelihoods. Watershed management challenges: Improved productivity, resources and livelihoods.

Wani, P., Chandrapalaiah, S., & Dart, P. J. (1985). Response of Pearl millet Cultivars to Inoculation with Nitrogen Fixing Bacteria. *Experimental Agriculture*, *21*(2), 175–182. doi:10.1017/S001447970001245X

Wani, P., Rupela, O. P., & Lee, K. K. (1995). Sustainable agriculture in the semi – arid tropics through biological nitrogen fixation in grain legumes. *Plant and Soil*, *174*(1-2), 29–49. doi:10.1007/BF00032240

Wani, S. A., Chand, S., & Ali, T. (2013). Potential use of *Azotobacter chroococcum* in crop production: An overview. *Current Agriculture Research Journal*, *1*(1), 35–38. doi:10.12944/CARJ.1.1.04

Warad, H. C., Ghosh, S. C., & Thanachayanont, C. (2004). Highly luminescent manganese doped ZnS quantum dots for biological labelling. *Proceeding of International Conference on Smart Materials/Intelligent Materials*, 203-206.

Ware, D. H., Jaiswal, P., Ni, J., Yap, I. V., Pan, X., Clark, K. Y., Teytelman, L., Schmidt, S. C., Zhao, W., Chang, K., Cartinhour, S., Stein, L. D., & McCouch, S. R. (2002). Gramene, a tool for grass genomics. *Plant Physiology*, *130*(4), 1606–1613. doi:10.1104/pp.015248 PMID:12481044

Wasaya, Ahmad, & Hassan. (2013). Enhancing crop productivity through wheat (*triticum aestivum* l.) - fenugreek intercropping system. *The Journal of Animal & Plant Sciences*, *23*(1), 210–215.

Wäspi, U., Blanc, D., Winkler, T., Ruedi, P., & Dudler, R. (1998). Syringolin, a novel peptide elicitor from Pseudomonas syringae pv. syringae that induces resistance to Pyricularia oryzae in rice. *Molecular Plant-Microbe Interactions*, *11*(8), 727–733. doi:10.1094/MPMI.1998.11.8.727

Wäspi, U., Schweizer, P., & Dudler, R. (2001). Syringolin reprograms wheat to undergo hypersensitive cell death in a compatible interaction with powdery mildew. *The Plant Cell*, *13*(1), 153–161. doi:10.1105/tpc.13.1.153 PMID:11158536

Watanabe, J. (1989). Increasing fertility with crop residues and green manures. *Noko To Engei*, *64*, 223–228.

Watershed management programmes. (n.d.). https://vikaspedia.in/social-welfare/rural-poverty-alleviation-1/schemes-department-of-land-resources/watershed-management-programmes

Webb, P., Stordalen, G. A., Singh, S., Wijesinha-Betton, R., Shetty, P., & Lartey, A. (2018). Hunger and malnutrition in the 21st century. *BMJ (Clinical Research Ed.)*, *361*, k2238. doi:10.1136/bmj.k2238 PMID:29898884

Weidmann, G., & Kilcher, L. (2011). *African organic agriculture training*. FiBL.

Wei, L., Zhang, D., Xiang, F., & Zhang, Z. (2009). Differentially expressed miRNAs potentially involved in the regulation of defense mechanism to drought stress in maize seedlings. *International Journal of Plant Sciences*, *170*(8), 979–989. doi:10.1086/605122

Wei, W., & Skye, H. M. (2021). Residential net-zero energy buildings: Review and perspective. *Renewable & Sustainable Energy Reviews*, *142*, 110859. doi:10.1016/j.rser.2021.110859 PMID:34413697

Welch, R. M., & Graham, R. D. (2002). Breeding crops for enhanced micronutrient content. *Plant and Soil*, *245*(1), 205–214. doi:10.1023/A:1020668100330

Wellen, K. E., & Hotamisligil, G. S. (2005). Inflammation, stress, and diabetes. *The Journal of Clinical Investigation*, *115*(5), 1111–1119.

Wenger, K., Kayser, A., Gupta, S. K., Furrer, G., & Schulin, R. (2002). Comparison of NTA and elemental sulfur as potential soil amendments in phytoremediation. *Soil & Sediment Contamination*, *11*(5), 655–672. doi:10.1080/20025891107023

Westerink, J., Plomp, M., Ottburg, F., Zanen, M., & Schrijver, R. (2018). Farmers for Nature: the ultimate nature-inclusive agriculture? *Alterra-rapport-Wageningen University and Research Centre*, (2858).

Wetser, K., Sudirjo, E., Buisman, C. J., & Strik, D. P. (2015). Electricity generation by a plant microbial fuel cell with an integrated oxygen reducing biocathode. *Applied Energy*, *137*, 151–157. doi:10.1016/j.apenergy.2014.10.006

Wheeler, T. R., Craufurd, P. Q., Ellis, R. H., Porter, J. R., & Prasad, P. V. V. (2000). Temperature variability and the yield of annual crops. *Agriculture, Ecosystems & Environment*, *82*, 159–167.

White, C. (2020). Why Regenerative Agriculture? *American Journal of Economics and Sociology*, *79*(3), 799–812.

Whitelaw, M. A. (1999). Growth promotion of plants inoculated with phosphate-solubilizing fungi. *Advances in Agronomy*, *69*, 99–151. doi:10.1016/S0065-2113(08)60948-7

White, P. J., & Broadley, M. R. (2009). Biofortification of crops with seven mineral elements often lacking in human diets—Iron, zinc, copper, calcium, magnesium, selenium and iodine. *The New Phytologist*, *182*(1), 49–84. doi:10.1111/j.1469-8137.2008.02738.x PMID:19192191

Whitfield, L., Richards, A. J., & Rimmer, D. L. (2003). Effects of mycorrhizal colonization on *Thymus polytrichus* from heavy metalcontaminated sites in north England. *Mycorrhiza*, *14*, 47–54. PMID:14564560

Whiting, S. N., De Souza, M., & Terry, N. (2001). Rhizosphere bacteria mobilize Zn for hyperaccumulator by *Thlaspi caerulescens*. *Environmental Science & Technology*, *35*(15), 3144–3150. doi:10.1021/es001938v PMID:11505990

WHO. (2015). *World Health Organization and Secretariat of the Convention of Biological Diversity Connecting Global Priorities: Biodiversity and Human Health: A Status of Knowledge Review*. WHO.

Wiesel, L., Newton, A. C., Elliott, I., Booty, D., Gilroy, E. M., Birch, P. R. J., & Hein, I. (2014). Molecular effects of resistance elicitors from biological origin and their potential for crop protection. *Frontiers in Plant Science*, *5*, 655. doi:10.3389/fpls.2014.00655 PMID:25484886

Wieshammer, G., Unterbrunner, R., Garcia, T. B., Zivkovic, M. F., Puschenreiter, M., & Wenzel, W. W. (2007). Phytoextraction of Cd and Zn from agricultural soils by *Salix* ssp and intercropping of *Salix caprea* and *Arabidopsis halleri. Plant and Soil*, *298*(1-2), 255–264. doi:10.100711104-007-9363-9

Wightman, B., Ha, I., & Ruvkun, G. (1993). Post transcriptional regulation of the heterochronic gene lin-4 mediates temporal formation in *C. elegans. Cell*, *75*(5), 855–862. doi:10.1016/0092-8674(93)90530-4 PMID:8252622

Wilhelm, M., Davin, E. L., & Seneviratne, S. I. (2015). Climate engineering of vegetated land for hot extremes mitigation: An Earth system model sensitivity study. *Journal of Geophysical Research, D, Atmospheres*, *120*(7), 2612–2623.

Williams, D. A. (1971). A test for differences between treatment means when several dose levels are compared with a zero dose control. *Biometrics*, *27*(1), 103–117. doi:10.2307/2528930 PMID:5547548

Williams, D. A. (1972). The comparison of several dose levels with a zero dose control. *Biometrics*, *28*(2), 510–531. doi:10.2307/2556164 PMID:5037867

Williams, M. M. II, Walsh, D. B., & Boydston, R. A. (2004). Integrating arthropod herbivory and reduced herbicide use for weed management. *Weed Science*, *52*(6), 1018–1025. doi:10.1614/WS-04-034R

Wilson, M. A., Tran, N. H., Milev, A. S., Kannangara, G. S. K., Volk, H., & Lu, G. Q. M. (2008). Nanomaterials in soils. *Geoderma*, *146*(1-2), 291–302. doi:10.1016/j.geoderma.2008.06.004

Winer, B. J. (1971). *Statistical principles in experimental design* (2nd ed.). McGraw Hill.

Wirth, J., Wirth, J., Aeschlimann, B., & Yakandawala, N. (2009). Rice endosperm iron biofortification by targeted and synergistic action of nicotianamine synthase and ferritin. *Plant Biotechnology Journal*, *7*(7), 631–644. doi:10.1111/j.1467-7652.2009.00430.x PMID:19702755

Wiwanitkit, V. (2012). Database and tools for nutrigenomics: A brief summary. *J Med Nutr Nutraceut.*, *1*, 87–91.

Wong, N. H., Chen, Y., Ong, C. L., & Sia, A. (2003). Investigation of Thermal Benefits of Rooftop Garden in the Tropical Environment. *Building and Environment*, *38*(2), 261–270. doi:10.1016/S0360-1323(02)00066-5

Wong, N. H., Cheong, D. K. W., Yan, H., Soh, J., Ong, C. L., & Sia, A. (2002). The effects of rooftop garden on energy consumption of a commercial building in Singapore. *Energy and Building*, *35*(4), 353–364. doi:10.1016/S0378-7788(02)00108-1

Wong, N. H., Tay, S. F., Wong, R., Ong, C. L., & Sia, A. (2003). Life cycle cost analysis of rooftop gardens in Singapore. *Building and Environment*, *38*(3), 499–509. doi:10.1016/S0360-1323(02)00131-2

Wrobel-Kwiatkowska, M., Kropiwnicki, M., Zebrowski, J., Beopoulos, A., Dyminska, L., Hanuza, J., & Rymowicz, W. (2019). Effect of mcl-PHA synthesis in flax on plant mechanical properties and cell wall composition. *Transgenic Research*, *28*(1), 77–90. doi:10.100711248-018-0105-y PMID:30484148

Wrobel-Kwiatkowska, M., Skorkowska-Telichowska, K., Dyminska, L., Mączka, M., Hanuza, J., & Szopa, J. (2009). Biochemical, mechanical, and spectroscopic analyses of genetically engineered flax fibers producing bioplastic (poly-β-hydroxybutyrate). *Biotechnology Progress*, *25*(5), 1489–1498. doi:10.1002/btpr.194 PMID:19572280

Wrobel-Kwiatkowska, M., Zebrowski, J., Starzycki, M., Oszmianski, J., & Szopa, J. (2007). Engineering of PHB synthesis causes improved elastic properties of flax fibers. *Biotechnology Progress*, *23*(1), 269–277. doi:10.1021/bp0601948 PMID:17269698

Wrobel-Kwiatkowska, M., Zebrowski, J., & Szopa, J. (2004). Polyhydroxybutyrate synthesis in transgenic flax. *Journal of Biotechnology*, *107*(1), 41–54. doi:10.1016/j.jbiotec.2003.10.005 PMID:14687970

Wu, J. G., Ren, C., & Li, H. Z. (1997). The effects of undecomposed organic materials on rice yield in paddy soils. *Jilin Nongye Daxue Xuebao*, *19*, 68–73.

Wu, L. H., Li, Z., Han, C. L., Liu, L., Teng, Y., Sun, X. H., Pan, C., Huang, Y. J., Luo, Y. M., & Christie, P. (2012). Phytoremediation of soil contaminated with cadmium, copper and polychlorinated biphenyls. *International Journal of Phytoremediation*, *14*(6), 570–584. doi:10.1080/15226514.2011.619227 PMID:22908627

Wu, L., & Liu, M. (2008). Preparation and properties of chitosan coated NPK compound fertilizer with controlled release and water-retention. *Carbohydrate Polymers*, *72*(2), 240–247. doi:10.1016/j.carbpol.2007.08.020

WWEC. (2010). *National Statistics on Water and Wastewater Treatment*. WWEC.

Xiang, S., Fruehauf, J., & Li, C. J. (2006). Short hairpin RNA-expressing bacteria elicit RNA interference in mammals. *Nature Biotechnology*, *24*(6), 697–702. doi:10.1038/nbt1211 PMID:16699500

Xin, M., Wang, Y., Yao, Y., Xie, C., Peng, H., Ni, Z., & Sun, Q. (2010). Diverse set of microRNAs are responsive to powdery mildew infection and heat stress in wheat (*Triticum aestivum* L.). *BMC Plant Biology*, *10*(1), 123. doi:10.1186/1471-2229-10-123 PMID:20573268

Xu, B., Chen, W., Zhang, G., Wang, J., Ping, W., Luo, L., & Chen, J. (2020). How to achieve green growth in China's agricultural sector. *Journal of Cleaner Production*, *271*, 122770. doi:10.1016/j.jclepro.2020.122770

Yadav & Abburi. (2001). *Atlas of Cropping Systems in India*. PDCSR Bulletin No. 2001-2, Project Directorate for Cropping Systems Research.

Yadav, A.K. (2015). *Organic agriculture (Concept, Scenario, Principles and Practices)*. National Project on Organic Farming, Dept. of Agriculture and Cooperation, Govt of India.

Yadav, A. K., Dash, P., Mohanty, A., Abbassi, R., & Mishra, B. K. (2012). Performance assessment of innovative constructed wetland-microbial fuel cell for electricity production and dye removal. *Ecological Engineering*, *47*, 126–131. doi:10.1016/j.ecoleng.2012.06.029

Yadav, A. K., Jena, S., Acharya, B. C., & Mishra, B. K. (2012). Removal of azo dye in innovative constructed wetlands: Influence of iron scrap and sulfate reducing bacterial enrichment. *Ecological Engineering*, *49*, 53–58. doi:10.1016/j.ecoleng.2012.08.032

Yadava, D. K., Hossain, F., & Mohapatra, T. (2018). Nutritional security through crop biofortification in India: Status & future prospects. *The Indian Journal of Medical Research*, *148*, 1–11. PMID:30666987

Yadav, S. K., Babu, S., Yadav, M. K., Singh, K., Yadav, G. S., & Pal, S. (2013). A Review of Organic Farming for Sustainable Agriculture in Northern India. *International Journal of Agronomy*, 1–8.

Yadav, S. P. (2002). Performance of effective microorganisms (EM) on growth and yields of selected vegetables. *Nature Farming & Environment*, *3*, 35–38.

Yamamura, K., & Kiritani, K. (1998). A simple method to estimate the potential increase in the number of generations under global warming in temperate zones. *Applied Entomology and Zoology*, *33*(2), 289–298. doi:10.1303/aez.33.289

Yang, F. L., Li, S. G., Zhu, F., & Lei, C. L. (2009). Structural characterization of nanoparticles loaded with garlic essential oil and their insecticidal activity against *Triboliumcastaneum* (Herbst) (*Coleoptera: Tenebrionidae*). *Journal of Agricultural and Food Chemistry*, *57*(21), 10156–10162. doi:10.1021/jf9023118 PMID:19835357

Yang, F., Tian, J., Fang, H., Gao, Y., Xu, M., Lou, Y., ... Kuzyakov, Y. (2019). Functional soil organic matter fractions, microbial community, and enzyme activities in a mollisol under 35 years manure and mineral fertilization. *Journal of Soil Science and Plant Nutrition*, *19*(2), 430–439.

Yang, L., Kai, L., Yong, L., Yang, W. W., Fuzhong, Z. P., Jian, Z., Chen, L., Gao, S., & Zhang, L. (2016). Cadmium contamination of soil and crops is affected by intercropping and rotation systems in the lower reaches of the Minjiang River in south-western China. *Environmental Geochemistry and Health*, *38*(3), 811–820. doi:10.100710653-015-9762-4 PMID:26323960

Yan, S., Ren, B., Zeng, B., & Shen, J. (2020). Improving RNAi efficiency for pest control in crop species. *BioTechniques*, *68*(5), 283–290. doi:10.2144/btn-2019-0171 PMID:32202134

Yin, C., Jurgenson, J. E., & Hulbert, S. H. (2011). Development of a host-induced RNAi system in the wheat stripe rust fungus *Puccinia striiformis* f. sp. *tritici*. *Molecular Plant-Microbe Interactions*, *24*(5), 554–561. doi:10.1094/MPMI-10-10-0229 PMID:21190437

Yirefu, F., Tana, T., Tafesse, A., & Zekarias, Y. (2012). Competitive ability of sugarcane (*Saccharum officinarum* L.) cultivars to weed interference in sugarcane plantations of Ethiopia. *Crop Protection (Guildford, Surrey)*, *32*, 138–143. doi:10.1016/j.cropro.2011.10.016

Yokoo, T., Matsumoto, K., Ooba, T., Morimoto, K., & Taguchi, S. (2015). Enhanced poly (3-hydroxybutyrate) production in transgenic tobacco BY-2 cells using engineered acetoacetyl-CoA reductase. *Bioscience, Biotechnology, and Biochemistry*, *79*(6), 986–988. doi:10.1080/09168451.2014.1002448 PMID:25647430

Youssef, M. M., & Ali, M. S. (1998). Management of *Meloidogyne incognita* infecting cowpea by using some native blue green algae. *Anzeiger für Schädlingskunde, Pflanzenschutz. Umweltschutz*, *71*, 15–16.

Youssef, M. M., & Eissa, M. F. (2014). Biofertilizers and their role in management of plant parasitic nematodes. A review. *E3 Journal of Biotechnology & Pharmaceutical Research*, *5*, 1–6.

Yuan, D., Bassie, L., & Sabalza, M. (2011). Potential impact of plant biotechnology on the millennium development goals. *Plant Cell Reports*, *30*(3), 249–265.

Yu, B., Lydiate, D. J., Young, L. W., Schäfer, U. A., & Hannoufa, A. (2007). Enhancing the carotenoid content of *Brassica napus* seeds by downregulating lycopene epsilon cyclase. *Transgenic Research*, *17*(4), 573–585. doi:10.100711248-007-9131-x PMID:17851775

Yu, G. Y., & Muehlbauer, G. J. (2001). Benzothiadiazole-induced gene expression in wheat spikes does not provide resistance to Fusarium head blight. *Physiological and Molecular Plant Pathology*, *59*(3), 129–136. doi:10.1006/pmpp.2001.0348

Yunlong, C., & Smit, B. (1994). Sustainability in agriculture: A general review. *Agriculture, Ecosystems & Environment*, *49*(3), 299–307. doi:10.1016/0167-8809(94)90059-0

Yunus, A. M. M., Parveez, G. K. A., & Ho, C. L. (2008). Transgenic plants producing polyhydroxyalkanoates. *Asia-Pacific Journal of Molecular Biology and Biotechnology*, *16*(1), 1–10.

Yuranova, T., Laub, D., & Kiwi, J. (2007). Synthesis, activity and characterization of textiles showing self-cleaning activity under daylight irradiation. *Catalysis Today*, *122*(1-2), 109–117. doi:10.1016/j.cattod.2007.01.040

Zaidi, S., Usmani, S., Singh, B. R., & Musarrat, J. (2006). Significance of *Bacillus subtilis* strain SJ-101 as a bioinoculant for concurrent plant growth promotion and nickel accumulation in *Brassica juncea*. *Chemosphere*, *64*(6), 991–997. doi:10.1016/j.chemosphere.2005.12.057 PMID:16487570

Zerbino, D. R., & Birney, E. (2008). Velvet: Algorithms for de novo short read assembly using de Bruijn graphs. *Genome Research, 18*(5), 821–829. doi:10.1101/gr.074492.107 PMID:18349386

Zhang, H., Meng, W., & Zhao, F. (2020, April). Study on the exploitation, utilization and cooperation of geothermal energy between Iceland and China. *IOP Conference Series. Earth and Environmental Science, 480*(1), 012010. doi:10.1088/1755-1315/480/1/012010

Zhang, J., Xu, Y., Huan, Q., & Chong, K. (2009). Deep sequencing of *Brachypodium* small RNAs at the global genome level identifies microRNAs involved in cold stress response. *BMC Genomics, 10*(1), 449. doi:10.1186/1471-2164-10-449 PMID:19772667

Zhang, P., Guo, Z., Ullah, S., Melagraki, G., Afantitis, A., & Lynch, I. (2021). Nanotechnology and artificial intelligence to enable sustainable and precision agriculture. *Nature Plants, 7*(7), 864–876. Advance online publication. doi:10.103841477-021-00946-6 PMID:34168318

Zhang, Q., Han, L., Jing, H., Blom, D. A., Lin, Y., Xin, H. L., & Wang, H. (2016). Facet control of gold nanorods. *ACS Nano, 10*(2), 2960–2974. doi:10.1021/acsnano.6b00258 PMID:26795706

Zhang, S., Schisler, D. A., Boehm, M. J., & Slininger, P. J. (2007). Utilization of chemical inducers of resistance and *Cryptococcus flavescens* OH 182.9 to reduce *Fusarium* head blight under greenhouse conditions. *Biological Control, 42*(3), 308–315. doi:10.1016/j.biocontrol.2007.05.020

Zhang, S., Yang, X. L., Li, H., Song, H. L., Wang, R. C., & Dai, Z. Q. (2017). Degradation of sulfamethoxazole in bioelectrochemical system with power supplied by constructed wetland-coupled microbial fuel cells. *Bioresource Technology, 244*, 345–352. doi:10.1016/j.biortech.2017.07.143 PMID:28780269

Zhang, T., Jin, Y., Zhao, J.-H., Gao, F., Zhou, B.-J., Fang, Y.-Y., & Guo, H.-S. (2016). Host-induced gene silencing of the target gene in fungal cells confers effective resistance to the cotton wilt disease pathogen *Verticillium dahliae. Molecular Plant, 9*(6), 939–942. doi:10.1016/j.molp.2016.02.008 PMID:26925819

Zhang, W., Han, D. Y., Dick, W. A., Davis, K. R., & Hoitink, H. A. J. (1998). Compost and compost water extract-induced systemic acquired resistance in cucumber and *Arabidopsis. Phytopathology, 88*(5), 450–454. doi:10.1094/PHYTO.1998.88.5.450 PMID:18944926

Zheng, L., Hong, F., Lu, S., & Liu, C. (2005). Effect of nano-$TiO_{(2)}$ on strength of naturally aged seeds and growth of spinach. *Biological Trace Element Research, 104*(1), 83–91. doi:10.1385/BTER:104:1:083 PMID:15851835

Zheng, Z., Zhou, Y., Li, X., Liu, S., & Tang, Z. (2011). Highly-sensitive organophosphorous pesticide biosensors based on nanostructured films of acetylcholinesterase and CdTe quantum dots. *Biosensors & Bioelectronics, 26*(6), 3081–3085. doi:10.1016/j.bios.2010.12.021 PMID:21196108

Zhong, S. H., Liu, J. Z., Jin, H., Lin, L., Li, Q., Chen, Y., Yuan, Y.-X., Wang, Z.-Y., Huang, H., Qi, Y.-J., Chen, X.-Y., Vaucheret, H., Chory, J., Li, J., & He, Z.-H. (2013). Warm temperatures induce transgenerational epigenetic release of RNA silencing by inhibiting siRNA biogenesis in *Arabidopsis. Proceedings of the National Academy of Sciences of the United States of America, 110*(22), 9171–9176. doi:10.1073/pnas.1219655110 PMID:23686579

Zhou, F., Menke, F. L. H., Yoshioka, K., Moder, W., Shirano, Y., & Klessig, D. F. (2004). High humidity suppresses ssi4-mediated cell death and disease resistance upstream of MAP kinase activation, H_2O_2 production and defense gene expression. *The Plant Journal, 39*(6), 920–932. doi:10.1111/j.1365-313X.2004.02180.x PMID:15341634

Zhu, K.Y., & Palli, S.R. (2020). Mechanisms, applications, and challenges of insect RNA interference. *Annu. Rev. Entomol., 65*, 14.1–14.19.

Zhuang, P., Ye, Z. H., Lan, C. Y., Xie, Z. W., & Shu, W. S. (2005). Chemically assisted phytoextraction of heavy metal contaminated soils using three plant species. *Plant and Soil*, *276*(1-2), 153–162. doi:10.100711104-005-3901-0

Zhu, Y. L., Pilon-Smits, E. A. H., Tarun, A. S., Weber, S. U., Jouanin, L., & Terry, N. (1999). Cadmium tolerance and accumulation in Indian mustard is enhanced by overexpressing g-glutamylcysteine synthetase. *Plant Physiology*, *121*(4), 1169–1177. doi:10.1104/pp.121.4.1169 PMID:10594104

Zoheir, Sadeghi, Mashhadi. (2008). Allelopathic Effects of sunflower (*Helianthus annuus*) on Germination and Growth of Wild Barley (*Hordeum spontaneum*). *Agricultural Technology (Thailand)*, *4*(1), 219–229.

Zularisam, A. W., Zahirah, Z. S., Zakaria, I., Syukiri, M. M., Anwar, A., & Sakinah, M. (2010). Production of biofertilizer from vermicomposting process of municipal sewage sludge. *Journal of Applied Sciences*, *10*(7), 580–584. doi:10.3923/jas.2010.580.584

About the Contributors

R. S. Sengar is Professor and Head, Dept. of Ag. Biotechnology, College of Agriculture in Sardar Vallabhbhai Patel University of Agriculture & Technology, Meerut. Dr. Sengar has 24 years teaching, research and extension experience to his credit. He remained associated with the teaching for a period of about 10 years at G.B. Pant University of Agri. & Tech. Pantnagar, U.S. Nagar (Uttarakhand). Prof. Sengar possesses a career of leadership in improving agrarian livelihood, food, nutrition and ecological security in cutting-edge research and technology development, in higher education and human resources development and in policy and programme formulation, execution and appraisal each pursued in national and global capacities He has supervised 11 Ph.D. students and published more than 118 research papers in India and foreign journals of repute. Dr. Sengar contributed more than two hundred abstracts/ papers to different conferences/symposiums/workshop organized at National and International forums. He also participated in several international programmes seminar/conferences as chairperson of technical sessions, group leader, invited speaker and delegate. He has published 897 popular articles in Hindi and English languages in important journals and magazines. He has also published 16 books for scientific community and farmers. Dr. Sengar has delivered more than 217 talks at All India Radio and Television at Delhi, Lucknow, Rampur and Bareilly centers. Dr. Sengar is life member of several professional societies. He is a member of the editorial boards and review committees of few journals. He is also Chief Editor of an International Journal of Biological Science ‘Biotech Today’ and Founder of ‘Society of Green World for Sustainable Environment’. He has been the Principal and Co-Investigator of 17 projects financed by various government departments. Wide exposure to international programmes and also in handling international projects. He has widely travelled and has visited United Kingdom, Nepal and Dubai under various programme. Dr. Sengar recipient of the ‘Best writer’ award from Vishwa Agro Marketing and Communication Kota Rajasthan, ‘Kunwar Saxena Bahadur SRDA’ award from society for Recent Development in Agriculture, Meerut, ‘Man of the Year’ & ‘Research Board of Advisors’ awards from American Biographical Institute, Inc. U.S.A., ‘Aryabhat’ award from Vigyan Bharti, New Delhi. Dr. Sengar is also received Gold medal 2011 from Hi-tech Horticulture Society, Meerut and Dr. J.C. Edward Medal 2012 from Bioved Research Society, Allahabad and “Scientist of the Year – 2013” from The Academy of Environmental Biology, Lucknow; Life Time Achievement Award – 2014 from Dr. Babasaheb Ambedkar National Institute of Social Sciences, Dr. Ambedkar Nagar (Mhow) M.P.; Dr. R.B. Lal Vishisht Sikshak Puruskar- 2015 from Uttar Pradesh Academy of Agricultural Sciences, Lucknow, Farm and Food Award-2016 from Farm and Food, New Delhi and Young Achiever Award – 2016 from Society of Plant Research, Bareilly. He is fellow of the Society of Plant Research, Bareilly; Society for recent development in agriculture, Meerut; Gramin Vikas Avam Siksha Prasar Samiti, Agra and Technical Society for Basic and Allied Sciences, New Delhi. Dr. Sengar has made significant contribution in the fields of Plant physiology, Biotechnology and Crop production and management systems besides providing leadership in shaping up national level research institutions and agricultural universities. Dr. Sengar has assisted several developing countries in enhancing their human resource, research and technology development capacities, in augmenting productivity, sustainability to enhance farmer's income and to alleviate hunger and poverty, crisis.

Reshu Chaudhary, M.Sc. Biochemistry and Ph.D. from Chaudhary Charan Singh University Meerut, is presently working as Women Scientist, Department of Agricultural Biotechnology, College of Agriculture, Sardar Vallabhbhai Patel University of Agriculture & Technology, Meerut. She has a Research experience of nearly 6 years. She is a good researcher and published 3 Books and more than 43 scientific papers/ articles and research paper in different national & International Journals/ Seminars of repute. Apart from this she is also an Editor of International Journal of Biological Sciences 'Biotech Today' published under the Society of Green World for sustainable Environment along with being an active reviewer at many reputed journals. Dr. Chaudhary is recipient of "Young Scientist Award– 2012" from Bioved Research Society, Allahabad and "Fellow Award – 2016" from The Society for Plant Research, Bareilly. She is working for the sustainable development of different crops from last two years.

* * *

Rekha Dixit is working as a Professor and Head, Department of Biochemistry and Physiology, College of Biotechnology, Sardar Vallabhbhai Patel University of Agriculture and Technology.

Elhoucine Essefi obtained his PhD in geology from the National Engineering School of Sfax in 2013 titled 'Wet Aeolian Sedimentology and Sequence Stratigraphy within the Terrestrial Analogues in Eastern Tunisia: Implications for Wet Aeolian Sedimentology and Sequence Stratigraphy on Mars'. In 2009, obtained the Master degree from the Faculty of Sciences of Sfax titled 'Multidisciplinary study of Sidi El Hani Saline Environment: the Geological History and the Climatic Variability'. In 2003, he obtained the 'Maitrise' degree in Earth Sciences from the Faculty of Sciences of Sfax. In 2011, he obtained the License in Linguistics and Anglophone Civilization from the Faculty of Arts and Humanities of Sfax. In 2015, he obtained the Master's degree in Specialized English titled 'Suggested solutions to some problems of Tunisian novice researchers with academic writing in English: case studies of a Research Article (Gallela et al., 2009) and a Master Thesis (Essefi, 2009)'.

Laxman Singh Gautam has been working in the Department of Zoology in S. S. V. college Hapur since January 2016. Prior to this Dr. Gautam was working in J. V. College Baraut (Baghpat) as Assistant Professor in the Department of Zoology from August 2010. Dr. Gautam interested in toxicology and fish and fisheries and published more than 15 research papers in National and International journals. Dr. Gautam worked in various positions in various committees in college.

Aakansha Goswami is currently working as Assistant Professor in The Department of Biotechnology, Keral Verma Subharti College of Science, Swami Vivekanand Subharti University (NAAC 'A' Accredited), situated in Meerut city, Uttar Pradesh. She has achieved her Doctorate degree in Biotechnology and has more than 8 years academic and research experience. She has Published 04 text books, 12 Book chapters and 50 research publications and reviews in peer reviewed National as well as International

Journals. She has received 10 Awards at national and International level. She has guided more than 05 students at post graduation level for dissertation. Her active areas of research are Plant Biotechnology, tissue culture, molecular biology, microbiology and animal Biotechnology etc. She has actively participated in Seminar/ Symposia and Conferences at national and International Level. She has the 03 foreign visits like- Dubai, Malaysia and Singapore. She has organised the many conferences and Seminar as Co-organizing Secretary and Joint Organizing Secretary at national and International level.

Soumaya Hajji is an Engineer in Geology from National Engineering School in Sfax (ENIS), Tunisia in 2000. She obtained her PhD in Modeling in Hydrogeology in 2015. She obtained her HDR in Environmental Hydrogeology and geo-Informatic in 2021. She is a researcher at geochemistry and hydrogeology, modelling and geostatistics at 'Water-Energy-Environment' Research Laboratory (LR3E) in Engineering School Sfax, Tunisia. Her research activities deal with geochemistry, modeling, analysis, optimisation, simulation of groundwater resources, groundwater quality and. She is an Associate Professor in Geology Departement in Faculty of Sciences of Sfax, Tunisia.

Amit Kumar is working as an assistant Professor Department of Biotechnology, Keral Verma Subhati College of Science, Swami Vivekanand Subharti University, Meerut.

Ashok Kumar is working as a UGC-BSR fellow at the School of Biotechnology, Banaras Hindu University, Varanasi. He has more than 40 years of teaching and research experience. He has published more than 100 research papers in reputed journals. His area of research interest includes molecular microbiology, environmental biotechnology, and clinical microbiology.

Devendra Kumar is presently working as Assistant Professor in the Department of Biotechnology at Keral Verma Subharti College of Science, Swami Vivekanand Subharti University, Meerut. He had graduated from Dr Bhim Rao Ambedkar University, Agra in 2013 and post-graduated in M.Sc. Biotechnology in 2016 from U.P. Pt. Deen Dayal Upadhyaya Pashu Chikitsa Vigyan Vishwavidyalaya Evam Go-Anusandhan Sansthan (DUVASU), Mathura. He achieved his Ph.D. in Agricultural Biotechnology from Sardar Vallabhbhai Patel University of Agriculture and Technology, Meerut in 2020. He also cleared National Eligibility Test (ICAR_NET) conducted by ASRB, New Delhi. He has won Research Excellence Award, Young Scientist Award and many Certificate of Expertise and Certificate of Appreciations from prestigious societies in various conferences and workshops for his outstanding contribution and recognition in the field of Agricultural Biotechnology. Dr. Devendra is a scholastic researcher in his academic career with Hands-on-Expertise on plant tissue culture, hydroponics, qPCR, plant hybridization and various molecular biology techniques. He had published 8 international and 6 national research papers in reputed journals of Springer, Elsevier, Taylor & Francis, Willey Online Library as well as NAAS rated journals. He has also published many Hindi and English popular articles in reputed magazines. He had given many expert talks and participated in many FDPs, trainings and worked as organizing secretary, member of various committees of conferences, workshops, Hands-on-Training courses. He has reviewed for many international journals of repute including NAAS scored and well as Taylor & Francis group. He is the life time member of many scientific professional societies.

Jay Kumar is currently working as Research Associate at CSIR- Central Drug Research Institute, Lucknow. He has completed his Ph.D. in Biotechnology from the School of Biotechnology, Banaras Hindu University, Varanasi (U.P.), and M.Sc. in Molecular Biology and Biotechnology from Tezpur University, Tezpur (Assam). He has published more than 5 research articles in journals of international repute. His area of interest includes microbial biotechnology and genetic engineering.

Permod Kumar is currently working as an Assistant Professor in The Department of Botany, Keral Verma Subharti College of Science, Swami Vivekanand Subharti University (NAAC 'A' Accredited), situated in the City Meerut of State Uttar Pradesh.

Nisha Malik is presently working as an Assistant Professor in the Department of Biotechnology, Keral Verma Subharti College of Science, Swami Vivekanand Subharti University, (NAAC 'A' Accredited), Meerut. Her area of interest is Plant Tissue Culture, Molecular biology, and Antioxidant Studies for sustainable agriculture development.

Pragati Misra has completed Master of Science in Plant Physiology in 2000 and Ph. D with major in Plant Physiology and minors in Biotechnology and Biochemistry from G.B. Pant University of Agriculture and Technology in the year 2004. Dr. Misra has qualified CSIR-NETJRF and ICAR NET in year 2001 and 2003 respectively. At present she is working as an Assistant Professor in the Department of Molecular and Cellular Engineering, SHIATS since December, 2007. She has research and teaching experience of 16 years and having 48 research publications in National and International Journals with Impact Factor. She has Supervised/guided more than 50 undergraduate, 29 post graduate dissertations and supervised 10 doctoral students.

Manjul Mungali was born on 10 January 1984 in Distract Almora, Uttarakhand. The Author did his primary education from UP board, UG from Kumaun university Nainital. The Author obtained his PG from G.B.Pant university of Ag & Tech., Pantnagar, Uttarakhand. The author completed his PhD from Kumaun University Nainital. The Author have more than seven years of teaching and research experience. The author wrote International and national research paper, book and book chapter. The author is also awardees of one patent in biosensor. The specialization area of author is biotechnology, nanotechnology, and biosensors.

Vivek Mungali is Professor & Head at the Department of Agronomy, College of Agriculture, Sardar Vallabhbhai Patel University of Agriculture & Technology, Meerut-250110. Area of specialization is soil fertility & weed management.

Vinay Panwar is Professor and Head, Department of Zoology, Keral Varma Subharti College of Science, Swami Vivekanand Subharti University, Meerut. He has supervised 09 Ph.D., 02 M.Phil. and 15 M.Sc. students and published more than 20 research papers in India and foreign journals of repute. Dr. Panwar has developed and edited several academic programmes. He edited more than 50 books for Monad and Swami Vivekanand University, developed and launched more than 15 new programmes at Monad University and Swami Vivekanand Subharti University. Dr. Panwar is an internationally acclaimed Scientist, Consultant, Manager and Academician innovator with vast experience in the area of higher academics and professional education. He has over 25 years of experience in national teaching, lecturing, research, management and administration. He served as; Professor and Dean, Board of Studies, active role in UGC inspection and active role in various committee at Monad and Swami Vivekanand Subharti University.

J. P. N. Rai is presently working as the full time professor, Department of Environmental Sciences, Gobind Ballabh Pant University of Agriculture & Technology, Pantnagar, Uttarakhand, India, besides, chairing the department for eight years as Head of the department. Dr. Rai has supervised more than 20 Ph. D. thesis on the subject, besides teaching PG and Ph.D courses entitled Biodegradation and recycling of wastes, Waste treatment design and Environmental waste utilization. He has more than two hundred publications to his credit, including research papers, review articles & book chapters in scientific journals of national and international repute, apart from being a versatile reviewer of several journals. He has enough teaching and research experience of thirty seven years in the field of Environmental Science in general and Environmental Biotechnology in particular. He has been successfully completed more than five projects in various fields of ecotechnology, bioenergy production, bioremediation of waste water treatment, terrestrial sites, nanobiomaterial synthesis and applications. Currently he has actively engaged in the research of environmental waste degradation and applications, wastewater treatment and purification through photo-degradation and biological degradation employing bioresources.

Pramod W. Ramteke served as Professor in Biotechnology and Dean of Post Graduate Studies at Sam Higginbottom University of Agriculture, Technology and Sciences, Prayagraj, India. He is elected Fellow of National Academy of Agricultural Sciences, Royal Society of Biology, Academy of Microbiological Sciences (AMI), Biotech Research Society of India, National Academy of Biological Sciences and many other scientific societies. Also, recipient of several awards including BHU Centennial award, Excellence in Science Award (SCON), J. C. Bose Gold Medal, Dr. J. C. Edward Medal, Prof. K. S. Bilgramy Memorial Award, Prof. K. V. Shastri Gold Medal and V. S. Chauhan Gold Medal.

Anju Rani is currently working as an Associate Professor in The Department of Botany, Keral Verma Subharti College of Science, Swami Vivekanand Subharti University (NAAC 'A' Accredited), situated in the City Meerut of State Uttar Pradesh. She has achieved her Doctorate degree in 'Botany, Plant Pathology' and has more than 14 years academic and research experience. She has Published 4 books, 10 book chapters and more than 52 research publications and reviews in peer reviewed National as well as International Journals. She has received Young Women Scientist Award in the International Conference and Excellence in Research (Scientist) Award in 2017 for her contributions to these. She has published one book on nanotechnology entitled "Applications of Nanotechnology: An Introduction" in 2017. Received Best Plant Pathologist -2019 Award in the international conference on 'Global Initiatives in Agricultural and Applied Sciences for Eco Friendly Environment (GAAFES-2019)' held on 1st & 2nd December 2019, UGC-HRDC Hall, Kumaun University, Nainital, UK, India. Her active areas of research are Microbiology, Biotechnology, Mycology and Plant Pathology, Genetics and Plant Breeding, Nanotechnology.

Sanjay Rawal is working in the field of Agronomy with special reference to Organic farming, Crop based systems, precision agriculture, mechanization in potato. The research interest includes potato crop management with focus on nutrition management. Working in the field of Agronomy with special reference to Organic farming, Crop based systems, precision agriculture, mechanization in potato. The research interest includes potato crop management with focus on nutrition management.

Naved Sabir was born in Sikandra Rao (Then in Dist. Aligarh), UP in 1964. He obtained his B.Sc., M.Sc. and M.Phil. from A.M.U., Aligarh and Ph.D. from Indian Agricultural Research Institute (IARI), New Delhi. Dr Sabir, joined ARS in 1989 and has been working mainly in horticultural IPM, Nematode Management, Good Agricultural Practices etc. Presently, he is a Principal Scientist at CPCT, IARI, New Delhi. He has additional training courses in Food Safety Management Systems as well as in GAP. During the last 15 years Dr Sabir has engaged himself in the area of protected cultivation and has worked in more than 5 Nat. and Int. projects developing several technologies and published material. He is a Faculty of PG School of the institute also. He is the author/editor of several books in the area of IPM, GAP and Protected Cultivation. He has published more than 35 research papers in Nat. and Int. Journals, including two books on protected floriculture for vocational courses of NCERT and more than 25 Technical Bulletins. He has been an editorial board member in several journals and annual reports. He is the Founder and Executive Member of Indian Society for Protected Cultivation and life member of many scientific societies and has organized/co-organized three national level seminars and brainstorming sessions. Dr Sabir was Country Representative on GlobalGAP to Philippines Development Agency in 2014. He has travelled widely across India and several nations in official and personal capacities. He has been a member in several committees at institutional and national level and is a tech. expert and member in DST-TIFAC GOI, NIPHM, BIS, GOI and National Horticultural Board (NHB), Min. of Agri., GoI. Dr Sabir is very popular among farmers and has delivered hundreds of lectures across country.

Suman Sanju is M.Sc. (Botany)Gold medalist, M. Phil in Botany(Plant Physiology). To explore her love for science, she acquired Ph.D in Botany on RNAi technology. She is former Project Assistant at CSIR IHBT, Palampur. Then she has worked at CPRI, Shimla as Senior Research Fellow and is also a former SRF at CSK HPKV, Palampur. She has done research on various aspects of science including Heavy metal toxicity, RNAi Technology and mutations in plant breeding. She has published research and review articles on the related topics. She is a botany teacher also worked as an Assistant professor at M.C.M. D.A.V. College, Kangra and presently working as assistant professor at Abhialshi P.G. institute of sciences, Ner chowk, Mandi. She aspires to continue to explore her passion for science through research and teaching.

Shweta Saraswat is Assistant Professor in Department of Environmental Sciences, G. B. Pant University of Agri. & Technology, Pantnagar, Uttarakhand, India and former Post Doctoral fellow in various projects funded by Department of Science & Technology, Indian Council of Medical Research and Ministry of Environment, Forest and Climate Change, Government of India. She has ten years of research and teaching experience to her credit. Besides, she has published fifteen research papers including review articles and book chapters in reputed scientific journals in the field of bioremediation of terrestrial and aquatic systems, green technology development, environmental waste utilization, and bioenergy production. Recently, She is working on integrating nanotechnology with bio-based systems for sensing, analyzing, and treating the contaminated wastewater in an efficient and sustainable manner.

Pradeep Kumar Shukla is presently working as Assistant Professor in the Department of Biological Sciences at Sam Higginbottom University of Agricultural Technology and Sciences, Prayagraj since October, 2009. He has completed M.Sc. Plant Physiology and doctoral degree in Plant Physiology with minors in Biotechnology and Biochemistry from College of Basic Sciences and Humanities, G.B. Pant University of Agriculture & Technology, Pantnagar, U.S. Nagar, Uttarakhand. He has qualified ICAR NET. He has teaching and research experience of 17 years and 16 years respectively, in Plant Physiology & Plant Biotechnology. His area of research is Molecular Plant Physiology of Abiotic stresses and Plant Nanotechnology. Dr. Shukla has around 90 National and International publications and written 13 book chapters. He has attended and presented research paper in several national and international conferences and trainings and has been an active member of the organizing committee for various events. He has guided several undergraduate, postgraduate and students for their doctoral research. He has already organized 2 National Conferences and 2 workshops. He has been a member of several professional bodies and societies. He has written many laboratory manuals and edited journals and proceedings.

Ashu Singh was born in Allahabad, India. She received the B.Tech. degree in Genetic Engineering from the Allahabad Agricultural Deemed University, Allahabad, India, in 2008, and the M.Tech. degree in Biotechnology Engineering from the Sardar Vallabhbhai Patel University of Agriculture and Technology (State Agriculture University) Modipuram, Meerut, in 2011 and finally her PhD. from SHUATS Prayagraj in Plant biotechnology. In 2013, she joined the Department of Agriculture Biotechnology, Sardar Vallabhbhai Patel University of Agriculture and Technology (State Agriculture University) Modipuram, Meerut, as a Women Scientist (WOS-A). Since April 2013, she has been with the Department of Agriculture Biotechnology, College of Agriculture, where she is working on varietal improvement of aromatic rice. Her current research interests include molecular characterization and gene mapping of salinity and drought resistance of aroma rice (Oryza sativa L.). Miss Singh is a National Woman Scientist Fellow of the Department of Science and Technology New Delhi, India. She has a long experience of eight trainings of more than 3 months at different reputed institutes and research laboratories with technical expertise in molecular breeding and also has made significant contribution in the vast arena of stress physiology, biotechnology, and basmati rice production while exploring the applicability of research in enhancing productivity. She also holds Diploma in "Industry Programme in Bioinformatics" (BII, Noida). She has submitted research study on "Importance of Protein Folding for Protein Structure." Apart from this she is also an Associate Editor at Biotech Today journal published under the Society of Green World for Sustainable Environment along with being an active reviewer of many reputed journals. Miss Singh contributed 12 research papers in reputed indexed journals worldwide and attended several conferences/symposiums/workshops organized at national and international forums.

Ashutosh Singh is associated in teaching and research at Rani Lakshmi Bai Central Agricultural University, Jhansi since four years. He has more than ten years of research experience in the field of molecular breeding for various biotic and abiotic stresses in Basmati Rice and Pyramiding/incorporation of major gene(s) for bacterial blight and blast. He has published more than 32 papers in International and National journals, 15 popular articles in reputed magazine, five books and 8 book chapters. Dr. Singh is the recipient of CSIR–NMITLI SRF fellowship and qualified ARS–NET in agricultural biotechnology conducted by ICAR–ASRB. He has also received various scientific honors in the field of agricultural sciences like young scientist, young rice breeder and young teacher award.

Bir Pal Singh was Director of Central Potato Research Institute, Shimla. A plant pathologist, by profession he has research experience of over 32 years in Potato. He has made significant contribution in late blight management and had developed ten potato varieties in partnership with other scientists. He has also developed "JHULSACAST" and mathematical model of applying need-based fungicides for managing late blight. He has also worked on PVY resistant Triplex parental lines development and standardization of aeroponic system of Potato seed Production. There are 167 research papers, 11 technical bulletins and 7 books to his credit. He has also been member of several recognized societies and committees. Dr. Singh is a recipient of several awards which include Hari Om Ashram trust award (20022003) of ICAR, New Delhi and Prof M. J. Narsimhan medal by Indian Phytopathological society (1991).

Divya Singh is currently pursuing PhD from School of Biotechnology, Banaras Hindu University, Varanasi, India. She is a Gold Medalist in M.Sc. (Biotechnology) from Maharishi Dayanand Saraswati University, Ajmer. She has cleared several prestigious examinations of national repute. She is also awarded DST-INSPIRE JRF Fellowship by the Department of Science and Technology, Government of India. She has published research papers and book chapters in various international journals and books. Her area of interest is microbial biotechnology.

Narendra Singh is presently working as Professor and Head Department of Agronomy. Having additional responsibilities of Associate Director Extension, Convener Vice Chancellor Technical Cell and PIO of BUAT, Banda. Earlier work as Head KVK, Chitrakoor for 4 years. Having total 27+ years of experience of Extension Teaching and Research. Published 2 books, 20+ research papers, 4 book chapers,15 articles and 100+ extension literatures in reputed research journals and books.

Aditi Thakur is working as a Scientist (Biotechnology) in Dr. Y.S Parmar University of Horticulture and Forestry, College of Horticulture and Forestry, Neri, Hamirpur (H.P). She is doctorate in Biotechnology from the Central Potato Research institute, Shimla and Banasthali University, Rajasthan and has worked in the area of host delivered gene silencing approach through amiRNAs for the management of late blight disease. She has also worked in prestigious institutes i.e CSIR lab, IHBT Palampur as well as in Plant Pathology Division, IARI, PUSA, New Delhi. She has almost 8 years of experience and actively engaged in research and teaching of undergraduate as well as graduate classes. Her area of expertise includes proteomics, Gene silencing, Plant genetic engineering, Plant -pathogen interactions. She has published and presented many papers in leading national as well as international journals.

Saritha Vara has 17 years of teaching and research experience in the area of Environmental Sustainability and has also published around 80 research papers on addressing challenges of environmental issues and providing sustainable solutions.

Index

D

E

F

G

H

I

J

L

Y

www.ingramcontent.com/pod-product-compliance
Lightning Source LLC
LaVergne TN
LVHW080358011026
849161LV00009B/2332